PARALLELOGRAM

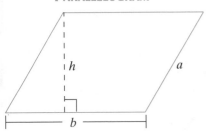

Perimeter: $P = 2a + 2b$
Area: $A = bh$

CIRCLE

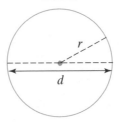

Circumference: $C = \pi d$
$C = 2\pi r$
Area: $A = \pi r^2$

RECTANGULAR SOLID

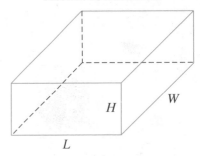

Volume: $V = LWH$
Surface Area: $A = 2HW + 2LW + 2LH$

CUBE

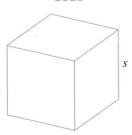

Volume: $V = s^3$
Surface Area: $A = 6s^2$

CONE

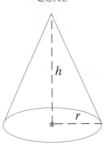

Volume: $V = \dfrac{1}{3}\pi r^2 h$
Surface Area: $A = \pi r \sqrt{r^2 + h^2}$

RIGHT CIRCULAR CYLINDER

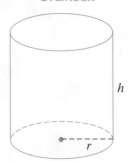

Volume: $V = \pi r^2 h$
Surface Area:
$A = 2\pi rh + 2\pi r^2$

OTHER FORMULAS

Distance: $d = rt$ (r = rate, t = time)

Temperature: $F = \dfrac{9}{5}C + 32 \qquad C = \dfrac{5}{9}(F - 32)$

Simple Interest: $I = Prt$
   (P = principal, r = annual interest rate, t = time in years)

Compound Interest: $A = P\left(1 + \dfrac{r}{n}\right)^{nt}$
   (P = principal, r = annual interest rate, t = time in years, n = number of compoundings per year)

BEGINNING ALGEBRA

SECOND EDITION

K. ELAYN MARTIN-GAY

University of New Orleans

This Annotated Instructor's Edition is exactly like your student's text, but it contains all exercise answers. Where possible the answer is displayed on the same page as the exercise. Exercise answers that require graphical solutions, and both answers and suggestions for instructional strategies for the group activities, can be found in the Instructor's Answers section in Appendix F.

The Student Edition of the text contains answers to selected exercises. Located in the back of the Student Edition are answers to odd-numbered exercises, all chapter test exercises, and all cumulative review exercises including text references. It does not contain answers to even-numbered exercises, group activities, or instructor's notes.

Prentice Hall
Upper Saddle River
New Jersey 07458

Sponsoring Editors, Melissa Acuña and Ann Marie Jones
Editor in Chief, Jerome Grant
Editorial Director, Tim Bozik
Production Editor, Barbara Mack
Managing Editor, Linda Behrens
Assistant Vice President of Production and Manufacturing, David W. Riccardi
Development Editor, Laurie Golson
Editor in Chief of Development, Ray Mullaney
Marketing Manager, Jolene Howard
Creative Director, Paula Maylahn
Art Director, Amy Rosen
Assistant to Art Director, Rod Hernandez
Art Manager, Gus Vibal
Interior Design, Geri Davis, The Davis Group, Inc.
Cover Design, Bruce Kenselaar
Photo Editor, Lori Morris-Nantz
Photo Research, Rhoda Sidney
Manufacturing Buyer, Alan Fischer
Manufacturing Manager, Trudy Pisciotti
Supplements Editor/Editorial Assistant, April Thrower
Cover Photo, Art Wolfe/Tony Stone Images

Photo credits appear on page P-1, which constitutes a continuation of the copyright page.

Printed in the United States of America

10 9 8 7 6 5 4 3 2 1

ISBN 0-13-568460-9

Prentice-Hall International (UK) Limited, *London*
Prentice-Hall of Australia Pty. Limited, *Sydney*
Prentice-Hall Canada Inc., *Toronto*
Prentice-Hall Hispanoamericana, S.A., *Mexico City*
Prentice-Hall of India Private Limited, *New Delhi*
Prentice-Hall of Japan, Inc., *Tokyo*
Simon & Schuster Asia Pte. Ltd., *Singapore*
Editora Prentice-Hall do Brasil, Ltda., *Rio de Janeiro*

To my mother, Barbara M. Miller,
and her husband, Leo Miller,
and to the memory
of my father, Robert J. Martin

CONTENTS

3

Graphing

163

4

Exponents and Polynomials

227

5

Factoring Polynomials

281

PREFACE

ABOUT THIS BOOK

This book was written to provide a solid foundation in algebra for students who might have had no previous experience in algebra. Specific care has been taken to prepare students to go on to their next course in mathematics, and to help students to succeed in nonmathematical courses that require a grasp of algebraic fundamentals. I have tried to achieve this by writing a user-friendly text that is keyed to objectives and contains many worked-out examples. The basic concepts of graphing are introduced early, and applications, data interpretation, and geometric concepts are emphasized and integrated throughout the book.

The many factors that contributed to the success of the first edition have been retained. In preparing this edition, I considered the comments and suggestions of colleagues throughout the country and of the many users of the first edition. The AMATYC Crossroads Document and the NCTM Standards (plus Addenda), together with advances in technology, also influenced the careful reexamination of every section of the text. All of these inputs helped to update the presentation, enhancing the content and pedagogical value.

KEY PEDAGOGICAL FEATURES IN THE SECOND EDITION

Readability and Connections Many reviewers of this edition as well as users of the previous edition have commented favorably on the readability and clear, organized presentation. I have tried to make the writing style as clear as possible while still retaining the mathematical integrity of the content. When a new topic is presented, an effort has been made to relate the new ideas to those the students may already know. Constant reinforcement and connections within problem-solving strategies, data interpretation, geometry, patterns, graphs, and situations from everyday life can help students gradually master both new and old information.

Problem-Solving Process This is formally introduced in Chapter 2, with a new six-step process that is integrated throughout the text. The six steps are *Understand,*

Assign, Illustrate, Translate, Complete, and *Interpret.* The repeated use of these steps in a variety of examples shows their wide applicability. Reinforcing the steps can increase students' comfort level and confidence in tackling problems.

Applications and Connections This book contains a wealth of practical applications found throughout the book in worked-out examples and exercise sets. The applications help to motivate students and strengthen their understanding of mathematics in the real world. They help show connections to a wide range of areas such as biology, environmental issues, consumer applications, allied health, business, entertainment, history, art, literature, finance, sports, and music, as well as to related mathematical areas such as geometry. Many involve interesting real-life data. Sources for data include newspapers, magazines, government publications, and reference books. Opportunities for obtaining your own real data are also included.

Group Activities Each chapter opens with a photograph and description of a real-life situation. At the close of the chapter, students can work cooperatively to apply the algebraic and critical thinking skills they have learned to make decisions and answer the Group Activity that is related to the chapter opening situation. The Group Activity is a multi-part, often hands-on, problem. These new situations, designed for student involvement and interaction, allow for a variety of teaching and learning styles. Answers and tips for instuctional strategies for Group Activities are available in the Annotated Instructor's Edition. In addition, there are opportunities for group activities within section exercise sets.

Reminder Reminders, formerly Helpful Hint boxes, contain practical advice on problem solving. Reminders appear in the context of material in the chapter and give students extra help in understanding and working problems. They are highlighted in a box for quick reference.

Exercise Sets Each exercise set is divided into two parts. Both parts contain graded problems. The first part is carefully keyed to worked examples in the text. Once a student has gained confidence in a skill, the second part contains exercises not keyed to examples. There are ample exercises throughout the book, including end-of-chapter reviews, tests, and cumulative reviews. In addition, each exercise set contains one or more of the following features.

Mental Mathematics These problems are found at the beginning of an exercise set. They are mental warmups that reinforce concepts found in the accompanying section and increase students' confidence before they tackle an exercise set. By relying on their own mental skills, students increase not only their confidence in themselves, but also their number sense and estimation ability.

Conceptual and Writing Exercises These exercises, now found in almost every exercise set, are keyed with the icon ▢ . These exercises require students to show an understanding of a concept learned in the corresponding section. This is accomplished by asking students questions that require them to use two or more concepts together. Some require students to stop, think, and explain in their own words the concept(s) used in the exercises they have just completed. Guidelines recommended by the American Mathematical Association of Two Year Colleges (AMATYC) and other professional groups recommend incorporating writing in mathematics courses to reinforce concepts.

Data and Graphical Interpretation There is increased emphasis on data interpretation in exercises via tables and graphs. The ability to interpret data and read and create a variety of types of graphs is developed gradually so students become comfortable with it. In addition, a new appendix on mean, median, and mode together with exercises is included.

Scientific Calculator Explorations and Exercises Scientific Calculator Explorations, although optional, contain examples and exercises to reinforce concepts or motivate discovery learning. This feature is placed appropriately throughout the text to instruct students on the proper use of the calculator.

Additional exercises building on the skills developed in the Explorations may be found in exercise sets throughout the text, and are marked with an icon ▦ .

Graphing Calculator Explorations and Exercises For graphing calculators or computer graphing utilities, these new Explorations are integrated appropriately throughout the text. Entirely optional, they contain examples and exercises to reinforce concepts, help interpret graphs, or motivate discovery learning.

Additional new exercises building on the skills developed in the Explorations may be found in exercise sets throughout the text, and are marked with an icon ▦ .

Review Exercises Formerly called Skill Review, these exercises are found at the end of each section after Chapter 1. These problems are keyed to earlier sections and review concepts learned earlier in the text that are needed in the next section or in the next chapter. These exercises show the connections between earlier topics and later material.

A Look Ahead These are examples and problems similar to those found in a next algebra course. "A Look Ahead" is presented as a natural extension of the material and contains an example followed by advanced exercises. I strongly suggest that any student who plans to take another algebra course work these problems.

Graphics The text contains numerous graphics, models, and illustrations to visually clarify and reinforce concepts. These include new bar charts, line graphs, calculator screens, application illustrations, and geometric figures. The inside front cover of the text includes a quick reference to geometric figures and formulas, and the inside back cover now includes a summary of common graphs.

Chapter Highlights Found at the end of each chapter, the new Chapter Highlights contain key definitions, concepts, and examples to help students understand and retain what they have learned.

Chapter Review and Test The end of each chapter contains a review of topics introduced in the chapter. These review problems are keyed to sections. The chapter test is not keyed to sections.

Cumulative Review Each chapter after the first contains a cumulative review. Each problem contained in the cumulative review is actually an earlier worked example in the text that is referenced in the back of the book along with the answer. Students who need to see a complete worked-out solution, with explanation, can do so by turning to the appropriate example in the text.

Functional Use of Color and Design Elements of the text are highlighted with color or design to make it easier for students to read and study.

Videotape and Software Icons At the beginning of each section, videotape and software icons are displayed. These icons help reinforce that these learning aids are available should students wish to use them to help them review concepts and skills at their own pace. These items have direct correlation to the text and emphasize the text's methods of solution.

KEY CONTENT FEATURES IN THE SECOND EDITION

Overview In addition to the traditional topics in beginning algebra courses, this text contains a strong emphasis on problem solving, and geometric concepts and reading and interpreting graphs and data are integrated throughout. The geometry concepts covered are those most important to a student's understanding of algebra, and I have included many applications and exercises devoted to this topic. Also, geometric figures and a review of angles, lines, and special triangles are covered in the appendices. I have also integrated reading and interpreting line and bar graphs throughout the text. Not only does this naturally lead to the rectangular coordinate system and beyond, but it gives students practice at interpreting real data. Students are also given the opportunity to see how today's technology can be of help. Exercises are a critical part of student learning, and particular care was taken in writing these.

Increased Emphasis on Data Interpretation There is an increased emphasis on data interpretation via tables and graphs that begins in the first section of the book and continues throughout the text. The ability to interpret data and a variety of types of graphs including bar, line, and circle graphs is developed gradually so students become comfortable with it. In addition, a new appendix on mean, median, and mode is included.

Early and Intuitive Introduction to Graphing As bar and line graphs are gradually introduced in Chapters 1 and 2, an emphasis is slowly placed on the notion of paired data. This leads naturally to the concepts of ordered pair and the rectangular coordinate system introduced in Chapter 3. Chapter 3 is devoted to graphing and concepts of graphing linear equations such as slope and intercepts. These concepts are reinforced throughout exercise sets in subsequent chapters, helping prepare students for more work with equations of lines in Chapter 7.

Increased Emphasis on Problem Solving Building on the strengths of the first edition, a special emphasis and strong commitment is given to contemporary and practical applications of algebra. Real data was drawn from a variety of sources including magazines, newspapers, government publications, and reference books. Generating and using personal real data is also encouraged.

Increased Opportunities to Use Technology Optional calculator as well as graphing calculator explorations are integrated appropriately throughout the text.

New Examples Additional detailed step-by-step examples were added where needed. Many of these reflect real life. Examples are used in two ways. Often there are numbered, formal examples, and occasionally an example or application is used to introduce a topic or informally discuss the topic.

New Exercises A significant amount of time was spent on the exercise sets. New exercises and additional examples help address a wide range of student learning styles and abilities. New kinds of exercises include group activities, conceptual and writing exercises, multi-part exercises, optional graphing calculator exercises, and data analysis from tables and graphs. In addition, the mental math, drill, and word problems were refined and enhanced.

SUPPLEMENTS FOR THE INSTRUCTOR

PRINTED SUPPLEMENTS

Annotated Instructor's Edition (ISBN 0-13-568460-9)

- Answers to exercises on the same text page or in Instructor's Answers section
- Instructor's Answers section contains answers to exercises requiring graphical solutions
- Instructor's Answers section also contains answers and pedagogical suggestions for group activities
- Notes to the Instructor

Instructor's Solutions Manual (ISBN 0-13-568379-3)

- Solutions to even-numbered exercises, chapter tests, and cumulative review exercises
- Graphics computer-generated for clarity
- Answers checked for accuracy

Test Item File (ISBN 0-13-568411-0)

- Six forms (A, B, C, D, E, and F) of Chapter Tests
  —three forms contain multiple-choice items
  —three forms contain free-response items

- Two forms of Cumulative Review Tests
  —every two chapters
- Final Exams
  —four forms with free-response scrambled items
  —four forms with multiple-choice scrambled items
- Answers to all items

MEDIA SUPPLEMENTS

TestPro2 Computerized Testing (Sample Disk IBM, ISBN 0-13-258104-3; Sample Disk Mac, ISBN 0-13-258112-4; IBM, ISBN 0-13-258112-4; Mac, ISBN 0-13-568429-3)

- Comprehensive text-specific testing
- Generates test questions and drill worksheets from algorithms keyed to the text learning objectives
- Edit or add your own questions
- Compatible with Scantron or possible other scanners

USING INTERNET AND WEB BROWSER

Using the Internet and a Web browser, such as Netscape, can add to your mathematical resources. Below is a list of some of the sites that may be worth your or your students' visit.

- Prentice Hall Home Page   http://www/prenhall.com
- The Mathematical Association of America   http://www/maa.org
- The American Mathematical Society   http://www/ams.org
- The National Council of Teachers of Mathematics   http://www/nctm.org
- The Census Bureau   http://www.census.gov
- Texas Instruments   http://www.ti.com/calc
- The Math Archives   http://archives.math.utk.edu

INTERNET GUIDE

- Contact your local Prentice Hall representative.

SUPPLEMENTS FOR THE STUDENT

PRINTED SUPPLEMENTS

Student Solutions Manual (ISBN 0-13-568387-4)

- Detailed step-by-step solutions to odd-numbered text and review exercises
- Solutions to all chapter practice tests and cumulative review exercises
- Solution methods reflect those emphasized in the text
- Ask your bookstore about ordering

Student Study Guide (ISBN 0-13-568403-X)

- Additional step-by-step worked out examples and exercises
- Practice tests and final examination
- Solution methods reflect those emphasized in the text
- Includes study skills and note-taking suggestions
- Ask your bookstore about ordering

New York Times Supplement

- A free newspaper from Prentice Hall and *The New York Times*
- Interesting and current articles on mathematics
- Invites talking and writing about mathematics
- Created new each year

MEDIA SUPPLEMENTS

Videotape Series (Sample Video, ISBN 0-13-258146-9;
Video Series, ISBN 0-13-568395-5)

- Specifically keyed to the textbook by section
- Presentation and step-by-step examples by the textbook author, an award-winning teacher
- Comprehensive coverage

MathPro Tutorial Software (Sample Disk IBM, ISBN 0-13-258120-5;
Sample Disk Mac, ISBN 0-13-258138-8;
IBM Network-User, ISBN 0-13-568445-5;
IBM Single-User, ISBN 0-13-578543-X;
Mac, ISBN 0-13-568452-8)

- Text-specific tutorial exercises
- Interactive feedback
- Unlimited practice Warmup Exercises
- Graded and recorded Practice Problems
- New user interface, glossary, and expressions editor for ease of use and flexibility

ACKNOWLEDGMENTS

First, as usual, I would like to thank my husband Clayton for his constant encouragement. I would also like to thank my children, Eric and Bryan, for continuing to eat my burnt bacon and cookies even though they now realize that they are "extra crispy."

I would also like to thank my extended family for their invaluable help and wonderful sense of humor. Their contributions are too numerous to list. They are Peter, Karen, Michael, Christopher, Matthew, and Jessica Callac; Stuart, Earline, Melissa, and Mandy Martin; Mark, Sabrina, and Madison Martin; Leo and Barbara Miller; and Jewett Gay.

A special thank you to all the users of the first edition of this text and for their suggestions for improvements that were incorporated into the second edition. I would also like to thank the following reviewers of this text:

Dennis Carrie *Golden West College*
Celeste Carter *Richland College*
Carol Cheshire *Macon College*
Gregory Davis *University of Wisconsin—Green Bay*
Margarita Fresquez *Palo Alto College*
Terry Y. Fung *Kean College of New Jersey*
Gail Gonyo *Adirondack Community College*
Ethel H. Guinyard *Tallahassee Community College*
Pat Hirschy *Asnuntuck Community Technical College*
Robert Horvath *El Camino College*
Robert Malena *Community College of Allegheny County—South College*
Christopher G. McNally *Tallahassco Community College*
Lois Miller *Golden West College*
Wayne L. Miller *Lee College*
Sofya Nayer *Borough of Manhattan Community College*
Dorothy Pennington *Tallahassee Community College*
Dianne Phelps *Sullivan County Community College*
Karen Schwitters *Seminole Community College*
Kenneth Shabell *Riverside Community College*
Edelma Simes *Phillips County Community College*
Lyndon Weberg *University of Wisconsin*
Cathleen M. Zucco *Le Moyne College*

Cheryl Roberts and Phyllis Barnidge did an excellent job of providing answers and solutions, and contributing to the overall accuracy of the book. Laurie Golson was invaluable for her many suggestions during the development of the second edition. I very much appreciated the writers and accuracy checkers of the supplements to accompany this text as well as Emily Keaton's contributions. Last, but by no means least, a special thanks to the staff at Prentice Hall for their support and assistance: Melissa Acuña, Ann Marie Jones, Barbara Mack, Linda Behrens, Alan Fischer, Amy Rosen, Gus Vibal, Paula Maylahn, April Thrower, Evan Girard, Jolene Howard, Gary June, Jerome Grant, and Tim Bozik.

K. Elayn Martin-Gay

ABOUT THE AUTHOR

K. Elayn Martin-Gay has taught mathematics at the University of New Orleans for over 16 years and has received numerous teaching awards, including the local University Alumni Association's Award for Excellence in Teaching.

Over the years, Elayn has developed a videotaped lecture series to help her students understand algebra better. This highly successful video material is the basis for the four-book series, *Prealgebra, Beginning Algebra, Intermediate Algebra,* and *Introductory and Intermediate Algebra,* a combined approach.

HOW TO USE THE TEXT:
A GUIDE FOR STUDENTS

Beginning Algebra, Second Edition has been designed as just one of the tools in a fully integrated learning package to help you develop beginning algebra skills. Our goal is to encourage your success and mastery of the mathematical concepts introduced in this text. Take a few moments now to see how this text will help you excel.

CHAPTER

3

3.1 THE RECTANGULAR COORDINATE SYSTEM

3.2 GRAPHING LINEAR EQUATIONS

3.3 INTERCEPTS

3.4 SLOPE

3.5 GRAPHING LINEAR INEQUALITIES

GRAPHING

FINANCIAL ANALYSIS

Investment analysts must investigate a company's financial data, such as sales, profit margin, debt, and assets, to evaluate whether investing in it is a wise choice. One way to analyze such data is to graph the data and identify trends in it visually over time. Another way to analyze such data is to find algebraically the rate at which it is changing over time.

IN THE CHAPTER GROUP ACTIVITY ON PAGE 215, YOU WILL HAVE THE OPPORTUNITY TO ANALYZE THE SALES OF SEVERAL COMPANIES IN THE AEROSPACE INDUSTRY.

The photo application at the opening of every chapter and **applications throughout** offer real-world scenarios that connect mathematics to your life. In addition, at the end of the chapter, a group activity or discovery-based project further shows the chapter's applicability.

Page 163

APPLY THE PROBLEM-SOLVING PROCESS

As you study, **make connections**–this text's organization can help you. There are features in this text designed to help you relate material you are learning to previously mastererd material. Math topics are tied to real life as often as possible. Key learning objectives are introduced and easily identifiable in every section.

Save time by having a plan. Follow the **six-step process**, and you will find yourself successfully solving a wide range of problems.

PROBLEM-SOLVING STEPS

1. UNDERSTAND the problem. During this step don't work with variables (except for known formulas), but simply become comfortable with the problem. Some ways of accomplishing this are listed below.
   - Read and reread the problem.
   - Construct a drawing.
   - Look up an unknown formula.
   - Propose a solution and check. Pay careful attention to how to check your proposed solution. This will help later when writing an equation to model the problem.

2. ASSIGN a variable to an unknown in the problem. Use this variable to represent any other unknown quantities.

3. ILLUSTRATE the problem. A diagram or chart using the assigned variables can often help to visualize the known facts.

4. TRANSLATE the problem into a mathematical model. This is often an equation.

5. COMPLETE the work. This often means to solve the equation.

6. INTERPRET the results. *Check* the proposed solution in the stated problem and *state* your conclusion.

Page 113

REMINDER   Be careful when identifying the base of an exponential expression.

$$(-3)^2 \qquad\qquad -3^2 \qquad\qquad 2 \cdot 3^2$$

Base is -3 Base is 3   Base is 3

$$(-3)^2 = (-3)(-3) = 9 \qquad -3^2 = -(3 \cdot 3) = -9 \qquad 2 \cdot 3^2 = 2 \cdot 3 \cdot 3 = 18$$

Page 229

"Reminders" contain practical advice, and provide extra help in understanding and working problems.

VISUALIZE . . . SEE THE CONCEPTS!

Graphing is introduced early and intuitively. Fully integrated in all appropriate sections with an increased emphasis on **data interpretation** via tables and graphs, this concept is developed gradually throughout the text. Knowing how to use data and graphs is a valuable skill in the workplace as well as in other courses.

The following bar graph shows average miles per gallon for cars during the years shown.

Real Data is integrated throughout the text, drawn from familiar sources such as magazines and newspapers.

73. In 1988, cars averaged how many miles per gallon of fuel?

74. In 1991, cars averaged how many miles per gallon of fuel?

75. Find the increase of average miles per gallon for automobiles for the years 1985 to 1992.

76. Do you notice any trends from this graph?

77. What line segment has the greatest slope?

78. What line segment has the least slope?

79. What year showed the greatest increase in miles per gallon?

80. What year showed the least increase in miles per gallon?

81. Find the slope of a line parallel to the line passing through $(-7, -5)$ and $(-2, -6)$.

Page 207

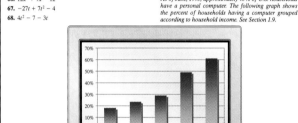

66. $12x^2 + 7x - 12$
67. $-27t + 7t^2 - 4$
68. $4t^2 - 7 - 3t$

As of June, 1995, approximately 40% of U.S. households have a personal computer. The following graph shows the percent of households having a computer grouped according to household income. See Section 1.9.

Page 303

Many graphics, models and illustrations provide **visual reinforcement.**

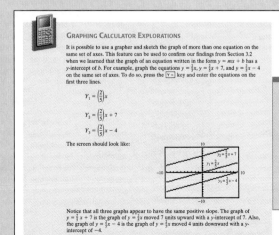

GRAPHING CALCULATOR EXPLORATIONS

It is possible to use a grapher and sketch the graph of more than one equation on the same set of axes. This feature can be used to confirm our findings from Section 3.2 when we learned that the graph of an equation written in the form $y = mx + b$ has a y-intercept of b. For example, graph the equations $y = \frac{2}{5}x$, $y = \frac{2}{5}x + 7$, and $y = \frac{2}{5}x - 4$ on the same set of axes. To do so, press the $\boxed{Y=}$ key and enter the equations on the first three lines.

$$Y_1 = \left(\frac{2}{5}\right)x$$

$$Y_2 = \left(\frac{2}{5}\right)x + 7$$

$$Y_3 = \left(\frac{2}{5}\right)x - 4$$

The screen should look like:

Notice that all three graphs appear to have the same positive slope. The graph of $y = \frac{2}{5}x + 7$ is the graph of $y = \frac{2}{5}x$ moved 7 units upward with a y-intercept of 7. Also, the graph of $y = \frac{2}{5}x - 4$ is the graph of $y = \frac{2}{5}x$ moved 4 units downward with a y-intercept of -4.

Graph the equations on the same set of axes. Describe the similarities and differences in their graphs.

Scientific and Graphing Calculator explorations and exercises are woven into the appropriate sections to reinforce concepts and **motivate discovery-based learning.**

Page 204

CHECK YOUR UNDERSTANDING. EXPAND IT. EXPLORE!

Good exercise sets are essential to the make-up of a solid beginning algebra textbook. The exercises you will find in this textbook are designed to help you understand skills and concepts as well as challenge and motivate you. Note, too, the Highlights, Test, Review, and Cumulative Review found at the end of each chapter.

MENTAL MATH

Decide whether a line with the given slope is upward, downward, horizontal or vertical.

1. $m = \dfrac{7}{6}$ **2.** $m = -3$ **3.** $m = 0$ **4.** m is undefined.

Confidence building **Mental Math** problems are in most sections.

Page 204

Build your confidence with the beginning exercises; the first part of each exercise set is keyed to already worked examples. Then try the remaining exercises.

Conceptual and Writing Exercises bring together two or more concepts and often require "in your own words" written explanation.

7. The area of the larger rectangle below is $x(x + 3)$. Find another expression for this area by finding the sum of the areas of the smaller rectangles.

8. Write an expression for the area of the larger rectangle below in two different ways.

Find the following products. See Examples 2 and 3.

9. $(a + 7)(a - 2)$ **10.** $(y + 5)(y + 7)$
11. $(2y - 4)^2$ **12.** $(6x - 7)^2$
13. $(5x - 9y)(6x - 5y)$ **14.** $(3x - 7y)(7x + 2y)$
15. $(2x^2 - 5)^2$ **16.** $(x^2 - 4)^2$

17. The area of the figure below is $(x + z)(x + 3)$. Find another expression for this area by finding the sum of the areas of the smaller rectangles.

24. $(3 + b)(2 - 5b - 3b^2)$

Find the following products. See Example 5.

25. $(x + 2)^3$ **26.** $(y - 1)^3$
27. $(2y - 3)^3$ **28.** $(3x + 4)^3$

Find the following products. Use the vertical multiplication method. See Example 6.

29. $(x + 3)(2x^2 + 4x - 1)$
30. $(2x - 5)(3x^2 - 4x + 7)$
31. $(x^2 + 5x - 7)(x^2 - 7x - 9)$
32. $(3x^2 - x + 2)(x^2 + 2x + 1)$

33. Evaluate each of the following.
   a. $(2 + 3)^2$; $2^2 + 3^2$
   b. $(8 + 10)^2$; $8^2 + 10^2$
   Does $(a + b)^2 = a^2 + b^2$ no matter what the values of a and b are? Why or why not?

34. Perform the indicated operations. Explain the difference between the two expressions.
   a. $(3x + 5) + (3x + 7)$
   b. $(3x + 5)(3x + 7)$

Page 250

A Look Ahead

EXAMPLE
Find the product $[(a + b) - 2][(a + b) + 2]$.

Solution:
If we think of $(a + b)$ as one term, we can think of $[(a + b) - 2][(a + b) + 2]$ as the product of the sum and difference of two terms.

$$[(a + b) - 2][(a + b) + 2] = (a + b)^2 - 2^2$$

Next, square $(a + b)$. $= a^2 + 2ab + b^2 - 4$

Find each product.

87. $[(x + y) - 3][(x + y) + 3]$
88. $[(a + c) - 5][(a + c) + 5]$
89. $[(a - 3) + b][(a - 3) - b]$
90. $[(x - 2) + y][(x - 2) - y]$
91. $[(2x + 1) - y][(2x + 1) + y]$
92. $[(3x + 2) - z][(3x + 2) + z]$

A Look Ahead examples and problems are similar to those found in the *next* algebra course and include more advanced exercises.

Page 257

Find the slope of the line through the given points.

87. (2.1, 6.7) and (−8.3, 9.3)

88. (−3.8, 1.2) and (−2.2, 4.5)

89. (2.3, 0.2) and (7.9, 5.1)

90. (14.3, −10.1) and (9.8, −2.9)

91. Find the slope and the *y*-intercept of the line defined by $x + 3y = 6$. Next, solve the equation $x + 3y = 6$ for *y*. Compare the slope and the *y*-intercept with the equation solved for *y*. Write down any observations.

92. Find the slope and the *y*-intercept of the line defined by $5x + 2y = 7$. Next, solve the equation $5x + 2y = 7$ for *y*. Compare the slope and the *y*-intercept with the equation solved for *y*. Write down any observations.

93. The graph of $y = \frac{1}{2}x$ has a slope of $\frac{1}{2}$. The graph of $y = 3x$ has a slope of 3. The graph of $y = 5x$ has a slope of 5. Graph all three equations on a single coordinate system. As slope becomes larger, how

Page 208

Scientific and Graphing Calculator exercises are found within most exercise sets.

Review Exercises

Solve each linear inequality. See Section 2.9.

95. $-3x \le -9$ **96.** $-x > -16$

97. $\dfrac{x - 6}{2} < 3$ **98.** $\dfrac{2x + 1}{3} \ge -1$

Graph each linear equation in two variables. See Sections 3.2 and 3.3.

99. $x - y = 6$ **100.** $x = -2y$

101. $y = 3x$ **102.** $5x + 3y = 15$

103. $x = -2$ **104.** $y = 5$

Page 208

Review Exercises review concepts learned earlier that are needed in the next section or next chapter.

GROUP ACTIVITY

FINANCIAL ANALYSIS

OPTIONAL MATERIALS:
• Financial magazines
• Annual reports

The table below gives the sales in millions of dollars for the leading U.S. businesses in the aerospace industry for the years 1993 and 1994. You have been asked to analyze the performances of these companies and, based on this information alone, make an investment recommendation.

Assuming that the trends in the sales are li ear, graph the line represented by the orde pairs for each company. Describe the tren shown by each graph.

2. Find the slope of the line for each compan

3. Which of the lines, if any, have negative slop What does that mean in this context? Whic the lines, if any, have zero slopes? What doe that mean in this context? Which of the line any, are parallel? What does that mean in th context? Which, if any, of the lines are perp dicular? What does that mean in this contex

AEROSPACE SALES (IN MILLIONS OF DOLLARS)

COMPANY	1993	1994
Boeing	25,300	21,900
United Technologies	20,700	21,200
McDonnell-Douglas	14,500	13,200

Page 217

There is an opportunity to **explore** an exercise that relates to the chapter-opening photo as a group activity or discovery-based project.

STUDENT RESOURCES

Seek out these items to match your personal learning style.

3.2 | GRAPHING LINEAR EQUATIONS

OBJECTIVES

1. Identify linear equations.
2. Graph a linear equation by finding and plotting ordered pair solutions.

In the previous section, we found that equations in two variables may have more than one solution. For example, both (6, 0) and (2, −2) are solutions of the equation $x − 2y = 6$. In fact, this equation has an infinite number of solutions. Other solutions include (0, −3), (4, −1), (−2, −4), and (8, 1). If we graph these solutions, notice that a pattern appears.

Page 175

Text-specific videos hosted by the award-winning teacher and author of *Beginning Algebra, Second Edition,* cover each objective in every chapter section as a supplementary review.

MathPro Tutorial Software, developed around the content of *Beginning Algebra, Second Edition,* provides interactive warm-up and graded algorithmic practice problems with step-by-step worked solutions.

ALSO AVAILABLE:

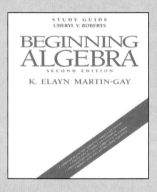

New York Times/Themes of the Times
Newspaper-format supplement –
*ask your professor about
this exciting free supplement!*

REVIEW OF REAL NUMBERS

CREATING AND INTERPRETING GRAPHS

Companies often rely on market research to investigate the types of consumers who buy their products and their competitors' products. One way of gathering this type of data is through the use of surveys. The raw data collected from such surveys can be difficult to interpret without some kind of organization. Graphs serve to organize data visually. They also enable a user to interpret the data represented by the graph quickly.

IN THE CHAPTER GROUP ACTIVITY ON PAGE 62, YOU WILL HAVE THE OPPORTUNITY TO CONDUCT A BRIEF SURVEY ON BREAKFAST CEREAL CONSUMPTION AND CREATE GRAPHS TO REPRESENT THE RESULTS.

The power of mathematics is its flexibility. We apply numbers to almost every aspect of our lives, from an ordinary trip to the grocery store to a rocket launched into space. The power of algebra is its generality. Using letters to represent numbers, we tie together the trip to the grocery store and the launched rocket.

In this chapter we review the basic symbols and words—the language—of arithmetic and introduce using variables in place of numbers. This is our starting place in the study of algebra.

1.1   SYMBOLS AND SETS OF NUMBERS

O B J E C T I V E S

 1   Identify natural and whole numbers, and picture them on a number line.

2   Define the meaning of the symbols $=$, $\neq$, $<$, $>$, $\leq$, and $\geq$.

3   Translate sentences into mathematical statements.

4   Identify integers, rational numbers, irrational numbers, and real numbers.

5   Find the absolute value of a real number.

TAPE BA 1.1

1   We begin with a review of the set of natural numbers and the set of whole numbers and how we use symbols to compare these numbers. A **set** is a collection of objects, each of which is called a **member** or **element** of the set. A pair of brace symbols { } encloses the list of elements and is translated as "the set of" or "the set containing."

> **NATURAL NUMBERS**
>
> The set of **natural numbers** is $\{1, 2, 3, 4, 5, 6, \ldots\}$.

> **WHOLE NUMBERS**
>
> The set of **whole numbers** is $\{0, 1, 2, 3, 4, \ldots\}$.

The three dots (an ellipsis) at the end of the list of elements of a set means that the list continues in the same manner indefinitely.

These numbers can be pictured on a **number line.** We will use the number line often to help us visualize objects and relationships. Visualizing mathematical concepts is an important skill and tool, and later we will develop and explore other visualizing tools.

To draw a number line, first draw a line. Choose a point on the line and label it 0. To the right of 0, label any other point 1. Being careful to use the same distance as from 0 to 1, mark off equally spaced distances. Label these points 2, 3, 4, 5, and so on. Since the whole numbers continue indefinitely, it is not possible to show every whole number on the number line. The arrow at the right end of the line indicates that the pattern continues indefinitely.

2 Picturing whole numbers on a number line helps us to see the order of the numbers. Symbols can be used to describe concisely in writing the order that we see.

> The **equal symbol** = means "is equal to."
> The symbol ≠ means "is not equal to."

These symbols may be used to form a **mathematical statement.** The statement might be true or it might be false. The two statements below are both true.

> 2 = 2   states that "two is equal to two"
>
> 2 ≠ 6   states that "two is not equal to six"

If two numbers are not equal, then one number is larger than the other. The symbol > means "is greater than." The symbol < means "is less than." For example,

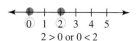

2 > 0 or 0 < 2

> 2 > 0 states that "two is greater than zero"
>
> 3 < 5 states that "three is less than five"

On the number line, we see that a number **to the right of** another number is **larger.** Similarly, a number **to the left of** another number is smaller. For example, 3 is to the left of 5 on the number line, which means that 3 is less than 5, or 3 < 5. Similarly, 2 is to the right of 0 on the number line, which means 2 is greater than 0, or 2 > 0. Since 0 is to the left of 2, we can also say that 0 is less than 2, or 0 < 2. The symbols ≠, <, and > are called **inequality symbols.**

3 < 5

> REMINDER   Notice that 2 > 0 has exactly the same meaning as 0 < 2. Switching the order of the numbers and reversing the "direction of the inequality symbol" does not change the meaning of the statement.
>
> 5 > 3 has the same meaning as 3 < 5.
>
> Also notice that, when the statement is true, the inequality arrow points to the smaller number.

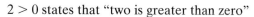

EXAMPLE 1   Insert <, >, or = in the space between the paired numbers to make each statement true.

a. 2   3   **b.** 7   4   **c.** 72   27

Solution: **a.** 2 < 3 since 2 is to the left of 3 on the number line.

b. 7 > 4 since 7 is to the right of 4 on the number line.

c. 72 > 27 since 72 is to the right of 27 on the number line.

Two other symbols are used to compare numbers. The symbol $\le$ means "is less than or equal to." The symbol $\ge$ means "is greater than or equal to." For example,

$7 \le 10$ states that "seven is less than or equal to ten"

This statement is true since $7 < 10$ is true. If either $7 < 10$ or $7 = 10$ is true, then $7 \le 10$ is true.

$3 \ge 3$ states that "three is greater than or equal to three"

This statement is true since $3 = 3$ is true. If either $3 > 3$ or $3 = 3$ is true, then $3 \ge 3$ is true.

The statement $6 \ge 10$ is false since neither $6 > 10$ nor $6 = 10$ is true.

The symbols $\le$ and $\ge$ are also called **inequality symbols.**

EXAMPLE 2   Tell whether each statement is true or false.

a. $8 \ge 8$ **b.** $8 \le 8$ **c.** $23 \le 0$ **d.** $23 \ge 0$

Solution: **a.** True, since $8 = 8$ is true.   **b.** True, since $8 = 8$ is true.

c. False, since neither $23 < 0$ nor $23 = 0$ is true.   **d.** True, since $23 > 0$ is true.

---

3   Now, let's use the symbols discussed above to translate sentences into mathematical statements.

EXAMPLE 3   Translate each sentence into a mathematical statement.

a. Nine is less than or equal to eleven.

b. Eight is greater than one.

c. Three is not equal to four.

Solution: **a.**

Nine	is less than or equal to	eleven
9	$\le$	11

b.

Eight	is greater than	one
8	$>$	1

c.

Three	is not equal to	four
3	$\ne$	4

---

4   Whole numbers are not sufficient to describe many situations in the real world. For example, quantities smaller than zero must sometimes be represented, such as temperatures less than 0 degrees.

We can picture numbers less than zero on the number line as follows:

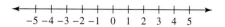

Numbers less than 0 are to the left of 0 and are labeled -1, -2, -3, and so on. A $-$ sign, such as the one in -1, tells us that the number is to the left of 0 on the number line. In words, -1 is read "negative one." A $+$ sign or no sign tells us that a number lies to the right of 0 on the number line. For example, 3 and $+3$ both mean positive three.

The numbers we have pictured are called the set of **integers.** Integers to the left of 0 are called **negative integers;** integers to the right of 0 are called **positive integers.** The integer 0 is neither positive nor negative.

INTEGERS

The set of **integers** is $\{\ldots, -3, -2, -1, 0, 1, 2, 3, \ldots\}$.

Notice the ellipses (three dots) to the left and to the right of the list for the integers. This indicates that the positive integers and the negative integers continue indefinitely.

A problem with integers in real-life settings arises when quantities are smaller than some integer but greater than the next smallest integer. On the number line, these quantities may be visualized by points between integers. Some of these quantities between integers can be represented as a quotient of integers. For example,

The point on the number line halfway between 0 and 1 can be represented by $\frac{1}{2}$, a quotient of integers.

The point on the number line halfway between 0 and -1 can be represented by $-\frac{1}{2}$. Other quotients of integers and their graphs are shown.

The set of numbers, each of which can be represented as a quotient of integers, is called the set of **rational numbers.** Notice that every integer is also a rational number since each integer can be expressed as a quotient of integers. For example, the integer 5 is also a rational number since $5 = \frac{5}{1}$.

RATIONAL NUMBERS

The set of **rational numbers** is the set of all numbers that can be expressed as a quotient of integers.

The number line also contains points that cannot be expressed as quotients of integers. These numbers are called **irrational numbers** because they cannot be represented by rational numbers. For example, $\sqrt{2}$ and π are irrational numbers.

IRRATIONAL NUMBERS

The set of **irrational numbers** is the set of all numbers that correspond to points on the number line but that are not rational numbers. That is, an irrational number is a number that cannot be expressed as a quotient of integers.

Rational numbers and irrational numbers can be written as decimal numbers. The decimal equivalent of a rational number will either terminate or repeat in a pattern. For example, upon dividing we find that

$$\frac{3}{4} = 0.75 \text{ (decimal number terminates or ends) and}$$

$$\frac{2}{3} = 0.66666\ldots \text{ (decimal number repeats in a pattern)}$$

The decimal representation of an irrational number will neither terminate nor repeat. (For further review of decimals, see the appendix.)

The set of numbers, each of which corresponds to a point on the number line, is called the set of **real numbers.** One and only one point on the number line corresponds to each real number.

REAL NUMBERS

The set of **real numbers** is the set of all numbers each of which corresponds to a point on the number line.

On the following number line, we see that real numbers can be positive, negative, or 0. Numbers to the left of 0 are called **negative numbers;** numbers to the right of 0 are called **positive numbers.** Positive and negative numbers are also called **signed numbers.**

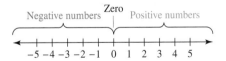

Several different sets of numbers have been discussed in this section. The following diagram shows the relationships among these sets of real numbers.

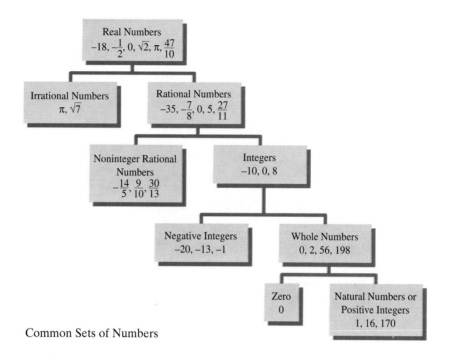

Common Sets of Numbers

EXAMPLE 4   Given the set $\left\{-2, 0, \frac{1}{4}, 112, -3, 11, \sqrt{2}\right\}$, list the numbers in this set that belong to the set of:

a. Natural numbers   **b.** Whole numbers   **c.** Integers

d. Rational numbers   **e.** Irrational numbers   **f.** Real numbers

Solution: **a.** The natural numbers are 11 and 112.

b. The whole numbers are 0, 11, and 112.

c. The integers are -3, -2, 0, 11, and 112.

d. Recall that integers are rational numbers also. The rational numbers are -3, -2, 0, $\frac{1}{4}$, 11, and 112.

e. The irrational number is $\sqrt{2}$.

f. The real numbers are all numbers in the given set.

We can now extend the meaning and use of inequality symbols such as $<$ and $>$ to apply to all real numbers.

ORDER PROPERTY FOR REAL NUMBERS

Given any two real numbers a and b, $a < b$ if a is to the left of b on the number line. Similarly, $a > b$ if a is to the right of b on the number line.

EXAMPLE 5 Insert $<$, $>$, or $=$ in the space between the paired numbers to make each statement true.

a. -1 0 **b.** 7 $\frac{14}{2}$ **c.** -5 -6

Solution: **a.** $-1 < 0$ since -1 is to the left of 0 on the number line.

b. $7 = \frac{14}{2}$ since $\frac{14}{2}$ simplifies to 7.

c. $-5 > -6$ since -5 is to the right of -6 on the number line.

The number line not only gives us a picture of the real numbers, it also helps us visualize the distance between numbers. The distance between a real number a and 0 is given a special name called the **absolute value** of a. "The absolute value of a" is written in symbols as $|a|$.

ABSOLUTE VALUE

The absolute value of a real number a, denoted by $|a|$, is the distance between a and 0 on a number line.

For example, $|3| = 3$ and $|-3| = 3$ since both 3 and -3 are a distance of 3 units from 0 on the number line.

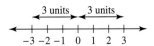

R E M I N D E R   Since $|a|$ is a distance, $|a|$ is always either positive or 0, never negative. That is, **for any real number a, $|a| \geq 0$.**

EXAMPLE 6 Find the absolute value of each number.

a. $|4|$ **b.** $|-5|$ **c.** $|0|$

Solution: **a.** $|4| = 4$ since 4 is 4 units from 0 on the number line.

b. $|-5| = 5$ since -5 is 5 units from 0 on the number line.

c. $|0| = 0$ since 0 is 0 units from 0 on the number line.

EXAMPLE 7 Insert $<$, $>$, or $=$ in the appropriate space to make the statement true.

a. $|0|$ 2 **b.** $|-5|$ 5 **c.** $|-3|$ $|-2|$ **d.** $|5|$ $|6|$ **e.** $|-7|$ $|6|$

Solution: **a.** $|0| < 2$ since $|0| = 0$ and $0 < 2$.

b. $|-5| = 5$.

c. $|-3| > |-2|$ since $3 > 2$.

d. $|5| < |6|$ since $5 < 6$.

e. $|-7| > |6|$ since $7 > 6$.

EXERCISE SET 1.1

Insert <, >, or = in the space between the paired numbers to make each statement true. See Example 1.

1. $4 < 10$

2. $8 > 5$

3. $7 > 3$

4. $9 < 15$

5. $6.26 = 6.26$

6. $2.13 > 1.13$

7. $0 < 7$

8. $20 > 0$

9. The freezing point of water is 32° Fahrenheit. The boiling point of water is 212° Fahrenheit. Write an inequality statement using < or > comparing the numbers 32 and 212. $32 < 212$

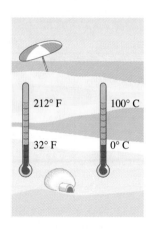

10. The freezing point of water is 0° Celsius. The boiling point of water is 100° Celsius. Write an inequality statement using < or > comparing the numbers 0 and 100. $0 < 100$

Are the following statements true or false? See Example 2.

11. $11 \leq 11$ true

12. $4 \geq 7$ false

13. $10 > 11$ false

14. $17 > 16$ true

15. $3 + 8 \geq 3(8)$ false

16. $8 \cdot 8 \leq 8 \cdot 7$ false

17. $7 > 0$ true

18. $4 < 7$ true

19. An angle measuring 30° is shown and an angle measuring 45° is shown. Use the inequality symbol $\leq$ or $\geq$ to write a statement comparing the numbers 30 and 45. $30 \leq 45$

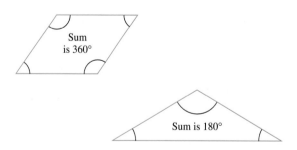

20. The sum of the measures of the angles of a triangle is 180°. The sum of the measures of the angles of a parallelogram is 360°. Use the inequality symbol $\leq$ or $\geq$ to write a statement comparing the numbers 360 and 180. $360 \geq 180$

Sum is 360°

Sum is 180°

Write each sentence as a mathematical statement. See Example 3.

21. Eight is less than twelve. $8 < 12$

22. Fifteen is greater than five. $15 > 5$

23. Five is greater than or equal to four. $5 \geq 4$

24. Negative ten is less than or equal to thirty-seven.

25. Fifteen is not equal to negative two. $15 \neq -2$

26. Negative seven is not equal to seven. $-7 \neq 7$

24. $-10 \leq 37$

The graph below is called a bar graph. This particular graph shows the first three quiz scores for Bill Seggerson in his anatomy class. Each bar represents a different quiz and the height of each bar represents Bill's score for that particular quiz.

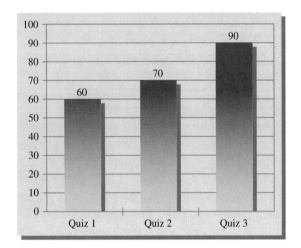

27. What is Bill's highest quiz score?   90

28. What is Bill's lowest quiz score?   60

29. Write an inequality statement using ≤ or ≥ comparing the scores for Quiz 2 and Quiz 3.   70 ≤ 90

30. Do you notice any trends shown by this bar graph?   His quiz scores are improving.

Tell which set or sets each number belongs to: natural numbers, whole numbers, integers, rational numbers, irrational numbers, and real numbers. See Example 4.

31. 0   whole, integers, rational, real

32. $\frac{1}{4}$ rational, real

33. −2   integers, rational, real

34. $-\frac{1}{2}$ rational, real

35. 6

36. 5

37. $\frac{2}{3}$ rational, real

38. $\sqrt{3}$ irrational, real

Tell whether each statement is true or false.

39. Every rational number is also an integer.   false

40. Every negative number is also a rational number.

41. Every natural number is positive.   true

42. Every rational number is also a real number.   true

43. 0 is a real number.   true

35. natural, whole, integers, rational, real

44. Every real number is also a rational number.   false

45. Every whole number is an integer.   true

46. $\frac{1}{2}$ is an integer.   false

Insert <, >, or = in the appropriate space to make a true statement. See Examples 5–7.

47. $-10 > -100$

48. $-200 < -20$

49. $32 > 5.2$

50. $7 > -7$

51. $\frac{18}{3} < \frac{24}{3}$

52. $\frac{8}{2} = \frac{12}{3}$

53. $-51 < -50$

54. $|-20| > -200$

55. $|-5| > -4$

56. $0 = |0|$

57. $|-1| = |1|$

58. $\left|\frac{2}{5}\right| = \left|-\frac{2}{5}\right|$

59. $|-2| < |-3|$

60. $-500 < |-50|$

61. $|0| < |-8|$

62. $|-12| = \frac{24}{2}$

The apparent magnitude of a star is the measure of its brightness as seen by someone on Earth. The smaller the apparent magnitude, the brighter the star. Below, the apparent magnitudes of some stars are listed.

STAR	APPARENT MAGNITUDE	STAR	APPARENT MAGNITUDE
Alpha Centauri	0	Spica	0.98
Sirius	−1.46	Rigel	0.12
Vega	0.03	Regulus	1.35
Antares	0.96	Canopus	−0.72
Sun	−26.7	Hadar	0.61

63. The apparent magnitude of the sun is −26.7. The apparent magnitude of the star Alpha Centauri is 0. Write an inequality statement comparing the numbers 0 and −26.7.   0 > −26.7

64. The apparent magnitude of Antares is 0.96. The apparent magnitude of Spica is 0.98. Write an inequality statement comparing the numbers 0.96 and 0.98.   0.96 < 0.98

65. Which is brighter, the sun or Alpha Centauri?   sun

66. Which is dimmer, Antares or Spica?   Spica

67. Which star listed is the brightest?   sun

68. Which star listed is the dimmest?   Regulus

36. natural, whole, integers, rational, real

40. false

Tell whether each statement is true or false.

69. $5 < 6$ true

70. $7 > 8$ false

71. $-5 < -6$ false

72. $-7 > -8$ true

73. $|-5| < |-6|$ true

74. $|-7| > |-8|$ false

75. $|-5| \geq |5|$ true

76. $|-3| < |0|$ false

77. $-3 > 2$ false

78. $-5 < 5$ true

79. $|8| = |-8|$ true

80. $|9| = |-9|$ true

81. $|0| > |-4|$ false

82. $|0| \leq |0|$ true

Rewrite the following inequalities so that the inequality symbol points in the opposite direction and the resulting statement has the same meaning as the given one.

83. $25 \geq 20$ $20 \leq 25$ **84.** $-13 \leq 13$ $13 \geq -13$

85. $0 < 6$ $6 > 0$ **86.** $5 > 3$ $3 < 5$

87. $-10 > -12$ $-12 < -10$ **88.** $-4 < -2$ $-2 > -4$

89. In your own words, explain how to find the absolute value of a number.   answers may vary

90. Give an example of a real-life situation that can be described with integers but not with whole numbers.   answers may vary

1.2   FRACTIONS

TAPE BA 1.2

OBJECTIVES

1 Write fractions in simplest form.

2 Multiply and divide fractions.

3 Add and subtract fractions.

A quotient of two integers such as $\frac{2}{9}$ is called a **fraction.** In the fraction $\frac{2}{9}$, the top number, 2, is called the **numerator** and the bottom number, 9, is called the **denominator.**

A fraction may be used to refer to part of a whole. For example, $\frac{2}{9}$ of the circle below is shaded. The denominator 9 tells us how many equal parts the whole circle is divided into and the numerator 2 tells us how many equal parts are shaded.

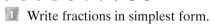

$\frac{2}{9}$ of the circle
is shaded.

To simplify fractions, we can factor the numerator and the denominator. In the statement $3 \cdot 5 = 15$, 3 and 5 are called **factors** and 15 is the **product.** (The raised dot symbol indicates multiplication.)

$$3 \quad \cdot \quad 5 \quad = \quad 15$$
$$\uparrow \qquad \uparrow \qquad \quad \uparrow$$
$$\text{factor} \quad \text{factor} \quad \text{product}$$

To **factor** 15 means to write it as a product. The number 15 can be factored as $3 \cdot 5$ or as $1 \cdot 15$.

A fraction is said to be **simplified** or in **lowest terms** when the numerator and the denominator have no factors in common other than 1. For example, the fraction $\frac{5}{11}$ is in lowest terms since 5 and 11 have no common factors other than 1.

To help us simplify fractions, we write the numerator and the denominator as a product of **prime numbers.** A prime number is a whole number, other than 1, whose only factors are 1 and itself. The first few prime numbers are

2, 3, 5, 7, 11, 13, 17, 19, 23, 29, and so on.

EXAMPLE 1   Write each of the following numbers as a product of primes.

a. 40   **b.** 63

Solution: **a.** First, write 40 as the product of any two whole numbers.

$$40 = 4 \cdot 10$$

Next, factor each of these numbers. Continue this process until all of the factors are prime numbers.

$$40 = 4 \cdot 10$$
$$= 2 \cdot 2 \cdot 2 \cdot 5$$

All the factors are now prime numbers. Then 40 written as a product of primes is

$$40 = 2 \cdot 2 \cdot 2 \cdot 5$$

b. $63 = 9 \cdot 7$
$$= 3 \cdot 3 \cdot 7$$

To use prime factors to write a fraction in lowest terms, apply the fundamental principle of fractions.

FUNDAMENTAL PRINCIPLE OF FRACTIONS

If $\frac{a}{b}$ is a fraction and c is a nonzero real number, then

$$\frac{a \cdot c}{b \cdot c} = \frac{a}{b}$$

EXAMPLE 2   Write each fraction in lowest terms.

a. $\frac{42}{49}$ **b.** $\frac{11}{27}$ **c.** $\frac{88}{20}$

Solution: **a.** Write the numerator and the denominator as products of primes; then apply the fundamental principle to the common factor 7.

$$\frac{42}{49} = \frac{2 \cdot 3 \cdot \boxed{7}}{7 \cdot \boxed{7}} = \frac{2 \cdot 3}{7} = \frac{6}{7}$$

b. $\dfrac{11}{27} = \dfrac{11}{3 \cdot 3 \cdot 3}$

There are no common factors other than 1, so $\frac{11}{27}$ is already in lowest terms.

c. $\dfrac{88}{20} = \dfrac{\boxed{2} \cdot \boxed{2} \cdot 2 \cdot 11}{\boxed{2} \cdot \boxed{2} \cdot 5} = \dfrac{22}{5}$

2   To multiply two fractions, multiply numerator times numerator to obtain the numerator of the product; multiply denominator times denominator to obtain the denominator of the product.

MULTIPLYING FRACTIONS

$$\frac{a}{b} \cdot \frac{c}{d} = \frac{a \cdot c}{b \cdot d}, \qquad \text{if } b \neq 0 \text{ and } d \neq 0$$

EXAMPLE 3   Find the product of $\dfrac{2}{15}$ and $\dfrac{5}{13}$. Write the product in lowest terms.

Solution: $\dfrac{2}{15} \cdot \dfrac{5}{13} = \dfrac{2 \cdot 5}{15 \cdot 13}$ Multiply numerators.
                                   Multiply denominators.

Next, simplify the product by dividing the numerator and the denominator by any common factors.

$$= \frac{2 \cdot \boxed{5}}{3 \cdot \boxed{5} \cdot 13}$$

$$= \frac{2}{39}$$

Before dividing fractions, we first define **reciprocals.** Two fractions are reciprocals of each other if their product is 1. For example $\frac{2}{3}$ and $\frac{3}{2}$ are reciprocals since $\frac{2}{3} \cdot \frac{3}{2} = 1$. Also, the reciprocal of 5 is $\frac{1}{5}$ since $5 \cdot \frac{1}{5} = \frac{5}{1} \cdot \frac{1}{5} = 1$.

To divide fractions, multiply the first fraction by the reciprocal of the second fraction.

DIVIDING FRACTIONS

$$\frac{a}{b} \div \frac{c}{d} = \frac{a}{b} \cdot \frac{d}{c}, \qquad \text{if } b \neq 0, d \neq 0, \text{ and } c \neq 0$$

EXAMPLE 4   Find each quotient. Write all answers in lowest terms.

a. $\dfrac{4}{5} \div \dfrac{5}{16}$ 　　　　　　**b.** $\dfrac{7}{10} \div 14$ 　　　　　**c.** $\dfrac{3}{8} \div \dfrac{3}{10}$

Solution:　**a.** $\dfrac{4}{5} \div \dfrac{5}{16} = \dfrac{4}{5} \cdot \dfrac{16}{5} = \dfrac{4 \cdot 16}{5 \cdot 5} = \dfrac{64}{25}$

b. $\dfrac{7}{10} \div 14 = \dfrac{7}{10} \div \dfrac{14}{1} = \dfrac{7}{10} \cdot \dfrac{1}{14} = \dfrac{7 \cdot 1}{2 \cdot 5 \cdot 2 \cdot 7} = \dfrac{1}{20}.$

c. $\dfrac{3}{8} \div \dfrac{3}{10} = \dfrac{3}{8} \cdot \dfrac{10}{3} = \dfrac{3 \cdot 2 \cdot 5}{2 \cdot 2 \cdot 2 \cdot 3} = \dfrac{5}{4}$

3   To add or subtract fractions with the same denominator, combine numerators and place the sum or difference over the common denominator.

ADDING AND SUBTRACTING FRACTIONS WITH THE SAME DENOMINATOR

$$\frac{a}{b} + \frac{c}{b} = \frac{a + c}{b}, \qquad \text{if } b \neq 0$$

$$\frac{a}{b} - \frac{c}{b} = \frac{a - c}{b}, \qquad \text{if } b \neq 0$$

EXAMPLE 5   Add or subtract as indicated. Write each result in lowest terms.

a. $\dfrac{2}{7} + \dfrac{4}{7}$ 　　　**b.** $\dfrac{3}{10} + \dfrac{2}{10}$ 　　　**c.** $\dfrac{9}{7} - \dfrac{2}{7}$ 　　　**d.** $\dfrac{5}{3} - \dfrac{1}{3}$

Solution:　**a.** $\dfrac{2}{7} + \dfrac{4}{7} = \dfrac{2 + 4}{7} = \dfrac{6}{7}$

b. $\dfrac{3}{10} + \dfrac{2}{10} = \dfrac{3 + 2}{10} = \dfrac{5}{10} = \dfrac{5}{2 \cdot 5} = \dfrac{1}{2}$

c. $\dfrac{9}{7} - \dfrac{2}{7} = \dfrac{9 - 2}{7} = \dfrac{7}{7} = 1$

d. $\dfrac{5}{3} - \dfrac{1}{3} = \dfrac{5 - 1}{3} = \dfrac{4}{3}$

To add or subtract fractions without the same denominator, first write the fractions as **equivalent fractions** with a common denominator. Equivalent fractions are fractions that represent the same quantity. For example, $\frac{3}{4}$ and $\frac{12}{16}$ are equivalent

Whole

$$\frac{3}{4} = \frac{12}{16}$$

fractions since they represent the same portion of a whole, as the diagram shows. Count the larger squares and the shaded portion is $\frac{3}{4}$. Count the smaller squares and the shaded portion is $\frac{12}{16}$. Thus, $\frac{3}{4} = \frac{12}{16}$.

We can write equivalent fractions by multiplying a given fraction by 1, as shown in the next example. Multiplying a number by 1 does not change the value of the number.

EXAMPLE 6   Write $\frac{2}{5}$ as an equivalent fraction with a denominator of 20.

Solution: Since $5 \cdot 4 = 20$, multiply the fraction by $\frac{4}{4}$. Multiplying by $\frac{4}{4} = 1$ does not change the value of the fraction.

$$\frac{2}{5} = \frac{2}{5} \cdot \frac{4}{4} = \frac{2 \cdot 4}{5 \cdot 4} = \frac{8}{20}$$

EXAMPLE 7   Add or subtract as indicated. Write each answer in lowest terms.

a. $\frac{2}{5} + \frac{1}{4}$ **b.** $\frac{1}{2} + \frac{17}{22} - \frac{2}{11}$ **c.** $3\frac{1}{6} - 1\frac{11}{12}$

Solution: **a.** Fractions must have a common denominator before they can be added or subtracted. Since 20 is the smallest number that both 5 and 4 divide into evenly, 20 is the **least common denominator.** Write both fractions as equivalent fractions with denominators of 20. Since

$$\frac{2}{5} \cdot \frac{4}{4} = \frac{2 \cdot 4}{5 \cdot 4} = \frac{8}{20} \quad \text{and} \quad \frac{1}{4} \cdot \frac{5}{5} = \frac{1 \cdot 5}{4 \cdot 5} = \frac{5}{20}$$

then

$$\frac{2}{5} + \frac{1}{4} = \frac{8}{20} + \frac{5}{20} = \frac{13}{20}$$

b. The least common denominator for denominators 2, 22, and 11 is 22. First, write each fraction as an equivalent fraction with a denominator of 22. Then add or subtract from left to right.

$$\frac{1}{2} = \frac{1}{2} \cdot \frac{11}{11} = \frac{11}{22}, \qquad \frac{17}{22} = \frac{17}{22}, \qquad \text{and} \qquad \frac{2}{11} = \frac{2}{11} \cdot \frac{2}{2} = \frac{4}{22}$$

Then

$$\frac{1}{2} + \frac{17}{22} - \frac{2}{11} = \frac{11}{22} + \frac{17}{22} - \frac{4}{22} = \frac{24}{22} = \frac{12}{11}$$

c. To find $3\frac{1}{6} - 1\frac{11}{12}$, first rewrite each mixed number as follows:

$$3\frac{1}{6} = 3 + \frac{1}{6} = \frac{18}{6} + \frac{1}{6} = \frac{19}{6}$$

$$1\frac{11}{12} = 1 + \frac{11}{12} = \frac{12}{12} + \frac{11}{12} = \frac{23}{12}$$

Then

$$3\frac{1}{6} - 1\frac{11}{12} = \frac{19}{6} - \frac{23}{12} = \frac{38}{12} - \frac{23}{12} = \frac{15}{12} = \frac{5}{4}$$

EXERCISE SET 1.2

Represent the shaded part of each geometric figure by a fraction.

1. $\frac{3}{8}$

2. $\frac{1}{4}$

3. $\frac{5}{7}$

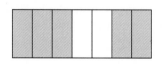

4. $\frac{2}{5}$

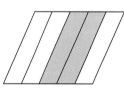

Write each of the following numbers as a product of primes. See Example 1.

5. 20 $2 \cdot 2 \cdot 5$
6. 56 $2 \cdot 2 \cdot 2 \cdot 7$
7. 75 $3 \cdot 5 \cdot 5$
8. 32 $2 \cdot 2 \cdot 2 \cdot 2 \cdot 2$
9. 45 $3 \cdot 3 \cdot 5$
10. 24 $2 \cdot 2 \cdot 2 \cdot 3$

Write the following fractions in lowest terms. See Example 2.

11. $\frac{2}{4}$ $\frac{1}{2}$
12. $\frac{3}{6}$ $\frac{1}{2}$
13. $\frac{10}{15}$ $\frac{2}{3}$
14. $\frac{15}{20}$ $\frac{3}{4}$
15. $\frac{3}{7}$ $\frac{3}{7}$
16. $\frac{5}{9}$ $\frac{5}{9}$

17. $\frac{18}{30}$ $\frac{3}{5}$
18. $\frac{42}{45}$ $\frac{14}{15}$

Multiply or divide as indicated. Write the answer in lowest terms. See Examples 3 and 4.

19. $\frac{1}{2} \cdot \frac{3}{4}$ $\frac{3}{8}$
20. $\frac{10}{6} \cdot \frac{3}{5}$ 1
21. $\frac{2}{3} \cdot \frac{3}{4}$ $\frac{1}{2}$
22. $\frac{7}{8} \cdot \frac{3}{21}$ $\frac{1}{8}$
23. $\frac{1}{2} \div \frac{7}{12}$ $\frac{6}{7}$
24. $\frac{7}{12} \div \frac{1}{2}$ $\frac{7}{6}$
25. $\frac{3}{4} \div \frac{1}{20}$ 15
26. $\frac{3}{5} \div \frac{9}{10}$ $\frac{2}{3}$
27. $\frac{7}{10} \cdot \frac{5}{21}$ $\frac{1}{6}$
28. $\frac{3}{35} \cdot \frac{10}{63}$ $\frac{2}{147}$
29. $2\frac{7}{9} \cdot \frac{1}{3}$ $\frac{25}{27}$
30. $\frac{1}{4} \cdot 5\frac{5}{6}$ $\frac{35}{24}$

The area of a plane figure is a measure of the amount of surface of the figure. Find the area of each figure below. (The area of a rectangle is the product of its length and width. The area of a triangle is $\frac{1}{2}$ the product of its base and height.)

31.

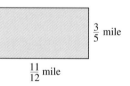

$\frac{3}{5}$ mile

$\frac{11}{12}$ mile

32.

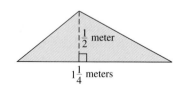

$\frac{1}{2}$ meter

$1\frac{1}{4}$ meters

31. $\frac{11}{20}$ sq. mile

32. $\frac{5}{16}$ sq. meter

Add or subtract as indicated. Write the answer in lowest terms. See Example 5.

33. $\dfrac{4}{5} - \dfrac{1}{5}$ $\dfrac{3}{5}$

34. $\dfrac{6}{7} - \dfrac{1}{7}$ $\dfrac{5}{7}$

35. $\dfrac{4}{5} + \dfrac{1}{5}$ 1

36. $\dfrac{6}{7} + \dfrac{1}{7}$ 1

37. $\dfrac{17}{21} - \dfrac{10}{21}$ $\dfrac{1}{3}$

38. $\dfrac{18}{35} - \dfrac{11}{35}$ $\dfrac{7}{35}$

39. $\dfrac{23}{105} + \dfrac{4}{105}$ $\dfrac{9}{35}$

40. $\dfrac{13}{132} + \dfrac{35}{132}$ $\dfrac{4}{11}$

Write each of the following fractions as an equivalent fraction with the given denominator. See Example 6.

41. $\dfrac{7}{10}$ with a denominator of 30 $\dfrac{21}{30}$

42. $\dfrac{2}{3}$ with a denominator of 9 $\dfrac{6}{9}$

43. $\dfrac{2}{9}$ with a denominator of 18 $\dfrac{4}{18}$

44. $\dfrac{8}{7}$ with a denominator of 56 $\dfrac{64}{56}$

45. $\dfrac{4}{5}$ with a denominator of 20 $\dfrac{16}{20}$

46. $\dfrac{4}{5}$ with a denominator of 25 $\dfrac{20}{25}$

Add or subtract as indicated. Write the answer in lowest terms. See Example 7.

47. $\dfrac{2}{3} + \dfrac{3}{7}$ $\dfrac{23}{21}$

48. $\dfrac{3}{4} + \dfrac{1}{6}$ $\dfrac{11}{12}$

49. $2\dfrac{13}{15} - 1\dfrac{1}{5}$ $1\dfrac{2}{3}$

50. $5\dfrac{2}{9} - 3\dfrac{1}{6}$ $2\dfrac{1}{18}$

51. $\dfrac{5}{22} - \dfrac{5}{33}$ $\dfrac{5}{66}$

52. $\dfrac{7}{10} - \dfrac{8}{15}$ $\dfrac{1}{6}$

53. $\dfrac{12}{5} - 1$ $\dfrac{7}{5}$

54. $2 - \dfrac{3}{8}$ $\dfrac{13}{8}$

Each circle below represents a whole, or 1. Determine the unknown part of the circle.

55. $\dfrac{2}{10}$

56. $\dfrac{6}{11}$

57. $\dfrac{3}{8}$

58. 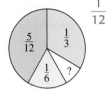 $\dfrac{1}{12}$

Perform the following operations. Write answers in lowest terms.

59. $\dfrac{10}{21} + \dfrac{5}{21}$ $\dfrac{5}{7}$

60. $\dfrac{11}{35} + \dfrac{3}{35}$ $\dfrac{2}{5}$

61. $\dfrac{10}{3} - \dfrac{5}{21}$ $\dfrac{65}{21}$

62. $\dfrac{11}{7} - \dfrac{3}{35}$ $\dfrac{52}{35}$

63. $\dfrac{2}{3} \cdot \dfrac{3}{5}$ $\dfrac{2}{5}$

64. $\dfrac{2}{3} \div \dfrac{3}{4}$ $\dfrac{8}{9}$

65. $\dfrac{3}{4} \div \dfrac{7}{12}$ $\dfrac{9}{7}$

66. $\dfrac{3}{5} + \dfrac{2}{3}$ $\dfrac{19}{15}$

67. $\dfrac{5}{12} + \dfrac{4}{12}$ $\dfrac{3}{4}$

68. $\dfrac{2}{7} + \dfrac{4}{7}$ $\dfrac{6}{7}$

69. $5 + \dfrac{2}{3}$ $\dfrac{17}{3}$

70. $7 + \dfrac{1}{10}$ $\dfrac{71}{10}$

71. $\dfrac{7}{8} \div 3\dfrac{1}{4}$ $\dfrac{7}{26}$

72. $3 \div \dfrac{3}{4}$ 4

73. $\dfrac{7}{18} \div \dfrac{14}{36}$ 1

74. $4\dfrac{3}{7} \div \dfrac{31}{7}$ 1

75. $\dfrac{23}{105} - \dfrac{2}{105}$ $\dfrac{1}{5}$

76. $\dfrac{57}{132} - \dfrac{13}{132}$ $\dfrac{1}{3}$

77. $1\dfrac{1}{2} + 3\dfrac{2}{3}$ $\dfrac{31}{6}$

78. $2\dfrac{3}{5} - 4\dfrac{7}{10}$ $\dfrac{73}{10}$

79. $\dfrac{2}{3} - \dfrac{5}{9} + \dfrac{5}{6}$ $\dfrac{17}{18}$

80. $\dfrac{8}{11} - \dfrac{1}{4} + \dfrac{1}{2}$ $\dfrac{43}{44}$

The perimeter of a plane figure is the total distance around the figure. Find the perimeter of each figure below.

❏ **81.**

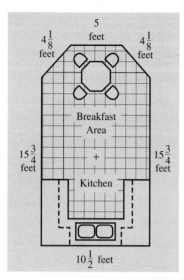

$55\frac{1}{4}$ feet

❏ **82.**

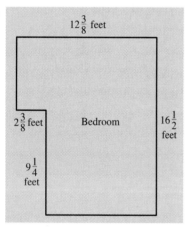

$57\frac{3}{4}$ feet

83. One of the New Orleans Saints guards was told to lose 20 pounds during the summer. So far, the guard has lost $11\frac{1}{2}$ pounds. How many more pounds must he lose? $8\frac{1}{2}$ lbs.

84. In an English class at Cartez College, Stuart was told to read a 340-page book. So far, he has read $\frac{2}{5}$ of it. How many pages does he *have left to read*?

❏ **85.** In your own words, explain how to add two fractions with different denominators.

❏ **86.** In your own words, explain how to multiply two fractions.   answers may vary

The following trail chart is given to visitors at the Lakeview Forest Preserve.

TRAIL NAME	DISTANCE (MILES)
Robin Path	$3\frac{1}{2}$
Red Falls	$5\frac{1}{2}$
Green Way	$2\frac{1}{8}$
Autumn Walk	$1\frac{3}{4}$

87. How much longer is Red Falls Trail than Green Way Trail?

88. Find the total distance traveled by someone who hiked along all four trails.

Most of the water on Earth is in the form of oceans. Only a small part is fresh water. The graph below is called a circle graph or pie chart. This particular circle graph shows the distribution of fresh water.

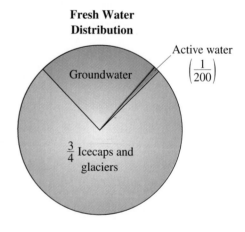

❏ **89.** What fractional part of fresh water is icecaps and glaciers?

❏ **90.** What fractional part of fresh water is active water?

❏ **91.** What fractional part of fresh water is groundwater?

❏ **92.** What fractional part of fresh water is groundwater or icecaps and glaciers?

84. 204 pages   **85.** answers may vary   **87.** $3\frac{3}{8}$ miles   **88.** $12\frac{7}{8}$ miles   **89.** $\frac{3}{4}$ **90.** $\frac{1}{200}$ **91.** $\frac{49}{200}$
92. $\frac{199}{200}$

1.3 | EXPONENTS AND ORDER OF OPERATIONS

O B J E C T I V E S

1. Use and simplify exponential expressions.
2. Define and use the order of operations.
3. Translate word statements to symbols.

TAPE BA 1.3

1

Frequently in algebra, products occur that contain repeated multiplication of the same factor. For example, the volume of a cube whose sides each measure 2 centimeters is $(2 \cdot 2 \cdot 2)$ cubic centimeters. We may use **exponential notation** to write such products in a more compact form. For example,

$$2 \cdot 2 \cdot 2 \quad \text{may be written as} \quad 2^3.$$

The 2 in 2^3 is called the **base;** it is the repeated factor. The 3 in 2^3 is called the **exponent** and is the number of times the base is used as a factor. The expression 2^3 is called an **exponential expression.**

Volume is $(2 \cdot 2 \cdot 2)$ cubic centimeters.

$$\overset{\text{exponent}}{2^3} = 2 \cdot 2 \cdot 2 = 8$$
$$\text{base} \longrightarrow 2 \text{ is a factor } 3 \text{ times}$$

EXAMPLE 1   Evaluate the following:

a. 3^2 [read as "3 squared" or as "3 to the second power"]
b. 5^3 [read as "5 cubed" or as "5 to the third power"]
c. 2^4 [read as "2 to the fourth power"]
d. 7^1 **e.** $\left(\dfrac{3}{7}\right)^2$

Solution: **a.** $3^2 = 3 \cdot 3 = 9$
b. $5^3 = 5 \cdot 5 \cdot 5 = 125$
c. $2^4 = 2 \cdot 2 \cdot 2 \cdot 2 = 16$
d. $7^1 = 7$
e. $\left(\dfrac{3}{7}\right)^2 = \left(\dfrac{3}{7}\right)\left(\dfrac{3}{7}\right) = \dfrac{9}{49}$

> R E M I N D E R $2^3 \neq 2 \cdot 3$ since 2^3 indicates repeated **multiplication** of the same factor.
> $2^3 = 2 \cdot 2 \cdot 2 = 8$, whereas $2 \cdot 3 = 6$.

2

Using symbols for mathematical operations is a great convenience. The more operation symbols presented in an expression, the more careful we must be when performing the indicated operation. For example, in the expression $2 + 3 \cdot 7$, do we add first or multiply first? To eliminate confusion, **grouping symbols** are used.

Examples of grouping symbols are parentheses ( ), brackets [ ], braces { }, and the fraction bar. If we wish $2 + 3 \cdot 7$ to be simplified by adding first, enclose $2 + 3$ in parentheses.

$$(2 + 3) \cdot 7 = 5 \cdot 7 = 35$$

If we wish to multiply first, $3 \cdot 7$ may be enclosed in parentheses.

$$2 + (3 \cdot 7) = 2 + 21 = 23$$

To eliminate confusion when no grouping symbols are present, use the following agreed upon order of operations.

ORDER OF OPERATIONS

Simplify expressions using the following order. If grouping symbols such as parentheses are present, simplify expressions within those first, starting with the innermost set. If fraction bars are present, simplify the numerator and the denominator separately.

1. Evaluate exponential expressions.

2. Perform multiplications or divisions in order from left to right.

3. Perform additions or subtractions in order from left to right.

Now simplify $2 + 3 \cdot 7$. There are no grouping symbols and no exponents, so we multiply and then add.

$$2 + 3 \cdot 7 = 2 + 21 \qquad \text{Multiply.}$$
$$= 23 \qquad \text{Add.}$$

EXAMPLE 2 Simplify each expression.

a. $6 \div 3 + 5^2$ **b.** $\dfrac{2(12 + 3)}{|-15|}$ **c.** $3 \cdot 10 - 7 \div 7$ **d.** $3 \cdot 4^2$ **e.** $\dfrac{3}{2} \cdot \dfrac{1}{2} - \dfrac{1}{2}$

Solution: **a.** Evaluate 5^2 first.

$$6 \div 3 + 5^2 = 6 \div 3 + 25$$

Next divide, then add.

$$6 \div 3 + 25 = 2 + 25 \qquad \text{Divide.}$$
$$= 27 \qquad \text{Add.}$$

b. First, simplify the numerator and the denominator separately.

$$\frac{2(12 + 3)}{|-15|} = \frac{2(15)}{15} \qquad \text{Simplify numerator and denominator separately.}$$

$$= \frac{30}{15}$$

$$= 2 \qquad \text{Simplify.}$$

c. Multiply and divide from left to right. Then subtract.

$$3 \cdot 10 - 7 \div 7 = 30 - 1$$
$$= 29 \qquad \text{Subtract.}$$

d. In this example, only the 4 is squared. The factor of 3 is not part of the base because no grouping symbol includes it as part of the base.

$$3 \cdot 4^2 = 3 \cdot 16 \qquad \text{Evaluate the exponential expression.}$$
$$= 48 \qquad \text{Multiply.}$$

e. The order of operations applies to operations with fractions in exactly the same way as it applies to operations with whole numbers.

$$\frac{3}{2} \cdot \frac{1}{2} - \frac{1}{2} = \frac{3}{4} - \frac{1}{2} \qquad \text{Multiply.}$$
$$= \frac{3}{4} - \frac{2}{4} \qquad \text{The least common denominator is 4.}$$
$$= \frac{1}{4} \qquad \text{Subtract.}$$

R E M I N D E R   Be careful when evaluating an exponential expression. In $3 \cdot 4^2$, the exponent 2 applies only to the base 4. In $(3 \cdot 4)^2$, we multiply first because of parentheses, so the exponent 2 applies to the product $3 \cdot 4$.

$$3 \cdot 4^2 = 3 \cdot 16 = 48 \qquad (3 \cdot 4)^2 = (12)^2 = 144$$

Expressions that include many grouping symbols can be confusing. When simplifying these expressions, keep in mind that grouping symbols separate the expression into distinct parts. Each is then simplified separately.

EXAMPLE 3   Simplify $\dfrac{3 + |4 - 3| + 2^2}{6 - 3}$.

Solution: The fraction bar serves as a grouping symbol and separates the numerator and denominator. Simplify each separately. Also, the absolute value bars here serve as a grouping symbol. We begin in the numerator by simplifying within the absolute value bars.

$$\frac{3 + |4 - 3| + 2^2}{6 - 3} = \frac{3 + |1| + 2^2}{6 - 3} \qquad \text{Simplify the expression inside the absolute value bars.}$$
$$= \frac{3 + 1 + 2^2}{3} \qquad \text{Find the absolute value and simplify the denominator.}$$
$$= \frac{3 + 1 + 4}{3} \qquad \text{Evaluate the exponential expression.}$$
$$= \frac{8}{3} \qquad \text{Simplify the numerator.}$$

EXAMPLE 4   Simplify $3[4(5 + 2) - 10]$.

Solution: Notice that both parentheses and brackets are used as grouping symbols. Start with the innermost set of grouping symbols.

$$3[4(5 + 2) - 10] = 3[4(7) - 10]$$ Simplify the expression in parentheses.
$$= 3[28 - 10]$$ Multiply 4 and 7.
$$= 3[18]$$ Subtract inside the brackets.
$$= 54$$ Multiply.

EXAMPLE 5   Simplify $\dfrac{8 + 2 \cdot 3}{2^2 - 1}$.

Solution: $\dfrac{8 + 2 \cdot 3}{2^2 - 1} = \dfrac{8 + 6}{4 - 1} = \dfrac{14}{3}$

3   Oftentimes solving problems involves the ability to translate word phrases and sentences into symbols. Below is a list of key words and phrases to help us translate.

ADDITION (+)	SUBTRACTION (−)	MULTIPLICATION (·)	DIVISION (÷)	EQUALITY (=)
sum	difference of	product	quotient	equals
plus	minus	times	divide	gives
added to	subtracted from	multiply	into	is/was
more than	less than	twice	ratio	yields
increased by	decreased by	of	divided by	amounts to
total	less			represents
				is the same as

EXAMPLE 6   Translate each word statement into symbols.

   a. Seventeen plus nine is twenty-six.

   b. Fourteen decreased by four equals ten.

   c. The product of two and three amounts to thirty-six divided by six.

Solution: **a.** In words:

Seventeen	plus	nine	is	twenty-six
Translate: 17	+	9	=	26

   b. In words:

Fourteen	decreased by	four	equals	ten
Translate: 14	−	4	=	10

   c. In words:

The product of two and three	amounts to	thirty-six	divided by	six
Translate: $2 \cdot 3$	=	36	÷	6

SCIENTIFIC CALCULATOR EXPLORATIONS

EXPONENTS

To evaluate exponential expressions on a scientific calculator, find the key marked $\boxed{y^x}$.
To evaluate, for example, 3^5, press the following keys: $\boxed{3}$ $\boxed{y^x}$ $\boxed{5}$ $\boxed{=}$. The display
should read $\boxed{243}$.

ORDER OF OPERATIONS

Some calculators follow the order of operations, and others do not. To see whether or
not your calculator has the order of operations built in, use your calculator to find
$2 + 3 \cdot 4$. To do this, press the following sequence of keys:

$\boxed{2}$ $\boxed{+}$ $\boxed{3}$ $\boxed{\times}$ $\boxed{4}$ $\boxed{=}$.

The correct answer is 14 because the order of operations is to multiply before we
add. If the calculator displays $\boxed{14}$, then it has order of operations built in.
Even if the order of operations is built in, parentheses must sometimes be inserted.
For example, to simplify $\frac{5}{12 - 7}$, press the keys

$\boxed{5}$ $\boxed{\div}$ $\boxed{(}$ $\boxed{1}$ $\boxed{2}$ $\boxed{-}$ $\boxed{7}$ $\boxed{)}$ $\boxed{=}$.

The display should read $\boxed{1}$.

Use a calculator to evaluate each expression.

1. 5^3 125

2. 7^4 2401

3. 9^5 59,049

4. 8^6 262,144

5. $2(20 - 5)$ 30

6. $3(14 - 7) + 21$ 42

7. $24(862 - 455) + 89$ 9857

8. $99 + (401 + 962)$ 1462

9. $\dfrac{4623 + 129}{36 - 34}$ 2376

10. $\dfrac{956 - 452}{89 - 86}$ 168

EXERCISE SET 1.3

Evaluate. See Example 1.

1. 3^5 243

2. 5^3 125

3. 3^3 27

4. 4^4 256

5. 1^5 1

6. 1^8 1

7. 5^1 5

8. 8^1 8

9. $\left(\dfrac{1}{5}\right)^3$ $\dfrac{1}{125}$

10. $\left(\dfrac{6}{11}\right)^2$ $\dfrac{36}{121}$

11. $\left(\dfrac{2}{3}\right)^4$ $\dfrac{16}{81}$

12. $\left(\dfrac{1}{2}\right)^5$ $\dfrac{1}{32}$

13. 7^2 49

14. 9^2 81

15. 2^5 32

16. 2^6 64

17. 0^3 0

18. 0^2 0

19. 4^2 16

20. 2^4 16

21. $(1.2)^2$ 1.44

22. $(0.07)^2$ 0.0049

23. The area of a square whose sides each measure 5
meters is $(5 \cdot 5)$ square meters. Write this area
using exponential notation. 5^2 square meters

24. The volume of a solid is a measure of the space it encloses. The volume of a sphere whose radius is 5 meters is $\left(\frac{4}{3}\pi \cdot 5 \cdot 5 \cdot 5\right)$ cubic meters. Write this volume using exponential notation.

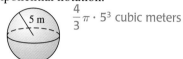

$\frac{4}{3}\pi \cdot 5^3$ cubic meters

Simplify each expression. See Examples 2 through 5.

25. $5 + 6 \cdot 2$ 17
26. $8 + 5 \cdot 3$ 23
27. $4 \cdot 8 - 6 \cdot 2$ 20
28. $12 \cdot 5 - 3 \cdot 6$ 42
29. $2(8 - 3)$ 10
30. $5(6 - 2)$ 20
31. $2 + (5 - 2) + 4^2$ 21
32. $6 - 2 \cdot 2 + 2^5$ 34
33. $5 \cdot 3^2$ 45
34. $2 \cdot 5^2$ 50
35. $\frac{1}{4} \cdot \frac{2}{3} - \frac{1}{6}$ 0
36. $\frac{3}{4} \cdot \frac{1}{2} + \frac{2}{3}$ $\frac{25}{24}$
37. $\frac{6 - 4}{9 - 2}$ $\frac{2}{7}$
38. $\frac{8 - 5}{24 - 20}$ $\frac{3}{4}$
39. $2[5 + 2(8 - 3)]$ 30
40. $3[4 + 3(6 - 4)]$ 30
41. $\frac{3 + 3(5 + 3)}{3^2 + 1}$ $\frac{27}{10}$
42. $\frac{3 + 6(8 - 5)}{4^2 + 2}$ $\frac{7}{6}$
43. $\frac{6 + |8 - 2| + 3^2}{18 - 3}$ $\frac{7}{5}$
44. $\frac{16 + |13 - 5| + 4^2}{17 - 5}$ $\frac{10}{3}$

45. Are parentheses necessary in the expression $2 + (3 \cdot 5)$? Explain your answer.

46. Are parentheses necessary in the expression $(2 + 3) \cdot 5$? Explain your answer.

Translate each word statement to symbols. See Example 6.

47. Ten added to twelve is equal to the product of eleven and two. $10 + 12 = 11 \cdot 2$

48. One hundred is the product of four and the sum of twenty and five. $100 = 4 \cdot (20 + 5)$

49. The sum of five and six is greater than ten.

50. The difference of fifteen and seven is less than nine. $15 - 7 < 9$

51. The product of three and five is greater than twelve. $3 \cdot 5 > 12$ **52.** $4 \div 7 < 30$

52. The quotient of four and seven is less than thirty.

Insert $<$, $>$, or $=$ in the space provided to make the following true.

53. 8^1 $8(1)$ $=$
54. 9^1 $9(1)$ $=$
55. 2^3 $2 \cdot 3$ $>$
56. 2^4 $2 \cdot 4$ $>$
57. $\left(\frac{2}{3}\right)^3$ $\frac{1}{9} + \frac{1}{27}$ $>$
58. $\left(\frac{3}{10}\right)^2$ $\frac{1}{10} \cdot \frac{2}{10}$ $>$

59. $(0.6)^2$ 36 $<$ **60.** $(0.3)^3$ 27 $<$

61. Why is 8^2 usually read as "eight squared"? (*Hint:* What is the area of the **square** below?)   answers may vary

8 inches

62. Why is 4^3 usually read as "four cubed"? (*Hint:* What is the volume of the **cube** below?)   answers may vary

4 cm

Simplify each expression.

63. $\frac{19 - 3 \cdot 5}{6 - 4}$ 2
64. $\frac{4 \cdot 3 + 2}{4 + 3 \cdot 2}$ $\frac{7}{5}$
65. $\frac{|6 - 2| + 3}{8 + 2 \cdot 5}$ $\frac{7}{18}$
66. $\frac{15 - |3 - 1|}{12 - 3 \cdot 2}$ $\frac{13}{6}$
67. $\frac{3(2 + 5)}{6 + 2}$ $\frac{21}{8}$
68. $\frac{4(8 - 3)}{6 + 3}$ $\frac{20}{9}$
69. $5[3(0.2 + 0.1) + 4]$ 24.5   **70.** $4[5(2.1 + 4.3) - 8]$ 96

71. Insert parentheses so that the following expression simplifies to 32.
$20 - 4 \cdot 4 \div 2$ $(20 - 4) \cdot 4 \div 2$

72. Insert parentheses so that the following expression simplifies to 28.
$2 \cdot 5 + 3^2$ $2 \cdot (5 + 3^2)$

Translate each word statement into symbols.

73. Three cubed is the same as the sum of twenty and seven. $3^3 = 20 + 7$

74. Fifteen divided by the sum of two and three is three. $15 \div (2 + 3) = 3$

75. One increased by two equals the quotient of nine and three. $1 + 2 = 9 \div 3$

76. Four subtracted from eight is equal to two squared. $8 - 4 = 2^2$

77. Nine is less than or equal to the product of eleven and two. $9 \le 11 \cdot 2$

78. Eight is greater than one subtracted from seven.

79. Three is not equal to four divided by two.

80. The difference of sixteen and four is greater than ten. $16 - 4 > 10$

45. No; multiplication comes before addition in order of operations.   **49.** $5 + 6 > 10$
46. Yes; if the parentheses weren't there, the multiplication would be done first.   **78.** $8 > 7 - 1$ **79.** $3 \ne 4 \div 2$

1.4 | INTRODUCTION TO VARIABLE EXPRESSIONS AND EQUATIONS

O B J E C T I V E S

TAPE BA 1.4

1. Evaluate algebraic expressions, given replacement values for variables.
2. Translate word phrases into algebraic expressions.
3. Define equation and solution or root of an equation.
4. Translate word statements into equations.

---

1 In algebra, we use symbols, usually letters such as x, y, or z, to represent unknown numbers. A symbol that is used to represent a number is called a **variable.** An **algebraic expression** is a collection of numbers, variables, operation symbols, and grouping symbols. For example,

$$2x, \quad -3, \quad 2x - 10, \quad 5(p^2 + 1), \quad \text{and} \quad \frac{3y^2 - 6y + 1}{5}$$

are algebraic expressions. The expression $2x$ means $2 \cdot x$. Also, $5(p^2 + 1)$ means $5 \cdot (p^2 + 1)$ and $3y^2$ means $3 \cdot y^2$. If we give a specific value to a variable, we can **evaluate an algebraic expression.** To evaluate an algebraic expression means to find its numerical value once we know the value of the variables. Make sure the order of operations is followed when evaluating an expression.

EXAMPLE 1   Evaluate each expression if $x = 3$ and $y = 2$.

a. $2x - y$ **b.** $\dfrac{3x}{2y}$ **c.** $\dfrac{x}{y} + \dfrac{y}{2}$ **d.** $x^2 - y^2$

Solution: **a.** Replace x with 3 and y with 2.

$$
\begin{aligned}
2x - y &= 2(3) - 2 && \text{Let } x = 3 \text{ and } y = 2.\\
&= 6 - 2 && \text{Multiply.}\\
&= 4 && \text{Subtract.}
\end{aligned}
$$

b. $\dfrac{3x}{2y} = \dfrac{3 \cdot 3}{2 \cdot 2} = \dfrac{9}{4}$ Let $x = 3$ and $y = 2$.

c. Replace x with 3 and y with 2. Then simplify.

$$\frac{x}{y} + \frac{y}{2} = \frac{3}{2} + \frac{2}{2} = \frac{5}{2}$$

d. Replace x with 3 and y with 2.

$$x^2 - y^2 = 3^2 - 2^2 = 9 - 4 = 5$$

---

2 Now that we can represent an unknown number by a variable, let's practice translating phrases into algebraic expressions.

EXAMPLE 2   Write an algebraic expression that represents each of the following phrases. Let the variable x represent the unknown number.

    a. The sum of a number and 3

    b. The product of a 3 and a number

    c. Twice a number

    d. 10 decreased by a number

    e. 5 times a number increased by 7

Solution: **a.** $x + 3$ since "sum" means to add.

    b. $3 \cdot x$ and $3x$ are both ways to denote the product of 3 and x.

    c. $2 \cdot x$ or $2x$.

    d. $10 - x$ because "decreased by" means to subtract.

    e. $\underbrace{5x}$ $+ 7$
        5 times
        a number

An **equation** is a mathematical statement that two expressions have equal value. The equal symbol "=" is used to equate the two expressions. For example, $3 + 2 = 5$, $7x = 35$, $\frac{2(x - 1)}{3} = 0$, and $I = PRT$ are all equations.

> R E M I N D E R   An equation contains the equal symbol "=". An algebraic expression does not.

When an equation contains a variable, deciding which values of the variable make an equation a true statement is called **solving** an equation for the variable. A **solution** or **root** of an equation is a value for the variable that makes the equation true. For example, 3 is a solution of the equation $x + 4 = 7$, because if x is replaced by 3 the statement is true.

$$x + 4 = 7$$
$$\downarrow$$
$$3 + 4 = 7 \qquad \text{Replace } x \text{ with 3.}$$
$$7 = 7 \qquad \text{True.}$$

Similarly, 1 is not a solution of the equation $x + 4 = 7$, because $1 + 4 = 7$ is **not** a true statement. The **solution set** of an equation is the set of all solutions of the equation. The solution set of the equation $x + 4 = 7$ is $\{3\}$.

EXAMPLE 3   Decide whether 2 is a solution of $3x + 10 = 8x$.

Solution: Replace x with 2 and see if a true statement results.

$$3x + 10 = 8x \qquad \text{Original equation.}$$
$$3(2) + 10 = 8(2) \qquad \text{Replace } x \text{ with 2.}$$
$$6 + 10 = 16 \qquad \text{Simplify each side.}$$
$$16 = 16 \qquad \text{True.}$$

Since we arrived at a true statement after replacing x with 2 and simplifying both sides of the equation, 2 is a solution of the equation.

 We now practice translating sentences into equations.

EXAMPLE 4 Write each sentence as an equation. Let x represent the unknown number.

a. The quotient of 15 and a number is 4.

b. Three subtracted from 12 is a number.

c. Four times a number added to 17 is 21.

Solution: **a.** In words: | The quotient of 15 and a number | is | 4 |

Translate: $\qquad\qquad \dfrac{15}{x} \qquad\qquad = \quad 4$

b. In words: | Three subtracted **from** 12 | is | a number |

Translate: $\qquad\qquad 12 - 3 \qquad\qquad = \qquad x$

Care must be taken when the operation is subtraction. The expression $3 - 12$ would be incorrect. Notice that $3 - 12 \neq 12 - 3$.

c. In words: | 4 times a number | added to | 17 | is | 21 |

Translate: $\qquad\quad 4x \qquad\qquad\quad + \qquad 17 \ = \ 21$

EXERCISE SET 1.4

Evaluate each expression if x = 1, y = 3, and z = 5. See Example 1.

1. $3x - 2$ 1

2. $6y - 8$ 10

3. $|2x + 3y|$ 11

4. $|5z - 2y|$ 19

5. $xy + z$ 8

6. $yz - x$ 14

7. $5y^2$ 45

8. $2z^2$ 50

9. $|y^2 + z^2|$ 34

10. $|10y - 3z|$ 15

Neglecting air resistance, the expression $16t^2$ gives the distance in feet an object will fall in t seconds.

11. Complete the chart below. To evaluate $16t^2$, remember to first find t^2, then multiply by 16.

TIME t (IN SECONDS)	DISTANCE $16t^2$ (IN FEET)
1	16
2	64
3	144
4	256

12. Does an object fall the same distance *during* each second? Why or why not? (See Exercise 11.)

No, the expression $16t^2$ is not linear.

Write each phrase as an algebraic expression. Let x represent the unknown number. See Example 2.

13. Fifteen more than a number. $x + 15$

14. One-half times a number. $\frac{1}{2}x$

15. Five subtracted from a number. $x - 5$

16. The quotient of a number and 9. $\frac{x}{9}$

17. Three times a number increased by 22. $3x + 22$

18. The product of 8 and a number. $8x$

Decide whether the given number is a solution of the given equation. See Example 3.

19. $3x - 6 = 9$; 5 solution

20. $2x + 7 = 3x$; 6 not a solution

21. $2x + 6 = 5x - 1$; 0 not a solution

22. $4x + 2 = x + 8$; 2 solution

23. $x^2 + 2x + 1 = 0$; 1 not a solution

24. $x^2 + 2x - 35 = 0$; 5 solution

25. $2x - 5 = 5$; 8 not a solution

26. $3x - 10 = 8$: 6 solution

27. $x + 6 = x + 6$; 2 solution

28. $x + 6 = x + 6$; 10 solution

29. $x = 5x + 15$; 0 not a solution

30. $4 = 1 - x$; 1 not a solution

Write each of the following sentences as an equation. Use x to represent the unknown number. See Example 4.

31. The sum of 5 and a number is 20. $5 + x = 20$

32. Twice a number is 17. $2x = 17$

33. Thirteen minus three times a number is 13.

34. Seven subtracted from a number is 0. $x - 7 = 0$

35. The quotient of 12 and a number is $\frac{1}{2}$. $\frac{12}{x} = \frac{1}{2}$

36. The sum of 8 and twice a number is 42.

Evaluate each expression if x = 2, y = 6, and z = 3.

37. $3x + 8y$ 54

38. $|4z - 2y|$ 0

39. $\frac{4x}{3y}$ $\frac{4}{9}$

40. $\frac{6z}{5x}$ $\frac{9}{5}$

33. $13 - 3x = 13$ **36.** $8 + 2x = 42$

41. $\frac{y}{x} + \frac{y}{x}$ 6

42. $\frac{9}{z} + \frac{4z}{y}$ 5

43. $3x^2 + y$ 18

44. $y + 4z^2$ 42

45. $x + yz$ 20

46. $z + xy$ 15

Evaluate each expression if x = 12, y = 8, and z = 4.

47. $\frac{x}{z} + 3y$ 27

48. $\frac{y}{z} + 8x$ 98

49. $3z + z^2$ 28

50. $7y + y^2$ 120

51. $x^2 - 3y + x$ 132

52. $y^2 - 3x + y$ 36

53. $\frac{2x + z}{3y - z}$ $\frac{7}{5}$

54. $\frac{3x - y}{4z + x}$ 1

55. $\frac{x^2 + z}{y^2 + 2z}$ $\frac{37}{18}$

56. $\frac{y^2 + x}{x^2 + 3y}$ $\frac{19}{42}$

57. In your own words, explain the difference between an expression and an equation. answers may vary

58. Determine whether each is an expression or an equation.

    a. $3x^2 - 26$ expression

    b. $3x^2 - 26 = 1$ equation

    c. $2x - 5 = 7x - 5$ equation

    d. $9y + x - 8$ expression

Write each of the following as an algebraic expression or an equation. **61.** $2x - 10 = 18$ **63.** $20 - 30x$

59. A number divided by 13 $\frac{x}{13}$

60. Seven times the sum of a number and 19 $7(x + 19)$

61. Ten subtracted from twice a number is 18.

62. A number subtracted from four times that number is 75.6. $4x - x = 75.6$

63. Twenty less the product of 30 and a number

64. Twelve decreased by a number $12 - x$

65. A number times 0.02 equals 1.76. $0.02x = 1.76$

66. The product of $\frac{3}{4}$ and the sum of a number and 1 equals 9. $\frac{3}{4}(x + 1) = 9$

67. A number subtracted from 19 is three times that number. $19 - x = 3x$

68. Three times a number added to $1\frac{11}{12}$ equals the number increased by 2. $1\frac{11}{12} + 3x = x + 2$

Solve the following.

69. The perimeter of a figure is the distance around the figure. The expression $2l + 2w$ represents the perimeter of a rectangle when l is its length and w is its width. Find the perimeter of the following rectangle by substituting 8 for l and 6 for w.   28 meters

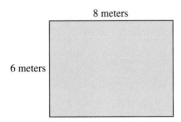

8 meters

6 meters

70. The expression $a + b + c$ represents the perimeter of a triangle when a, b, and c are the lengths of its sides. Find the perimeter of the following triangle.

$\frac{11}{14}$ yd.

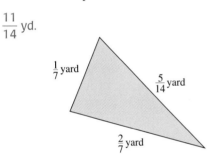

$\frac{1}{7}$ yard

$\frac{5}{14}$ yard

$\frac{2}{7}$ yard

71. The area of a figure is the total enclosed surface of the figure. Area is measured in square units. The expression lw represents the area of a rectangle when l is its length and w is its width. Find the area of the following rectangular-shaped lot.   12,000 sq. ft.

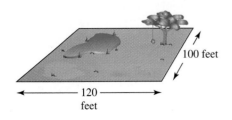

100 feet

120 feet

72. A trapezoid is a four-sided figure with exactly one pair of parallel sides. The expression $\frac{(B + b)h}{2}$ represents its area, when B and b are the lengths of the two parallel sides and h is the height between these sides. Find the area if $B = 15$ inches, $b = 7$ inches, and $h = 5$ inches.   55 sq. in.

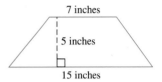

7 inches

5 inches

15 inches

73. The expression $\frac{I}{PT}$ represents the rate of interest being charged if a loan of P dollars for T years required I dollars in interest to be paid. Find the interest rate if a \$650 loan for 3 years to buy a used IBM personal computer requires \$126.75 in interest to be paid.   6.5%

74. The expression $\frac{d}{t}$ represents the average speed r in miles per hour if a distance of d miles is traveled in t hours. Find the rate to the nearest whole number if the distance between Dallas, Texas, and Kaw City, Oklahoma, is 432 miles, and it takes Barbara Goss 8.5 hours to drive the distance.   51 mph

75. Peter Callac earns a base salary plus a commission on all sales he makes at St. Joe Brick Company. The expression $B + RS$ represents his gross income, when B is the base income, R is the commission rate, and S is the amount sold. Find Peter's income if $B = \$300$, $R = 8\%$, $S = \$500$.   \$340

Evaluate each expression if $x = 5$ and $y = 1$.

76. $2(x - 3)^2 + 5(y + 2)^2$ 53

77. $7(x - 4)^2 - 5(y - 1)^2$ 7

78. $\dfrac{5x}{2} + \dfrac{7y}{4}$ $\dfrac{57}{4}$

79. $\dfrac{3xy}{6} - \dfrac{y^3}{12}$ $\dfrac{29}{12}$

80. $\dfrac{4(2x - 3) - 5(y + 2)}{(x + 1)(y + 1)}$ $\dfrac{13}{12}$

81. $\dfrac{3[(x + 1) + (2y + 5)]}{(x + 1)^2}$ $\dfrac{13}{12}$

1.5 ADDING REAL NUMBERS

TAPE BA 1.5

O B J E C T I V E S

1 Add real numbers with the same sign.

2 Add real numbers with unlike signs.

3 Solve problems that involve addition of real numbers.

4 Find the opposite of a number.

1 Real numbers can be added, subtracted, multiplied, divided, and raised to powers, just as whole numbers can. We use the number line to help picture the addition of real numbers.

On a number line, a positive number can be represented anywhere by an arrow of appropriate length pointing to the right. A negative number can be represented by an arrow pointing to the left. For example, to find the sum $3 + 2$ on a number line, start at 0 and draw an arrow representing 3. This arrow should be three units long pointing to the right, since 3 is positive. From the tip of this arrow, draw another arrow representing 2. The tip of the second arrow ends at their sum, 5.

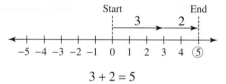

$$3 + 2 = 5$$

To find the sum $-1 + (-2)$, start at 0 and draw an arrow representing -1. This arrow is one unit long pointing to the left. At the tip of this arrow, draw an arrow representing -2. The tip of the second arrow ends at their sum, -3.

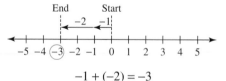

$$-1 + (-2) = -3$$

Using a number line each time we add two numbers can be time consuming. Instead, we can notice patterns in the previous examples and write rules for adding signed numbers. When adding two numbers with the same sign, notice that the sign of the sum is the same as the sign of the addends.

TO ADD TWO NUMBERS WITH THE SAME SIGN

Step 1.   Find the sum of their absolute values.

Step 2.   Use their common sign as the sign of the sum.

Add $(-7) + (-6)$.

Step 1.   Find the sum of their absolute values.

$$|-7| = 7, \quad |-6| = 6, \quad \text{and} \quad 7 + 6 = 13$$

Step 2.   Use their common sign as the sign of the sum. This means that

$$(-7) + (-6) = -13$$

↑— common sign

Thinking of signed numbers as money earned or lost might help make addition more meaningful. Earnings can be thought of as positive numbers. If $1 is earned and later another $3 is earned, the total amount earned is $4. $1 + 3 = 4$.

On the other hand, losses can be thought of as negative numbers. If $1 is lost and later another $3 is lost, a total of $4 is lost. $(-1) + (-3) = -4$.

EXAMPLE 1   Find each sum.

       a. $-3 + (-7)$ **b.** $5 + (+12)$ **c.** $(-1) + (-20)$ **d.** $-2 + (-10)$

Solution: **a.** $-3 + (-7) = -10$ **b.** $5 + (+12) = 17$ **c.** $(-1) + (-20) = -21$

       d. $-2 + (-10) = -12$

Adding numbers whose signs are not the same can be pictured on the number line, also. To find the sum $-4 + 6$, begin at 0 and draw an arrow representing -4. This arrow is 4 units long and pointing to the left. At the tip of this arrow, draw an arrow representing 6. The tip of the second arrow ends at their sum, 2.

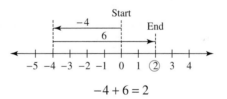

$$-4 + 6 = 2$$

Using temperature as an example, if the thermometer registers 4 degrees below 0 degrees and then rises 6 degrees, the new temperature is 2 degrees above 0 degrees. Thus, it is reasonable that $-4 + 6 = 2$. Likewise, if the temperature is 4 degrees above 0 degrees and then falls 6 degrees, the new temperature is 2 degrees below 0 degrees. On the number line, we see that $4 + (-6) = -2$.

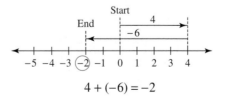

$$4 + (-6) = -2$$

Once again, we can observe a pattern: when adding two numbers with different signs, the sign of the sum is the same as the sign of the addend whose absolute value is larger.

> **TO ADD TWO NUMBERS WITH DIFFERENT SIGNS**
>
> *Step 1.* Find the difference of the larger absolute value and the smaller absolute value.
>
> *Step 2.* Use the sign of the addend whose absolute value is larger as the sign of the sum.

EXAMPLE 2 Find each sum.

a. $3 + (-7)$ **b.** $(-2) + (10)$ **c.** $0.2 + (-0.5)$

Solution: **a.** Since $|-7|$ is larger than $|3|$, their sum has the same sign as -7. That is, the sum is negative. Then $|-7| = 7$, $|3| = 3$, and $7 - 3 = 4$. The sum is -4.

$$3 + (-7) = -4$$

b. Since $|10|$ is larger than $|-2|$, the sign of the sum is the same as the sign of 10, positive. Then $|10| = 10$, $|-2| = 2$, and $10 - 2 = 8$. The sum is $+8$ or 8.

$$(-2) + 10 = 8$$

c. $0.2 + (-0.5) = -0.3$

EXAMPLE 3 Find the sums.

a. $3 + (-7) + (-8)$ **b.** $[7 + (-10)] + [-2 + (-4)]$

Solution: **a.** Perform the additions from left to right.

$$3 + (-7) + (-8) = -4 + (-8) \qquad \text{Adding numbers with different signs.}$$
$$= -12 \qquad \text{Adding numbers with like signs.}$$

b. Simplify inside brackets first.

$$[7 + (-10)] + [-2 + (-4)] = [-3] + [-6]$$
$$= -9 \qquad \text{Add.}$$

 Signed numbers are used in everyday life. Stock market returns show gains and losses as positive and negative numbers. Temperatures in cold climates often dip into the negative range, commonly referred to as "below zero" temperatures. Bank statements report deposits and withdrawals as positive and negative numbers.

EXAMPLE 4 During three days, a share of Lamplighter's International stock recorded the following gains and losses:

Monday	Tuesday	Wednesday
a gain of \$2	a loss of \$1	a loss of \$3

Find the overall gain or loss for the stock for the three days.

Solution: The overall gain or loss is the sum of the gains and losses.

In words: gain plus loss plus loss

Translate: 2   + (-1) + (-3) $= -2$

The overall loss is $2.

4

To help us subtract real numbers in the next section, we first review the concept of opposites. The graph of 4 and -4 is shown on the number line below.

Notice that 4 and -4 lie on opposite sides of 0, and each is 4 units away from 0.

This relationship between -4 and $+4$ is an important one. Such numbers are known as **opposites** or **additive inverses** of each other.

OPPOSITES OR ADDITIVE INVERSES

Two numbers that are the same distance from 0 but lie on opposite sides of 0 are called opposites or additive inverses of each other.

The opposite of 10 is -10.

The opposite of -3 is 3.

The opposite of $\frac{1}{2}$ is $-\frac{1}{2}$.

Notice that the sum of a number and its opposite is 0.

$$10 + (-10) = 0$$
$$-3 + 3 = 0$$
$$\frac{1}{2} + \left(-\frac{1}{2}\right) = 0$$

In general, **if a is a number, we write the opposite or additive inverse of a as $-a$.**

SUM OF OPPOSITES

The sum of a number a and its opposite $-a$ is 0.

$$a + (-a) = 0$$

Since the opposite of a is $-a$, this means that we write the opposite of -3 as $-(-3)$. But we said above that the opposite of -3 is 3. This can be true only if $-(-3) = 3$.

If a is a number, then $-(-a) = a$.

For example, $-(-10) = 10$, $-\left(-\frac{1}{2}\right) = \frac{1}{2}$, and $-(-2x) = 2x$.

EXAMPLE 5   Find the opposite or additive inverse of each number.

        a. 5   **b.** 0   **c.** -6

Solution: **a.** The opposite of 5 is -5. Notice that 5 and -5 are on opposite sides of 0 when plotted on a number line and are equal distances away.

        b. The opposite of 0 is 0 since $0 + 0 = 0$.

        c. The opposite of -6 is 6.

EXAMPLE 6   Simplify the following.

        a. $-(-6)$ **b.** $-|-6|$

Solution: **a.** $-(-6) = 6$

        b. $-|-6| = -6$

EXERCISE SET 1.5

Find the following sums. See Examples 1 through 3.

1. $6 + 3$ 9   **2.** $9 + (-12)$ -3

3. $-6 + (-8)$ -14 **4.** $-6 + (-14)$ -10

5. $-8 + (-7)$ -15 **6.** $6 + (-4)$ 2

7. $-52 + 36$ -16 **8.** $-94 + 27$ -67

9. $6 + (-4) + 9$ 11   **10.** $-18 + |-53|$ 35

11. $2\frac{3}{4} + \left(-\frac{1}{8}\right)$ $2\frac{5}{8}$ **12.** $4\frac{4}{5} + \left(-\frac{3}{10}\right)$ $4\frac{1}{2}$

Find the additive inverse or the opposite. See Example 5.

13. 6 -6 **14.** 4 -4 **15.** -2 2

16. -8 8   **17.** 0   0   **18.** $-\frac{1}{4}$ $\frac{1}{4}$

19. $|-6|$ -6 **20.** $|-11|$ -11

21. In your own words, explain how to find the opposite of a number.   answers may vary

22. In your own words, explain why 0 is the only number that is its own opposite.   answers may vary

Simplify the following. See Example 6.

23. $-|-2|$ -2 **24.** $-(-3)$ 3   **25.** $-|0|$ 0

26. $\left|-\frac{2}{3}\right|$ $\frac{2}{3}$ **27.** $-\left|-\frac{2}{3}\right|$ $-\frac{2}{3}$ **28.** $-(-7)$ 7

Find the sums.   **43.** -59

29. $-3 + (-5)$ -8 **30.** $-7 + (-4)$ -11

31. $-9 + (-3)$ -12 **32.** $+8 + (-6)$ 2

33. $+9 + (-3)$ 6   **34.** $-9 + (+4)$ -5

35. $-7 + (+3)$ -4 **36.** $-5 + (+9)$ 4

37. $-3 + (+10)$ 7   **38.** $3 + (-8)$ -5

39. $8 + 4$ 12   **40.** $12 + (+2)$ 14

41. $-15 + 9 + (-2)$ -8 **42.** $-9 + 15 + (-5)$ 1

43. $-21 + (-16) + (-22)$ **44.** $-14 + |-16|$ 2

45. $|-8| + (-16)$ -8 **46.** $|-6| + (-61)$ -55

47. $-\frac{7}{16} + \frac{1}{4}$ $-\frac{3}{16}$ **48.** $-\frac{5}{9} + \frac{1}{3}$ $-\frac{2}{9}$

49. $-\frac{7}{10} + \left(-\frac{3}{5}\right)$ $-\frac{13}{10}$ **50.** $-33 + (-14)$ -47

51. $27 + (-46)$ -19 **52.** $53 + (-37)$ 16

53. $-18 + 49$ 31   **54.** $-26 + 14$ (-12)

55. $126 + (-67)$ 59   **56.** $-\frac{5}{6} + \left(-\frac{2}{3}\right)$ $-\frac{3}{2}$

57. $6.3 + (-8.4)$ -2.1 **58.** $9.2 + (-11.4)$ -2.2

59. $117 + (-79)$ 38   **60.** $-114 + (-88)$ -202

61. $-214 + (-86)$ -300 **62.** $18 + (-6) + 4$ 16

63. $-23 + 16 + (-2)$ -9 **64.** $-14 + (-3) + 11$ -6

65. $-9.6 + (-3.5)$ -13.1 **66.** $-6.7 + (-7.6)$ -14.3

67. $|5 + (-10)|$ 5 **68.** $[-3 + 5] + (-11)$ -9

69. $[-2 + (-7)] + [-11 + (-4)]$ -24

70. $8 + [-5 + 5] + (-12)$ -4

71. $|7 + (-10)| + |-16|$ 19

72. $\left| -\dfrac{3}{10} + \dfrac{1}{10} \right| + \left| \dfrac{2}{5} - \dfrac{1}{10} \right|$ $\dfrac{1}{2}$

The following bar graph shows the daily low temperatures for a week in Sioux Falls, South Dakota.

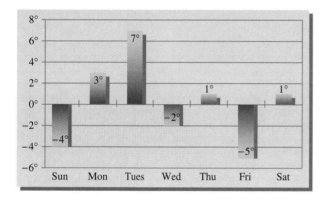

☐ 73. On what day of the week was the graphed temperature the highest?   Tues.

☐ 74. On what day of the week was the graphed temperature the lowest?   Fri.

☐ 75. What is the highest temperature shown on the graph? $7°$

☐ 76. What is the lowest temperature shown on the graph? $-5°$

☐ 77. Find the average daily low temperature for Sunday through Thursday. (*Hint:* To find the average of the five temperatures, find their sum and divide by 5.) $1°$

Solve the following. See Example 4.

78. The low temperature in Anoka, Minnesota, was -15 degrees last night. During the day it rose only 9 degrees. Find the high temperature for the day. $-6°$

79. On January 2, 1943, the temperature was -4 degrees at 7:30 A.M. in Spearfish, South Dakota. Incredibly, it got 49 degrees warmer in the next 2 minutes. To what temperature did it rise by 7:32?

79. 45°   **81.** 146 ft.   **83.** $-\dfrac{3}{8}$ point

80. The lowest elevation on Earth is -1312 feet (that is, 1312 feet below sea level) at the Dead Sea. If you are standing 658 feet above the Dead Sea, what is your elevation? -654 ft.

81. The lowest point in Africa is -512 feet at Lake Assal in Djibouti. If you are standing at a point 658 feet above Lake Assal, what is your elevation?

82. In checking the stock market results, Alexis discovers our stock posted changes of $-1\frac{5}{8}$ points and $-2\frac{1}{2}$ points over the last two days. What is the combined change? $-4\frac{1}{8}$ points

83. Yesterday our stock posted a change of $-1\frac{1}{4}$ points, but today it showed a gain of $+\frac{7}{8}$ point. Find the overall change for the two days.

84. In golf, scores that are under par for the entire round are shown as negative scores; positive scores are shown for scores that are over par. In two rounds in the United States Open, Arnold Palmer had overall scores of -6 and $+4$. What was his total overall score? -2

Decide whether the given number is a solution of the given equation.

85. $x + 9 = 5; -4$ **86.** $7 = -x + 3; 10$

87. $y + (-3) = -7; -1$ **88.** $1 = y + 7; -6$

☐ 89. Explain why adding a negative number to another negative number produces a negative sum.

☐ 90. When a positive and a negative number are added, sometimes the sum is positive, sometimes it is zero, and sometimes it is negative. Explain why this happens.   answers may vary

85. solution   **86.** not a solution   **87.** not a solution   **88.** solution   **89.** answers may vary

1.6 SUBTRACTING REAL NUMBERS

TAPE BA 1.6

O B J E C T I V E S

1. Subtract real numbers.
2. Add and subtract real numbers.
3. Evaluate algebraic expressions using real numbers.
4. Solve problems that involve subtraction of real numbers.

Now that addition of signed numbers has been discussed, we can explore subtraction. We know that $9 - 7 = 2$. Notice that $9 + (-7) = 2$, also. This means that

$$9 - 7 = 9 + (-7)$$

In general, the difference of two numbers a and b is the same as the sum of a and the opposite of b.

SUBTRACTING TWO REAL NUMBERS

If a and b are real numbers, then $a - b = a + (-b)$.

In other words, to find the difference of two numbers, add the first number to the opposite of the second number.

EXAMPLE 1 Find each difference.

a. $-13 - (+4)$ **b.** $5 - (-6)$ **c.** $3 - 6$ **d.** $-1 - (-7)$

Solution:

a. $-13 - (+4) = -13 + (-4)$ Add the opposite of $+4$, which is -4.

opposite

$= -17$

b. $5 - (-6) = 5 + (6)$ Add the opposite of -6, which is 6.

opposite

$= 11$

c. $3 - 6 = 3 + (-6)$ Add 3 to the opposite of 6, which is -6.
$= -3$

d. $-1 - (-7) = -1 + (7) = 6$

EXAMPLE 2   Subtract 8 from -4.

Solution: Be careful when interpreting. The order of numbers in subtraction is important. 8 is to be subtracted **from** -4.

$$-4 - 8 = -4 + (-8) = -12$$

2   Expressions containing both sums and differences make good use of grouping symbols. In simplifying these expressions, remember to rewrite differences as sums and follow the standard order of operations.

EXAMPLE 3   Simplify each expression.

a. $-3 + [(-2 - 5) - 2]$ **b.** $2^3 - |10| + [-6 - (-5)]$

Solution: **a.** Start with the innermost sets of parentheses. Rewrite $-2 - 5$ as a sum.

$$\begin{aligned}
-3 + [(-2 - 5) - 2] &= -3 + [(-2 + (-5)) - 2] \\
&= -3 + [(-7) - 2] & \text{Add: } -2+(-5). \\
&= -3 + [-7 + (-2)] & \text{Write } -7-2 \text{ as a sum.} \\
&= -3 + [-9] & \text{Add.} \\
&= -12 & \text{Add.}
\end{aligned}$$

b. Start simplifying the expression inside the brackets by writing $-6 - (-5)$ as a sum.

$$\begin{aligned}
2^3 - |10| + [-6 - (-5)] &= 2^3 - |10| + [-6 + 5] \\
&= 2^3 - 10 + [-1] & \text{Add. Write } |10| \text{ as 10.} \\
&= 8 - 10 + (-1) & \text{Evaluate } 2^3. \\
&= 8 + (-10) + (-1) & \text{Write } 8-10 \text{ as a sum.} \\
&= -2 + (-1) & \text{Add.} \\
&= -3 & \text{Add.}
\end{aligned}$$

3   Knowing how to evaluate expressions for given replacement values is helpful when checking solutions of equations and when solving problems whose unknowns satisfy given expressions. The next example illustrates this.

EXAMPLE 4   If $x = 2$ and $y = -5$, evaluate the following expressions.

a. $\dfrac{x - y}{12 + x}$ **b.** $x^2 - y$

Solution: **a.** Replace x with 2 and y with -5. Be sure to put parentheses around -5 to separate signs. Then simplify the resulting expression.

$$\frac{x - y}{12 + x} = \frac{2 - (-5)}{12 + 2} = \frac{2 + 5}{14} = \frac{7}{14} = \frac{1}{2}$$

b. Replace the x with 2 and y with -5 and simplify.

$$x^2 - y = 2^2 - (-5) = 4 - (-5) = 4 + 5 = 9$$

▨ **4**

Another use of signed numbers is in recording altitudes above and below sea level, as shown in the next example.

EXAMPLE 5   The lowest point in North America is in Death Valley, at an elevation of 282 feet below sea level. Nearby, Mount Whitney reaches 14,494 feet, the highest point in the United States outside Alaska. How much of a variation in elevation is there between these two extremes?

Solution: To find the variation in elevation between the two heights, find the difference of the high point and the low point.

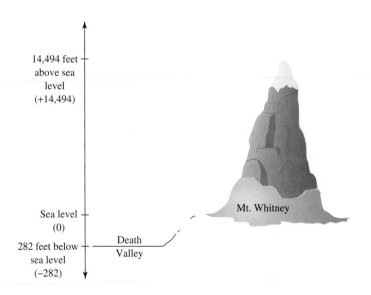

In words: high point   minus   low point

Translate: 14,494   −   (−282)   = 14,494 + 282

= 14,776 feet

Thus, the variation in elevation is 14,776 feet.

EXAMPLE 6   At 6:00 P.M., the temperature at the Winter Olympics was 14 degrees; by morning the temperature dropped to -23 degrees. Find the overall change in temperature.

Solution: To find the overall change, find the difference of the morning temperature and the 6:00 P.M. temperature.

In words: | morning temp. | minus | 6:00 P.M. temp. |

Translate: $-23°$ $-$ $14°$ $= -23° + (-14°) = -37°$

Thus, the overall change is $-37°$. That is, the temperature dropped by 37 degrees.

EXERCISE SET 1.6

Find each difference. See Example 1.

1. $-6 - 4$ -10

2. $-12 - 8$ -20

3. $4 - 9$ -5

4. $8 - 11$ -3

5. $16 - (-3)$ 19

6. $12 - (-5)$ 17

7. $\dfrac{1}{2} - \dfrac{1}{3}$ $\dfrac{1}{6}$

8. $\dfrac{3}{4} - \dfrac{7}{8}$ $-\dfrac{1}{8}$

9. $-16 - (-18)$ 2

10. $-20 - (-48)$ 28

Perform the operation. See Example 2.

11. Subtract -5 from 8. 13

12. Subtract 3 from -2. -5

13. Subtract -1 from -6. -5

14. Subtract 17 from 1. -16

Simplify each expression. (Remember the order of operations.) See Example 3.

15. $-6 - (2 - 11)$ 3

16. $-9 - (3 - 8)$ -4

17. $3^3 - 8 \cdot 9$ -45

18. $2^3 - 6 \cdot 3$ -10

19. $2 - 3(8 - 6)$ -4

20. $4 - 6(7 - 3)$ -20

21. $|-3| + 2^2 + [-4 - (-6)]$ 9

22. $|-2| + 6^2 + (-3 - 8)$ 27

🖋 **23.** If a and b are positive numbers, then is $a - b$ always positive, always negative, or sometimes positive and sometimes negative?

🖋 **24.** If a and b are negative numbers, then is $a - b$ always positive, always negative, or sometimes positive and sometimes negative?

If $x = -6$, $y = -3$, and $t = 2$, evaluate each expression. See Example 4.

25. $x - y$ -3

26. $3t - x$ 12

27. $x + t - 12$ -16

28. $y - 3t - 8$ -17

29. $\dfrac{x - (-12)}{y + 6}$ 2

30. $\dfrac{x - (-9)}{t + 1}$ 1

31. $x - t^2$ -10

32. $y - t^3$ -11

33. $\dfrac{4t - y}{3t}$ $\dfrac{11}{6}$

34. $\dfrac{5t - x}{5t}$ $\dfrac{8}{5}$

Simplify each expression.

35. $-6 - 5$ -11

36. $-8 - 4$ -12

37. $7 - (-4)$ 11

38. $3 - (-6)$ 9

39. $-6 - (-11)$ 5

40. $-4 - (-16)$ 12

41. $6\dfrac{2}{5} - \dfrac{7}{10}$ $5\dfrac{7}{10}$

42. $-\dfrac{3}{5} - \dfrac{5}{6}$ $-\dfrac{43}{30}$

43. $-\dfrac{3}{4} - \dfrac{1}{9}$ $-\dfrac{31}{36}$

44. $-6.1 - (-5.3)$ -0.8

45. $-2.6 - (-6.7)$ 4.1

46. $6 - 11 - (-8)$ 3

47. $4 - (-6) - 9$ 1

48. $-6 - 14 - (-3)$ -17

49. $-11 - (-8) - 4$ -7

50. $-13 - 27$ -40

51. $16 - (-21)$ 37

52. $15 - (-33)$ 48

53. $-44 - (-27)$ -17

54. $-36 - (-51)$ 15

55. $4\dfrac{2}{3} - \dfrac{5}{12}$ $4\dfrac{1}{4}$

56. $-\dfrac{1}{6} - \left(-\dfrac{3}{4}\right)$ $\dfrac{7}{12}$

57. $-\dfrac{1}{10} - \left(-\dfrac{7}{8}\right)$ $\dfrac{31}{40}$

58. $8.3 - 11.2$ -2.9

59. $9.7 - 16.1$ -6.4

60. $8.3 - (-0.62)$ 8.92

61. $4.3 - (-0.87)$ 5.17

62. $3 - (6[3 - (-2)] - 8)$

63. $4 - (3[9 - (-8)] - 11)$ -36

64. $3 - 6 \cdot 4^2$ -93

65. $2 - 3 \cdot 5^2$ -73

66. $(3 - 6) + 4^2$ 13

67. $(2 - 3) + 5^2$ 24

68. $3 - (6 \cdot 4)^2$ -573

69. $2 - (3 \cdot 5)^2$ -223

70. $-2 + [(8 - 11) - (-2 - 9)]$ 6

71. $-5 + [(4 - 15) - (-6) - 8]$ -18

23. sometimes positive and sometimes negative   **24.** sometimes positive and sometimes negative   **62.** -19

72. Subtract 8 from 7. −1

73. Subtract 9 from −4. −13

74. Decrease −8 by 15. −23

75. Decrease 11 by −14. 25

Solve the following. See Examples 5 and 6.

76. Within 24 hours in 1916, the temperature in Browning, Montana, fell from 44 degrees to −56 degrees. How large a drop in temperature was this? 100°

77. Much of New Orleans is just barely above sea level. If George descends 12 feet from an elevation of 5 feet above sea level, what is his new elevation? 7 feet below sea level

78. In a series of plays, the San Francisco 49ers gain 2 yards, lose 5 yards, and then lose another 20 yards. What is their total gain or loss of yardage?

79. In some card games, it is possible to have a negative score. Lavonne Schultz currently has a score of 15 points. She then loses 24 points. What is her new score? −9

80. Aristotle died in the year −322 (or 322 B.C.). When was he born, if he was 62 years old when he died? 384 B.C.

81. Augustus Caesar died in A.D. 14 in his 77th year. When was he born? 63 B.C.

82. Tyson Industries stock posted a loss of $1\frac{5}{8}$ points yesterday. If it drops another $\frac{3}{4}$ points today, find its overall change for the two days. $-2\frac{3}{8}$ points

83. A commercial jet liner hits an air pocket and drops 250 feet. After climbing 120 feet, it drops another 178 feet. What is its overall vertical change? −308 ft.

78. 23 yd. loss

The following bar graph is from an earlier section and shows the daily low temperatures for a week in Sioux Falls, South Dakota.

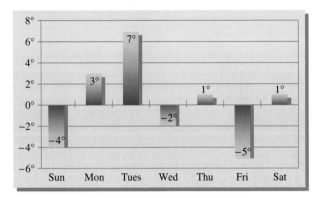

84. Use the bar graph to record the daily increases and decreases in the low temperatures from the previous day.

DAY	DAILY INCREASE OR DECREASE
Monday	7°
Tuesday	4°
Wednesday	−9°
Thursday	3°
Friday	−6°
Saturday	6°

85. Which day of the week had the greatest increase in temperature? Monday

86. Which day of the week had the greatest decrease in temperature? Wednesday

If x = −5, y = 4, and t = 10, simplify each expression.

87. $x - y$ −9

88. $y - x$ 9

89. $|x| + 2t - 8y$ −7

90. $|x + t - 7y|$ 23

91. $\dfrac{9 - x}{y + 6}$ $\dfrac{7}{5}$

92. $\dfrac{15 - x}{y + 2}$ $\dfrac{10}{3}$

93. $y^2 - x$ 21

94. $t^2 - x$ 105

95. $\dfrac{|x - (-10)|}{2t}$ $\dfrac{1}{4}$

96. $\dfrac{|5y - x|}{6t}$ $\dfrac{5}{12}$

97. In your own words, explain why $5 - 8$ simplifies to a negative number. answers may vary

98. Explain why $6 - 11$ is the same as $6 + (-11)$.

98. answers may vary

Decide whether the given number is a solution of the given equation.

99. $x - 9 = 5$; -4 not a solution

100. $x - 10 = -7$; 3 solution

101. $-x + 6 = -x - 1$; -2 not a solution

102. $-x - 6 = -x - 1$; -10 not a solution

103. $-x - 13 = -15$; 2 solution

104. $4 = 1 - x$; 5 not a solution

1.7   MULTIPLYING AND DIVIDING REAL NUMBERS

TAPE BA 1.7

O B J E C T I V E S

 Multiply and divide real numbers.

2 Find the values of algebraic expressions.

1 In this section, we discover patterns for multiplying and dividing real numbers. To discover sign rules for multiplication, recall that multiplication is repeated addition. Thus $3 \cdot 2$ means that 2 is an addend 3 times. That is,

$$2 + 2 + 2 = 3 \cdot 2$$

which equals 6. Similarly, $3 \cdot (-2)$ means -2 is an addend 3 times. That is,

$$(-2) + (-2) + (-2) = 3 \cdot (-2)$$

Since $(-2) + (-2) + (-2) = -6$, then $3 \cdot (-2) = -6$. This suggests that the product of a positive number and a negative number is a negative number.

What about the product of two negative numbers? To find out, consider the following pattern.

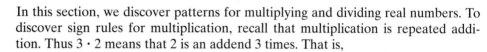

Factor decreases by 1 each time

$$
\left.
\begin{array}{l}
-3 \cdot 2 = -6 \\
-3 \cdot 1 = -3 \\
-3 \cdot 0 = 0
\end{array}
\right\}
\text{ Product increases by 3 each time.}
$$

This pattern continues as

Factor decreases by 1 each time

$$
\left.
\begin{array}{l}
-3 \cdot -1 = 3 \\
-3 \cdot -2 = 6
\end{array}
\right\}
\text{ Product increases by 3 each time.}
$$

This suggests that the product of two negative numbers is a positive number.

MULTIPLYING REAL NUMBERS

1. The product of two numbers with the same sign is a positive number.

2. The product of two numbers with different signs is a negative number.

EXAMPLE 1 Find the product.

  a. $(-6)(4)$ **b.** $2(-1)$ **c.** $(-5)(-10)$

Solution: **a.** $(-6)(4) = -24$ **b.** $2(-1) = -2$ **c.** $(-5)(-10) = 50$

We know that every whole number multiplied by zero equals zero. This remains true for signed numbers.

ZERO AS A FACTOR

If b is a real number, then $b \cdot 0 = 0$. Also, $0 \cdot b = 0$.

EXAMPLE 2 Perform the indicated operations.

  a. $(7)(0)(-6)$ **b.** $(-2)(-3)(-4)$ **c.** $(-1)(5)(-9)$ **d.** $(-2)^3$
  e. $(-4)(-11)-(5)(-2)$

Solution: **a.** By the order of operations, we multiply from left to right. Notice that, because one of the factors is 0, the product is 0.

$$(7)(0)(-6) = 0(-6) = 0$$

  b. Multiply two factors at a time, from left to right.

$$(-2)(-3)(-4) = (6)(-4) \qquad \text{Multiply } (-2)(-3).$$
$$= -24$$

  c. Multiply from left to right.

$$(-1)(5)(-9) = (-5)(-9) \qquad \text{Multiply } (-1)(5).$$
$$= 45$$

  d. The exponent 3 means 3 factors of the base -2.

$$(-2)^3 = (-2)(-2)(-2)$$
$$= -8 \qquad \text{Multiply.}$$

  e. Follow the rules for order of operation.

$$(-4)(-11)-(5)(-2) = 44 - (-10) \qquad \text{Find the products.}$$
$$= 44 + 10 \qquad \text{Add 44 to the opposite of } -10.$$
$$= 54 \qquad \text{Add.}$$

Multiplying signed decimals or fractions is carried out exactly the same way as multiplying by integers.

EXAMPLE 3   Find each product.

a. $(-1.2)(0.05)$

b. $\dfrac{2}{3} \cdot -\dfrac{7}{10}$

Solution: **a.** The product of two numbers with different signs is negative.

$$(-1.2)(0.05) = -[(1.2)(0.05)]$$
$$= -0.06$$

b. $\dfrac{2}{3} \cdot -\dfrac{7}{10} = -\dfrac{2 \cdot 7}{3 \cdot 10} = -\dfrac{2 \cdot 7}{3 \cdot 2 \cdot 5} = -\dfrac{7}{15}$

Just as every difference of two numbers $a - b$ can be written as the sum $a + (-b)$, so too every quotient of two numbers can be written as a product. For example, the quotient $6 \div 3$ can be written as $6 \cdot \frac{1}{3}$. Recall that the pair of numbers 3 and $\frac{1}{3}$ has a special relationship. Their product is 1 and they are called reciprocals or **multiplicative inverses** of each other.

> ### RECIPROCALS OR MULTIPLICATIVE INVERSES
> Two numbers whose product is 1 are called reciprocals or multiplicative inverses of each other.

Notice that **0 has no multiplicative inverse** since 0 multiplied by any number is never 1 but always 0.

EXAMPLE 4   Find the multiplicative inverse of each number.

a. 22

b. $\dfrac{3}{16}$

c. -10

d. $-\dfrac{9}{13}$

Solution: **a.** The multiplicative inverse of 22 is $\frac{1}{22}$ since $22 \cdot \frac{1}{22} = 1$.

b. The multiplicative inverse of $\frac{3}{16}$ is $\frac{16}{3}$ since $\frac{3}{16} \cdot \frac{16}{3} = 1$.

c. The multiplicative inverse of -10 is $-\frac{1}{10}$.

d. The multiplicative inverse of $-\frac{9}{13}$ is $-\frac{13}{9}$.

We may now write a quotient as an equivalent product.

> ### QUOTIENT OF TWO REAL NUMBERS
> If a and b are real numbers and b is not 0, then
> $$\frac{a}{b} = a \cdot \frac{1}{b}$$

In other words, the quotient of two real numbers is the product of the first number and the multiplicative inverse or reciprocal of the second number.

EXAMPLE 5   Use the definition of the quotient of two numbers to find each quotient.

  a. $-18 \div 3$ **b.** $\dfrac{-14}{-2}$ **c.** $\dfrac{20}{-4}$

Solution: **a.** $-18 \div 3 = -18 \cdot \dfrac{1}{3} = -6$ **b.** $\dfrac{-14}{-2} = -14 \cdot -\dfrac{1}{2} = 7$

  c. $\dfrac{20}{-4} = 20 \cdot -\dfrac{1}{4} = -5$

Since the quotient $a \div b$ can be written as the product $a \cdot \dfrac{1}{b}$, it follows that sign patterns for dividing two real numbers are the same as sign patterns for multiplying two real numbers.

MULTIPLYING AND DIVIDING REAL NUMBERS

1. The product or quotient of two numbers with the same sign is a positive number.

2. The product or quotient of two numbers with different signs is a negative number.

EXAMPLE 6   Find each quotient.

  a. $\dfrac{-24}{-4}$ **b.** $\dfrac{-36}{3}$ **c.** $\dfrac{2}{3} \div \left(-\dfrac{5}{4}\right)$

Solution: **a.** $\dfrac{-24}{-4} = 6$

  b. $\dfrac{-36}{3} = -12$

  c. $\dfrac{2}{3} \div \left(-\dfrac{5}{4}\right) = \dfrac{2}{3} \cdot \left(-\dfrac{4}{5}\right) = -\dfrac{8}{15}$

The definition of the quotient of two real numbers does not allow for division by 0 because 0 does not have a multiplicative inverse. There is no number we can multiply 0 by to get 1. How then do we interpret $\dfrac{3}{0}$? We say that division by 0 is not allowed or not defined and that $\dfrac{3}{0}$ does not represent a real number. The denominator of a fraction can never be 0.

Can the numerator of a fraction be 0? Can we divide 0 by a number? Yes. For example,

$$\dfrac{0}{3} = 0 \cdot \dfrac{1}{3} = 0$$

In general, the quotient of 0 and any nonzero number is 0.

ZERO AS A DIVISOR OR DIVIDEND

1. The quotient of any nonzero real number and 0 is undefined. In symbols, if $a \neq 0$, $\dfrac{a}{0}$ is **undefined.**

2. The quotient of 0 and any real number except 0 is 0. In symbols, if $a \neq 0$, $\dfrac{0}{a} = 0$.

EXAMPLE 7   Perform the indicated operations.

   a. $\dfrac{1}{0}$ **b.** $\dfrac{0}{-3}$ **c.** $\dfrac{0(-8)}{2}$

Solution: **a.** $\dfrac{1}{0}$ is undefined   **b.** $\dfrac{0}{-3} = 0$ **c.** $\dfrac{0(-8)}{2} = \dfrac{0}{2} = 0$

Notice that $\dfrac{12}{-2} = -6, \dfrac{12}{2} = -6$ and $\dfrac{-12}{2} = -6$. This means that

$$\frac{12}{-2} = -\frac{12}{2} = \frac{-12}{2}$$

In words, a single negative sign in a fraction can be written in the denominator, in the numerator, or in front of the fraction without changing the value of the fraction. Thus,

$$\frac{1}{-7} = \frac{-1}{7} = -\frac{1}{7}$$

In general, if a and b are real numbers, $b \neq 0$, $\dfrac{a}{-b} = \dfrac{-a}{b} = -\dfrac{a}{b}$.

Examples combining basic arithmetic operations along with the principles of order of operations help us to review these concepts.

EXAMPLE 8   Simplify each expression.

   a. $\dfrac{(-12)(-3) + 4}{-7 - (-2)}$

   b. $\dfrac{2(-3)^2 - 20}{-5 + 4}$

Solution: **a.** First, simplify the numerator and denominator separately, then divide.

$$\frac{(-12)(-3) + 4}{-7 - (-2)} = \frac{36 + 4}{-7 + 2}$$

$$= \frac{40}{-5}$$

$$= -8 \qquad \text{Divide.}$$

b. Simplify the numerator and denominator separately, then divide.

$$\frac{2(-3)^2 - 20}{-5 + 4} = \frac{2 \cdot 9 - 20}{-5 + 4} = \frac{18 - 20}{-5 + 4} = \frac{-2}{-1} = 2$$

2

EXAMPLE 9   If $x = -2$ and $y = -4$, evaluate each expression.

a. $5x - y$ **b.** $x^3 - y^2$ **c.** $\dfrac{3x}{2y}$

Solution: **a.** Replace x with -2 and y with -4 and simplify.

$$5x - y = 5(-2) - (-4) = -10 - (-4) = -10 + 4 = -6$$

b. Replace x with -2 and y with -4.

$$\begin{aligned} x^3 - y^2 &= (-2)^3 - (-4)^2 && \text{Substitute the given values for the variables.} \\ &= -8 - (16) && \text{Evaluate exponential expressions.} \\ &= -8 + (-16) && \text{Write as a sum.} \\ &= -24 && \text{Add.} \end{aligned}$$

c. Replace x with -2 and y with -4 and simplify.

$$\frac{3x}{2y} = \frac{3(-2)}{2(-4)} = \frac{-6}{-8} = \frac{3}{4}$$

SCIENTIFIC CALCULATOR EXPLORATIONS

ENTERING NEGATIVE NUMBERS

To enter a negative number on a calculator, find a key marked +/−. (On some calculators, this key is marked CHS for "change sign.") To enter −8, for example, press the keys 8 +/−. The display will read ⌐ −8 ⌐.

OPERATIONS WITH REAL NUMBERS

To evaluate $-2(7 - 9) - 20$ on a calculator, press the keys

2 +/− × (7 − 9) − 2 0 = .

The display will read ⌐ −16 ⌐.

Use a calculator to simplify each expression.

1. $-38(26 - 27)$ 38

2. $-59(-8) + 1726$ 2198

3. $134 + 25(68 - 91)$ −441

4. $45(32) - 8(218)$ −304

5. $\dfrac{-50(294)}{175 - 265}$ $163\dfrac{1}{3}$

6. $\dfrac{-444 - 444.8}{-181 - 324}$ 1.76

7. $9^5 - 4550$ 54,499

8. $5^8 - 6259$ 384,366

9. $(-125)^2$ Be careful.   15,625

10. -125^2 Be careful.   −15,625

Exercise Set 1.7

Find the following products. See Examples 1 through 3.

1. $(-3)(+4)$ -12

2. $(+8)(-2)$ -16

3. $-6(-7)$ 42

4. $(-3)(-8)$ 24

5. $(-2)(-5)(0)$ 0

6. $(7)(0)(-3)$ 0

7. $2(-9)$ -18

8. $(-5)(3)$ -15

9. $\left(-\dfrac{3}{4}\right)\left(\dfrac{8}{9}\right)$ $-\dfrac{2}{3}$

10. $\left(\dfrac{5}{6}\right)\left(-\dfrac{3}{10}\right)$ $-\dfrac{1}{4}$

11. $\left(-1\dfrac{1}{5}\right)\left(-1\dfrac{2}{3}\right)$ 2

12. $\left(-\dfrac{5}{6}\right)\left(-\dfrac{3}{10}\right)$ $\dfrac{1}{4}$

13. $(-1)(2)(-3)(-5)$ -30

14. $(-2)(-3)(-4)(-2)$

15. $(2)(-1)(-3)(5)(3)$ 90

48

16. $(3)(-5)(-2)(-1)(-2)$ 60

17. $(-4)^2$ 16

18. $(-3)^3$ -27

Decide whether each statement is true or false.

19. The product of three negative integers is negative.

20. The product of three positive integers is positive.

21. The product of four negative integers is negative.

22. The product of four positive integers is positive.

19. true **20.** true **21.** false **22.** true

Find the multiplicative inverse or reciprocal of each number. See Example 4.

23. 9 $\dfrac{1}{9}$

24. 100 $\dfrac{1}{100}$

25. $\dfrac{2}{3}$ $\dfrac{3}{2}$

26. $\dfrac{1}{7}$ 7

27. -14 $-\dfrac{1}{14}$

28. $-\dfrac{3}{11}$ $-\dfrac{11}{3}$

29. Find any real numbers that are their own reciprocal.   1, −1

30. Explain why 0 has no reciprocal. answers may vary

Find the following quotients. If the quotient is undefined, state so. See Examples 5–7.

31. $\dfrac{18}{-2}$ -9

32. $-\dfrac{14}{7}$ -2

33. $\dfrac{-12}{-4}$ 3

34. $-\dfrac{20}{5}$ -4

35. $\dfrac{-45}{-9}$ 5

36. $\dfrac{30}{-2}$ -15

37. $\dfrac{0}{-3}$ 0

38. $-\dfrac{4}{0}$ undefined

39. $-\dfrac{3}{0}$

40. $\dfrac{0}{-4}$ 0

39. undefined

70. 12

Simplify the following. See Example 8.

41. $\dfrac{-6^2 + 4}{-2}$ 16

42. $\dfrac{3^2 + 4}{5}$ $\dfrac{13}{5}$

43. $\dfrac{8 + (-4)^2}{4 - 12}$ -3

44. $\dfrac{6 + (-2)^2}{4 - 9}$ -2

45. $\dfrac{22 + (3)(-2)}{-5 - 2}$ $-\dfrac{16}{7}$

46. $\dfrac{-20 + (-4)(3)}{1 - 5}$ 8

If $x = -5$ and $y = -3$, evaluate each expression. See Example 9.

47. $3x + 2y$ -21

48. $4x + 5y$ -35

49. $2x^2 - y^2$ 41

50. $x^2 - 2y^2$ 7

51. $x^3 + 3y$ -134

52. $y^3 + 3x$ -42

53. $\dfrac{2x - 5}{y - 2}$ 3

54. $\dfrac{2y - 12}{x - 4}$ 2

55. $\dfrac{6 - y}{x - 4}$ -1

56. $\dfrac{4 - 2x}{y + 3}$ undefined

Perform indicated operations. If the expression is undefined, state so.

57. $(-6)(-2)$ 12

58. $5(-3)$ -15

59. $(-7)(2)$ -14

60. $(-3)(-9)$ 27

61. $\dfrac{18}{-3}$ -6

62. $\dfrac{-16}{-4}$ 4

63. $-\dfrac{6}{0}$ undefined

64. $-\dfrac{16}{2}$ -8

65. $-\dfrac{15}{-3}$ 5

66. $\dfrac{48}{-12}$ -4

67. $\dfrac{0}{-7}$ 0

68. $-\dfrac{48}{-8}$ 6

69. $(-6)(3)(-2)(-1)$ -36

70. $(-3)(-2)(-1)(-2)$

71. $(-5)^3$ -125

72. $(-2)^5$ -32

73. $(-4)^2$ 16

74. $(-6)^2$ 36

75. -4^2 -16

76. -6^2 -36

77. $\dfrac{-3 - 5^2}{2(-7)}$ 2

78. $\dfrac{-2 - 4^2}{3(-6)}$ 1

79. $\dfrac{6 - 2(-3)}{4 - 3(-2)}$ $\dfrac{6}{5}$

80. $\dfrac{8 - 3(-2)}{2 - 5(-4)}$ $\dfrac{7}{11}$

81. $\dfrac{-3 - 2(-9)}{-15 - 3(-4)}$ -5

82. $\dfrac{-4 - 8(-2)}{-9 - 2(-3)}$ -4

83. $-3(2 - 8)$ 18

84. $-4(3 - 9)$ 24

85. $6(3 - 8)$ $\quad -30$ **86.** $4(8 - 11)$ $\quad -12$

87. $-3[(2 - 8) - (-6 - 8)]$ $\quad -24$

88. $-2[(3 - 5) - (2 - 9)]$ $\quad -10$

89. $\left(\frac{2}{5}\right)\left(-1\frac{1}{4}\right)$ $\quad -\frac{1}{2}$ **90.** $\left(-4\frac{2}{3}\right)\left(-\frac{8}{21}\right)$ $\quad \frac{16}{9}$

91. $(1.82)(-4.6)$ $\quad -8.372$ **92.** $(-3.6)(-0.61)$ $\quad 2.196$

93. $-22.4 \div (-1.6)$ $\quad 14$ **94.** $12.24 \div (-2.4)$ $\quad -5.1$

If q is a negative number, r is a negative number, and t is a positive number, determine whether each expression simplifies to a positive or negative number. If it is not possible to determine, state so.

95. $\frac{q}{r \cdot t}$ positive   **96.** $q^2 \cdot r \cdot t$ negative

97. $q + t$ **98.** $t + r$

99. $t(q + r)$ negative   **100.** $r(q - t)$ positive

Write each of the following as an expression and evaluate.

101. The sum of -2 and the quotient of -15 and 3

102. The sum of 1 and the product of -8 and -5

103. Twice the sum of -5 and -3

104. 7 subtracted from the quotient of 0 and 5

Decide whether the given number is a solution of the given equation.

105. $-5x = -35; 7$ **106.** $2x = x - 1; -4$

107. $\frac{x}{-10} = 2; -20$ **108.** $\frac{45}{x} = -15; -3$

109. $-3x - 5 = -20; 5$ **110.** $2x + 4 = x + 8; -4$

111. The following graph shows Trader's stock consistently decreasing in value by $1.50 per share per day. If this trend continues, when will the stock be worth $20 per share?   Sat., Oct. 2

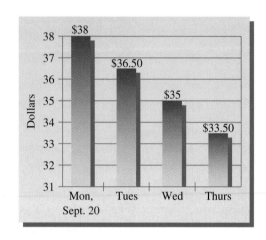

112. Explain why the product of an even number of negative numbers is a positive number.

113. If a and b are any real numbers, is the statement $a \cdot b = b \cdot a$ always true? Why or why not?

114. If a and b are any real numbers, is the statement $a - b = b - a$ always true? Why or why not?

97. not possible to determine   **98.** not possible to determine   **101.** $-2 + \frac{-15}{3}; -7$ **102.** $1 + (-8)(-5); 41$

103. $2[-5 + (-3)]; -16$ **104.** $\frac{0}{5} - 7; -7$ **105.** solution   **106.** not a solution   **107.** solution   **108.** solution

1.8 PROPERTIES OF REAL NUMBERS

TAPE BA 1.8

O B J E C T I V E S

1. Identify the commutative property.
2. Identify the associative property.
3. Identify the distributive property.
4. Identify the additive and multiplicative identities.
5. Identify the additive and multiplicative inverse properties.

Specialized terms occur in every area of study. A biologist or ecologist will often be concerned with "symbiotic relationships." "Torque" is a major concern to the physicist. An economist or business analyst might be interested in "marginal revenue." A nurse needs to be familiar with "hemostasis." Mathematics, too, has its own specialized terms and principles.

109. solution **110.** not a solution **112.** answers may vary **113.** answers may vary   **114.** answers may vary

This section introduces the basic properties of the real number system. Throughout this section, a, b, and c represent real numbers.

When we add, subtract, multiply, or divide two real numbers (except for division by zero), the result is a real number. This is guaranteed by the closure properties.

CLOSURE PROPERTIES

If a and b are real numbers then $a + b$, $a - b$, and ab are real numbers. Also, $\dfrac{a}{b}$, $b \neq 0$, is a real number.

---

1 Next, we look at the commutative properties. These properties state that the order in which any two real numbers are added or multiplied does not change their sum or product.

COMMUTATIVE PROPERTIES

Addition: $a + b = b + a$

Multiplication: $a \cdot b = b \cdot a$

For example, if we let $a = 3$ and $b = 5$, then the commutative properties guarantee that

$$3 + 5 = 5 + 3 \qquad \text{and} \qquad 3 \cdot 5 = 5 \cdot 3$$

R E M I N D E R   Is subtraction also commutative? Try an example. Is $3 - 2 = 2 - 3$? **No!** The left side of this statement equals 1; the right side equals -1. There is no commutative property of subtraction. Similarly, there is no commutative property for division. For example, $\frac{10}{2}$ does not equal $\frac{2}{10}$.

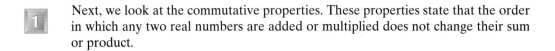

EXAMPLE 1   If $a = -2$, and $b = 5$, show that:

a. $a + b = b + a$ **b.** $a \cdot b = b \cdot a$

Solution: **a.** Replace a with -2 and b with 5. Then

$$a + b = b + a$$

becomes

$$-2 + 5 = 5 + (-2)$$

or

$$3 = 3$$

Since both sides represent the same number, $-2 + 5 = 5 + (-2)$ is a true statement.

b. Replace a with -2 and b with 5. Then

$$a \cdot b = b \cdot a$$

becomes

$$-2 \cdot 5 = 5 \cdot (-2)$$

or

$$-10 = -10$$

Since both sides represent the same number, the statement is true.   ▬▬▬▬

2 When adding or multiplying three numbers, does it matter how we group the numbers? This question is answered by the associative properties. These properties state that when adding or multiplying three numbers, any two adjacent numbers may be grouped together without changing their sum or product.

> **ASSOCIATIVE PROPERTIES**
> **Addition:** $(a + b) + c = a + (b + c)$
> **Multiplication:** $(a \cdot b) \cdot c = a \cdot (b \cdot c)$

Illustrate these properties by working the following example.

EXAMPLE 2   If $a = -3$, $b = 2$, and $c = 4$, show that:

a. $(a + b) + c = a + (b + c)$ **b.** $(a \cdot b) \cdot c = a \cdot (b \cdot c)$

Solution: Replace a with -3, b with 2, and c with 4.

a. $(a + b) + c = a + (b + c)$

$(-3 + 2) + 4 = -3 + (2 + 4)$ Replace a with -3, b with 2, and c with 4.

$-1 + 4 = -3 + 6$ Simplify inside parentheses.

$3 = 3$ Add.

b. $(a \cdot b) \cdot c = a \cdot (b \cdot c)$

$(-3 \cdot 2) \cdot 4 = -3 \cdot (2 \cdot 4)$ Replace a with -3, b with 2, and c with 4.

$-6 \cdot 4 = -3 \cdot 8$ Simplify inside parentheses.

$-24 = -24$ Multiply.   ▬▬▬▬

3 The **distributive property of multiplication over addition** is used repeatedly throughout algebra. It is useful because it allows us to write a product as a sum or a sum as a product.

DISTRIBUTIVE PROPERTY OF MULTIPLICATION OVER ADDITION

$a(b + c) = ab + ac$

Since multiplication is commutative, the distributive property can also be written

$$(b + c)a = ba + ca$$

The truth of this property can be illustrated by letting $a = 3$, $b = 2$, and $c = 5$.

$$a(b + c) = ab + ac$$

becomes

$$3(2 + 5) = 3 \cdot 2 + 3 \cdot 5$$
$$3 \cdot 7 = 6 + 15$$
$$21 = 21$$

Notice in this example that 3 is "being distributed to" each addend inside the parentheses. That is, 3 is multiplied by each addend.

The distributive property can be extended so that factors can be distributed to more than two addends in parentheses. For example,

$$3(x + y + 2) = 3(x) + 3(y) + 3(2)$$
$$= 3x + 3y + 6$$

EXAMPLE 3   Use the distributive property to write each expression without parentheses.

a. $2(x + y)$ **b.** $-5(-3 + z)$ **c.** $5(x + y - z)$ **d.** $-1(2 - y)$

e. $-(3 + x - w)$

Solution: **a.** $2(x + y) = 2 \cdot x + 2 \cdot y$
$$= 2x + 2y$$

b. $-5(-3 + z) = -5(-3) + (-5) \cdot z$
$$= 15 - 5z$$

c. $5(x + y - z) = 5 \cdot x + 5 \cdot y + 5(-z)$
$$= 5x + 5y - 5z$$

d. $-1(2 - y) = (-1)(2) + (-1)(-y)$
$$= -2 + y$$

e. $-(3 + x - w) = -1(3 + x - w)$
$$= (-1)(3) + (-1)(x) + (-1)(-w)$$
$$= -3 - x + w$$

Notice in the last example that $-(3 + x - w)$ is rewritten as $-\mathbf{1}(3 + x - w)$.

EXAMPLE 4   Use the distributive property to write each sum as a product.

a. $8 \cdot 2 + 8 \cdot x$ **b.** $7s + 7t$

Solution: **a.** $8 \cdot 2 + 8 \cdot x = 8(2 + x)$
b. $7s + 7t = 7(s + t)$

Next, we look at the **identity properties.** These properties guarantee that two special numbers exist. These numbers are called the **identity element for addition** and the **identity element for multiplication.**

IDENTITIES FOR ADDITION AND MULTIPLICATION

0 is the identity element for addition.

$$a + 0 = a \quad \text{and} \quad 0 + a = a$$

1 is the identity element for multiplication.

$$a \cdot 1 = a \quad \text{and} \quad 1 \cdot a = a$$

Notice that 0 is the only number that can be added to any real number with the result that the sum is the same real number. Also, 1 is the only number that can be multiplied by any other number with the result that the product is the same real number.

Additive inverses or **opposites** were introduced in Section 1.5. Two numbers are called additive inverses or opposites if their sum is 0. The additive inverse or opposite of 6 is -6 because $6 + (-6) = 0$. The additive inverse or opposite of -5 is 5 because $-5 + 5 = 0$.

Reciprocals or **multiplicative inverses** were introduced in Section 1.2. Two nonzero numbers are called reciprocals or multiplicative inverses if their product is 1. The reciprocal or multiplicative inverse of $\frac{2}{3}$ is $\frac{3}{2}$ because $\frac{2}{3} \cdot \frac{3}{2} = 1$. Likewise, the reciprocal of -5 is $-\frac{1}{5}$ because $-5\left(-\frac{1}{5}\right) = 1$.

ADDITIVE OR MULTIPLICATIVE INVERSES

The numbers a and $-a$ are additive inverses or opposites of each other because their sum is 0; that is,

$$a + (-a) = 0$$

The numbers b and $\frac{1}{b}$ (for $b \neq 0$) are called reciprocals or multiplicative inverses of each other because their product is 1; that is,

$$b \cdot \frac{1}{b} = 1$$

EXAMPLE 5   Find the additive inverse or opposite of each number.

    a. -3 **b.** 5   **c.** 0   **d.** $|-2|$

Solution: **a.** The additive inverse of -3 is 3 because $-3 + 3 = 0$.

    b. The additive inverse of 5 is -5 because $5 + (-5) = 0$.

    c. The additive inverse of 0 is 0 because $0 + 0 = 0$.

    d. $|-2| = 2$. The additive inverse of 2 (and $|-2|$) is -2.

EXAMPLE 6   Find the multiplicative inverse or reciprocal of each number.

    a. 7   **b.** $\dfrac{-1}{9}$

Solution: **a.** The multiplicative inverse of 7 is $\dfrac{1}{7}$ because $7 \cdot \dfrac{1}{7} = 1$.

    b. The multiplicative inverse of $\dfrac{-1}{9}$ is $\dfrac{9}{-1}$, or -9, because $\left(\dfrac{-1}{9}\right)(-9) = 1$.

EXAMPLE 7   Name the property illustrated by each true statement.

    a. $2 \cdot 3 = 3 \cdot 2$

    b. $3(x + 5) = 3x + 15$

    c. $2 + (4 + 8) = (2 + 4) + 8$

Solution: **a.** The commutative property of multiplication

    b. The distributive property

    c. The associative property of addition

EXAMPLE 8   Use the indicated property and the given expression to write a true statement.

    a. $2 + 9$; the commutative property of addition

    b. $(5 \cdot 8) \cdot 9$; the associative property of multiplication

    c. $x + 0$; the additive identity property

Solution: **a.** $2 + 9 = 9 + 2$

    b. $(5 \cdot 8) \cdot 9 = 5 \cdot (8 \cdot 9)$

    c. $x + 0 = x$

1. commutative property of multiplication **3.** associative property of addition **7.** associative property of multiplication
8. multiplicative inverse **10.** associative property of addition **12.** commutative property of addition

EXERCISE SET 1.8

13. associative property of multiplication **22.** $-10r - 55$ **23.** $5x + 20m + 10$
24. $24y + 8z - 48$ **25.** $-4 + 8m - 4n$ **26.** $-16 - 8p - 20$ **29.** $-r + 3 + 7p$

Name the properties illustrated by each true statement. See Examples 1, 2, and 7.

1. $3 \cdot 5 = 5 \cdot 3$

2. $4(3 + 8) = 4 \cdot 3 + 4 \cdot 8$ distributive property

3. $2 + (8 + 5) = (2 + 8) + 5$

4. $4 + 9 = 9 + 4$ commutative property of addition

5. $9(3 + 7) = 9 \cdot 3 + 9 \cdot 7$ distributive property

6. $1 \cdot 9 = 9$ multiplicative identity property

7. $(4 \cdot 8) \cdot 9 = 4 \cdot (8 \cdot 9)$ **8.** $6 \cdot \dfrac{1}{6} = 1$

9. $0 + 6 = 6$ identity property of addition

10. $(4 + 9) + 6 = 4 + (9 + 6)$

11. $-4(3 + 7) = -4 \cdot 3 + (-4) \cdot 7$ distributive property

12. $11 + 6 = 6 + 11$

13. $-4 \cdot (8 \cdot 3) = (-4 \cdot 8) \cdot 3$

14. $10 + 0 = 10$ identity property of addition

15. Write an example that shows that division is not commutative. answers may vary

16. Write an example that shows that subtraction is not commutative. answers may vary

Use the distributive property to write each expression without parentheses. See Example 3.

17. $3(6 + x)$ $18 + 3x$ **18.** $2(x - 5)$ $2x - 10$

19. $-2(y - z)$ $-2y + 2z$ **20.** $-3(z - y)$ $-3z + 3y$

21. $-7(3y - 5)$ $-21y + 35$ **22.** $-5(2r + 11)$

23. $5(x + 4m + 2)$ **24.** $8(3y + z - 6)$

25. $-4(1 - 2m + n)$ **26.** $-4(4 + 2p + 5)$

27. $-(5x + 2)$ $-5x - 2$ **28.** $-(9r + 5)$ $-9r - 5$

29. $-(r - 3 - 7p)$ **30.** $-(-q - 2 + 6r)$

Use the distributive property to write each sum as a product. See Example 4.

31. $4 \cdot 1 + 4 \cdot y$ $4(1 + y)$ **32.** $14 \cdot z + 14 \cdot 5$

33. $11x + 11y$ $11(x + y)$ **34.** $9a + 9b$ $9(a + b)$

35. $(-1) \cdot 5 + (-1) \cdot x$ **36.** $(-3)a + (-3)y$

Find the additive inverse or opposite of each of the following numbers. See Example 5.

37. 16 -16 **38.** 14 -14

39. -8 8 **40.** -3 3

41. $|-9|$ -9

42. $|11|$ -11

43. $\dfrac{2}{3}$ $-\dfrac{2}{3}$

44. $-\dfrac{7}{8}$ $\dfrac{7}{8}$

45. $-(-1.2)$ -1.2

46. $-(7.9)$ 7.9

47. $-|2|$ 2

48. $-|-9|$ 9

Find the multiplicative inverse or reciprocal of each of the following numbers. See Example 6.

49. $\dfrac{2}{3}$ $\dfrac{3}{2}$

50. $\dfrac{3}{4}$ $\dfrac{4}{3}$

51. $-\dfrac{5}{6}$ $-\dfrac{6}{5}$

52. $-\dfrac{7}{8}$ $-\dfrac{8}{7}$

53. 6 $\dfrac{1}{6}$

54. 3 $\dfrac{1}{3}$

55. -2 $-\dfrac{1}{2}$

56. -5 $-\dfrac{1}{5}$

57. $-\left|-\dfrac{3}{5}\right|$ $-\dfrac{5}{3}$

58. $-\left|-\dfrac{2}{5}\right|$ $-\dfrac{5}{2}$

59. $3\dfrac{5}{6}$ $\dfrac{6}{23}$

60. $2\dfrac{3}{5}$ $\dfrac{5}{13}$

Write the additive inverse and the multiplicative inverse of each expression. Assume that the value of each expression is not 0.

61. x $-x; \dfrac{1}{x}$

62. y $-y; \dfrac{1}{y}$

63. $-3z$

64. $5a$

65. $a + b$ $-a - b; \dfrac{1}{a + b}$

66. $x - y$ $-x + y; \dfrac{1}{x - y}$

Use the indicated property and the given expression to write a true statement. See Example 8.

67. $\dfrac{2}{3} \cdot \dfrac{3}{2}$; multiplicative inverse property $\dfrac{2}{3} \cdot \dfrac{3}{2} = 1$

68. $8 + 16$; commutative property of addition $8 + 16$

69. $(-4)(-3)$; commutative property of multiplication

70. $-4 + 4$; additive inverse property $-4 + 4 = 0$

71. $3 + (8 + 9)$; associative property of addition

72. $(4 \cdot 3) \cdot 9$; associative property of multiplication

73. $y + 0$; additive identity property $y + 0 = y$

74. $1 \cdot x$; multiplicative identity property $(1 \cdot x) = x$

75. $x(a + b)$; distributive property $x(a + b) = xa + xb$

76. $(m + n)y$; distributive property

30. $q + 2 - 6r$ **32.** $14(z + 5)$ **35.** $-1(5 + x)$ **36.** $-3(a + y)$ **63.** $3z; -\dfrac{1}{3z}$ **64.** $-5a; \dfrac{1}{5a}$
69. $(-4)(-3) = (-3)(-4)$ **71.** $3 + (8 + 9) = (3 + 8) + 9$ **72.** $(4 \cdot 3) \cdot 9 = 4 \cdot (3 \cdot 9)$ **76.** $(m + n)y = my + ny$

77. $a(b + c) = (b + c)a$ **78.** $x + 2 = 2 + x$ **79. a.** distributive property   **79. b.** commutative property of addition

77. $a(b + c)$; commutative property of multiplication

78. $x + 2$; commutative property of addition

Name the property illustrated by each step.

79. a. $7(2 + x) + 5 = 14 + 7x + 5$

b. $= 7x + 14 + 5$

      $= 7x + 19$

80. a. $-10 + 3(y + 4) = -10 + 3y + 12$

b. $= -10 + 12 + 3y$

      $= 2 + 3y$

81. a. $\triangle + (\square + \bigcirc) = (\square + \bigcirc) + \triangle$

b. $= (\bigcirc + \square) + \triangle$

c. $= \bigcirc + (\square + \triangle)$

82. a. $(x + y) + z = x + (y + z)$

b. $= (y + z) + x$

c. $= (z + y) + x$

83. Explain why 0 is called the identity element for addition.   answers may vary

84. Explain why 1 is called the identity element for multiplication.   answers may vary

80. a. distributive property   **80. b.** commutative property of addition   **81. a.** commutative property of addition
81. b. commutative property of addition   **81. c.** associative property of addition   **82. a.** associative property of addition
82. b. commutative property of addition   **82. c.** commutative property of addition

1.9 | READING GRAPHS

TAPE BA 1.9

O B J E C T I V E S

1. Read bar graphs.
2. Read line graphs.

In today's world, where the exchange of information is required to be fast and entertaining, graphs are becoming increasingly popular. They provide a quick way of making comparisons, drawing conclusions, and approximating quantities. Thus far, we have practiced reading bar graphs and circle graphs or pie charts. In this section we continue our study of bar graphs and we introduce line graphs.

A bar graph consists of a series of bars arranged vertically or horizontally. The bar graph on the next page shows a comparison of the rates charged by selected electricity companies. The names of the companies are listed horizontally and a bar is shown for each company. Corresponding to the height of the bar for each company is a number along a vertical axis. These vertical numbers are cents charged for each kilowatt-hour of electricity used.

EXAMPLE 1   The bar graph on the next page shows the cents charged per kilowatt-hour for selected electricity companies.

a. Which company charges the highest rate?

b. Which company charges the lowest rate?

c. Approximate the electricity rate charged by the first four companies listed.

d. Approximate the difference in the rates charged by the companies in parts (a) and (b).

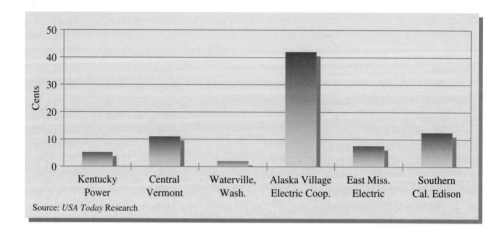

Source: *USA Today* Research

Solution: **a.** The tallest bar corresponds to the company that charges the highest rate. Alaska Village Electric Cooperative charges the highest rate.

b. The shortest bar corresponds to the company that charges the lowest rate. Waterville, Washington charges the lowest rate.

c. To approximate the rate charged by Kentucky Power, go to the top of the bar that corresponds to this company. From the top of the bar, move horizontally to the left until the vertical axis is reached.

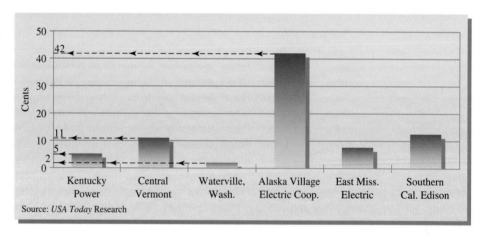

Source: *USA Today* Research

The height of the bar is approximately halfway between the 0 and 10 marks. We therefore conclude that

Kentucky Power charges approximately 5¢ per kilowatt-hour.

Central Vermont charges approximately 11¢ per kilowatt-hour.

Waterville, Washington charges approximately 2¢ per kilowatt-hour.

Alaska Village Electric charges approximately 42¢ per kilowatt-hour.

d. The difference in rates for Alaska Village Electric Cooperative and Waterville, Washington is approximately 42¢ − 2¢ or 40¢.

As mentioned earlier, a bar graph can consist of vertically arranged bars as in the graph above or horizontally arranged bars as in the next graph.

EXAMPLE 2   The following bar graph shows Disney's top six animated films before 1995 and the amount of money they generated at theaters.

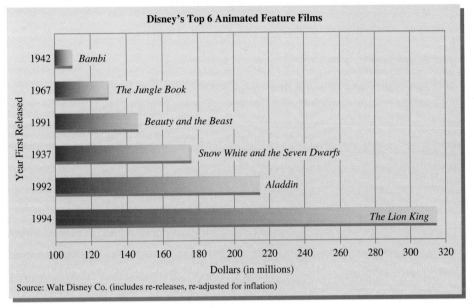

a. Find the film shown that generated the most income for Disney and approximate the income.

b. How much more money did the film *Aladdin* make than the film *Beauty and the Beast?*

Solution: **a.** Since these bars are arranged horizontally, look for the longest bar, which is the bar representing the film *The Lion King.* To approximate the income from this film, from the right edge of this bar move vertically downward to the dollars axis. This film generated approximately 315 million dollars, or $315,000,000, the most income for Disney.

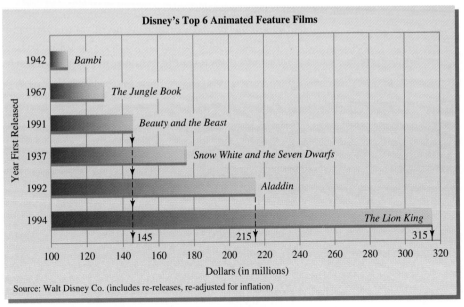

b. *Aladdin* generated approximately 215 million dollars. *Beauty and the Beast* generated approximately 145 million dollars. To find how much more money *Aladdin* generated than *Beauty and the Beast,* subtract: 215 − 145 = 70 million dollars, or $70,000,000.

The next graph is called a **line graph.**

EXAMPLE 3   The line graph below shows the relationship between two sets of measurements: the distance driven in a 14-foot U-Haul truck in one day and the total cost of renting this truck for that day. Notice that the horizontal axis is labeled Distance and the vertical axis is labeled Total Cost.

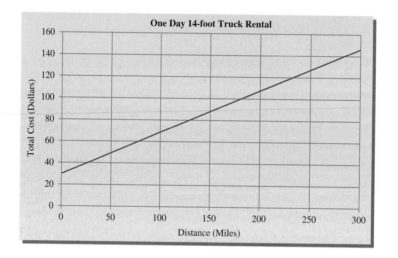

a. Find the total cost of renting the truck if 100 miles are driven.

b. Find the number of miles driven if the total cost of renting is $140.

Solution: **a.** Find the number 100 on the horizontal scale and move vertically upward until the line is reached. From this point on the line, move horizontally to the left until the vertical scale is reached. The total cost of renting the truck if 100 miles are driven is approximately $70.

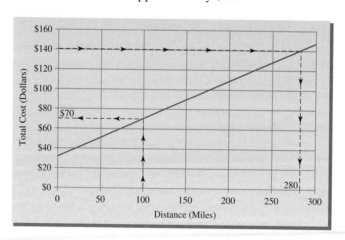

b. Find the number 140 on the vertical scale and move horizontally to the right until the line is reached. From this point on the line, move vertically downward until the horizontal scale is reached. The truck is driven approximately 280 miles.

From the previous example, we can see that graphing provides a quick way to approximate quantities. In Chapter 3 we show how we can use equations to find exact answers to the questions posed in Example 3. The next graph is another example of a line graph. It is also sometimes called a **broken line graph.**

EXAMPLE 4   This line graph shows the relationship between time spent smoking a cigarette and pulse rate. Time is recorded along the horizontal axis in minutes, with 0 minutes being the moment a smoker lights a cigarette. Pulse is recorded along the vertical axis in heartbeats per minute.

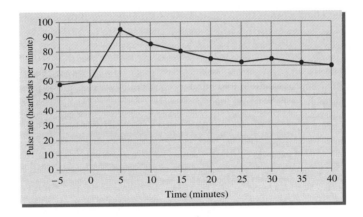

a. What is the pulse rate 15 minutes after lighting a cigarette?

b. When is the pulse rate the lowest?

c. When does the pulse rate show the greatest change?

Solution: **a.** Locate the number 15 along the time axis and move vertically upward until the line is reached. From this point on the line, move horizontally to the left until the pulse rate axis is reached. Read the number of beats per minute. The pulse rate is 80 beats per minute 15 minutes after lighting a cigarette.

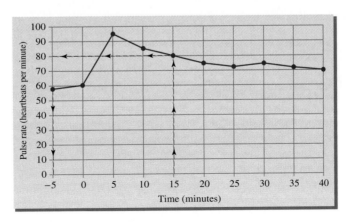

b. Find the lowest point of the line graph, which represents the lowest pulse rate. From this point, move vertically downward to the time axis. The pulse rate is the lowest at −5 minutes, which means 5 minutes *before* lighting a cigarette.

c. The pulse rate shows the greatest change during the 5 minutes between 0 and 5. Notice that the line graph is *steepest* between 0 and 5 minutes.

EXERCISE SET 1.9

Use the bar graph in Example 1 to answer the following.

1. Approximate the electricity rate charged by East Miss. Electric.   7¢

2. Approximate the electricity rate charged by Southern Cal. Edison.   12¢

3. Which companies shown charge more than 12¢ per kilowatt-hour?   Alaska Village Electric Coop.

4. Which companies shown charge less than 10¢ per kilowatt hour?

5. Find the rate charged by the company shown that is the nearest to your home.   answers may vary

Use the bar graph in Example 2 to answer the following.

6. Approximate the income generated by the film *Bambi*. $110 million

7. Approximate the income generated by the film *The Jungle Book*. $130 million

8. Approximate the income generated by the film *Snow White and the Seven Dwarfs*. $176 million

9. How much more money did the film *Snow White and the Seven Dwarfs* generate than *Bambi*?

10. Before 1990, which Disney film generated the most income?   *Snow White and the Seven Dwarfs*

11. After 1990, which Disney Film generated the most income?   *The Lion King*

Use the line graph in Example 3 to answer the following.

12. Find the total cost of renting the truck if 50 miles are driven. $50

13. Find the total cost of renting the truck if 230 miles are driven. $125

14. Find the number of miles driven if the total cost of renting is $80.   125 miles

15. Find the number of miles driven if the total cost of renting is $50.   50 miles

Use the line graph in Example 4 to answer the following.

16. Approximate the pulse rate 5 minutes before lighting a cigarette.   58 heartbeats per minute

17. Approximate the pulse rate 10 minutes after lighting a cigarette.   85 heartbeats per minute

18. Find the difference in pulse rate between 5 minutes before and 10 minutes after lighting a cigarette.

19. What is the highest pulse rate shown on the graph?

The following bar graph shows the team in each sport that has gone the longest time without being in a playoff.

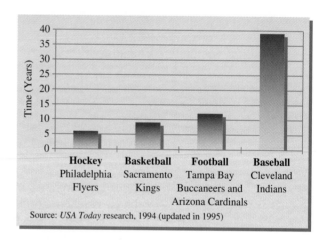

Source: *USA Today* research, 1994 (updated in 1995)

20. Which team for the sports shown has gone the longest time without being in a playoff?

21. Which hockey team has gone the longest without being in a playoff?   Philadelphia Flyers

22. Why are 2 football teams listed?

23. Approximate the greatest number of years that a basketball team has gone without being in the playoffs.   9 years

4. Kentucky Power; Waterville, Wash.; East Miss. Electric

19. 95 heartbeats per minute   **20.** Cleveland Indians being in a playoff.

9. $66 million   **18.** 27 heartbeats per minute

22. They have gone the same number of years without

24. How many more years has a football team gone without being in the playoffs than a basketball team? 3 years

25. How many more years has a football team gone without being in the playoffs than a hockey team?

The line graph below shows the average cost of newsprint per metric ton since 1984.

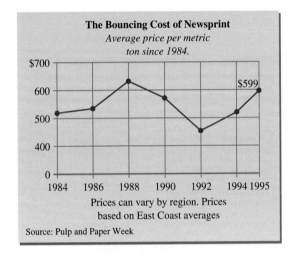

The Bouncing Cost of Newsprint
Average price per metric ton since 1984.

Prices can vary by region. Prices based on East Coast averages

Source: Pulp and Paper Week

26. During what year was the cost of newsprint less than $500 per metric ton? 1992

27. What year shown had the highest cost of newsprint? Estimate this cost. 1988; $633

28. What year shown had the lowest cost of newsprint? Estimate this cost. 1992; $444

29. Estimate the cost of newsprint in 1986. $533

30. In April 1995, the *Houston Post* newspaper closed. One reason given for this close is shown by a trend on this line graph. What is that trend?
rising costs after 1992

Geographic locations can be described by a gridwork of lines called latitudes and longitudes as shown below. For example, the location of Houston, Texas, can be described by latitude 30° north and longitude 95° west.

31. Using latitude and longitude, describe the location of New Orleans, Louisiana.

32. Using latitude and longitude, describe the location of Denver, Colorado.

33. Use an atlas and describe the location of your hometown. answers may vary

34. Give another name for 0° latitude. equator

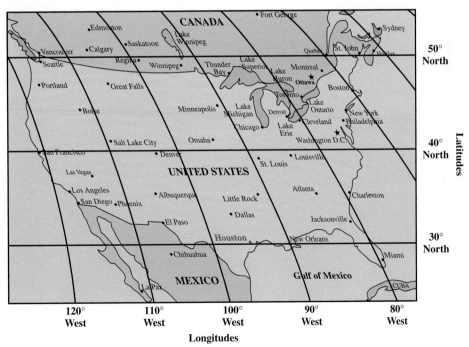

25. A football team has gone 6 years more without being in a playoff than a hockey team. **31.** latitude 30° north, longitude 90° west **32.** latitude 39° north, longitude 104° west

GROUP ACTIVITY

CREATING AND INTERPRETING GRAPHS

MATERIALS:

- Compass or circle template
- Grapher with bar graph and circle graph capabilities (optional)

Conduct the following survey during class and record the results.

a. Into which of the following categories does your age fall?

age ≤ 20	20 < age ≤ 25
25 < age ≤ 30	30 < age ≤ 35
35 < age ≤ 40	age > 40

b. What is your gender?

female   male

c. Which of the following brands of breakfast cereal did you eat today? (Choose only one.)

Kellogg's	Post
General Mills	Quaker
Ralston	Other

Did not eat breakfast cereal today

1. For each survey question, tally the results for each category. Present the results in a table.

2. For each survey question, find the fraction of the total number of responses that fall in each answer category.

3. Using a compass (or circle template), create a circle graph for each set of responses to questions (a)–(c). Label the graph.

4. Create a bar graph for each set of responses to questions (a)–(c). Label the graph.

5. Study your graphs and discuss what you may conclude from them. What do the graphs tell you about your survey respondents? Write a paragraph summarizing your group's conclusions. For any of the survey question results, do you prefer one type of graph over the other? Why?

6. (Optional) Use a grapher to create the circle graphs and bar graphs in Questions 3 and 4. Compare them to your own graphs. Are there any differences? Explain.

See App. F for Group Activity answers and suggestions.

CHAPTER 1   HIGHLIGHTS

DEFINITIONS AND CONCEPTS	EXAMPLES

SECTION 1.1   SYMBOLS AND SETS OF NUMBERS

A **set** is called a collection of objects, called **elements,** enclosed in braces.	$\{a, c, e\}$

Natural Numbers: $\{1, 2, 3, 4, \ldots\}$ **Whole Numbers:** $\{0, 1, 2, 3, 4, \ldots\}$ **Integers:** $\{\ldots, -3, -2, -1, 0, 1, 2, 3, \ldots\}$ **Rational Numbers:** {real numbers that can be expressed as a quotient of integers} **Irrational Numbers:** {real numbers that cannot be expressed as a quotient of integers} **Real Numbers:** {all numbers that correspond to a point on the number line}	Given the set $\left\{-3.4, \sqrt{3}, 0, \frac{2}{3}, 5, -4\right\}$ list the numbers that belong to the set of Natural numbers 5 Whole numbers 0, 5 Integers $-4, 0, 5$ Rational numbers $-3.4, 0, \frac{2}{3}, 5, -4$ Irrational numbers $\sqrt{3}$ Real numbers $-3.4, \sqrt{3}, 0, \frac{2}{3}, 5, -4$

A line used to picture numbers is called a **number line.**	

The **absolute value** of a real number a denoted by $|a|$ is the distance between a and 0 on the number line.	$|5| = 5 \qquad |0| = 0 \qquad |-2| = 2$

Symbols: $=$ is equal to $\neq$ is not equal to $>$ is greater than $<$ is less than $\leq$ is less than or equal to $\geq$ is greater than or equal to	$-7 = -7$ $3 \neq -3$ $4 > 1$ $1 < 4$ $6 \leq 6$ $18 \geq -\dfrac{1}{3}$

Order Property for Real Numbers For any two real numbers a and b, a is less than b if a is to the left of b on the number line.	 $-3 < 0 \qquad 0 > -3 \qquad 0 < 2.5 \qquad 2.5 > 0$

SECTION 1.2   FRACTIONS

A quotient of two integers is called a **fraction.** The **numerator** of a fraction is the top number. The **denominator** of a fraction is the bottom number.	$\dfrac{13}{17}\quad \begin{array}{l} \leftarrow \text{ numerator} \\ \leftarrow \text{ denominator} \end{array}$

If $a \cdot b = c$, then a and b are **factors** and c is the **product.**	$\underset{\text{factor}}{7} \quad \cdot \quad \underset{\text{factor}}{9} \quad = \quad \underset{\text{product}}{63}$

(continued)

DEFINITIONS AND CONCEPTS	EXAMPLES
SECTION 1.2 FRACTIONS	
A fraction is in **lowest terms** when the numerator and the denominator have no factors in common other than 1.	$\dfrac{13}{17}$ is in lowest terms.
To write a fraction in lowest terms, factor the numerator and the denominator; then apply the fundamental property.	Write in lowest terms. $\dfrac{6}{14} = \dfrac{2 \cdot 3}{2 \cdot 7} = \dfrac{3}{7}$
Two fractions are **reciprocals** if their product is 1. The reciprocal of $\frac{a}{b}$ is $\frac{b}{a}$.	The reciprocal of $\frac{6}{25}$ is $\frac{25}{6}$
To multiply fractions, numerator times numerator is the numerator of the product and denominator times denominator is the denominator of the product. **To divide fractions,** multiply the first fraction by the reciprocal of the second fraction. **To add fractions with the same denominator,** add the numerators and place the sum over the common denominator. **To subtract fractions with the same denominator,** subtract the numerators and place the difference over the common denominator.	Perform the indicated operations. $\dfrac{2}{5} \cdot \dfrac{3}{7} = \dfrac{6}{35}$ $\dfrac{5}{9} \div \dfrac{2}{7} = \dfrac{5}{9} \cdot \dfrac{7}{2} = \dfrac{35}{18}$ $\dfrac{5}{11} + \dfrac{3}{11} = \dfrac{8}{11}$ $\dfrac{13}{15} - \dfrac{3}{15} = \dfrac{10}{15} = \dfrac{2}{3}$
Fractions that represent the same quantity are called **equivalent fractions.**	$\dfrac{1}{5} = \dfrac{1 \cdot 4}{5 \cdot 4} = \dfrac{4}{20}$ $\dfrac{1}{5}$ and $\dfrac{4}{20}$ are equivalent fractions.
SECTION 1.3 EXPONENTS AND ORDER OF OPERATIONS	
The expression a^n is an **exponential expression.** The number a is called the **base;** it is the repeated factor. The number n is called the **exponent;** it is the number of times that the base is a factor.	$4^3 = 4 \cdot 4 \cdot 4 = 64$ $7^2 = 7 \cdot 7 = 49$
Order of Operations Simplify expressions in the following order. If grouping symbols are present, simplify expressions within those first, starting with the innermost set. Also, simplify the numerator and the denominator of a fraction separately. **1.** Simplify exponential expressions. **2.** Multiply or divide in order from left to right. **3.** Add or subtract in order from left to right.	$\dfrac{8^2 + 5(7 - 3)}{3 \cdot 7} = \dfrac{8^2 + 5(4)}{21}$ $= \dfrac{64 + 5(4)}{21}$ $= \dfrac{64 + 20}{21}$ $= \dfrac{84}{21}$ $= 4$

(continued)

DEFINITIONS AND CONCEPTS	EXAMPLES
SECTION 1.4 INTRODUCTION TO VARIABLE EXPRESSIONS AND EQUATIONS	

A symbol used to represent a number is called a **variable.**	Examples of variables are:
An **algebraic expression** is a collection of numbers, variables, operation symbols, and grouping symbols.	$$q, x, z$$ Examples of algebraic expressions are:
To **evaluate an algebraic expression** containing a variable, substitute a given number for the variable and simplify.	$$5x, 2(y - 6), \frac{q^2 - 3q + 1}{6}$$ Evaluate $x^2 - y^2$ if $x = 5$ and $y = 3$. $$x^2 - y^2 = (5)^2 - 3^2$$ $$= 25 - 9$$ $$= 16$$
A mathematical statement that two expressions are equal is called an **equation.**	Equations:
A **solution** or **root** of an equation is a value for the variable that makes the equation a true statement.	$$3x - 9 = 20$$ $$A = \pi r^2$$
	Determine whether 4 is a solution of $5x + 7 = 27$. $$5x + 7 = 27$$ $$5(4) + 7 = 27$$ $$20 + 7 = 27$$ $$27 = 27 \qquad \text{True}$$ 4 is a solution.

SECTION 1.5 ADDING REAL NUMBERS	

To Add Two Numbers with the Same Sign	Add.
1. Add their absolute values.	$$10 + 7 = 17$$
2. Use their common sign as the sign of the sum.	$$-3 + (-8) = -11$$
To Add Two Numbers with Different Signs	
1. Subtract their absolute values.	$$-25 + 5 = -20$$
2. Use the sign of the number whose absolute value is larger as the sign of the sum.	$$14 + (-9) = 5$$

Two numbers that are the same distance from 0 but lie on opposite sides of 0 are called **opposites** or **additive inverses.** The opposite of a number a is denoted by $-a$.	The opposite of -7 is 7. The opposite of 123 is -123.
The sum of a number a and its opposite, $-a$, is 0. $$a + (-a) = 0$$ If a is a number, then $-(-a) = a$	$$-4 + 4 = 0$$ $$12 + (-12) = 0$$ $$-(-8) = 8$$ $$-(-14) = 14$$

(continued)

DEFINITIONS AND CONCEPTS	EXAMPLES

SECTION 1.6 SUBTRACTING REAL NUMBERS

To subtract two numbers a and b, add the first number a to the opposite of the second number b.

$$a - b = a + (-b)$$

Subtract.

$$3 - (-44) = 3 + 44 = 47$$
$$-5 - 22 = -5 + (-22) = -27$$
$$-30 - (-30) = -30 + 30 = 0$$

SECTION 1.7 MULTIPLYING AND DIVIDING REAL NUMBERS

Quotient of two real numbers

$$\frac{a}{b} = a \cdot \frac{1}{b}$$

Multiply or divide.

$$\frac{42}{2} = 42 \cdot \frac{1}{2} = 21$$

Multiplying and Dividing Real Numbers

The product or quotient of two numbers with the same sign is a positive number. The product or quotient of two numbers with different signs is a negative number.

$$7 \cdot 8 = 56 \qquad -7 \cdot (-8) = 56$$
$$-2 \cdot 4 = -8 \qquad 2 \cdot (-4) = -8$$
$$\frac{90}{10} = 9 \qquad \frac{-90}{-10} = 9$$

Products and Quotients Involving Zero

The product of 0 and any number is 0.

$$b \cdot 0 = 0 \text{ and } 0 \cdot b = 0$$

$$\frac{42}{-6} = -7 \qquad \frac{-42}{6} = -7$$

$$-4 \cdot 0 = 0 \qquad 0 \cdot \left(-\frac{3}{4}\right) = 0$$

The quotient of a nonzero number and 0 is undefined.

$$\frac{b}{0} \text{ is undefined}$$

$$\frac{-85}{0} \text{ is undefined.}$$

The quotient of 0 and any nonzero number is 0.

$$\frac{0}{b} = 0$$

$$\frac{0}{18} = 0 \qquad \frac{0}{-47} = 0$$

SECTION 1.8 PROPERTIES OF REAL NUMBERS

Commutative Properties

Addition: $a + b = b + a$

Multiplication: $a \cdot b = b \cdot a$

Associative Properties

Addition: $(a + b) + c = a + (b + c)$

Multiplication: $(a \cdot b) \cdot c = a \cdot (b \cdot c)$

Two numbers whose product is 1 are called **multiplicative inverses** or **reciprocals.** The reciprocal of a nonzero number a is $\frac{1}{a}$ because $a \cdot \frac{1}{a} = 1$.

$$3 + (-7) = -7 + 3$$
$$-8 \cdot 5 = 5 \cdot (-8)$$

$$(5 + 10) + 20 = 5 + (10 + 20)$$
$$(-3 \cdot 2) \cdot 11 = -3 \cdot (2 \cdot 11)$$

The reciprocal of 3 is $\frac{1}{3}$.

The reciprocal of $-\frac{2}{5}$ is $-\frac{5}{2}$.

(continued)

DEFINITIONS AND CONCEPTS	EXAMPLES
SECTION 1.8 PROPERTIES OF REAL NUMBERS	

Distributive Property

$a(+ c) = a \cdot b + a \cdot c$

Identities

$a + 0 = a \qquad 0 + a = a$

$a \cdot 1 = a \qquad 1 \cdot a = a$

Inverses

Addition or opposite: $\quad a + (-a) = 0$

Multiplication or reciprocal: $\quad b \cdot \frac{1}{b} = 1$

$5(6 + 10) = 5 \cdot 6 + 5 \cdot 10$

$-2(3 + x) = -2 \cdot 3 + (-2)(x)$

$5 + 0 = 5 \qquad 0 + (-2) = -2$

$-14 \cdot 1 = -14 \qquad 1 \cdot 27 = 27$

$7 + (-7) = 0$

$3 \cdot \dfrac{1}{3} = 1$

SECTION 1.9   READING GRAPHS

To find the value on the vertical axis representing a location on a graph, move horizontally from the location on the graph until the vertical axis is reached. To find the value on the horizontal axis representing a location on a graph, move vertically from the location on the graph until the horizontal axis is reached.

The broken line graph to the right shows the average public classroom teachers' salaries for the school year ending in the years shown.

Estimate the average public teacher's salary for the school year ending in 1989. The average salary is approximately $29,500.

Find the earliest year that the average salary rose above $32,000. The year was 1991.

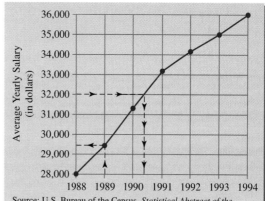

Source: U.S. Bureau of the Census, *Statistical Abstract of the United States: 1994* (114th edition) Washington, D.C., 1994

CHAPTER 1 REVIEW

(1.1) *Insert* $<$*,* $>$*, or* $=$ *in the appropriate space to make the following statements true.*

1. $8 < 10$

2. $7 > 2$

3. $-4 > -5$

4. $\dfrac{12}{2} > -8$

5. $|-7| < |-8|$

6. $|-9| > -9$

7. $-|-1| = -1$

8. $|-14| = -(-14)$

9. $1.2 > 1.02$

10. $-\dfrac{3}{2} < -\dfrac{3}{4}$

Translate each statement into symbols. **11.** $4 \geq -3$

11. Four is greater than or equal to negative three.

12. Six is not equal to five. $6 \neq 5$

13. 0.03 is less than 0.3. $0.03 < 0.3$

15. a. $\{1, 3\}$ **b.** $\{0, 1, 3\}$ **c.** $\{-6, 0, 1, 3\}$ **d.** $\{-6, 0, 1, 1\frac{1}{2}, 3, 9.62\}$ **e.** $\{\pi\}$ **f.** $\{-6, 0, 1, 1\frac{1}{2}, 3, 9.62, \pi\}$

14. Lions and hyenas were featured in the Disney film *The Lion King*. For short distances, lions can run at a rate of 50 miles per hour whereas hyenas can run at a rate of 40 miles per hour. Write an inequality statement comparing the numbers 50 and 40.
$50 > 40$

Given the following sets of numbers, list the numbers in each set that also belong to the set of:

a. Natural numbers   **b.** Whole numbers

c. Integers   **d.** Rational numbers

e. Irrational numbers   **f.** Real numbers

15. $\left\{-6, 0, 1, 1\frac{1}{2}, 3, \pi, 9.62\right\}$

16. $\left\{-3, -1.6, 2, 5, \frac{11}{2}, 15.1, \sqrt{5}, 2\pi\right\}$

The following chart shows the gains and losses in dollars of Density Oil and Gas stock for a particular week.

DAY	GAIN OR LOSS IN DOLLARS
Monday	+1
Tuesday	−2
Wednesday	+5
Thursday	+1
Friday	−4

17. Which day showed the greatest loss?   Friday

18. Which day showed the greatest gain?   Wednesday

(1.2) *Write the number as a product of prime factors.*

19. 36 $2 \cdot 2 \cdot 3 \cdot 3$

20. 120 $2 \cdot 2 \cdot 2 \cdot 3 \cdot 5$

Perform the indicated operations. Write results in lowest terms.

21. $\frac{8}{15} \cdot \frac{27}{30}$ $\frac{12}{25}$

22. $\frac{7}{8} \div \frac{21}{32}$ $\frac{4}{3}$

23. $\frac{7}{15} + \frac{5}{6}$ $\frac{13}{10}$

24. $\frac{3}{4} - \frac{3}{20}$ $\frac{3}{5}$

25. $2\frac{3}{4} + 6\frac{5}{8}$ $9\frac{3}{8}$

26. $7\frac{1}{6} - 2\frac{2}{3}$ $4\frac{1}{2}$

27. $5 \div \frac{1}{3}$ 15

28. $2 \cdot 8\frac{3}{4}$ $17\frac{1}{2}$

29. Determine the unknown part of the given circle. $\frac{7}{12}$

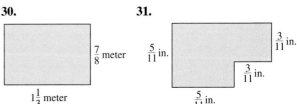

Find the area and the perimeter of each figure.

30.

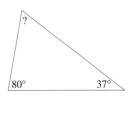

$\frac{7}{8}$ meter

$1\frac{1}{3}$ meter

31.

$\frac{5}{11}$ in.

$\frac{3}{11}$ in.

$\frac{3}{11}$ in.

$\frac{5}{11}$ in.

32. A trim carpenter needs a piece of quarter round molding $6\frac{1}{8}$ feet long for a bathroom. She finds a piece $7\frac{1}{2}$ feet long. How long a piece does she need to cut from the $7\frac{1}{2}$-foot-long molding in order to use it in the bathroom? $1\frac{3}{8}$ ft.

(1.3) *Simplify each expression.*

33. $6 \cdot 3^2 + 2 \cdot 8$ 70

34. $68 - 5 \cdot 2^3$ 28

35. $3(1 + 2 \cdot 5) + 4$ 37

36. $8 + 3(2 \cdot 6 - 1)$ 41

37. $\frac{4 + |6 - 2| + 8^2}{4 + 6 \cdot 4}$ $\frac{18}{7}$

38. $5[3(2 + 5) - 5]$ −70

Translate each word statement to symbols.

39. The difference of twenty and twelve is equal to the product of two and four. $20 - 12 = 2 \cdot 4$

40. The quotient of nine and two is greater than negative five. $\frac{9}{2} > -5$

(1.4) *Evaluate each expression if $x = 6$, $y = 2$, $z = 8$.*

41. $2x + 3y$ 18

42. $x(y + 2z)$ 108

43. $\frac{x}{y} + \frac{z}{2y}$ 5

44. $x^2 - 3y^2$ 24

45. The expression $180 - a - b$ represents the measure of the unknown angle of the given triangle. Replace a with 37 and b with 80 to find the measure of the unknown angle.   63°

16. a. $\{2, 5\}$ **b.** $\{2, 5\}$ **c.** $\{-3, 2, 5\}$ **d.** $\left\{-3, -1.6, 2, 5, \frac{11}{2}, 15.1\right\}$ **e.** $\{\sqrt{5}, 2\pi\}$ **f.** $\left\{-3, -1.6, 2, 5, \frac{11}{2}, 15.1, \sqrt{5}, 2\pi\right\}$

30. $A = 1\frac{1}{6}$ sq. meters $P = 4\frac{5}{12}$ meters   **31.** $A = \frac{34}{121}$ sq. in. $P = 2\frac{4}{11}$ in.

Decide whether the given number is a solution to the given equation.

46. $7x - 3 = 18$; 3   solution

47. $3x^2 + 4 = x - 1$; 1   not a solution

(1.5) *Find the additive inverse or the opposite.*

48. -9 9

49. $\dfrac{2}{3}$ $\dfrac{-2}{3}$

50. $|-2|$ -2

51. $-|-7|$ 7

Find the following sums.

52. $-15 + 4$ -11

53. $-6 + (-11)$ -17

54. $\dfrac{1}{16} + \left(-\dfrac{1}{4}\right)$ $-\dfrac{3}{16}$

55. $-8 + |-3|$ -5

56. $-4.6 + (-9.3)$ -13.9

57. $-2.8 + 6.7$ 3.9

(1.6) *Perform the indicated operations.*

58. $6 - 20$ -14

59. $-3.1 - 8.4$ -11.5

60. $-6 - (-11)$ 5

61. $4 - 15$ -11

62. $-21 - 16 + 3(8 - 2)$

63. $\dfrac{11 - (-9) + 6(8 - 2)}{2 + 3 \cdot 4}$

62. -19 **63.** 4

If $x = 3$, $y = -6$, and $z = -9$, evaluate each expression.

64. $2x^2 - y + z$ 15

65. $\dfrac{y - x + 5x}{2x}$ 1

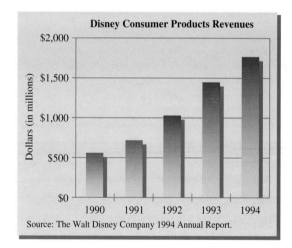 **66.** At the beginning of the week the price of Density Oil and Gas stock from Exercises 17 and 18 is $50 per share. Find the price of a share of stock at the end of the week. $51

Find the multiplicative inverse or reciprocal.

67. -6 $-\dfrac{1}{6}$

68. $\dfrac{3}{5}$ $\dfrac{5}{3}$

(1.7) *Simplify each expression.*

69. $6(-8)$ -48

70. $(-2)(-14)$ 28

71. $\dfrac{-18}{-6}$ 3

72. $\dfrac{42}{-3}$ -14

73. $-3(-6)(-2)$ -36

74. $(-4)(-3)(0)(-6)$ 0

75. $\dfrac{4 \cdot (-3) + (-8)}{2 + (-2)}$

76. $\dfrac{3(-2)^2 - 5}{-14}$ $-\dfrac{1}{2}$

75. undefined   **77.** commutative property of addition
84. associative property of multiplication

(1.8) *Name the property illustrated.*

77. $-6 + 5 = 5 + (-6)$

78. $6 \cdot 1 = 6$ multiplicative identity property

79. $3(8 - 5) = 3 \cdot 8 + 3 \cdot (-5)$ distributive property

80. $4 + (-4) = 0$ additive inverse property

81. $2 + (3 + 9) = (2 + 3) + 9$

82. $2 \cdot 8 = 8 \cdot 2$ commutative property of multiplication

83. $6(8 + 5) = 6 \cdot 8 + 6 \cdot 5$ distributive property

84. $(3 \cdot 8) \cdot 4 = 3 \cdot (8 \cdot 4)$

85. $4 \cdot \dfrac{1}{4} = 1$ multiplicative inverse

86. $8 + 0 = 8$ additive identity property

87. $4(8 + 3) = 4(3 + 8)$ commutative property of addition

(1.9) *Use the graph below showing Disney's consumer products revenues to answer the exercises.*

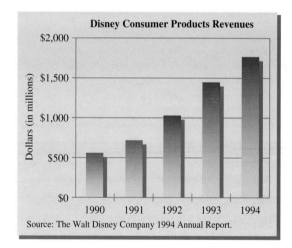

88. Approximate Disney's consumer products revenue in 1994. $1,800 million

89. Approximate the increase in consumer products revenue in 1992. $400 million

90. What year shows the greatest revenue?   1994

91. What trend is shown by this graph?   revenue is increasing

81. associative property of addition

27. associative property of addition   **28.** commutative property of multiplication

CHAPTER 1 TEST

Translate the statement into symbols.

1. The absolute value of negative seven is greater than five. $|-7| > 5$

2. The sum of nine and five is greater than or equal to four. $(9 + 5) \geq 4$

Simplify the expression.

3. $-13 + 8$ -5

4. $-13 - (-2)$ -11

5. $6 \cdot 3 - 8 \cdot 4$ -14

6. $(13)(-3)$ -39

7. $(-6)(-2)$ 12

8. $\dfrac{|-16|}{-8}$ -2

9. $\dfrac{-8}{0}$ undefined

10. $\dfrac{|-6| + 2}{5 - 6}$ -8

11. $\dfrac{1}{2} - \dfrac{5}{6}$ $-\dfrac{1}{3}$

12. $-1\dfrac{1}{8} + 5\dfrac{3}{4}$ $4\dfrac{5}{8}$

13. $-\dfrac{3}{6} + \dfrac{15}{8}$ $\dfrac{51}{40}$

14. $3(-4)^2 - 80$ -32

15. $6[5 + 2(3 - 8) - 3]$ -48

16. $\dfrac{-12 + 3 \cdot 8}{4}$ 3

17. $\dfrac{(-2)(0)(-3)}{-6}$ 0

Insert $<$, $>$, or $=$ in the appropriate space to make each of the following statements true.

18. $-3 > -7$

19. $4 > -8$

20. $|-3| > 2$

21. $|-2| = -1 - (-3)$

22. Given $\left\{-5, -1, \dfrac{1}{4}, 0, 1, 7, 11.6, \sqrt{7}, 3\pi\right\}$, list the numbers in this set that also belong to the set of:

   a. Natural numbers $\{1, 7\}$

   b. Whole numbers $\{0, 1, 7\}$

   c. Integers $\{-5, -1, 0, 1, 7\}$

   d. Rational numbers $\left\{-5, -1, \dfrac{1}{4}, 0, 1, 7, 11.6\right\}$

   e. Irrational numbers $\{\sqrt{7}, 3\pi\}$

   f. Real numbers $\left\{-5, -1, \dfrac{1}{4}, 0, 1, 7, 11.6, \sqrt{7}, 3\pi\right\}$

If $x = 6$, $y = -2$, and $z = -3$, evaluate each expression.

23. $x^2 + y^2$ 40

24. $x + yz$ 12

25. $2 + 3x - y$ 22

26. $\dfrac{y + z - 1}{x}$ -1

Identify the property illustrated by each expression.

27. $8 + (9 + 3) = (8 + 9) + 3$

28. $6 \cdot 8 = 8 \cdot 6$

29. $-6(2 + 4) = -6 \cdot 2 + (-6) \cdot 4$ distributive property

30. $\dfrac{1}{6}(6) = 1$ multiplicative inverse

31. Find the opposite of -9. 9

32. Find the reciprocal of $-\dfrac{1}{3}$. -3

The New Orleans Saints were 22 yards from the goal when the following series of gains and losses occurred.

	GAINS AND LOSSES IN YARDS
First Down	5
Second Down	−10
Third Down	−2
Fourth Down	29

33. During which down did the greatest loss of yardage occur?   second down

34. Was a touchdown scored?   yes

35. The temperature at the Winter Olympics was a frigid 14 degrees below zero in the morning, but by noon it had risen 31 degrees. What was the temperature at noon? $17°$

36. Jean Avarez decided to sell 280 shares of stock, which decreased in value by $1.50 per share yesterday. How much money did she lose?   loss of $420

Intel is a semiconductor manufacturer that makes almost one-third of the world's computer chips. (You may have seen the slogan "Intel Inside" in commercials on television.) The line graph to the right shows Intel's net revenues in billions of dollars. Use this figure to answer the questions below.

37. Estimate Intel's revenue in 1993. $8 billion

38. Estimate Intel's revenue in 1989. $3 billion

39. Find the increase in Intel's revenue from 1993 to 1995. $5.5 billion

40. What year shows the greatest increase in revenue?   1994

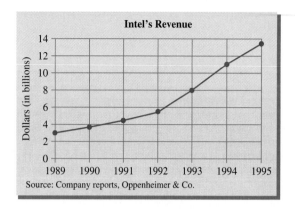

Source: Company reports, Oppenheimer & Co.

EQUATIONS, INEQUALITIES, AND PROBLEM SOLVING

CALCULATING PRICE PER UNIT

Business people, such as caterers and printers, provide services as well as products to their customers. They must often calculate a price per unit (item, person, hour, etc.) to charge their customers that covers the cost of all materials as well as payment for their time and labor. As a first step in calculating price per unit, it is often helpful to start by figuring the business person's actual cost per unit of the materials, which can be described by an algebraic formula.

IN THE CHAPTER GROUP ACTIVITY ON PAGE 149, YOU WILL HAVE THE OPPORTUNITY TO HELP A CATERER FIGURE A PRICE PER PERSON FOR FRUIT SALAD TO QUOTE TO A PROSPECTIVE CUSTOMER.

Much of mathematics relates to deciding which statements are true and which are false. When a statement, such as an equation, contains variables, it is usually not possible to decide whether the equation is true or false until the variable has been replaced by a value. For example, the statement $x + 7 = 15$ is an equation stating that the sum $x + 7$ has the same value as 15. Is this statement true or false? It is false for some values of x and true for just one value of x, namely 8. Our purpose in this chapter is to learn ways of deciding which values make an equation or an inequality true. To begin, we spend a bit more time learning about algebraic expressions.

2.1 SIMPLIFYING ALGEBRAIC EXPRESSIONS

OBJECTIVES

1. Identify terms, like terms, and unlike terms.
2. Combine like terms.
3. Use the distributive property to remove parentheses.
4. Write word phrases as algebraic expressions.

TAPE BA 2.1

As we explore in this section, an expression such as $3x + 2x$ is not as simple as possible, because—even without replacing x by a value—we can perform the indicated addition.

Before we practice simplifying expressions, some new language is presented. A **term** is a number or the product of a number and variables raised to powers. For example,

$$-y, \quad 2x^3, \quad -5, \quad 3xz^2, \quad \frac{2}{y}, \quad 0.8z$$

are terms. The **numerical coefficient** of a term is the numerical factor. The numerical coefficient of $3x$ is 3. Recall that $3x$ means $3 \cdot x$.

TERM	NUMERICAL COEFFICIENT
$3x$	3
$\dfrac{y^3}{5}$	$\dfrac{1}{5}$ since $\dfrac{y^3}{5}$ means $\dfrac{1}{5} \cdot y^3$
$-0.7ab^3c^5$	-0.7
z	1
$-y$	-1
-5	-5

> REMINDER   The term $-y$ means $-1y$ and thus has a numerical coefficient of -1. The term z means $1z$ and thus has a numerical coefficient of 1.

EXAMPLE 1   Identify the numerical coefficient.

a. $-3y$ **b.** $22z^4$ **c.** y **d.** $-x$ **e.** $\dfrac{x}{7}$

Solution: **a.** The numerical coefficient of $-3y$ is -3.
b. The numerical coefficient of $22z^4$ is 22.
c. The numerical coefficient of y is 1, since y is $1y$.
d. The numerical coefficient of $-x$ is -1, since $-x$ is $-1x$.
e. The numerical coefficient of $\frac{x}{7}$ is $\frac{1}{7}$, since $\frac{x}{7}$ is $\frac{1}{7} \cdot x$.

Terms with the same variables raised to exactly the same powers are called **like terms.**

LIKE TERMS	UNLIKE TERMS	
$3x, 2x$	$5x, 5x^2$	Why? Same variable x, but different powers x and x^2
$-6x^2y, 2x^2y, 4x^2y$	$7y, 3z, 8x^2$	Why? Different variables
$2ab^2c^3, ac^3b^2$	$6abc^3, 6ab^2$	Why? Different variables and different powers

Each variable and its exponent must match exactly in like terms, but like terms need not have the same numerical coefficients, nor do their factors need to be in the same order. For example, $2x^2y$ and $-yx^2$ are like terms.

EXAMPLE 2   Tell whether the terms are like or unlike.

a. $-x^2, 3x^3$ **b.** $4x^2y, x^2y, -2x^2y$ **c.** $-2yz, -3zy$ **d.** $-x^4, x^4$

Solution: **a.** Unlike terms, since the exponents on x are not the same.
b. Like terms, since each variable and its exponent match.
c. Like terms, since $zy = yz$ by the commutative property.
d. Like terms.

2 An algebraic expression containing the sum or difference of like terms can be simplified by applying the distributive property. For example, by the distributive property, we rewrite the sum of the like terms $3x + 2x$ as

$$3x + 2x = (3 + 2)x = 5x$$

Also,

$$-y^2 + 5y^2 = (-1 + 5)y^2 = 4y^2$$

Simplifying the sum or difference of like terms is called **combining like terms.**

EXAMPLE 3   Simplify the following by combining like terms.

a. $7x - 3x$ b. $10y^2 + y^2$ c. $8x^2 + 2x - 3x$

Solution: a. $7x - 3x = (7 - 3)x = 4x$
b. $10y^2 + y^2 = (10 + 1)y^2 = 11y^2$
c. $8x^2 + 2x - 3x = 8x^2 + (2 - 3)x = 8x^2 - x$

EXAMPLE 4   Simplify each expression by combining like terms.

a. $2x + 3x + 5 + 2$ b. $-5a - 3 + a + 2$ c. $4y - 3y^2$ d. $2.3x + 5x - 6$

Solution: Use the distributive property to combine the numerical coefficients of like terms.

a. $2x + 3x + 5 + 2 = (2 + 3)x + (5 + 2)$
$$= 5x + 7$$

b. $-5a - 3 + a + 2 = -5a + 1a + (-3 + 2)$
$$= (-5 + 1)a + (-3 + 2)$$
$$= -4a - 1$$

c. $4y - 3y^2$ These two terms cannot be combined because they are unlike terms.
d. $2.3x + 5x - 6 = (2.3 + 5)x - 6$
$$= 7.3x - 6$$

The examples above suggest the following:

> To **combine like terms,** add the numerical coefficients and multiply the
> result by the common variable factors.

3   Simplifying expressions makes frequent use of the distributive property to remove
parentheses.

EXAMPLE 5   Find each product by using the distributive property to remove parentheses.

a. $5(x + 2)$ b. $-2(y + 0.3z - 1)$ c. $-(x + y - 2z + 6)$

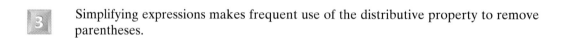

Solution: a. $5(x + 2) = 5(x) + 5(2)$ Apply the distributive
property.

$$= 5x + 10$$ Multiply.

b. $-2(y + 0.3z - 1) = -2(y) + (-2)(0.3z) + (-2)(-1)$ Apply the distributive property

$$= -2y - 0.6z + 2$$ Multiply.

c. $-(x + y - 2z + 6) = -1(x + y - 2z + 6)$ Distribute -1 over each term.

$$= -1(x) - 1(y) - 1(-2z) - 1(6)$$

$$= -x - y + 2z - 6$$

R E M I N D E R   If a "$-$" sign precedes parentheses, the sign of each term inside the parentheses is changed when the distributive property is applied to remove parentheses.

Examples:

$$-(2x + 1) = -2x - 1$$ $$-(x - 2y) = -x + 2y$$

$$-(-5x + y - z) = 5x - y + z$$ $$-(-3x - 4y - 1) = 3x + 4y + 1$$

To simplify an expression containing parentheses, we use the distributive property to remove parentheses and then the distributive property to combine any like terms.

EXAMPLE 6   Simplify the following expressions.

a. $3(2x - 5) + 1$ **b.** $8 - (7x + 2) + 3x$ **c.** $-2(4x + 7) - (3x - 1)$

Solution: **a.** $3(2x - 5) + 1 = 6x - 15 + 1$ Apply the distributive property.

$$= 6x - 14$$ Combine like terms.

b. $8 - (7x + 2) + 3x = 8 - 7x - 2 + 3x$ Apply the distributive property.

$$= -7x + 3x + 8 - 2$$

$$= -4x + 6$$ Combine like terms.

c. $-2(4x + 7) - (3x - 1) = -8x - 14 - 3x + 1$ Apply the distributive property.

$$= -11x - 13$$ Combine like terms.

EXAMPLE 7   Subtract $4x - 2$ from $2x - 3$.

Solution: "Subtract $4x - 2$ **from** $2x - 3$" translates to $(2x - 3) - (4x - 2)$. Next, simplify the algebraic expression.

$$(2x - 3) - (4x - 2) = 2x - 3 - 4x + 2$$ Apply the distributive property.

$$= -2x - 1$$ Combine like terms.

4

Next, we practice writing word phrases as algebraic expressions.

EXAMPLE 8 Write the following phrases as algebraic expressions and simplify if possible. Let x represent the unknown number.

a. Twice a number, added to 6.

b. The difference of a number and 4, divided by 7.

c. Five added to 3 times the sum of a number and 1.

Solution: **a.**

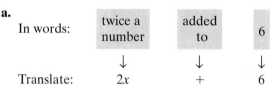

b.

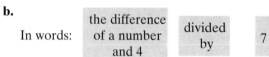

c.

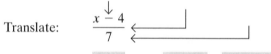

Next, we simplify this expression.

$$5 + 3(x + 1) = 5 + 3x + 3$$
$$= 8 + 3x$$

MENTAL MATH

Identify the numerical coefficient of each term. See Example 1.

1. $-7y$ -7

2. $3x$ 3

3. x 1

4. $-y$ -1

5. $17x^2y$ 17

6. $1.2xyz$ 1.2

Indicate whether the following lists of terms are like or unlike. See Example 2.

7. $5y, -y$ like

8. $-2x^2y, 6xy$ unlike

9. $2z, 3z^2$ unlike

10. $ab^2, -7ab^2$ like

11. $8wz, \frac{1}{7}zw$ like

12. $7.4p^3q^2, 6.2p^3q^2r$ unlike

EXERCISE SET 2.1

5. $-7b - 9$ **6.** $3g - 2$ **7.** $-m - 6$ **8.** $-3a - 2$

Simplify each expression by combining any like terms.
See Examples 3 and 4.

1. $7y + 8y$ $15y$
2. $5x - 2x$ $3x$

3. $8w - w + 6w$ $13w$
4. $c - 7c + 2c$ $-4c$

5. $3b - 5 - 10b - 4$
6. $6g + 5 - 3g - 7$

7. $m - 4m + 2m - 6$
8. $a + 3a - 2 - 7a$

11. $7d - 11$ **12.** $9z + 48$ **13.** $-3x + 2y - 1$

Simplify each expression. Use the distributive property
to remove any parentheses. See Examples 5 and 6.

9. $5(y - 4)$ $5y - 20$
10. $7(r - 3)$ $7r - 21$

11. $7(d - 3) + 10$
12. $9(z + 7) - 15$

13. $-(3x - 2y + 1)$
14. $-(y + 5z - 7)$

15. $5(x + 2) - (3x - 4)$
16. $4(2x - 3) - 2(x + 1)$

17. In your own words, explain how to combine like terms.   answers may vary

18. Do like terms contain the same numerical coefficients? Explain your answer.   answers may vary

14. $-y - 5z + 7$ **15.** $2x + 14$ **16.** $6x - 14$

Write each of the following as an algebraic expression.
Simplify if possible. See Example 7.

19. Add $6x + 7$ to $4x - 10$. $10x - 3$

20. Add $3y - 5$ to $y + 16$. $4y + 11$

21. Subtract $7x + 1$ from $3x - 8$. $-4x - 9$

22. Subtract $4x - 7$ from $12 + x$. $19 - 3x$

24. $\dfrac{x - 2}{5}$ **25.** $\dfrac{3}{4}x + 12$

Write each of the following phrases as an algebraic ex-
pression and simplify if possible. Let x represent the un-
known number. See Example 8.

23. Twice a number decreased by four. $2x - 4$

24. The difference of a number and two, divided by five.

25. Three-fourths of a number increased by twelve.

26. Eight more than triple the number. $3x + 8$

27. The sum of -2 and 5 times a number, added to 7 times a number. $-2 + 12x$

28. The sum of 3 times a number and 10, **subtracted from** 9 times a number. $6x - 10$

31. $4x - 3$ **33.** $8x - 53$ **34.** $5r - 56$ **36.** -11

Simplify each expression.

29. $7x^2 + 8x^2 - 10x^2$ $5x^2$ **30.** $8x + x - 11x$ $-2x$

31. $6x - 5x + x - 3 + 2x$

32. $8h + 13h - 6 + 7h - h$ $27h - 6$

33. $-5 + 8(x - 6)$
34. $-6 + 5(r - 10)$

35. $5g - 3 - 5 - 5g$ -8
36. $8p + 4 - 8p - 15$

37. $6.2x - 4 + x - 1.2$
38. $7.9y - 0.7 - y + 0.2$

39. $2k - k - 6$ $k - 6$
40. $7c - 8 - c$ $6c - 8$

41. $0.5(m + 2) + 0.4m$
42. $0.2(k + 8) - 0.1k$

43. $-4(3y - 4)$ $-12y + 16$ **44.** $-3(2x + 5)$

45. $3(2x - 5) - 5(x - 4)$
46. $2(6x - 1) - (x - 7)$

47. $3.4m - 4 - 3.4m$ -7

48. $2.8w - 0.9 - 0.5 - 2.8w$ -1.4

49. $6x + 0.5 - 4.3x - 0.4x + 3$

50. $0.4y - 6.7 + y - 0.3 - 2.6y$

51. $-2(3x - 4) + 7x - 6$
52. $8y - 2 - 3(y + 4)$

53. $-9x + 4x + 18 - 10x$
54. $5y - 14 + 7y - 20y$

55. $5k - (3k - 10)$
56. $-11c - (4 - 2c)$

57. $(3x + 4) - (6x - 1)$
58. $(8 - 5y) - (4 + 3y)$

59. Recall that the perimeter of a figure is the total distance around the figure. Given the following rectangle, express the perimeter as an algebraic expression containing the variable x. $(18x - 2)$ ft.

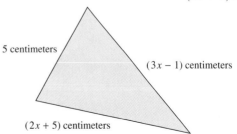

5x feet

$(4x - 1)$ feet

$(4x - 1)$ feet

5x feet

60. Given the following triangle, express its perimeter as an algebraic expression containing the variable x. $(5x + 9)$ cm

5 centimeters

$(3x - 1)$ centimeters

$(2x + 5)$ centimeters

37. $7.2x - 5.2$ **38.** $6.9y - 0.5$ **41.** $0.9m + 1$ **42.** $0.1k + 1.6$ **44.** $-6x - 15$ **45.** $x + 5$ **46.** $11x + 5$
47. -11 **49.** $1.3x + 3.5$ **50.** $-1.2y - 7$ **51.** $x + 2$ **52.** $5y - 14$ **53.** $-15x + 18$ **54.** $-8y - 14$
55. $2k + 10$ **56.** $-9c - 4$ **57.** $-3x + 5$
58. $4 - 8y$

Write each of the following as an algebraic expression. Simplify if possible.

61. Subtract $5m - 6$ from $m - 9$. $-4m - 3$

62. Subtract $m - 3$ from $2m - 6$. $m - 3$

63. Eight times the sum of a number and six. $8(x + 6)$

64. Five less than four times the number. $4x - 5$

65. Double a number minus the sum of the number and ten. $x - 10$

66. Half a number minus the product of the number and eight. $-7.5x$

67. Seven multiplied by the quotient of a number and six.

68. The product of a number and ten, less twenty.

Given the following, determine whether each scale is balanced or not.

1 cone balances 1 cube

1 cylinder balances 2 cubes

69.

70.

71.

72.

73. To convert from feet to inches, we multiply by 12. For example, the number of inches in 2 feet is $12 \cdot 2$ inches. If one board has a length of $(x + 2)$ *feet* and a second board has a length of $3x - 1$ *inches,* express their total length in inches as an algebraic expression. $(15x + 23)$ in.

74. The value of 7 nickels is $5 \cdot 7$ cents. Likewise, the value of x nickels is $5x$ cents. If the money box in a drink machine contains x *nickels,* $3x$ *dimes,* and $30x - 1$ *quarters,* express their total value in cents as an algebraic expression. $(785x - 25)$ ¢

Review Exercises

Evaluate the following expressions for the given values. See Section 1.7.

75. If $x = -1$ and $y = 3$, find $y - x^2$. 2

76. If $g = 0$ and $h = -4$, find $gh - h^2$. -16

77. If $a = 2$ and $b = -5$, find $a - b^2$. -23

78. If $x = -3$, find $x^3 - x^2 + 4$. -32

79. If $y = -5$ and $z = 0$, find $yz - y^2$. -25

80. If $x = -2$, find $x^3 - x^2 - x$. -10

A Look Ahead

EXAMPLE

Simplify $-3xy + 2x^2y - (2xy - 1)$.

Solution:

$$-3xy + 2x^2y - (2xy - 1) = -3xy + 2x^2y - 2xy + 1$$
$$= -5xy + 2x^2y + 1$$

Simplify each expression.

81. $5b^2c^3 + 8b^3c^2 - 7b^3c^2$ $5b^2c^3 + b^3c^2$

82. $4m^4p^2 + m^4p^2 - 5m^2p^4$ $5m^4p^2 - 5m^2p^4$

83. $3x - (2x^2 - 6x) + 7x^2$ $5x^2 + 9x$

84. $9y^2 - (6xy^2 - 5y^2) - 8xy^2$ $14y^2 - 14xy^2$

85. $-(2x^2y + 3z) + 3z - 5x^2y$ $-7x^2y$

86. $-(7c^3d - 8c) - 5c - 4c^3d$ $-11c^3d + 3c$

67. $\dfrac{7x}{6}$ **68.** $10x - 20$ **69.** balanced   **70.** not balanced   **71.** balanced   **72.** balanced

2.2 | THE ADDITION PROPERTY OF EQUALITY

O B J E C T I V E S

1. Define linear equation in one variable and equivalent equations.
2. Use the addition property of equality to solve linear equations.
3. Write word phrases as algebraic expressions.

Recall from Section 1.4 that an equation is a statement that two expressions have the same value. Also, a value of the variable that makes an equation a true statement is called a solution or root of the equation. The process of finding the solution of an equation is called **solving** the equation for the variable. In this section we concentrate on solving **linear equations** in one variable.

> **LINEAR EQUATION IN ONE VARIABLE**
>
> A linear equation in one variable can be written in the form
>
> $$ax + b = c$$
>
> where a, b, and c are real numbers and $a \neq 0$.

Evaluating a linear equation for a given value of the variable, as we did in Section 1.4, can tell us whether that value is a solution, but we can't rely on evaluating an equation as our method of solving it.

Instead, to solve a linear equation in x, we write a series of simpler equations, all *equivalent* to the original equation, so that the final equation has the form

$$x = \textbf{number} \qquad \text{or} \qquad \textbf{number} = x$$

Equivalent equations are equations that have the same solution. This means that the "number" above is the solution to the original equation.

The first property of equality that helps us write simpler equivalent equations is the **addition property of equality.**

> **ADDITION PROPERTY OF EQUALITY**
>
> If a, b, and c are real numbers, then
>
> $$a = b \qquad \text{and} \qquad a + c = b + c$$
>
> are equivalent equations.

This property guarantees that adding the same number to both sides of an equation does not change the solution of the equation. Since subtraction is defined in terms of addition, we may also **subtract the same number from both sides** without changing the solution.

A good way to picture a true equation is as a balanced scale. Since it is balanced, each side of the scale weighs the same amount.

If the same weight is added to or subtracted from each side, the scale remains balanced.

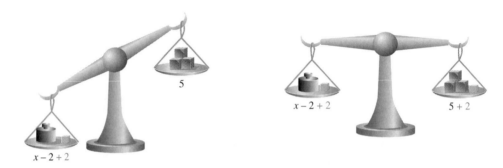

We use the addition property of equality to write equivalent equations until the variable is **isolated** (by itself on one side of the equation) and the equation looks like "x = number" or "number = x."

EXAMPLE 1   Solve $x - 7 = 10$ for x.

Solution: To solve for x, isolate x on one side of the equation. To do this, add 7 to both sides of the equation.

$$x - 7 = 10$$
$$x - 7 + 7 = 10 + 7 \qquad \text{Add 7 to both sides.}$$
$$x = 17 \qquad \text{Simplify.}$$

The solution of the equation $x = 17$ is obviously 17. Since we are writing equivalent equations, the solution of the equation $x - 7 = 10$ is also 17.

To check, replace x with 17 in the original equation.

$$x - 7 = 10$$
$$17 - 7 = 10 \qquad \text{Replace } x \text{ with 17 in the original equation.}$$
$$10 = 10 \qquad \text{True.}$$

Since the statement is true, 17 is the solution and the solution set is $\{17\}$. ▬▬

EXAMPLE 2   Solve $y + 0.6 = -1.0$.

Solution: To solve for y, subtract 0.6 from both sides of the equation.

$$y + 0.6 = -1.0$$
$$y + 0.6 - 0.6 = -1.0 - 0.6 \qquad \text{Subtract 0.6 from both sides.}$$
$$y = -1.6 \qquad \text{Combine like terms.}$$

To check the proposed solution, -1.6, replace y with -1.6 in the original equation.

Check:

$$y + 0.6 = -1.0$$
$$-1.6 + 0.6 = -1.0 \qquad \text{Replace } y \text{ with } -1.6 \text{ in the original equation.}$$
$$-1.0 = -1.0 \qquad \text{True.}$$

The solution set is $\{-1.6\}$.

EXAMPLE 3   Solve $5t - 5 = 6t + 2$ for t.

Solution: To solve for t, we first want all terms containing t on one side of the equation and all other terms on the other side of the equation. To do this, first subtract $5t$ from both sides of the equation.

$$5t - 5 = 6t + 2$$
$$5t - 5 - 5t = 6t + 2 - 5t \qquad \text{Subtract } 5t \text{ from both sides.}$$
$$-5 = t + 2 \qquad \text{Combine like terms.}$$

Next, subtract 2 from both sides and the variable t will be isolated.

$$-5 = t + 2$$
$$-5 - 2 = t + 2 - 2 \qquad \text{Subtract 2 from both sides.}$$
$$-7 = t$$

Check the solution, -7, in the original equation. The solution set is $\{-7\}$.

> **R E M I N D E R**   We may isolate the variable on either side of the equation. $-7 = t$ is equivalent to $t = -7$.

Many times, it is best to simplify one or both sides of an equation before applying the addition property of equality.

EXAMPLE 4   Solve $2x + 3x - 5 + 7 = 10x + 3 - 6x - 4$ for x.

Solution: First, simplify both sides of the equation.

$$2x + 3x - 5 + 7 = 10x + 3 - 6x - 4$$

$$5x + 2 = 4x - 1 \qquad \text{Combine like terms on each side of the equation.}$$

$$5x + 2 - 4x = 4x - 1 - 4x \qquad \text{Subtract } 4x \text{ from both sides.}$$

$$x + 2 = -1 \qquad \text{Combine like terms.}$$

$$x + 2 - 2 = -1 - 2 \qquad \text{Subtract 2 from both sides.}$$

$$x = -3 \qquad \text{Combine like terms.}$$

Check by replacing x with -3 in the original equation.

$$2x + 3x - 5 + 7 = 10x + 3 - 6x - 4$$

$$2(-3) + 3(-3) - 5 + 7 = 10(-3) + 3 - 6(-3) - 4 \qquad \text{Replace } x \text{ with } -3.$$

$$-6 - 9 - 5 + 7 = -30 + 3 + 18 - 4 \qquad \text{Multiply.}$$

$$-13 = -13 \qquad \text{True.}$$

The solution set is $\{-3\}$.

If an equation contains parentheses, use the distributive property to remove them.

EXAMPLE 5 Solve $-5(2a - 1) - (-11a + 6) = 7$ for a.

Solution:
$$-5(2a - 1) - (-11a + 6) = 7$$

$$-10a + 5 + 11a - 6 = 7 \qquad \text{Apply the distributive property.}$$

$$a - 1 = 7 \qquad \text{Combine like terms.}$$

$$a - 1 + 1 = 7 + 1 \qquad \text{Add 1 to both sides to isolate } a$$

$$a = 8 \qquad \text{Combine like terms.}$$

Check to see that 8 is the solution or the solution set is $\{8\}$.

When solving equations, we may sometimes encounter an equation such as

$$-x = 5$$

This equation is not solved for x because x is not isolated. To solve this equation for x, recall that

"$-$" can be read as "the opposite of"

We can read the equation $-x = 5$, then, as "the opposite of x is 5." If the opposite of x is 5, this means that x is the opposite of 5 or -5.

In summary,

$$-x = 5 \qquad \text{and} \qquad x = -5$$

are equivalent equations and $x = -5$ is solved for x.

EXAMPLE 6   Solve $3 - x = 7$ for x.

Solution: First, subtract 3 from both sides.

$$3 - x = 7$$
$$3 - x - 3 = 7 - 3 \qquad \text{Subtract 3 from both sides.}$$
$$-x = 4 \qquad \text{Simplify.}$$

Since the opposite of x is 4, this means that x is the opposite of 4, or -4.

$$x = -4$$

Check by replacing x with -4 in the original equation.

$$3 - x = 7$$
$$3 - (-4) = 7 \qquad \text{Replace } x \text{ with } -4.$$
$$7 = 7 \qquad \text{True.}$$

The solution set is $\{-4\}$.

 Next, we practice writing algebraic expressions.

EXAMPLE 7   **a.** The sum of two numbers is 8. If one number is 3, find the other number.
   b. The sum of two numbers is 8. If one number is x, write an expression representing the other number.

Solution: **a.** If the sum of two numbers is 8 and one number is 3, we find the other number by subtracting 3 from 8. The other number is $8 - 3$ or 5.

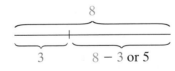

   b. If the sum of two numbers is 8 and one number is x, we find the other number by subtracting x from 8. The other number is represented by $8 - x$.

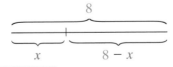

MENTAL MATH

Solve each equation mentally. See Examples 1 and 2.

1. $x + 4 = 6$ $x = 2$
2. $x + 7 = 10$ $x = 3$
3. $n + 18 = 30$ $n = 12$
4. $z + 22 = 40$ $z = 18$
5. $b - 11 = 6$ $b = 17$
6. $d - 16 = 5$ $d = 21$

EXERCISE SET 2.2

Solve each equation. See Examples 1 and 2.

1. $x + 11 = -2$ $x = -13$ **2.** $y - 5 = -9$ $y = -4$

3. $5y + 14 = 4y$ $y = -14$ **4.** $8x - 7 = 9x$ $x = -7$

5. $8x = 7x - 8$ $x = -8$ **6.** $x = 2x + 3$ $x = -3$

7. $x - 2 = -4$ $x = -2$ **8.** $y + 7 = 5$ $y = -2$

9. $\dfrac{1}{2} + f = \dfrac{3}{4}$ $f = \dfrac{1}{4}$ **10.** $c + \dfrac{1}{4} = \dfrac{3}{8}$ $c = \dfrac{1}{8}$

Solve each equation. See Examples 3 and 4.

11. $3x - 6 = 2x + 5$ $x = 11$ **12.** $7y + 2 = 6y + 2$

13. $3t - t - 7 = t - 7$ **14.** $4c + 8 - c = 8 + 2c$

15. $7x + 2x = 8x - 3$ **16.** $3n + 2n = 7 + 4n$

17. $2y + 10 = y$ $y = -10$ **18.** $4x - 4 = 3x$ $x = 4$

19. $y + 0.8 = 9.7$ $y = 8.9$ **20.** $w + 0.9 = 3.6$

21. $5b - 0.7 = 6b$ **22.** $8n + 1.5 = 9n$

23. $5x - 6 = 6x - 5$ **24.** $2x + 7 = x - 10$

25. $7t - 12 = 6t$ $t = 12$ **26.** $9m + 14 = 8m$

27. $y - 5y + 0.6 = 0.8 - 5y$ $y = 0.2$

28. $6z + z - 0.9 = 6z + 0.9$ $z = 1.8$

29. In your own words, explain what is meant by the solution of an equation.   answers may vary

30. In your own words, explain how to check a solution of an equation.   answers may vary

Solve each equation. See Examples 5 and 6.

31. $2(x - 4) = x + 3$ **32.** $3(y + 7) = 2y - 5$

33. $7(6 + w) = 6(2 + w)$ **34.** $6(5 + c) = 5(c - 4)$

35. $10 - (2x - 4) = 7 - 3x$ $x = -7$

36. $15 - (6 - 7k) = 2 + 6k$ $k = -7$

37. $-5(n - 2) = 8 - 4n$ **38.** $-4(z - 3) = 2 - 3z$

39. $-3(x - 4) = -4x$ $x = -12$

40. $-2(x - 1) = -3x$ $x = -2$

41. $3(n - 5) - (6 - 2n) = 4n$ $n = 21$

42. $5(3 + z) - (8z + 9) = -4z$ $z = -6$

43. $-2(t - 1) - 3t = 8 - 4t$ $t = -6$

44. $-4(r + 5) - 7r = 12 - 10r$ $r = -32$

45. $4y - 6(y + 4) = 1 - y$ $y = -25$

46. $-7k + 3(k - 1) = 6 - 5k$ $k = 9$

47. $7(m - 2) - 6(m + 1) = -20$ $m = 0$

48. $-4(x - 1) - 5(2 - x) = -6$ $x = 0$

49. $0.8t + 0.2(t - 0.4) = 1.75$ $t = 1.83$

50. $0.6v + 0.4(0.3 + v) = 2.34$ $v = 2.22$

See Example 7.

51. Two numbers have a sum of 20. If one number is p, express the other number in terms of p. $20 - p$

52. Two numbers have a sum of 13. If one number is y, express the other number in terms of y. $13 - y$

53. A 10-foot board is cut into two pieces. If one piece is x feet long, express the other length in terms of x. $(10 - x)$ ft.

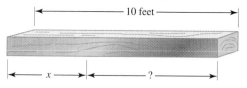

54. A 5-foot piece of string is cut into two pieces. If one piece is x feet long, express the other length in terms of x. $(5 - x)$ ft.

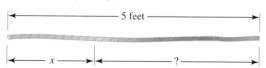

55. Two angles are *supplementary* if their sum is $180°$. If one angle measures $x°$, express the measure of its supplement in terms of x. $(180 - x)°$

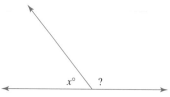

56. Two angles are *complementary* if their sum is $90°$. If one angle measures $x°$, express the measure of its complement in terms of x. $(90 - x)°$

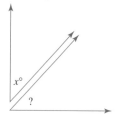

12. $y = 0$ **13.** $t = 0$ **14.** $c = 0$ **15.** $x = -3$ **16.** $n = 7$ **20.** $w = 2.7$ **21.** $b = -0.7$
22. $n = 1.5$ **23.** $x = -1$ **24.** $x = -17$ **26.** $m = -14$ **31.** $x = 11$ **32.** $y = -26$
33. $w = -30$ **34.** $c = -50$ **37.** $n = 2$ **38.** $z = 10$

57. The sum of the angles of a triangle is 180°. If one angle of a triangle measures $x°$ and a second angle measures $(2x + 7)°$, express the measure of the third angle in terms of x. Simplify the expression.

$(173 - 3x)°$

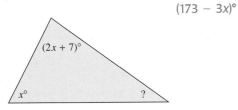

58. A quadrilateral is a four-sided figure like the one shown next whose angle sum is 360°. If one angle measures $x°$, a second angle measures $3x°$, and a third angle measures $5x°$, express the measure of the fourth angle in terms of x. Simplify the expression.

$(360 - 9x)°$

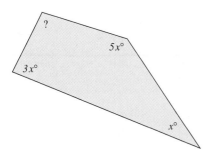

59. In a mayoral election, April Catarella received 284 more votes than Charles Pecot. If Charles received n votes, how many votes did April receive?

60. The length of the top of a computer desk is $1\frac{1}{2}$ feet longer than its width. If its width measures m feet, express its length as an algebraic expression in m.

Use a calculator to determine whether the given value is a solution of the given equation.

61. $1.23x - 0.06 = 2.6x - 0.1285$; $x = 0.05$ solution

62. $8.13 + 5.85y = 20.05y - 8.91$; $y = 1.2$ solution

63. $3(a + 4.6) = 5a + 2.5$; $a = 6.3$ not a solution

64. $7(z - 1.7) + 9.5 = 5(z + 3.2) - 9.2$; $z = 4.8$
64. not a solution

Review Exercises

Find the multiplicative inverse or reciprocal of each. See Section 1.7.

65. $\dfrac{5}{8}$ $\dfrac{8}{5}$ **66.** $\dfrac{7}{6}$ $\dfrac{6}{7}$ **67.** 2 $\dfrac{1}{2}$

68. 5 $\dfrac{1}{5}$ **69.** $-\dfrac{1}{9}$ -9 **70.** $-\dfrac{3}{5}$ $-\dfrac{5}{3}$

Bryan, a 4-year old child, was admitted to Slidell Memorial Hospital with pneumonia and a fever of 104°. His temperature was taken every hour and the results are recorded on the bar graph shown. (See Section 1.9.)

71. How many hours after arrival at the hospital was Bryan's fever the highest?   5 hours

72. What was Bryan's highest recorded temperature?

73. How many hours after arrival at the hospital was Bryan's temperature 101°?   2 hours

74. When did Bryan's temperature show the greatest increase?   5 hours after arrival

75. When did Bryan's temperature show the greatest decrease?   6 hours and 7 hours after arrival

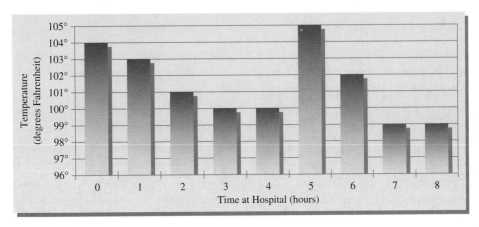

59. $(n + 284)$ votes   **60.** $\left(m + 1\frac{1}{2}\right)$ ft.   **72.** 105° F

TAPE BA 2.3

2.3 | THE MULTIPLICATION PROPERTY OF EQUALITY

O B J E C T I V E S

1. Use the multiplication property of equality to solve linear equations.
2. Use both the addition and multiplication properties of equality to solve linear equations.
3. Write word phrases as algebraic expressions.

As useful as the addition property of equality is, it cannot help us solve every type of linear equation in one variable. For example, adding or subtracting a value on both sides of the equation does not help solve

$$\frac{5}{2}x = 15.$$

Instead, we apply another important property of equality, the **multiplication property of equality.**

MULTIPLICATION PROPERTY OF EQUALITY

If a, b, and c are real numbers and $c \neq 0$, then

$$a = b \quad \text{and} \quad ac = bc$$

are equivalent equations.

This property guarantees that multiplying both sides of an equation by the same nonzero number does not change the solution of the equation. Since division is defined in terms of multiplication, we may also divide both sides of the equation by the same nonzero number without changing the solution.

EXAMPLE 1   Solve for x: $\frac{5}{2}x = 15$.

Solution: To isolate x, multiply both sides of the equation by the reciprocal of $\frac{5}{2}$, which is $\frac{2}{5}$.

$$\frac{5}{2}x = 15$$

$$\frac{2}{5} \cdot \frac{5}{2}x = \frac{2}{5} \cdot 15 \qquad \text{Multiply both sides by } \frac{2}{5}.$$

$$\left(\frac{2}{5} \cdot \frac{5}{2}\right)x = \frac{2}{5} \cdot 15 \qquad \text{Apply the associative property.}$$

$$1x = 6 \qquad \text{Simplify.}$$

or

$$x = 6$$

To check, replace x with 6 in the original equation.

$$\frac{5}{2}x = 15$$

$$\frac{5}{2}(6) = 15 \qquad \text{Replace } x \text{ with 6.}$$

$$15 = 15 \qquad \text{True.}$$

The solution set is $\{6\}$.

Why did we multiply both sides by the reciprocal of the coefficient of x? Multiplying the coefficient $\frac{5}{2}$ by $\frac{2}{5}$ leaves a coefficient of 1 and thus isolates the variable.

In general, multiplying by the reciprocal of the variable's coefficient is a way to isolate a variable. Multiplying by the reciprocal of a number is, of course, the same as dividing by the number.

EXAMPLE 2   Solve $-3x = 33$ for x.

Solution: Recall that $-3x$ means $-3 \cdot x$. To isolate x, divide both sides by the coefficient of x, that is, -3.

$$-3x = 33$$

$$\frac{-3x}{-3} = \frac{33}{-3} \qquad \text{Divide both sides by } -3.$$

$$1x = -11 \qquad \text{Simplify.}$$

$$x = -11$$

To check, replace x with -11 in the original equation.

$$-3x = 33$$

$$-3(-11) = 33 \qquad \text{Replace } x \text{ with } -11 \text{ in the original equation.}$$

$$33 = 33 \qquad \text{True.}$$

The solution set is $\{-11\}$.

EXAMPLE 3   Solve $\dfrac{y}{7} = 20$ for y.

Solution: Recall that $\dfrac{y}{7} = \dfrac{1}{7}y$. To isolate y, multiply both sides of the equation by 7, the reciprocal of $\dfrac{1}{7}$.

$$\frac{y}{7} = 20$$

$$7 \cdot \frac{y}{7} = 7 \cdot 20 \qquad \text{Multiply each side by 7.}$$

$$y = 140 \qquad \text{Simplify.}$$

To check, replace y with 140 in the original equation.

$$\frac{y}{7} = 20 \qquad \text{Original equation.}$$

$$\frac{140}{7} = 20 \qquad \text{Replace } y \text{ with 140.}$$

$$20 = 20 \qquad \text{True.}$$

The solution set is {140}.

EXAMPLE 4   Solve $-\dfrac{2}{3}x = -5$ for x.

Solution: To isolate x, multiply both sides of the equation by $-\dfrac{3}{2}$, the reciprocal of the coefficient of x.

$$-\frac{2}{3}x = -5$$

$$-\frac{3}{2} \cdot -\frac{2}{3}x = -\frac{3}{2} \cdot -5 \qquad \text{Multiply both sides by the reciprocal of } -\frac{2}{3}.$$

$$x = \frac{15}{2} \qquad \text{Simplify.}$$

The solution set is $\left\{\dfrac{15}{2}\right\}$. Check this solution in the original equation.

2   We are now ready to combine the skills learned in the last section with the skills learned from this section to solve equations by applying more than one property.

EXAMPLE 5   Solve $-z - 4 = 6$ for z.

Solution: First, isolate $-z$, the term containing the variable. To do so, add 4 to both sides of the equation.

$$-z - 4 + 4 = 6 + 4 \qquad \text{Add 4 to both sides.}$$

$$-z = 10 \qquad \text{Simplify.}$$

Next, recall that $-z$ means $-1 \cdot z$. To isolate z, either multiply or divide both sides of the equation by -1. In this example, we divide.

$$-z = 10$$

$$\frac{-z}{-1} = \frac{10}{-1} \qquad \text{Divide both sides by the coefficient } -1.$$

$$z = -10 \qquad \text{Simplify.}$$

Check: $-z - 4 = 6$

$-(-10) - 4 = 6$ Replace z with -10.

$10 - 4 = 6$

$6 = 6$ True.

Since this is a true statement, -10 is the solution and the solution set is $\{-10\}$.

EXAMPLE 6   Solve $5x - 2 = 18$ for x.

Solution: First, isolate $5x$, the term containing the variable. To do so, add 2 to both sides of the equation.

$$5x - 2 = 18$$
$$5x - 2 + 2 = 18 + 2 \qquad \text{Add 2 to both sides.}$$
$$5x = 20 \qquad \text{Simplify.}$$

Having isolated the variable term, we now use the multiplication property of equality to achieve a coefficient of 1.

$$\frac{5x}{5} = \frac{20}{5} \qquad \text{Divide both sides by 5.}$$
$$x = 4 \qquad \text{Simplify.}$$

The solution is 4. As usual, we can check this solution by replacing x with 4 in the original equation.

$$5x - 2 = 18$$
$$5(4) - 2 = 18 \qquad \text{Replace } x \text{ with 4.}$$
$$20 - 2 = 18 \qquad \text{Simplify.}$$
$$18 = 18 \qquad \text{True.}$$

The solution set is $\{4\}$.

EXAMPLE 7   Solve for a: $2a + 5a - 10 + 7 = 5a - 13$.

Solution: First, simplify both sides of the equation by combining like terms.

$$2a + 5a - 10 + 7 = 5a - 13$$
$$7a - 3 = 5a - 13 \qquad \text{Combine like terms.}$$
$$7a - 3 - 5a = 5a - 13 - 5a \qquad \text{Subtract } 5a \text{ from both sides.}$$

$$2a - 3 = -13 \qquad \text{Combine like terms.}$$
$$2a - 3 + 3 = -13 + 3 \qquad \text{Add 3 to both sides.}$$
$$2a = -10 \qquad \text{Simplify.}$$
$$\frac{2a}{2} = \frac{-10}{2} \qquad \text{Divide both sides by 2.}$$
$$a = -5 \qquad \text{Simplify.}$$

To check, replace a with -5 in the original equation. The solution set is $\{-5\}$.

3   Next, we continue to sharpen our problem-solving skills by writing algebraic expressions.

EXAMPLE 8   If x is the first of three consecutive integers, express the sum of the three integers in terms of x. Simplify if possible.

Solution: An example of three consecutive integers is 7,   8,   9. The second consecutive integer is always 1 more than the first, and the third consecutive integer is 2 more than the first. If x is the first of three consecutive integers, the three consecutive integers are

$$x, \quad x + 1, \quad x + 2$$

Their sum is

In words: first integer $+$ second integer $+$ third integer

Translate: $x \quad + \quad (x + 1) \quad + \quad (x + 2)$

which simplifies to $3x + 3$.

   The exercise set mentions consecutive even and odd integers.

An example of three consecutive **even** integers is 14, 16, and 18. Notice that each integer is *two more* than the previous integer.

An example of three consecutive **odd** integers is 5, 7, and 9. Notice that each integer is *again two more* than the previous integer.

In general, if x is the first integer, we have the following:

     Three consecutive integers: $x, x + 1, x + 2$
     Three consecutive odd integers: $x, x + 2, x + 4$, if x is odd
     Three consecutive even integers: $x, x + 2, x + 4$, if x is even

Solve the following equations mentally. See Example 1.

1. $3a = 27$ $a = 9$

2. $9c = 54$ $c = 6$

3. $5b = 10$ $b = 2$

4. $7t = 14$ $t = 2$

5. $6x = -30$ $x = -5$

6. $8r = -64$ $r = -8$

EXERCISE SET 2.3

Solve the following equations. See Examples 1 through 4.

1. $-5x = 20$ $x = -4$

2. $-7x = -49$ $x = 7$

3. $3x = 0$ $x = 0$

4. $-2x = 0$ $x = 0$

5. $-x = -12$ $x = 12$

6. $-y = 8$ $y = -8$

7. $\frac{2}{3}x = -8$ $x = -12$

8. $\frac{3}{4}n = -15$ $n = -20$

9. $\frac{1}{6}d = \frac{1}{2}$ $d = 3$

10. $\frac{1}{8}v = \frac{1}{4}$ $v = 2$

11. $\frac{a}{-2} = 1$ $a = -2$

12. $\frac{d}{15} = 2$ $d = 30$

13. $\frac{k}{7} = 0$ $k = 0$

14. $\frac{f}{-5} = 0$ $f = 0$

Solve the following equations. See Examples 5 and 6.

15. $2x - 4 = 16$ $x = 10$

16. $3x - 1 = 26$ $x = 9$

17. $-5x + 2 = 22$ $x = -4$

18. $7x + 4 = -24$

19. $6x + 10 = -20$ $x = -5$

20. $-10y + 15 = 5$

Solve the following equations. See Example 7.

21. $-4y + 10 = -6y - 2$

22. $-3z + 1 = -2z + 4$

23. $9x - 8 = 10 + 15x$

24. $15t - 5 = 7 + 12t$

25. $2x - 7 = 6x - 27$

26. $3 + 8y = 3y - 2$

27. $6 - 2x + 8 = 10$

28. $-5 - 6y + 6 = 19$

29. $-3a + 6 + 5a = 7a - 8a$ $a = -2$

30. $4b - 8 - b = 10b - 3b$ $b = -2$

31. The equation $3x + 6 = 2x + 10 + x - 4$ is true for all real numbers. Substitute a few real numbers for x to see that this is so and then try solving the equation.   answers may vary

32. The equation $6x + 2 - 2x = 4x + 1$ has no solution. Try solving this equation for x and see what happens.   answers may vary

33. From the results of Exercises 31 and 32, when do you think an equation has all real numbers as its solution set?   answers may vary

34. From the results of Exercises 31 and 32, when do you think an equation has no solution?

34. answers may vary

Solve the following equations.

35. $-3w = 18$ $w = -6$

36. $5j = -45$ $j = -9$

37. $-0.2z = -0.8$ $z = 4$

38. $-0.1m = 3.6$

39. $-h = -\frac{3}{4}$ $h = \frac{3}{4}$

40. $-b = \frac{4}{7}$ $b = -\frac{4}{7}$

41. $6a + 3 = 3$ $a = 0$

42. $8t + 5 = 5$ $t = 0$

43. $5 - 0.3k = 5$ $k = 0$

44. $2 + 0.4p = 2$ $p = 0$

45. $2x + \frac{1}{2} = \frac{7}{2}$ $x = \frac{3}{2}$

46. $3n - \frac{1}{3} = \frac{8}{3}$ $n = 1$

47. $\frac{x}{3} - 2 = 5$ $x = 21$

48. $\frac{b}{4} + 1 = 7$ $b = 24$

49. $10 = 2x - 1$ $x = \frac{11}{2}$

50. $12 = 3j - 4$ $j = \frac{16}{3}$

51. $4 - 12x = 7$ $x = -\frac{1}{4}$

52. $24 + 20b = 10$

53. $-\frac{2}{3}x = \frac{5}{9}$ $x = -\frac{5}{6}$

54. $-\frac{3}{8}y = -\frac{1}{16}$ $y = \frac{1}{6}$

55. $10 = -6n + 16$ $n = 1$

56. $-5 = -2m + 7$

57. $z - 5z = 7z - 9 - z$ $z = \frac{9}{10}$

58. $t - 6t = -13 + t - 3t$ $t = \frac{13}{3}$

59. $5x + 20 = 8 - x$

60. $6t - 18 = t - 3$

18. $x = -4$

20. $y = 1$

21. $y = -6$

22. $z = -3$

23. $x = -3$

24. $t = 4$

25. $x = 5$

26. $y = -1$

27. $x = 2$

28. $y = -3$

38. $m = -36$

52. $b = -\frac{7}{10}$

56. $m = 6$

59. $x = -2$

60. $t = 3$

61. $5y - y = 2y - 14$ **62.** $k - 4k = 20 - 5k$

63. $6z - 8 - z + 3 = 0$ **64.** $4a + 1 + a - 11 = 0$

65. $10 - n - 2 = 2n + 2$ $n = 2$

66. $4 - 10 - 5c = 3c - 12$ $c = \dfrac{3}{4}$

67. $0.4x - 0.6x - 5 = 1$ **68.** $0.4x - 0.9x - 6 = 19$

Write each algebraic expression described. Simplify if possible. See Example 8.

69. If x represents the first odd integer, express the next odd integer in terms of x. $x + 2$

70. If x represents the first even integer, express the next even integer in terms of x. $x + 2$

71. If x represents the first of two consecutive even integers, express the sum of the two integers in terms of x. $2x + 2$

72. If x represents the first of two consecutive odd integers, express the sum of the two integers in terms of x. $2x + 2$

73. If x is the first of three consecutive integers, express the sum of the first integer and the third integer as an algebraic expression containing the variable x. $2x + 2$

74. If x is the first of two consecutive integers, express the sum of 20 and the second consecutive integer as an algebraic expression containing the variable x. $x + 21$

75. Express the sum of three odd consecutive integers as an algebraic expression in x. Let x be the first odd integer. $3x + 6$

76. If x is the first of four consecutive even integers, write their sum as an algebraic expression in x. $4x + 12$

Solve each equation.

77. $-3.6x = 10.62$ **78.** $4.95y = -31.185$

79. $7x - 5.06 = -4.92$ $x = 0.02$

80. $0.06y + 2.63 = 2.5562$ $y = -1.23$

Review Exercises

Simplify each expression. See Section 2.1.

81. $5x + 2(x - 6)$ $7x - 12$ **82.** $-7y + 2y - 3(y + 1)$

83. $6(2z + 4) + 20$ **84.** $-(3a - 3) + 2a - 6$

85. $-(x - 1) + x$ 1 **86.** $8(z - 6) + 7z - 1$

Insert $<$, $>$, or $=$ in the appropriate space to make each statement true. See Sections 1.5 and 1.7.

87. $(-3)^2 > -3^2$ **88.** $(-2)^4 > -2^4$

89. $(-2)^3 = -2^3$ **90.** $(-4)^3 = -4^3$

91. $-|-6| < 6$ **92.** $-|-0.7| = -0.7$

2.4 │ SOLVING LINEAR EQUATIONS

O B J E C T I V E S

1. Apply the general strategy for solving a linear equation.
2. Solve equations containing fractions.
3. Solve equations containing decimals.
4. Recognize identities and equations with no solution.
5. Write sentences as equations and solve.

TAPE BA 2.4

We now present a general strategy for solving linear equations. One new piece of strategy is a suggestion to "clear an equation of fractions" as a first step. Doing so makes the equation more manageable, since operating on integers is more convenient than operating on fractions.

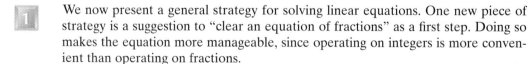

61. $y = -7$ **62.** $k = 10$ **63.** $z = 1$ **64.** $a = 2$

67. $x = -30$ **68.** $x = -50$ **77.** $x = -2.95$ **78.** $y = -6.3$ **82.** $-8y - 3$ **83.** $12z + 44$

84. $-a - 3$ **86.** $15z - 49$

> **To Solve Linear Equations in One Variable**
>
> *Step 1.*   Clear the equation of fractions by multiplying both sides of the equation by the lowest common denominator (LCD) of all denominators in the equation.
>
> *Step 2.*   Remove any grouping symbols, such as parentheses, by using the distributive property.
>
> *Step 3.*   Simplify each side of the equation by combining like terms.
>
> *Step 4.*   Write the equation with variable terms on one side and numbers on the other side by using the addition property of equality.
>
> *Step 5.*   Isolate the variable by using the multiplication property of equality.
>
> *Step 6.*   Check the solution by substituting it in the original equation.

EXAMPLE 1   Solve $4(2x - 3) + 7 = 3x + 5$.

Solution: There are no fractions to clear, so begin with step 2.

$$4(2x - 3) + 7 = 3x + 5$$

Step 2. $8x - 12 + 7 = 3x + 5$ Apply the distributive property.

Step 3. $8x - 5 = 3x + 5$ Combine like terms.

Step 4.   Get variable terms on the same side of the equation by subtracting $3x$ from both sides; then add 5 to both sides.

$$8x - 5 - 3x = 3x + 5 - 3x$$ Subtract $3x$ from both sides.

$$5x - 5 = 5$$ Simplify.

$$5x - 5 + 5 = 5 + 5$$ Add 5 to both sides.

$$5x = 10$$ Simplify.

Step 5.   Use the multiplication property of equality to isolate x.

$$\frac{5x}{5} = \frac{10}{5}$$ Divide both sides by 5.

$$x = 2$$ Simplify.

Step 6.   To check, replace x with 2 in the original equation.

Check: $4(2x - 3) + 7 = 3x + 5$

$4[2(2) - 3] + 7 = 3(2) + 5$ Replace x with 2.

$4(4 - 3) + 7 = 6 + 5$

$4(1) + 7 = 11$

$4 + 7 = 11$

$11 = 11$ True.

The solution set is {2}.

EXAMPLE 2 Solve $8(2 - t) = -5t$.

Solution: First, apply the distributive property.

$$8(2 - t) = -5t$$

Step 2. $\qquad 16 - 8t = -5t$ $\qquad$ Use the distributive property.

Step 4. $\qquad 16 - 8t + 8t = -5t + 8t$ $\qquad$ Add $8t$ to both sides.

$\qquad\qquad 16 = 3t$ $\qquad$ Combine like terms.

Step 5. $\qquad \dfrac{16}{3} = \dfrac{3t}{3}$ $\qquad$ Divide both sides by 3.

$\qquad\qquad \dfrac{16}{3} = t$ $\qquad$ Simplify.

Step 6. Check to see that the solution is $\dfrac{16}{3}$.

$$8(2 - t) = -5t$$

$$8\left(2 - \dfrac{16}{3}\right) = -5\left(\dfrac{16}{3}\right) \qquad \text{Replace } t \text{ with } \dfrac{16}{3}.$$

$$8\left(\dfrac{6}{3} - \dfrac{16}{3}\right) = -\dfrac{80}{3} \qquad \text{The LCD is 3.}$$

$$8\left(-\dfrac{10}{3}\right) = -\dfrac{80}{3} \qquad \text{Subtract fractions.}$$

$$-\dfrac{80}{3} = -\dfrac{80}{3} \qquad \text{True.}$$

The solution set is $\left\{\dfrac{16}{3}\right\}$.

[2] Next, we solve an equation containing fractions.

EXAMPLE 3 Solve for x: $\dfrac{x}{2} - 1 = \dfrac{2}{3}x - 3$.

Solution: This equation contains fractions, so we begin by clearing fractions. To do this, multiply both sides of the equation by the LCD of 2 and 3, which is 6.

$$\dfrac{x}{2} - 1 = \dfrac{2}{3}x - 3$$

Step 1. $\qquad 6\left(\dfrac{x}{2} - 1\right) = 6\left(\dfrac{2}{3}x - 3\right)$ $\qquad$ Multiply both sides by 6.

Step 2. $\qquad 6\left(\dfrac{x}{2}\right) - 6(1) = 6\left(\dfrac{2}{3}x\right) - 6(3)$ $\qquad$ Apply the distributive property.

$\qquad\qquad 3x - 6 = 4x - 18$ $\qquad$ Simplify.

There are no longer grouping symbols and no like terms on either side of the equation, so we continue with step 4.

$$3x - 6 = 4x - 18$$

Step 4. $3x - 6 - 3x = 4x - 18 - 3x$ Subtract $3x$ from both sides.

$$-6 = x - 18$$ Simplify.

$$-6 + 18 = x - 18 + 18$$ Add 18 to both sides.

$$12 = x$$ Simplify.

Step 5.   The variable is isolated so there is no need to apply the multiplication property of equality.

Step 6.   The equation is solved for x and the solution is 12. To check, replace x with 12 in the original equation.

$$\frac{x}{2} - 1 = \frac{2}{3}x - 3$$ Original equation.

$$\frac{12}{2} - 1 = \frac{2}{3} \cdot 12 - 3$$ Replace x with 12.

$$6 - 1 = 8 - 3$$ Simplify.

$$5 = 5$$ True.

The solution set is {12}.

───────

EXAMPLE 4   Solve $\dfrac{2(a + 3)}{3} = 6a + 2$.

Solution: Clear the equation of fractions first.

$$\frac{2(a + 3)}{3} = 6a + 2$$

Step 1. $3 \cdot \dfrac{2(a + 3)}{3} = 3(6a + 2)$ Clear fraction by multiplying both sides by the LCD 3.

Step 2.   Next, use the distributive property and remove parentheses.

$$2a + 6 = 18a + 6$$ Apply the distributive property.

Step 4. $2a + 6 - 6 = 18a + 6 - 6$ Subtract 6 from both sides.

$$2a = 18a$$

$$2a - 18a = 18a - 18a$$ Subtract $18a$ from both sides.

$$-16a = 0$$

Step 5. $\dfrac{-16a}{-16} = \dfrac{0}{-16}$ Divide both sides by -16.

$$a = 0$$ Write the fraction in simplest form.

Step 6.   To check, replace a with 0 in the original equation. The solution set is {0}.

3

Oftentimes, especially when solving a problem having to do with money, we encounter an equation containing decimals. An equation such as this may be cleared of decimals by multiplying both sides by an appropriate power of 10 as shown in the next example.

EXAMPLE 5   Solve $.25x + .10(x - 3) = .05(x + 18)$

Solution: First, clear this equation of decimals by multiplying both sides of the equation by 100. Recall that multiplying a decimal number by 100 has the effect of moving the decimal point 2 places to the right.

$$.25x + .10(x - 3) = .05(x + 18)$$

Step 1. $.25x + .10(x - 3) = .05(x + 18)$ Multiply both sides by 100.

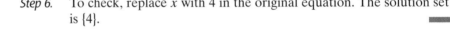

$$25x + 10(x - 3) = 5(x + 18)$$

Step 2. $25x + 10x - 30 = 5x + 90$ Apply the distributive property.

Step 3. $35x - 30 = 5x + 90$ Combine like terms.

Step 4. $35x - 30 - 5x = 5x + 90 - 5x$ Subtract $5x$.

$$30x - 30 = 90$$ Combine like terms.

$$30x - 30 + 30 = 90 + 30$$ Add 30.

$$30x = 120$$ Combine like terms.

Step 5. $\dfrac{30x}{30} = \dfrac{120}{30}$ Divide by 30.

$$x = 4$$

Step 6.   To check, replace x with 4 in the original equation. The solution set is $\{4\}$.

4

So far, each equation that we have solved has had a single solution.

Not every equation in one variable has a single solution. Some equations have no solution, while others have an infinite number of solutions. For example,

$$x + 5 = x + 7$$

has no solution since, no matter which **real number** we replace x with, the equation is false.

real number + 5 = same **real number** + 7   **FALSE**

On the other hand,

$$x + 6 = x + 6$$

has infinitely many solutions since x can be replaced by any real number and the equation is always true.

real number + 6 = same **real number** + 6   **TRUE**

The equation $x + 6 = x + 6$ is called an **identity.** The next few examples illustrate equations like these.

EXAMPLE 6   Solve $-2(x - 5) + 10 = -3(x + 2) + x$.

Solution:

$$-2(x - 5) + 10 = -3(x + 2) + x$$

$-2x + 10 + 10 = -3x - 6 + x$	Distribute on both sides.
$-2x + 20 = -2x - 6$	Combine like terms.
$-2x + 20 + 2x = -2x - 6 + 2x$	Add $2x$ to both sides.
$20 = -6$	Combine like terms.

Notice that no value for x makes $20 = -6$ a true equation. We conclude that there is **no solution** to this equation. Its solution set is written as either $\{\ \}$ or $\varnothing$.

EXAMPLE 7   Solve $3(x - 4) = 3x - 12$.

Solution:

| $3(x - 4) = 3x - 12$ | |
| $3x - 12 = 3x - 12$ | Apply the distributive property. |

The left side of the equation is now identical to the right side. Every real number may be substituted for x and a true statement will result. We arrive at the same conclusion if we continue.

$3x - 12 = 3x - 12$	
$3x - 12 + 12 = 3x - 12 + 12$	Add 12 to both sides.
$3x = 3x$	Combine like terms.
$0 = 0$	Subtract $3x$ from both sides.

Again, one side of the equation is identical to the other side. Thus, $3(x - 4) = 3x - 12$ is an **identity** and every real number is a solution. The solution set may be written as

$\{x \mid x$ is a real number$\}$. This is read as

$\uparrow$ $\uparrow$ ⌣

"the   such
set of   that x is a real number"
all x

---

5   We can apply our equation-solving skills to solving problems written in words.
   Many times, writing an equation that describes or models a problem involves a direct translation from a word sentence to an equation.

EXAMPLE 8 Twice a number added to seven is the same as three subtracted from the number. Find the number.

Solution: Translate the sentence into an equation and solve.

In words:	twice a number	added to	seven	is the same as	three subtracted from the number
Translate:	$2x$	$+$	7	$=$	$x - 3$

To solve, begin by subtracting x on both sides to isolate the variable term.

$$2x + 7 = x - 3$$

$2x + 7 - x = x - 3 - x$ Subtract x from both sides.

$x + 7 = -3$ Combine like terms.

$x + 7 - 7 = -3 - 7$ Subtract 7 from both sides.

$x = -10$ Combine like terms.

Check the solution in the problem as it was originally stated. To do so, replace "number" in the sentence with -10. Twice "-10" added to 7 is the same as 3 subtracted from "-10."

$$2(-10) + 7 = -10 - 3$$
$$-13 = -13$$

The unknown number is -10.

R E M I N D E R When checking solutions, go back to the original stated problem, rather than to your equation in case errors have been made in translating to an equation.

Scientific Calculator Explorations

Checking Equations
We can use a calculator to check possible solutions of equations. To do this, replace the variable by the possible solution and evaluate both sides of the equation separately.

Equation: $3x - 4 = 2(x + 6)$ Solution: $x = 16$

$3x - 4 = 2(x + 6)$ Original equation.

$3(16) - 4 \stackrel{?}{=} 2(16 + 6)$ Replace x with 16.

(continued)

Now evaluate each side with your calculator.

Evaluate left side: $\boxed{3}$ $\boxed{\times}$ $\boxed{16}$ $\boxed{-}$ $\boxed{4}$ $\boxed{=}$ Display: $\boxed{44}$

Evaluate right side: $\boxed{2}$ $\boxed{(}$ $\boxed{16}$ $\boxed{+}$ $\boxed{6}$ $\boxed{)}$ $\boxed{=}$ Display: $\boxed{44}$

Since the left side equals the right side, the equation checks.

Use a calculator to check the possible solutions to each of the following equations.

1. $2x = 48 + 6x; x = -12$ solution

2. $-3x - 7 = 3x - 1; x = -1$ solution

3. $5x - 2.6 = 2(x + 0.8); x = 4.4$ not a solution

4. $-1.6x - 3.9 = -6.9x - 25.6; x = 5$ not a solution

5. $\dfrac{564x}{4} = 200x - 11(649); x = 121$ solution

6. $20(x - 39) = 5x - 432; x = 23.2$ solution

EXERCISE SET 2.4

Solve each equation. See Examples 1 and 2.

1. $-2(3x - 4) = 2x$ $x = 1$ **2.** $-(5x - 1) = 9$

3. $4(2n - 1) = (6n + 4) + 1$

4. $3(4y + 2) = 2(1 + 6y) + 8$ no solution

5. $5(2x - 1) - 2(3x) = 4$

6. $3(2 - 5x) + 4(6x) = 12$

7. $6(x - 3) + 10 = -8$ **8.** $-4(2 + n) + 9 = 1$

Solve each equation. See Examples 3 through 5.

9. $\dfrac{3}{4}x - \dfrac{1}{2} = 1$ $x = 2$ **10.** $\dfrac{2}{3}x + \dfrac{5}{3} = \dfrac{5}{3}$ $x = 0$

11. $x + \dfrac{5}{4} = \dfrac{3}{4}x$ $x = -5$ **12.** $\dfrac{7}{8}x + \dfrac{1}{4} = \dfrac{3}{4}x$ $x = -2$

13. $\dfrac{x}{2} - 1 = \dfrac{x}{5} + 2$ $x = 10$ **14.** $\dfrac{x}{5} - 2 = \dfrac{x}{3}$ $x = -15$

15. $\dfrac{6(3 - z)}{5} = -z$ $z = 18$ **16.** $\dfrac{4(5 - w)}{3} = -w$

17. $\dfrac{2(x + 1)}{4} = 3x - 2$ $x = 1$

18. $\dfrac{3(y + 3)}{5} = 2y + 6$ $y = -3$

2. $x = -\dfrac{8}{5}$ **3.** $n = \dfrac{9}{2}$ **5.** $x = \dfrac{9}{4}$ **6.** $x = \dfrac{2}{3}$

7. $x = 0$ **8.** $n = 0$ **16.** $w = 20$

19. $.50x + .15(70) = .25(142)$ $x = 50$

20. $.40x + .06(30) = .20(49)$ $x = 20$

21. $.12(y - 6) + .06y = .08y - .07(10)$ $y = 0.2$

22. $.60(z - 300) + .05z = .70z - .41(500)$ $z = 500$

Solve each equation. See Examples 6 and 7.

23. $5x - 5 = 2(x + 1) + 3x - 7$ all real numbers

24. $3(2x - 1) + 5 = 6x + 2$ all real numbers

25. $\dfrac{x}{4} + 1 = \dfrac{x}{4}$ no solution   **26.** $\dfrac{x}{3} - 2 = \dfrac{x}{3}$ no solution

27. $3x - 7 = 3(x + 1)$ **28.** $2(x - 5) = 2x + 10$

29. Explain the difference between simplifying an expression and solving an equation.

30. When solving an equation, if the final equivalent equation is $0 = 5$, what can we conclude? If the final equivalent equation is $-2 = -2$, what can we conclude?   answers may vary

31. On your own, construct an equation for which every real number is a solution.

32. On your own, construct an equation that has no solution.   answers may vary

27. no solution   **28.** no solution
29. answers may vary   **31.** answers may vary

33. $x = 4$ **34.** $y = 5$ **35.** $y = -4$ **36.** $n = 5$ **37.** $x = 3$ **38.** $x = 5$ **39.** $y = -2$ **40.** $y = -1$ **45.** no solution

Solve each equation.

33. $4x + 3 = 2x + 11$

34. $6y - 8 = 3y + 7$

35. $-2y - 10 = 5y + 18$

36. $7n + 5 = 10n - 10$

37. $.6x - .1 = .5x + .2$

38. $.2x - .1 = .6x - 2.1$

39. $2y + 2 = y$

40. $7y + 4 = -3$

41. $3(5c - 1) - 2 = 13c + 3$ $c = 4$

42. $4(3t + 4) - 20 = 3 + 5t$ $t = 1$

43. $x + \dfrac{7}{6} = 2x - \dfrac{7}{6}$ $x = \dfrac{7}{3}$ **44.** $\dfrac{5}{2}x - 1 = x + \dfrac{1}{4}$ $x = \dfrac{5}{6}$

45. $2(x - 5) = 7 + 2x$

46. $-3(1 - 3x) = 9x - 3$

47. $\dfrac{2(z + 3)}{3} = 5 - z$

48. $\dfrac{3(w + 2)}{4} = 2w + 3$

49. $\dfrac{4(y - 1)}{5} = -3y$

50. $\dfrac{5(1 - x)}{6} = -4x$

51. $8 - 2(a - 1) = 7 + a$ $a = 1$

52. $5 - 6(2 + b) = b - 14$ $b = 1$

53. $2(x + 3) - 5 = 5x - 3(1 + x)$ no solution

54. $4(2 + x) + 1 = 7x - 3(x - 2)$ no solution

55. $\dfrac{5x - 7}{3} = x$ $x = \dfrac{7}{2}$ **56.** $\dfrac{7n + 3}{5} = -n$

57. $\dfrac{9 + 5v}{2} = 2v - 4$ **58.** $\dfrac{6 - c}{2} = 5c - 8$

59. $-3(t - 5) + 2t = 5t - 4$

60. $-(4a - 7) - 5a = 10 + a$

61. $.02(6t - 3) = .05(t - 2) + .02$

62. $.03(m + 7) = .02(5 - m) + .03$ $m = -1.6$

63. $.06 - .01(x + 1) = -.02(2 - x)$ $x = 3$

64. $-.01(5x + 4) = .04 - .01(x + 4)$ $x = -1$

65. $\dfrac{3(x - 5)}{2} = \dfrac{2(x + 5)}{3}$ **66.** $\dfrac{5(x - 1)}{4} = \dfrac{3(x + 1)}{2}$

67. $1000(7x - 10) = 50(412 + 100x)$ $x = 15.3$

68. $10{,}000(x + 4) = 100(16 + 7x)$

69. $.035x + 5.112 = .010x + 5.107$ $x = -0.2$

70. $.127x - 2.685 = .027x - 2.38$ $x = 3.05$

Write each of the following as equations. Then solve.
See Example 8.

71. The sum of twice a number and $\frac{1}{5}$ is equal to the difference between three times a number and $\frac{4}{5}$. Find the number.

72. The sum of four times a number and $\frac{2}{3}$ is equal to

the difference of five times the number and $\frac{5}{6}$. Find the number.

73. The sum of twice a number and 7 is equal to the sum of a number and 6. Find the number.

74. The difference of three times a number and 1 is the same as twice a number. Find the number.

75. Three times a number minus 6 is equal to two times a number plus 8. Find the number.

76. The sum of 4 times a number and -2 is equal to the sum of 5 times a number and -2. Find the number. $4x - 2 = 5x - 2;\ 0$

77. One-third of a number is five-sixths. Find the number.

78. Seven-eighths of a number is one-half. Find the number.

79. The difference of a number and four is twice the number. Find the number.

80. The sum of double a number and six is four times the number. Find the number. $2x + 6 = 4x;\ 3$

81. If the quotient of a number and 4 is added to $\frac{1}{2}$, the result is $\frac{3}{4}$. Find the number.

82. If $\frac{3}{4}$ is added to three times a number, the result is $\frac{1}{2}$ subtracted from twice a number. Find the number.

83. The perimeter of a geometric figure is the sum of the lengths of its sides. If the perimeter of the following pentagon (five-sided figure) is 28 centimeters, find the length of each side. $x = 4$ cm, $2x = 8$ cm

x centimeters

x centimeters   *x centimeters*

2x centimeters   *2x centimeters*

84. The perimeter of the following triangle is 35 meters. Find the length of each side.

$x = 6$ meters; $2x + 1 = 13$ meters; $3x - 2 = 16$ meters

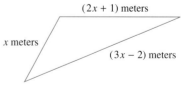

(2x + 1) meters

x meters

(3x − 2) meters

85. Five times a number subtracted from ten is triple the number. Find the number.

46. all real numbers **47.** $z = \frac{9}{5}$ **48.** $w = -\frac{6}{5}$ **49.** $y = \frac{4}{19}$ **50.** $x = -\frac{5}{19}$ **56.** $n = -\frac{1}{4}$ **57.** $v = -17$

58. $c = 2$ **59.** $t = \frac{19}{6}$ **60.** $a = -\frac{3}{10}$ **61.** $t = -\frac{2}{7}$ **65.** $x = 13$ **66.** $x = -11$ **68.** $x = -4\frac{4}{31}$

71. $2x + \frac{1}{5} = 3x - \frac{4}{5}$; 1 **72.** $4x + \frac{2}{3} = 5x - \frac{5}{6}$; $\frac{3}{2}$ **73.** $2x + 7 = x + 6$; −1 **74.** $3x - 1 = 2x$; 1 **75.** $3x - 6 = 2x + 8$; 14

86. Nine is equal to ten subtracted from double a number. Find the number.

The graph below is called a 3-dimensional bar graph. It shows the five most common names of cities, towns, or villages in the United States.

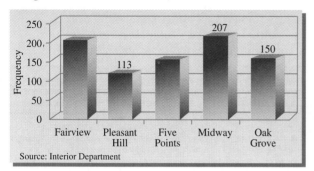

Source: Interior Department

87. What is the most popular name of a city, town, or village in the United States? Midway

88. How many more cities, towns, or villages are named Oak Grove than named Pleasant Hill? 37

89. Let x represent "the number of towns, cities, or villages named Five Points" and use the information given to determine the unknown number. "The number of towns, cities, or villages named Five Points" added to 55 is equal to twice "the number of towns, cities, or villages named Five Points" minus 90. Check your answer by noticing the height of the bar representing Five Points. Is your answer reasonable? 145

90. Let x represent "the number of towns, cities, or villages named Fairview" and use the information given to determine the unknown number. Three times "the number of towns, cities, or villages named Fairview" added to 24 is equal to 168 subtracted from 4 times "the number of towns, cities, or villages named Fairview." Check your answer by noticing the height of the bar representing Fairview. Is your answer reasonable? 192

Review Exercises

Evaluate. See Section 1.7.

91. $|2^3 - 3^2| - |5 - 7|$ −1

92. $|5^2 - 2^2| + |9 \div (-3)|$ 24

93. $\dfrac{5}{4 + 3 \cdot 7}$ $\dfrac{1}{5}$ **94.** $\dfrac{8}{24 - 8 \cdot 2}$ 1

See Section 2.1.

95. A plot of land is in the shape of a triangle. If one side is x meters, a second side is $2x - 3$ meters and a third side is $3x - 5$ meters, express the perimeter of the lot as a simplified expression in x.

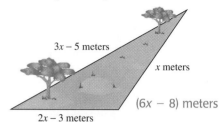

96. A portion of a board has length x feet. The other part has length $7x - 9$ feet. Express the total length of the board as a simplified expression in x.

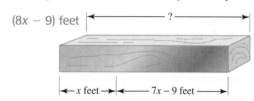

A Look Ahead

EXAMPLE

Solve $t(t + 4) = t^2 - 2t + 10$.

Solution:
$$t(t + 4) = t^2 - 2t + 10$$
$$t^2 + 4t = t^2 - 2t + 10$$
$$t^2 + 4t - t^2 = t^2 - 2t + 10 - t^2$$
$$4t = -2t + 10$$
$$4t + 2t = -2t + 10 + 2t$$
$$6t = 10$$
$$\frac{6t}{6} = \frac{10}{6}$$
$$t = \frac{5}{3}$$

Solve each equation.

97. $x(x - 3) = x^2 + 5x + 7$ $x = -\dfrac{7}{8}$

98. $t^2 - 6t = t(8 + t)$ $t = 0$

99. $2z(z + 6) = 2z^2 + 12z - 8$ no solution

100. $y^2 - 4y + 10 = y(y - 5)$ $y = -10$

101. $n(3 + n) = n^2 + 4n$ $n = 0$

102. $3c^2 - 8c + 2 = c(3c - 8)$ no solution

77. $\frac{1}{3}x = \frac{5}{6}$; $\frac{5}{2} = 2\frac{1}{2}$ **78.** $\frac{7}{8}x = \frac{1}{2}$; $\frac{4}{7}$ **79.** $x - 4 = 2x$; −4 **81.** $\frac{x}{4} + \frac{1}{2} = \frac{3}{4}$; 1 **82.** $\frac{3}{4} + 3x = 2x - \frac{1}{2}$; $-\frac{5}{4}$

85. $\frac{5}{4}$ **86.** $\frac{19}{2} = 9\frac{1}{2}$

2.5 | AN INTRODUCTION TO PROBLEM SOLVING

OBJECTIVE

TAPE BA 2.5

1. Apply the steps for problem solving

In previous sections, you practiced writing word phrases and sentences as algebraic expressions and equations to help prepare for problem solving. We now use these translations to help write equations that model a problem. The problem-solving steps given next may be helpful.

PROBLEM-SOLVING STEPS

1. **UNDERSTAND** the problem. During this step don't work with variables, but simply become comfortable with the problem. Some ways of accomplishing this are listed below.
   - Read and reread the problem.
   - Construct a drawing.
   - Propose a solution and check. Pay careful attention to how you check your proposed solution. This will help later when writing an equation to model the problem.

2. **ASSIGN** a variable to an unknown in the problem. Use this variable to represent any other unknown quantities.

3. **ILLUSTRATE** the problem. A diagram or chart using the assigned variables can often help visualize the known facts.

4. **TRANSLATE** the problem into a mathematical model. This is often an equation.

5. **COMPLETE** the work. This often means to solve the equation.

6. **INTERPRET** the results: *Check* the proposed solution in the stated problem and *state* your conclusion.

EXAMPLE 1   A 10-foot board is to be cut into two pieces so that the longer piece is 4 times the shorter. Find the length of each piece.

Solution: 1. UNDERSTAND the problem. To do so, read and reread the problem, construct a drawing, and propose a solution. For example, if 3 feet represents the length of the shorter piece, then $4(3) = 12$ feet is the length of the longer piece, since it is 4 times the length of the shorter piece.

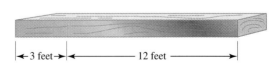

|←3 feet→|←——— 12 feet ———→|

This guess gives a total board length of 3 feet + 12 feet = 15 feet, too long. The purpose of guessing a solution is not to guess correctly, but to help better understand the problem and how to model it. At this point, we have a better understanding of the problem. Although it may be possible now to guess the solution, for purposes of practicing problem-solving skills, we continue with the problem-solving steps.

2. ASSIGN a variable. Use this variable to represent any other known quantities. If we let

   x = length of shorter piece,

   then $4x$ = length of longer piece.

3. ILLUSTRATE the problem. Draw a picture of the board and label the picture with the assigned variables.

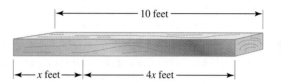

4. TRANSLATE the problem. First, write the equation in words.

In words:	length of shorter piece	added to	length of longer	equals	total length of board
Translate:	x	$+$	$4x$	$=$	10

5. COMPLETE the work. Here, we solve the equation.

   $x + 4x = 10$

   $5x = 10$ Combine like terms.

   $\dfrac{5x}{5} = \dfrac{10}{5}$ Divide both sides by 5.

   $x = 2$

6. INTERPRET the results. First, *check* the solution in the stated problem. If the shorter piece of board is 2 feet, the longer piece is 4 · (2 feet) = 8 feet and the sum of the two pieces is 2 feet + 8 feet = 10 feet. Next, *state* the conclusions. The shorter piece of board is 2 feet and the longer piece of board is 8 feet.

EXAMPLE 2   In 1996, Congress had 8 more Republican senators than Democratic. If the total number of senators is 100, how many senators of each party were there?

Solution: 1. UNDERSTAND the problem. Read and reread the problem. Let's guess that there are 40 Democratic senators. Since there are 8 more Republicans than Democrats, there must be $40 + 8 = 48$ Republicans. The total number of Democrats and Republicans is then $40 + 48 = 88$. This is an incorrect guess, since the total should be 100, but we now have a better understanding of the problem. This was the purpose for guessing.

2. ASSIGN a variable. Let

$$x = \text{number of Democrats,}$$

then

$$x + 8 = \text{number of Republicans.}$$

3. ILLUSTRATE the problem. No diagram or chart is needed.

4. TRANSLATE the problem. First write the equation in words.

In words:	number of Democrats	added to	number of Republicans	equals	100
Translate:	x	$+$	$(x + 8)$	$=$	100

5. COMPLETE the work. We solve the equation.

$$x + (x + 8) = 100$$
$$2x + 8 = 100 \qquad \text{Combine like terms.}$$
$$2x + 8 - 8 = 100 - 8 \qquad \text{Subtract 8 from both sides.}$$
$$2x = 92$$
$$\frac{2x}{2} = \frac{92}{2} \qquad \text{Divide both sides by 2.}$$
$$x = 46$$

6. INTERPRET the results.
   Check: If there are 46 Democratic senators, then there are $46 + 8 = 54$ Republican senators. The total number of senators is then $46 + 54 = 100$. The results check.
   State: In 1996, there were 46 Democratic and 54 Republican senators.

Much of problem solving involves a direct translation from a sentence to an equation.

EXAMPLE 3   Twice the sum of a number and 4 is the same as four times the number decreased by 12. Find the number.

Solution: 1. UNDERSTAND. Read and reread the problem. Propose a solution and check.

2. ASSIGN. Let

$$x = \text{the unknown number.}$$

3. ILLUSTRATE. No illustration is needed.

4. TRANSLATE.

In words:	twice	sum of a number and 4	is the same as	four times the number	decreased by	12
Translate:	2	$(x + 4)$	=	$4x$	−	12

5. COMPLETE. Now solve the equation.

$$2(x + 4) = 4x - 12$$
$$2x + 8 = 4x - 12 \qquad \text{Apply the distributive property.}$$
$$2x + 8 - 4x = 4x - 12 - 4x \qquad \text{Subtract } 4x \text{ from both sides.}$$
$$-2x + 8 = -12$$
$$-2x + 8 - 8 = -12 - 8 \qquad \text{Subtract 8 from both sides.}$$
$$-2x = -20$$
$$\frac{-2x}{-2} = \frac{-20}{-2} \qquad \text{Divide both sides by } -2.$$
$$x = 10$$

6. INTERPRET.

Check: Check this solution in the problem as it was originally stated. To do so, replace "number" with 10. Twice the sum of "10" and 4 is 28, which is the same as 4 times "10" decreased by 12.

State: The number is 10.

EXAMPLE 4

A local cellular phone company charges Elaine Chapoton $50 per month and $0.36 per minute of phone use in her usage category. If Elaine was charged $99.68 for a month's cellular phone use, determine the number of whole minutes of phone use.

Solution:

1. UNDERSTAND. Read and reread the problem. Next, guess an answer. Let's guess 70 minutes, and pay careful attention as to how we check this guess. For 70 minutes of use, Elaine's phone bill will be $50 plus $0.36 per minute of use. This is $50 + 0.36(70) = $75.20, less than $99.68. We now understand the problem and know that the number of minutes is greater than 70.

2. ASSIGN. Let

x represent the unknown in this problem,

or let x = number of minutes.

3. ILLUSTRATE. No diagram is needed.

4. TRANSLATE.

In words:	$50	added to	minute charge	is equal to	$99.68
Translate:	50	+	$.36x$	=	99.68

5. COMPLETE.

$$50 + 0.36x = 99.68$$
$$50 + 0.36x - 50 = 99.68 - 50$$
$$0.36x = 49.68$$
$$\frac{0.36x}{0.36} = \frac{49.68}{0.36}$$
$$x = 138$$

6. INTERPRET.

Check: If Elaine spends 138 minutes on her cellular phone, her bill is $50 + $0.36(138) = $99.68.

State: Elaine spent 138 minutes on her cellular phone this month.

EXERCISE SET 2.5

Solve the following. See Examples 1 and 2.

1. The governor of New York makes twice as much money as the governor of Nebraska. If the total of their salaries is $195,000, find the salary of each.

2. In the 1992 Summer Olympics, the Unified Team, consisting of athletes from 12 former Soviet republics, won 8 more gold medals than the United

In the '92 Olympics, Michael Marsh won the 200-m run in 20.01 seconds

States Team. If the total number of gold medals for both is 82, find the number of gold medals that each team won.

3. A 40-inch board is to be cut into three pieces so that the second piece is twice as long as the first piece and the third piece is 5 times as long as the first piece. If *x* represents the length of the first piece, find the lengths of all three pieces.

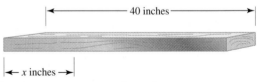

4. A 21-foot beam is to be divided so that the longer piece is 1 foot more than 3 times the shorter piece. If *x* represents the length of the shorter piece, find the lengths of both pieces.

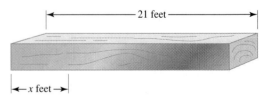

Solve. See Example 3.

5. The product of twice a number and three is the same as the difference of five times the number and $\frac{3}{4}$. Find the number. $-\frac{3}{4}$

1. Nebraska: $65,000; New York: $130,000 **2.** Unified Team: 45 gold medals; U.S.: 37 gold medals
3. 1st: 5 in.; 2nd: 10 in.; 3rd: 25 in. **4.** shorter piece: 5 ft.; longer piece: 16 ft.

6. If the difference of a number and four is doubled, the result is $\frac{1}{4}$ less than the number. Find the number. $\frac{31}{4}$

7. If the sum of a number and five is tripled, the result is one less than twice the number. Find the number. -16

8. Twice the sum of a number and six equals three times the sum of the number and four. Find the number.   0

Solve. See Example 4.

9. A car rental agency advertised renting a Buick Century for $24.95 per day and $0.29 per mile. If you rent this car for 2 days, how many whole miles can be driven on a $100 budget?   172 miles

10. A plumber gave an estimate for the renovation of a kitchen. Her hourly pay is $27 per hour and the plumber's parts will cost $80. If her total estimate is $404, how many hours does she expect this job to take?   12 hours

Solve.

11. A 17-foot piece of string is cut into two pieces so that one piece is 2 feet longer than twice the shorter piece. If the shorter piece is x feet long, find the lengths of both pieces.

12. In a recent election in Florida for a seat in the United States House of Representatives, Corrine Brown received 16,950 more votes than Marc Little. If the total number of votes was 110,740 find the number of votes for each candidate.

13. Two angles are supplementary if their sum is 180°. One angle measures three times the measure of a smaller angle. If x represents the measure of the smaller angle and these two angles are supplementary, find the measure of each angle.   45° and 135°

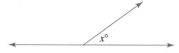

14. Two angles are complementary if their sum is 90°. Given the measures of the complementary angles shown, find the measure of each angle.

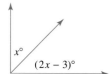

15. On June 20, 1994, John Paxson sank a 3-point shot with 3.9 seconds left to give the Chicago Bulls their third straight National Basketball Association championship. The opposing team was the Phoenix Suns. If the final score of the game was 2 consecutive integers whose sum is 197, find each final score.   Bulls: 99; Suns: 98

16. To make an international telephone call, you need the code for the country you are calling. The codes for Mali Republic, Côte d'Ivoire, and Niger are three consecutive odd integers whose sum is 675. Find the code for each country.

17. Determine whether there are two consecutive odd integers such that 7 times the first exceeds 5 times the second by 54.   There are none.

18. The sum of three consecutive integers is 13 more than twice the smallest integer. Find the integers.

19. Twice the difference of a number and 8 is equal to three times the sum of a number and 3. Find the number. -25

20. Five times the sum of a number and -1 is the same as 6 times a number. Find the number. -5

21. Find two consecutive odd integers such that twice the larger is 15 more than three times the smaller.

22. Find three consecutive even integers whose sum is negative 114. $-40, -38, -36$

23. On December 7, 1995, a probe launched from the robot explorer called Galileo entered the atmosphere of Jupiter at 100,000 miles per hour. The diameter of the probe is 19 inches less than twice its height. If the sum of the height and the diameter is 83 inches, find each dimension.

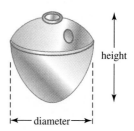

24. Over the past few years the satellite Voyager II has passed by the planets Saturn, Uranus, and Neptune, continually updating information about these planets, including the number of moons for each.

11. Shorter piece: 5 ft.; longer piece: 12 ft.
12. Corrine Brown: 63,845
   Marc Little: 46,895
14. 31° and 59°

16. Mali Republic: 223; Côte d'Ivoire: 225; Niger: 227
18. 10, 11, 12
21. $-11, -9$
23. height: 34 in.; diameter: 49 in.

Uranus is now believed to have 7 more moons than Neptune. Also, Saturn is now believed to have three times the number of moons of Neptune. If the total number of moons for these planets is 47, find the number of moons for each planet.

25. A woman's $15,000 estate is to be divided so that her husband receives twice as much as her son. If x represents the amount of money that her son receives, find the amount of money that her husband receives and the amount of money that her son receives. son: $5000 husband: $10,000

26. The sum of two consecutive integers is 31. What are the numbers? 15, 16

27. The flag of Brazil contains a parallelogram. One angle of the parallelogram is 15° less than twice the measure of the angle next to it. Find the measure of each angle of the parallelogram. (*Hint:* Recall that opposite angles of a parallelogram have the same measure and that the sum of the angles is 360°.) 65°, 115°

28. The flag of Equatorial Guinea contains an isosceles triangle. (Recall that an isosceles triangle con-

24. Saturn: 24 moons
Uranus: 15 moons
Neptune: 8 moons

tains two angles with the same measure.) If the measure of the third angle of the triangle is 30° more than twice the measure of either of the other two angles, find the measure of each angle of the triangle. (*Hint:* Recall that the sum of the measure of the angles of a triangle is 180°.) 37.5°, 37.5°, 105°

29. The measures of the angles of a triangle are 3 consecutive even integers. Find the measure of each angle. 58°, 60°, 62°

30. The golden rectangle is a rectangle whose length is approximately 1.6 times its width. The early Greeks thought that a rectangle with these dimensions was the most pleasing to the eye and examples of the golden rectangle are found in many early works of art. For example, the Parthenon in Athens contains many examples of golden rectangles. Mike Hallahan would like to plant a rectangular garden in the shape of a golden rectangle. If he has 78 feet of fencing available, find the dimensions of the garden. 15 ft. by 24 ft.

The length-width rectangle approximates the golden rectangle as well as the width-height rectangle.

31. Dr. Dorothy Smith gave the students in her geometry class at the University of New Orleans the following question. Is it possible to construct a triangle such that the second angle of the triangle has a measure that is twice the measure of the first angle and the measure of the third angle is 3 times the measure of the first? If so, find the measure of each angle. (*Hint:* Recall that the sum of the measures of the angles of a triangle is 180°.)

32. Give an example of how you recently solved a problem using mathematics.   answers may vary

33. In your own words, explain why a solution of a word problem should be checked using the original wording of the problem and not the equation written from the wording.   answers may vary

 Recall from Exercise 30 that the golden rectangle is a rectangle whose length is approximately 1.6 times its width.

34. It is thought that for about 75% of adults, a rectangle in the shape of the golden rectangle is the most pleasing to the eye. Draw 3 rectangles, one in the shape of the golden rectangle, and poll your class. Do the results agree with the percentage given above?   answers may vary

35. Examples of golden rectangles can be found today in architecture and manufacturing packaging. Find an example of a golden rectangle in your home. A few suggestions: the front face of a book, the floor of a room, the front of a box of food.

36. Measure the dimensions of each rectangle and decide which one best approximates the shape of the golden rectangle.   c

b.

c.

Review Exercises

Perform indicated operations. See Sections 1.5 and 1.6.

37. $3 + (-7)$ -4

38. $-2 + (-8)$ -10

39. $4 - 10$ -6

40. $-11 + 2$ -9

41. $-5 - (-1)$ -4

42. $-12 - 3$ -15

Translate each sentence into an equation. See Sections 1.4 and 2.5.

43. Half of the difference of a number and one is thirty-seven. $\frac{1}{2}(x - 1) = 37$

44. Five times the opposite of a number is the number plus sixty. $5(-x) = x + 60$

45. If three times the sum of a number and 2 is divided by 5, the quotient is 0. $\frac{3(x + 2)}{5} = 0$

46. If the sum of a number and 9 is subtracted from 50, the result is 0. $50 - (x + 9) = 0$

a.

31. 30°, 60°, 90°   **35.** answers may vary

2.6 | FORMULAS AND PROBLEM SOLVING

TAPE
BA 2.6

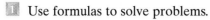

O B J E C T I V E S

1. Use formulas to solve problems.
2. Solve a formula or equation for one of its variables.

An equation that describes a known relationship among quantities, such as distance, time, volume, weight, and money is called a **formula.** These quantities are represented by letters and are thus variables of the formula. Here are some common formulas and their meanings.

$A = lw$
Area of a rectangle = length · width

$I = PRT$
Simple Interest = Principal · Rate · Time

$P = a + b + c$
Perimeter of a triangle = side a + side b + side c

$d = rt$
distance = rate · time

$V = lwh$
Volume of a rectangular solid = length · width · height

$F = \left(\dfrac{9}{5}\right)C + 32$

degrees Fahrenheit = $\left(\dfrac{9}{5}\right)$ · degrees Celsius + 32

Formulas are valuable tools because they allow us to calculate measurements as long as we know certain other measurements. For example, if we know we traveled a distance of 100 miles at a rate of 40 miles per hour, we can replace the variables d and r in the formula $d = rt$ and find our time, t.

$$d = rt \qquad \text{Formula.}$$
$$100 = 40t \qquad \text{Replace } d \text{ with } 100 \text{ and } r \text{ with } 40.$$

This is a linear equation in one variable, t. To solve for t, divide both sides of the equation by 40.

$$\frac{100}{40} = \frac{40t}{40} \qquad \text{Divide both sides by 40.}$$

$$\frac{5}{2} = t \qquad \text{Simplify.}$$

The time traveled is $\frac{5}{2}$ hours or $2\frac{1}{2}$ hours.

In this section we solve problems that can be modeled by known formulas. We use the same problem-solving steps that were introduced in the previous section. These steps have been slightly revised to include formulas.

PROBLEM-SOLVING STEPS

1. UNDERSTAND the problem. During this step don't work with variables (except for known formulas), but simply become comfortable with the problem. Some ways of accomplishing this are listed below.

- Read and reread the problem.
- Construct a drawing.
- Look up an unknown formula.
- Propose a solution and check. Pay careful attention to how to check your proposed solution. This will help later when writing an equation to model the problem.

2. ASSIGN a variable to an unknown in the problem. Use this variable to represent any other unknown quantities.

3. ILLUSTRATE the problem. A diagram or chart using the assigned variables can often help to visualize the known facts.

4. TRANSLATE the problem into a mathematical model. This is often an equation.

5. COMPLETE the work. This often means to solve the equation.

6. INTERPRET the results. *Check* the proposed solution in the stated problem and *state* your conclusion.

EXAMPLE 1   A glacier is a giant mass of rocks and ice that flows downhill like a river. Portage Glacier in Alaska is about 6 miles, or 31,680 feet, long and moves 400 feet per year. Icebergs are created when the front end of the glacier flows into Portage Lake. How long does it take for ice at the head (beginning) of the glacier to reach the lake?

Solution: **1.** UNDERSTAND. Read and reread the problem. The appropriate formula needed to solve this problem is the distance formula, $d = rt$. To become familiar with this formula, let's find the distance that ice traveling at a rate of 400 feet per year travels in 100 years. To do so, we let t time be 100 years, r rate be the given 400 feet per year, and substitute these values into the formula $d = rt$. We then have that distance $d = 400(100) = 40,000$ feet. Since we are interested in finding how long it takes ice to travel 31,680 feet, we now know that it is less than 100 years.

2. ASSIGN. Since we are using the formula $d = rt$, we let

t = the time in years for ice to reach the lake,

r = rate or speed of ice, and

d = distance from beginning of glacier to lake.

3. ILLUSTRATE. No other illustration needed.

4. TRANSLATE. To translate to an equation, we use the formula $d = rt$ and let distance $d = 31,680$ feet and rate $r = 400$ feet per year.

Formula: $d \;=\; r \cdot t$
Substitute: $31,680 = 400 \cdot t$ Let $d = 31,680$ and $r = 400$.

5. COMPLETE. Solve the equation for t. To solve for t, divide both sides by 400.

$$\frac{31,680}{400} = \frac{400 \cdot t}{400} \qquad \text{Divide both sides by 400.}$$

$$79.2 = t \qquad \text{Simplify.}$$

6. INTERPRET.
Check: To check, substitute 79.2 for t, and 400 for r in the distance formula and check to see that the distance is 31,680 feet.
State: It takes 79.2 years for the ice at the head of Portage Glacier to reach the lake.

EXAMPLE 2   Charles Pecot can afford enough fencing to enclose a rectangular garden with a perimeter of 140 feet. If the width of his garden is to be 30 feet, find the length.

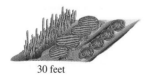

30 feet

Solution: **1.** UNDERSTAND. Read and reread the problem. The formula needed to solve this problem is the formula for the perimeter of a rectangle, $P = 2l + 2w$. Before continuing, become familiar with this formula.

2. ASSIGN. Let

l = the length of the rectangular garden

w = the width of the rectangular garden, and

P = perimeter of the garden.

3. ILLUSTRATE.

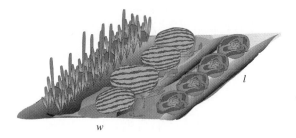

4. TRANSLATE. To translate to an equation, we use the formula $P = 2l + 2w$ and let perimeter $P = 140$ feet and width $w = 30$ feet.

Formula: $P = 2l + 2w$

Substitute: $140 = 2l + 2(30)$ Let $P = 140$ and $w = 30$.

5. COMPLETE.

$$140 = 2l + 2(30)$$
$$140 = 2l + 60$$
$$140 - 60 = 2l + 60 - 60 \quad \text{Subtract 60 from both sides.}$$
$$80 = 2l \quad\quad\quad\quad \text{Combine like terms.}$$
$$40 = l \quad\quad\quad\quad \text{Divide both sides by 2.}$$

6. INTERPRET. To *check*, substitute 40 for l and 30 for w in the perimeter formula and check to see that the perimeter is 140 feet.
   State: The length of the rectangular garden is 40 feet.   ▬▬▬▬

EXAMPLE 3   The average maximum temperature for January in Algerias, Algeria, is 59° Fahrenheit. Find the equivalent temperature in degrees Celsius.

Solution: **1.** UNDERSTAND. Read and reread the problem. A formula that can be used to solve this problem is the formula for converting degrees Celsius to degrees Fahrenheit, $F = \frac{9}{5}C + 32$. Before continuing, become familiar with this formula.

2. ASSIGN. Let

> C = temperature in degrees Celsius, and
>
> F = temperature in degrees Fahrenheit.

3. ILLUSTRATE. No illustration is needed.

4. TRANSLATE. To translate to an equation, we use the formula $F = \frac{9}{5}C + 32$ and let degrees Fahrenheit $F = 59$.

> Formula: $\qquad F = \frac{9}{5}C + 32$
>
> Substitute: $\quad 59 = \frac{9}{5}C + 32 \qquad$ Let $F = 59$.

5. COMPLETE.

$$59 = \frac{9}{5}C + 32$$

$$59 - 32 = \frac{9}{5}C + 32 - 32 \qquad \text{Subtract 32 from both sides.}$$

$$27 = \frac{9}{5}C \qquad \text{Combine like terms.}$$

$$\frac{5}{9} \cdot 27 = \frac{5}{9} \cdot \frac{9}{5}C \qquad \text{Multiply both sides by } \tfrac{5}{9}.$$

$$15 = C \qquad \text{Combine like terms.}$$

6. INTERPRET

> *Check:* To check, replace C with 15 and F with 59 in the formula and see that a true statement results.
>
> *State:* Thus, 59° Fahrenheit is equivalent to 15° Celsius. ▬▬▬

2 We say that the formula $F = \frac{9}{5}C + 32$ is solved for F because F is isolated on one side of the equation and the other side of the equation contains no F's. Suppose that we need to convert many Fahrenheit temperatures to equivalent degrees Celsius. In this case, it is easier to perform this task by solving the formula $F = \frac{9}{5}C + 32$ for C. (See Example 7.) For this reason, it is important to be able to solve an equation for any one of its specified variables. For example, the formula $d = rt$ is solved for d in terms of r and t. We can also solve $d = rt$ for t in terms of d and r. To solve for t, divide both sides of the equation by r.

$$d = rt$$
$$\frac{d}{r} = \frac{rt}{r} \qquad \text{Divide both sides by } r.$$
$$\frac{d}{r} = t \qquad \text{Simplify.}$$

To solve a formula or an equation for a specified variable, we use the same steps as for solving a linear equation. These steps are listed next.

TO SOLVE EQUATIONS FOR A SPECIFIED VARIABLE

Step 1.   Clear the equation of fractions by multiplying each side of the equation by the lowest common denominator.

Step 2.   Remove grouping symbols such as parentheses by using the distributive property.

Step 3.   Simplify each side of the equation by combining like terms.

Step 4.   Write the equation with terms containing the specified variable on one side and all other terms on the other side by using the addition property of equality.

Step 5.   Isolate the specified variable by using the distributive property and the multiplication property of equality.

EXAMPLE 4   Solve $V = lwh$ for l.

Solution: This formula is used to find the volume of a box. To solve for l, divide both sides by wh.

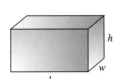

$$V = lwh$$

$$\frac{V}{wh} = \frac{lwh}{wh} \qquad \text{Divide both sides by } wh.$$

$$\frac{V}{wh} = l \qquad \text{Simplify.}$$

Since we have isolated l on one side of the equation, we have solved for l in terms of V, w, and h. Remember that it does not matter on which side of the equation we isolate the variable.

EXAMPLE 5   Solve $y = mx + b$ for x.

Solution: First, isolate mx by subtracting b from both sides.

$$y = mx + b$$

$$y - b = mx + b - b \qquad \text{Subtract } b \text{ from both sides.}$$

$$y - b = mx \qquad \text{Combine like terms.}$$

Next, solve for x by dividing both sides by m.

$$\frac{y - b}{m} = \frac{mx}{m}$$

$$\frac{y - b}{m} = x \qquad \text{Simplify.}$$

EXAMPLE 6 Solve $P = 2l + 2w$ for w.

Solution: This formula relates the perimeter of a rectangle to its length and width. To solve for w, begin by subtracting $2l$ from both sides.

$$P = 2l + 2w$$
$$P - 2l = 2l + 2w - 2l \qquad \text{Subtract } 2l \text{ from both sides.}$$
$$P - 2l = 2w \qquad \text{Combine like terms.}$$
$$\frac{P - 2l}{2} = \frac{2w}{2} \qquad \text{Divide both sides by 2.}$$
$$\frac{P - 2l}{2} = w \qquad \text{Simplify.}$$

The next example has an equation containing a fraction. We will first clear the equation of fractions and then solve for the specified variable.

EXAMPLE 7 Solve $F = \frac{9}{5}C + 32$ for C.

Solution:
$$F = \frac{9}{5}C + 32$$
$$5(F) = 5\left(\frac{9}{5}C + 32\right) \qquad \text{Clear the fraction by multiplying both sides by the LCD.}$$
$$5F = 9C + 160 \qquad \text{Distribute the 5.}$$
$$5F - 160 = 9C + 160 - 160 \qquad \text{Subtract 160 from both sides.}$$
$$5F - 160 = 9C \qquad \text{Combine like terms.}$$
$$\frac{5F - 160}{9} = \frac{9C}{9} \qquad \text{Divide both sides by 9.}$$
$$\frac{5F - 160}{9} = C \qquad \text{Simplify.}$$

EXERCISE SET 2.6

Substitute the given values into the given formulas and solve for the unknown variable. See Examples 1 through 7.

1. $A = bh$; $A = 45$, $b = 15$ $h = 3$

2. $D = rt$; $D = 195$, $t = 3$ $r = 65$

3. $S = 4lw + 2wh$; $S = 102$, $l = 7$, $w = 3$ $h = 3$

4. $V = lwh$; $l = 14$, $w = 8$, $h = 3$ $V = 336$

5. $C = 2\pi r$; $C = 15.7$ (use the approximation 3.14 for π) $r = 2.5$

6. $A = \pi r^2$; $r = 4.5$ (use the approximation 3.14 for π) $A = 63.585$

7. $I = PRT$; $I = 3750$, $P = 25,000$, $R = .05$ $T = 3$

8. $I = PRT$; $I = 1,056,000$, $R = .055$, $T = 6$ $P = 3,200,000$

9. $A = \frac{1}{2}(B + b)h; A = 180, B = 11, b = 7$ $h = 20$

10. $A = \frac{1}{2}(B + b)h; A = 60, B = 7, b = 3$ $h = 12$

11. $P = a + b + c; P = 30, a = 8, b = 10$ $c = 12$

12. $V = \frac{1}{3}Ah; V = 45, h = 5$ $A = 27$

13. $V = \frac{1}{3}\pi r^2 h; V = 565.2, r = 6$ (use the approximation 3.14 for π) $h = 15$

14. $V = \frac{4}{3}\pi r^3; r = 3$ (use the approximation 3.14 for π)
$V = 113.04$

Solve each formula for the specified variable. See Examples 4 through 7.

15. $f = 5gh$ for h **16.** $C = 2\pi r$ for r

17. $V = LWH$ for W **18.** $T = mnr$ for n

19. $3x + y = 7$ for y **20.** $-x + y = 13$ for y

21. $A = p + PRT$ for R **22.** $A = p + PRT$ for T

23. $V = \frac{1}{3}Ah$ for A $A = \frac{3V}{h}$ **24.** $D = \frac{1}{4}fk$ for k $k = \frac{4D}{f}$

25. $P = a + b + c$ for a $a = P - b - c$

26. $PR = s_1 + s_2 + s_3 + s_4$ for s_3 $s_3 = PR - s_1 - s_2 - s_4$

27. $S = 2\pi rh + 2\pi r^2$ for h **28.** $S = 4lw + 2wh$ for h

29. The formula $V = LWH$ is used to find the volume of a box. If the length of a box is doubled, the width is doubled, and the height is doubled, how does this affect the volume?

30. The formula $A = bh$ is used to find the area of a parallelogram. If the base of a parallelogram is doubled and its height is doubled, how does this affect the area?   It multiplies the area by 4.

29. It multiplies the volume by 8.

*Delta Airlines awards "Frequency Flyer" miles equal to the number of miles traveled rounded **up** to the nearest thousand. Use this for Exercises 31 and 32.*

31. A 5.5-hour nonstop flight from Orlando, Florida, to San Francisco, California, averages 470 mph. Find the "Frequency Flyer" miles earned on this flight.
3000 miles

32. A 45-minute $\left(\frac{3}{4}\text{-hour}\right)$ nonstop flight from New Orleans, Louisiana, to Houston, Texas, averages

500 mph. Find the "Frequent Flyer" miles earned on this flight.   1000 miles

Solve by using a known formula.

33. The distance from the sun to the Earth is approximately 93,000,000 miles. If light travels at a rate of 186,000 miles per second, how long does it take light from the sun to reach us?   500 sec. or $8\frac{1}{3}$ min.

34. Light travels at a rate of 186,000 miles per second. If our moon is 238,860 miles from the Earth, how long does it take light from the moon to reach us? (Round to the nearest tenth of a second.)   1.3 sec.

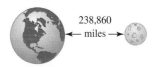

238,860
← miles →

35. Find how much rope is needed to wrap around the Earth at the equator, if the radius of the Earth is 4000 miles. (*Hint:* Use 3.14 for π and the formula for circumference.)   25,120 miles

36. If the length of a rectangularly shaped garden is 6 meters and its width is 4.5 meters, find the amount of fencing required.   21 meters

4.5 meters

6 meters

37. Dry Ice is a name given to solidified carbon dioxide. At $-78.5°$ Celsius it changes directly from a solid to a gas. Convert this temperature to Fahrenheit. $-109.3°$F

15. $h = \frac{f}{5g}$ **16.** $r = \frac{C}{2\pi}$ **17.** $W = \frac{V}{LH}$ **18.** $n = \frac{T}{mr}$ **19.** $y = 7 - 3x$ **20.** $y = 13 + x$

21. $R = \frac{A - p}{PT}$ **22.** $T = \frac{A - p}{PR}$ **27.** $h = \frac{S - 2\pi r^2}{2\pi r}$ **28.** $h = \frac{S - 4lw}{2w}$

38. Lightning bolts can reach a temperature of 50,000° Fahrenheit. Convert this to degrees Celsius.   27,760°C

39. Convert Nome, Alaska's 14°F high temperature to Celsius.   −10°C

40. Convert Paris, France's low temperature of −5°C to Fahrenheit.   23°

41. The SR-71 is a top secret spy plane. It is capable of traveling from Rochester, New York, to San Francisco, California, a distance of approximately 3000 miles, in $1\frac{1}{2}$ hours. Find the rate of the SR-71.

42. A limousine built in 1968 for the president cost $500,000 and weighed 5.5 tons. This Lincoln Continental Executive could travel at 50 miles per hour with all of its tires shot away. At this rate, how long would it take to travel from Charleston, Virginia, to Washington, D.C., a distance of 375 miles?   7.5 hours

The Dante II is a spider-like robot that is used to map the depths of an active Alaskan volcano.

43. The dimensions of the Dante II are 10 feet long by 8 feet wide by 10 feet high. Find the volume of the smallest box needed to store this robot.

44. The Dante II traveled 600 feet into an active Alaskan volcano in $3\frac{1}{3}$ hours. Find the traveling rate of Dante II in feet per minute. (*Hint:* First convert $3\frac{1}{3}$ hours to minutes.)   3 ft. per minute

41. 2000 mph   **43.** 800 cubic feet

45. Piranha fish require 1.5 cubic feet of water per fish to maintain a healthy environment. Find the maximum number of piranhas you could put in a tank measuring 8 feet by 3 feet by 6 feet.   96 piranhas

46. Maria's Pizza sells one 16-inch cheese pizza or two 10-inch cheese pizzas for $9.99. Determine which size gives more pizza.   one 16-in. pizza

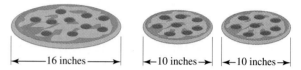

47. Find how long it takes Tran Nguyen to drive 135 miles on I-10 if he merges onto I-10 at 10 A.M. and drives nonstop with his cruise control set on 60 mph.   2.25 hours

48. Beaumont, Texas, is about 150 miles from Toledo Bend. If Leo Miller leaves Beaumont at 4 A.M. and averages 45 mph, when should he arrive at Toledo Bend?   7:20 A.M.

49. Find the temperature at which the Celsius measurement and Fahrenheit measurement are the same number.   −40°

50. Find how many goldfish you can put in a cylindrical tank whose diameter is 8 meters and whose height is 3 meters, if each goldfish needs 2 cubic meters of water.   75 goldfish

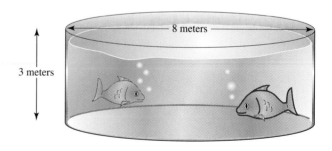

51. Bolts of lightning can travel at 270,000 miles per second. How many times can a lightning bolt travel around the world in one second? (See Exercise 35. Round to the nearest tenth.)   10.8

52. A glacier is a giant mass of rocks and ice that flows downhill like a river. Exit Glacier, near Seward, Alaska moves at a rate of 20 inches a day. Find the distance in feet the glacier moves in a year. (Assume 365 days a year. Round to 2 decimal places.)   608.33 ft.

53. 33,493,333,333 cubic miles

53. On July 16, 1994, the Shoemaker-Levy 9 comet collided with Jupiter. The impact of the largest fragment of the comet, a massive chunk of rock and ice, created a fireball with a radius of 2000 miles. Find the volume of this spherical fireball. (Use 3.14 for π. Round to the nearest whole cubic mile.)

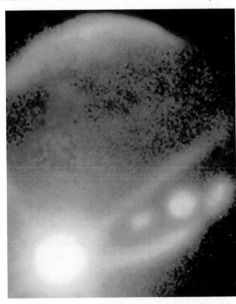

54. The fireball from the largest fragment of the comet (see Exercise 53) immediately collapsed, as it was pulled down by gravity. As it fell, it cooled to approximately $-350°$ Fahrenheit. Convert this to degrees Celsius. $-212\frac{2}{9}°$C

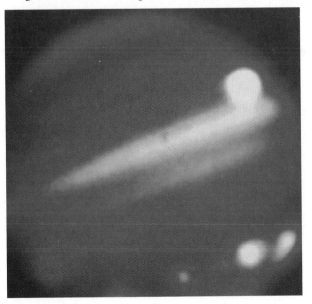

55. Stalactites join stalagmites to form columns. A column found at Natural Bridge Caverns near San Antonio, Texas, rises 15 feet and has a *diameter* of only 2 inches. Find the volume of this column in cubic inches. (*Hint:* Use the formula for volume of a cylinder and use 3.14 for π.)   565.2 cubic inches

56. A lawn is in the shape of a trapezoid with a height of 60 feet and bases of 70 feet and 130 feet. How many bags of fertilizer must be purchased to cover the lawn if each bag covers 4000 square feet?

2 bags

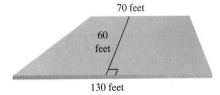

70 feet

60 feet

130 feet

57. Normal room temperature is about 78°F. Write this temperature as degrees Celsius. $25\frac{5}{9}°$C

58. Flying fish do not fly, but actually glide. (See photo at the top of next page.) They have been known to travel a distance of 1300 feet at a rate of 20 miles per hour. How many seconds did it take to travel

Color image made by the Hubble Space Telescope of the impact site of Fragment G of comet.

this distance? (*Hint:* First convert miles per hour to feet per second. Recall that 1 mile = 5280 feet. Round to the nearest tenth of a second.)   44.3 sec.

59. The X-30 is a new "space plane" being developed that will skim the edge of space at 4000 miles per hour. Neglecting altitude, if the circumference of the earth is approximately 25,000 miles, how long will it take for the X-30 to travel around the Earth?   6.25 hrs

60. In the United States, the longest hang glider flight was a 303-mile, $8\frac{1}{2}$ hour flight from New Mexico to Kansas. What was the average rate during this flight? $35\frac{11}{17}$ mph

61. If the area of a right-triangularly shaped sail is 20 square feet and its base is 5 feet, find the height of the sail.   8 ft.

Review Exercises

Write the following phrases as algebraic expressions. See Section 2.1.

62. Nine divided by the sum of a number and 5.

63. Half the product of a number and five. $\frac{1}{2}(5x)$

64. Three times the sum of a number and four.

65. One-third of the quotient of a number and six.

66. Double the sum of ten and four times the number.

67. Twice a number divided by three times the number.

68. Triple the difference of a number and twelve.

69. A number minus the sum of the number and six. $x - (x + 6)$

2.7 PERCENT AND PROBLEM SOLVING

OBJECTIVES

1. Write percents as decimals and decimals as percents.
2. Find percents of numbers.
3. Read and interpret graphs containing percents.
4. Solve percent equations.
5. Solve problems containing percents.

TAPE
BA 2.7

1️⃣ Much of today's statistics is given in terms of percent: a basketball player's free throw percent, current interest rates, stock market trends, and nutrition labeling, just to name a few. In this section, we explore percent and applications involving percents.

If 37 percent of all households in the United States have a home computer,

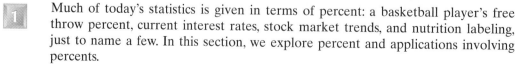

62. $\dfrac{9}{x + 5}$ **64.** $3(x + 4)$ **65.** $\dfrac{1}{3}\left(\dfrac{x}{6}\right)$ **66.** $2(10 + 4x)$ **67.** $\dfrac{2x}{3x}$ **68.** $3(x - 12)$

what does this percent mean? It means that 37 households out of every 100 households have a home computer. In other words,

the word **percent** means **per hundred** so that

37 **percent** means 37 **per hundred** or $\dfrac{37}{100}$.

We use the symbol % to denote percent, and we can write a percent as a fraction or a decimal.

$$37\% = \frac{37}{100} = 0.37$$

$$8\% = \frac{8}{100} = 0.08$$

$$100\% = \frac{100}{100} = 1.00$$

This suggests the following:

To Write a Percent as a Decimal

Drop the percent symbol and move the decimal point two places to the left.

EXAMPLE 1   Write each percent as a decimal.

a. 35%   **b.** 89.5%   **c.** 150%

Solution: **a.** 35% = 0.35   **b.** 89.5% = 0.895   **c.** 150% = 1.5

To write a decimal as a percent, we reverse the procedure above.

To Write a Decimal as a Percent

Move the decimal point two places to the right and attach the percent symbol, %.

EXAMPLE 2   Write each number as a percent.

a. 0.73   **b.** 1.39   **c.** $\dfrac{1}{4}$

Solution: **a.** 0.73 = 73%   **b.** 1.39 = 139%

c. First, write $\frac{1}{4}$ as a decimal.

$$\frac{1}{4} = 0.25 = 25\%.$$

2   To find a percent of a number, recall that the word "of" means multiply.

EXAMPLE 3   Find 72% of 200.

Solution: To find 72% of 200, we multiply.

$$72\% \text{ of } 200 = 72\%(200)$$
$$= 0.72(200)$$
$$= 144.$$

Thus, 72% of 200 is 144.

3   As mentioned earlier, percents are often used in statistics. Recall that the graph below is called a circle graph or a pie chart. The circle or pie represents a whole, or 100%. Each circle is divided into sectors (shaped like pieces of a pie) which represent various parts of the whole 100%.

EXAMPLE 4   The circle graph below shows how much money homeowners in the United States spend annually on maintaining their homes. Use this graph to answer the questions below.

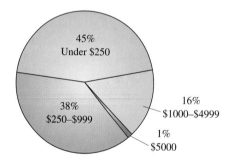

Yearly Home Maintenance in the U.S.

a. What percent of homeowners spend under $250 on yearly home maintenance?

b. What percent of homeowners spend less than $1000 per year on maintenance?

c. How many of the 22,000 homeowners in a town called Fairview might we expect to spend under $250 a year on home maintenance?

d. Find the number of degrees in the 16% sector.

Solution: **a.** From the circle graph, we see that 45% of homeowners spend under $250 per year on home maintenance.

  b. From the circle graph, we know that 45% of homeowners spend under $250 per year and 38% of homeowners spend $250–$999 per year, so that the sum 45% + 38% or 83% of homeowners spend less than $1000 per year.

c. Since 45% of homeowners spend under $250 per year on maintenance, we find 45% of 22,000.

$$45\% \text{ of } 22{,}000 = 0.45(22{,}000)$$
$$= 9900$$

We might then expect that 9900 homeowners in Fairview spend under $250 per year on home maintenance.

d. To find the number of degrees in the 16% sector, recall that the number of degrees around a circle is 360°. Thus, to find the number of degrees in the 16% sector, we find 16% of 360°.

57.6°

$$16\% \text{ of } 360° = 0.16(360°)$$
$$= 57.6°$$

The 16% sector contains 57.6°.

4

Next, we practice writing sentences as percent equations. Since these problems involve direct translations, all of our previous problem-solving steps are not needed.

EXAMPLE 5   The number 63 is what percent of 72?

Solution: **1. UNDERSTAND.** Read and reread the problem. Next, let's guess a solution. Suppose we guess that 63 is 80% of 72. We may check our guess by finding 80% of 72. 80% of 72 = 0.80(72) = 57.6. Close, but not 63. At this point, though, we have a better understanding of the problem, we know the correct answer is close to and greater than 80%, and we know how to check our proposed solution later.

2. ASSIGN. Let x = the unknown percent.

4. TRANSLATE. Recall that "is" means "equals" and "of" signifies multiplying. Translate the sentence directly.

In words: | The number 63 | is | what percent | of | 72 ? |

Translate: 63   = x ·   72

5. COMPLETE.

$$63 = 72x$$
$$0.875 = x \qquad \text{Divide both sides by 72.}$$
$$87.5\% = x \qquad \text{Write as a percent.}$$

6. INTERPRET.
Check by verifying that 87.5% of 72 is 63.
State the results. The number 63 is 87.5% of 72.

EXAMPLE 6   The number 120 is 15% of what number?

Solution: **1. UNDERSTAND.** Read and reread the problem. Guess a solution and check your guess.

2. ASSIGN. Let x = the unknown number.

4. TRANSLATE.

In words: | The number 120 | is | 15% | of | what number? |

Translate: 120   =   15%   · x

5. COMPLETE.

$$120 = 0.15x$$

$$800 = x \qquad \text{Divide both sides by 0.15.}$$

6. INTERPRET.

Check the proposed solution of 800 by finding 15% of 800 and verifying that the result is 120.

State: Thus, 120 is 15% of 800.

5. Percent increase or percent decrease is a common way to describe how some measurement has increased or decreased. For example, crime increased by 8%, teachers received a 5.5% increase in salary, or a company decreased its employees by 10%. The next example is a review of percent increase.

EXAMPLE 7

The cost of a hand-tossed pepperoni pizza at Domino's recently increased from $5.80 to $7.03. Find the percent increase.

Solution:

1. UNDERSTAND. Read and reread the problem. Let's guess that the percent increase is 10%. To see if this is the case, we find 10% of $5.80 to find the *increase* in price. Then we add this increase to $5.80 to find the *new price*. In other words, 10%($5.80) = 0.10($5.80) = $0.58, the *increase* in price. The new price then would be $5.80 + $0.58 = $6.38, less than the actual new price of $7.03. We now know that the increase is greater than 10% and we know how to check our proposed solution.

2. ASSIGN. Let x = the percent increase.

4. TRANSLATE. First, find the **increase,** and then the **percent increase.** The increase in price is found by:

In words: increase = | new price | − | old price | or

Translate: increase = $7.03   − $5.80
                      = $1.23

Next, find the percent increase. The percent increase or percent decrease is always a percent of the original number or in this case, the old price.

In words: | increase | is | what percent increase | of | old price |

Translate: $1.23   = x · $5.80

5. COMPLETE.

$$1.23 = 5.80x$$

$$0.212 \approx x \qquad \text{Divide both sides by 5.80 and round to 3 decimal places.}$$

$$21.2\% \approx x \qquad \text{Write as a percent.}$$

6. INTERPRET.

Check the proposed solution.

State: The percent increase in price is approximately 21.2%.

EXERCISE SET 2.7

Write each percent as a decimal. See Example 1.

1. 120%   1.2

2. 73%   0.73

3. 22.5%   0.225

4. 4.2%   0.042

5. 0.12%   0.0012

6. 0.86%   0.0086

Write each number as a percent. See Example 2.

7. 0.75   75%

8. 0.3   30%

9. 2   200%

10. 5.1   510%

11. $\frac{1}{8}$ 12.5%

12. $\frac{3}{5}$ 60%

Use the home maintenance graph to answer each question. See Example 4.

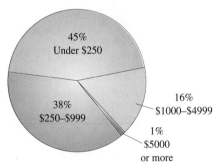

Yearly Home Maintenance in the U.S.

45% Under $250

38% $250–$999

16% $1000–$4999

1% $5000 or more

Source: Census Bureau

13. What percent of homeowners spend $250–$999 on yearly home maintenance?   38%

14. What percent of homeowners spend $5000 or more on yearly home maintenance?   1%

15. What percent of homeowners spend $250–$4999 on yearly home maintenance?   54%

16. What percent of homeowners spend $250 or more on yearly home maintenance?   55%

17. Find the number of degrees in the 38% sector.

18. Find the number of degrees in the 1% sector.   3.6°

19. How many homeowners in your town might you expect to spend $250–$999 on yearly home maintenance?   answers may vary

20. How many homeowners in your town might you expect to spend $5000 or more on yearly home maintenance?   answers may vary

Solve the following. See Examples 3, 5, and 6.

21. What number is 16% of 70?   11.2

22. What number is 88% of 1000?   880

23. The number 28.6 is what percent of 52?   55%

24. The number 87.2 is what percent of 436?   20%

25. The number 45 is 25% of what number?   180

26. The number 126 is 35% of what number?   360

27. Find 23% of 20.   4.6   **28.** Find 140% of 86.

29. The number 40 is 80% of what number?   50

30. The number 56.25 is 45% of what number?   125

31. The number 144 is what percent of 480?   30%

32. The number 42 is what percent of 35?   120%

Solve. See Example 7. Many applications in this exercise set may be solved more efficiently with the use of a calculator.

33. Dillard's advertised a 25% off sale. If a London Fog coat originally sold for $156, find the decrease and the sale price. $39 decrease; $117 sale price

34. Time Saver increased the price of a $0.75 cola by 15%. Find the increase and the new price.

35. Hallahan's Construction Company increased their estimate for building a new house from $95,500 to $110,000. Find the percent increase.

36. By buying in quantity, the Cannon family was able to decrease their weekly food bill from $150 a week to $130 a week. Find the percent decrease.

17. 136.8°   **28.** 120.4   **34.** $0.11 increase; $0.86 new price   **35.** 15.2% increase   **36.** 13.3% decrease

37. At this writing, the women's world record for throwing a disc (like a Frisbee) is held by Anni Kreml of the USA. Her throw was 447.2 feet. The men's record is held by Niclas Bergehamn of Sweden. His throw was 44.8% further than Anni's. Find the length of his throw. (Round to the nearest tenth of a foot.) (*Source:* World Flying Disc Federation.)

38. Scoville units are used to measure the hotness of a pepper. An alkaloid, capsaicin, is the ingredient that makes a pepper hot and liquid chromatography measures the amount of capsaicin in parts per million. The jalapeno measures around 5000 Scoville units, while the hottest pepper, the habanero, measures around 3000% of the measure of the jalapeno. Find the measure of the habanero pepper.   150,000 Scoville units

43. Do the percents shown in the graph below have a sum of 100%? Why or why not?

44. Survey your algebra class and find what percent of the class has used over-the-counter drugs for each of the categories listed. Draw a bar graph of the results.   answers may vary

45. Iceberg lettuce is grown and shipped to stores for about 40 cents a head, and consumers purchase it for about 70 cents a head. Find the percent increase.   75% increase

46. The lettuce consumption per capita in 1968 was about 21.5 pounds, and in 1992 the consumption rose to 26.1 pounds. Find the percent increase. (Round to the nearest tenth of a percent.)

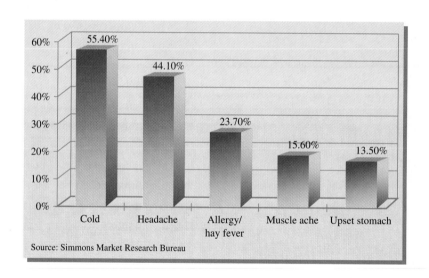

Source: Simmons Market Research Bureau

The above graph shows the percent of people in a survey who have used various over-the-counter drugs in a twelve-month period.

39. What percent of those surveyed used over-the-counter drugs to combat the common cold?

40. What percent of those surveyed used over-the-counter drugs to combat an upset stomach?

41. If 230 people were surveyed, how many of these used over-the-counter drugs for allergies?

42. The city of Chattanooga has a population of approximately 152,000. How many of these people would you expect to have used over-the-counter drugs for relief of a headache?   67,032 people

47. A recent study showed that 26% of men have dozed off at their place of work. If you currently employ 121 men, how many of these men might you expect to have dozed off at work? (*Source:* Better Sleep Council.)   31 men

48. A recent study showed that women and girls spend 41% of the household clothes budget. If a family spent $2000 last year on clothing, how much might have been spent on clothing for women and girls? (*Source:* The Interep Radio Store.) $820

49. The table on the next page shows where lightning strikes. Use this table to draw a circle graph or pie chart of this information.

37. 647.5 ft.   **39.** 55.40%   **40.** 13.50%   **41.** 54 people   **43.** No, many people use several medications.
46. 21.4% increase

FIELDS, BALLPARKS	**UNDER TREES**	**BODIES OF WATER**	**GOLF COURSES**	**NEAR HEAVY EQUIPMENT**	**TELEPHONE POLES**	**OTHER**
28%	17%	13%	4%	6%	1%	31%

50. During a recent 5-year period, bank fees for bounced checks have risen from $12.62 to $15.65. Find the percent increase. (Round to the nearest whole percent.) If inflation over the same period has been 16%, do you think the increase in bank fees is fair? (*Source:* Federal Reserve.)   24%; no

51. During a recent 5-year period, the bank fee for depositing a bad check has risen from $5.38 to $6.08. Find the percent increase. (Round to the nearest whole percent.) Given the inflation rate of 16%, do you think that this increase in bank fees is fair?

52. The first Barbie doll was introduced in March 1959 and cost $3. This same 1959 Barbie doll now costs up to $5000. Find the percent increase rounded to the nearest whole percent.   166, 567%

53. The ACT Assessment is a college entrance exam taken by about 60% of college-bound students. The national average score was 20.7 in 1993 and rose to 20.8 in 1994. Find the percent increase. (Round to the nearest hundredth of a percent.)

The double bar graph below shows selected services and products and the percent of supermarkets offering each.

54. What percent of supermarkets offered plastic grocery bags in 1990?   88%

55. What percent of supermarkets accepted credit cards in 1992?   39%

56. Suppose that there are 20 supermarkets in your city. How many of these 20 supermarkets might have offered customers plastic grocery bags in 1992?   18

57. How many more supermarkets (in percent) offered reusable grocery bags in 1992 than in 1990?

58. Do you notice any trends shown in this graph?

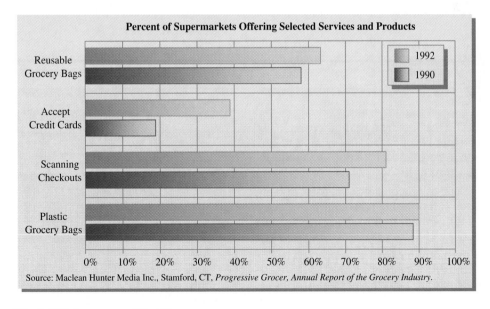

Percent of Supermarkets Offering Selected Services and Products

Source: Maclean Hunter Media Inc., Stamford, CT, *Progressive Grocer, Annual Report of the Grocery Industry.*

51. 13%; yes   **53.** 0.48% increase   **57.** 5%
58. More supermarkets were offering services in 1992 than in 1990.

Standardized nutrition labels like the ones below have been displayed on food items since 1994. The percent column on the right shows the percent of daily values based on a 2000-calorie diet shown at the bottom of the label. For example, a serving of this food contains 4 grams of total fat, where the recommended daily fat based on a 2000 calorie diet is 65 grams of fat. This means that $\frac{4}{65}$ or approximately 6% (as shown) of your daily recommended fat is taken in by eating a serving of this food.

Nutrition Facts

Serving Size 18 crackers (31g)
Servings Per Container About 9

Amount Per Serving

Calories 130 Calories from Fat 35

	% Daily Value*
Total Fat 4g	**6%**
Saturated Fat 0.5g	**3%**
Polyunsaturated Fat 0g	
Monounsaturated Fat 1.5g	
Cholesterol 0mg	**0%**
Sodium 230mg	**x**
Total Carbohydrate 23g	**y**
Dietary Fiber 2g	**8%**
Sugars 3g	
Protein 2g	

Vitamin A 0%	•	Vitamin C 0%
Calcium 2%	•	Iron 6%

*Percent Daily Values are based on a 2,000 calorie diet. Your daily values may be higher or lower depending on your calorie needs.

	Calories:	2,000	2,500
Total Fat	Less than	65g	80g
Sat Fat	Less than	20g	25g
Cholesterol	Less than	300mg	300mg
Sodium	Less than	2400mg	2400mg
Total Carbohydrate		300g	375g
Dietary Fiber		25g	30g

59. Based on a 2000-calorie diet, what percent of daily values of sodium is contained in a serving of this food? In other words, find *x*. (Round to the nearest tenth of a percent.)   9.6%

60. Based on a 2000-calorie diet, what percent of daily values of total carbohydrate is contained in a serving of this food? In other words, find *y*. (Round to the nearest tenth of a percent.)   7.7%

61. Notice on the nutrition label that one serving of this food contains 130 calories and 35 of these calories are from fat. Find the percent of calories

from fat. (Round to the nearest tenth of a percent.) It is recommended that no more than 30% of calorie intake come from fat. Does this food satisfy this recommendation?   26.9%; yes

Below is a nutrition label for a particular food.

NUTRITIONAL INFORMATION PER SERVING

Serving Size: 9.8 oz.   **Servings Per Container: 1**

Calories	280	Polyunsaturated Fat	1g
Protein	12g	Saturated Fat	3g
Carbohydrate	45g	Cholesterol	20mg
Fat	6g	Sodium	520mg
Percent of Calories from Fat	**?**	Potassium	220mg

62. If fat contains approximately 9 calories per gram, find the percent of calories from fat in one serving of this food. (Round to the nearest tenth of a percent.)   19.3%

63. If protein contains approximately 4 calories per gram, find the percent of calories from protein from one serving of this food. (Round to the nearest tenth of a percent.)   17.1%

64. Find a food that contains more than 30% of its calories per serving from fat. Analyze the nutrition label and verify that the percents shown are correct.   answers may vary

Review Exercises

Evaluate the following expressions for the given values. See Section 1.7.

65. $2a + b - c$; $a = 5$, $b = -1$, and $c = 3$ 6

66. $-3a + 2c - b$; $a = -2$, $b = 6$, and $c = -7$ -14

67. $4ab - 3bc$; $a = -5$, $b = -8$, and $c = 2$ 208

68. $ab + 6bc$; $a = 0$, $b = -1$, and $c = 9$ -54

69. $n^2 - m^2$; $n = -3$ and $m = -8$ -55

70. $2n^2 + 3m^2$; $n = -2$ and $m = 7$ 155

Solve. See Sections 2.6 and 2.7.

71. Find how much interest $2000 earns in 3 years in a savings account paying 4% simple interest annually. $240

72. Find the amount of principal that must be invested in a CD paying 5% simple interest annually to earn $125 in $2\frac{1}{2}$ years. $1000

73. Find the distance traveled if driving at a speed of 57 miles per hour for 3.5 hours.   199.5 miles

74. The distance from Kansas City, Missouri, to Duluth, Minnesota, is approximately 590 miles. How long does it take to travel from Kansas City to Duluth at an average speed of 55 miles per hour? (Round to the nearest tenth of an hour.)
10.7 hours

2.8 | FURTHER PROBLEM SOLVING

O B J E C T I V E S

TAPE
BA 2.8

1. Solve problems involving geometry concepts.
2. Solve problems involving distance.
3. Solve problems involving mixtures.
4. Solve problems involving interest.

This section is devoted to solving problems in the categories listed. The same problem-solving steps used in previous sections are also followed in this section. They are listed below for review.

PROBLEM-SOLVING STEPS

1. UNDERSTAND the problem. During this step don't work with variables (except for known formulas), but simply become comfortable with the problem. Some ways of accomplishing this are listed below.
   - Read and reread the problem.
   - Construct a drawing.
   - Look up an unknown formula.
   - Propose a solution and check. Pay careful attention to how to check your proposed solution. This will help later when writing an equation to model the problem.

2. ASSIGN a variable to an unknown in the problem. Use this variable to represent any other unknown quantities.

3. ILLUSTRATE the problem. A diagram or chart using the assigned variables can often help visualize the known facts.

4. TRANSLATE the problem into a mathematical model. This is often an equation.

5. COMPLETE the work. This often means to solve the equation.

6. INTERPRET the results. *Check* the proposed solution in the stated problem and *state* your conclusion.

EXAMPLE 1 The length of a rectangular road sign is 2 feet less than three times its width. Find the dimensions if the perimeter is 28 feet.

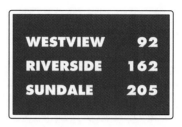

Solution:

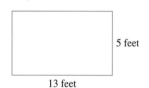

13 feet

1. UNDERSTAND. Read and reread the problem. Recall that the formula for the perimeter of a rectangle is $P = 2l + 2w$. Draw a rectangle and guess the solution. If the width of the rectangular sign is 5 feet, its length is 2 feet less than 3 times the width or 3(5 feet) − 2 feet = 13 feet. The perimeter P of the rectangle is then 2(13 feet) + 2(5 feet) = 36 feet, too much. We now know that the width is less than 5 feet.

2. ASSIGN. Let

$$w = \text{the width of the rectangular sign; then}$$
$$3w - 2 = \text{the length of the sign.}$$

3. ILLUSTRATE. Draw a rectangle and label it with the assigned variables.

$3w - 2$

4. TRANSLATE.

Formula: $\quad P = 2l \quad\quad + 2w$ or
Substitute: $\quad 28 = 2(3w - 2) + 2w.$

5. COMPLETE. $\quad 28 = 2(3w - 2) + 2w$

$$28 = 6w - 4 + 2w \quad\quad \text{Apply the distributive property.}$$
$$28 = 8w - 4$$
$$28 + 4 = 8w - 4 + 4 \quad\quad \text{Add 4 to both sides.}$$
$$32 = 8w$$
$$\frac{32}{8} = \frac{8w}{8} \quad\quad \text{Divide both sides by 8.}$$
$$4 = w$$

6. INTERPRET.

Check: If the width of the sign is 4 feet, the length of the sign is 3(4 feet) − 2 feet = 10 feet. This gives a perimeter of $P = 2(4 \text{ feet}) + 2(10 \text{ feet}) = 28$ feet, the correct perimeter.

State: The width of the sign is 4 feet and the length of the sign is 10 feet.

2

EXAMPLE 2   Marie Antonio, a bicycling enthusiast, rode her 10-speed at an average speed of 18 miles per hour on level roads and then slowed down to an average of 10 miles per hour on the hilly roads of the trip. If she covered a distance of 98 miles, how long did the entire trip take if traveling the level roads took the same time as traveling the hilly roads?

Solution: **1.** UNDERSTAND the problem. To do so, read and reread the problem. The formula $d = r \cdot t$ is needed. At this time, let's guess a solution. Suppose that she spent 2 hours traveling on the level roads. This means that she also spent 2 hours traveling on the hilly roads, since the times spent were the same. What is her total distance? Her distance on the level road is rate $\cdot$ time $= 18(2) = 36$ miles. Her distance on the hilly roads is rate $\cdot$ time $= 10(2) = 20$ miles. This gives a total distance of 36 miles $+$ 20 miles $=$ 56 miles, not the correct distance of 98 miles. Remember that the purpose of guessing a solution is not to guess correctly (although this may happen) but to help better understand the problem and how to model it with an equation.

2. ASSIGN a variable to an unknown in the problem. We are looking for the length of the entire trip so begin by letting

$x =$ the time spent on level roads.

Because the same amount of time is spent on hilly roads, then also

$x =$ the time spent on hilly roads.

3. ILLUSTRATE the problem. We summarize the information from the problem on the following chart. Fill in the rates given, the variables used to represent the times, and use the formula $d = r \cdot t$ to fill in the distance column.

	RATE · TIME = DISTANCE		
LEVEL	18	x	$18x$
HILLY	10	x	$10x$

4. TRANSLATE the problem into a mathematical model. Since the entire trip covered 98 miles, we have that

In words: total distance $=$ level distance $+$ hilly distance

Translate: 98 $=$ $18x$ $+$ $10x$

5. COMPLETE the work by solving the equation.

$$98 = 28x$$

$$\frac{98}{28} = \frac{28x}{28}$$

$$3.5 = x$$

6. INTERPRET the results.

Check: Recall that *x* represents the time spent on the level portion of the trip and also the time spent on the hilly portion. If Marie rides for 3.5 hours at 18 mph, her distance is 18(3.5) = 63 miles. If Marie rides for 3.5 hours at 10 mph, her distance is 10(3.5) = 35 miles. The total distance is 63 miles + 35 miles = 98 miles, the required distance.

State: The time of the entire trip is then 3.5 hours + 3.5 hours or 7 hours.

 Mixture problems involve two or more different quantities being combined to form a new mixture. These applications range from Dow Chemical's need to form a chemical mixture of a required strength to Planter's Peanut Company's need to find the correct mixture of peanuts and cashews, given taste and price constraints.

EXAMPLE 3   A chemist working on his doctoral degree at Massachusetts Institute of Technology needs 12 liters of a 50% acid solution for a lab experiment. The stockroom has only 40% and 70% solutions. How much of each solution should be mixed together to form 12 liters of a 50% solution?

Solution: **1.** UNDERSTAND. First, read and reread the problem a few times. Next, guess a solution. Suppose that we need 7 liters of the 40% solution. Then we need 12 − 7 = 5 liters of the 70% solution. To see if this is indeed the solution, find the amount of pure acid in 7 liters of the 40% solution, in 5 liters of the 70% solution, and in 12 liters of a 50% solution, the required amount and strength.

number of liters	×	acid strength	=	amount of pure acid
7 liters	×	40%	=	7(0.40) or 2.8 liters
5 liters	×	70%	=	5(0.70) or 3.5 liters
12 liters	×	50%	=	12(0.50) or 6 liters

Since 2.8 liters + 3.5 liters = 6.3 liters and not 6, our guess is incorrect, but we have gained some invaluable insight into how to model and check this problem.

2. ASSIGN. Let

$$x = \text{number of liters of 40\% solution; then}$$
$$12 - x = \text{number of liters of 70\% solution.}$$

3. ILLUSTRATE. The following table summarizes the information given. Recall that the amount of acid in each solution is found by multiplying the acid strength of each solution by the number of liters.

| | **ACID** | | |
	No. of liters · Strength = Amount of Acid		
40% SOLUTION	x	40%	$0.40x$
70% SOLUTION	$12 - x$	70%	$0.70(12 - x)$
50% SOLUTION NEEDED	12	50%	$0.50(12)$

TRANSLATE. The amount of acid in the final solution is the sum of the amounts of acid in the two beginning solutions.

In words:	acid in 40% solution	+	acid in 70% solution	=	acid in 50% mixture
Translate:	$0.40x$	+	$0.70(12 - x)$	=	$0.50(12)$

5. COMPLETE.

$$0.40x + 0.70(12 - x) = 0.50(12)$$

$$0.4x + 8.4 - 0.7x = 6 \qquad \text{Apply the distributive property.}$$

$$-0.3x + 8.4 = 6 \qquad \text{Combine like terms.}$$

$$-0.3x = -2.4 \qquad \text{Subtract 8.4 from both sides.}$$

$$x = 8 \qquad \text{Divide both sides by } -0.3.$$

6. INTERPRET.

Check: To check, recall how we checked our guess.
State: If 8 liters of the 40% solution are mixed with $12 - 8$ or 4 liters of the 70% solution, the result is 12 liters of a 50% solution. ▬▬▬

The next example is an investment problem.

EXAMPLE 4   Rajiv Puri invested part of his $20,000 inheritance in a mutual funds account that pays 7% simple interest yearly and the rest in a certificate of deposit that pays 9% simple interest yearly. At the end of one year, Rajiv's investments earned $1550. Find the amount he invested at each rate.

Solution: **1.** UNDERSTAND: Read and reread the problem. Next, guess a solution. Suppose that Rajiv invested $8000 in the 7% fund and the rest, $12,000, in the fund paying 9%. To check, find his interest after one year. Recall the formula, $I = PRT$, so the interest from the 7% fund = $8000(0.07)(1) = $560. The interest from the 9% fund = $12,000(0.09)(1) = $1080. The sum of the interests is $560 + $1080 = $1640. Our guess is incorrect, since the sum of the interests is not $1550, but we now have a better understanding of the problem.

2. **ASSIGN.** Let

x = amount of money in the account paying 7%. The rest of the money is \$20,000 less x or

$20,000 - x$ = amount of money in the account paying 9%.

3. **ILLUSTRATE.** We apply the simple interest formula $I = PRT$ and organize our information in the following chart. Since there are two different rates of interest and two different amounts invested, we apply the formula twice.

	PRINCIPAL ·	**RATE** ·	**TIME** =	**INTEREST**
7% FUND	x	0.07	1	$x(0.07)(1)$ or $0.07x$
9% FUND	$20,000 - x$	0.09	1	$(20,000 - x)(0.09)(1)$ or $0.09(20,000 - x)$
TOTAL	20,000			1550

4. **TRANSLATE.** The total interest earned, \$1550, is the sum of the interest earned at 7% and the interest earned at 9%.

In words:

interest at 7%	+	interest at 9%	=	total interest

Translate: $\quad\quad 0.07x \quad + \ 0.09(20,000 - x) = \quad\quad 1550$

5. **COMPLETE.**

$$0.07x + 0.09(20,000 - x) = 1550$$
$$0.07x + 1800 - 0.09x = 1550 \quad \text{Apply the distributive property.}$$
$$1800 - 0.02x = 1550 \quad \text{Combine like terms.}$$
$$-0.02x = -250 \quad \text{Subtract 1800 from both sides.}$$
$$x = 12,500 \quad \text{Divide both sides by } -0.02.$$

6. **INTERPRET.**

Check: If $x = 12,500$, then $20,000 - x = 20,000 - 12,500$ or 7500. These solutions are reasonable, since their sum is \$20,000 as required. The annual interest on \$12,500 at 7% is \$875; the annual interest on \$7500 at 9% is \$675, and \$875 + \$675 = \$1550.

State: The amount invested at 7% is \$12,500. The amount invested at 9% is \$7500.

EXERCISE SET 2.8

Solve each word problem. See Example 1.

1. An architect designs a rectangular flower garden such that the width is exactly two-thirds of the length. If 260 feet of antique picket fencing are to be used, find the dimensions of the garden.

2. If the length of a rectangular parking lot is 10 me-

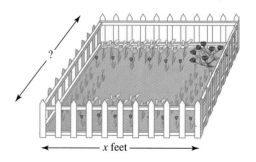

1. length: 78 ft; width: 52 ft.

ters less than twice its width, and the perimeter is 400 meters, find the length of the parking lot.

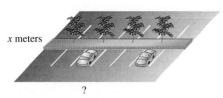

x meters

?

3. A flower bed is in the shape of a triangle with one side twice the length of the shortest side, and the third side is 30 feet more than the length of the shortest side. Find the dimensions if the perimeter is 102 feet.   18 ft., 36 ft., 48 ft.

4. The perimeter of a yield sign in the shape of an isosceles triangle is 22 feet. If the shortest side is 2 feet less than the other two sides, find the length of the shortest side. (*Hint:* An isosceles triangle has two sides the same length.)   6 ft.

?

Yield

x feet   *x* feet

Solve. See Example 2.

5. A jet plane traveling at 500 mph overtakes a propeller plane traveling at 200 mph that had a 2-hour head start. How far from the starting point are the planes? $666\frac{2}{3}$ miles

6. How long will it take a bus traveling at 60 miles per hour to overtake a car traveling at 40 mph if the car had a 1.5-hour head start?   3 hours

7. The Jones family drove to Disneyland at 50 miles per hour and returned on the same route at 40 mph. Find the distance to Disneyland if the total driving time was 7.2 hours.   160 miles

8. A bus traveled on a level road for 3 hours at an average speed 20 miles per hour faster than it traveled on a winding road. The time spent on the winding road was 4 hours. Find the average speed on the level road if the entire trip was 305 miles.

Solve. See Example 3.

9. How much pure acid should be mixed with 2 gallons of a 40% acid solution in order to get a 70% acid solution?   2 gal

10. How many cubic centimeters of a 25% antibiotic solution should be added to 10 cubic centimeters of a 60% antibiotic solution in order to get a 30% antibiotic solution?   60 cc

11. Planter's Peanut Company wants to mix 20 pounds of peanuts worth $3 a pound with cashews worth $5 a pound in order to make an experimental mix worth $3.50 a pound. How many pounds of cashews should be added to the peanuts? $6\frac{2}{3}$ lbs.

12. Community Coffee Company wants a new flavor of Cajun coffee. How many pounds of coffee worth $7 a pound should be added to 14 pounds of coffee worth $4 a pound to get a mixture worth $5 a pound?   7 lbs.

13. Is it possible to mix a 30% antifreeze solution with a 50% antifreeze solution and obtain a 70% antifreeze solution? Why or why not?   no

14. A trail mix is made by combining peanuts worth $3 a pound, raisins worth $2 a pound, and M & M's worth $4 a pound. Would it make good business sense to sell the trail mix for $1.98 a pound? Why or why not?   no

15. Zoya invested part of her $25,000 advance at 8% annual simple interest and the rest at 9% annual simple interest. If her total yearly interest from both accounts was $2135, find the amount invested at each rate. $11,500 @8%; $13,500 @9%

16. Shirley invested some money at 9% annual simple interest and $250 more than that amount at 10% annual simple interest. If her total yearly interest was $101, how much was invested at each rate?

17. Michael invested part of his $10,000 bonus in a fund that paid an 11% profit and invested the rest in stock that suffered a 4% loss. Find the amount of each investment if his overall net profit was $650. $7000 @ 11% profit; $3000 @ 4% loss

18. Bruce invested a sum of money at 10% annual simple interest and invested twice that amount at 12% annual simple interest. If his total yearly income from both investments was $2890, how much was invested at each rate?

2. 130 meters   **8.** 55 mph   **16.** $400 @ 9%; $650 @ 10%
18. $8500 @ 10%; $17,000 @ 12%

19. square's side length: 8 in.; triangle's side length: 13 in.
Solve.

19. The perimeter of an equilateral triangle is 7 inches more than the perimeter of a square, and the side of the triangle is 5 inches longer than the side of the square. Find the side of the triangle.

20. A square animal pen and a pen shaped like an equilateral triangle have equal perimeters. Find the length of the sides of each pen if the sides of the triangular pen are fifteen less than twice a side of the square pen. (*Hint:* An equilateral triangle has three sides the same length.)

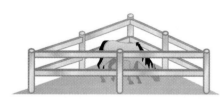

21. How can $54,000 be invested, part at 8% annual simple interest and the remainder at 10% annual simple interest, so that the interest earned by the two accounts will be equal?

22. Kathleen and Cade Williams leave simultaneously from the same point hiking in opposite directions, Kathleen walking at 4 miles per hour and Cade at 5 mph. How long can they talk on their walkie-talkies if the walkie-talkies have a 20-mile radius?

23. Ms. Mills invested her $20,000 bonus in two accounts. She took a 4% loss on one investment and made a 12% profit on another investment, but ended up breaking even. How much was invested in each account? $5,000 @ 12%; $15,000 @ 4%

24. Alan and Dave Schaferkötter leave from the same point driving in opposite directions, Alan driving at 55 miles per hour and Dave at 65 mph. Alan has a one-hour head start. How long will they be able to talk on their car phones if the phones have a 250-mile range?   2 hours and $37\frac{1}{2}$ min.

25. If $3000 is invested at 6% annual simple interest, how much should be invested at 9% annual simple interest so that the total yearly income from both investments is $585? $4500

26. How much of an alloy that is 20% copper should be mixed with 200 ounces of an alloy that is 50% copper in order to get an alloy that is 30% copper?

27. Trudy Waterbury, a financial planner, invested a certain amount of money at 9% annual simple interest, twice that amount at 10% annual simple interest, and three times that amount at 11% annual simple interest. Find the amount invested at each rate if her total yearly income from the investments was $2790.

28. How much water should be added to 30 gallons of a solution that is 70% antifreeze in order to get a mixture that is 60% antifreeze?   5 gal.

29. April Thrower spent $32.25 to take her daughter's birthday party guests to the movies. Adult tickets cost $5.75 and children tickets cost $3.00. If 8 persons were at the party, how many adult tickets were bought?   3 adult tickets

30. Nedra and Latonya Dominguez are 12 miles apart hiking toward each other. How long will it take them to meet if Nedra walks at 3 miles per hour and Latonya walks 1 mph faster? $1\frac{5}{7}$ hours

31. Two hikers are 11 miles apart and walking toward each other. They meet in 2 hours. Find the rate of each hiker if one hiker walks 1.1 miles per hour faster than the other.   2.2 mph; 3.3 mph

32. On a 255-mile trip, Gary Alessandrini traveled at an average speed of 70 miles per hour, got a speeding ticket, and then traveled at 60 mph for the remainder of the trip. If the entire trip took 4.5 hours and the speeding ticket stop took 30 minutes, how long did Gary speed before getting stopped? $1\frac{1}{2}$ hours

33. Mark Martin can row upstream at 5 miles per hour and downstream at 11 mph. If Mark starts rowing upstream until he gets tired and then rows downstream to his starting point, how far did Mark row if the entire trip took 4 hours?   27.5 miles

   *To "break even" in a manufacturing business, revenue R (income) **must equal** the cost C of production, or R = C.*

34. The cost C to produce x number of skateboards is given by $C = 100 + 20x$. The skateboards are sold

20. square's side length: 22.5 units; triangle's side length: 30 units   **21.** $30,000 @ 8%; $24,000 @ 10%

22. $2\frac{2}{9}$ hours   **26.** 400 oz.   **27.** $4500 @ 9%; $9000 @ 10%; $13,500 @ 11%

wholesale for $24 each, so revenue R is given by $R = 24x$. Find how many skateboards the manufacturer needs to produce and sell to break even. (*Hint:* Set the expression for R equal to the expression for C, then solve for x.)   25 skateboards

35. The revenue R from selling x number of computer boards is given by $R = 60x$, and the cost C of producing them is given by $C = 50x + 5000$. Find how many boards must be sold to break even. Find how much money is needed to produce the break-even number of boards.   500; $30,000

36. The cost C of producing x number of paperback books is given by $C = 4.50x + 2400$. Income R from these books is given by $R = 7.50x$. Find how many books should be produced and sold to break even.   800 books

37. Find the break-even quantity for a company that makes x number of computer monitors at a cost C given by $C = 870 + 70x$ and receives revenue R given by $R = 105x$.   25 monitors

38. Problems 34 through 37 involve finding the break-even point for manufacturing. Discuss what happens if a company makes and sells fewer products than the break-even point. Discuss what happens if more products than the break-even point are made and sold.   answers may vary

Review Exercises

Perform the indicated operations. See Section 1.6.

39. $3 + (-7)$ -4 **40.** $(-2) + (-8)$ -10

41. $\dfrac{3}{4} - \dfrac{3}{16}$ $\dfrac{9}{16}$ **42.** $-11 + 2.9$ -8.1

43. $-5 - (-1)$ -4 **44.** $-12 - 3$ -15

Place $<$, $>$, or $=$ in the appropriate space to make each a true statement. See Sections 1.1 and 1.7.

45. $-5 \; > \; -7$ **46.** $\dfrac{12}{3} \; = \; 2^2$

47. $|-5| \; = \; -(-5)$ **48.** $-3^3 \; = \; (-3)^3$

2.9 | SOLVING LINEAR INEQUALITIES

O B J E C T I V E S

TAPE
BA 2.9

1. Define linear inequality in one variable.
2. Graph solution sets on a number line.
3. Solve linear inequalities.
4. Solve compound inequalities.
5. Solve inequality applications.

In Chapter 1, we reviewed these inequality symbols and their meanings:

$<$ means "is less than" $\leq$ means "is less than or equal to"

$>$ means "is greater than" $\geq$ means "is greater than or equal to"

A linear inequality is similar to a linear equation except that the equality symbol is replaced with an inequality symbol.

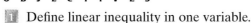
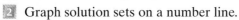

LINEAR EQUATIONS	LINEAR INEQUALITIES
$x = 3$	$x < 3$
$5n - 6 = 14$	$5n - 6 \geq 14$
$12 = 7 - 3y$	$12 \leq 7 - 3y$
$\dfrac{x}{4} - 6 = 1$	$\dfrac{x}{4} - 6 > 1$

Linear Inequality in One Variable

A linear inequality in one variable is an inequality that can be written in the form

$$ax + b < c$$

where a, b, and c are real numbers and a is not 0.

This definition and all other definitions, properties, and steps in this section also hold true for the inequality symbols, $>$, $\geq$, and $\leq$.

A **solution of an inequality** is a value of the variable that makes the inequality a true statement. The solution set is the set of all solutions. In the inequality $x < 3$, replacing x with any number less than 3, that is, to the left of 3 on the number line, makes the resulting inequality true. This means that any number less than 3 is a solution of the inequality $x < 3$. Since there are infinitely many such numbers, we cannot list all the solutions of the inequality. We *can* use set notation and write

$$\{\ x\ \ \ \ |\ \ \ \ \ x < 3\ \}. \text{ Recall that this is read}$$

$$\uparrow\qquad\uparrow\qquad\overbrace{\qquad}$$

the   such
set of   that x is less than 3.
all x

We can also picture the solution set on a number line. To do so, shade the portion of the number line corresponding to numbers less than 3.

Recall that all the numbers less than 3 lie to the left of 3 on the number line. An open circle about the point representing 3 indicates that 3 **is not** a solution of the inequality: 3 **is not** less than 3. The shaded arrow indicates that the solutions of $x < 3$ continue indefinitely to the left of 3.

Picturing the solutions of an inequality on a number line is called **graphing** the solutions or graphing the inequality, and the picture is called the **graph** of the inequality.

To graph $x \leq 3$, shade the numbers to the left of 3 and place a closed circle on the point representing 3. The closed circle indicates that 3 **is** a solution: 3 **is** less than or equal to 3.

EXAMPLE 1   Graph $x \geq -1$.

Solution: We place a closed circle at -1 since the inequality symbol is $\geq$ and -1 is greater than or equal to -1. Then shade to the right of -1.

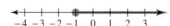

Inequalities containing one inequality symbol are called **simple inequalities,** while inequalities containing two inequality symbols are called **compound inequalities.** A compound inequality is really two simple inequalities in one. The compound inequality

$$3 < x < 5 \quad \text{means} \quad 3 < x \text{ and } x < 5$$

This can be read "x is greater than 3 and less than 5."

A solution of a compound inequality is a value that is a solution of both of the simple inequalities that make up the compound inequality. For example,

$$4\frac{1}{2} \text{ is a solution of } 3 < x < 5 \text{ since } 3 < 4\frac{1}{2} \text{ and } 4\frac{1}{2} < 5.$$

To graph $3 < x < 5$, place open circles at both 3 and 5 and shade between.

EXAMPLE 2   Graph $2 < x \le 4$.

Solution: Graph all numbers greater than 2 and less than or equal to 4. Place an open circle at 2, a closed circle at 4, and shade between.

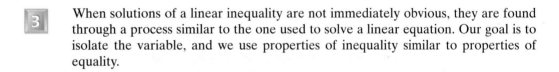

When solutions of a linear inequality are not immediately obvious, they are found through a process similar to the one used to solve a linear equation. Our goal is to isolate the variable, and we use properties of inequality similar to properties of equality.

> **ADDITION PROPERTY OF INEQUALITY**
> If a, b, and c are real numbers, then
> $$a < b \qquad \text{and} \qquad a + c < b + c$$
> are equivalent inequalities.

This property also holds true for subtracting values, since subtraction is defined in terms of addition. In other words, adding or subtracting the same quantity from both sides of an inequality does not change the solutions of the inequality.

EXAMPLE 3   Solve $x + 4 \le -6$ for x. Graph the solution set.

Solution: To solve for x, subtract 4 from both sides of the inequality.

$$x + 4 \le -6 \qquad \text{Original inequality.}$$
$$x + 4 - 4 \le -6 - 4 \qquad \text{Subtract 4 from both sides.}$$

$$x \leq -10 \qquad \text{Simplify.}$$

The solution set is $\{x \mid x \leq -10\}$.

An important difference between linear equations and linear inequalities is shown when we multiply or divide both sides of an inequality by a nonzero real number. For example, start with the true statement $6 < 8$ and multiply both sides by 2. As we see below, the resulting inequality is also true.

$$6 < 8 \qquad \text{True.}$$
$$2(6) < 2(8) \qquad \text{Multiply both sides by 2.}$$
$$12 < 16 \qquad \text{True.}$$

But if we start with the same true statement $6 < 8$ and multiply both sides by -2, the resulting inequality is not a true statement.

$$6 < 8 \qquad \text{True.}$$
$$-2(6) < -2(8) \qquad \text{Multiply both sides by } -2.$$
$$-12 < -16 \qquad \text{False.}$$

Notice, however, that if we reverse the direction of the inequality symbol, the resulting inequality is true.

$$-12 < -16 \qquad \text{False.}$$
$$-12 > -16 \qquad \text{True.}$$

This demonstrates the multiplication property of inequality.

MULTIPLICATION PROPERTY OF INEQUALITY

1. If a, b, and c are real numbers, and c is **positive,** then

$$a < b \qquad \text{and} \qquad ac < bc$$

are equivalent inequalities.

2. If a, b, and c are real numbers, and c is **negative,** then

$$a < b \qquad \text{and} \qquad ac > bc$$

are equivalent inequalities.

Because division is defined in terms of multiplication, this property also holds true when dividing both sides of an inequality by a nonzero number: If we multiply or divide both sides of an inequality by a negative number, **the direction of the inequality sign must be reversed for the inequalities to remain equivalent.**

R E M I N D E R   Whenever both sides of an inequality are multiplied or divided by a negative number, the direction of the inequality symbol **must be** reversed to form an equivalent inequality.

EXAMPLE 4   Solve $-2x \leq -4$, and graph the solution set.

Solution: Remember to reverse the direction of the inequality symbol when dividing by a negative number.

$$-2x \leq -4$$

$$\frac{-2x}{-2} \geq \frac{-4}{-2} \qquad \begin{array}{l} \text{Divide both sides by } -2 \text{ and} \\ \text{reverse the inequality sign.} \end{array}$$

$$x \geq 2 \qquad \text{Simplify.}$$

The solution set $\{x \mid x \geq 2\}$ is graphed as shown.

EXAMPLE 5   Solve $2x < -4$, and graph the solution set.

Solution: $$2x < -4$$

$$\frac{2x}{2} < \frac{-4}{2} \qquad \begin{array}{l} \text{Divide both sides by } 2. \\ \text{Do not reverse the inequality sign.} \end{array}$$

$$x < -2 \qquad \text{Simplify.}$$

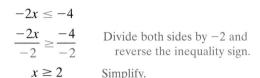

The graph of $\{x \mid x < -2\}$ is shown.

Follow these steps to solve linear inequalities.

TO SOLVE LINEAR INEQUALITIES IN ONE VARIABLE

Step 1.   Clear the inequality of fractions by multiplying both sides of the inequality by the lowest common denominator (LCD) of all fractions in the inequality.

Step 2.   Remove grouping symbols such as parentheses by using the distributive property.

Step 3.   Simplify each side of the inequality by combining like terms.

Step 4.   Write the inequality with variable terms on one side and numbers on the other side by using the addition property of inequality.

Step 5.   Isolate the variable by using the multiplication property of inequality.

Don't forget that if both sides of an inequality are multiplied or divided by a negative number, the direction of the inequality sign must be reversed.

EXAMPLE 6   Solve $-4x + 7 \geq -9$, and graph the solution set.

Solution:

$$-4x + 7 \geq -9$$

$$-4x + 7 - 7 \geq -9 - 7 \qquad \text{Subtract 7 from both sides.}$$

$$-4x \geq -16 \qquad \text{Simplify.}$$

$$\frac{-4x}{-4} \leq \frac{-16}{-4} \qquad \text{Divide both sides by } -4 \text{ and reverse the direction of the inequality sign.}$$

$$x \leq 4 \qquad \text{Simplify.}$$

The graph of $\{x \mid x \leq 4\}$ is shown.

EXAMPLE 7 Solve $2x + 7 \leq x - 11$, and graph the solution set.

Solution:

$$2x + 7 \leq x - 11$$

$$2x + 7 - x \leq x - 11 - x \qquad \text{Subtract } x \text{ from both sides.}$$

$$x + 7 \leq -11 \qquad \text{Combine like terms.}$$

$$x + 7 - 7 \leq -11 - 7 \qquad \text{Subtract 7 from both sides.}$$

$$x \leq -18 \qquad \text{Combine like terms.}$$

The graph of the solution set $\{x \mid x \leq -18\}$ is shown.

EXAMPLE 8 Solve $-5x + 7 < 2(x - 3)$, and graph the solution set.

Solution:

$$-5x + 7 < 2(x - 3)$$

$$-5x + 7 < 2x - 6 \qquad \text{Apply the distributive property.}$$

$$-5x + 7 - 2x < 2x - 6 - 2x \qquad \text{Subtract } 2x \text{ from both sides.}$$

$$-7x + 7 < -6 \qquad \text{Combine like terms.}$$

$$-7x + 7 - 7 < -6 - 7 \qquad \text{Subtract 7 from both sides.}$$

$$-7x < -13 \qquad \text{Combine like terms.}$$

$$\frac{-7x}{-7} > \frac{-13}{-7} \qquad \text{Divide both sides by } -7 \text{ and reverse the direction of the inequality sign.}$$

$$x > \frac{13}{7} \qquad \text{Simplify.}$$

The graph of the solution set $\left\{x \mid x > \frac{13}{7}\right\}$ is shown.

EXAMPLE 9 Solve $2(x - 3) - 5 \leq 3(x + 2) - 18$, and graph the solution set.

Solution:

$$2(x - 3) - 5 \leq 3(x + 2) - 18$$

$$2x - 6 - 5 \leq 3x + 6 - 18 \qquad \text{Apply the distributive property.}$$

$$2x - 11 \leq 3x - 12 \qquad \text{Combine like terms.}$$

$$-x - 11 \leq -12 \qquad \text{Subtract } 3x \text{ from both sides.}$$
$$-x \leq -1 \qquad \text{Add 11 to both sides.}$$
$$\frac{-x}{-1} \geq \frac{-1}{-1} \qquad \text{Divide both sides by } -1 \text{ and reverse the direction of the inequality sign.}$$
$$x \geq 1 \qquad \text{Simplify.}$$

The graph of the solution set $\{x \mid x \geq 1\}$ is shown.

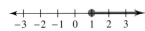

 When we solve a simple inequality, we isolate the variable on one side of the inequality. When we solve a compound inequality, we isolate the variable in the middle part of the inequality. Also, when solving a compound inequality, we must perform the same operation to all **three** parts of the inequality: left, middle, and right.

EXAMPLE 10   Solve $-1 \leq 2x - 3 < 5$, and graph the solution set.

Solution:
$$-1 \leq 2x - 3 < 5$$
$$-1 + 3 \leq 2x - 3 + 3 < 5 + 3 \qquad \text{Add 3 to all three parts.}$$
$$2 \leq 2x < 8 \qquad \text{Combine like terms.}$$
$$\frac{2}{2} \leq \frac{2x}{2} < \frac{8}{2} \qquad \text{Divide all three parts by 2.}$$
$$1 \leq x < 4 \qquad \text{Simplify.}$$

The solution set $\{x \mid 1 \leq x < 4\}$ is graphed.

EXAMPLE 11   Solve $3 \leq \dfrac{3x}{2} + 4 \leq 5$, and graph the solution set.

Solution:
$$3 \leq \frac{3x}{2} + 4 \leq 5$$
$$2(3) \leq 2\left(\frac{3x}{2} + 4\right) \leq 2(5) \qquad \text{Multiply all three parts by 2 to clear the fraction.}$$
$$6 \leq 3x + 8 \leq 10 \qquad \text{Distribute.}$$
$$-2 \leq 3x \leq 2 \qquad \text{Subtract 8 from all three parts.}$$
$$\frac{-2}{3} \leq \frac{3x}{3} \leq \frac{2}{3} \qquad \text{Divide all three parts by 3.}$$
$$\frac{-2}{3} \leq x \leq \frac{2}{3} \qquad \text{Simplify.}$$

The graph of the solution set $\left\{x \mid -\frac{2}{3} \leq x \leq \frac{2}{3}\right\}$ is shown.

5 Problems containing words such as "at least," "at most," "between," "no more than," and "no less than" usually indicate that an inequality should be solved instead of an equation. In solving applications involving linear inequalities, use the same procedure you use to solve applications involving linear equations.

EXAMPLE 12

Marie Chase and Jonathan Edwards are having their wedding reception at the Gallery reception hall. They may spend at most $1000 for the reception. If the reception hall charges a $100 cleanup fee plus $14 per person, find the greatest number of people that they can invite and still stay within their budget.

Solution:

1. UNDERSTAND. Read and reread the problem. Next, guess a solution. If 50 people attend the reception, the cost is $100 + $14(50) = $100 + $700 = $800.

2. ASSIGN. Let x = the number of people who attend the reception.

4. TRANSLATE.

In words:	cleanup fee	+	cost per person	must be less than or equal to	$1000
Translate:	100	+	14x	≤	1000

5. COMPLETE.

$$100 + 14x \leq 1000$$
$$14x \leq 900 \qquad \text{Subtract 100 from both sides.}$$
$$x \leq 64\frac{2}{7} \qquad \text{Divide both sides by 14.}$$

6. INTERPRET.

Check: Since x represents the number of people, we round down to the nearest whole, or 64. Notice that if 64 people attend, the cost is $100 + $14(64) = $996. If 65 people attend, the cost is $100 + $14(65) = $1010, which is more than the given $1000.

State: Marie Chase and Jonathan Edwards can invite at most 64 people to the reception.

MENTAL MATH

Solve each of the following inequalities.

1. $5x > 10$ $x > 2$ **2.** $4x < 20$ $x < 5$ **3.** $2x \geq 16$ $x \geq 8$ **4.** $9x \leq 63$ $x \leq 7$

EXERCISE SET 2.9

For answer graphs please see page 162.

Graph each on a number line. See Examples 1 and 2.

1. $x \le -1$

2. $y < 0$

3. $x > \dfrac{1}{2}$

4. $z \ge -\dfrac{2}{3}$

5. $-1 < x < 3$

6. $2 \le y \le 3$

7. $0 \le y < 2$

8. $-1 \le x \le 4$

Solve each inequality and graph the solution set. See Examples 3 through 5.

9. $2x < -6$

10. $3x > -9$

11. $x - 2 \ge -7$

12. $x + 4 \le 1$

13. $-8x \le 16$

14. $-5x < 20$

Solve each inequality and graph the solution set. See Examples 6 and 7.

15. $3x - 5 > 2x - 8$

16. $3 - 7x \ge 10 - 8x$

17. $4x - 1 \le 5x - 2x$

18. $7x + 3 < 9x - 3x$

19. $x - 7 < 3(x + 1)$

20. $3x + 9 \ge 5(x - 1)$

Solve each inequality and graph the solution set. See Examples 8 and 9.

21. $-6x + 2 \ge 2(5 - x)$

22. $-7x + 4 > 3(4 - x)$

23. $4(3x - 1) \le 5(2x - 4)$

24. $3(5x - 4) \le 4(3x - 2)$

25. $3(x + 2) - 6 > -2(x - 3) + 14$

26. $7(x - 2) + x \le -4(5 - x) - 12$

Solve each inequality, and then graph the solution set. See Examples 10 and 11.

27. $-3 < 3x < 6$

28. $-5 < 2x < -2$

29. $2 \le 3x - 10 \le 5$

30. $4 \le 5x - 6 \le 19$

31. $-4 < 2(x - 3) < 4$

32. $0 < 4(x + 5) < 8$

□ 33. Explain how solving a linear inequality is similar to solving a linear equation.   answers may vary

□ 34. Explain how solving a linear inequality is different from solving a linear equation.   answers may vary

Solve the following inequalities. Graph each solution set.

35. $-2x \le -40$

36. $-7x > 21$

37. $-9 + x > 7$

38. $y - 4 \le 1$

62. 1st side $\le$ 15 in.; 2nd side $\le$ 60 in.

39. $3x - 7 < 6x + 2$

40. $2x - 1 \ge 4x - 5$

41. $5x - 7x \le x + 2$

42. $4 - x < 8x + 2x$

43. $\dfrac{3}{4}x > 2$

44. $\dfrac{5}{6}x \ge -8$

45. $3(x - 5) < 2(2x - 1)$

46. $5(x + 4) < 4(2x + 3)$

47. $4(2x + 1) > 4$

48. $6(2 - x) \ge 12$

49. $-5x + 4 \le -4(x - 1)$

50. $-6x + 2 < -3(x + 4)$

51. $-2 < 3x - 5 < 7$

52. $1 < 4 + 2x \le 7$

53. $-2(x - 4) - 3x < -(4x + 1) + 2x$

54. $-5(1 - x) + x \le -(6 - 2x) + 6$

55. $-3x + 6 \ge 2x + 6$

56. $-(x - 4) < 4$

57. $-6 < 3(x - 2) < 8$

58. $-5 \le 2(x + 4) < 8$

Solve the following. See Example 12.

59. Six more than twice a number is greater than negative fourteen. Find all numbers that make this statement true. $x > -10$

60. Five times a number increased by one is less than or equal to ten. Find all such numbers. $x \le \dfrac{9}{5}$

61. The perimeter of a rectangle is to be no greater than 100 centimeters and the width must be 15 centimeters. Find the maximum length of the rectangle.   35 cm

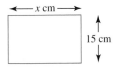

62. One side of a triangle is four times as long as another side, and the third side is 12 inches long. If the perimeter can be no longer than 87 inches, find the maximum lengths of the other two sides.

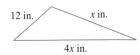

63. The temperatures in Ohio range from $-39°C$ to $45°C$. Use a compound inequality to convert these temperatures to Fahrenheit temperatures. (*Hint:* Use $C = \dfrac{5}{9}(F - 32)$.) $-38.2° \le F \le 113°$

64. $89.3 \le x \le 100$

64. Mario Lipco has scores of 85, 95, and 92 on his algebra tests. Use a compound inequality to find the range of scores he can make on his final exam in order to receive an A in the course. The final exam counts as three tests, and an A is received if the final course average is from 90 to 100. (*Hint:* The average of a list of numbers is their sum divided by the number of numbers in the list.)

65. The formula $C = 3.14d$ can be used to approximate the circumference of a circle given its diameter. Waldo Manufacturing manufactures and sells a certain washer with an outside circumference of 3 centimeters. The company has decided that a washer whose actual circumference is in the interval $2.9 \le C \le 3.1$ centimeters is acceptable. Use a compound inequality and find the corresponding interval for diameters of these washers. (Round to 3 decimal places.) $0.924 \le d \le 0.987$

66. Bunnie Supplies manufactures plastic Easter eggs that open. The company has determined that if the circumference of the opening of each part of the egg is in the interval $118 \le C \le 122$ millimeters, the eggs will open and close comfortably. Use a compound inequality and find the corresponding interval for diameters of these openings. (Round to 2 decimal places.) $37.58 \le d \le 38.85$

67. Twice a number increased by one is between negative five and seven. Find all such numbers.

68. Half a number decreased by four is between two and three. Find all such numbers. $12 < x < 14$

69. A financial planner has a client with $15,000 to invest. If he invests $10,000 in a certificate of deposit paying 11% annual simple interest, at what rate does the remainder of the money need to be in-

67. $-3 < x < 3$ **70.** $x \ge 61,000$

vested so that the two investments together yield at least $1600 in yearly interest? 10%

70. Alex earns $600 per month plus 4% of all his sales over $1000. Find the minimum sales that will allow Alex to earn at least $3000 per month.

71. Ben Holladay bowled 146 and 201 in his first two games. What must he bowl in his third game to have an average of at least 180? 193

72. On an NBA team the two forwards measure 6'8" and 6'6" and the two guards measure 6'0" and 5'9" tall. How tall a center should they hire if they wish to have a starting team average height of at least 6'5"? $x \ge 7'2"$

Review Exercises

Evaluate the following. See Section 1.3.

73. $(2)^3$ 8   **74.** $(3)^3$ 27   **75.** $(1)^{12}$ 1

76. 0^5 0   **77.** $\left(\dfrac{4}{7}\right)^2$ $\dfrac{16}{49}$ **78.** $\left(\dfrac{2}{3}\right)^3$ $\dfrac{8}{27}$

This broken line graph shows the enrollment of people (members) in a Health Maintenance Organization (HMO). The height of each dot corresponds to the number of members (in millions).

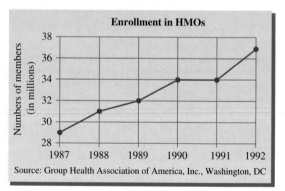

Enrollment in HMOs

Source: Group Health Association of America, Inc., Washington, DC

79. How many people were enrolled in Health Maintenance Organizations in 1989? 32 million

80. How many people were enrolled in Health Maintenance Organizations in 1992? 37 million

81. Which year shows the greatest increase in number of members? 1992

82. In what year were there 31,000,000 members of HMOs? 1988

A Look Ahead

EXAMPLE

Solve $x(x - 6) > x^2 - 5x + 6$, and then graph the solution.

Solution:

$$x(x - 6) > x^2 - 5x + 6$$
$$x^2 - 6x > x^2 - 5x + 6$$
$$x^2 - 6x - x^2 > x^2 - 5x + 6 - x^2$$
$$-6x > -5x + 6$$
$$-x > 6$$
$$\frac{-x}{-1} < \frac{6}{-1}$$
$$x < -6$$

The solution set $\{ x \mid x < -6 \}$ is graphed as shown.

Solve each inequality and then graph the solution set.

83. $x(x + 4) > x^2 - 2x + 6$

84. $x(x - 3) \geq x^2 - 5x - 8$

85. $x^2 + 6x - 10 < x(x - 10)$

86. $x^2 - 4x + 8 < x(x + 8)$

87. $x(2x - 3) \leq 2x^2 - 5x$

88. $x(4x + 1) < 4x^2 - 3x$

83.

85.

87.

84.

86.

88.

GROUP ACTIVITY

CALCULATING PRICE PER UNIT

MATERIALS:
- Grocery store circulars
- Supermarket newspaper advertisements

Sabrina owns a catering business, and she has been asked to quote a price per person for catering fruit salad at an afternoon reception for 20 people. (*Price per person* is the amount per per-son Sabrina actually charges her customer.) The recipe she plans to use is given in the table.

1. Study the recipe in the figure and give a de-scription of the general process necessary to fig-ure the fruit salad cost per person. (*Cost per person* is the amount Sabrina pays, per person, for her raw materials.)

FRUIT SALAD RECIPE
2   pounds green grapes, halved
3   pounds watermelon, cubed
1   pint fresh blueberries
1.25 pounds nectarines or mangos, sliced
1   6 ounce can frozen pineapple juice concentrate, thawed
Combine ingredients, and serve chilled. Serves 10 guests.

(continued)

2. Assign variables to the unknowns and translate your verbal description from Question 1 into an algebraic formula for the fruit salad cost per person.

3. Collect actual prices for the fruit salad ingredients from grocery store circulars, newspaper advertisements, or trips to a supermarket. Organize the ingredient prices you have found in a table. Use this real data and the algebraic formula from Question 2 to compute an estimate of the fruit salad cost per person. (*Note:* If you are unable to find prices for the exact ingredi-

ents listed in the recipe, you may substitute a new ingredient and use the new ingredient's price. In your calculations, make a note of any substitutions you are making.)

4. What *additional* amount should Sabrina charge per person if she hopes to receive a total of $90 (including both cost of raw materials and profit) for the catering job? What percent increase does the price Sabrina ultimately charges per person represent over her actual cost per person?

See App. F for Group Activity answers and suggestions.

CHAPTER 2   HIGHLIGHTS

DEFINITIONS AND CONCEPTS	EXAMPLES
SECTION 2.1 SIMPLIFYING ALGEBRAIC EXPRESSIONS	

The **numerical coefficient** of a **term** is its numerical factor.

TERM	NUMERICAL COEFFICIENT
$-7y$	-7
x	1
$\frac{1}{5}a^2b$	$\frac{1}{5}$

Terms with the same variables raised to exactly the same powers are **like terms.**

LIKE TERMS	UNLIKE TERMS
$12x, -x$	$3y, 3y^2$
$-2xy, 5yx$	$7a^2b, -2ab^2$

To combine like terms, add the numerical coefficients and multiply the result by the common variable factor.

To remove parentheses, apply the distributive property.

$$9y + 3y = 12y$$
$$-4z^2 + 5z^2 - 6z^2 = -5z^2$$

$$-4(x + 7) + 10(3x - 1)$$
$$= -4x - 28 + 30x - 10$$
$$= 26x - 38$$

(continued)

DEFINITIONS AND CONCEPTS	EXAMPLES

SECTION 2.2 THE ADDITION PROPERTY OF EQUALITY

A **linear equation in one variable** can be written in the form $ax + b = c$ where a, b, and c are real numbers and $a \neq 0$.

LINEAR EQUATIONS

$$-3x + 7 = 2$$
$$3(x - 1) = -8(x + 5) + 4$$

Equivalent equations are equations that have the same solution.

$x - 7 = 10$ and $x = 17$
are equivalent equations.

ADDITION PROPERTY OF EQUALITY

Adding the same number to or subtracting the same number from both sides of an equation does not change its solution.

$$y + 9 = 3$$
$$y + 9 - 9 = 3 - 9$$
$$y = -6$$

SECTION 2.3 THE MULTIPLICATION PROPERTY OF EQUALITY

MULTIPLICATION PROPERTY OF EQUALITY

Multiplying both sides or dividing both sides of an equation by the same nonzero number does not change its solution.

$$\frac{2}{3}a = 18$$
$$\frac{3}{2}\left(\frac{2}{3}a\right) = \frac{3}{2}(18)$$
$$a = 27$$

SECTION 2.4 SOLVING LINEAR EQUATIONS

TO SOLVE LINEAR EQUATIONS

Solve: $\dfrac{5(-2x + 9)}{6} + 3 = \dfrac{1}{2}$

1. Clear the equation of fractions.

1. $6 \cdot \dfrac{5(-2x + 9)}{6} + 6 \cdot 3 = 6 \cdot \dfrac{1}{2}$

$$5(-2x + 9) + 18 = 3$$

2. Remove any grouping symbols such as parentheses.

2. $-10x + 45 + 18 = 3$ Distributive property.

3. Simplify each side by combining like terms.

3. $-10x + 63 = 3$ Combine like terms.

4. Write variable terms on one side and numbers on the other side using the addition property of equality.

4. $-10x + 63 - 63 = 3 - 63$ Subtract 63.
$$-10x = -60$$

5. Isolate the variable using the multiplication property of equality.

5. $\dfrac{-10x}{-10} = \dfrac{-60}{-10}$ Divide by -10.
$$x = 6$$

(continued)

Definitions and Concepts	**Examples**

Section 2.4 Solving Linear Equations

6. Check by substituting in the original equation.

6.
$$\frac{5(-2x+9)}{6} + 3 = \frac{1}{2}$$

$$\frac{5(-2 \cdot 6 + 9)}{6} + 3 = \frac{1}{2}$$

$$\frac{5(-3)}{6} + 3 = \frac{1}{2}$$

$$-\frac{5}{2} + \frac{6}{2} = \frac{1}{2}$$

$$\frac{1}{2} = \frac{1}{2} \quad \text{True.}$$

Section 2.5 An Introduction to Problem Solving

Problem-Solving Steps

The height of the Hudson volcano in Chili is twice the height of the Kiska volcano in the Aleutian Islands. If the sum of their heights is 12,870 feet, find the height of each.

1. UNDERSTAND the problem.

1. Read and reread the problem. Guess a solution and check your guess.

2. ASSIGN a variable.

2. Let x be the height of the Kiska volcano. Then $2x$ is the height of the Hudson volcano.

3. ILLUSTRATE the problem.

3.

$x \; \rvert$ $2x \; \rvert$
Kiska   Hudson

4. TRANSLATE the problem.

4. In words:

height of Kiska	added to	height of Hudson	is	12,870
Translate: x	$+$	$2x$	$=$	12,870

5. COMPLETE by solving.

5. $x + 2x = 12{,}870$
$3x = 12{,}870$
$x = 4290$

6. INTERPRET the results.

6. *Check:* If x is 4290 then $2x$ is 2(4290) or 8580. Their sum is 4290 + 8580 or 12,870, the required amount.

State: Kiska volcano is 4290 feet high and Hudson volcano is 8580 feet high.

Section 2.6 Formulas and Problem Solving

Formulas

An equation that describes a known relationship among quantities is called a **formula.**

$A = lw$ (area of a rectangle)

$I = PRT$ (simple interest)

(continued)

DEFINITIONS AND CONCEPTS	EXAMPLES

SECTION 2.6 FORMULAS AND PROBLEM SOLVING

To solve a formula for a specified variable, use the same steps as for solving a linear equation. Treat the specified variable as the only variable of the equation.

Solve $P = 2l + 2w$ for l.

$$P = 2l + 2w$$

$$P - 2w = 2l + 2w - 2w \qquad \text{Subtract } 2w.$$

$$P - 2w = 2l$$

$$\frac{P - 2w}{2} = \frac{2l}{2} \qquad \text{Divide by 2.}$$

$$\frac{P - 2w}{2} = l \qquad \text{Simplify.}$$

If all values for the variables in a formula are known except for one, this unknown value may be found by substituting in the known values and solving.

If $d = 182$ miles and $r = 52$ miles per hour in the formula $d = r \cdot t$, find t.

$$d = r \cdot t$$

$$182 = 52 \cdot t \qquad \text{Let } d = 182 \text{ and } r = 52.$$

$$3.5 = t$$

The time is 3.5 hours.

SECTION 2.7 PERCENT AND PROBLEM SOLVING

The word **percent** means **per hundred.** The symbol % is used to denote percent.

To write a percent as a decimal, drop the percent symbol and move the decimal point two places to the left.

$$49\% = \frac{49}{100}, \; 1\% = \frac{1}{100}$$

$$85.\% = .85, \quad 3.5\% = .035$$

To write a decimal as a percent, move the decimal point two places to the right and attach the percent symbol, %.

$$.35 = 35\%, \; 10.1 = 1010\%$$

$$\frac{1}{8} = 0.125 = 12.5\%$$

Use the same problem-solving steps to solve a problem containing percents.

1. UNDERSTAND.

2. ASSIGN.

4. TRANSLATE.

5. COMPLETE.

32% of what number is 36.8?

1. Read and reread. Guess a solution and check.

2. Let x = the unknown number.

4. In words:

32%	of	what number	is	36.8

Translate: $32\% \qquad \cdot \qquad x \qquad = \qquad 36.8$

5. Solve $32\% \cdot x = 36.8$

$$.32x = 36.8$$

$$\frac{.32x}{.32} = \frac{36.8}{.32} \qquad \text{Divide by .32.}$$

$$x = 115 \qquad \text{Simplify.}$$

(continued)

DEFINITIONS AND CONCEPTS	EXAMPLES

SECTION 2.7 PERCENT AND PROBLEM SOLVING

6. INTERPRET.	**6.** *Check:* 32% of 115 is .32(115) = 36.8.
	State: The unknown number is 115.

SECTION 2.8 FURTHER PROBLEM SOLVING

	How many liters of a 20% acid solution must be mixed with a 50% acid solution in order to obtain 12 liters of a 30% solution?
1. UNDERSTAND.	**1.** Read and reread. Guess a solution and check.
2. ASSIGN.	**2.** Let x = number of liters of 20% solution.
	Then $12 - x$ = number of liters of 50% solution.
3. ILLUSTRATE.	**3.**

	NO. OF LITERS ·	ACID STRENGTH	= AMOUNT OF ACID
20% SOLUTION	x	20%	$0.20x$
50% SOLUTION	$12 - x$	50%	$0.50(12 - x)$
30% SOLUTION NEEDED	12	30%	$0.30(12)$

4. TRANSLATE.	**4.** In words: acid in 20% solution + acid in 50% solution = acid in 30% solution
	Translate: $0.20x + 0.50(12 - x) = 0.30(12)$
5. COMPLETE.	**5.** Solve $0.20x + 0.50(12 - x) = 0.30(12)$
	$0.20x + 6 - 0.50x = 3.6$ Apply the distributive property.
	$-0.30x + 6 = 3.6$
	$-0.30x = -2.4$ Subtract 6.
	$x = 8$ Divide by -0.30.
6. INTERPRET.	**6.** *Check,* then *state.*
	If 8 liters of a 20% acid solution are mixed with $12 - 8$ or 4 liters of a 50% acid solution, the result is 12 liters of a 30% solution.

SECTION 2.9 SOLVING LINEAR INEQUALITIES

A **linear inequality in one variable** is an inequality that can be written in one of the forms:

$ax + b < c \qquad ax + b \leq c$

$ax + b > c \qquad ax + b \geq c$

where a, b, and c are real numbers and a is not 0.

LINEAR INEQUALITIES

$2x + 3 < 6 \qquad\qquad 5(x - 6) \geq 10$

$\dfrac{x - 2}{5} > \dfrac{5x + 7}{2} \qquad \dfrac{-(x + 8)}{9} \leq \dfrac{-2x}{11}$

DEFINITIONS AND CONCEPTS	EXAMPLES

SECTION 2.9 SOLVING LINEAR INEQUALITIES

ADDITION PROPERTY OF INEQUALITY

Adding the same number to or subtracting the same number from both sides of an inequality does not change the solutions.

$$y + 4 \leq -1$$
$$y + 4 - 4 \leq -1 - 4 \qquad \text{Subtract 4.}$$
$$y \leq -5$$

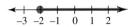

-6 -5 -4 -3 -2 -1 0 1 2

MULTIPLICATION PROPERTY OF INEQUALITY

Multiplying or dividing both sides of an inequality by the same *positive number* does not change its solutions.

$$\frac{1}{3}x > -2$$
$$3\left(\frac{1}{3}x\right) > 3 \cdot -2 \qquad \text{Multiply by 3.}$$
$$x > -6$$

-6 -4 -2 0 2

Multiplying or dividing both sides of an inequality by the same **negative number and reversing the direction of the inequality symbol** does not change its solutions.

$$-2x \leq 4$$
$$\frac{-2x}{-2} \geq \frac{4}{-2} \qquad \text{Divide by } -2, \text{ reverse inequality symbol.}$$
$$x \geq -2$$

-3 -2 -1 0 1 2

TO SOLVE LINEAR INEQUALITIES

1. Clear the equation of fractions.
2. Remove grouping symbols.
3. Simplify each side by combining like terms.
4. Write variable terms on one side and numbers on the other side using the addition property of inequality.
5. Isolate the variable using the multiplication property of inequality.

Solve: $3(x + 2) \leq -2 + 8$

1. No fractions to clear. $3(x + 2) \leq -2 + 8$
2. $3x + 6 \leq -2 + 8 \qquad$ Distributive property
3. $3x + 6 \leq 6 \qquad$ Combine like terms.
4. $3x + 6 - 6 \leq 6 - 6 \qquad$ Subtract 6.
   $3x \leq 0$
5. $\dfrac{3x}{3} \leq \dfrac{0}{3} \qquad$ Divide by 3.
   $x \leq 0$

-2 -1 0 1 2

COMPOUND INEQUALITIES

Inequalities containing two inequality symbols are called **compound inequalities.**

$$-2 < x < 6$$
$$5 \leq 3(x - 6) < \frac{20}{3}$$

(continued)

DEFINITIONS AND CONCEPTS	**EXAMPLES**
SECTION 2.9 SOLVING LINEAR INEQUALITIES	

To solve a compound inequality, isolate the variable in the middle part of the inequality. Perform the same operation to all three parts of the inequality: left, middle, right.

Solve: $-2 < 3x + 1 < 7$

$$-2 - 1 < 3x + 1 - 1 < 7 - 1 \qquad \text{Subtract 1.}$$
$$-3 < 3x < 6$$

$$\frac{-3}{3} < \frac{3x}{3} < \frac{6}{3} \qquad \text{Divide by 3.}$$
$$-1 < x < 2$$

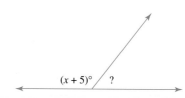

CHAPTER 2 REVIEW

(2.1) *Simplify the following expressions.*

1. $5x - x + 2x$ $6x$

2. $0.2z - 4.6x - 7.4z$ $-4.6x - 7.2z$

3. $\frac{1}{2}x + 3 + \frac{7}{2}x - 5$ $4x - 2$

4. $\frac{4}{5}y + 1 + \frac{6}{5}y + 2$ $2y + 3$

5. $2(n - 4) + n - 10$ $3n - 18$

6. $3(w + 2) - (12 - w)$ $4w - 6$

7. Subtract $7x - 2$ from $x + 5$ $-6x + 7$

8. Subtract $1.4y - 3$ from $y - 0.7$ $-0.4y + 2.3$

Write each of the following as algebraic expressions.

9. Three times a number decreased by 7. $3x - 7$

10. Twice the sum of a number and 2.8 added to 3 times a number. $2(x + 2.8) + 3x$

(2.2) *Solve the following.*

11. $8x + 4 = 9x$ $x = 4$

12. $5y - 3 = 6y$ $y = -3$

13. $3x - 5 = 4x + 1$ $x = -6$

14. $2x - 6 = x - 6$ $x = 0$

15. $4(x + 3) = 3(1 + x)$ $x = -9$

16. $6(3 + n) = 5(n - 1)$ $n = -23$

Write each as an algebraic expression.

17. The sum of two numbers is 10. If one number is x, express the other number in terms of x. $10 - x$

18. Mandy is 5 inches taller than Melissa. If x inches represents the height of Mandy, express Melissa's height in terms of x. $x - 5$

19. If one angle measures $(x + 5)°$, express the measure of its supplement in terms of x. $(175 - x)°$

(2.3) *Solve each equation.*

20. $\frac{3}{4}x = -9$ $x = -12$

21. $\frac{x}{6} = \frac{2}{3}$ $x = 4$

22. $-3x + 1 = 19$ $x = -6$

23. $5x + 25 = 20$ $x = -1$

24. $5x - 6 + x = 9 + 4x - 1$ $x = 7$

25. $8 - y + 4y = 7 - y - 3$ $y = -1$

26. Express the sum of three even consecutive integers as an expression in x. Let x be the first even integer. $3x + 6$

(2.4) *Solve the following.*

27. $\frac{2}{7}x - \frac{5}{7} = 1$ $x = 6$

28. $\frac{5}{3}x + 4 = \frac{2}{3}x$ $x = -4$

29. $-(5x + 1) = -7x + 3$ $x = 2$

30. $-4(2x + 1) = -5x + 5$ $x = -3$

31. $-6(2x - 5) = -3(9 + 4x)$ no solution

32. $3(8y - 1) = 6(5 + 4y)$ no solution

33. $\frac{3(2 - z)}{5} = z$ $z = \frac{3}{4}$

34. $\frac{4(n + 2)}{5} = -n$ $n = -\frac{8}{9}$

35. $5(2n - 3) - 1 = 4(6 + 2n)$ $n = 20$

36. $-2(4y - 3) + 4 = 3(5 - y)$ $y = -1$

37. $9z - z + 1 = 6(z - 1) + 7$ $z = 0$

38. $5t - 3 - t = 3(t + 4) - 15$ $t = 0$

39. $-n + 10 = 2(3n - 5)$ $n = \frac{20}{7}$

40. $-9 - 5a = 3(6a - 1)$ $a = -\frac{6}{23}$

41. $\frac{5(c + 1)}{6} = 2c - 3$ $c = \frac{23}{7}$

42. $\frac{2(8 - a)}{3} = 4 - 4a$ $a = -\frac{2}{5}$

43. $200(70x - 3560) = -179(150x - 19{,}300)$ $x = 102$

44. $1.72y - .04y = 0.42$ $y = 0.25$

45. The quotient of a number and 3 is the same as the difference of the number and two. Find the number.   3

46. Double the sum of a number and six is the opposite of the number. Find the number. -4

(2.5) *Solve each of the following.*

47. The height of the Eiffel Tower is 68 feet more than three times a side of its square base. If the sum of these two dimensions is 1380 feet, find the height of the Eiffel Tower.   1052 ft.

48. A 12-foot board is to be divided into two pieces so that one piece is twice as long as the other. If x represents the length of the shorter piece, find the length of each piece.   4 ft.; 8 ft.

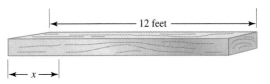
12 feet

x

49. One area code in Ohio is 34 more than three times another area code used in Ohio. If the sum of these area codes is 1262, find the two area codes. 307; 955

50. Find three consecutive even integers whose sum is negative 114. $-40, -38, -36$

(2.6) *Substitute the given values into the given formulas and solve for the unknown variable.*

51. $P = 2l + 2w$; $P = 46, l = 14$ $w = 9$

52. $V = lwh$; $V = 192, l = 8, w = 6$ $h = 4$

Solve each of the following for the indicated variable.

53. $y = mx + b$ for m $m = \frac{y - b}{x}$

54. $r = vst - 5$ for s

55. $2y - 5x = 7$ for x $x = \frac{2y - 7}{5}$

56. $3x - 6y = -2$ for y

57. $C = \pi D$ for π $\pi = \frac{C}{D}$

58. $C = 2\pi r$ for π

59. A swimming pool holds 900 cubic meters of water. If its length is 20 meters and its height is 3 meters, find its width.   15 meters

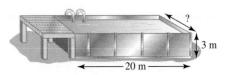
?

3 m

20 m

60. The highest temperature on record in Rome, Italy, is 104° Fahrenheit. Convert this temperature to degrees Celsius.   40°C

61. A charity 10K race is given annually to benefit a local hospice organization. How long will it take to run/walk a 10K race (10 kilometers or 10,000 meters) if your average pace is 125 **meters** per minute?   1 hour and 20 min.

(2.7) *Solve.*

62. Find 12% of 250.   30

54. $s = \frac{r + 9}{vt}$ **56.** $y = \frac{2 + 3x}{6}$ **58.** $\pi = \frac{c}{2r}$

72. no; some business travelers have chosen more than one category   **74.** $35,000 @ 8.5%; $15,000 @ 10.5%

63. Find 110% of 85.   93.5

64. The number 9 is what percent of 45?   20%

65. The number 59.5 is what percent of 85?   70%

66. The number 137.5 is 125% of what number?   110

67. The number 768 is 60% of what number?   1280

68. The state of Mississippi has the highest phoneless rate in the United States, 12.6% of households. If a city in Mississippi has 50,000 households, how many of these would you expect to be phoneless?   6300

The graph below shows how business travelers relax when in their hotel rooms.

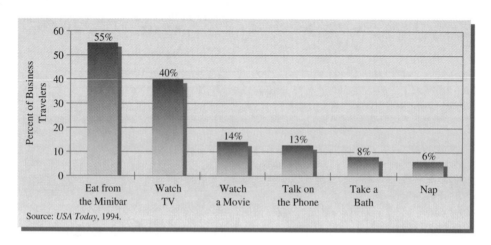

Source: *USA Today*, 1994.

69. What percent of business travelers surveyed relax by taking a nap?   6%

70. What is the most popular way to relax according to the survey?   Eat from the Minibar

71. If a hotel in New York currently has 300 business travelers, how many might you expect to relax by watching TV?   120 travelers

72. Do the percents in the graph above have a sum of 100%? Why or why not?

73. The number of employees at Arnold's Box Manufacturers just decreased from 210 to 180. Find the percent decrease. Round to the nearest tenth of a percent.   14.3%

(2.8) Solve each of the following.

74. A $50,000 retirement pension is to be invested into two accounts: a money market fund that pays 8.5% and a certificate of deposit that pays 10.5%. How much should be invested at each rate in order to provide a yearly interest income of $4550?

75. A pay phone is holding its maximum number of 500 coins consisting of nickels, dimes, and quarters. The number of quarters is twice the number of dimes. If the value of all the coins is $88.00, how many nickels were in the pay phone?   80 nickels

76. How long will it take an Amtrak passenger train to catch up to a freight train if their speeds are 60 and 45 miles per hour and the freight train had an hour and a half head start?   4.5 hours

77. Fabio Casartelli, from Italy, won a gold medal in cycling during the 1992 Summer Olympics. Suppose he rides a bicycle up a mountain trail at 8 miles per hour and down the same trail at 12 mph. Find the round-trip distance traveled if the total travel time was 5 hours.   48 miles

(2.9) Solve and graph the solution of each of the following inequalities.

78. $x \leq -2$

79. $x > 0$

80. $-1 < x < 1$

81. $0.5 \leq y < 1.5$

82. $-2x \geq -20$

83. $-3x > 12$

84. $5x - 7 > 8x + 5$

85. $x + 4 \geq 6x - 16$

86. $2 \leq 3x - 4 < 6$

87. $-3 < 4x - 1 < 2$

88. $-2(x - 5) > 2(3x - 2)$ **89.** $4(2x - 5) \leq 5x - 1$

90. Tina earns $175 per week plus a 5% commission on all her sales. Find the minimum amount of sales to ensure that she earns at least $300 per week.

91. Ellen shot rounds of 76, 82, and 79 golfing. What must she shoot on her next round so that her average will be below 80?   score must be less than 83

78.–89. For problems 78–89, see p. 161.   **82.** $x \leq 10$ **83.** $x < -4$ **84.** $x < -4$ **85.** $x \leq 4$ **86.** $2 \leq x < \dfrac{10}{3}$

87. $-\dfrac{1}{2} < x < \dfrac{3}{4}$ **88.** $x < \dfrac{7}{4}$ **89.** $x \leq \dfrac{19}{3}$ **90.** $x \geq \$2500$

CHAPTER 2 TEST

Simplify each of the following expressions.

1. $2y - 6 - y - 4$ $y - 10$

2. $2.7x + 6.1 + 3.2x - 4.9$ $5.9x + 1.2$

3. $4(x - 2) - 3(2x - 6)$ $-2x + 10$

4. $-5(y + 1) + 2(3 - 5y)$ $-15y + 1$

Solve each of the following equations.

5. $-\dfrac{4}{5}x = 4$ $x = -5$

6. $4(n - 5) = -(4 - 2n)$ $n = 8$

7. $5y - 7 + y = -(y + 3y)$ $y = \dfrac{7}{10}$

8. $4z + 1 - z = 1 + z$ $z = 0$

9. $\dfrac{2(x + 6)}{3} = x - 5$ $x = 27$

10. $\dfrac{4(y - 1)}{5} = 2y + 3$ $y = -\dfrac{19}{6}$

11. $\dfrac{1}{2} - x + \dfrac{3}{2} = x - 4$ $x = 3$

12. $\dfrac{1}{3}(y + 3) = 4y$ $y = \dfrac{3}{11}$

13. $-.3(x - 4) + x = .5(3 - x)$ $x = 0.25$

14. $-4(a + 1) - 3a = -7(2a - 3)$ $a = \dfrac{25}{7}$

Solve each of the following applications.

15. A number increased by two-thirds of the number is 35. Find the number.   21

16. A gallon of water seal covers 200 square feet. How many gallons are needed to paint two coats of water seal on a deck that measures 20 feet by 35 feet?   7 gal.

35 feet   20 feet

17. Sedric Angell invested an amount of money in Amoxil stock that earned an annual 10% return, and then he invested twice the original amount in

17. $8500 @ 10%; $17,000 @ 12%

IBM stock that earned an annual 12% return. If his total return from both investments was $2890, find how much he invested in each stock.

18. Two trains leave Los Angeles simultaneously traveling on the same track in opposite directions at speeds of 50 and 64 miles per hour. How long will it take before they are 285 miles apart? $2\frac{1}{2}$ hours

19. Find the value of x if $y = -14$, $m = -2$, and $b = -2$ in the formula $y = mx + b$. $x = 6$

Solve each of the following equations for the indicated variable.

20. $V = \pi r^2 h$ for h $h = \dfrac{V}{\pi r^2}$

21. $3x - 4y = 10$ for y $y = \dfrac{3x - 10}{4}$

Solve and graph each of the following inequalities.

22. $3x - 5 > 7x + 3$ $x < -2$ See p. 161

23. $x + 6 > 4x - 6$ $x < 4$

24. $-2 < 3x + 1 < 8$ $-1 < x < \dfrac{7}{3}$

25. $0 < 4x - 7 < 9$ $\dfrac{7}{4} < x < 4$

26. $\dfrac{2(5x + 1)}{3} > 2$ $x > \dfrac{2}{5}$

The following graph shows the source of income for charities.

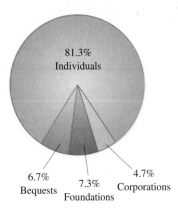

81.3% Individuals

6.7% Bequests   7.3% Foundations   4.7% Corporations

27. What percent of charity income comes from individuals?   81.3%

28. If the total annual income for charities is $126.2 billion, find the amount that comes from corporations. $5.9314 billion

29. Find the number of degrees in the Bequests sector. 24.12°

1. *(Sec. 1.1, Ex. 2)*   **10.** *(Sec. 2.1, Ex. 2)*

CHAPTER 2 CUMULATIVE REVIEW

1. Tell whether each statement is true or false.

   a. $8 \geq 8$ True   **b.** $8 \leq 8$ True

   c. $23 \leq 0$ False   **d.** $23 \geq 0$ True

2. Insert $<$, $>$, or $=$ in the appropriate space to make the statement true.   *(Sec. 1.1, Ex. 7)*

   a. $|0| \; < \; 2$ **b.** $|-5| \; = \; 5$

   c. $|-3| \; > \; |-2|$ **d.** $|5| \; < \; |6|$

   e. $|-7| \; > \; |6|$

3. Find the product of $\frac{2}{15}$ and $\frac{5}{13}$. Write the product in lowest terms. $\frac{2}{39}$ *(Sec. 1.2, Ex. 3)*

7. Find each product.   *(Sec. 1.7, Ex. 3)*

   a. $(-1.2)(0.05)$ -0.06 **b.** $\dfrac{2}{3} \cdot -\dfrac{7}{10}$ $-\dfrac{7}{15}$

8. Find the additive inverse or opposite of each number.   *(Sec. 1.8, Ex. 5)*

   a. -3 3   **b.** 5 -5 **c.** 0 0   **d.** $|-2|$ -2

9. The line graph below shows the relationship between two sets of measurements: the distance driven in a 14-foot U-Haul truck in one day and the total costs of renting this truck for that day. Notice that the horizontal axis is labeled Distance and the vertical axis is labeled Total Cost.   *(Sec. 1.9, Ex. 3)*

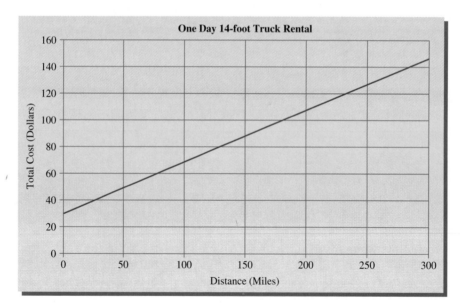

4. Simplify $\dfrac{3 + |4 - 3| + 2^2}{6 - 3}$. $\dfrac{8}{3}$ *(Sec. 1.3, Ex. 3)*

5. Find each sum.   *(Sec. 1.5, Ex. 2)*

   a. $3 + (-7)$ -4 **b.** $(-2) + (10)$ 8

   c. $0.2 + (-0.5)$ -0.3

6. Simplify each expression.   *(Sec. 1.6, Ex. 3)*

   a. $-3 + [(-2 - 5) - 2]$ -12

   b. $2^3 - |10| + [-6 - (-5)]$ -3

   a. Find the total cost of renting the truck if 100 miles are driven. $70

   b. Find the number of miles driven if the total cost of renting is $140.   280 miles

10. Tell whether the terms are like or unlike.

   a. $-x^2, 3x^3$ unlike   **b.** $4x^2y, x^2y, -2x^2y$ like

   c. $-2yz, -3zy$ like   **d.** $-x^4, x^4$ like

18. *(Sec. 2.7, Ex. 5)*   **19.** *(Sec. 2.8, Ex. 2)*

11. Subtract $4x - 2$ from $2x - 3$. $-2x - 1$ *(Sec. 2.1, Ex. 7)*

12. Solve $x - 7 = 10$ for x. $x = 17$ *(Sec. 2.2, Ex. 1)*

13. Solve $5x - 2 = 18$ for x. $x = 4$ *(Sec. 2.3, Ex. 6)*

14. Solve $\dfrac{2(a + 3)}{3} = 6a + 2$. $a = 0$ *(Sec 2.4, Ex. 4)*

15. In 1996, Congress had 8 more Republican senators than Democratic. If the number of senators is 100, how many senators of each party were there?

16. A glacier is a giant mass of rocks and ice that flows downhill like a river. Portage Glacier in Alaska is about 6 miles, or 31,680 feet, long and moves 400 feet per year. Icebergs are created when the front end of the glacier flows into Portage Lake. How long does it take for ice at the head (beginning) of the glacier to reach the lake?

15. 54 Republicans; 46 Democrats *(Sec. 2.5, Ex. 2)*   **16.** 79.2 years *(Sec. 2.6, Ex. 1)*

17. Write each percent as a decimal.   *(Sec. 2.7, Ex. 1)*
   a. 35%   0.35   **b.** 89.5%   0.895   **c.** 150%   1.5

18. The number 63 is what percent of 72?   87.5%

19. Marie Antonio, a bicycling enthusiast, rode her 10-speed at an average speed of 18 miles per hour on level roads and then slowed down to an average of 10 miles per hour on the hilly roads of the trip. If she covered a distance of 98 miles, how long did the entire trip take if traveling the level roads took the same time as traveling the hilly roads?   7 hours

20. Graph $2 < x \le 4$.   *(Sec. 2.9, Ex. 2)*

21. Solve $2(x - 3) - 5 \le 3(x + 2) - 18$, and graph the solution set. $x \ge 1$ *(Sec. 2.9, Ex. 9)*

The following are answer graphs for Chapter Review

78.

79. $-2\ -1\ 0\ 1\ 2$

80. $-2\ -1\ 0\ 1\ 2$

81. $-2\ -1\ 0\ 1\ 2\ 3\ 4$

82. $x \le 10$ $6\ 7\ 8\ 9\ 10\ 11$

83. $x < -4$ $-6\ -5\ -4\ -3\ -2$

84. $x < -4$ $-6\ -5\ -4\ -3\ -2\ -1\ 0$

85. $x \le 4$ $0\ 1\ 2\ 3\ 4\ 5$

86. $2 \le x < \dfrac{10}{3}$ $0\ 1\ 2\ 3\ 4$

87. $-\dfrac{1}{2} < x < \dfrac{3}{4}$ $-1\ 0\ 1$

88. $x < \dfrac{7}{4}$ $-2\ -1\ 0\ 1\ 2$

89. $x \le \dfrac{19}{3}$ $4\ 5\ 6\ 7\ 8$

The following are answer graphs for Chapter Tests

22. $x < -2$ $-4\ -3\ -2\ -1\ 0$

23. $x < 4$ $0\ 1\ 2\ 3\ 4\ 5$

24. $-1 < x < \dfrac{7}{3}$ $-1\ 0\ 1\ 2\ 3$

25. $\dfrac{7}{4} < x < 4$ $1\ 2\ 3\ 4\ 5\ 6$

26. $x > \dfrac{2}{5}$ $-1\ 0\ 1\ 2\ 3$

The following are answer graphs for Exercise Set 2.9

1. −4 −3 −2 −1 0 1 2

2. −2 −1 0 1 2 3

3. $\frac{1}{2}$; −1 0 1 2

4. $-\frac{2}{3}$; −3 −2 −1 0 1 2

5. −2 −1 0 1 2 3 4

6. 0 1 2 3 4

7. −1 0 1 2 3 4 5

8. −2 −1 0 1 2 3 4

9. $x < -3$; −5 −4 −3 −2 −1 0

10. $x > -3$; −5 −4 −3 −2 −1 0 1

11. $x \geq -5$; −8 −7 −6 −5 −4 −3 −2

12. $x \leq -3$; −5 −4 −3 −2 −1 0 1

13. $x \geq -2$; −5 −4 −3 −2 −1 0 1

14. $x > -4$; −6 −5 −4 −3 −2 −1 0

15. $x > -3$; −4 −3 −2 −1 0

16. $x \geq 7$; 4 5 6 7 8 9

17. $x \leq 1$; −1 0 1 2 3 4

18. $x < -3$; −4 −3 −2 −1 0

19. $x > -5$; −6 −5 −4 −3 −2 −1

20. $x \leq 7$; 3 4 5 6 7 8

21. $x \leq -2$; −5 −4 −3 −2 −1 0

22. $x < -2$; −4 −3 −2 −1 0

23. $x \leq -8$; −11 −10 −9 −8 −7 −6

24. $x \leq \frac{4}{3}$; −2 −1 0 1 2

25. $x > 4$; 0 1 2 3 4 5

26. $x > \frac{2}{5}$; −1 0 1 2 3

27. $-1 < x < 2$; −4 −3 −2 −1 0 1 2 3 4 5

28. $-\frac{5}{2} < x < -1$; $-\frac{5}{2}$; −5 −4 −3 −2 −1 0 1 2

29. $4 \leq x \leq 5$; 3 4 5 6 7

30. $2 \leq x \leq 5$; −1 0 1 2 3 4 5 6 7 8

31. $1 < x < 5$; −2 −1 0 1 2 3 4 5

32. $-5 < x < -3$; −6 −5 −4 −3 −2 −1

35. $x \geq 20$; 8 12 16 20 24 28 32

36. $x < -3$; −6 −5 −4 −3 −2 −1 0

37. $x > 16$; 4 8 12 16 20 24 28

38. $y \leq 5$; 2 3 4 5 6 7 8

39. $x > -3$; −6 −5 −4 −3 −2 −1 0

40. $x \leq 2$; −1 0 1 2 3 4 5

41. $x \geq -\frac{2}{3}$; $-\frac{2}{3}$; −3 −2 −1 0 1 2 3

42. $x > \frac{4}{11}$; $\frac{4}{11}$; −2 −1 0 1 2 3 4

43. $x > \frac{8}{3}$; $\frac{8}{3}$; −3 −2 −1 0 1 2 3 4 5

44. $x \geq -\frac{48}{5}$; $-\frac{48}{5}$; −12 −11 −10 −9 −8 −7

45. $x > -13$; −16 −15 −14 −13 −12 −11 −10

46. $x > \frac{8}{3}$; $\frac{8}{3}$; 0 1 2 3 4 5

47. $x > 0$; −3 −2 −1 0 1 2 3

48. $x \leq 0$; −3 −2 −1 0 1 2 3

49. $x \geq 0$; −3 −2 −1 0 1 2 3

50. $x > \frac{14}{3}$; $\frac{14}{3}$; 2 3 4 5 6 7

51. $1 < x < 4$; −2 −1 0 1 2 3 4 5 6 7

52. $-\frac{3}{2} < x \leq \frac{3}{2}$; $-\frac{3}{2}$; $\frac{3}{2}$; −4 −3 −2 −1 0 1 2 3 4

53. $x > 3$; 0 1 2 3 4 5 6

54. $x \leq \frac{5}{4}$; $\frac{5}{4}$; −1 0 1 2 3 4

55. $x \leq 0$; −3 −2 −1 0 1 2 3

56. $x > 0$; −3 −2 −1 0 1 2 3

57. $0 < x < \frac{14}{3}$; $\frac{14}{3}$; −3 −2 −1 0 1 2 3 4 5 6 7

58. $-\frac{13}{2} \leq x < 0$; $-\frac{13}{2}$; −8 −7 −6 −5 −4 −3 −2 −1 0

GRAPHING

FINANCIAL ANALYSIS

Investment analysts must investigate a company's financial data, such as sales, profit margin, debt, and assets, to evaluate whether investing in it is a wise choice. One way to analyze such data is to graph the data and identify trends in it visually over time. Another way to analyze such data is to find algebraically the rate at which it is changing over time.

IN THE CHAPTER GROUP ACTIVITY ON PAGE 217, YOU WILL HAVE THE OPPORTUNITY TO ANALYZE THE SALES OF SEVERAL COMPANIES IN THE AEROSPACE INDUSTRY.

I n Chapter 2 we learned to solve and graph the solutions of linear equations and inequalities in one variable. Now we define and present techniques for solving and graphing linear equations and inequalities in two variables.

3.1 | THE RECTANGULAR COORDINATE SYSTEM

O B J E C T I V E S

1. Define the rectangular coordinate system.
2. Plot ordered pairs of numbers.
3. Determine whether an ordered pair is a solution of an equation in two variables.
4. Find the missing coordinate of an ordered pair solution, given one coordinate of the pair.

TAPE
BA 3.1

In Section 1.9, we learned how to read graphs. Example 4 in Section 1.9 presented the graph below showing the relationship between time spent smoking a cigarette and pulse rate. Notice in this graph that there are two numbers associated with each point of the graph. For example, we discussed earlier that 15 minutes after "lighting up," the pulse rate is 80 beats per minute. If we agree to write the time first and the pulse rate second, we can say there is a point on the graph corresponding to the **ordered pair** of numbers (15, 80). A few more ordered pairs are listed alongside their corresponding points.

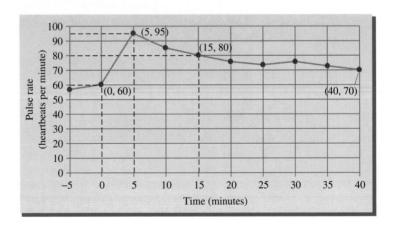

In general, we use this same ordered pair idea to describe the location of a point in a plane (such as a piece of paper). We start with a horizontal and a vertical axis. Each axis is a number line, and for the sake of consistency we construct our axes to intersect at the 0 coordinate of both. This point of intersection is called the

origin. Notice that these two number lines or axes divide the plane into four regions called **quadrants.** The quadrants are usually numbered with Roman numerals as shown. The axes are not considered to be in any quadrant.

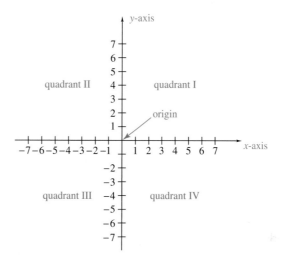

It is helpful to label axes, so we label the horizontal axis the **x-axis** and the vertical axis the **y-axis.** We call the system described above the **rectangular coordinate system.**

Just as with the pulse rate graph, we can then describe the locations of points by ordered pairs of numbers. We list the horizontal **x-axis** measurement first and the vertical **y-axis** measurement second.

The location of the point shown below can be described by the ordered pair of numbers (3, 2). Here, the x-value or **x-coordinate** is 3 and the y-value or **y-coordinate** is 2.

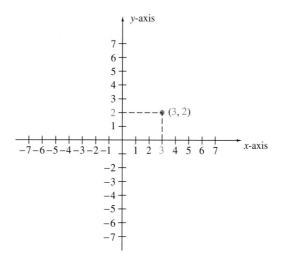

Does the order in which the coordinates are listed matter? Yes! Notice that the point corresponding to the ordered pair (2, 3) is in a different location than the point corresponding to (3, 2). These two ordered pairs of numbers describe two different points of the plane.

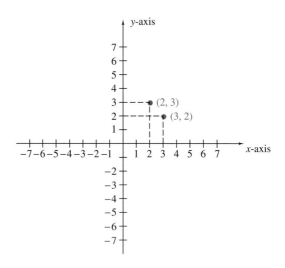

2 Given an ordered pair, how do we **graph** or **plot** it? That is, how do we find its corresponding point? To see, let's graph the ordered pair $(-2, 5)$. Start at the origin and move 2 units in the negative x-direction; from there, move 5 units in the positive y-direction. The ending location is the location of the point corresponding to the ordered pair $(-2, 5)$.

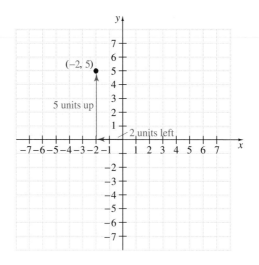

Here are some more ordered pairs that have been plotted.

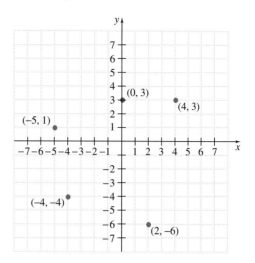

Keep in mind that **each ordered pair corresponds to exactly one point in the real plane and that each point in the plane corresponds to exactly one ordered pair.** Because of this correspondence, we may refer to the point corresponding to the ordered pair (2, 5) as simply the point (2, 5), for example.

EXAMPLE 1   On a single coordinate system, plot the ordered pairs. State in which quadrant, if any, each point lies.

a. $(3, 2)$ **b.** $(-2, -4)$ **c.** $(1, -2)$ **d.** $(-5, 3)$

e. $(0, 0)$ **f.** $(0, 2)$ **g.** $(-5, 0)$ **h.** $\left(0, -1\frac{1}{2}\right)$

Solution:

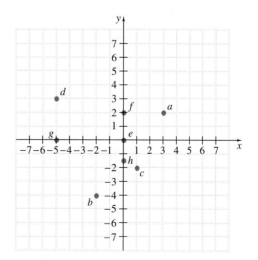

Point **a** lies in quadrant I.

Point **d** lies in quadrant II.

Point **b** lies in quadrant III.

Point **c** lies in quadrant IV.

Points **e, f, g,** and **h** lie on an axis, so they are not in any quadrant.

Notice that the y-coordinate of any point on the x-axis is 0. For example, the coordinates of point g are $(-5, 0)$. Also, the x-coordinate of any point on the y-axis is 0. For example, the coordinates of point f are $(0, 2)$.

3   Let's see how we can use ordered pairs to record solutions of equations containing two variables. An equation in one variable such as $x + 1 = 5$ has one solution, which is 4: the number 4 is the value of the variable x that makes the equation true.

An equation in two variables, such as $2x + y = 8$, has solutions consisting of two values, one for x and one for y. For example, $x = 3$ and $y = 2$ is a solution of $2x + y = 8$ because, if x is replaced with 3 and y with 2, we get a true statement.

$$2x + y = 8$$
$$2(3) + 2 = 8$$
$$8 = 8 \qquad \text{True.}$$

The solution $x = 3$ and $y = 2$ can be written as $(3, 2)$, an **ordered pair** of numbers. The first number, 3, is the x-value and the second number, 2, is the y-value.

In general, an ordered pair is a **solution** of an equation in two variables if replacing the variables by the values of the ordered pair results in a true statement.

EXAMPLE 2   Determine whether each ordered pair is a solution of the equation $x - 2y = 6$.

a. $(6, 0)$ **b.** $(0, 3)$ **c.** $(2, -2)$

Solution: **a.** Let $x = 6$ and $y = 0$ in the equation $x - 2y = 6$.

$$x - 2y = 6$$
$$6 - 2(0) = 6 \qquad \text{Replace } x \text{ with 6 and } y \text{ with 0.}$$
$$6 - 0 = 6 \qquad \text{Simplify.}$$
$$6 = 6 \qquad \text{True.}$$

$(6, 0)$ is a solution, since $6 = 6$ is a true statement.

b. Let $x = 0$ and $y = 3$.

$$x - 2y = 6$$
$$0 - 2(3) = 6 \qquad \text{Replace } x \text{ with 0 and } y \text{ with 3.}$$
$$0 - 6 = 6$$
$$-6 = 6 \qquad \text{False.}$$

(0, 3) is *not* a solution, since $-6 = 6$ is a false statement.

c. Let $x = 2$ and $y = -2$ in the equation.

$$x - 2y = 6$$
$$2 - 2(-2) = 6 \qquad \text{Replace } x \text{ with 2 and } y \text{ with } -2.$$
$$2 + 4 = 6$$
$$6 = 6 \qquad \text{True.}$$

$(2, -2)$ is a solution, since $6 = 6$ is a true statement.

4 If one value of an ordered pair solution of an equation is known, the other value can be determined. To find the unknown value, replace one variable in the equation by its known value. Doing so results in an equation with just one variable that can be solved for the variable using the methods of Chapter 2.

EXAMPLE 3 Complete the following ordered pair solutions for the equation $3x + y = 12$.

a. $(0, \ )$ **b.** $(\ , 6)$ **c.** $(-1, \ )$

Solution: **a.** In the ordered pair $(0, \ )$, the x-value is 0. Let $x = 0$ in the equation and solve for y.

$$3x + y = 12$$
$$3(0) + y = 12 \qquad \text{Replace } x \text{ with 0.}$$
$$0 + y = 12$$
$$y = 12$$

The completed ordered pair is $(0, 12)$.

b. In the ordered pair $(\ , 6)$, the y-value is 6. Let $y = 6$ in the equation and solve for x.

$$3x + y = 12$$
$$3x + 6 = 12 \qquad \text{Replace } y \text{ with 6.}$$
$$3x = 6 \qquad \text{Subtract 6 from both sides.}$$
$$x = 2 \qquad \text{Divide both sides by 3.}$$

The ordered pair is $(2, 6)$.

c. In the ordered pair $(-1, \ )$, the x-value is -1. Let $x = -1$ in the equation and solve for y.

$$3x + y = 12$$
$$3(-1) + y = 12 \qquad \text{Replace } x \text{ with } -1.$$
$$-3 + y = 12$$
$$y = 15 \qquad \text{Add 3 to both sides.}$$

The ordered pair is $(-1, 15)$.

Solutions of equations in two variables can also be recorded in a **table of values,** as shown in the next example.

EXAMPLE 4   Complete the table for the equation $y = 3x$.

x	y
a. -1	
b.	0
c.	-9

Solution: **a.** Replace x with -1 in the equation and solve for y.

$$y = 3x$$
$$y = 3(-1) \quad \text{Let } x = -1.$$
$$y = -3$$

The ordered pair is $(-1, -3)$.

b. Replace y with 0 in the equation and solve for x.

$$y = 3x$$
$$0 = 3x \quad \text{Let } y = 0.$$
$$0 = x \quad \text{Divide both sides by 3.}$$

The completed ordered pair is $(0, 0)$.

c. Replace y with -9 in the equation and solve for x.

$$y = 3x$$
$$-9 = 3x \quad \text{Let } y = -9.$$
$$-3 = x \quad \text{Divide both sides by 3.}$$

x	y
-1	-3
0	0
-3	-9

The completed ordered pair is $(-3, -9)$. The completed table is shown to the right.

EXAMPLE 5   Complete the table for the equation $y = 3$.

x	y
-2	
0	
-5	

Solution: The equation $y = 3$ is the same as $0x + y = 3$. No matter what value we replace x by, y always equals 3. The completed table is:

x	y
-2	3
0	3
-5	3

By now, you have noticed that equations in two variables often have more than one solution. We discuss this more in the next section.

A table showing ordered pair solutions may be written vertically or horizontally as shown in the next example.

EXAMPLE 6 A small business purchased a computer for $2000. The business predicts that the computer will be used for 5 years and the value in dollars y of the computer in x years is $y = -300x + 2000$. Complete the table.

x	0	1	2	3	4	5
y						

Solution: To find the value of y when x is 0, replace x with 0 in the equation. We use this same procedure to find y when x is 1 and when x is 2.

WHEN $x = 0$,

$y = -300x + 2000$
$y = -300 \cdot 0 + 2000$
$y = 0 + 2000$
$y = 2000$

WHEN $x = 1$,

$y = -300x + 2000$
$y = -300\,(1) + 2000$
$y = -300 + 2000$
$y = 1700$

WHEN $x = 2$,

$y = -300x + 2000$
$y = -300 \cdot 2 + 2000$
$y = -600 + 2000$
$y = 1400$

We have the ordered pairs (0, 2000), (1, 1700), and (2, 1400). This means that in 0 years the value of the computer is $2000, in 1 year the value of the computer is $1700, and in 2 years the value is $1400. Complete the table of values.

WHEN $x = 3$,

$y = -300x + 2000$
$y = -300 \cdot 3 + 2000$
$y = -900 + 2000$
$y = 1100$

WHEN $x = 4$,

$y = -300x + 2000$
$y = -300 \cdot 4 + 2000$
$y = -1200 + 2000$
$y = 800$

WHEN $x = 5$,

$y = -300x + 2000$
$y = -300 \cdot 5 + 2000$
$y = -1500 + 2000$
$y = 500$

The completed table is

x	0	1	2	3	4	5
y	2000	1700	1400	1100	800	500

The ordered pair solutions recorded in the completed table for the example above are graphed on the following page. Notice that the graph gives a visual picture of the decrease in value of the computer.

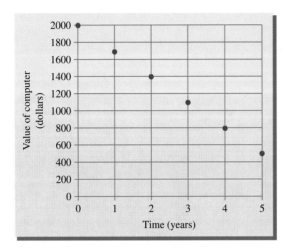

x	y
0	2000
1	1700
2	1400
3	1100
4	800
5	500

MENTAL MATH

Give two ordered pair solutions for each of the following linear equations.

1. $x + y = 10$ answers may vary; Ex. (5, 5), (7, 3)

2. $x + y = 6$ answers may vary; Ex. (0, 6), (6, 0)

3. $x = 3$ answers may vary; Ex. (3, 5), (3, 0)

4. $y = -2$ answers may vary; Ex. (0, −2), (1, −2)

EXERCISE SET 3.1

Plot the ordered pairs. State in which quadrant, if any, each point lies. See Example 1. graphical answers in App. F

1. (1, 5)   **2.** (−5, −2)   **3.** (−6, 0)

4. (0, −1)   **5.** (2, −4)   **6.** (−1, 4)

7. $\left(4\frac{3}{4}, 0\right)$ **8.** $\left(0, \frac{7}{8}\right)$ **9.** (0, 0)

10. (5, 0)   **11.** (0, 4)   **12.** (−3, −3)

13. When is the graph of the ordered pair (a, b) the same as the graph of the ordered pair (b, a)? $a = b$

14. In your own words, describe how to plot an ordered pair.   answers may vary

15. Find the perimeter of the rectangle whose vertices are the points with coordinates (−1, 5), (3, 5), (3, −4), and (−1, −4).   26 units

16. Find the area of the rectangle whose vertices are the points with coordinates (5, 2), (5, −6), (0, −6), and (0, 2).   40 sq. units

Determine whether each ordered pair is a solution of the given linear equation. See Example 2.

17. $2x + y = 7$; (3, 1), (7, 0), (0, 7)   yes; no; yes

18. $x - y = 6$; (5, −1), (7, 1), (0, −6)   yes; yes; yes

19. $y = -5x$; (−1, −5), (0, 0), (2, −10)   no; yes; yes

20. $x = 2y$; (0, 0), (2, 1), (−2, −1)   yes; yes; yes

21. $x = 5$; $(4, 5)$, $(5, 4)$, $(5, 0)$ no; yes; yes

22. $y = 2$; $(-2, 2)$, $(2, 2)$, $(0, 2)$ yes; yes; yes

23. $x + 2y = 9$; $(5, 2)$, $(0, 9)$ yes; no

24. $3x + y = 8$; $(2, 3)$, $(0, 8)$ no; yes

25. $2x - y = 11$; $(3, -4)$, $(9, 8)$ no; no

26. $x - 4y = 14$; $(2, -3)$, $(14, 6)$ yes; no

27. $x = \frac{1}{3}y$; $(0, 0)$, $(3, 9)$ yes; yes

28. $y = -\frac{1}{2}x$; $(0, 0)$, $(4, 2)$ yes; no

29. $y = -2$; $(-2, -2)$, $(5, -2)$ yes; yes

30. $x = 4$; $(4, 0)$, $(4, 4)$ yes; yes

Find the x- and y-coordinates of the following labeled points.

31.

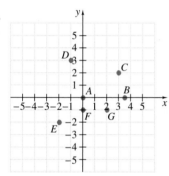

A, $(0, 0)$;
B, $(3\frac{1}{2}, 0)$;
C, $(3, 2)$;
D, $(-1, 3)$;
E, $(-2, -2)$;
F, $(0, -1)$;
G, $(2, -1)$

32.

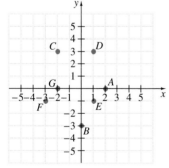

A, $(2, 0)$;
B, $(0, -3)$;
C, $(-2, 3)$;
D, $(1, 3)$;
E, $(1, -1)$;
F, $(-3, -1)$;
G, $(-2, 0)$

Determine the quadrant or quadrants in which the points described below lie.

33. The first coordinate is positive and the second coordinate is negative.   quadrant IV

34. Both coordinates are negative.   quadrant III

35. The first coordinate is negative.   quadrants II or III

36. The second coordinate is positive. quadrants I or II

Complete each ordered pair so that it is a solution of the given linear equation. See Examples 3 through 5.

37. $x - 4y = 4$; $(\ , -2)$, $(4, \ )$ $(-4, -2)$, $(4, 0)$

38. $x - 5y = -1$; $(\ , -2)$, $(4, \ )$ $(-11, -2)$, $(4, 1)$

39. $3x + y = 9$; $(0, \ )$, $(\ , 0)$ $(0, 9)$, $(3, 0)$

40. $x + 5y = 15$; $(0, \ )$, $(\ , 0)$ $(0, 3)$, $(15, 0)$

41. $y = -7$; $(11, \ )$, $(\ , -7)$

42. $x = \frac{1}{2}$; $(\ , 0)$, $\left(\frac{1}{2}, \ \right)$

Complete the table of values for each given linear equation; then plot each solution. Use a single coordinate system for each equation. See Examples 4 through 6.

43. $x + 3y = 6$

x	y
0	
	0
	1

$(0, 2)$
$(6, 0)$
$(3, 1)$
See App. F.

44. $2x + y = 4$

x	y
0	
	0
	2

$(0, 4)$
$(2, 0)$
$(1, 2)$
See App. F.

45. $2x - y = 12$

x	y
0	
	-2
-3	

$(0, -12)$
$(5, -2)$
$(-3, -18)$
See App. F.

46. $-5x + y = 10$

x	y
	0
	5
2	

$(-2, 0)$
$(-1, 5)$
$(2, 20)$
See App. F.

47. $2x + 7y = 5$

x	y
0	
	0
	1

$\left(0, \frac{5}{7}\right)$
$\left(\frac{5}{2}, 0\right)$
$(-1, 1)$
See App. F.

41. $(11, -7)$; answers may vary, Ex. $(2, -7)$ **42.** $\left(\frac{1}{2}, 0\right)$; answers may vary, Ex. $\left(\frac{1}{2}, 4\right)$

48. $x - 6y = 3$

x	y
0	
1	
	−1

$\left(0, -\frac{1}{2}\right)$
$\left(1, -\frac{1}{3}\right)$
$(-3, -1)$
See App. F.

49. $x = 3$

x	y
	0
	−0.5
	$\frac{1}{4}$

$(3, 0)$
$(3, -0.5)$
$\left(3, \frac{1}{4}\right)$
See App. F.

50. $y = -1$

x	y
−2	
0	
−1	

$(-2, -1)$
$(0, -1)$
$(-1, -1)$
See App. F.

51. $x = -5y$

x	y
	0
	1
10	

$(0, 0)$
$(-5, 1)$
$(10, -2)$
See App. F.

52. $y = -3x$

x	y
0	
−2	
	9

$(0, 0)$
$(-2, 6)$
$(-3, 9)$
See App. F.

53. Discuss any similarities in the graphs of the ordered pair solutions for Exercises 43–52.

54. Explain why equations in two variables have more than one solution.   answers may vary

55. The cost in dollars y of producing x computer desks is given by $y = 80x + 5000$.

  a. Complete the following table and graph the results.

x	100	200	300
y	13,000	21,000	29,000

  b. Find the number of computer desks that can be produced for $8600. (*Hint:* Find x when $y = 8600$.)   45 desks

56. The hourly wage y of an employee at a certain production company is given by $y = 0.25x + 9$ where x is the number of units produced in an hour.

  a. Complete the table and graph the results.   See App. F.

x	0	1	5	10
y	9	9.25	10.25	11.50

  b. Find the number of units that must be produced each hour to earn an hourly wage of $12.25. (*Hint:* Find x when $y = 12.25$.)   13 units

The graph below shows Walt Disney Company's annual revenues.

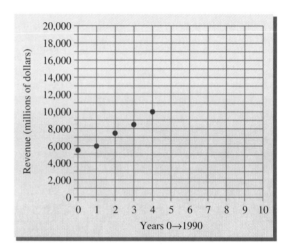

57. Estimate the increase in revenues for years 1, 2, 3, and 4.

58. Use a straight edge or ruler and this graph to predict Disney's revenue in the year 2000.
approximately $18,000 million

Review Exercises

Solve each equation for y. See Section 2.4.

59. $x + y = 5$ $y = 5 - x$ **60.** $x - y = 3$ $y = x - 3$

61. $2x + 4y = 5$ **62.** $5x + 2y = 7$

63. $10x = -5y$ $y = -2x$ **64.** $4y = -8x$ $y = -2x$

65. $x - 3y = 6$ **66.** $2x - 9y = -20$

53. answers may vary   **57.** year 1: $500 million; year 2: $1500 million; year 3: $1000 million; year 4: $1500 million

61. $y = -\frac{1}{2}x + \frac{5}{4}$ **62.** $y = -\frac{5}{2}x + \frac{7}{2}$ **65.** $y = \frac{1}{3}x - 2$ **66.** $y = \frac{2}{9}x + \frac{20}{9}$

3.2 | GRAPHING LINEAR EQUATIONS

O B J E C T I V E S

1 Identify linear equations.

2 Graph a linear equation by finding and plotting ordered pair solutions.

TAPE
BA 3.2

In the previous section, we found that equations in two variables may have more than one solution. For example, both $(6, 0)$ and $(2, -2)$ are solutions of the equation $x - 2y = 6$. In fact, this equation has an infinite number of solutions. Other solutions include $(0, -3)$, $(4, -1)$, $(-2, -4)$, and $(8, 1)$. If we graph these solutions, notice that a pattern appears.

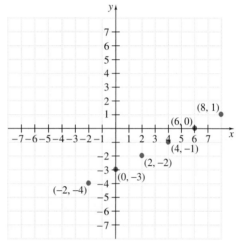

These solutions all appear to lie on the same line, which has been filled in below. It can be shown that every ordered pair solution of the equation corresponds to a point on this line, and every point on this line corresponds to an ordered pair solution. Thus, we say that this line is the graph of the equation $x - 2y = 6$.

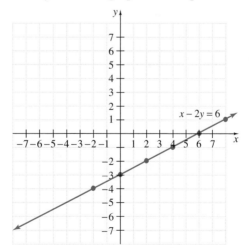

The equation $x - 2y = 6$ is called a **linear equation in two variables** and **the graph of every linear equation in two variables is a line.**

LINEAR EQUATION IN TWO VARIABLES

A linear equation in two variables is an equation that can be written in the form

$$Ax + By = C$$

where A, B, and C are real numbers and A and B are not both 0.

The form $Ax + By = C$ is called **standard form.**

EXAMPLES OF LINEAR EQUATIONS IN TWO VARIABLES

$$2x + y = 8 \qquad -2x = 7y \qquad y = \frac{1}{3}x + 2 \qquad y = 7$$

Before we graph linear equations in two variables, let's practice identifying these equations.

EXAMPLE 1   Identify the linear equations in two variables.

a. $x - 1.5y = -1.6$ **b.** $y = -2x$ **c.** $x + y^2 = 9$ **d.** $x = 5$

Solution: **a.** This is a linear equation in two variables because it is written in the form $Ax + By = C$ with $A = 1$, $B = -1.5$, and $C = -1.6$.
b. This is a linear equation in two variables because it can be written in the form $Ax + By = C$.

$$y = -2x$$
$$2x + y = 0 \qquad \text{Add } 2x \text{ to both sides.}$$

c. This is *not* a linear equation in two variables because y is squared.
d. This is a linear equation in two variables because it can be written in the form $Ax + By = C$.

$$x = 5$$
$$x + 0y = 5 \qquad \text{Add } 0 \cdot y.$$

---

2   From geometry, we know that a straight line is determined by just two points. Graphing a linear equation in two variables, then, requires that we find just two of its infinitely many solutions. Once we do so, we plot the solution points and draw the line connecting the points. Usually, we find a third solution as well, as a check.

EXAMPLE 2   Graph the linear equation $2x + y = 5$.

Solution: Find three ordered pair solutions of $2x + y = 5$. To do this, choose a value for one variable, x or y, and solve for the other variable. For example, let $x = 1$. Then $2x + y = 5$ becomes

$$2x + y = 5$$
$$2(\mathbf{1}) + y = 5 \qquad \text{Replace } x \text{ with 1.}$$
$$2 + y = 5 \qquad \text{Multiply.}$$
$$y = \mathbf{3} \qquad \text{Subtract 2 from both sides.}$$

Since $y = 3$ when $x = 1$, the ordered pair $(1, 3)$ is a solution of $2x + y = 5$. Next, let $x = 0$.

$$2x + y = 5$$
$$2(\mathbf{0}) + y = 5 \qquad \text{Replace } x \text{ with 0.}$$
$$0 + y = 5$$
$$y = \mathbf{5}$$

The ordered pair $(0, 5)$ is a second solution.

The two solutions found so far allow us to draw the straight line that is the graph of all solutions of $2x + y = 5$. However, we find a third ordered pair as a check. Let $y = -1$.

$$2x + y = 5$$
$$2x + (\mathbf{-1}) = 5 \qquad \text{Replace } y \text{ with } -1.$$
$$2x - 1 = 5$$
$$2x = 6 \qquad \text{Add 1 to both sides.}$$
$$x = \mathbf{3} \qquad \text{Divide both sides by 2.}$$

The third solution is $(3, -1)$. These three ordered pair solutions are listed in table form as shown. The graph of $2x + y = 5$ is the line through the three points.

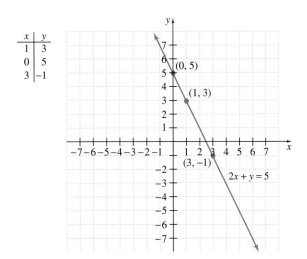

x	y
1	3
0	5
3	−1

EXAMPLE 3   Graph the linear equation $-5x + 3y = 15$.

Solution: Find three ordered pair solutions of $-5x + 3y = 15$.

LET $x = 0$.
$$-5x + 3y = 15$$
$$-5 \cdot 0 + 3y = 15$$
$$0 + 3y = 15$$
$$3y = 15$$
$$y = 5$$

LET $y = 0$.
$$-5x + 3y = 15$$
$$-5x + 3 \cdot 0 = 15$$
$$-5x + 0 = 15$$
$$-5x = 15$$
$$x = -3$$

LET $x = -2$.
$$-5x + 3y = 15$$
$$-5(-2) + 3y = 15$$
$$10 + 3y = 15$$
$$3y = 5$$
$$y = \frac{5}{3}$$

The ordered pairs are $(0, 5)$, $(-3, 0)$, and $\left(-2, \frac{5}{3}\right)$. The graph of $-5x + 3y = 15$ is the line through the three points.

x	y
0	5
-3	0
-2	$\frac{5}{3} = 1\frac{2}{3}$

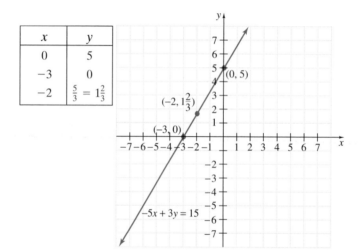

EXAMPLE 4   Graph the linear equation $y = 3x$.

Solution: To graph this linear equation, we find three ordered pair solutions. Since this equation is solved for y, choose three x values.

If $x = 2$, $y = 3 \cdot 2 = 6$.

If $x = 0$, $y = 3 \cdot 0 = 0$.

If $x = -3$, $y = 3 \cdot -3 = -9$.

x	y
2	6
0	0
-3	-9

Next, graph the ordered pair solutions listed in the table above and draw a line through the plotted points. The line is the graph of $y = 3x$. Every point on the graph represents an ordered pair solution of the equation and every ordered pair solution is a point on this line.

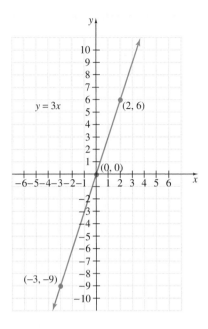

EXAMPLE 5   Graph the linear equation $y = -\frac{1}{3}x$.

Solution: Find three ordered pair solutions, graph the solutions, and draw a line through the plotted solutions. To avoid fractions, choose x values that are multiples of 3 to substitute in the equation.

If $x = 6$, then $y = -\dfrac{1}{3} \cdot 6 = -2$.

If $x = 0$, then $y = -\dfrac{1}{3} \cdot 0 = 0$.

If $x = -3$, then $y = -\dfrac{1}{3} \cdot -3 = 1$.

x	y
6	−2
0	0
−3	1

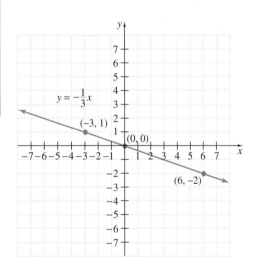

Let's compare the graphs in Examples 4 and 5. The graph of $y = 3x$ tilts upward (as we follow the line from left to right) and the graph of $y = -\frac{1}{3}x$ tilts downward (as we follow the line from left to right). Also notice that both lines go through the origin or that $(0, 0)$ is an ordered pair solution of both equations. This

m is a constant. The graph of an equation in this form goes through the origin $(0, 0)$ because when x is 0, $y = mx$ becomes $y = m \cdot 0 = 0$.

EXAMPLE 6   Graph the linear equation $y = 3x + 6$ and compare this graph with the graph of $y = 3x$ in Example 4.

Solution: Find ordered pair solutions, graph the solutions, and draw a line through the plotted solutions. We choose x values and substitute in the equation $y = 3x + 6$.

If $x = -3$, then $y = 3(-3) + 6 = -3$.
If $x = 0$, then $y = 3(0) + 6 = 6$.
If $x = 1$, then $y = 3(1) + 6 = 9$.

x	y
-3	-3
0	6
1	9

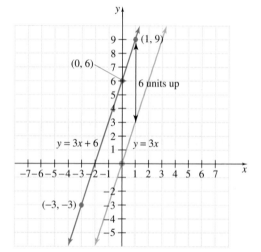

The most startling similarity is that both graphs appear to have the same upward tilt as we move from left to right. Also, the graph of $y = 3x$ crosses the y-axis at the origin, while the graph of $y = 3x + 6$ crosses the y-axis at 6. In fact, the graph of $y = 3x + 6$ is the same as the graph of $y = 3x$ moved vertically upward 6 units.

Notice above that the graph of $y = 3x + 6$ crosses the y-axis at 6. This happens because when $x = 0$, $y = 3x + 6$ becomes $y = 3 \cdot 0 + 6 = 6$. The graph contains the point $(0, 6)$, which is on the y-axis.

In general, if a linear equation in two variables is solved for y, we say that it is written in the form $y = mx + b$. The graph of this equation contains the point $(0, b)$ because when $x = 0$, $y = mx + b$ is $y = m \cdot 0 + b = b$.

The graph of $y = mx + b$ crosses the y-axis at b.

GRAPHING CALCULATOR EXPLORATIONS

In this section, we begin a study of graphing calculators and graphing software packages for computers. These graphers use the same point plotting technique that was introduced in this section. The advantage of this graphing technology is, of course, that graphing calculators and computers can find and plot ordered pair solutions much faster than we can. Note, however, that the features described in these boxes may not be available on all graphing calculators.

The rectangular screen where a portion of the rectangular coordinate system is displayed is called a **window.** We call it a **standard window** for graphing when both the x- and y-axes show coordinates between -10 and 10. This information is often displayed in the window menu on a graphing calculator as

Xmin $= -10$

Xmax $= 10$

  Xscl $= 1$ The scale on the x-axis is one unit per tick mark.

Ymin $= -10$

Ymax $= 10$

  Yscl $= 1$ The scale on the y-axis is one unit per tick mark.

To use a graphing calculator to graph the equation $y = 2x + 3$, press the Y = key and enter the keystrokes 2 x + 3 . The top row should now read $Y_1 = 2x + 3$. Next press the GRAPH key, and the display should look like this:

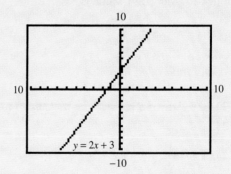

Use a standard window and graph the following linear equations. (Unless otherwise stated, use a standard window when graphing.)   graphical answers in App. F

1. $y = -3x + 7$

2. $y = -x + 5$

3. $y = \dfrac{1}{4}x - 2$

4. $y = \dfrac{2}{3}x - 1$

5. $y = 2.5x - 7.9$

6. $y = -1.3x + 5.2$

7. $y = -\dfrac{3}{10}x + \dfrac{32}{5}$

8. $y = \dfrac{2}{9}x - \dfrac{22}{3}$

EXERCISE SET 3.2

Determine whether each equation is a linear equation in two variables. See Example 1.

1. $-x = 3y + 10$ yes
2. $y = x - 15$ yes
3. $x = y$ yes
4. $x = y^3$ no
5. $x^2 + 2y = 0$ no
6. $0.01x - 0.2y = 8.8$ yes
7. $y = -1$ yes
8. $x = 25$ yes

Graph each linear equation. See Examples 2 through 5.

9. $x + y = 4$
10. $x + y = 7$ graphical answers in App. F
11. $x - y = -2$
12. $-x + y = 6$
13. $x - 2y = 4$
14. $-x + 5y = 5$
15. $y = 6x + 3$
16. $y = -2x + 7$

Write each statement as an equation in two variables. Then graph the equation. graphical answers in App. F

17. The y-value is 5 more than the x-value. $y = x + 5$
18. The y-value is twice the x-value. $y = 2x$
19. Two times the x-value added to three times the y-value is 6. $2x + 3y = 6$
20. Five times the x-value added to twice the y-value is -10. $5x + 2y = -10$

Graph each pair of linear equations on the same set of axes. Discuss how the graphs are similar and how they are different. See Example 6. graphical answers in App. F

21. $y = 5x; y = 5x + 4$
22. $y = 2x; y = 2x + 5$
23. $y = -2x; y = -2x - 3$
24. $y = x; y = x - 7$
25. $y = \frac{1}{2}x; y = \frac{1}{2}x + 2$
26. $y = -\frac{1}{4}x; y = -\frac{1}{4}x + 3$

Graph each linear equation. graphical answers in App. F

27. $x - 2y = -6$
28. $-x + 2y = 5$
29. $y = 6x$
30. $x = -2y$
31. $3y - 10 = 5x$
32. $-2x + 7 = 2y$
33. $x + 3y = 9$
34. $2x + y = -2$
35. $y - x = -1$
36. $x - y = 5$
37. $x = -3y$
38. $y = -x$

39. $5x - y = 10$
40. $7x - y = 2$
41. $y = \frac{1}{2}x + 2$
42. $y = -\frac{1}{5}x - 1$

43. Graph the nonlinear equation $y = x^2$ by completing the table shown. Plot the ordered pairs and connect them with a smooth curve.

x	y
0	0
1	1
-1	1
2	4
-2	4

44. Graph the nonlinear equation $y = |x|$ by completing the table shown. Plot the ordered pairs and connect them. This curve is "V" shaped.

$$y = |x|$$

x	y
0	0
1	1
-1	1
2	2
-2	2

The graph of $y = 5x$ is below as well as Figures A–D. For Exercises 45 through 48, match each equation with its graph.

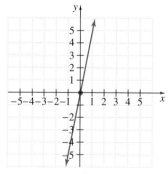

$y = 5x$

A.

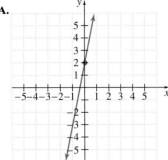

B.

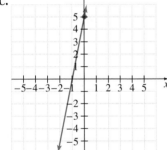

C.

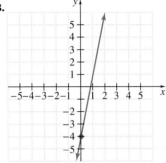

D.

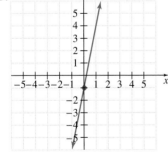

45. $y = 5x + 5$ C   **46.** $y = 5x - 4$ B

47. $y = 5x - 1$ D   **48.** $y = 5x + 2$ A

Recall that if an equation is written in the form $y = mx + b$, its graph crosses the y-axis at b. Use this to match each graph with its corresponding equation in Exercises 49 through 54.

A.

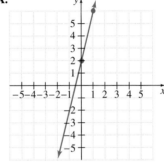

B.

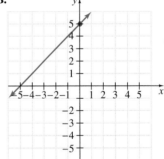

C.

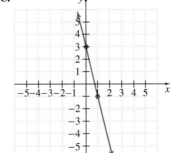

D.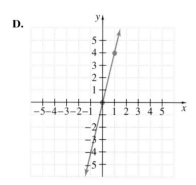

E.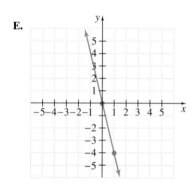

F.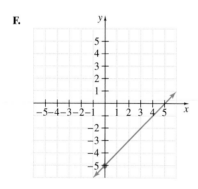

49. $y = 4x$ D   **50.** $y = -4x$ E

51. $y = 4x + 2$ A   **52.** $y = -4x + 3$ C

53. $y = x + 5$ B   **54.** $y = x - 5$ F

55. Explain how to find ordered pair solutions of linear equations in two variables.   answers may vary

56. $x + y = 12; y = 9$ **57.** yes; answers may vary   **58.** $x + y = 25; x = 5$

56. The perimeter of the trapezoid below is 22 centimeters. Write a linear equation in two variables for the perimeter. Find y if x is 3 cm.

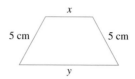

57. If (a, b) is an ordered pair solution of $x + y = 5$, is (b, a) also a solution? Explain why or why not.

58. The perimeter of the rectangle below is 50 miles. Write a linear equation in two variables for this perimeter. Use this equation to find x when y is 20.

Review Exercises

59. The coordinates of three vertices of a rectangle are $(-2, 5)$, $(4, 5)$, and $(-2, -1)$. Find the coordinates of the fourth vertex. See Section 3.1. $(4, -1)$

60. The coordinates of two vertices of square are $(-3, -1)$ and $(2, -1)$. Find the coordinates of two pairs of points possible for the third and fourth vertices. See Section 3.1. $(-3, 4), (2, 4); (-3, -5), (2, -5)$

Solve the following equations. See Section 2.4.

61. $3(x - 2) + 5x = 6x - 16$ $x = -5$

62. $5 + 7(x + 1) = 12 + 10x$ $x = 0$

63. $3x + \dfrac{2}{5} = \dfrac{1}{10}$ $x = -\dfrac{1}{10}$

64. $\dfrac{1}{6} + 2x = \dfrac{2}{3}$ $x = \dfrac{1}{4}$

Complete each table. See Section 3.1.

65. $x - y = -3$

x	y	
0		$(0, 3)$
	0	$(-3, 0)$

66. $y - x = 5$

x	y	
0		$(0, 5)$
	0	$(-5, 0)$

67. $y = 2x$

x	y
0	
	0

(0, 0)
(0, 0)

68. $x = -3y$

x	y
0	
	0

(0, 0)
(0, 0)

3.3 | INTERCEPTS

O B J E C T I V E S

TAPE
BA 3.3

1. Identify intercepts of a graph.
2. Graph a line given its intercepts.
3. Graph a linear equation by finding and plotting intercepts.
4. Identify and graph vertical and horizontal lines.

1

In this section, we graph linear equations in two variables by identifying intercepts. For example, the graph of $y = 4x - 8$ is shown below. Notice that this graph crosses the y-axis at the point $(0, -8)$. The y-coordinate of this point, -8, is called the **y-intercept.** Likewise, the graph crosses the x-axis at $(2, 0)$ and the x-coordinate, 2, is called the **x-intercept.**

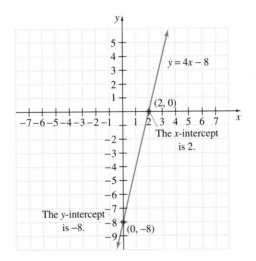

In general, an **intercept point** of a graph is a point where the graph intersects an axis. When an intercept point is on the x-axis, the x-coordinate of the point is called an x-intercept. When an intercept point is on the y-axis, the y-coordinate of the point is called a y-intercept.

EXAMPLE 1   Identify the *x*- and *y*-intercepts and the intercept points.

a.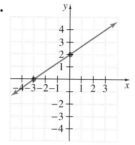

b.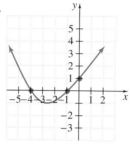

c.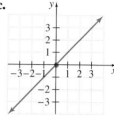

d.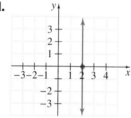

e.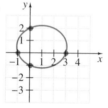

Solution: **a.** The graph crosses the *x*-axis at −3, so the *x*-intercept is −3. The graph crosses the *y*-axis at 2, so the *y*-intercept is 2. The intercept points are (−3, 0) and (0, 2).

b. The graph crosses the *x*-axis at −4 and −1, so the *x*-intercepts are −4 and −1. The graph crosses the *y*-axis at 1, so the *y*-intercept is 1. The intercept points are (−4, 0), (−1, 0), and (0, 1).

c. The *x*-intercept is 0 and the *y*-intercept is 0. The intercept point is (0, 0).

d. The *x*-intercept is 2. There are no *y*-intercepts. The intercept point is (2, 0).

e. The *x*-intercepts are −1 and 3. The *y*-intercepts are −1 and 2. The intercept points are (−1, 0), (3, 0), (0, −1), and (0, 2).

2   Since a line is determined by two points, we can usually graph a line when we know its *x*-intercept and its *y*-intercept. For example, if the *x*-intercept is 3 and the *y*-intercept is −5, we graph the line containing the points (3, 0) and (0, −5).

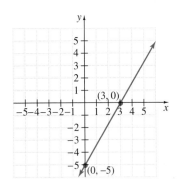

EXAMPLE 2   Graph the line with x-intercept 1 and y-intercept -1.

Solution: To graph the line, we identify and graph the intercept points. The intercept points are $(1, 0)$ and $(0, -1)$. Plot these points and draw a line through them.

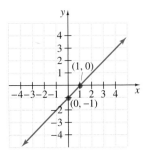

3   Given an equation of a line, intercept points are usually easy to find since one co-ordinate is 0.

One way to find the y-intercept of a line, given its equation, is to let $x = 0$, since a point on the y-axis has an x-coordinate of 0. To find the x-intercept of a line, let $y = 0$, since a point on the x-axis has a y-coordinate of 0.

> **FINDING X- AND Y-INTERCEPTS**
>
> To find the x-intercept, let $y = 0$ and solve for x.
> To find the y-intercept, let $x = 0$ and solve for y.

EXAMPLE 3   Graph $x - 3y = 6$ by finding and plotting intercept points.

Solution: Let $y = 0$ to find the x-intercept and let $x = 0$ to find the y-intercept.

$$
\begin{array}{ll}
\text{Let } y = 0 & \text{Let } x = 0 \\
x - 3y = 6 & x - 3y = 6 \\
x - 3(0) = 6 & 0 - 3y = 6 \\
x - 0 = 6 & -3y = 6 \\
x = 6 & y = -2
\end{array}
$$

The x-intercept is 6 and the y-intercept is -2. We find a third ordered pair solution to check our work. If we let $y = -1$, then $x = 3$. Plot the points $(6, 0)$, $(0, -2)$, and $(3, -1)$. The graph of $x - 3y = 6$ is the line drawn through these points, as shown.

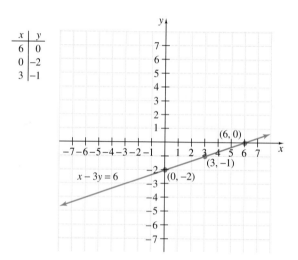

x	y
6	0
0	-2
3	-1

EXAMPLE 4   Graph $x = -2y$ by plotting intercept points.

Solution: Let $y = 0$ to find the x-intercept and $x = 0$ to find the y-intercept.

$$
\begin{array}{llll}
\text{If } y = 0 & \text{then} & \text{If } x = 0 & \text{then} \\
x = -2y & & & \\
x = -2(0) & \text{or} & 0 = -2y & \text{or} \\
x = 0 & & 0 = y &
\end{array}
$$

Both the x-intercept and y-intercept are 0. In other words, when $x = 0$, then $y = 0$, which gives the ordered pair $(0, 0)$. Also, when $y = 0$, then $x = 0$, which gives the same ordered pair $(0, 0)$. This happens when the graph passes through the origin. Since two points are needed to determine a line, we must find at least one more ordered pair that satisfies $x = -2y$. Let $y = -1$ to find a second ordered pair solution and let $y = 1$ as a checkpoint.

If $y = -1$ then	If $y = 1$ then
$x = -2(-1)$ or	$x = -2(1)$ or
$x = 2$	$x = -2$

The ordered pairs are $(0, 0)$, $(2, -1)$, and $(-2, 1)$. Plot these points to graph $x = -2y$.

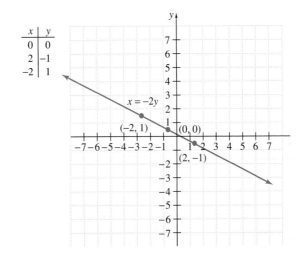

EXAMPLE 5   Graph $4x = 3y - 9$.

Solution: Find the x- and y-intercepts, and then choose $x = 2$ to find a third checkpoint.

If $y = 0$ then	If $x = 0$ then	If $x = 2$ then
$4x = 3(0) - 9$ or	$4 \cdot 0 = 3y - 9$ or	$4(2) = 3y - 9$ or
$4x = -9$	$9 = 3y$	$8 = 3y - 9$
Solve for x.	Solve for y.	Solve for y.
$x = -\dfrac{9}{4}$ or $-2\dfrac{1}{4}$	$3 = y$	$17 = 3y$
		$\dfrac{17}{3} = y$ or $y = 5\dfrac{2}{3}$

The ordered pairs are $\left(-2\dfrac{1}{4}, 0\right)$, $(0, 3)$, and $\left(2, 5\dfrac{2}{3}\right)$. The equation $4x = 3y - 9$ is graphed as follows.

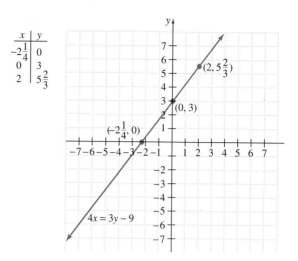

x	y
$-2\frac{1}{4}$	0
0	3
2	$5\frac{2}{3}$

4 The equation $x = c$, where c is a real number constant, is a linear equation in two variables because it can be written in the form $x + 0y = c$. The graph of this equation is a vertical line as shown in the next example.

EXAMPLE 6 Graph $x = 2$.

Solution: The equation $x = 2$ can be written as $x + 0y = 2$. For any y-value chosen, notice that x is 2. No other value for x satisfies $x + 0y = 2$. Any ordered pair whose x-coordinate is 2 is a solution of $x + 0y = 2$. We will use the ordered pair solutions $(2, 3)$, $(2, 0)$, and $(2, -3)$ to graph $x = 2$.

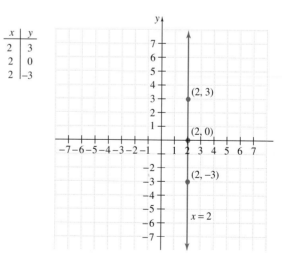

x	y
2	3
2	0
2	-3

The graph is a vertical line with x-intercept 2. Note that this graph has no y-intercept because x is never 0.

VERTICAL LINES

The graph of $x = c$, where c is a real number, is a vertical line with x-intercept c.

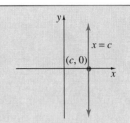

EXAMPLE 7   Graph $y = -3$.

Solution: The equation $y = -3$ can be written as $0x + y = -3$. For any x-value chosen, y is -3. If we choose 4, 1, and -2 as x-values, the ordered pair solutions are $(4, -3)$, $(1, -3)$, and $(-2, -3)$. Use these ordered pairs to graph $y = -3$. The graph is a horizontal line with y-intercept -3 and no x-intercept.

x	y
4	-3
1	-3
-2	-3

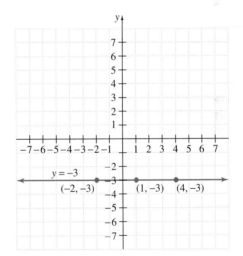

HORIZONTAL LINES

The graph of $y = c$, where c is a real number, is a horizontal line with y-intercept c.

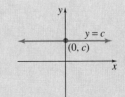

GRAPHING CALCULATOR EXPLORATIONS

You may have noticed that to use the [Y =] key on a grapher to graph an equation, the equation must be solved for y. For example, to graph $2x + 3y = 7$, we solve this equation for y.

$$2x + 3y = 7$$

$$3y = -2x + 7 \qquad \text{Subtract } 2x \text{ from both sides.}$$

$$\frac{3y}{3} = -\frac{2x}{3} + \frac{7}{3} \qquad \text{Divide both sides by 3.}$$

$$y = -\frac{2}{3}x + \frac{7}{3} \qquad \text{Simplify.}$$

To graph $2x + 3y = 7$ or $y = -\frac{2}{3}x + \frac{7}{3}$, press the [Y =] key and enter

$$Y_1 = -\frac{2}{3}x + \frac{7}{3}$$

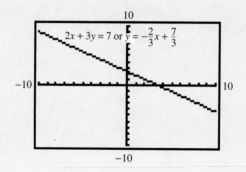

Graph each linear equation.   graphical answers in App. F

1. $x = 3.78y$ **2.** $-2.61y = x$

3. $3x + 7y = 21$ **4.** $-4x + 6y = 12$

5. $-2.2x + 6.8y = 15.5$ **6.** $5.9x - 0.8y = -10.4$

EXERCISE SET 3.3

Identify the intercepts and intercept points. See Example 1.

1.

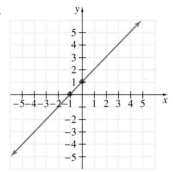

1. $x = -1$; $y = 1$; $(-1, 0)$; $(0, 1)$

2.

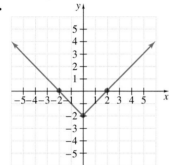

2. $x = -2$; $x = 2$; $y = -2$; $(-2, 0)$; $(2, 0)$; $(0, -2)$

3.

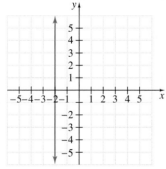

3. $x = -2$; $(-2, 0)$

4.

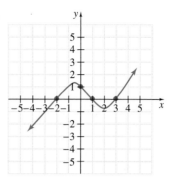

4. $x = -2$; $x = 1$; $x = 3$; $y = 1$;
$(-2, 0)$; $(1, 0)$; $(3, 0)$; $(0, 1)$

5.

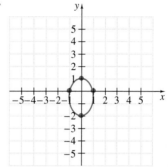

5. $x = -1$; $x = 1$; $y = 1$; $y = -2$; $(-1, 0)$;
$(1, 0)$; $(0, 1)$; $(0, -2)$

6.

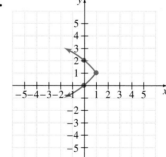

6. $x = 0$; $y = 0$; $y = 2$; $(0, 0)$; $(0, 2)$

⟐ 7. What is the greatest number of intercepts for a line?

⟐ 8. What is the least number of intercepts for a line?   1

⟐ 9. What is the least number of intercepts for a circle?   0

⟐ 10. What is the greatest number of intercepts for a circle? 4

7. infinite

Graph each line with given x- and y-intercept. See Example 2. graphical answers in App. F

11. *x*-intercept: 5
   y-intercept: -3

12. *x*-intercept: 0
   y-intercept: 4

13. *x*-intercept: 7/2
   y-intercept: 0

14. *x*-intercept: $-1/2$
   y-intercept: -9

Graph each linear equation by finding x- and y-intercepts. See Examples 3 through 5. graphical answers in App. F

15. $x - y = 3$

16. $x - y = -4$

17. $x = 5y$

18. $2x = y$

19. $-x + 2y = 6$

20. $x - 2y = -8$

21. $2x - 4y = 8$

22. $2x + 3y = 6$

Graph each linear equation. See Examples 6 and 7. graphical answers in App. F

23. $x = -1$

24. $y = 5$

25. $y = 0$

26. $x = 0$

27. $y + 7 = 0$

28. $x - 2 = 0$

*Two lines in the same plane that do not intersect are called **parallel lines**.* graphical answers in App. F

29. Draw a line parallel to the line $x = 5$ that intersects the *x*-axis at 1. What is the equation of this line?

30. Draw a line parallel to the line $y = -1$ that intersects the *y*-axis at -4. What is the equation of this line?

Graph each linear equation. graphical answers in App. F

31. $x + 2y = 8$

32. $x - 3y = 3$

33. $x - 7 = 3y$

34. $y - 3x = 2$

35. $x = -3$

36. $y = 3$

37. $3x + 5y = 7$

38. $3x - 2y = 5$

39. $x = y$

40. $x = -y$

41. $x + 8y = 8$

42. $x - 3y = 9$

43. $5 = 6x - y$

44. $4 = x - 3y$

45. $-x + 10y = 11$

46. $-x + 9 = -y$

47. $y = 1$

48. $x = 1$

49. $x = 2y$

50. $y = -2x$

51. $x + 3 = 0$

52. $y - 6 = 0$

53. $x = 4y - \dfrac{1}{3}$

54. $y = -3x + \dfrac{3}{4}$

55. $2x + 3y = 6$

56. $4x + y = 5$

For Exercises 57 through 62, match each equation with its graph.

A.
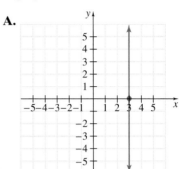

B.
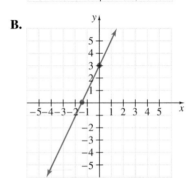

C.
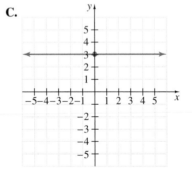

D.
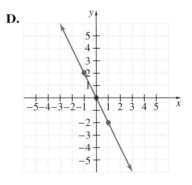

E.

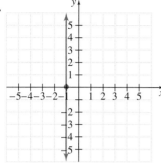

F.

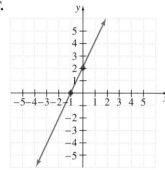

a particular office chair and 6 hours to manufacture an office desk. A total of 1200 hours is available to produce office chairs and desks of this style. The linear equation that models this situation is $3x + 6y = 1200$, where x represents the number of chairs produced and y the number of desks manufactured.

a. Complete the ordered pair solution (0,) of this equation. Describe the manufacturing situation that corresponds to this solution.

b. Complete the ordered pair solution (, 0) of this equation. Describe the manufacturing situation that corresponds to this solution.

c. Use the ordered pairs found above and graph the equation $3x + 6y = 1200$. See App. F.

d. If 50 desks are manufactured, find the greatest number of chairs that they can make. 300 chairs

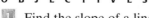 **65.** Discuss whether a vertical line ever has a y-intercept.

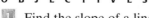 **66.** Explain why it is a good idea to use three points to graph a linear equation. answers may vary

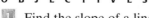 **67.** Discuss whether a horizontal line ever has an x-intercept. answers may vary

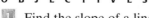 **68.** Explain how to find intercepts. answers may vary

57. $y = 3$ C

58. $y = 2x + 2$ F

59. $x = -1$ E

60. $x = 3$ A

61. $y = 2x + 3$ B

62. $y = -2x$ D

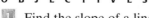 **63.** A computer purchased for $4000 loses value or depreciates in the amount of $800 per year. The equation $y = 4000 - 800x$ models the depreciated value of the computer, where y is the depreciated value after x years. Graph this equation for x values from 0 to 5.

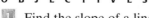 **64.** The production supervisor at Alexandra's Office Products finds that it takes 3 hours to manufacture

Review Exercises

Simplify.

69. $\dfrac{-6 - 3}{2 - 8}$ $\dfrac{3}{2}$

70. $\dfrac{4 - 5}{-1 - 0}$ 1

71. $\dfrac{-8 - (-2)}{-3 - (-2)}$ 6

72. $\dfrac{12 - 3}{10 - 9}$ 9

73. $\dfrac{0 - 6}{5 - 0}$ $-\dfrac{6}{5}$

74. $\dfrac{2 - 2}{3 - 5}$ 0

63. See App. F. **64. a.** (0, 200); answers may vary **64. b.** (400, 0); answers may vary **65.** answers may vary

---

3.4 | Slope

TAPE
BA 3.4

O B J E C T I V E S

1 Find the slope of a line given two points of the line.

2 Graph a line given its slope and a point of the line.

3 Find the slope of a line given its equation.

4 Find the slopes of horizontal and vertical lines.

5 Compare the shapes of parallel and perpendicular lines.

1 Thus far, much of this chapter has been devoted to graphing lines. You have probably noticed by now that a key feature of a line is its slant or steepness. In mathematics, the slant or steepness of a line is formally known as its **slope.** We measure the slope of a line by the ratio of vertical change to the corresponding horizontal change as we move along the line.

On the line below, for example, suppose that we begin at the point (1, 2) and move to the point (4, 6). The vertical change is the change in y-coordinates: $6 - 2$ or 4 units. The corresponding horizontal change is the change in x-coordinates: $4 - 1 = 3$ units. The ratio of these changes is

$$\text{slope} = \frac{\text{change in } y \text{ (vertical change)}}{\text{change in } x \text{ (horizontal change)}} = \frac{4}{3}$$

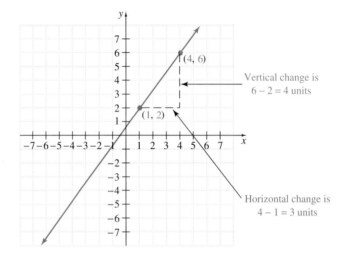

The slope of this line, then, is $\frac{4}{3}$: for every 4 units of change in y-coordinates, there is a corresponding change of 3 units in x-coordinates. You may wonder if the slope depends on what two points of a line are chosen. The answer is no, it makes no difference at all which two points we choose: the ratio of change is constant, and this is why we call this constant ratio slope. Slope, or steepness of a line, after all, is the same everywhere on a line.

To find the slope of a line, then, choose two points of the line. Label the two x-coordinates of two points, x_1 and x_2 (read "x sub one" and "x sub two"), and label the corresponding y-coordinates y_1 and y_2.

The vertical change or **rise** between these points is the difference in the y-coordinates: $y_2 - y_1$. The horizontal change or **run** between the points is the difference of the x-coordinates: $x_2 - x_1$. The slope of the line is the ratio of $y_2 - y_1$ to $x_2 - x_1$ and we traditionally use the letter m to denote slope.

$$m = \frac{y_2 - y_1}{x_2 - x_1}$$

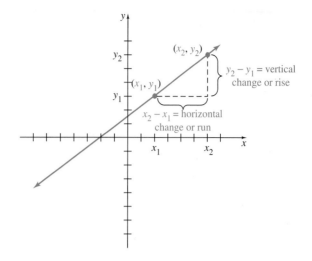

SLOPE OF A LINE

The slope m of the line through points (x_1, y_1) and (x_2, y_2), is given by

$$m = \frac{\text{rise}}{\text{run}} = \frac{y_2 - y_1}{x_2 - x_1}, \qquad \text{as long as } x_2 \neq x_1.$$

EXAMPLE 1   Find the slope of the line through $(-1, 5)$ and $(2, -3)$. Graph the line.

Solution: If we let (x_1, y_1) be $(-1, 5)$, then $x_1 = -1$ and $y_1 = 5$. Also, let (x_2, y_2) be point $(2, -3)$ so that $x_2 = 2$ and $y_2 = -3$. Then, by the definition of slope,

$$m = \frac{y_2 - y_1}{x_2 - x_1}$$

$$= \frac{-3 - 5}{2 - (-1)}$$

$$= \frac{-8}{3} = -\frac{8}{3}$$

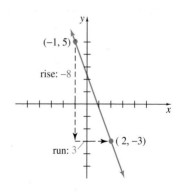

The slope of the line is $-\dfrac{8}{3}$.

In Example 1, we could just as well have identified (x_1, y_1) with $(2, -3)$ and (x_2, y_2) with $(-1, 5)$. It makes no difference which point is called (x_1, y_1) or (x_2, y_2).

> REMINDER When finding the slope of a line through two given points, it makes no difference which given point is called (x_1, y_1) and which is called (x_2, y_2). However, once an x-coordinate is called x_1, make sure its corresponding y-coordinate is called y_1.

EXAMPLE 2 Find the slope of the line through $(-1, -2)$ and $(2, 4)$. Graph the line.

Solution: Let (x_1, y_1) be $(2, 4)$ and let (x_2, y_2) be $(-1, -2)$.

$$m = \frac{y_2 - y_1}{x_2 - x_1}$$

$$= \frac{-2 - 4}{-1 - 2}$$

$$= \frac{-6}{-3} = 2$$

The slope is 2.

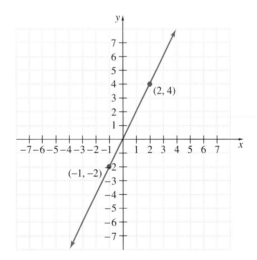

Notice that the slope of the line in Example 1 is negative, whereas the slope of the line in Example 2 is positive. Let your eye follow the line with negative slope from left to right and notice that the line "goes down." Following the line with positive slope from left to right, notice that the line "goes up." This is true in general.

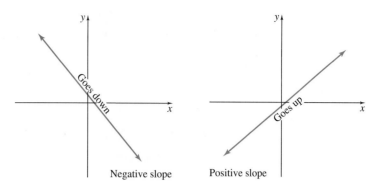

Negative slope   Positive slope

2 From geometry, we know that a line is uniquely determined by 2 points. A line is also uniquely determined by 1 point and a slope.

EXAMPLE 3   Graph the line through $(-1, 5)$ with slope -2.

Solution: To graph the line, we need two points. One point is $(-1, 5)$, and we use the slope -2, which can be written as $\frac{-2}{1}$, to find another point.

$$m = \frac{\text{rise}}{\text{run}} = \frac{-2}{1}$$

To find another point, start at $(-1, 5)$ and move vertically two units down, since the numerator of the slope is -2; then move horizontally 1 unit to the right since the denominator of the slope is 1. We stop at the point $(0, 3)$. The line through $(-1, 5)$ and $(0, 3)$ has the required slope of -2.

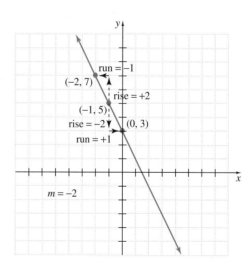

The slope -2 can also be written as $\frac{2}{-1}$, so to find another point we could start at $(-1, 5)$ and move two units up and then one unit left. We would stop at the point $(-2, 7)$. The line through $(-1, 5)$ and $(-2, 7)$ has the required slope and is the same line as shown previously through $(-1, 5)$ and $(0, 3)$.

 As we have seen, the slope of a line is defined by two points on the line. Thus, if we know the equation of a line, we can find its slope by finding two of its points.

EXAMPLE 4   Find the slope of the line whose equation is $-2x + 3y = 12$.

Solution: To find the slope of the line defined by $-2x + 3y = 12$, find any two points on the line. Let's find and use intercept points as our two points.

IF $x = 0$	IF $y = 0$
$-2 \cdot 0 + 3y = 12$	$-2x + 3 \cdot 0 = 12$
$3y = 12$	$-2x = 12$
$y = 4$	$x = -6$

If $x = 0$, the corresponding y-value is 4. The y-intercept point is $(0, 4)$.

If $y = 0$, the corresponding x-value is -6. The x-intercept point is $(-6, 0)$.

Use the points $(0, 4)$ and $(-6, 0)$ to find the slope. Let (x_1, y_1) be $(0, 4)$ and (x_2, y_2) be $(-6, 0)$. Then

$$m = \frac{y_2 - y_1}{x_2 - x_1} = \frac{0 - 4}{-6 - 0} = \frac{-4}{-6} = \frac{2}{3}$$

The slope is $\dfrac{2}{3}$.

 Next, we find the slopes of horizontal and vertical lines.

EXAMPLE 5   Find the slope of the line $y = -1$.

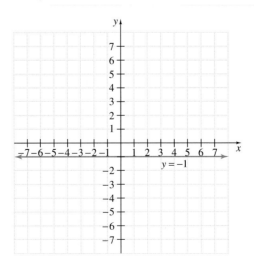

Solution: Recall that $y = -1$ is a horizontal line with y-intercept -1. To find the slope, find two ordered pair solutions of $y = -1$. Solutions of $y = -1$ must have a y-value of -1.

Let $(x_1, y_1) = (2, -1)$ and $(x_2, y_2) = (-3, -1)$. Then

$$m = \frac{y_2 - y_1}{x_2 - x_1} = \frac{-1 - (-1)}{-3 - 2} = \frac{0}{-5} = 0$$

The slope of the line $y = -1$ is 0. Since the y-values will have a difference of 0 for all horizontal lines, we can say that all **horizontal lines have a slope of 0.** ▬▬▬▬

EXAMPLE 6   Find the slope of the line $x = 5$.

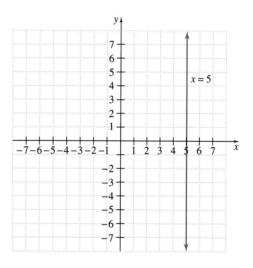

Solution: Recall that the graph of $x = 5$ is a vertical line with x-intercept 5.

To find the slope, find two ordered pair solutions of $x = 5$. Solutions of $x = 5$ must have an x-value of 5.

Let $(x_1, y_1) = (5, 0)$ and $(x_2, y_2) = (5, 4)$. Then

$$m = \frac{y_2 - y_1}{x_x - x_1} = \frac{4 - 0}{5 - 5} = \frac{4}{0}$$

Since $\frac{4}{0}$ is undefined, we say the slope of the vertical line $x = 5$ is undefined. Since the x-values will have a difference of 0 for all vertical lines, we can say that all **vertical lines have undefined slope.** ▬▬▬▬

REMINDER   Slope of 0 and undefined slope are not the same. Vertical lines have undefined slope or no slope, while horizontal lines have a slope of 0.

Here is a general review of slope.

SUMMARY OF SLOPE

Slope m of the line through (x_1, y_1) and (x_2, y_2) is given by the equation
$m = \dfrac{y_2 - y_1}{x_2 - x_1}$.

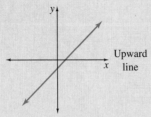

Positive slope: $m > 0$
Upward line

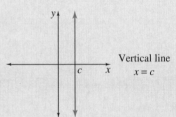

Negative slope: $m < 0$
Downward line

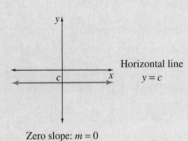

Zero slope: $m = 0$
Horizontal line $y = c$

No slope or undefined slope
Vertical line $x = c$

5 Two lines in the same plane are parallel if they do not intersect. Slopes of lines can help us determine whether lines are parallel. Parallel lines have the same steepness, so it follows that they have the same slope.

PARALLEL LINES

Nonvertical parallel lines have the same slope.

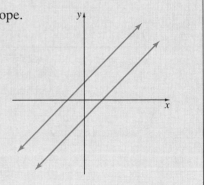

How do the slopes of perpendicular lines compare? Two lines that intersect at right angles are said to be **perpendicular.** The product of the slopes of two perpendicular lines is -1.

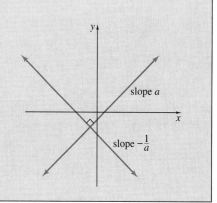

PERPENDICULAR LINES

If the product of the slopes of two lines is -1, then the lines are perpendicular.

(Two nonvertical lines are perpendicular if the slope of one is the negative reciprocal of the slope of the other.)

slope a

slope $-\dfrac{1}{a}$

EXAMPLE 7   Is the line passing through the points $(-6, 0)$ and $(-2, 3)$ parallel to the line passing through the points $(5, 4)$ and $(7, 5)$?

Solution: To see if these lines are parallel, we find and compare slopes. The line passing through the points $(-6, 0)$ and $(-2, 3)$ has slope

$$m = \frac{3 - 0}{-2 - (-6)} = \frac{3}{4}$$

The line passing through the points $(5, 4)$ and $(7, 5)$ has slope

$$m = \frac{5 - 4}{7 - 5} = \frac{1}{2}$$

Since the slopes are not the same, these lines are not parallel.

EXAMPLE 8   Find the slope of a line perpendicular to the line passing through the points $(-1, 7)$ and $(2, 2)$.

Solution: First, let's find the slope of the line through $(-1, 7)$ and $(2, 2)$. This line has slope

$$m = \frac{2 - 7}{2 - (-1)} = \frac{-5}{3}$$

The slope of every line perpendicular to the given line has a slope equal to the negative reciprocal of

$$-\frac{5}{3} \qquad \text{or} \qquad -\left(-\frac{3}{5}\right) = \frac{3}{5}$$

↑   ↑

negative   reciprocal

The slope of a line perpendicular to the given line has slope $\dfrac{3}{5}$.

GRAPHING CALCULATOR EXPLORATIONS

It is possible to use a grapher and sketch the graph of more than one equation on the same set of axes. This feature can be used to confirm our findings from Section 3.2 when we learned that the graph of an equation written in the form $y = mx + b$ has a y-intercept of b. For example, graph the equations $y = \frac{2}{5}x$, $y = \frac{2}{5}x + 7$, and $y = \frac{2}{5}x - 4$ on the same set of axes. To do so, press the $\boxed{Y =}$ key and enter the equations on the first three lines.

$$Y_1 = \left(\frac{2}{5}\right)x$$

$$Y_2 = \left(\frac{2}{5}\right)x + 7$$

$$Y_3 = \left(\frac{2}{5}\right)x - 4$$

The screen should look like:

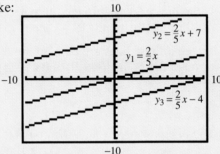

Notice that all three graphs appear to have the same positive slope. The graph of $y = \frac{2}{5}x + 7$ is the graph of $y = \frac{2}{5}x$ moved 7 units upward with a y-intercept of 7. Also, the graph of $y = \frac{2}{5}x - 4$ is the graph of $y = \frac{2}{5}x$ moved 4 units downward with a y-intercept of -4.

Graph the equations on the same set of axes. Describe the similarities and differences in their graphs. graphical answers in App. F

1. $y = 3.8x$, $y = 3.8x - 3$, $y = 3.8x + 9$ **2.** $y = -4.9x$, $y = -4.9x + 1$, $y = -4.9x + 8$

3. $y = \frac{1}{4}x$; $y = \frac{1}{4}x + 5$, $y = \frac{1}{4}x - 8$ **4.** $y = -\frac{3}{4}x$, $y = -\frac{3}{4}x - 5$, $y = -\frac{3}{4}x + 6$

MENTAL MATH

Decide whether a line with the given slope is upward, downward, horizontal or vertical.

1. $m = \frac{7}{6}$ upward   **2.** $m = -3$ downward   **3.** $m = 0$ horizontal   **4.** m is undefined.   vertical

EXERCISE SET 3.4

Find the slope of each line if it exists. See Example 1.

1.

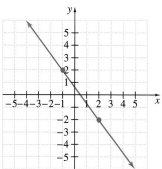

$m = -\dfrac{4}{3}$

2.

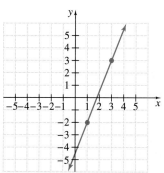

$m = \dfrac{5}{2}$

3.

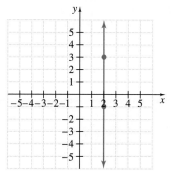

undefined slope

4.

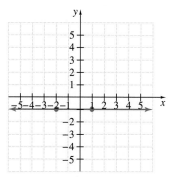

$m = 0$

5.

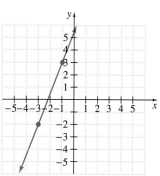

$m = \dfrac{5}{2}$

6.

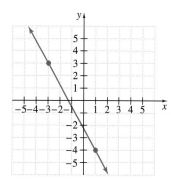

$m = \dfrac{7}{4}$

Find the slope of the line that goes through the given points. See Examples 1 and 2. **13.** undefined

7. $(0, 0)$ and $(7, 8)$ $\quad\frac{8}{7}$ **8.** $(-1, 5)$ and $(0, 0)$ $\quad -5$

9. $(-1, 5)$ and $(6, -2)$ -1 **10.** $(-1, 9)$ and $(-3, 4)$ $\quad\frac{5}{2}$

11. $(1, 4)$ and $(5, 3)$ $\quad -\frac{1}{4}$ **12.** $(3, 1)$ and $(2, 6)$ $\quad -5$

13. $(-4, 3)$ and $(-4, 5)$ **14.** $(6, -6)$ and $(6, 2)$

15. $(-2, 8)$ and $(1, 6)$ $\quad -\frac{2}{3}$ **16.** $(4, -3)$ and $(2, 2)$ $\quad -\frac{5}{2}$

17. $(1, 0)$ and $(1, 1)$ **18.** $(0, 13)$ and $(-4, 13)$ $\quad 0$

19. $(5, -11)$ and $(1, -11)$ **20.** $(5, 4)$ and $(0, 5)$ $\quad -\frac{1}{5}$

For each graph, determine which line has the greater slope.

21.

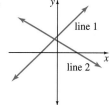

line 1

22.

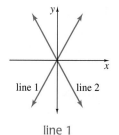

line 1

14. undefined   **17.** undefined   **19.** 0

23.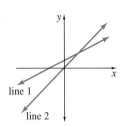

24.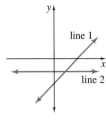

line 2   line 1

 Match each line with its slope.

A. $m = 0$ **B.** no slope   **C.** $m = 3$

D. $m = 1$ **E.** $m = -\dfrac{1}{2}$ **F.** $m = -\dfrac{3}{4}$

25.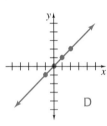

D

26.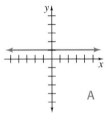

A

27.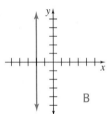

B

28.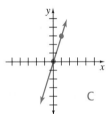

C

29.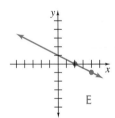

E

30.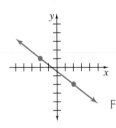

F

Graph each line passing through the given point with the given slope. See Example 3.   graphical answers in App. F

31. Through $(2, 3)$, with slope $\dfrac{1}{4}$.

32. Through $(-1, 5)$, with slope $\dfrac{2}{3}$.

33. Through $(0, -4)$, with slope 3.

34. Through $(3, 1)$, with slope 4.

35. Through $(-2, -7)$, with slope -1.

36. Through $(-4, -3)$, with slope -2.

37. Through $(-3, 4)$, with slope $-\dfrac{3}{5}$.

38. Through $(6, -2)$, with slope $-\dfrac{1}{3}$.

Find the slope of each line. See Examples 4–6.

39. $y = 5x - 2$ $m = 5$ **40.** $y = -2x + 6$ $m = -2$

41. $2x + y = 7$ $m = -2$ **42.** $-5x + y = 10$ $m = 5$

43. $x = 1$ undefined slope   **44.** $y = -2$ $m = 0$

45. $y = -3$ $m = 0$ **46.** $x = 5$ undefined slope

47. $2x - 3y = 10$ $m = \dfrac{2}{3}$ **48.** $-3x - 4y = 6$ $m = -\dfrac{3}{4}$

49. $x = 2y$ $m = \dfrac{1}{2}$ **50.** $x = -4y$ $m = -\dfrac{1}{4}$

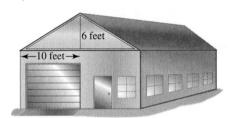 *The pitch of a roof is its slope. Find the pitch of each roof shown.*

51. $\dfrac{3}{5}$

6 feet   10 feet

52. $\dfrac{1}{2}$

1   2

53. Find x so that the pitch of the roof is $\dfrac{1}{3}$. $x = 6$

x 18 feet

54. Find x so that the pitch of the roof is $\frac{2}{5}$. $x = 20$

The grade of a road is its slope written as a percent. Find the grade of the road shown.

55.

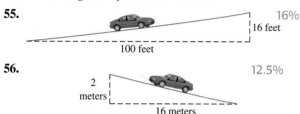

16%

16 feet

100 feet

56.

2 meters

16 meters

12.5%

Find the slope of the line a. parallel and b. perpendicular to the line through each pair of points. See Examples 7 and 8.

57. $(-3, -3)$ and $(0, 0)$ **a.** 1 **b.** -1

58. $(6, -2)$ and $(1, 4)$ **a.** $-\frac{6}{5}$ **b.** $\frac{5}{6}$

59. $(-8, -4)$ and $(3, 5)$ **a.** $\frac{9}{11}$ **b.** $-\frac{11}{9}$

60. $(6, -1)$ and $(-4, -10)$ **a.** $\frac{9}{10}$ **b.** $-\frac{10}{9}$

Determine whether the lines through each pair of points are parallel, perpendicular, or neither. See Examples 7 and 8.

61. $(0, 6)$ and $(-2, 0)$
$(0, 5)$ and $(1, 8)$

62. $(1, -1)$ and $(-1, -11)$
$(-2, -8)$ and $(0, 2)$

63. $(2, 6)$ and $(-2, 8)$
$(0, 3)$ and $(1, 5)$

64. $(-1, 7)$ and $(1, 10)$
$(0, 3)$ and $(1, 5)$

65. $(3, 6)$ and $(7, 8)$
$(0, 6)$ and $(2, 7)$

66. $(4, 2)$ and $(6, 6)$
$(0, -2)$ and $(1, 0)$

67. $(2, -3)$ and $(6, -5)$
$(5, -2)$ and $(-3, -4)$

68. $(-1, -5)$ and $(4, 4)$
$(10, 8)$ and $(-7, 4)$

69. $(-4, -3)$ and $(-1, 0)$
$(4, -4)$ and $(0, 0)$

70. $(2, -2)$ and $(4, -8)$
$(0, 7)$ and $(3, 8)$

71. Show that a triangle with vertices at the points $(2, 5)$, $(-4, 4)$, and $(-3, 0)$ is a right triangle.

72. Show that the quadrilateral with vertices $(1, 3)$, $(2, 1)$, $(-4, 0)$, and $(-3, -2)$ is a parallelogram.

The following bar graph shows average miles per gallon for cars during the years shown.

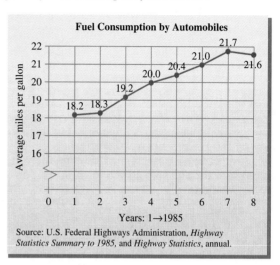

Fuel Consumption by Automobiles

Average miles per gallon

21.7, 21.0, 20.4, 20.0, 19.2, 18.2, 18.3, 21.6

Years: 1→1985

Source: U.S. Federal Highways Administration, *Highway Statistics Summary to 1985*, and *Highway Statistics*, annual.

73. In 1988, cars averaged how many miles per gallon of fuel? 20 mpg

74. In 1991, cars averaged how many miles per gallon of fuel? 21.7 mpg

75. Find the increase of average miles per gallon for automobiles for the years 1985 to 1992. 3.4 mpg

76. Do you notice any trends from this graph?

77. What line segment has the greatest slope?

78. What line segment has the least slope?

79. What year showed the greatest increase in miles per gallon? 1987

80. What year showed the least increase in miles per gallon? 1992

81. Find the slope of a line parallel to the line passing through $(-7, -5)$ and $(-2, -6)$.

82. Find the slope of a line parallel to the line passing through the origin and $(-2, 10)$. -5

83. Find the slope of a line perpendicular to the line passing through the origin and $(1, -3)$.

84. Find the slope of the line perpendicular to the line passing through $(-1, 3)$ and $(2, -8)$.

85. Find the slope of a line parallel to the line $y = x$. 1

86. Find the slope of a line parallel to the line $x + 2y = 6$. $-\frac{1}{2}$

Find the slope of the line through the given points.

87. $(2.1, 6.7)$ and $(-8.3, 9.3)$ -0.25

88. $(-3.8, 1.2)$ and $(-2.2, 4.5)$ 2.0625

89. $(2.3, 0.2)$ and $(7.9, 5.1)$ 0.875

90. $(14.3, -10.1)$ and $(9.8, -2.9)$ -1.6

91. Find the slope and the *y*-intercept of the line defined by $x + 3y = 6$. Next, solve the equation $x + 3y = 6$ for *y*. Compare the slope and the *y*-intercept with the equation solved for *y*. Write down any observations.

92. Find the slope and the *y*-intercept of the line defined by $5x + 2y = 7$. Next, solve the equation $5x + 2y = 7$ for *y*. Compare the slope and the *y*-intercept with the equation solved for *y*. Write down any observations.

93. The graph of $y = \frac{1}{2}x$ has a slope of $\frac{1}{2}$. The graph of $y = 3x$ has a slope of 3. The graph of $y = 5x$ has a slope of 5. Graph all three equations on a single coordinate system. As slope becomes larger, how does the steepness of the line change?

94. The graph of $y = -\frac{1}{3}x + 2$ has a slope of $-\frac{1}{3}$. The graph of $y = -2x + 2$ has a slope of -2. The graph of $y = -4x + 2$ has a slope of -4. Graph all three equations on a single coordinate system. As the absolute value of the slope becomes larger, how does the steepness of the line change?   See App. F.; the line becomes steeper.

Review Exercises

Solve each linear inequality. See Section 2.9.

95. $-3x \le -9$ $x \ge 3$

96. $-x > -16$ $x < 16$

97. $\dfrac{x - 6}{2} < 3$ $x < 12$

98. $\dfrac{2x + 1}{3} \ge -1$ $x \ge -2$

Graph each linear equation in two variables. See Sections 3.2 and 3.3.   graphical answers in App. F

99. $x - y = 6$

100. $x = -2y$

101. $y = 3x$

102. $5x + 3y = 15$

103. $x = -2$

104. $y = 5$

91. $m = -\dfrac{1}{3}$; $b = 2$; answers may vary   **92.** $m = -\dfrac{5}{2}$; $b = \dfrac{7}{2}$; answers may vary

93. See App. F.; the line becomes steeper.

3.5 GRAPHING LINEAR INEQUALITIES

OBJECTIVES

TAPE
BA 3.5

1 Determine whether an ordered pair is a solution of a linear inequality in two variables.

2 Graph a linear inequality in two variables.

1 Recall that a linear equation in two variables is an equation that can be written in the form $Ax + By = C$ where *A*, *B*, and *C* are real numbers and *A* and *B* are not both 0. The definition of a linear inequality is the same except that the equal sign is replaced with an inequality sign.

A **linear inequality in two variables** is an inequality that can be written in one of the forms:

$$Ax + By < C \qquad Ax + By \le C$$
$$Ax + By > C \qquad Ax + By \ge C$$

where A, B, and C are real numbers and A and B are not both 0. Just as for linear equations in x and y, an ordered pair is a **solution** of an inequality in x and y if replacing the variables by coordinates of the ordered pair results in a true statement.

EXAMPLE 1   Determine whether each ordered pair is a solution of the equation $2x - y < 6$.

a. $(5, -1)$ **b.** $(2, 7)$ **c.** $(0, -6)$

Solution: **a.** Replace x with 5 and y with -1 and see if a true statement results.

$$2x - y < 6$$
$$2(5) - (-1) < 6 \qquad \text{Replace } x \text{ with 5 and } y \text{ with } -1.$$
$$10 + 1 < 6$$
$$11 < 6 \qquad \text{False.}$$

The ordered pair $(5, -1)$ is not a solution since $11 < 6$ is a false statement.

b. Replace x with 2 and y with 7 and see if a true statement results.

$$2x - y < 6$$
$$2(2) - 7 < 6 \qquad \text{Replace } x \text{ with 2 and } y \text{ with 7.}$$
$$4 - 7 < 6$$
$$-3 < 6 \qquad \text{True.}$$

The ordered pair $(2, 7)$ is a solution since $-3 < 6$ is a true statement.

c. Replace x with 0 and y with -6 and see if a true statement results.

$$2x - y < 6$$
$$2(0) - (-6) < 6 \qquad \text{Replace } x \text{ with 0 and } y \text{ with } -6$$
$$0 + 6 < 6$$
$$6 < 6 \qquad \text{False.}$$

The ordered pair $(0, -6)$ is not a solution since $6 < 6$ is a false statement.

2   The linear equation $x + y = 5$ is graphed next. Recall that all points on the line correspond to ordered pairs that satisfy the equation $x + y = 5$. It can be shown that all the points above the line $x + y = 5$ have coordinates that satisfy the inequality $x + y > 5$. Similarly, all points below the line have coordinates that satisfy the inequality $x + y < 5$.

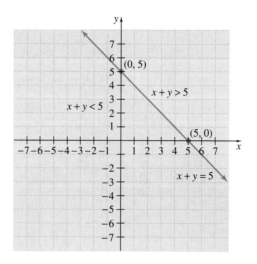

The region above the line and the region below the line are called **half-planes.** Every line divides the plane (similar to a sheet of paper extending indefinitely in all directions) into two half-planes; the line is called the **boundary.**

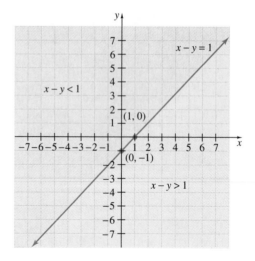

Similarly, the half-plane above the boundary $x - y = 1$ is the graph of the inequality $x - y < 1$. The half-plane below the boundary $x - y = 1$ is the graph of the inequality $x - y > 1$. Since the inequality $x - y \leq 1$ means

$$x - y = 1 \text{ or } x - y < 1$$

the graph of $x - y \leq 1$ is the half-plane $x - y < 1$ along with the boundary line $x - y = 1$.

The steps to graph a linear inequality are given next.

TO GRAPH A LINEAR INEQUALITY IN TWO VARIABLES

Step 1.   Graph the boundary line found by replacing the inequality sign with an equal sign. If the inequality sign is > or <, graph a dashed boundary line indicating that the points on the line are not solutions of the inequality. If the inequality sign is ≥ or ≤, graph a solid boundary line indicating that the points on the line are solutions of the inequality.

Step 2.   Choose a point, not on the boundary line, as a test point. Substitute the coordinates of this test point into the original inequality.

Step 3.   If a true statement is obtained in step 2, shade the half-plane that contains the test point. If a false statement is obtained, shade the half-plane that does not contain the test point.

EXAMPLE 2   Graph $x + y < 7$.

Solution: First, graph the boundary line by graphing the equation $x + y = 7$. Graph this boundary as a dashed line because the inequality sign is <, and thus the points on the line are not solutions of the inequality $x + y < 7$.

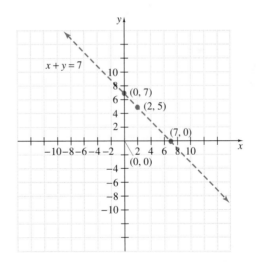

Next, choose a test point, being careful not to choose a point on the boundary line. We choose $(0, 0)$. Substitute the coordinates of $(0, 0)$ into $x + y < 7$.

$x + y < 7$ Original inequality.

$0 + 0 < 7$ Replace x with 0 and y with 0.

$0 < 7$ True.

Since the result is a true statement, $(0, 0)$ is a solution of $x + y < 7$, and every point in the same half-plane as $(0, 0)$ is also a solution. To indicate this, shade the entire half-plane containing $(0, 0)$, as shown.

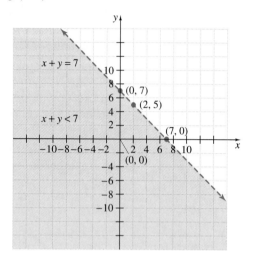

EXAMPLE 3   Graph $2x - y \geq 3$.

Solution: Graph the boundary line by graphing $2x - y = 3$. Draw this line as a solid line since the inequality sign is $\geq$, and thus the points on the line are solutions of $2x - y \geq 3$. Once again, $(0, 0)$ is a convenient test point since it is not on the boundary line.

Substitute 0 for x and 0 for y into the **original inequality.**

$$2x - y \geq 3$$
$$2(0) - 0 \geq 3 \qquad \text{Let } x = 0 \text{ and } y = 0.$$
$$0 \geq 3 \qquad \text{False.}$$

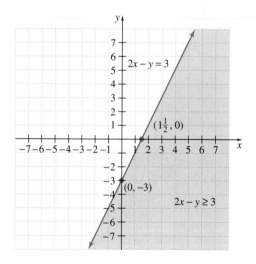

Since the statement is false, no point in the half-plane containing $(0, 0)$ is a solution. Shade the half-plane that does not contain $(0, 0)$. Every point in the shaded half-plane and every point on the boundary line satisfies $2x - y \geq 3$.   ▬▬▬▬

REMINDER   When graphing an inequality, make sure the test point is substituted in the **original inequality.** For example, when graphing $x + y < 3$, test $(0, 0)$ in $x + y < 3$, not $x + y = 3$. Since $(0, 0)$ is a solution of $x + y < 3$, shade the half-plane containing $(0, 0)$ as shown.

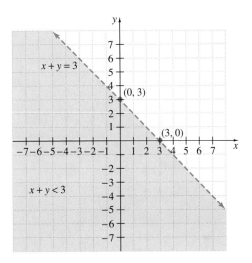

▬▬▬▬

EXAMPLE 4   Graph $x > 2y$.

Solution: Find the boundary line by graphing $x = 2y$. The boundary line is a dashed line since the inequality symbol is $>$. We cannot use $(0, 0)$ as a test point because it is a point on the boundary line. Choose $(0, 2)$ as the test point.

$$x > 2y$$
$$0 > 2(2) \qquad \text{Let } x = 0 \text{ and } y = 2.$$
$$0 > 4 \qquad \text{False.}$$

Since the statement is false, shade the half-plane that does not contain the test point $(0, 2)$.

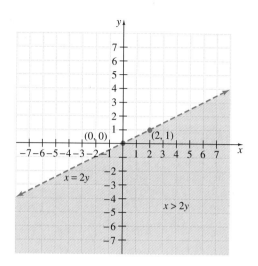

EXAMPLE 5   Graph $5x + 4y \leq 20$.

Solution: Graph the solid boundary line $5x + 4y = 20$. Choose $(0, 0)$ as the test point.

$$5x + 4y \leq 20$$
$$5(0) + 4(0) \leq 20 \qquad \text{Let } x = 0 \text{ and } y = 0.$$
$$0 \leq 20 \qquad \text{True.}$$

Shade the half-plane that contains $(0, 0)$. The graph of $5x + 4y \leq 20$ is given next.

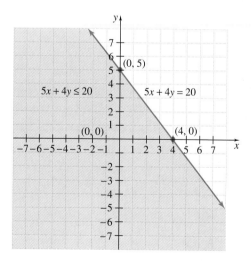

EXAMPLE 6   Graph $y > 3$.

Solution: Graph the dashed boundary line $y = 3$. Recall that the graph of $y = 3$ is a horizontal line with y-intercept 3. Choose $(0, 0)$ as the test point.

$$y > 3$$
$$0 > 3 \qquad \text{Let } y = 0.$$
$$0 > 3 \qquad \text{False.}$$

Shade the half-plane that does not contain $(0, 0)$. The graph of $y > 3$ is shown next.

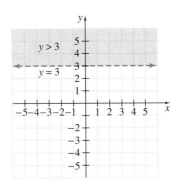

MENTAL MATH

State whether the graph of each inequality includes its corresponding boundary line.

1. $y \geq x + 4$ yes
2. $x - y > -7$ no
3. $y \geq x$ yes
4. $x > 0$ no

Decide whether (0, 0) is a solution of each given inequality.

5. $x + y > -5$ yes
6. $2x + 3y < 10$ yes
7. $x - y \leq -1$ no
8. $\dfrac{2}{3}x + \dfrac{5}{6}y > 4$ no

EXERCISE SET 3.5

Determine which ordered pairs given are solutions of the linear inequality in two variables. See Example 1.

1. $x - y > 3$; $(0, 3)$, $(2, -1)$, $(5, 1)$ no; no; yes

2. $y - x < -2$; $(2, 1)$, $(5, -1)$, $(3, 7)$ no; yes; no

3. $3x - 5y \leq -4$; $(2, 3)$, $(-1, -1)$, $(4, 0)$ yes; no; no

4. $2x + y \geq 10$; $(0, 11)$, $(-1, -4)$, $(5, 0)$ yes; no; yes

5. $x < -y$; $(6, 6)$, $(0, 2)$, $(-5, 1)$ no; no; yes

6. $y > 3x$; $(0, 0)$, $(1, 4)$, $(-1, -4)$ no; yes; no

graphical answers in App. F
Graph each inequality. See Examples 2 through 6.

7. $x + y \leq 1$
8. $x + y \geq -2$
9. $2x + y > -4$
10. $x + 3y \leq 3$
11. $x + 6y \leq -6$
12. $7x + y > -14$
13. $2x + 5y > -10$
14. $5x + 2y \leq 10$
15. $x + 2y \leq 3$
16. $2x + 3y > -5$
17. $2x + 7y > 5$
18. $3x + 5y \leq -2$
19. $x - 2y \geq 3$
20. $4x + y \leq 2$

21. $5x + y < 3$

22. $x + 2y > -7$

23. $4x + y < 8$

24. $9x + 2y \geq -9$

25. $y \geq 2x$

26. $x < 5y$

27. $x \geq 0$

28. $y \leq 0$

29. $y \leq -3$

30. $x > -\dfrac{2}{3}$

31. $2x - 7y > 0$

32. $5x + 2y \leq 0$

33. $3x - 7y \geq 0$

34. $-2x - 9y > 0$

35. $x > y$

36. $x \leq -y$

37. $x - y \leq 6$

38. $x - y > 10$

39. $-\dfrac{1}{4}y + \dfrac{1}{3}x > 1$

40. $\dfrac{1}{2}x - \dfrac{1}{3}y \leq -1$

41. $-x < 0.4y$

42. $0.3x \geq 0.1y$

 43. Write an inequality whose solutions are all pairs of numbers x and y whose sum is at least 13. Graph the inequality. $x + y \geq 13$; see App. F.

44. Write an inequality whose solutions are all the pairs of numbers x and y whose sum is at most -4. Graph the inequality. $x + y \leq -4$; see App. F.

Match each inequality with its graph.

a. $x > 2$ **b.** $y < 2$ **c.** $y \leq 2x$ **d.** $y \leq -3x$

e. $2x + 3y < 6$ **f.** $3x + 2y > 6$

45.

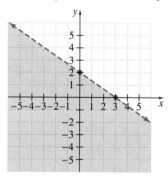

e

46.

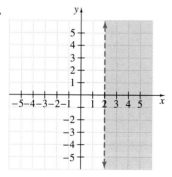

a

47.

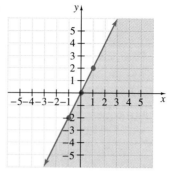

c

48.

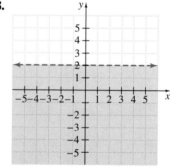

b

49.

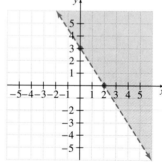

f

50.

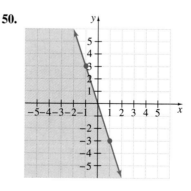

d

51. Explain why a point on the boundary line should not be chosen as the test point.   answers may vary

52. Describe the graph of a linear inequality.
answers may vary

Review Exercises

Evaluate. See Section 1.3.

53. 2^3 8

54. 3^4 81

55. $(-2)^5$ −32

56. -2^5 −32

57. $3 \cdot 4^2$ 48

58. $4 \cdot 3^3$ 108

Evaluate each expression for the given replacement value. See Section 1.7.

59. x^2 if x is −5   25

60. x^3 if x is −5   −125

61. $2x^3$ if x is −1   −2

62. $3x^2$ if x is −1   3

GROUP ACTIVITY

FINANCIAL ANALYSIS

OPTIONAL MATERIALS:
- Financial magazines
- Annual reports

The table below gives the sales in millions of dollars for the leading U.S. businesses in the aerospace industry for the years 1993 and 1994. You have been asked to analyze the performances of these companies and, based on this information alone, make an investment recommendation.

AEROSPACE SALES (IN MILLIONS OF DOLLARS)

COMPANY	1993	1994
Boeing	25,300	21,900
United Technologies	20,700	21,200
McDonnell-Douglas	14,500	13,200
Lockheed	13,100	13,100
Allied-Signal	11,800	12,800
Martin Marietta	9,400	9,900
Textron	8,700	9,700
General Dynamics	4,700	3,700

Source: *FORTUNE* magazine

1. Write the data for each company as two ordered pairs in the form (year, sales).

See App. F. for Group Activity answers and suggestions.

Assuming that the trends in the sales are linear, graph the line represented by the ordered pairs for each company. Describe the trend shown by each graph.

2. Find the slope of the line for each company.

3. Which of the lines, if any, have negative slopes? What does that mean in this context? Which of the lines, if any, have zero slopes? What does that mean in this context? Which of the lines, if any, are parallel? What does that mean in this context? Which, if any, of the lines are perpendicular? What does that mean in this context?

4. Of these aerospace industry companies, which one(s) would you recommend as an investment choice? Why?

5. Do you think it is wise to make a decision after looking at only 2 years of sales? What other factors do you think should be taken into consideration when making an investment choice?

6. (Optional) Use financial magazines and/or company annual reports to find sales or revenue information for two different years for two to four companies in the same industry. Analyze the sales and make an investment recommendation.

CHAPTER 3   HIGHLIGHTS

DEFINITIONS AND CONCEPTS	EXAMPLES

SECTION 3.1 THE RECTANGULAR COORDINATE SYSTEM

The **rectangular coordinate system** consists of a plane and a vertical and a horizontal number line intersecting at their 0 coordinate. The vertical number line is called the **y-axis** and the horizontal number line is called the **x-axis.** The point of intersection of the axes is called the **origin.**

To **plot** or **graph** an ordered pair means to find its corresponding point on a rectangular coordinate system.

To plot or graph an ordered pair such as $(3, -2)$, start at the origin. Move 3 units to the right and from there, 2 units down.

To plot or graph $(-3, 4)$ start at the origin. Move 3 units to the left and from there, 4 units up.

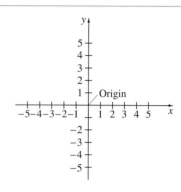

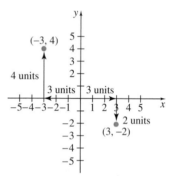

An ordered pair is a **solution** of an equation in two variables if replacing the variables by the coordinates of the ordered pair results in a true statement.

Determine whether $(-1, 5)$ is a solution to $2x + 3y = 13$.

$$2x + 3y = 13$$
$$2(-1) + 3 \cdot 5 = 13 \qquad \text{Let } x = -1, y = 5$$
$$-2 + 15 = 13$$
$$13 = 13 \qquad \text{True.}$$

If one coordinate of an ordered pair solution is known, the other value can be determined by substitution.

Complete the ordered pair $(0, \ )$ for the equation $x - 6y = 12$.

$$x - 6y = 12$$
$$0 - 6y = 12 \qquad \text{Let } x = 0.$$
$$\frac{-6y}{-6} = \frac{12}{-6} \qquad \text{Divide by } -6.$$
$$y = -2$$

The ordered pair solution is $(0, -2)$.

DEFINITIONS AND CONCEPTS	EXAMPLES

SECTION 3.2 GRAPHING LINEAR EQUATIONS

A **linear equation in two variables** is an equation that can be written in the form $Ax + By = C$ where A and B are not both 0. The form $Ax + By = C$ is called **standard form.**

To graph a linear equation in two variables, find three ordered pair solutions. Plot the solution points and draw the line connecting the points.

LINEAR EQUATIONS

$3x + 2y = -6$ $x = -5$

$y = 3$ $y = -x + 10$

$x + y = 10$ is in standard form.

Graph $x - 2y = 5$

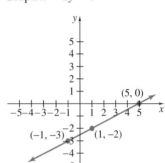

x	y
5	0
1	-2
-1	-3

SECTION 3.3 INTERCEPTS

An **intercept point** of a graph is a point where the graph intersects an axis. If a graph intersects the x-axis at a, then a is the **x-intercept** and the corresponding intercept point is $(a, 0)$. If a graph intersects the y-axis at b, then b is the **y-intercept** and the corresponding intercept point is $(0, b)$.

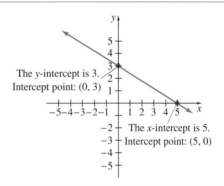

The y-intercept is 3.
Intercept point: $(0, 3)$

The x-intercept is 5.
Intercept point: $(5, 0)$

To find the x-intercept, let $y = 0$ and solve for x.

To find the y-intercept, let $x = 0$ and solve for y.

Graph $2x - 5y = -10$ by finding intercepts.

If $y = 0$, then

$$2x - 5 \cdot 0 = -10$$
$$2x = -10$$
$$\frac{2x}{2} = \frac{-10}{2}$$
$$x = -5$$

The x-intercept is -5.
Intercept point: $(-5, 0)$.

If $x = 0$, then

$$2 \cdot 0 - 5y = -10$$
$$-5y = -10$$
$$\frac{-5y}{-5} = \frac{-10}{-5}$$
$$y = 2$$

The y-intercept is 2.
Intercept point: $(0, 2)$.

(continued)

DEFINITIONS AND CONCEPTS	**EXAMPLES**

SECTION 3.3 INTERCEPTS

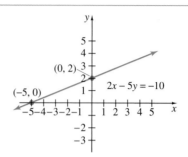

The graph of $x = c$ is a vertical line with x-intercept c.

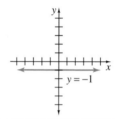

The graph of $y = c$ is a horizontal line with y-intercept c.

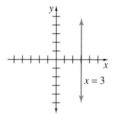

SECTION 3.4 SLOPE

The **slope** ***m*** of the line through points (x_1, y_1) and (x_2, y_2) is given by

$$m = \frac{y_2 - y_1}{x_2 - x_1} \quad \text{as long as } x_2 \neq x_1$$

A horizontal line has slope 0.

The slope of a vertical line is undefined.

Nonvertical parallel lines have the same slope.

The slope of the line through points $(-1, 6)$ and $(-5, 8)$ is

$$m = \frac{y_2 - y_1}{x_2 - x_1} = \frac{8 - 6}{-5 - (-1)} = \frac{2}{-4} = -\frac{1}{2}$$

The slope of the line $y = -5$ is 0.

The line $x = 3$ has undefined slope.

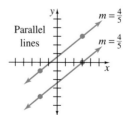

(continued)

Definitions and Concepts	Examples

Section 3.4 Slope

Two nonvertical lines are perpendicular if the slope of one is the negative reciprocal of the slope of the other.

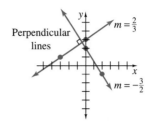

Perpendicular lines

$m = \frac{2}{3}$

$m = -\frac{3}{2}$

Section 3.5 Graphing Linear Inequalities

A **linear inequality in two variables** is an inequality that can be written in one of the forms:

$$Ax + By < C \qquad Ax + By \le C$$
$$Ax + By > C \qquad Ax + By \ge C$$

Linear Inequalities

$$2x - 5y < 6 \qquad x \ge -5$$
$$y > -8x \qquad y \le 2$$

To graph a linear inequality

1. Graph the boundary line by graphing the related equation. Draw the line solid if the inequality symbol is $\le$ or $\ge$. Draw the line dashed if the inequality symbol is $<$ or $>$.

2. Choose a test point not on the line. Substitute its coordinates into the original inequality.

3. If the resulting inequality is true, shade the half-plane that contains the test point. If the inequality is not true, shade the half-plane that does not contain the test point.

Graph $2x - y \le 4$.

1. Graph $2x - y = 4$. Draw a solid line because the inequality symbol is $\le$.

2. Check the test point $(0, 0)$ in the inequality $2x - y \le 4$.

$$2 \cdot 0 - 0 \le 4 \qquad \text{Let } x = 0 \text{ and } y = 0.$$
$$0 \le 4 \qquad \text{True.}$$

3. The inequality is true so we shade the half-plane containing $(0, 0)$.

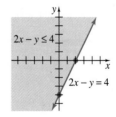

$2x - y \le 4$

$2x - y = 4$

CHAPTER 3 REVIEW

(3.1) *Plot the following ordered pairs on a Cartesian coordinate system.* graphical answers in App. F.

1. $(-7, 0)$

2. $\left(0, 4\frac{4}{5}\right)$

3. $(-2, -5)$

4. $(1, -3)$

5. $(0.7, 0.7)$

6. $(-6, 4)$

Determine whether each ordered pair is a solution of the given equation.

7. $7x - 8y = 56$; $(0, 56)$, $(8, 0)$ no; yes

8. $-2x + 5y = 10$; $(-5, 0)$, $(1, 1)$ yes; no

9. $x = 13$; $(13, 5)$, $(13, 13)$ **10.** $y = 2$; $(7, 2)$, $(2, 7)$

Complete the ordered pairs so that each is a solution of the given equation.

11. $-2 + y = 6x$; $(7, \ )$ **12.** $y = 3x + 5$; $(\ , -8)$

Complete the table of values for each given equation; then plot the ordered pairs. Use a single coordinate system for each exercise.

13. $9 = -3x + 4y$

x	y
	0
	3
9	

(−3, 0)
(1, 3)
(9, 9)
See App. F.

14. $y = 5$

x	y
7	
−7	
0	

(7, 5)
(−7, 5)
(0, 5)
See App. F.

15. $x = 2y$

x	y
	0
	5
	−5

(0, 0)
(10, 5)
(−10, −5)
See App. F.

16. The cost in dollars of producing x compact disk holders is given by $y = 5x + 2000$.

   a. Complete the following table.

x	1	100	1000
y	2005	2500	7000

   b. Find the number of compact disk holders that can be produced for $6430. 886 compact disks

9. yes; yes   **10.** yes; no   **11.** (7, 44)   **12.** $\left(-\dfrac{13}{3}, -8\right)$

(3.2) *Graph each linear equation.* graphical answers in App. F.

17. $x - y = 1$ **18.** $x + y = 6$

19. $x - 3y = 12$ **20.** $5x - y = -8$

21. $x = 3y$ **22.** $y = -2x$

23. $2x - 3y = 6$ **24.** $4x - 3y = 12$

(3.3) *Identify the intercepts and intercept points.*

25.

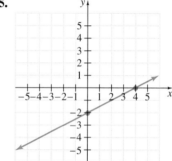

$x = 4$; $y = -2$;
(4, 0); (0, −2)

26.

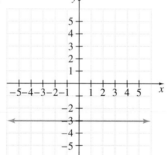

$y = -3$; (0, −3)

27.

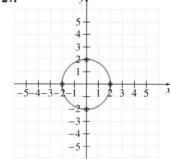

$x = -2$; $x = 2$;
$y = 2$; $y = -2$; (−2, 0);
(2, 0); (0, 2); (0, −2)

28.

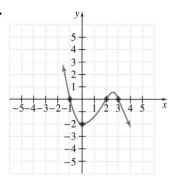

$x = -1; x = 2;$
$x = 3; y = -2;$
$(-1, 0); (2, 0);$
$(3, 0); (0, -2)$

graphical answers in App. F.

Graph each linear equation by finding its intercepts.

29. $x - 3y = 12$ **30.** $-4x + y = 8$

31. $y = -3$ **32.** $x = 5$

33. $y = -3x$ **34.** $x = 5y$

35. $x - 2 = 0$ **36.** $y + 6 = 0$

(3.4) Find the slope of each line.

37.

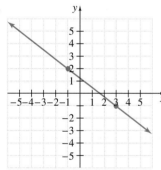

$m = -\dfrac{3}{4}$

38.

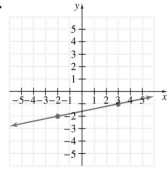

$m = \dfrac{1}{5}$

Match each line with its slope.

a.
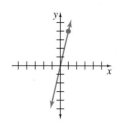

b.
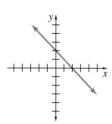

c.
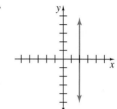

d.
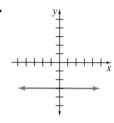

e.
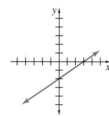

39. $m = 0$ d   **40.** $m = -1$ b

41. no slope   c   **42.** $m = 4$ a

43. $m = \dfrac{2}{3}$ e

Find the slope of the line that goes through the given points.

44. $(2, 5)$ and $(6, 8)$ $\dfrac{3}{4}$ **45.** $(4, 7)$ and $(1, 2)$ $\dfrac{5}{3}$

46. $(1, 3)$ and $(-2, -9)$ 4   **47.** $(-4, 1)$ and $(3, -6)$

Find the slope of each line.

48. $y = 3x + 7$ 3   **49.** $x - 2y = 4$ $\dfrac{1}{2}$

50. $y = -2$ 0   **51.** $x = 0$ undefined

Graph each line passing through the given point with the given slope.   graphical answers in App. F

52. through $(0, 3)$ with slope $\frac{1}{4}$

53. through $(-5, 3)$ with slope -3

Determine whether the lines through the pairs of points are parallel, perpendicular, or neither.

54. $(-3, 1)$ and $(1, -2)$ **55.** $(-7, 6)$ and $(0, 4)$
      $(2, 4)$ and $(6, 1)$ $(-9, -3)$ and $(1, 5)$

47. -1 **54.** parallel   **55.** neither

56. (9, 10) and (8, −7)   **57.** (−1, 3) and (3, −2)
     (−1, −3) and (2, −8)   (−2, −2) and (3, 2)

60. $x + 6y < 6$ **61.** $x + y > -2$

62. $y \geq -7$ **63.** $y \leq -4$

(3.5) *Graph the following inequalities.*   graphical answers in App. F.   **64.** $-x \leq y$ **65.** $x \geq -y$

58. $3x - 4y \leq 0$ **59.** $3x - 4y \geq 0$

56. neither   **57.** perpendicular

CHAPTER 3 TEST

Determine whether the ordered pairs are solutions of the equations.

1. $x - 2y = 3;\ (1, 1)$ no   **2.** $2x + 3y = 6;\ (0, -2)$
                                              no

Complete the ordered pair solution for the following equations.

3. $12y - 7x = 5;\ (1,\ )$ (1, 1)   **4.** $y = 17;\ (-4,\ )$
                                           (−4, 17)

Find the slopes of the following lines.

5.

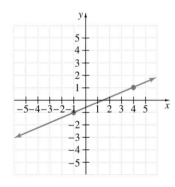

$m = \dfrac{2}{5}$

6.

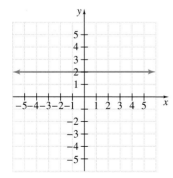

$m = 0$

7. Through $(6, -5)$ and $(-1, 2)$ −1

8. Through $(0, -8)$ and $(-1, -1)$ −7

9. $-3x + y = 5$ 3   **10.** $x = 6$ undefined

Determine whether the lines through the pairs of points are parallel, perpendicular, or neither.

11. $(-1, 3)$ and $(1, -3)$
     $(2, -1)$ and $(4, -7)$ parallel

12. $(-6, -6)$ and $(-1, -2)$
     $(-4, 3)$ and $(3, -3)$ neither

Graph the following.   graphical answers in App. F.

13. $2x + y = 8$ **14.** $-x + 4y = 5$

15. $x - y \geq -2$ **16.** $y \geq -4x$

17. $5x - 7y = 10$ **18.** $2x - 3y > -6$

19. $6x + y > -1$ **20.** $y = -1$

21. $x - 3 = 0$ **22.** $5x - 3y = 15$

Write each statement as an equation in two variables. Then graph the equation.   graphical answers in App. F.

23. The *y*-value is 1 more than twice the *x*-value. $y = 2x + 1$

24. The *x*-value added to four times the *y*-value is less than −4. $x + 4y < -4$

25. The perimeter of the parallelogram below is 42 meters. Write a linear equation in two variables for the perimeter. Use this equation to find *x* when *y* is 8. $x + 2y = 21;\ x = 5$ meters

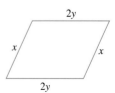

CHAPTER 3 CUMULATIVE REVIEW

1. Insert $<$, $>$, or $=$ in the space between the paired numbers to make each statement true. *(Sec. 1.1, Ex. 5)*

a. $-1\ <\ 0$ **b.** $7\ =\ \dfrac{14}{2}$ **c.** $-5\ >\ -6$

2. Write each fraction in lowest terms. *(Sec. 1.2, Ex. 2)*

a. $\dfrac{42}{49}$ $\dfrac{6}{7}$ **b.** $\dfrac{11}{27}$ $\dfrac{11}{27}$ **c.** $\dfrac{88}{20}$ $\dfrac{22}{5}$

3. Simplify $\dfrac{8+2\cdot3}{2^2-1}$. $\dfrac{14}{3}$ *(Sec. 1.3, Ex. 5)*

4. Write an algebraic expression that represents each of the following phrases. Let the variable x represent the unknown number. *(Sec. 1.4, Ex. 2)*

a. The sum of a number and 3 $x+3$

b. The product of 3 and a number $3x$

c. Twice a number $2x$

d. 10 decreased by a number $10-x$

e. 5 times a number increased by 7 $5x+7$

5. If $x=2$ and $y=-5$, evaluate the following expressions. *(Sec. 1.6, Ex. 4)*

a. $\dfrac{x-y}{12+x}$ $\dfrac{1}{2}$ **b.** x^2-y 9

6. Find each quotient. *(Sec. 1.7, Ex. 6)*

a. $\dfrac{-24}{-4}$ 6 **b.** $\dfrac{-36}{3}$ -12 **c.** $\dfrac{2}{3}\div\left(-\dfrac{5}{4}\right)$ $-\dfrac{8}{15}$

7. Find each product by using the distributive property to remove parentheses. **b.** $-2y-0.6z+2$

a. $5(x+2)$ $5x+10$ **b.** $-2(y+0.3z-1)$

c. $-(x+y-2z+6)$ $-x-y+2z-6$
(Sec. 2.1, Ex. 5)

8. Solve $-5(2a-1)-(-11a+6)=7$ for a. $a=8$
(Sec. 2.2, Ex. 5)

9. Solve $\dfrac{y}{7}=20$ for y. $y=140$ *(Sec. 2.3, Ex. 3)*

10. Solve $.25x+.10(x-3)=.05(x+18)$. $x=4$
(Sec. 2.4, Ex. 5)

11. Twice the sum of a number and 4 is the same as four times the number decreased by 12. Find the number. 10 *(Sec. 2.5, Ex. 3)*

12. Charles Pecot can afford enough fencing to enclose a rectangular garden with a perimeter of 140 feet. If the width of his garden is to be 30 feet, find the length. 40 ft. *(Sec. 2.6, Ex. 2)*

30 feet

13. Find 72% of 200. 144 *(Sec. 2.7, Ex. 3)*

14. Solve $-4x+7\geq-9$, and graph the solution set.
(Sec. 2.9, Ex. 6)

15. On a single coordinate system, plot the ordered pairs. State in which quadrant, if any, each point lies. *(Sec. 3.1, Ex. 1)*

a. $(3,2)$ **b.** $(-2,-4)$ **c.** $(1,-2)$ **d.** $(-5,3)$

e. $(0,0)$ **f.** $(0,2)$ **g.** $(-5,0)$ **h.** $\left(0,-1\dfrac{1}{2}\right)$

16. Graph the linear equation $2x+y=5$. *(Sec. 3.2, Ex. 2)*

17. Complete the table for the equation $y=3x$.

(Sec. 3.1, Ex. 4)

x	y
-1	-3
0	0
-3	-9

18. Graph $x=2$. See App. F *(Sec. 3.3, Ex. 6)*

19. Find the slope of the line through $(-1,5)$ and $(2,-3)$. Graph the line. *(Sec. 3.4, Ex. 1)*

20. Graph the line through $(-1,5)$ with slope -2.
(Sec. 3.4, Ex. 3)

21. Find the slope of the line $y=-1$ $m=0$
(Sec. 3.4, Ex. 5)

22. Graph $x+y<7$. See App. F *(Sec. 3.5, Ex. 2)*

14. $x\leq4$; 0 1 2 3 4 5 6

15. a. quadrant I **b.** quadrant III **c.** quadrant IV
d. quadrant II **e.** no quadrant; origin
f. no quadrant; y-axis **g.** no quadrant; x-axis
h. no quadrant; y-axis; see App. F

16. See App. F **19.** $-\dfrac{8}{3}$; See App. F

20. See App. F

EXPONENTS
AND
POLYNOMIALS

MAKING PREDICTIONS BASED ON HISTORICAL DATA

Accidental deaths include those due to motor vehicle accidents, poisonings, drownings, and fires. To help arrange for equipment/personnel resources and program funding for 911 service, it would be helpful for planners to be able to predict the number of accidental deaths that might occur in a given year. The first step in predicting such information is using the trends in historical data to model what the future might bring.

IN THE CHAPTER GROUP ACTIVITY ON PAGE 273, YOU WILL HAVE THE OPPORTUNITY TO INVESTIGATE TRENDS IN POLYNOMIALS REPRESENTING HISTORICAL DATA ON THE ANNUAL NUMBERS OF ACCIDENTAL DEATHS AND THEN USE THAT DATA AND RELATED POLYNOMIALS TO MAKE PREDICTIONS FOR THE FUTURE.

R ecall from Chapter 1 that an exponent is a shorthand notation for repeated factors. This chapter explores additional concepts about exponents and exponential expressions. An especially useful type of exponential expression is a polynomial. Polynomials model many real-world phenomena. Our goal in this chapter is to become proficient with operations on polynomials. ▬▬▬

4.1 | EXPONENTS

TAPE BA 4.1

O B J E C T I V E S

1 Evaluate exponential expressions.
2 Use the product rule for exponents.
3 Use the power rule for exponents.
4 Use the power rules for products and quotients.
5 Use the quotient rule for exponents, and define a number raised to the 0 power.

1 As we reviewed in Section 1.3, an exponent is a shorthand notation for repeated factors. For example, $2 \cdot 2 \cdot 2 \cdot 2 \cdot 2$ can be written as 2^5. The expression 2^5 is called an **exponential expression.** It is also called the fifth **power** of 2, or we say that 2 is **raised** to the fifth power.

$$5^6 = \underbrace{5 \cdot 5 \cdot 5 \cdot 5 \cdot 5 \cdot 5}_{6 \text{ factors; each factor is } 5} \qquad \text{and} \qquad (-3)^4 = \underbrace{(-3) \cdot (-3) \cdot (-3) \cdot (-3)}_{4 \text{ factors; each factor is } -3}$$

The **base** of an exponential expression is the repeated factor. The **exponent** is the number of times that the base is used as a factor.

$$5^6 \text{ exponent} \qquad (-3)^4 \text{ exponent}$$
$$\quad \text{base} \qquad\qquad\qquad \text{base}$$

DEFINITION OF a^n

If a is a real number and n is a positive integer, then a **raised to the nth power,** written a^n, is the product of n factors, each of which is a.

$$a^n = \underbrace{a \cdot a \cdot a \cdot a \cdot a \ldots a}_{n \text{ factors of } a}$$

▬▬▬

EXAMPLE 1   Evaluate each expression.

a. 2^3 **b.** 3^1 **c.** $(-4)^2$ **d.** -4^2 **e.** $\left(\frac{1}{2}\right)^4$ **f.** $4 \cdot 3^2$

Solution: **a.** $2^3 = 2 \cdot 2 \cdot 2 = 8$

b. To raise 3 to the first power means to use 3 as a factor only once. Therefore, $3^1 = 3$. Also, when no exponent is shown, the exponent is assumed to be 1.

c. $(-4)^2 = (-4)(-4) = 16$

d. $-4^2 = -(4 \cdot 4) = -16$

e. $\left(\dfrac{1}{2}\right)^4 = \dfrac{1}{2} \cdot \dfrac{1}{2} \cdot \dfrac{1}{2} \cdot \dfrac{1}{2} = \dfrac{1}{16}$

f. $4 \cdot 3^2 = 4 \cdot 9 = 36$

Notice how similar -4^2 is to $(-4)^2$ in the example above. The difference between the two is the parentheses. In $(-4)^2$, the parentheses tell us that the base, or repeated factor, is -4. In -4^2, only 4 is the base.

REMINDER   Be careful when identifying the base of an exponential expression.

$(-3)^2$	-3^2	$2 \cdot 3^2$
Base is -3	Base is 3	Base is 3
$(-3)^2 = (-3)(-3) = 9$	$-3^2 = -(3 \cdot 3) = -9$	$2 \cdot 3^2 = 2 \cdot 3 \cdot 3 = 18$

---

2   An exponent has the same meaning whether the base is a number or a variable. If x is a real number and n is a positive integer, then x^n is the product of n factors, each of which is x.

$$x^n = \underbrace{x \cdot x \cdot x \cdot x \cdot x \ldots x}_{n \text{ factors of } x}$$

EXAMPLE 2   Evaluate for the given value of x.

a. $2x^3$; x is 5

b. $\dfrac{9}{x^2}$; x is -3

Solution: **a.** If x is 5, $2x^3 = 2 \cdot 5^3$

$$= 2 \cdot (5 \cdot 5 \cdot 5)$$

$$= 2 \cdot 125$$

$$= 250$$

b. If x is -3, $\dfrac{9}{x^2} = \dfrac{9}{(-3)^2}$

$$= \dfrac{9}{(-3)(-3)}$$

$$= \dfrac{9}{9} = 1$$

Exponential expressions can be multiplied, divided, added, subtracted, and themselves raised to powers. By our definition of an exponent,

$$5^4 \cdot 5^3 = \underbrace{(5 \cdot 5 \cdot 5 \cdot 5)}_{4 \text{ factors of } 5} \cdot \underbrace{(5 \cdot 5 \cdot 5)}_{3 \text{ factors of } 5}$$

$$= \underbrace{5 \cdot 5 \cdot 5 \cdot 5 \cdot 5 \cdot 5 \cdot 5}_{7 \text{ factors of } 5}$$

$$= 5^7$$

Also,

$$x^2 \cdot x^3 = (x \cdot x) \cdot (x \cdot x \cdot x)$$

$$= x \cdot x \cdot x \cdot x \cdot x$$

$$= x^5$$

In both cases, notice that the result is exactly the same if the exponents are added.

$$5^4 \cdot 5^3 = 5^{4+3} = 5^7 \quad \text{and} \quad x^2 \cdot x^3 = x^{2+3} = x^5$$

This suggests the following **product rule for exponents.**

PRODUCT RULE FOR EXPONENTS

If m and n are positive integers and a is a real number, then

$$a^m \cdot a^n = a^{m+n}$$

In other words, to multiply two exponential expressions with a **common base,** keep the base and add the exponents.

EXAMPLE 3   Use the product rule to simplify.

a. $4^2 \cdot 4^5$ **b.** $x^2 \cdot x^5$ **c.** $y^3 \cdot y$ **d.** $y^3 \cdot y^2 \cdot y^7$ **e.** $(-5)^7 \cdot (-5)^8$

Solution: **a.** $4^2 \cdot 4^5 = 4^{2+5} = 4^7$

b. $x^2 \cdot x^5 = x^{2+5} = x^7$

c. $y^3 \cdot y = y^3 \cdot y^1$ Recall that if no exponent is written, it is assumed to be 1.

$$= y^{3+1}$$

$$= y^4$$

d. $y^3 \cdot y^2 \cdot y^7 = y^{3+2+7} = y^{12}$

e. $(-5)^7 \cdot (-5)^8 = (-5)^{7+8} = (-5)^{15}.$

We can simplify this expression further. Because $(-5)^{15}$ is the product of an odd number of negative numbers, the product is negative, so that

$$\underset{\underset{(-5)^{15}}{\uparrow}}{\overset{\overset{\text{odd number}}{\rule{1cm}{0.4pt}}}{}} \quad \text{can also be written as} \quad -5^{15}.$$

Both expressions have the same value.

EXAMPLE 4   Use the product rule to simplify $(2x^2)(-3x^5)$.

Solution: Recall that $2x^2$ means $2 \cdot x^2$ and $-3x^5$ means $-3 \cdot x^5$.

$$
\begin{aligned}
(2x^2)(-3x^5) &= 2 \cdot x^2 \cdot -3 \cdot x^5 && \text{Remove parentheses.} \\
&= 2 \cdot -3 \cdot x^2 \cdot x^5 && \text{Group factors with common bases.} \\
&= -6x^7 && \text{Simplify.}
\end{aligned}
$$

3   Exponential expressions can themselves be raised to powers. Let's try to discover a rule that simplifies an expression like $(x^2)^3$. By the definition of a^n,

$$(x^2)^3 = \underbrace{(x^2)(x^2)(x^2)}_{\text{3 factors of } x^2}$$

which can be simplified by the product rule for exponents.

$$(x^2)^3 = (x^2)(x^2)(x^2) = x^{2+2+2} = x^6$$

Notice that the result is exactly the same if we multiply the exponents.

$$(x^2)^3 = x^{2 \cdot 3} = x^6$$

The following property states this result.

> ### Power of a Power Rule for Exponents
> If m and n are positive integers and a is a real number, then
> $$(a^m)^n = a^{mn}$$

To raise a power to a power, keep the base and multiply the exponents.

EXAMPLE 5   Simplify each of the following expressions.

a. $(x^2)^5$ **b.** $(y^8)^2$ **c.** $[(-5)^3]^4$

Solution: **a.** $(x^2)^5 = x^{2 \cdot 5} = x^{10}$

b. $(y^8)^2 = y^{8 \cdot 2} = y^{16}$

c. $[(-5)^3]^4 = (-5)^{12}$. Because $(-5)^{12}$ is the product of an even number of negative numbers, their product is a positive number, so that

$$\underset{\underset{(-5)^{12}}{\downarrow}}{\overset{\text{even number}}{\rule{1.5cm}{0.4pt}}} \quad \text{can be written as} \quad 5^{12}$$

Both expressions have the same value.

When the base of an exponential expression is a product or quotient, the definition of a^n still applies. To simplify $(xy)^3$, for example,

$$(xy)^3 = \underbrace{(xy)(xy)(xy)}_{\text{3 factors of } xy}$$

$$= x \cdot x \cdot x \cdot y \cdot y \cdot y \qquad \text{Group factors with common bases.}$$

$$= x^3 y^3 \qquad \text{Simplify.}$$

Similarly, to simplify $\left(\dfrac{x}{y}\right)^3$:

$$\left(\frac{x}{y}\right)^3 = \underbrace{\left(\frac{x}{y}\right)\left(\frac{x}{y}\right)\left(\frac{x}{y}\right)}_{\text{3 factors of } \frac{x}{y}}$$

$$= \frac{x \cdot x \cdot x}{y \cdot y \cdot y} \qquad \text{Multiply fractions.}$$

$$= \frac{x^3}{y^3} \qquad \text{Simplify.}$$

Notice that the power of a product (or quotient) can be written as a product (or quotient) of powers.

$$(xy)^3 = x^3 y^3 \qquad \text{and} \qquad \left(\frac{x}{y}\right)^3 = \frac{x^3}{y^3}$$

In general, we have the following.

POWER OF A PRODUCT OR QUOTIENT RULE

If n is a positive integer and a, b, and c are real numbers, then

$$(ab)^n = a^n b^n \qquad \text{and} \qquad \left(\frac{a}{c}\right)^n = \frac{a^n}{c^n}$$

as long as c is not 0.

In other words, to raise a product to a power, raise each factor to the power. Also, to raise a quotient to a power, raise both the numerator and the denominator to the power.

EXAMPLE 6   Simplify each expression.

a. $(st)^4$ **b.** $\left(\dfrac{m}{n}\right)^7$ **c.** $(2a)^3$ **d.** $(-5x^2y^3z)^2$ **e.** $\left(\dfrac{2x^4}{3y^5}\right)^4$

Solution: **a.** $(st)^4 = s^4 \cdot t^4 = s^4t^4$ Use the power of a product rule.

b. $\left(\dfrac{m}{n}\right)^7 = \dfrac{m^7}{n^7}, n \neq 0$ Use the power of a quotient rule.

c. $(2a)^3 = 2^3 \cdot a^3 = 8a^3$ Use the power of a product rule.

d. $(-5x^2y^3z)^2 = (-5)^2 \cdot (x^2)^2 \cdot (y^3)^2 \cdot (z^1)^2$ Use the power of a product rule.

$= 25x^4y^6z^2$ Use the power rule for exponents.

e. $\left(\dfrac{2x^4}{3y^5}\right)^4 = \dfrac{2^4 \cdot (x^4)^4}{3^4 \cdot (y^5)^4}$ Use the power of a product or quotient rule.

$= \dfrac{16x^{16}}{81y^{20}}, y \neq 0$ Use the power rule for exponents.

---

5   Another pattern for simplifying exponential expressions involves quotients.

To simplify an expression like $\dfrac{x^5}{x^3}$, in which the numerator and the denominator have a common base, we can apply the fundamental principle of fractions and divide the numerator and the denominator by the common base factors. Assume for the remainder of this section that denominators are not 0.

$$\frac{x^5}{x^3} = \frac{x \cdot x \cdot x \cdot x \cdot x}{x \cdot x \cdot x}$$

$$= \frac{x \cdot x \cdot x \cdot x \cdot x}{x \cdot x \cdot x}$$

$$= x \cdot x$$

$$= x^2$$

Notice that the result is exactly the same if we subtract exponents of the common bases.

$$\frac{x^5}{x^3} = x^{5-3} = x^2$$

The quotient rule for exponents states this result in a general way.

> **QUOTIENT RULE FOR EXPONENTS**
>
> If m and n are positive integers and a is a real number, then
>
> $$\frac{a^m}{a^n} = a^{m-n}$$
>
> as long as a is not 0.

In other words, to divide one exponential expression by another with a common base, keep the base and subtract exponents.

EXAMPLE 7 Simplify each quotient.

    a. $\dfrac{x^5}{x^2}$ **b.** $\dfrac{4^7}{4^3}$ **c.** $\dfrac{(-3)^5}{(-3)^2}$ **d.** $\dfrac{2x^5y^2}{xy}$

Solution: **a.** $\dfrac{x^5}{x^2} = x^{5-2} = x^3$ Use the quotient rule.

    b. $\dfrac{4^7}{4^3} = 4^{7-3} = 4^4 = 256$ Use the quotient rule.

    c. $\dfrac{(-3)^5}{(-3)^2} = (-3)^3 = -27$

    d. Begin by grouping common bases.

$$\frac{2x^5y^2}{xy} = 2 \cdot \frac{x^5}{x^1} \cdot \frac{y^2}{y^1}$$

$$= 2 \cdot (x^{5-1}) \cdot (y^{2-1}) \qquad \text{Use the quotient rule.}$$

$$= 2x^4y^1 \qquad \text{or} \qquad 2x^4y$$

Let's look at one more case. To simplify $\dfrac{x^3}{x^3}$, we use the quotient rule and subtract exponents.

$$\frac{x^3}{x^3} = x^{3-3} = x^0$$

But our definition of a^n does not include the possibility that n might be 0, as in x^0. What is the meaning when 0 is an exponent? To find out, use another method to simplify $\dfrac{x^3}{x^3}$. Divide the numerator and denominator by common factors, by applying the fundamental principle.

$$\frac{x^3}{x^3} = \frac{x \cdot x \cdot x}{x \cdot x \cdot x} = 1$$

Since $\dfrac{x^3}{x^3} = x^0$ and $\dfrac{x^3}{x^3} = 1$, we conclude that $x^0 = 1$ as long as x is not 0.

> **ZERO EXPONENT**
>
> $a^0 = 1$, as long as a is not 0.

EXAMPLE 8   Simplify the following expressions.

  a. 3^0 **b.** $(ab)^0$ **c.** $(-5)^0$ **d.** -5^0

Solution: **a.** $3^0 = 1$

  b. Assume that neither a nor b is zero.

   $(ab)^0 = a^0 \cdot b^0 = 1 \cdot 1 = 1$

  c. $(-5)^0 = 1$ **d.** $-5^0 = -1 \cdot 5^0 = -1 \cdot 1 = -1$

> R E M I N D E R   These examples will remind you of the difference between adding and multiplying terms.
>
> | *Addition* | *Multiplication* |
> |---|---|
> | $5x^3 + 3x^3 = (5 + 3)x^3 = 8x^3$ | $(5x^3)(3x^3) = 5 \cdot 3 \cdot x^3 \cdot x^3 = 15x^{3+3} = 15x^6$ |
> | $7x + 4x^2 = 7x + 4x^2$ | $(7x)(4x^2) = 7 \cdot 4 \cdot x \cdot x^2 = 28x^{1+2} = 28x^3$ |

In the next example, exponential expressions are simplified using two or more of the exponent rules presented in this section.

EXAMPLE 9   Simplify the following.

  a. $\left(\dfrac{-5x^2}{y^3}\right)^2$ **b.** $\dfrac{(x^3)^4 x}{x^7}$ **c.** $\dfrac{(2x)^5}{x^3}$ **d.** $\dfrac{(a^2 b)^3}{a^3 b^2}$

Solution: **a.** Use the power of a product or quotient rule; then use the power of a power rule for exponents.

$$\left(\frac{-5x^2}{y^3}\right)^2 = \frac{(-5)^2 (x^2)^2}{(y^3)^2} = \frac{25x^4}{y^6}$$

  b. $\dfrac{(x^3)^4 x}{x^7} = \dfrac{x^{12} \cdot x}{x^7} = \dfrac{x^{12+1}}{x^7} = \dfrac{x^{13}}{x^7} = x^{13-7} = x^6$

  c. Use the power of a product or quotient rule; then use the quotient rule.

$$\frac{(2x)^5}{x^3} = \frac{2^5 \cdot x^5}{x^3} = 2^5 \cdot x^{5-3} = 32x^2$$

d. Begin by applying the power of a product rule to the numerator.

$$\frac{(a^2b)^3}{a^3b^2} = \frac{(a^2)^3 \cdot b^3}{a^3 \cdot b^2}$$

$$= \frac{a^6b^3}{a^3b^2} \qquad \text{Use the power of a power rule for exponents.}$$

$$= a^{6-3}b^{3-2} \qquad \text{Use the quotient rule.}$$

$$= a^3b^1 \quad \text{or} \quad a^3b$$

MENTAL MATH

State the bases and the exponents for each of the following expressions.

1. 3^2 base: 3; exponent: 2

2. 5^4 base: 5; exponent: 4

3. $(-3)^6$ base: -3; exponent: 6

4. -3^7 base: 3; exponent: 7

5. -4^2 base: 4; exponent: 2

6. $(-4)^3$ base: -4; exponent: 3

7. $5 \cdot 3^4$

8. $9 \cdot 7^6$

9. $5x^2$

10. $(5x)^2$ base: $5x$; exponent: 2

7. base: 5; exponent: 1; base: 3; exponent: 4
8. base: 9; exponent: 1; base: 7; exponent: 6
9. base: 5; exponent: 1; base: x; exponent: 2

EXERCISE SET 4.1

Evaluate each expression. See Example 1.

1. 7^2 49

2. -3^2 -9

3. $(-5)^1$ -5

4. $(-3)^2$ 9

5. -2^4 -16

6. -4^3 -64

7. $(-2)^4$ 16

8. $(-4)^3$ -64

9. $\left(\frac{1}{3}\right)^3$ $\frac{1}{27}$

10. $\left(-\frac{1}{9}\right)^2$ $\frac{1}{81}$

11. $7 \cdot 2^4$ 112

12. $9 \cdot 1^2$ 9

13. Explain why $(-5)^4 = 625$, while $-5^4 = -625$.

14. Explain why $5 \cdot 4^2 = 80$, while $(5 \cdot 4)^2 = 400$.

Evaluate each expression given the replacement values for x. See Example 2.

15. $x^2; x = -2$ 4

16. $x^3; x = -2$ -8

17. $5x^3; x = 3$ 135

18. $4x^2; x = -1$ 4

19. $2xy^2; x = 3$ and $y = 5$ 150

20. $-4x^2y^3; x = 2$ and $y = -1$ 16

21. $\frac{2z^4}{5}; z = -2$ $\frac{32}{5}$

22. $\frac{10}{3y^3}; y = 5$ $\frac{2}{75}$

23. The formula $V = x^3$ can be used to find the volume V of a cube with side length x. Find the volume of a cube with side length 7 meters. (Volume is measured in cubic units.)   343 cubic meters

x

24. The formula $S = 6x^2$ can be used to find the surface area S of a cube with side length x. Find the surface area of the cube with side length 5 meters. (Surface area is measured in square units.)

25. To find the amount of water that a swimming pool in the shape of a cube can hold, do we use the formula for volume of the cube or surface area of the cube? (See Exercises 23 and 24.)   volume

26. To find the amount of material needed to cover an ottoman in the shape of a cube, do we use the formula for volume of the cube or surface area of the cube? (See Exercises 23 and 24.)

13. answers may vary   **14.** answers may vary   **24.** 150 square meters   **26.** surface area

34. $-12x^{15}$ **36.** $18y^{17}$ sq. meters   **46.** $25y^2\pi$ sq. cm

Use the product rule to simplify each expression. Write the results using exponents. See Examples 3 and 4.

27. $x^2 \cdot x^5$ x^7 **28.** $y^2 \cdot y$ y^3

29. $(-3)^3 \cdot (-3)^9$ $(-3)^{12}$ **30.** $(-5)^7 \cdot (-5)^6$ $(-5)^{13}$

31. $(5y^4)(3y)$ $15y^5$ **32.** $(-2z^3)(-2z^2)$ $4z^5$

33. $(4z^{10})(-6z^7)(z^3)$ $-24z^{20}$**34.** $(12x^5)(-x^6)(x^4)$

❏ 35. The following rectangle has width $4x^2$ feet and length $5x^3$ feet. Find its area. $20x^5$ sq. ft.

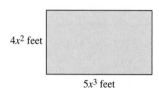

4x² feet

5x³ feet

❏ 36. The following parallelogram has base length $9y^7$ meters and height $2y^{10}$ meters. Find its area.

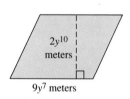

2y¹⁰ meters

9y⁷ meters

Use the power of a power rule and the power of a product or quotient rule to simplify each expression. See Examples 5 and 6.

37. $(pq)^7$ p^7q^7 **38.** $(4s)^3$ $64s^3$

39. $\left(\dfrac{m}{n}\right)^9$ $\dfrac{m^9}{n^9}$ **40.** $\left(\dfrac{xy}{7}\right)^2$ $\dfrac{x^2y^2}{49}$

41. $(x^2y^3)^5$ $x^{10}y^{15}$ **42.** $(a^4b)^7$ $a^{28}b^7$

❏ 43. $\left(\dfrac{-2xz}{y^5}\right)^2$ $\dfrac{4x^2z^2}{y^{10}}$ **44.** $\left(\dfrac{y^4}{-3z^3}\right)^3$ $\dfrac{y^{12}}{-27z^9}$

45. The square shown has sides of length $8z^5$ decimeters. Find its area. $64z^{10}$ sq. decimeters

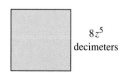

8z⁵
decimeters

❏ 46. Given the following circle with radius $5y$ centimeters, find its area. Do not approximate π.

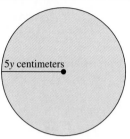

5y centimeters

❏ 47. The following vault is in the shape of a cube. If each side is $3y^4$ feet, find its volume.

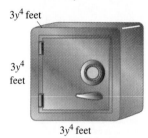

3y⁴ feet

3y⁴
feet

3y⁴ feet

❏ 48. The silo shown is in the shape of a cylinder. If its radius is $4x$ meters and its height is $5x^3$ meters, find its volume. Do not approximate π.

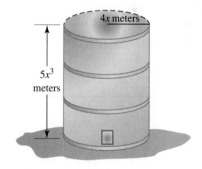

4x meters

5x³
meters

Use the quotient rule and simplify each expression. See Example 7.

49. $\dfrac{x^3}{x}$ x^2 **50.** $\dfrac{y^{10}}{y^9}$ y **51.** $\dfrac{(-2)^5}{(-2)^3}$ 4

52. $\dfrac{(-5)^{14}}{(-5)^{11}}$ -125 **53.** $\dfrac{p^7q^{20}}{pq^{15}}$ p^6q^5 **54.** $\dfrac{x^8y^6}{y^5}$ x^8y

55. $\dfrac{7x^2y^6}{14x^2y^3}$ $\dfrac{y^3}{2}$ **56.** $\dfrac{9a^4b^7}{3ab^2}$ $3a^3b^5$

47. $27y^{12}$ cubic ft.   **48.** $80x^5\pi$ cubic meters

Simplify the following. See Example 8.

57. $(2x)^0$ 1
58. $-4x^0$ -4
59. $-2x^0$ -2
60. $(4y)^0$ 1
61. $5^0 + y^0$ 2
62. $-3^0 + 4^0$ 0

63. In your own words, explain why $5^0 = 1$.

64. In your own words, explain when $(-3)^n$ is positive and when it is negative.   answers may vary

Simplify the following. See Example 9.

65. $\left(\dfrac{-3a^2}{b^3}\right)^3$ $\dfrac{-27a^6}{b^9}$
66. $\left(\dfrac{q^7}{-2p^5}\right)^5$ $\dfrac{q^{35}}{-32p^{25}}$

67. $\dfrac{(x^5)^7 \cdot x^8}{x^4}$ x^{39}
68. $\dfrac{y^{20}}{(y^2)^3 \cdot y^9}$ y^5

69. $\dfrac{(z^3)^6}{(5z)^4}$ $\dfrac{z^{14}}{625}$
70. $\dfrac{(3x)^4}{(x^2)^2}$ 81

71. $\dfrac{(6mn)^5}{mn^2}$ $7776m^4n^3$
72. $\dfrac{(6xy)^2}{9x^2y^2}$ 4

Simplify the following.

73. -5^2 -25
74. $(-5)^2$ 25

75. $\left(\dfrac{1}{4}\right)^3$ $\dfrac{1}{64}$
76. $\left(\dfrac{2}{3}\right)^3$ $\dfrac{8}{27}$

77. $(9xy)^2$ $81x^2y^2$
78. $(2ab)^5$ $32a^5b^5$

79. $(6b)^0$ 1
80. $(5ab)^0$ 1

81. $2^3 + 2^5$ 40
82. $7^2 - 7^0$ 48

83. b^4b^2 b^6
84. y^4y^1 y^5

85. $a^2a^3a^4$ a^9
86. $x^2x^{15}x^9$ x^{26}

87. $(2x^3)(-8x^4)$ $-16x^7$
88. $(3y^4)(-5y)$ $-15y^5$

89. $(4a)^3$ $64a^3$
90. $(2ab)^4$ $16a^4b^4$

91. $(-6xyz^3)^2$ $36x^2y^2z^6$
92. $(-3xy^2a^3b)^3$

93. $\left(\dfrac{3y^5}{6x^4}\right)^3$ $\dfrac{y^{15}}{8x^{12}}$
94. $\left(\dfrac{2ab}{6yz}\right)^4$ $\dfrac{a^4b^4}{81y^4z^4}$

95. $\dfrac{x^5}{x^4}$ x
96. $\dfrac{5x^9}{x^3}$ $5x^6$

97. $\dfrac{2x^3y^2z}{xyz}$ $2x^2y$
98. $\dfrac{x^{12}y^{13}}{x^5y^7}$ x^7y^6

99. $\dfrac{(3x^2y^5)^5}{x^3y}$ $243x^7y^{24}$
100. $\dfrac{(4a^2)^4}{a^4b}$ $\dfrac{256a^4}{b}$

Review Exercises

Simplify each expression by combining any like terms. Use the distributive property to remove any parenthesis. See Section 2.1.

101. $3x - 5x + 7$ $-2x + 7$
102. $7w + w - 2w$ $6w$

103. $y - 10 + y$ $2y - 10$
104. $-6z + 20 - 3z$

105. $7x + 2 - 8x - 6$
106. $10y - 14 - y - 14$

107. $2(x - 5) + 3(5 - x)$ $-x + 5$

108. $-3(w + 7) + 5(w + 1)$ $2w - 16$

A Look Ahead

EXAMPLE

Simplify $x^a \cdot x^{3a}$.

Solution:

Like bases, so add exponents.

$$x^a \cdot x^{3a} = x^{a+3a} = x^{4a}$$

Simplify each expression. Assume that variables represent positive integers. See the example.

109. $x^{5a}x^{4a}$ x^{9a}
110. $b^{9a}b^{4a}$ b^{13a}

111. $(a^b)^5$ a^{5b}
112. $(2a^{4b})^4$ $16a^{16b}$

113. $\dfrac{x^{9a}}{x^{4a}}$ x^{5a}
114. $\dfrac{y^{15b}}{y^{6b}}$ y^{9b}

115. $(x^ay^bz^c)^{5a}$ $x^{5a^2}y^{5ab}z^{5ac}$
116. $(9a^2b^3c^4d^5)^{ab}$ $9^{ab}a^{2ab}b^{3ab}c^{4ab}d^{5ab}$

63. answers may vary   **92.** $-27x^3y^6a^9b^3$ **104.** $-9z + 20$ **105.** $-x - 4$ **106.** $9y - 28$

4.2   ADDING AND SUBTRACTING POLYNOMIALS

TAPE BA 4.2

O B J E C T I V E S

1 Define monomial, binomial, trinomial, polynomial, and degree.

2 Find the value of a polynomial given replacement values for the variables.

3 Combine like terms.

4 Add and subtract polynomials.

In this section, we introduce a special algebraic expression called a polynomial. Let's first review some definitions presented in Section 2.1.

Recall that a term is a number or the product of a number and variables raised to powers. The terms of the expression $4x^2 + 3x$ are $4x^2$ and $3x$. The terms of the expression $9x^4 - 7x - 1$ are $9x^4$, $-7x$, and -1.

EXPRESSION	TERMS
$4x^2 + 3x$	$4x^2, 3x$
$9x^4 - 7x - 1$	$9x^4, -7x, -1$
$7y^3$	$7y^3$

The **numerical coefficient** of a term, or simply the **coefficient,** is the numerical factor of each term. If no numerical factor appears in the term, then the coefficient is understood to be 1. If the term is a number only, it is called a **constant** term or simply a **constant**.

TERM	COEFFICIENT
x^5	1
$3x^2$	3
$-4x$	-4
$-x^2y$	-1
3 (constant)	3

A **polynomial in x** is a finite sum of terms of the form ax^n, where a is a real number and n is a whole number. For example,

$$x^5 - 3x^3 + 2x^2 - 5x + 1$$

is a polynomial. Notice that this polynomial is written in **descending powers** of x because the powers of x decrease from left to right. (Recall that the term 1 can be thought of as $1x^0$.)

On the other hand,

$$x^{-5} + 2x - 3$$

is **not** a polynomial because it contains an exponent, -5, that is not a whole number. (We study negative exponents in Section 5 of this chapter.)

A **monomial** is a polynomial with exactly one term.

A **binomial** is a polynomial with exactly two terms.

A **trinomial** is a polynomial with exactly three terms.

The following are examples of monomials, binomials, and trinomials. Each of these examples is also a polynomial.

MONOMIALS	BINOMIALS	TRINOMIALS
ax^2	$x + y$	$x^2 + 4xy + y^2$
$-3z$	$3p + 2$	$x^5 + 7x^2 - x$
4	$4x^2 - 7$	$-q^4 + q^3r - 2q$

Each term of a polynomial has a **degree.**

DEGREE OF A TERM

The degree of a term is the sum of the exponents on the variables contained in the term.

EXAMPLE 1 Find the degree of each term.

    a. $-3x^2$ **b.** $5x^3yz$ **c.** 2

Solution: **a.** The exponent on x is 2, so the degree of the term is 2.

    b. $5x^3yz$ can be written as $5x^3y^1z^1$. The degree of the term is the sum of its exponents, so the degree is $3 + 1 + 1$ or 5.

    c. The constant, 2, can be written as $2x^0$ (since $x^0 = 1$). The degree of 2 or $2x^0$ is 0.

From the preceding, we can say that **the degree of a constant is 0.**

    Each polynomial also has a degree.

DEGREE OF A POLYNOMIAL

The degree of a polynomial is the greatest degree of any term of the polynomial.

EXAMPLE 2 Find the degree of each polynomial and tell whether the polynomial is a monomial, binomial, trinomial, or none of these.

    a. $-2t^2 + 3t + 6$ **b.** $15x - 10$ **c.** $7x + 3x^3 + 2x^2 - 1$ **d.** $7x^2y - 6xy$

Solution: **a.** The degree of the trinomial $-2t^2 + 3t + 6$ is 2, the greatest degree of any of its terms.

    b. The degree of the binomial $15x - 10$ is 1.

    c. The degree of the polynomial $7x + 3x^3 + 2x^2 - 1$ is 3.

    d. The degree of the binomial $7x^2y - 6xy$ is 3.

2   Polynomials have different values depending on replacement values for the variables.

EXAMPLE 3 Find the value of the polynomial $3x^2 - 2x + 1$ when $x = -2$.

Solution: Replace x with -2 and simplify.

$$3x^2 - 2x + 1 = 3(-2)^2 - 2(-2) + 1$$
$$= 3(4) + 4 + 1$$
$$= 12 + 4 + 1$$
$$= 17$$

Many physical phenomena can be modeled by polynomials.

EXAMPLE 4   The CN Tower in Toronto, Ontario, is 1821 feet tall and is the world's tallest self-supporting structure. An object is dropped from the top of this building. Neglecting air resistance, the height of the object at time t seconds is given by the polynomial $-16t^2 + 1821$. Find the height of the object when $t = 1$ second and when $t = 10$ seconds.

Solution: To find each height, we evaluate the polynomial when $t = 1$ and when $t = 10$.

$$-16t^2 + 1821 = -16(1)^2 + 1821 \qquad \text{Replace } t \text{ with 1.}$$
$$= -16(1) + 1821$$
$$= -16 + 1821$$
$$= 1805$$

The height of the object at 1 second is 1805 feet.

$$-16t^2 + 1821 = -16(10)^2 + 1821 \qquad \text{Replace } t \text{ with 10.}$$
$$= -16(100) + 1821$$
$$= -1600 + 1821$$
$$= 221$$

The height of the object at 10 seconds is 221 feet.

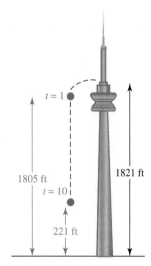

3   Polynomials with like terms can be simplified by combining like terms. Recall that like terms are terms that contain exactly the same variables raised to exactly the same powers.

LIKE TERMS

$5x^2, -7x^2$

$y, 2y$

$\dfrac{1}{2}a^2b, -a^2b$

Only like terms can be combined. We combine like terms by applying the distributive property.

EXAMPLE 5   Combine like terms.

        a. $-3x + 7x$ **b.** $11x^2 + 5 + 2x^2 - 7$

Solution: **a.** $-3x + 7x = (-3 + 7)x = 4x$

          b. $11x^2 + 5 + 2x^2 - 7 = 11x^2 + 2x^2 + 5 - 7$

$$= 13x^2 - 2 \qquad \text{Combine like terms.}$$

4   We now practice adding and subtracting polynomials.

TO ADD POLYNOMIALS

To add polynomials, combine all like terms.

EXAMPLE 6   Add $(-2x^2 + 5x - 1)$ and $(-2x^2 + x + 3)$.

Solution: $(-2x^2 + 5x - 1) + (-2x^2 + x + 3) = -2x^2 + 5x - 1 - 2x^2 + x + 3$

$$= (-2x^2 - 2x^2) + (5x + 1x) + (-1 + 3)$$

$$= -4x^2 + 6x + 2$$

EXAMPLE 7   Add: $(4x^3 - 6x^2 + 2x + 7) + (5x^2 - 2x)$.

Solution: $(4x^3 - 6x^2 + 2x + 7) + (5x^2 - 2x) = 4x^3 - 6x^2 + 2x + 7 + 5x^2 - 2x$

$$= 4x^3 + (-6x^2 + 5x^2) + (2x - 2x) + 7$$

$$= 4x^3 - x^2 + 7$$

Polynomials can be added vertically if we line up like terms underneath one another.

EXAMPLE 8 Add $(7y^3 - 2y^2 + 7)$ and $(6y^2 + 1)$ using the vertical format.

Solution: Vertically line up like terms and add.

$$\begin{array}{r} 7y^3 - 2y^2 + 7 \\ 6y^2 + 1 \\ \hline 7y^3 + 4y^2 + 8 \end{array}$$

To subtract one polynomial from another, recall the definition of subtraction. To subtract a number, we add its opposite: $a - b = a + (-b)$. To subtract a polynomial, we also add its opposite. Just as $-b$ is the opposite of b, $-(x^2 + 5)$ is the opposite of $(x^2 + 5)$.

EXAMPLE 9 Subtract: $(5x - 3) - (2x - 11)$.

Solution: From the definition of subtraction, we have

$$\begin{aligned} (5x - 3) - (2x - 11) &= (5x - 3) + [-(2x - 11)] &&\text{Add the opposite.} \\ &= (5x - 3) + (-2x + 11) &&\text{Apply the distributive} \\ &&&\text{property.} \\ &= 3x + 8 \end{aligned}$$

TO SUBTRACT POLYNOMIALS

To subtract two polynomials, change the signs of the terms of the polynomial being subtracted and then add.

EXAMPLE 10 Subtract: $(2x^3 + 8x^2 - 6x) - (2x^3 - x^2 + 1)$.

Solution: First, change the sign of each term of the second polynomial and then add.

$$\begin{aligned} (2x^3 + 8x^2 - 6x) - (2x^3 - x^2 + 1) &= (2x^3 + 8x^2 - 6x) + (-2x^3 + x^2 - 1) \\ &= 2x^3 - 2x^3 + 8x^2 + x^2 - 6x - 1 \\ &= 9x^2 - 6x - 1 &&\text{Combine like terms.} \end{aligned}$$

EXAMPLE 11 Subtract $(5y^2 + 2y - 6)$ from $(-3y^2 - 2y + 11)$ using the vertical format.

Solution: Arrange the polynomials in vertical format, lining up like terms.

$$\begin{array}{r} -3y^2 - 2y + 11 \\ -(5y^2 + 2y - 6) \\ \hline \end{array} \qquad \begin{array}{r} -3y^2 - 2y + 11 \\ -5y^2 - 2y + 6 \\ \hline -8y^2 - 4y + 17 \end{array}$$

EXAMPLE 12 Subtract $(5z - 7)$ from the sum of $(8z + 11)$ and $(9z - 2)$.

Solution: Notice that $(5z - 7)$ is to be subtracted **from** a sum. The translation is

$$[(8z + 11) + (9z - 2)] - (5z - 7)$$
$$= 8z + 11 + 9z - 2 - 5z + 7 \qquad \text{Remove grouping symbols.}$$
$$= 8z + 9z - 5z + 11 - 2 + 7 \qquad \text{Group like terms.}$$
$$= 12z + 16 \qquad \text{Combine like terms.}$$

MENTAL MATH

Combine like terms.

1. $-9y - 5y$ $-14y$

2. $6m^5 + 7m^5$ $13m^5$

3. $4y^3 + 3y^3$ $7y^3$

4. $21y^5 - 19y^5$ $2y^5$

5. $x + 6x$ $7x$

6. $7z - z$ $6z$

---

2. 2; trinomial **3.** 3; none of these **4.** 4; trinomial **5.** 6; trinomial **6.** 5; trinomial **9.** answers may vary
10. answers may vary **12.** answers may vary **14. (a)** -10; **(b)** -12 **15. (a)** -2; **(b)** 4 **17. (a)** -15; **(b)** -16

EXERCISE SET 4.2

Find the degree of each of the following polynomials and determine whether it is a monomial, binomial, trinomial, or none of these. See Examples 1 and 2.

1. $x + 2$ 1; binomial

2. $-6y + y^2 + 4$

3. $9m^3 - 5m^2 + 4m - 8$

4. $5a^2 + 3a^3 - 4a^4$

5. $12x^4y - x^2y^2 - 12x^2y^4$

6. $7r^2s^2 + 2r - 3s^5$

7. $3zx - 5x^2$ 2; binomial

8. $5y + 2$ 1; binomial

9. Describe how to find the degree of a term.

10. Describe how to find the degree of a polynomial.

11. Explain why xyz is a monomial while $x + y + z$ is a trinomial. answers may vary

12. Explain why the degree of the term $5y^3$ is 3 and the degree of the polynomial $2y + y + 2y$ is 1.

*Find the value of each polynomial when **(a)** $x = 0$ and **(b)** $x = -1$. See Examples 3 and 4.*

13. $x + 6$ **(a)** 6; **(b)** 5

14. $2x - 10$

15. $x^2 - 5x - 2$

16. $x^2 - 4$ **(a)** -4; **(b)** -3

17. x^3 -15

18. $-2x^3 + 3x^2$ -6

19. Find the height of the object in Example 4 when t is 10.8 seconds. Explain your result.

20. Approximate how long (to the nearest tenth of a second) before the object in Example 4 hits the ground. 10.7 sec.

Simplify each of the following by combining like terms. See Example 5.

21. $14x^2 + 9x^2$ $23x^2$

22. $18x^3 - 4x^3$ $14x^3$

23. $15x^2 - 3x^2$ $-y$

24. $12k^3 - 9k^3 + 11$

25. $8s - 5s + 4s$ $7s$

26. $5y + 7y - 6y$ $6y$

27. $0.1y^2 - 1.2y^2 + 6.7 - 1.9$ $-1.1y^2 + 4.8$

28. $7.6y + 3.2y^2 - 8y - 2.5y^2$ $0.7y^2 - 0.4y$

Perform the indicated operations. See Examples 6, 7, 9, and 10.

29. $(3x + 7) + (9x + 5)$

30. $(3x^2 + 7) + (3x^2 + 9)$

31. $(-7x + 5) + (-3x^2 + 7x + 5)$ $-3x^2 + 10$

32. $(3x - 8) + (4x^2 - 3x + 3)$ $4x^2 - 5$

33. $(2x + 5) - (3x - 9)$ $-x + 14$

34. $(5x^2 + 4) - (-2y^2 + 4)$ $5x^2 + 2y^2$

35. $3x - (5x - 9)$ $-2x + 9$

36. $4 - (-y - 4)$ $y + 8$

---

18. (a) -6; **(b)** -1 **19.** -45.24 ft; the object has reached the ground **23.** $12x^2 - y$ **24.** $3k^3 + 11$
29. $12x + 12$ **30.** $6x^2 + 16$

41. $(x^2 + 7x + 4)$ ft.   **43.** $(3y^2 + 4y + 11)$ meters

37. $(-5x^2 + 3) + (2x^2 + 1)$ $-3x^2 + 4$

38. $(-y - 2) + (3y + 5)$ $2y + 3$

39. $(2x^2 + 3x - 9) - (-4x + 7)$ $2x^2 + 7x - 16$

40. $(-7x^2 + 4x + 7) - (-8x + 2)$ $-7x^2 + 12x + 5$

41. Given the following triangle, find its perimeter.

42. Given the following quadrilateral, find its perimeter. $(2x^2 - 2x + 2)$ cm

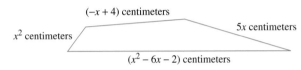

43. A wooden beam is $(4y^2 + 4y + 1)$ meters long. If a
piece $(y^2 - 10)$ meters is cut, express the length of
the remaining piece of beam as a polynomial in y.

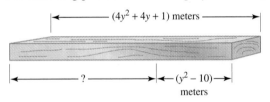

44. A piece of quarter-round molding is $(13x - 7)$
inches long. If a piece $(2x + 2)$ inches is removed,
express the length of the remaining piece of molding as a polynomial in x. $(11x - 9)$ in.

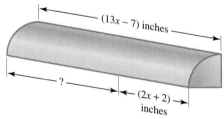

Perform the indicated operations. See Examples 8 and 11.

45. $\begin{array}{r} 3t^2 + 4 \\ + 5t^2 - 8 \\ \hline 8t^2 - 4 \end{array}$

46. $\begin{array}{r} 7x^3 + 3 \\ + 2x^3 + 1 \\ \hline 9x^3 + 4 \end{array}$

47. $\begin{array}{r} 4z^2 - 8z + 3 \\ - (6z^2 + 8z - 3) \\ \hline -2z^2 - 16z + 6 \end{array}$

48. $\begin{array}{r} 5u^5 - 4u^2 + 3u - 7 \\ - (3u^5 + 6u^2 - 8u + 2) \\ \hline 2u^5 - 10u^2 + 11u - 9 \end{array}$

49. $\begin{array}{r} 5x^3 - 4x^2 + 6x - 2 \\ - (3x^3 - 2x^2 - x - 4) \\ \hline 2x^3 - 2x^2 + 7x + 2 \end{array}$

50. $\begin{array}{r} 7a^2 - 9a + 6 \\ - (11a^2 - 4a + 2) \\ \hline -4a^2 - 5a + 4 \end{array}$

51. $\begin{array}{r} 10a^3 - 8a^2 + 9 \\ + 5a^3 + 9a^2 + 7 \\ \hline 15a^3 + a^2 + 16 \end{array}$

52. $\begin{array}{r} 2x^3 - 3x^2 + x - 4 \\ + 5x^3 + 2x^2 - 3x + 2 \\ \hline 7x^3 - x^2 - 2x - 2 \end{array}$

Perform the indicated operations. See Example 12.

53. Subtract $(19x^2 + 5)$ from $(81x^2 + 10)$. $62x^2 + 5$

54. Subtract $(2x + xy)$ from $(3x - 9xy)$. $x - 10xy$

55. Subtract $(2x + 2)$ from the sum of $(8x + 1)$ and
$(6x + 3)$. $12x + 2$

56. Subtract $(-12x - 3)$ from the sum of $(-5x - 7)$
and $(12x + 3)$. $19x - 1$

57. Subtract $(8x + 9)$ from $(9xy^2 + 7x - 18)$.

58. Subtract $(4x^2 + 7)$ from $(9x^3 + 9x^2 - 9)$.

Perform the indicated operations.

59. $-15x - (-4x)$ $-11x$ **60.** $16y - (-4y)$ $20y$

61. $2x - 5 + 5x - 8$ **62.** $x - 3 + 8x + 10$

63. $(-3y^2 - 4y) + (2y^2 + y - 1)$ $-y^2 - 3y - 1$

64. $(7x^2 + 2x - 9) + (-3^2 + 5)$ $7x^2 + 2x - 13$

65. $(-7y^2 + 5) - (-8y^2 + 12)$ $y^2 - 7$

66. $(4 + 5a) - (-a - 5)$ $6a + 9$

67. $(5x + 8) - (-2x^2 - 6x + 8)$ $2x^2 + 11x$

68. $(-6y^2 + 3y - 4) - (9y^2 - 3y)$ $-15y^2 + 6y - 4$

69. $(-8x^4 + 7x) + (-8x^4 + x + 9)$ $-16x^4 + 8x + 9$

70. $(6y^5 - 6y^3 + 4) + (-2y^5 - 8y^3 - 7)$

71. $(3x^2 + 5x - 8) + (5x^2 + 9x + 12) - (x^2 - 14)$

72. $(-a^2 + 1) - (a^2 - 3) + (5a^2 - 6a + 7)$

73. Subtract $4x$ from $7x - 3$. $3x - 3$

74. Subtract y from $y^2 - 4y + 1$. $y^2 - 5y + 1$

75. Subtract $(5x + 7)$ from $(7x^2 + 3x + 9)$.

76. Subtract $(5y^2 + 8y + 2)$ from $(7y^2 + 9y - 8)$.

77. Subtract $(4y^2 - 6y - 3)$ from the sum of $(8y^2 + 7)$
and $(6y + 9)$. $4y^2 + 12y + 19$

57. $9xy^2 - x - 27$ **58.** $9x^3 + 5x^2 - 16$ **61.** $7x - 13$ **62.** $9x + 7$ **70.** $4y^5 - 14y^3 - 3$
71. $7x^2 + 14x + 18$ **72.** $3a^2 - 6a + 11$ **75.** $7x^2 - 2x + 2$ **76.** $2y^2 + y - 10$

78. Subtract $(5y + 7x^2)$ from the sum of $(8y - x)$ and $(3 + 8x^2)$. $x^2 + 3y - x + 3$

79. Subtract $(-2x^2 + 4x - 12)$ from the sum of $(-x^2 - 2x)$ and $(5x^2 + x + 9)$. $6x^2 - 5x + 21$

80. Subtract $(4x^2 - 2x + 2)$ from the sum of $(x^2 + 7x + 1)$ and $(7x + 5)$. $-3x^2 + 16x + 4$

81. $[(1.2x^2 - 3x + 9.1) - (7.8x^2 - 3.1 + 8)] + (1.2x - 6)$ $-6.6x^2 - 1.8x - 1.8$

82. $[(7.9y^4 - 6.8y^3 + 3.3y) + (6.1y^3 - 5)] - (4.2y^4 + 1.1y - 1)$ $3.7y^4 - 0.7y^3 + 2.2y - 4$

 83. A rocket is fired upward from the ground with an initial velocity of 200 feet per second. Neglecting air resistance, the height of the rocket at any time t can be described by the polynomial $-16t^2 + 200t$. Find the height of the rocket at the given times.

    a. $t = 1$ second   184 ft.   **b.** $t = 5$ seconds   600 ft.

    c. $t = 7.6$ seconds   **d.** $t = 10.3$ seconds.

83. c. 595.84 ft.   **d.** 362.56 ft.   **84.** answers may vary   **86.** $(4xy + 2x^2)$ sq. units

84. Explain why the height of the rocket in Exercise 83 increases and then decreases as time passes.

85. Approximate (to the nearest tenth of a second) how long before the rocket in Exercise 83 hits the ground.   12.5 sec.

86. Write a polynomial that describes the surface area of the given figure in terms of its sides x and y. (Recall that the surface area of a solid is the sum of the areas of the faces or sides of the solid.)

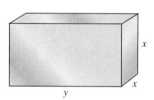

Review Exercises

Multiply. See Section 4.1.

87. $3x(2x)$ $6x^2$ **88.** $-7x(x)$ $-7x^2$

89. $(12x^3)(-x^5)$ $-12x^8$ **90.** $6r^3(7r^{10})$ $42r^{13}$

91. $10x^2(20xy^2)$ $200x^3y^2$ **92.** $-z^2y(11zy)$ $-11y^2z^3$
graphical answers in App. F

Graph each linear equation. See Sections 3.2 and 3.3

93. $2x - y = 6$ **94.** $3x - y = -6$

95. $x = -4$ **96.** $y = 2$

4.3 | Multiplying Polynomials

TAPE BA 4.3

O B J E C T I V E S

  1 Use the distributive property to multiply polynomials.

  2 Multiply polynomials vertically.

1 To multiply polynomials, we apply our knowledge of the rules and definitions of exponents.

    To multiply two monomials such as $(-5x^3)$ and $(-2x^4)$, use the associative and commutative properties and regroup. Remember that to multiply exponential expressions with a common base we add exponents.

$$(-5x^3)(-2x^4) = (-5)(-2)(x^3)(x^4) = 10x^7$$

    To multiply polynomials that are not monomials, use the distributive property.

EXAMPLE 1   Use the distributive property to find each product.

a. $5x(2x^3 + 6)$ **b.** $-3x^2(5x^2 + 6x - 1)$ **c.** $(3n^2 - 5n + 4)(2n)$

Solution: **a.** $5x(2x^3 + 6) = 5x(2x^3) + 5x(6)$ Use the distributive property.

$= 10x^4 + 30x$ Multiply.

b. $-3x^2(5x^2 + 6x - 1)$

$= (-3x^2)(5x^2) + (-3x^2)(6x) + (-3x^2)(-1)$ Use the distributive property.

$= -15x^4 - 18x^3 + 3x^2$ Multiply.

c. $(3n^2 - 5n + 4)(2n)$

$= (3n^2)(2n) + (-5n)(2n) + 4(2n)$ Use the distributive property.

$= 6n^3 - 10n^2 + 8n$ Multiply.

We also use the distributive property to multiply two binomials. To multiply $(x + 3)$ by $(x + 1)$, distribute the factor $(x + 1)$ first.

$(x + 3)(x + 1) = x(x + 1) + 3(x + 1)$ Distribute $(x + 1)$.

$= x(x) + x(1) + 3(x) + 3(1)$ Apply distributive property a second time.

$= x^2 + x + 3x + 3$ Multiply.

$= x^2 + 4x + 3$ Combine like terms.

This idea can be expanded so that we can multiply any two polynomials.

TO MULTIPLY TWO POLYNOMIALS

Multiply each term of the first polynomial by each term of the second polynomial, and then combine like terms.

EXAMPLE 2   Find the product: $(3x + 2)(2x - 5)$.

Solution: Multiply each term of the first binomial by each term of the second.

$(3x + 2)(2x - 5) = 3x(2x) + 3x(-5) + 2(2x) + 2(-5)$

$= 6x^2 - 15x + 4x - 10$ Multiply.

$= 6x^2 - 11x - 10$ Combine like terms.

EXAMPLE 3 Multiply: $(2x - y)^2$.

Solution: Recall that $a^2 = a \cdot a$, so $(2x - y)^2 = (2x - y)(2x - y)$. Multiply each term of the first polynomial by each term of the second.

$$(2x - y)(2x - y) = 2x(2x) + 2x(-y) + (-y)(2x) + (-y)(-y)$$
$$= 4x^2 - 2xy - 2xy + y^2 \qquad \text{Multiply.}$$
$$= 4x^2 - 4xy + y^2 \qquad \text{Combine like terms.}$$

EXAMPLE 4 Multiply: $(t + 2)$ by $(3t^2 - 4t + 2)$.

Solution: Multiply each term of the first polynomial by each term of the second.

$$(t + 2)(3t^2 - 4t + 2) = t(3t^2) + t(-4t) + t(2) + 2(3t^2) + 2(-4t) + 2(2)$$
$$= 3t^3 - 4t^2 + 2t + 6t^2 - 8t + 4$$
$$= 3t^3 + 2t^2 - 6t + 4 \qquad \text{Combine like terms.}$$

EXAMPLE 5 Multiply: $(3a + b)^3$.

Solution: Write $(3a + b)^3$ as $(3a + b)(3a + b)(3a + b)$.

$$(3a + b)(3a + b)(3a + b) = (9a^2 + 3ab + 3ab + b^2)(3a + b)$$
$$= (9a^2 + 6ab + b^2)(3a + b)$$
$$= (9a^2 + 6ab + b^2)3a + (9a^2 + 6ab + b^2)b$$
$$= 27a^3 + 18a^2b + 3ab^2 + 9a^2b + 6ab^2 + b^3$$
$$= 27a^3 + 27a^2b + 9ab^2 + b^3$$

2 Another convenient method for multiplying polynomials is to use a vertical format similar to the format used to multiply real numbers. We demonstrate this method by multiplying $(3y^2 - 4y + 1)$ by $(y + 2)$.

Step 1. Write the polynomials in a vertical format.

$$3y^2 - 4y + 1$$
$$\underline{\times \qquad y + 2}$$

Step 2.   Multiply 2 by each term of the top polynomial. Write the first **partial product** below the line.

$$3y^2 - 4y + 1$$
$$\times \qquad y + 2$$
$$6y^2 - 8y + 2 \qquad \text{Partial product.}$$

Step 3.   Multiply y by each term of the top polynomial. Write this partial product underneath the previous one, being careful to line up like terms.

$$3y^2 - 4y + 1$$
$$\times \qquad y + 2$$
$$6y^2 - 8y + 2 \qquad \text{Partial product.}$$
$$3y^3 - 4y^2 + y \qquad \text{Partial product.}$$

Step 4.   Combine like terms of the partial products.

$$3y^2 - 4y + 1$$
$$\times \qquad y + 2$$
$$6y^2 - 8y + 2$$
$$3y^3 - 4y^2 + y$$
$$\overline{3y^3 + 2y^2 - 7y + 2}$$

Thus, $(y + 2)(3y^2 - 4y + 1) = 3y^3 + 2y^2 - 7y + 2$.

EXAMPLE 6   Find the product of $(2x^2 - 3x + 4)$ and $(x^2 + 5x - 2)$ using the vertical format.

Solution: Multiply each term of the second polynomial by each term of the first polynomial.

$$2x^2 - 3x + 4$$
$$\times \qquad x^2 + 5x - 2$$
$$-4x^2 + 6x - 8 \qquad \text{Multiply } 2x^2 - 3x + 4 \text{ by } -2.$$
$$10x^3 - 15x^2 + 20x \qquad \text{Multiply } 2x^2 - 3x + 4 \text{ by } 5x.$$
$$2x^4 - 3x^3 + 4x^2 \qquad \text{Multiply } 2x^2 - 3x + 4 \text{ by } x^2.$$
$$\overline{2x^4 + 7x^3 - 15x^2 + 26x - 8} \qquad \text{Combine like terms.}$$

MENTAL MATH

Find the following products mentally.

1. $5x(2y)$ $10xy$

2. $7a(4b)$ $28ab$

3. $x^2 \cdot x^5$ x^7

4. $z \cdot z^4$ z^5

5. $6x(3x^2)$ $18x^3$

6. $5a^2(3a^2)$ $15a^4$

27. $8y^3 - 36y^2 + 54y - 27$ **28.** $27x^3 + 108x^2 + 144x + 64$
31. $x^4 - 2x^3 - 51x^2 + 4x + 63$ **32.** $3x^4 + 5x^3 + 3x^2 + 3x + 2$
33. No; answers may vary

EXERCISE SET 4.3

Find the following products. See Example 1.

1. $2a(2a - 4)$ $4a^2 - 8a$ **2.** $3a(2a + 7)$

3. $7x(x^2 + 2x - 1)$ **4.** $-5y(y^2 + y - 10)$

5. $3x^2(2x^2 - x)$ $6x^4 - 3x^3$ **6.** $-4y^2(5y - 6y^2)$

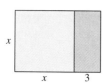
7. The area of the larger rectangle below is $x(x + 3)$. Find another expression for this area by finding the sum of the areas of the smaller rectangles.

2. $6a^2 + 21a$
3. $7x^3 + 14x^2 - 7x$
4. $-5y^3 - 5y^2 + 50y$
6. $-20y^3 + 24y^4$
7. $x^2 + 3x$

8. Write an expression for the area of the larger rectangle below in two different ways.

8. $x(1 + 2x); x + 2x^2$
9. $a^2 + 5a - 14$
10. $y^2 + 12y + 35$

Find the following products. See Examples 2 and 3.

9. $(a + 7)(a - 2)$ **10.** $(y + 5)(y + 7)$

11. $(2y - 4)^2$ **12.** $(6x - 7)^2$

13. $(5x - 9y)(6x - 5y)$ **14.** $(3x - 7y)(7x + 2y)$

15. $(2x^2 - 5)^2$ **16.** $(x^2 - 4)^2$

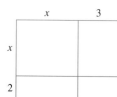
17. The area of the figure below is $(x + z)(x + 3)$. Find another expression for this area by finding the sum of the areas of the smaller rectangles.

11. $4y^2 - 16y + 16$
12. $36x^2 - 84x + 49$
13. $30x^2 - 79xy + 45y^2$
14. $21x^2 - 43xy - 14y^2$
15. $4x^4 - 20x^2 + 25$
16. $x^4 - 8x^2 + 16$
17. $x^2 + 5x + 6$

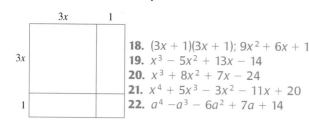
18. Write an expression for the area of the figure below in two different ways.

18. $(3x + 1)(3x + 1); 9x^2 + 6x + 1$
19. $x^3 - 5x^2 + 13x - 14$
20. $x^3 + 8x^2 + 7x - 24$
21. $x^4 + 5x^3 - 3x^2 - 11x + 20$
22. $a^4 - a^3 - 6a^2 + 7a + 14$

Find the following products. See Example 4.

19. $(x - 2)(x^2 - 3x + 7)$ **20.** $(x + 3)(x^2 + 5x - 8)$

21. $(x + 5)(x^3 - 3x + 4)$ **22.** $(a + 2)(a^3 - 3a^2 + 7)$

23. $(2a - 3)(5a^2 - 6a + 4)$ $10a^3 - 27a^2 + 26a - 12$

24. $(3 + b)(2 - 5b - 3b^2)$ $-3b^3 - 14b^2 - 13b + 6$

Find the following products. See Example 5.

25. $(x + 2)^3$ **26.** $(y - 1)^3$

27. $(2y - 3)^3$ **28.** $(3x + 4)^3$

25. $x^3 + 6x^2 + 12x + 8$ **26.** $y^3 - 3y^2 + 3y - 1$

Find the following products. Use the vertical multiplication method. See Example 6.

29. $(x + 3)(2x^2 + 4x - 1)$ $2x^3 + 10x^2 + 11x - 3$

30. $(2x - 5)(3x^2 - 4x + 7)$ $6x^3 - 23x^2 + 34x - 35$

31. $(x^2 + 5x - 7)(x^2 - 7x - 9)$

32. $(3x^2 - x + 2)(x^2 + 2x + 1)$

33. Evaluate each of the following.

a. $(2 + 3)^2; 2^2 + 3^2$ 25; 13

b. $(8 + 10)^2; 8^2 + 10^2$ 324; 164

Does $(a + b)^2 = a^2 + b^2$ no matter what the values of a and b are? Why or why not?

34. Perform the indicated operations. Explain the difference between the two expressions.

a. $(3x + 5) + (3x + 7)$ $6x + 12$; answers may vary

b. $(3x + 5)(3x + 7)$

34. b. $9x^2 + 36x + 35$; answers may vary

Find the following products.

35. $2a(a + 4)$ $2a^2 + 8a$ **36.** $-3a(2a + 7)$

37. $3x(2x^2 - 3x + 4)$ **38.** $-4x(5x^2 - 6x - 10)$

39. $(5x + 9y)(3x + 2y)$ **40.** $(5x - 5y)(2x - y)$

41. $(x + 2)(x^2 + 5x + 6)$ $x^3 + 7x^2 + 16x + 12$

42. $(x - 7)(x^2 - 15x + 56)$ $x^3 - 22x^2 + 161x - 392$

43. $(7x + 4)^2$ **44.** $(3x - 2)^2$

45. $-2a^2(3a^2 - 2a + 3)$ $-6a^4 + 4a^3 - 6a^2$

46. $-4b^2(3b^3 - 12b^2 - 6)$ $-12b^5 + 48b^4 + 24b^2$

47. $(x + 3)(x^2 + 7x + 12)$ $x^3 + 10x^2 + 33x + 36$

48. $(n + 1)(n^2 - 7n - 9)$ $n^3 - 6n^2 - 16n - 9$

49. $(a + 1)^3$ **50.** $(x - y)^3$

51. $(x + y)(x + y)$ **52.** $(x + 3)(7x + 1)$

36. $-6a^2 - 21a$ **37.** $6x^3 - 9x^2 + 12x$ **38.** $-20x^3 + 24x^2 + 40x$ **39.** $15x^2 + 37xy + 18y^2$
40. $10x^2 - 15xy + 5y^2$ **43.** $49x^2 + 56x + 16$ **44.** $9x^2 - 12x + 4$ **49.** $a^3 + 3a^2 + 3a + 1$
50. $x^3 - 3x^2y + 3xy^2 - y^3$ **51.** $x^2 + 2xy + y^2$ **52.** $7x^2 + 22x + 3$

53. $x^2 - 13x + 42$ **54.** $-12x^2 - 7x + 10$ **57.** $-4y^3 - 12y^2 + 44y$ **58.** $-10x^3 + 12x^2 - 2x$
59. $25x^2 - 1$ **60.** $6x^2 + xy - y^2$ **61.** $5x^3 - x^2 + 16x + 16$ **62.** $x^3 - 3x^2 + 5x - 6$

53. $(x - 7)(x - 6)$ **54.** $(4x + 5)(-3x + 2)$

55. $3a(a^2 + 2)$ $3a^3 + 6a$ **56.** $x^3(x + 12)$ $x^4 + 12x^3$

57. $-4y(y^2 + 3y - 11)$ **58.** $-2x(5x^2 - 6x + 1)$

59. $(5x + 1)(5x - 1)$ **60.** $(2x + y)(3x - y)$

61. $(5x + 4)(x^2 - x + 4)$ **62.** $(x - 2)(x^2 - x + 3)$

63. $(2x - 5)^3$ **64.** $(3y - 1)^3$

65. $(4x + 5)(8x^2 + 2x - 4)$ **66.** $(x + 7)(x^2 - 7x - 8)$

67. $(7xy - y)^2$ **68.** $(x + y)^2$ $x^2 + 2xy + y^2$

69. $(5y^2 - y + 3)(y^2 - 3y - 2)$

70. $(2x^2 + x - 1)(x^2 + 3x + 4)$

71. $(3x^2 + 2x - 4)(2x^2 - 4x + 3)$

72. $(a^2 + 3a - 2)(2a^2 - 5a - 1)$
$2a^4 + a^3 - 20a^2 + 7a + 2$
Express each of the following as polynomials.

73. Find the area of the following rectangle.
$(4x^2 - 25)$ sq. yds.

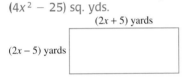
(2x + 5) yards
(2x − 5) yards

74. Find the area of the square field.

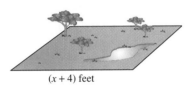
(x + 4) feet

75. Find the area of the following triangle.
$(6x^2 - 4x)$ sq. in.

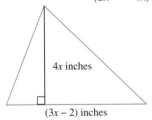
4x inches
(3x − 2) inches

76. Find the volume of the cube-shaped glass block.
$(y^3 - 3y^2 + 3y - 1)$ cu. meters

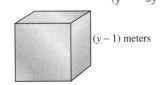
(y − 1) meters

77. Multiply the following polynomials.

a. $(a + b)(a - b)$ $a^2 - b^2$

b. $(2x + 3y)(2x - 3y)$ $4x^2 - 9y^2$

c. $(4x + 7)(4x - 7)$ $16x^2 - 49$

Can you make a general statement about all products of the form $(x + y)(x - y)$?
answers may vary

Review Exercises

Perform the indicated operation. See Section 4.1.

78. $(5x)^2$ $25x^2$ **79.** $(4p)^2$ $16p^2$

80. $(-3y^3)^2$ $9y^6$ **81.** $(-7m^2)^2$ $49m^4$

For income tax purposes, Rob Calcutta, the owner of Copy Services, uses a method called **straight-line depreciation** *to show the depreciated (or decreased) value of a copy machine he recently purchased. Rob assumes that he can use the machine for 7 years. The graph below shows the depreciated value of the machine over the years. See Sections 1.9 and 3.1.*

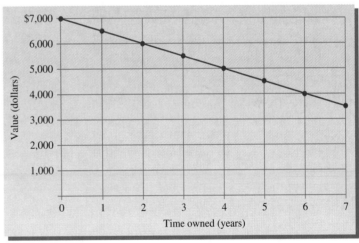

82. $7000   **84.** $500

82. What was the purchase price of the copy machine?

83. What is the depreciated value of the machine in 7 years? $3500

84. What loss in value occurred during the first year?

85. What loss in value occurred during the second year? $500

86. Why do you think this method of depreciating is called straight-line depreciation?   answers may vary

87. Why is the line tilted downward?

63. $8x^3 - 60x^2 + 150x - 125$ **64.** $27y^3 - 27y^2 + 9y - 1$ **65.** $32x^3 + 48x^2 - 6x - 20$ **66.** $x^3 - 57x - 56$
67. $49x^2y^2 - 14xy^2 + y^2$ **69.** $5y^4 - 16y^3 - 4y^2 - 7y - 6$ **70.** $2x^4 + 7x^3 + 10x^2 + x - 4$
71. $6x^4 - 8x^3 - 7x^2 + 22x - 12$ **74.** $(x^2 + 8x + 16)$ sq. ft.   **87.** There is a loss in value each year.

4.4 | SPECIAL PRODUCTS

TAPE BA 4.4

O B J E C T I V E S

 Multiply two binomials using the FOIL method.

2 Square a binomial.

3 Multiply the sum and difference of two terms.

In this section, we multiply binomials using special products. First, a special order for multiplying binomials called the FOIL order or method is introduced. This method is demonstrated by multiplying $(3x + 1)$ by $(2x + 5)$.

F stands for the product of the **First** terms. $(3x + 1)(2x + 5)$

$$(3x)(2x) = 6x^2 \qquad \textbf{F}$$

O stands for the product of the **Outer** terms. $(3x + 1)(2x + 5)$

$$(3x)(5) = 15x \qquad \textbf{O}$$

I stands for the product of the **Inner** terms. $(3x + 1)(2x + 5)$

$$(1)(2x) = 2x \qquad \textbf{I}$$

L stands for the product of the **Last** terms. $(3x + 1)(3x + 5)$

$$(1)(5) = 5 \qquad \textbf{L}$$

$$
\begin{array}{cccc}
\quad\ \textbf{F} & \textbf{O} & \textbf{I} & \textbf{L} \\
(3x + 1)(2x + 5) = 6x^2 & + 15x & + 2x & + 5 \\
\end{array}
$$
$$
= 6x^2 + 17x + 5 \qquad \text{Combine like terms.}
$$

EXAMPLE 1   Find $(x - 3)(x + 4)$ by the FOIL method.

Solution:
$$
\begin{array}{ccccc}
 & \textbf{F} & \textbf{O} & \textbf{I} & \textbf{L} \\
(x - 3)(x + 4) = (x)(x) & + (x)(4) & + (-3)(x) & + (-3)(4) \\
\end{array}
$$

$$= x^2 + 4x - 3x - 12$$
$$= x^2 + x - 12 \qquad \text{Collect like terms.}$$

EXAMPLE 2   Find $(5x - 7)(x - 2)$ by the FOIL method.

Solution:
$$
\begin{array}{ccccc}
 & \textbf{F} & \textbf{O} & \textbf{I} & \textbf{L} \\
(5x - 7)(x - 2) = 5x(x) & + 5x(-2) & + (-7)(x) & + (-7)(-2) \\
\end{array}
$$

$$= 5x^2 - 10x - 7x + 14$$
$$= 5x^2 - 17x + 14 \qquad \text{Collect like terms.}$$

EXAMPLE 3   Multiply $(y + 6)(2y - 1)$.

Solution:
$$\overset{\text{F}\quad\text{O}\quad\text{I}\quad\text{L}}{(y + 6)(2y - 1) = 2y^2 - 1y + 12y - 6}$$
$$= 2y^2 + 11y - 6$$

2   Now, try squaring a binomial using the FOIL method.

EXAMPLE 4   Multiply $(3y + 1)^2$.

Solution: $(3y + 1)^2 = (3y + 1)(3y + 1)$

$$\overset{\text{F}\qquad\text{O}\qquad\text{I}\qquad\text{L}}{= (3y)(3y) + (3y)(1) + 1(3y) + 1(1)}$$
$$= 9y^2 + 3y + 3y + 1$$
$$= 9y^2 + 6y + 1$$

Notice the pattern that appears in Example 4.

$(3y + 1)^2 = 9y^2 + 6y + 1$

$9y^2$ is the first term of the binomial squared. $(3y)^2 = 9y^2$.

$6y$ is 2 times the product of both terms of the binomial. $(2)(3y)(1) = 6y$.

1 is the second term of the binomial squared. $(1)^2 = 1$.

This pattern leads to the following, which can be used when squaring a binomial. We call these **special products.**

SQUARING A BINOMIAL
$$(a + b)^2 = a^2 + 2ab + b^2$$
$$(a - b)^2 = a^2 - 2ab + b^2$$

In other words, a binomial squared is equal to the square of the first term plus two times the product of the first and second terms plus the square of the second term. This product can be visualized geometrically. The area of the large square on the next page is $(a + b)^2$ square units because the length of each side is $(a + b)$ units. But the large square is also made up of 2 smaller squares and 2 rectangles. The area of the large square $(a + b)^2$ must equal the sum of the smaller areas. The sum of the smaller areas is $a^2 + ab + ab + b^2$ or $a^2 + 2ab + b^2$. Thus

$$(a + b)^2 = a^2 + 2ab + b^2$$

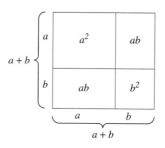

REMINDER
Notice that

$$(x + y)^2 \neq x^2 + y^2 \text{ and } (x - y)^2 \neq x^2 - y^2.$$
$$(x + y)^2 = (x + y)(x + y) = x^2 + 2xy + y^2 \text{ and not } x^2 + y^2.$$

Also,
$$(x - y)^2 = (x - y)(x - y) = x^2 - 2xy + y^2 \text{ and not } x^2 - y^2.$$

EXAMPLE 5   Use the preceding special products to square the following binomials.

a. $(t + 2)^2$ **b.** $(p - q)^2$ **c.** $(2x + 3y)^2$ **d.** $(5r - 7s)^2$

Solution: **a.** The first term is t and the second term is 2.

$$(t + 2)^2 = t^2 + 2(t)(2) + 2^2 = t^2 + 4t + 4$$

b. $(p - q)^2 = p^2 - 2(p)(q) + q^2 = p^2 - 2pq + q^2.$

c. $2x$ is the first term and $3y$ is the second term.

$$(2x + 3y)^2 = (2x)^2 + 2(2x)(3y) + (3y)^2 = 4x^2 + 12xy + 9y^2$$

d. $(5r - 7s)^2 = (5r)^2 - 2(5r)(7s) + (7s)^2 = 25r^2 - 70rs + 49s^2.$

3   Another special product is the product of the sum and difference of two terms, such as $(x + y)(x - y)$. Finding this product by the FOIL method, we see a pattern emerge.

$$
\begin{array}{c}
\text{L} \\
\text{F} \quad\quad \text{F} \quad \text{O} \quad \text{I} \quad \text{L} \\
(x + y)(x - y) = x^2 - xy + xy - y^2 \\
\text{I} \\
\text{O}
\end{array}
$$

$$= x^2 - y^2$$

Notice that middle two terms subtract out. This is because the **Outer** product is the opposite of the **Inner** product. Only the **difference of squares** remains.

MULTIPLYING THE SUM AND DIFFERENCE OF TWO TERMS

$$(a + b)(a - b) = a^2 - b^2$$

EXAMPLE 6   Find the following products using the preceding special product.

    a. $(2x + y)(2x - y)$ **b.** $(6t + 7)(6t - 7)$ **c.** $\left(c - \dfrac{1}{4}\right)\left(c + \dfrac{1}{4}\right)$

    d. $(2p - q)(2p + q)$

Solution: **a.** Using the preceding formula, $(2x + y)(2x - y) = (2x)^2 - y^2 = 4x^2 - y^2$.

    b. $(6t + 7)(6t - 7) = (6t)^2 - 7^2 = 36t^2 - 49$.

    c. $\left(c - \dfrac{1}{4}\right)\left(c + \dfrac{1}{4}\right) = c^2 - \left(\dfrac{1}{4}\right)^2 = c^2 - \dfrac{1}{16}$.

    d. $(2p - q)(2p + q) = (2p)^2 - q^2 = 4p^2 - q^2$.

EXAMPLE 7   Find $\left(x^2 - \dfrac{1}{3}y\right)\left(x^2 + \dfrac{1}{3}y\right)$.

Solution: $\left(x^2 - \dfrac{1}{3}y\right)\left(x^2 + \dfrac{1}{3}y\right) = (x^2)^2 - \left(\dfrac{1}{3}y\right)^2 = x^4 - \dfrac{1}{9}y^2$

---

1. $x^2 + 7x + 12$ **2.** $x^2 + 4x - 5$ **3.** $x^2 + 5x - 50$ **4.** $y^2 - 8y - 48$ **5.** $5x^2 + 4x - 12$
6. $6y^2 - 31y + 35$ **7.** $4y^2 - 25y + 6$ **8.** $2x^2 - 31x + 99$ **12.** $x^2 + 14x + 49$

EXERCISE SET 4.4

Find each product using the FOIL method. See Examples 1 through 3.

1. $(x + 3)(x + 4)$ **2.** $(x + 5)(x - 1)$

3. $(x - 5)(x + 10)$ **4.** $(y - 12)(y + 4)$

5. $(5x - 6)(x + 2)$ **6.** $(3y - 5)(2y - 7)$

7. $(y - 6)(4y - 1)$ **8.** $(2x - 9)(x - 11)$

9. $(2x + 5)(3x - 1)$ **10.** $(6x + 2)(x - 2)$

Find each product. See Examples 4 and 5.

11. $(x - 2)^2$ $x^2 - 4x + 4$ **12.** $(x + 7)^2$

13. $(2x - 1)^2$ $4x^2 - 4x + 1$ **14.** $(7x - 3)^2$

15. $(3a - 5)^2$ **16.** $(5a + 2)^2$

17. $(5x + 9)^2$ **18.** $(6s - 2)^2$

📝 **19.** Using your own words, explain how to square a binomial such as $(a + b)^2$.   answers may vary

📝 **20.** Explain how to find the product of two binomials using the FOIL method.   answers may vary

Find each product. See Examples 6 and 7.

21. $(a - 7)(a + 7)$ **22.** $(b + 3)(b - 3)$

23. $(3x - 1)(3x + 1)$ **24.** $(4x - 5)(4x + 5)$

25. $\left(3x - \dfrac{1}{2}\right)\left(3x + \dfrac{1}{2}\right)$ **26.** $\left(10x + \dfrac{2}{7}\right)\left(10x - \dfrac{2}{7}\right)$

27. $(9x + y)(9x - y)$ **28.** $(2x - y)(2x + y)$

---

9. $6x^2 + 13x - 5$ **10.** $6x^2 - 10x - 4$ **14.** $49x^2 - 42x + 9$ **15.** $9a^2 - 30a + 25$ **16.** $25a^2 + 20a + 4$
17. $25x^2 + 90x + 81$ **18.** $36s^2 - 24s + 4$ **21.** $a^2 - 49$ **22.** $b^2 - 9$ **23.** $9x^2 - 1$ **24.** $16x^2 - 25$
25. $9x^2 - \dfrac{1}{4}$ **26.** $100x^2 - \dfrac{4}{49}$ **27.** $81x^2 - y^2$ **28.** $4x^2 - y^2$

Find the area of each shaded region.

29. $(3x^2 - 9)$ sq. units

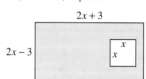

2x + 3

2x − 3

x
x

30. $\left(\dfrac{25}{2}a^2 - \dfrac{1}{2}b^2\right)$ sq. units

5a − b

5a + b

Find each product.

31. $(a + 5)(a + 4)$

32. $(a - 5)(a - 7)$

33. $(a + 7)^2$ $a^2 + 14a + 49$

34. $(b - 2)^2$ $b^2 - 4b + 4$

35. $(4a + 1)(3a - 1)$

36. $(6a + 7)(6a + 5)$

37. $(x + 2)(x - 2)$ $x^2 - 4$

38. $(x - 10)(x + 10)$

39. $(3a + 1)^2$ $9a^2 + 6a + 1$

40. $(4a - 2)^2$

41. $(x + y)(4x - y)$

42. $(3x + 2)(4x - 2)$

43. $(2a - 3)^2$ $4a^2 - 12a + 9$

44. $(5b - 4x)^2$

45. $(5x - 6z)(5x + 6z)$

46. $(11x - 7y)(11x + 7y)$

47. $(x - 3)(x - 5)$

48. $(a + 5b)(a + 6b)$

49. $\left(x - \dfrac{1}{3}\right)\left(x + \dfrac{1}{3}\right)$ $x^2 - \dfrac{1}{9}$

50. $\left(3x + \dfrac{1}{5}\right)\left(3x - \dfrac{1}{5}\right)$

51. $(a + 11)(a - 3)$

52. $(2x + 5)(x - 8)$

53. $(x - 2)^2$ $x^2 - 4x + 4$

54. $(3b + 7)^2$

55. $(3b + 7)(2b - 5)$

56. $(3y - 13)(y - 3)$

57. $(7p - 8)(7p + 8)$

58. $(3s - 4)(3s + 4)$

59. $\left(\dfrac{1}{3}a^2 - 7\right)\left(\dfrac{1}{3}a^2 + 7\right)$

60. $\left(\dfrac{2}{3}a - b^2\right)\left(\dfrac{2}{3}a - b^2\right)$

61. $(2r - 3s)(2r + 3s)$

62. $(6r - 2x)(6r + 2x)$

63. $(3x - 7y)^2$

64. $(4s - 2y)^2$

65. $(4x + 5)(4x - 5)$

66. $(3x + 5)(3x - 5)$

67. $(x + 4)(x + 4)$

68. $(3x + 2)(3x + 2)$

69. $\left(a - \dfrac{1}{2}y\right)\left(a + \dfrac{1}{2}y\right)$

70. $\left(\dfrac{a}{2} + 4y\right)\left(\dfrac{a}{2} - 4y\right)$

71. $\left(\dfrac{1}{5}x - y\right)\left(\dfrac{1}{5}x + y\right)$

72. $\left(\dfrac{y}{6} - 8\right)\left(\dfrac{y}{6} + 8\right)$

31. $a^2 + 9a + 20$

32. $a^2 - 12a + 35$

35. $12a^2 - a - 1$

36. $36a^2 + 72a + 35$

38. $x^2 - 100$

40. $16a^2 - 16a + 4$

41. $4x^2 + 3xy - y^2$

42. $12x^2 + 2x - 4$

44. $25b^2 - 40bx + 16x^2$

45. $25x^2 - 36z^2$

46. $121x^2 - 49y^2$

47. $x^2 - 8x + 15$

48. $a^2 + 11ab + 30b^2$

50. $9x^2 - \dfrac{1}{25}$

51. $a^2 + 8a - 33$

Express each of the following as a polynomial in x.

73. Find the area of the square rug shown if its side is $(2x + 1)$ feet. $(4x^2 + 4x + 1)$ sq. ft.

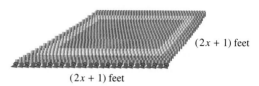

$(2x + 1)$ feet

$(2x + 1)$ feet

74. Find the area of the rectangular canvas if its length is $(3x - 2)$ inches and its width is $(x - 4)$ inches.

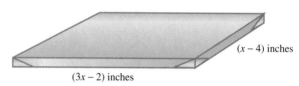

$(x - 4)$ inches

$(3x - 2)$ inches

75. Find the area of the shaded region.

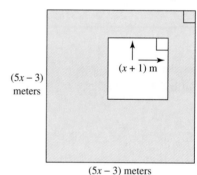

$(5x - 3)$ meters

$(x + 1)$ m

$(5x - 3)$ meters

76. Find the area of the shaded region.

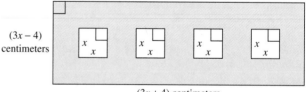

$(3x - 4)$ centimeters

x
x

x
x

x
x

x
x

$(3x + 4)$ centimeters

Review Exercises

Simplify each expression. See Section 4.1.

77. $\dfrac{50b^{10}}{70b^5}$ $\dfrac{5b^5}{7}$

78. $\dfrac{x^3y^6}{xy^2}$ x^2y^4

79. $\dfrac{8a^{17}b^5}{-4a^7b^{10}}$ $-\dfrac{2a^{10}}{b^5}$

80. $\dfrac{-6a^8y}{3a^4y}$ $-2a^4$

81. $\dfrac{2x^4y^{12}}{3x^4y^4}$ $\dfrac{2y^8}{3}$

82. $\dfrac{-48ab^6}{32ab^3}$ $-\dfrac{3b^3}{2}$

52. $2x^2 - 11x - 40$ **54.** $9b^2 + 42b + 49$ **55.** $6b^2 - b - 35$ **56.** $3y^2 - 22y + 39$ **57.** $49p^2 - 64$

58. $9s^2 - 16$ **59.** $\dfrac{1}{9}a^4 - 49$ **60.** $\dfrac{4}{9}a^2 - \dfrac{4}{3}ab^2 + b^4$ **61.** $4r^2 - 9s^2$ **62.** $36r^2 - 4x^2$ **63.** $9x^2 - 42xy + 49y^2$

Find the slope of each line. See Section 3.4.

83.

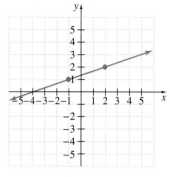

$\dfrac{1}{3}$

84.

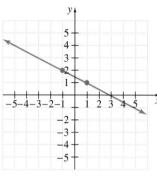

$-\dfrac{1}{2}$

85.

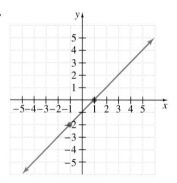

 1

86.

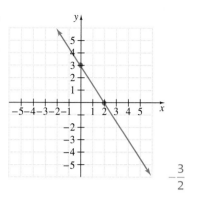

$-\dfrac{3}{2}$

A Look Ahead

EXAMPLE

Find the product $[(a + b) - 2][(a + b) + 2]$.

Solution:
If we think of $(a + b)$ as one term, we can think of $[(a + b) - 2][(a + b) + 2]$ as the product of the sum and difference of two terms.

$$[(a + b) - 2][(a + b) + 2] = (a + b)^2 - 2^2$$

Next, square $(a + b)$. $= a^2 + 2ab + b^2 - 4$

Find each product.

87. $[(x + y) - 3][(x + y) + 3]$ $x^2 + 2xy + y^2 - 9$
88. $[(a + c) - 5][(a + c) + 5]$ $a^2 + 2ac + c^2 - 25$
89. $[(a - 3) + b][(a - 3) - b]$ $a^2 - 6a + 9 - b^2$
90. $[(x - 2) + y][(x - 2) - y]$ $x^2 - 4x + 4 - y^2$
91. $[(2x + 1) - y][(2x + 1) + y]$ $4x^2 + 4x + 1 - y^2$
92. $[(3x + 2) - z][(3x + 2) + z]$ $9x^2 + 12x + 4 - z^2$

64. $16s^2 - 16sy + 4y^2$ **65.** $16x^2 - 25$ **66.** $9x^2 - 25$ **67.** $x^2 + 8x + 16$ **68.** $9x^2 + 12x + 4$ **69.** $a^2 - \dfrac{1}{4}y^2$

4.5   NEGATIVE EXPONENTS AND SCIENTIFIC NOTATION

O B J E C T I V E S

1. Evaluate numbers raised to negative integer powers.
2. Use all the rules and definitions for exponents to simplify exponential expressions.
3. Write numbers in scientific notation.
4. Convert numbers from scientific notation to standard form.

TAPE BA 4.5

70. $\dfrac{a^2}{4} - 16y^2$ **71.** $\dfrac{1}{25}x^2 - y^2$ **72.** $\dfrac{y^2}{36} - 64$ **74.** $(3x^2 - 14x + 8)$ sq. in.   **75.** $(24x^2 - 32x + 8)$ sq. meters
76. $(5x^2 - 16)$ square centimeters

Our work with exponential expressions so far has been limited to exponents that are positive integers or 0. Here we expand to give meaning to an expression like x^{-3}.

Suppose that we wish to simplify the expression $\dfrac{x^2}{x^5}$. If we use the quotient rule for exponents, we subtract exponents:

$$\frac{x^2}{x^5} = x^{2-5} = x^{-3}, \quad x \neq 0$$

But what does x^{-3} mean? Let's simplify $\dfrac{x^2}{x^5}$ using the definition of a^n.

$$\frac{x^2}{x^5} = \frac{x \cdot x}{x \cdot x \cdot x \cdot x \cdot x}$$

$$= \frac{x \cdot x}{x \cdot x \cdot x \cdot x \cdot x} \qquad \text{Divide numerator and denominator by common factors by applying the fundamental principle.}$$

$$= \frac{1}{x^3}$$

If the quotient rule is to hold true for negative exponents, then x^{-3} must equal $\dfrac{1}{x^3}$.

From this example, we state the definition for negative exponents.

NEGATIVE EXPONENTS

If a is a real number other than 0 and n is an integer, then

$$a^{-n} = \frac{1}{a^n}$$

EXAMPLE 1   Simplify by writing each expression with positive exponents only.

    a. 3^{-2} **b.** $2x^{-3}$ **c.** $\dfrac{1}{2^{-5}}$

Solution: **a.** $3^{-2} = \dfrac{1}{3^2} = \dfrac{1}{9}$ Use the definition of negative exponent.

          b. $2x^{-3} = 2 \cdot \dfrac{1}{x^3} = \dfrac{2}{x^3}$ Use the definition of negative exponent.

          Since there are no parentheses, notice that only x is the base for the exponent -3.

          c. $\dfrac{1}{2^{-5}} = \dfrac{1}{\dfrac{1}{2^5}}$ Use the definition of negative exponent.

               $= 1 \div \dfrac{1}{2^5}$

               $= 1 \cdot \dfrac{2^5}{1} = 2^5 \quad \text{or} \quad 32$

> REMINDER   A negative exponent does not affect the sign of its base.
>
> Remember: a negative exponent means a reciprocal. $a^{-n} = \frac{1}{a^n}$. For example,
>
> $$x^{-2} = \frac{1}{x^2}, \qquad 2^{-3} = \frac{1}{2^3} \text{ or } \frac{1}{8}$$
>
> $$\frac{1}{y^{-4}} = \frac{1}{\frac{1}{y^4}} = y^4, \qquad \frac{1}{5^{-2}} = 5^2 \text{ or } 25$$

EXAMPLE 2   Simplify each expression. Write results with positive exponents.

a. $\left(\frac{2}{3}\right)^{-4}$ 　　　　　　**b.** $2^{-1} + 4^{-1}$ 　　　　　　**c.** $(-2)^{-4}$

Solution: **a.** $\left(\frac{2}{3}\right)^{-4} = \frac{1}{\left(\frac{2}{3}\right)^4} = \frac{1}{\frac{2^4}{3^4}} = \frac{3^4}{2^4} = \frac{81}{16}$ **b.** $2^{-1} + 4^{-1} = \frac{1}{2} + \frac{1}{4} = \frac{2}{4} + \frac{1}{4} = \frac{3}{4}$

c. $(-2)^{-4} = \frac{1}{(-2)^4} = \frac{1}{(-2)(-2)(-2)(-2)} = \frac{1}{16}$

2   All the previously stated rules for exponents apply for negative exponents also. Here is a summary of the rules and definitions for exponents.

SUMMARY OF EXPONENT RULES

If m and n are integers and a, b, and c are real numbers, then:

Product rule for exponents: 　　　　$a^m \cdot a^n = a^{m+n}$

Power of a power rule for exponents: 　$(a^m)^n = a^{m \cdot n}$

Power of a product or quotient: 　　　$(ab)^n = a^n b^n$ and

$$\left(\frac{a}{c}\right)^n = \frac{a^n}{c^n}, c \neq 0$$

Quotient rule for exponents: 　　　　$\dfrac{a^m}{a^n} = a^{m-n}, a \neq 0$

Zero exponent: 　　　　　　　　　$a^0 = 1, a \neq 0$

Negative exponent: 　　　　　　　$a^{-n} = \dfrac{1}{a^n}, a \neq 0$

EXAMPLE 3   Simplify each expression. Write answers with positive exponents.

a. $\dfrac{y}{y^{-2}}$ 　　　　　　**b.** $\dfrac{p^{-4}}{q^{-9}}$ 　　　　　　**c.** $\dfrac{x^{-5}}{x^7}$

Solution: **a.** $\dfrac{y}{y^{-2}} = \dfrac{y^1}{y^{-2}} = y^{1-(-2)} = y^3$ **b.** $\dfrac{p^{-4}}{q^{-9}} = \dfrac{q^9}{p^4}$ **c.** $\dfrac{x^{-5}}{x^7} = x^{-5-7} = x^{-12} = \dfrac{1}{x^{12}}$

EXAMPLE 4 Simplify the following expressions. Write each result using positive exponents only.

a. $(2x^3)(5x)^{-2}$ **b.** $\left(\dfrac{3a^2}{b}\right)^{-3}$ **c.** $\dfrac{4^{-1}x^{-3}y}{4^{-3}x^2y^{-6}}$ **d.** $(y^{-3}z^{-6})^{-6}$ **e.** $\left(\dfrac{-2x^3y}{xy^{-1}}\right)^3$

Solution: **a.** $(2x^3)(5x)^{-2} = 2x^3 \cdot 5^{-2}x^{-2}$ Use the power of a power rule.

$= \dfrac{2x^{3+(-2)}}{5^2}$ Use the product and quotient rules and definition of negative exponent.

$= \dfrac{2x}{25}$

b. $\left(\dfrac{3a^2}{b}\right)^{-3} = \dfrac{3^{-3}a^{-6}}{b^{-3}} = \dfrac{b^3}{3^3a^6} = \dfrac{b^3}{27a^6}$

c. $\dfrac{4^{-1}x^{-3}y}{4^{-3}x^2y^{-6}} = 4^{-1-(-3)}x^{-3-2}y^{1-(-6)} = 4^2x^{-5}y^7 = \dfrac{4^2y^7}{x^5} = \dfrac{16y^7}{x^5}$

d. $(y^{-3}z^6)^{-6} = y^{18} \cdot z^{-36} = \dfrac{y^{18}}{z^{36}}$

e. $\left(\dfrac{-2x^3y}{xy^{-1}}\right)^3 = \dfrac{(-2)^3x^9y^3}{x^3y^{-3}} = \dfrac{-8x^9y^3}{x^3y^{-3}} = -8x^{9-3}y^{3-(-3)} = -8x^6y^6$

---

3 Both very large and very small numbers frequently occur in many fields of science. For example, the distance between the sun and the planet Pluto is approximately 5,906,000,000 kilometers, and the mass of a proton is approximately 0.000000000000000000000000165 gram. It can be tedious to write these numbers in this standard decimal notation, so **scientific notation** is used as a convenient shorthand for expressing very large and very small numbers.

SCIENTIFIC NOTATION

A positive number is written in scientific notation if it is written as the product of a number a, where $1 \le a < 10$, and an integer power r of 10:

$a \times 10^r$

The following numbers are written in scientific notation. The $\times$ sign for multiplication is used as part of the notation.

2.03×10^2 7.362×10^7

1×10^{-3} 8.1×10^{-5}

To write the distance between the sun and Pluto in scientific notation, begin by moving the decimal point to the left until we have a number between 1 and 10.

5906000000.

Next, count the number of places the decimal point is moved.

5906000000.

9 decimal places

We moved the decimal point 9 places **to the left.** This count is used as the power of 10.

$$5{,}906{,}000{,}000 = 5.906 \times 10^9$$

To express the mass of a proton in scientific notation, move the decimal until the number is between 1 and 10.

0.0000000000000000000000165

The decimal point was moved 24 places **to the right,** so the exponent on 10 is negative 24.

$$0.000\ 000\ 000\ 000\ 000\ 000\ 000\ 001\ 65 = 1.65 \times 10^{-24}$$

To Write a Number in Scientific Notation

Step 1.   Move the decimal point in the original number so that the new number has a value between 1 and 10.

Step 2.   Count the number of decimal places the decimal point is moved in step 1. If the decimal point is moved to the left, the count is positive. If the decimal point is moved to the right, the count is negative.

Step 3.   Multiply the new number in step 1 by 10 raised to an exponent equal to the count found in step 2.

EXAMPLE 5   Write the following numbers in scientific notation.

a. 367,000,000   **b.** 0.000003   **c.** 20,520,000,000   **d.** 0.00085

Solution: **a.** *Step 1*   Move the decimal point until the number is between 1 and 10.

367,000,000.

Step 2   The decimal point is moved to the left 8 places, so the count is positive 8.

Step 3 $367{,}000{,}000 = 3.67 \times 10^8$.

b. *Step 1*   Move the decimal point until the number is between 1 and 10.

0.000003

Step 2 The decimal point is moved 6 places to the right, so the count is −6.

Step 3 $0.000003 = 3.0 \times 10^{-6}$.

c. $20{,}520{,}000{,}000 = 2.052 \times 10^{10}$

d. $0.00085 = 8.5 \times 10^{-4}$

---

4 A number written in scientific notation can be rewritten in standard form. To write 8.63×10^3 in standard form, recall that $10^3 = 1000$.

$$8.63 \times 10^3 = 8.63(1000) = 8630$$

Notice that the exponent on the 10 is positive three and we moved the decimal point three places to the right.

To write 8.63×10^{-3} in standard form, recall that $10^{-3} = \dfrac{1}{10^3} = \dfrac{1}{1000}$.

$$8.63 \times 10^{-3} = 8.63\left(\frac{1}{1000}\right) = \frac{8.63}{1000} = 0.00863$$

The exponent on the 10 is negative three, and we moved the decimal to the left three places.

In general, **to write a scientific notation number in standard form,** move the decimal point the same number of places as the exponent on 10. If the exponent is positive, move the decimal point to the right; if the exponent is negative, move the decimal point to the left.

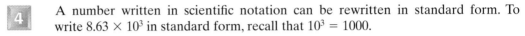

EXAMPLE 6 Write the following numbers in standard notation, without exponents.

    a. 1.02×10^5 **b.** 7.358×10^{-3} **c.** 8.4×10^7 **d.** 3.007×10^{-5}

Solution: **a.** Move the decimal point 5 places to the right and

$$1.02 \times 10^5 = 102{,}000.$$

    b. Move the decimal point 3 places to the left and

$$7.358 \times 10^{-3} = 0.007358$$

    c. $8.4 \times 10^7 = 84{,}000{,}000.$

        7 places to the right

    d. $3.007 \times 10^{-5} = 0.00003007$

        5 places to the left

Performing operations on numbers written in scientific notation makes use of the rules and definitions for exponents.

EXAMPLE 7   Perform the indicated operations. Write each answer in standard decimal notation.

    a. $(8 \times 10^{-6})(7 \times 10^3)$ **b.** $\dfrac{12 \times 10^2}{6 \times 10^{-3}}$

Solution: **a.** $(8 \times 10^{-6})(7 \times 10^3) = 8 \cdot 7 \cdot 10^{-6} \cdot 10^3$

$$= 56 \times 10^{-3}$$
$$= 0.056$$

    b. $\dfrac{12 \times 10^2}{6 \times 10^{-3}} = \dfrac{12}{6} \times 10^{2-(-3)} = 2 \times 10^5 = 200{,}000$

SCIENTIFIC CALCULATOR EXPLORATIONS

SCIENTIFIC NOTATION

To enter a number written in scientific notation on a scientific calculator, locate the key marked $\boxed{\text{EE}}$.

    To enter 3.1×10^7, press $\boxed{3.1}$ $\boxed{\text{EE}}$ $\boxed{7}$. The display should read $\boxed{3.1 \qquad 07}$.

Enter the following numbers written in scientific notation on your calculator.

1. 5.31×10^3 5.31 EE 03

2. -4.8×10^{14} −4.8 EE 14

3. 6.6×10^{-9} 6.6 EE −09

4. -9.9811×10^{-2} −9.9811 EE −2

Multiply the following on your calculator. Notice the form of the result.

5. $3{,}000{,}000 \times 5{,}000{,}000$ 1.5×10^{13}

6. $230{,}000 \times 1{,}000$ 2.3×10^8

Multiply the following on your calculator. Write the product in scientific notation.

7. $(3.26 \times 10^6)(2.5 \times 10^{13})$ 8.15×10^{19}

8. $(8.76 \times 10^{-4})(1.237 \times 10^9)$ 1.083612×10^6

MENTAL MATH

State each expression using positive exponents.

1. $5x^{-2}$ $\dfrac{5}{x^2}$ **2.** $3x^{-3}$ $\dfrac{3}{x^3}$ **3.** $\dfrac{1}{y^{-6}}$ y^6

4. $\dfrac{1}{x^{-3}}$ x^3 **5.** $\dfrac{4}{y^{-3}}$ $4y^3$ **6.** $\dfrac{16}{y^{-7}}$ $16y^7$

Exercise Set 4.5

Simplify each expression. Write results with positive exponents. See Examples 1 and 2.

1. 4^{-3} $\dfrac{1}{64}$

2. 6^{-2} $\dfrac{1}{36}$

3. $7x^{-3}$ $\dfrac{7}{x^3}$

4. $(7x)^{-3}$ $\dfrac{1}{343x^3}$

5. $\left(-\dfrac{1}{4}\right)^{-3}$ -64

6. $\left(-\dfrac{1}{8}\right)^{-2}$ 64

7. $3^{-1} + 2^{-1}$ $\dfrac{5}{6}$

8. $4^{-1} + 4^{-2}$ $\dfrac{5}{16}$

9. $\dfrac{1}{p^{-3}}$ p^3

10. $\dfrac{1}{q^{-5}}$ q^5

Simplify each expression. Write results with positive exponents. See Example 3.

11. $\dfrac{p^{-5}}{q^{-4}}$ $\dfrac{q^4}{p^5}$

12. $\dfrac{r^{-5}}{s^{-2}}$ $\dfrac{s^2}{r^5}$

13. $\dfrac{x^{-2}}{x}$ $\dfrac{1}{x^3}$

14. $\dfrac{y}{y^{-3}}$ y^4

15. $\dfrac{z^{-4}}{z^{-7}}$ z^3

16. $\dfrac{x^{-4}}{x^{-1}}$ $\dfrac{1}{x^3}$

17. It was stated earlier that, for an integer n,

$$x^{-n} = \dfrac{1}{x^n}, \quad x \neq 0$$

Explain why x may not equal 0.   answers may vary

18. If $a = \dfrac{1}{10}$, then find the value of a^{-2}.   100

Simplify the following. Write results with positive exponents only. See Example 4.

19. $(a^{-5}b^2)^{-6}$ $\dfrac{a^{30}}{b^{12}}$

20. $(4^{-1}x^5)^{-2}$ $\dfrac{16}{x^{10}}$

21. $\left(\dfrac{x^{-2}y^4}{x^3y^7}\right)^2$ $\dfrac{1}{x^{10}y^6}$

22. $\left(\dfrac{a^5b}{a^7b^{-2}}\right)^{-3}$ $\dfrac{a^6}{b^9}$

23. $\dfrac{4^2z^{-3}}{4^3z^{-5}}$ $\dfrac{z^2}{4}$

24. $\dfrac{3^{-1}x^4}{3^3x^{-7}}$ $\dfrac{x^{11}}{81}$

25. Explain why $(a^{-1})^3$ has the same value as $(a^3)^{-1}$.

26. Determine whether each statement is true or false.

   a. $5^{-1} < 5^{-2}$ false

   b. $\left(\dfrac{1}{5}\right)^{-1} < \left(\dfrac{1}{5}\right)^{-2}$ true

   c. $a^{-1} < a^{-2}$ for all nonzero numbers.   false

25. answers may vary

Simplify the following. Write results with positive exponents.

27. $(-3)^{-2}$ $\dfrac{1}{9}$

28. $(-2)^{-4}$ $\dfrac{1}{16}$

29. $\dfrac{-1}{p^{-4}}$ $-p^4$

30. $\dfrac{-1}{y^{-6}}$ $-y^6$

31. $-2^0 - 3^0$ -2

32. $5^0 + (-5)^0$ 2

33. $\dfrac{r}{r^{-3}r^{-2}}$ r^6

34. $\dfrac{p}{p^{-3}q^{-5}}$ p^4q^5

35. $(x^5y^3)^{-3}$ $\dfrac{1}{x^{15}y^9}$

36. $(z^5x^5)^{-3}$ $\dfrac{1}{z^{15}x^{15}}$

37. $2^0 + 3^{-1}$ $\dfrac{4}{3}$

38. $4^{-2} - 4^{-3}$ $\dfrac{3}{64}$

39. $\dfrac{2^{-3}x^{-4}}{2^2x}$ $\dfrac{1}{32x^5}$

40. $\dfrac{5^{-1}z^7}{5^{-2}z^9}$ $\dfrac{5}{z^2}$

41. $\dfrac{7ab^{-4}}{7^{-1}a^{-3}b^2}$ $\dfrac{49a^4}{b^6}$

42. $\dfrac{6^{-5}x^{-1}y^2}{6^{-2}x^{-4}y^4}$ $\dfrac{x^3}{216y^2}$

43. $\left(\dfrac{a^{-5}b}{ab^3}\right)^{-4}$ $a^{24}b^8$

44. $\left(\dfrac{r^{-2}s^{-3}}{r^{-4}s^{-3}}\right)^{-3}$ $\dfrac{1}{r^6}$

45. $\dfrac{(xy^3)^5}{(xy)^{-4}}$ x^9y^{19}

46. $\dfrac{(rs)^{-3}}{(r^2s^3)^2}$ $\dfrac{1}{r^7s^9}$

47. $\dfrac{(-2xy^{-3})^{-3}}{(xy^{-1})^{-1}}$ $-\dfrac{y^8}{8x^2}$

48. $\dfrac{(-3x^2y^2)^{-2}}{(xyz)^{-2}}$ $\dfrac{z^2}{9x^2y^2}$

49. Find the volume of the cube. $\dfrac{27}{x^6z^3}$ cu. in.

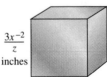

   $\dfrac{3x^{-2}}{z}$ inches

50. Find the area of the triangle. $\dfrac{10}{7x^4}$ sq. m

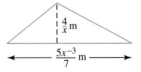

   $\dfrac{4}{x}$ m $\dfrac{5x^{-3}}{7}$ m

51. The product of a monomial and $7x^{-2}$ is $14x^5$. Find the monomial. $2x^7$

52. The product of an exponential expression and $-5y^3$ is $15y$. Find the expression. $-\dfrac{3}{y^2}$

Write each number in scientific notation. See Example 5.

53. 78,000 7.8×10^4

54. 9,300,000,000 9.3×10^9

55. 0.00000167 1.67 × 10⁻⁶ **56.** 0.00000017 1.7 × 10⁻⁷

57. 0.00635 6.35 × 10⁻³ **58.** 0.00194 1.94 × 10⁻³

59. 1,160,000 1.16 × 10⁶ **60.** 700,000 7.0 × 10⁵

61. The temperature at the interior of the Earth is 20,000,000 degrees Celsius. Write 20,000,000 in scientific notation. 2.0 × 10⁷

62. The half-life of a carbon isotope is 5000 years. Write 5000 in scientific notation. 5.0 × 10³

63. The distance between the earth and the sun is 93,000,000 miles. Write 93,000,000 in scientific notation. 9.3 × 10⁷

64. The population of the world is 5,506,000,000. Write 5,506,000,000 in scientific notation. 5.506 × 10⁹

Write each number in standard notation. See Example 6.

65. 7.86 × 10⁸ 786,000,000

66. 1.43 × 10⁷ 14,300,000

67. 8.673 × 10⁻¹⁰ 0.0000000008673

68. 9.056 × 10⁻⁴ 0.0009056

69. 3.3 × 10⁻² 0.033

70. 4.8 × 10⁻⁶ 0.0000048

71. 2.032 × 10⁴ 20,320

72. 9.07 × 10¹⁰ 90,700,000,000

73. One coulomb of electricity is 6.25 × 10¹⁸. Write this number in standard notation.

74. The mass of a hydrogen atom is 1.7 × 10⁻²⁴ grams. Write this number in standard notation.

75. The distance light travels in 1 year is 9.460 × 10¹² kilometers. Write this number in standard notation.

76. The population of the United States is 2.48 × 10⁸. Write this number in standard notation.

Evaluate the following expressions using exponential rules. Write the results in standard notation. See Example 7.

77. (1.2 × 10⁻³)(3 × 10⁻²) 0.000036

78. (2.5 × 10⁶)(2 × 10⁻⁶) 5

73. 6,250,000,000,000,000,000

74. 0.0000000000000000000000017

79. (4 × 10⁻¹⁰)(7 × 10⁻⁹) 0.0000000000000000028

80. (5 × 10⁶)(4 × 10⁻⁸) 0.2

81. $\dfrac{8 \times 10^{-1}}{16 \times 10^{5}}$ 0.0000005 **82.** $\dfrac{25 \times 10^{-4}}{5 \times 10^{-9}}$ 500,000

83. $\dfrac{1.4 \times 10^{-2}}{7 \times 10^{-8}}$ 200,000 **84.** $\dfrac{0.4 \times 10^{5}}{0.2 \times 10^{11}}$ 0.000002

85. The average amount of water flowing past the mouth of the Amazon River is 4.2 × 10⁶ cubic feet per second. How much water flows past in an hour? (1 hour equals 3600 seconds.) Write the result in scientific notation. 1.512 × 10¹⁰ cu. ft.

86. A beam of light travels 9.460 × 10¹² kilometers per year. How far does light travel in 10,000 years? Write the result in scientific notation. 9.46 × 10¹⁶ km

87. The total force (*F*) against the face of a dam that is 100 feet long by 20 feet high is given by the following:

$$F = \dfrac{(6.24 \times 10)(4 \times 10^{4})}{2}$$

Compute the force and express it in scientific notation. 1.248 × 10⁶

88. Suppose $1000 is invested at a rate of 9% and compounded monthly. The amount (*A*) after one year is given by

$$A = (1 \times 10^{3})(1.09381)$$

Compute this amount in standard notation. $1093.81

Simplify. Write results in standard notation.

89. (2.63 × 10¹²)(−1.5 × 10⁻¹⁰) −394.5

90. (6.785 × 10⁻⁴)(4.68 × 10¹⁰) 31,753,800

Light travels at a rate of 1.86 × 10⁵ miles per second. Use this information and the distance formula d = r · t to answer Exercises 91 and 92.

91. If the distance from the moon to the Earth is 238,857 miles, find how long it takes the reflected light of the moon to reach the Earth. (Round to the nearest tenth of a second.) 1.3 sec.

75. 9,460,000,000,000 km
76. 248,000,000

92. If the distance from the sun to the earth is 93,000,000 miles, find how long it takes the light of the sun to reach the earth. (Round to the nearest tenth of a second.)   500 sec.

Review Exercises

Simplify the following. See Section 4.1.

93. $\dfrac{5x^7}{3x^4}$ $\dfrac{5x^3}{3}$

94. $\dfrac{27y^{14}}{3y^7}$ $9y^7$

95. $\dfrac{15z^4y^3}{21zy}$ $\dfrac{5z^3y^2}{7}$

96. $\dfrac{18a^7b^{17}}{30a^7b}$ $\dfrac{3b^{16}}{5}$

Use the distributive property and multiply. See Section 4.3.

97. $\dfrac{1}{y}(5y^2 - 6y + 5)$ $5y - 6 + \dfrac{5}{y}$

98. $\dfrac{2}{x}(3x^5 + x^4 - 2)$ $6x^4 + 2x^3 - \dfrac{4}{x}$

99. $2x^2\left(10x - 6 + \dfrac{1}{x}\right)$ $20x^3 - 12x^2 + 2x$

100. $-5y^3\left(2y^2 - 4y + 2 - \dfrac{3}{y}\right)$
$-10y^5 + 20y^4 - 10y^3 + 15y^2$

A Look Ahead

EXAMPLE

Simplify the following expressions. Assume that the variable in the exponent represents an integer value.

a. $x^{m+1} \cdot x^m$ **b.** $(z^{2x+1})^x$ **c.** $\dfrac{y^{6a}}{y^{4a}}$

Solution:

a. $x^{m+1} \cdot x^m = x^{(m+1)+m} = x^{2m+1}$

b. $(z^{2x+1})^x = z^{(2x+1)x} = z^{2x^2+x}$

c. $\dfrac{y^{6a}}{y^{4a}} = y^{6a-4a} = y^{2a}$

Simplify each expression. Assume that variables represent positive integers. See the example.

101. $a^{-4m} \cdot a^{5m}$ a^m

102. $(x^{-3s})^3$ $\dfrac{1}{x^{9s}}$

103. $(3y^{2z})^3$ $27y^{6z}$

104. $a^{4m+1} \cdot a^4$ a^{4m+5}

105. $\dfrac{y^{4a}}{y^{-a}}$ y^{5a}

106. $\dfrac{y^{-6a}}{zy^{6a}}$ $\dfrac{1}{y^{12a}z}$

107. $(z^{3a+2})^{-2}$ $\dfrac{1}{z^{6a+4}}$

108. $(a^{4x-1})^{-1}$ $\dfrac{1}{a^{4x-1}}$

4.6 | DIVISION OF POLYNOMIALS

TAPE BA 4.6

OBJECTIVES

1 Divide a polynomial by a monomial.

2 Use long division to divide a polynomial by another polynomial.

 Although we didn't use the word monomial, in Sections 4.1 and 4.5 we divided monomials in developing rules and definitions for exponents. For example, $\frac{36a^2b}{6ab^3}$ can be thought of as the monomial $36a^2b$ divided by the monomial $6ab^3$.

EXAMPLE 1   Simplify by performing the indicated division of monomials. Write each answer with positive exponents.

a. $\dfrac{36a^2b}{6ab^3}$

b. $\dfrac{3x^4y^5}{9x^4y^2}$

Solution: To divide the monomials, use rules for exponents.

a. $\dfrac{36a^2b}{6ab^3} = \dfrac{6a}{b^2}$

b. $\dfrac{3x^4y^5}{9x^4y^2} = \dfrac{y^3}{3}$

To divide a polynomial by a monomial, recall addition of fractions. Fractions that have a common denominator are added by adding the numerators:

$$\frac{a}{c} + \frac{b}{c} = \frac{a + b}{c}$$

If we read this equation from right to left and let a, b, and c be monomials, $c \neq 0$, the following emerges.

TO DIVIDE A POLYNOMIAL BY A MONOMIAL

Divide each term of the polynomial by the monomial.

$$\frac{a + b}{c} = \frac{a}{c} + \frac{b}{c}, \quad c \neq 0$$

Throughout this section, we assume that denominators are not 0.

EXAMPLE 2   Divide $(6m^2 + 2m)$ by $2m$.

Solution: Begin by writing the quotient in fraction form. Then divide each term of the polynomial $6m^2 + 2m$ by the monomial $2m$.

$$\frac{6m^2 + 2m}{2m} = \frac{6m^2}{2m} + \frac{2m}{2m} = 3m + 1$$

To check, multiply the quotient $(3m + 1)$ by the divisor $2m$ and see that the product is the dividend $(6m^2 + 2m)$.

$$2m(3m + 1) = 2m(3m) + 2m(1) = 6m^2 + 2m$$

The quotient $(3m + 1)$ is correct.

EXAMPLE 3   Simplify the quotient $\dfrac{8x^2y^2 - 16xy + 2x}{4xy}$.

Solution: $\dfrac{8x^2y^2 - 16xy + 2x}{4xy} = \dfrac{8x^2y^2}{4xy} - \dfrac{16xy}{4xy} + \dfrac{2x}{4xy}$ Divide each term by $4xy$.

$$= 2xy - 4 + \frac{1}{2y}$$

Notice that the quotient is not a polynomial because of the term $\frac{1}{2y}$. This expression is called a rational expression and we will study rational expressions further in Chapter 6. Although the quotient of two polynomials is not always a polynomial, we may still check by multiplying.

$$4xy\left(2xy - 4 + \frac{1}{2y}\right) = 4xy(2xy) + 4xy(-4) + 4xy\left(\frac{1}{2y}\right)$$

$$= 8x^2y^2 - 16xy + 2x$$

EXAMPLE 4 Simplify $\dfrac{12x^5y^6 - 6x^2y^2 + 9x^3y^4 + 3}{6x^2y^2}$.

Solution: Divide each term by $6x^2y^2$.

$$\frac{12x^5y^6 - 6x^2y^2 + 9x^3y^4 + 3}{6x^2y^2} = \frac{12x^5y^6}{6x^2y^2} - \frac{6x^2y^2}{6x^2y^2} + \frac{9x^3y^4}{6x^2y^2} + \frac{3}{6x^2y^2}$$

$$= 2x^3y^4 - 1 + \frac{3xy^2}{2} + \frac{1}{2x^2y^2}$$

Check this result by multiplying.

2 To divide a polynomial by a polynomial other than a monomial, we use a process known as long division. Polynomial long division is similar to number long division, so we review long division by dividing 13 into 3660.

$$\begin{array}{r} 281 \\ 13\overline{)3660} \\ \underline{26}\downarrow \\ 106 \\ \underline{104}\downarrow \\ 20 \\ \underline{13} \\ 7 \end{array}$$

$2 \cdot 13 = 26$

Subtract and bring down the next digit in the dividend.

$8 \cdot 13 = 104$

Subtract and bring down the next digit in the dividend.

$1 \cdot 13 = 13$

Subtract. There are no more digits to bring down, so the remainder is 7.

The quotient is 281 R 7, which can be written as $281\frac{7}{13}$ ← remainder / ← divisor. Recall that division can be checked by multiplication. To check a division problem such as this one, we see that

$$13 \cdot 281 + 7 = 3660$$

Now we demonstrate long division of polynomials.

EXAMPLE 5 Divide $(x^2 + 7x + 12)$ by $(x + 3)$.

Solution: $(x^2 + 7x + 12)$ is the **dividend polynomial** and $(x + 3)$ is the **divisor polynomial.**

How many times does x
divide x^2? $\dfrac{x^2}{x} = x$.

$$\begin{array}{r} x \\ x + 3\overline{)x^2 + 7x + 12} \\ x^2 + 3x \\ \hline 4x + 12 \end{array}$$

To subtract, change the signs of these terms and add.

Multiply: $x(x + 3)$.

Subtract and bring down the next term.

Next, repeat this process.

$$
\begin{array}{r}
x + 4 \\
x + 3\overline{)x^2 + 7x + 12} \\
\underline{x^2 + 3x} \\
4x + 12 \\
\underline{4x + 12} \\
0
\end{array}
$$

How many times does x divide $4x$? $\dfrac{4x}{x} = 4.$

To subtract, change the signs of these terms and add. $\longrightarrow$

Multiply: $4(x + 3)$.

Subtract. The remainder is 0.

Then $(x^2 + 7x + 12)$ divided by $(x + 3)$ is $(x + 4)$, the **quotient polynomial.** To check, see that

$$\text{divisor} \cdot \text{quotient} + \text{remainder} = \text{dividend}$$

or

$$(x + 3) \cdot (x + 4) + 0 = x^2 + 7x + 12, \text{ the dividend, so the}$$
$$\text{division checks.}$$

EXAMPLE 6

Solution:

Divide $6x^2 + 10x - 5$ by $3x - 1$.

The divisor polynomial is $(3x - 1)$ and the dividend polynomial is $(6x^2 + 10x - 5)$.

$$
\begin{array}{r}
2x + 4 \\
3x - 1\overline{)6x^2 + 10x - 5} \\
\underline{6x^2 - 2x} \downarrow \\
12x - 5 \\
\underline{12x - 4} \\
-1
\end{array}
$$

$\dfrac{6x^2}{3x} = 2x$, so $2x$ is a term of the quotient.

$2x(3x - 1)$

Subtract and bring down the next term.

$\dfrac{12x}{3x} = 4, 4(3x - 1)$

Subtract. The remainder is -1.

Then $(6x^2 + 10x - 5)$ divided by $(3x - 1)$ is $(2x + 4)$ with a remainder of -1. This can be written as

$$\frac{6x^2 + 10x - 5}{3x - 1} = 2x + 4 + \frac{-1}{3x - 1} \left(\frac{\text{remainder}}{\text{divisor}} \right)$$

We call $(2x + 4)$ the quotient polynomial and -1 the **remainder polynomial.** To check, see that divisor $\cdot$ quotient $+$ remainder $=$ dividend.

$$(3x - 1)(2x + 4) + (-1) = (6x^2 + 12x - 2x - 4) - 1$$
$$= 6x^2 + 10x - 5$$

The division checks.

Notice that the division process is continued until the degree of the remainder polynomial is less than the degree of the divisor polynomial.

EXAMPLE 7 Divide: $\dfrac{4x^2 + 7 + 8x^3}{2x + 3}$.

Solution: Before we begin the division process, the dividend polynomial and the divisor polynomial should be written in descending order of exponents. Any missing powers are represented by a term whose coefficient is zero.

$$\frac{4x^2 + 7 + 8x^3}{2x + 3} = \frac{8x^3 + 4x^2 + 0x + 7}{2x + 3}$$

There is no x term, so include $0x$ as the missing power.

$$
\require{enclose}
\begin{array}{r}
4x^2 - 4x + 6 \\
2x + 3 \enclose{longdiv}{8x^3 + 4x^2 + 0x + 7} \\
\underline{8x^3 + 12x^2} \\
-8x^2 + 0x \\
\underline{-8x^2 - 12x} \\
12x + 7 \\
\underline{12x + 18} \\
-11
\end{array}
$$

Remainder polynomial.

Thus, $\dfrac{4x^2 + 7 + 8x^3}{2x + 3} = 4x^2 - 4x + 6 + \dfrac{-11}{2x + 3}$.

EXAMPLE 8 Divide: $\dfrac{2x^4 - x^3 + 3x^2 + x - 1}{x^2 + 1}$.

Solution: Before dividing, rewrite the divisor polynomial $(x^2 + 1)$ as $(x^2 + 0x + 1)$. The $0x$ term represents the missing x^1 term in the divisor.

$$
\require{enclose}
\begin{array}{r}
2x^2 - x + 1 \\
x^2 + 0x + 1 \enclose{longdiv}{2x^4 - x^3 + 3x^2 + x - 1} \\
\underline{2x^4 + 0x^3 + 2x^2} \\
-x^3 + x^2 + x \\
\underline{-x^3 - 0x^2 - x} \\
x^2 + 2x - 1 \\
\underline{x^2 + 0x + 1} \\
2x - 2
\end{array}
$$

Remainder polynomial.

Thus, $\dfrac{2x^4 - x^3 + 3x^2 + x - 1}{x^2 + 1} = 2x^2 - x + 1 + \dfrac{2x - 2}{x^2 + 1}$.

MENTAL MATH

Simplify each expression mentally.

1. $\dfrac{a^6}{a^4}$ a^2

2. $\dfrac{y^2}{y}$ y

3. $\dfrac{a^3}{a}$ a^2

4. $\dfrac{p^8}{p^3}$ p^5

5. $\dfrac{k^5}{k^2}$ k^3

6. $\dfrac{k^7}{k^5}$ k^2

10. $2m - \dfrac{27m^2}{7}$ **13.** $-3x^2 + x - \dfrac{4}{x^3}$ **14.** $3 - \dfrac{2}{a} + \dfrac{6}{a^2}$ **15.** $-1 + \dfrac{3}{2x} - \dfrac{7}{4x^4}$ **16.** $-4a^2 + 12 - \dfrac{5}{a}$

EXERCISE SET 4.6

17. $5x^3 - 3x + \dfrac{1}{x^2}$ **18.** $-2y + 2 + \dfrac{3}{y}$ **20.** $(6x^3 - 2x + 1)$ ft.   **25.** $2x + 1 + \dfrac{7}{x-4}$

Simplify each expression. See Example 1.

1. $\dfrac{8k^4}{2k}$ $4k^3$

2. $\dfrac{27r^4}{3r^6}$ $\dfrac{9}{r^2}$

3. $\dfrac{-6m^4}{-2m^3}$ $3m$

4. $\dfrac{15a^4}{-15a^5}$ $-\dfrac{1}{a}$

5. $\dfrac{-24a^6b}{6ab^2}$ $-\dfrac{4a^5}{b}$

6. $\dfrac{-5x^4y^5}{15x^4y^2}$ $-\dfrac{y^3}{3}$

7. $\dfrac{6x^2y^3}{-7xy^5}$ $-\dfrac{6x}{7y^2}$

8. $\dfrac{-8xa^2b}{-5xa^5b}$ $\dfrac{8}{5a^3}$

Perform each division. See Examples 2 through 4.

9. $\dfrac{15p^3 + 18p^2}{3p}$ $5p^2 + 6p$

10. $\dfrac{14m^2 - 27m^3}{7m}$

11. $\dfrac{-9x^4 + 18x^5}{6x^5}$ $-\dfrac{3}{2x} + 3$

12. $\dfrac{6x^5 + 3x^4}{3x^4}$ $2x + 1$

13. $\dfrac{-9x^5 + 3x^4 - 12}{3x^3}$

14. $\dfrac{6a^2 - 4a + 12}{2a^2}$

15. $\dfrac{4x^4 - 6x^3 + 7}{-4x^4}$

16. $\dfrac{-12a^3 + 36a - 15}{3a}$

17. $\dfrac{25x^5 - 15x^3 + 5}{5x^2}$

18. $\dfrac{-4y^2 + 4y + 6}{2y}$

19. The perimeter of a square is $(12x^3 + 4x - 16)$ feet. Find the length of its side. $(3x^3 + x - 4)$ ft.

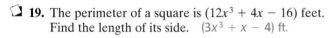

Perimeter is
$(12x^3 + 4x - 16)$
feet

20. The volume of the swimming pool shown is $(36x^5 - 12x^3 + 6x^2)$ cubic feet. If its height is $2x$ feet and its width is $3x$ feet, find its length.

3x feet

2x feet

Perform each division. See Examples 5 through 8.

21. $\dfrac{x^2 + 4x + 3}{x + 3}$ $x + 1$

22. $\dfrac{x^2 + 7x + 10}{x + 5}$ $x + 2$

23. $\dfrac{2x^2 + 13x + 15}{x + 5}$ $2x + 3$

24. $\dfrac{3x^2 + 8x + 4}{x + 2}$ $3x + 2$

25. $\dfrac{2x^2 - 7x + 3}{x - 4}$

26. $\dfrac{3x^2 - x - 4}{x - 1}$

27. $\dfrac{8x^2 + 6x - 27}{2x - 3}$ $4x + 9$

28. $\dfrac{18w^2 + 18w - 8}{3w + 4}$

29. $\dfrac{9a^3 - 3a^2 - 3a + 4}{3a + 2}$

30. $\dfrac{-x^3 - 6x^2 + 2x - 3}{x - 1}$

31. $\dfrac{2b^3 + 9b^2 + 6b - 4}{b + 4}$

32. $\dfrac{2x^3 + 3x^2 - 3x + 4}{x + 2}$

33. Explain how to check a polynomial long division result when the remainder is 0.

34. Explain how to check a polynomial long division result when the remainder is not 0.

35. The area of the following parallelogram is $(10x^2 + 31x + 15)$ square meters. If its base is $(5x + 3)$ meters, find its height. $(2x + 5)$ meters

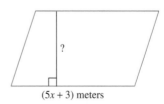

?

$(5x + 3)$ meters

36. The area of the top of the Ping-Pong table is $(49x^2 + 70x - 200)$ square inches. If its length is $(7x + 20)$ inches, find its width. $(7x - 10)$ in.

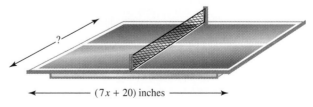

?

$(7x + 20)$ inches

26. $3x + 2 - \dfrac{2}{x - 1}$ **28.** $6w - 2$ **29.** $3a^2 - 3a + 1 + \dfrac{2}{3a + 2}$ **30.** $-x^2 - 7x - 5 - \dfrac{8}{x - 1}$

31. $2b^2 + b + 2 - \dfrac{12}{b + 4}$ **32.** $2x^2 - x - 1 + \dfrac{6}{x + 2}$ **33.** answers may vary   **34.** answers may vary

37. $\dfrac{4}{x} + \dfrac{1}{x^2} + \dfrac{9}{5x^3}$ **38.** $4x - 2 + \dfrac{3}{x} + \dfrac{1}{x^2}$

39. $5x - 2 + \dfrac{2}{x + 6}$ **41.** $x^2 - \dfrac{12x}{5} - 1$ **42.** $\dfrac{x}{2} + 3 + \dfrac{4}{x^2}$ **43.** $6x - 1 - \dfrac{1}{x + 3}$ **44.** $2x + 3 + \dfrac{23}{x - 6}$

Perform each division.

37. $\dfrac{20x^2 + 5x + 9}{5x^3}$

38. $\dfrac{8x^3 - 4x^2 + 6x + 2}{2x^2}$

39. $\dfrac{5x^2 + 28x - 10}{x + 6}$

40. $\dfrac{2x^2 + x - 15}{x + 3}$ $2x - 5$

41. $\dfrac{10x^3 - 24x^2 - 10x}{10x}$

42. $\dfrac{2x^3 + 12x^2 + 16}{4x^2}$

43. $\dfrac{6x^2 + 17x - 4}{x + 3}$

44. $\dfrac{2x^2 - 9x + 15}{x - 6}$

45. $\dfrac{12x^4 + 3x^2}{3x^2}$ $4x^2 + 1$

46. $\dfrac{15x^2 - 9x^5}{9x^5}$ $\dfrac{5}{3x^3} - 1$

47. $\dfrac{2x^3 + 2x^2 - 17x + 8}{x - 2}$ $2x^2 + 6x - 5 - \dfrac{2}{x - 2}$

48. $\dfrac{4x^3 + 11x^2 - 8x - 10}{x + 3}$ $4x^2 - x - 5 + \dfrac{5}{x + 3}$

49. $\dfrac{30x^2 - 17x + 2}{5x - 2}$ $6x - 1$ **50.** $\dfrac{4x^2 - 13x - 12}{4x + 3}$ $x - 4$

51. $\dfrac{3x^4 - 9x^3 + 12}{-3x}$

52. $\dfrac{8y^6 - 3y^2 - 4y}{4y}$

53. $\dfrac{8x^2 + 10x + 1}{2x + 1}$

54. $\dfrac{3x^2 + 17x + 7}{3x + 2}$

55. $\dfrac{4x^2 - 81}{2x - 9}$ $2x + 9$

56. $\dfrac{16x^2 - 36}{4x + 6}$ $4x - 6$

57. $\dfrac{4x^3 + 12x^2 + x - 12}{2x + 3}$

58. $\dfrac{6x^2 + 11x - 10}{3x - 2}$

59. $\dfrac{x^3 - 27}{x - 3}$ $x^2 + 3x + 9$ **60.** $\dfrac{x^3 + 64}{x + 4}$ $x^2 - 4x + 16$

61. $\dfrac{x^3 + 1}{x + 1}$ $x^2 - x + 1$ **62.** $\dfrac{x^5 + x^2}{x^2 + x}$

63. $\dfrac{1 - 3x^2}{x + 2}$

64. $\dfrac{7 - 5x^2}{x + 3}$

65. $\dfrac{-4b + 4b^2 - 5}{2b - 1}$

66. $\dfrac{-3y + 2y^2 - 15}{2y + 5}$

Review Exercises

Multiply each expression. See Section 4.3.

67. $2a(a^2 + 1)$ $2a^3 + 2a$ **68.** $-4a(3a^2 - 4)$

69. $2x(x^2 + 7x - 5)$ **70.** $4y(y^2 - 8y - 4)$

71. $-3xy(xy^2 + 7x^2y + 8)$ $-3x^2y^3 - 21x^3y^2 - 24xy$

72. $-9xy(4xyz + 7xy^2z + 2)$

73. $9ab(ab^2c + 4bc - 8)$ $9a^2b^3c + 36ab^2c - 72ab$

74. $-7sr(6s^2r + 9sr^2 + 9rs + 8)$

Use the bar graph below to answer exercises 75 through 78. See Section 1.9.

75. Which album has sold the most copies?   *Thriller*

76. Estimate how many more copies the album *Rumours* has sold than the album *Boston*.

77. Which album(s) shown has sold the least copies?

78. If Michael Jackson made \$7 for every *Thriller* album sold, what was his income for this album?

68. $-12a^3 + 16a$ **69.** $2x^3 + 14x^2 - 10x$ **70.** $4y^3 - 32y^2 - 16y$ **72.** $-36x^2y^2z - 63x^2y^3z - 18xy$
74. $-42s^3r^2 - 63s^2r^3 - 63s^2r^2 - 56sr$ **76.** 2 million **77.** *Born in the U.S.A.; Eagles Greatest Hits* **78.** \$168,000,000

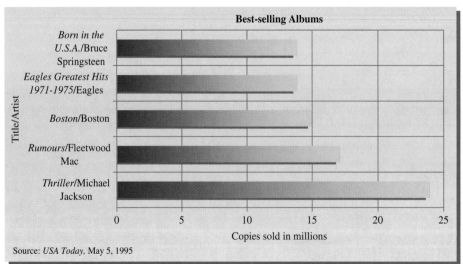

Best-selling Albums

Title/Artist (vertical axis):
Born in the U.S.A./Bruce Springsteen
Eagles Greatest Hits 1971-1975/Eagles
Boston/Boston
Rumours/Fleetwood Mac
Thriller/Michael Jackson

Copies sold in millions (horizontal axis): 0, 5, 10, 15, 20, 25

Source: *USA Today*, May 5, 1995

51. $-x^3 + 3x^2 - \dfrac{4}{x}$ **52.** $2y^5 - \dfrac{3y}{4} - 1$ **53.** $4x + 3 - \dfrac{2}{2x + 1}$ **54.** $x + 5 - \dfrac{3}{3x + 2}$ **57.** $2x^2 + 3x - 4$ **58.** $2x + 5$

62. $x^3 - x^2 + x$ **63.** $-3x + 6 - \dfrac{11}{x + 2}$ **64.** $-5x + 15 - \dfrac{38}{x + 3}$ **65.** $2b - 1 - \dfrac{6}{2b - 1}$ **66.** $y - 4 + \dfrac{5}{2y + 5}$

GROUP ACTIVITY

MAKING PREDICTIONS BASED ON HISTORICAL DATA

MATERIALS:
- Calculator
- Grapher with bar graph capabilities (optional)

According to data from the National Safety Council, a polynomial that represents the annual number of deaths due to motor vehicle accidents over the period 1990–1994 is

$$46{,}426 - 4381x + 893x^2,$$

where $x = 0$ represents 1990. A polynomial, also based on data from the National Safety Council, that represents the annual number of deaths due to all other types of accidents over the period 1990–1994 is

$$33{,}139 - 2540x + 730x^2,$$

where $x = 0$ represents 1990.

1. Use the polynomials to complete the following table showing the number of accidental deaths per year over the period 1990–1994 by evaluating each polynomial at the given values of x. What trends do you notice in the data?

YEAR	x	NUMBER OF ACCIDENTAL DEATHS DUE TO MOTOR VEHICLE ACCIDENTS	NUMBER OF ACCIDENTAL DEATHS DUE TO ALL OTHER TYPES OF ACCIDENTS
1990	0		
1991	1		
1992	2		
1993	3		
1994	4		

2. Use the given polynomials to find a polynomial that represents the total number of accidental deaths per year. Evaluate this new polynomial to find the total number of accidental deaths for the years 1990–1994.

3. Numerically verify the values you calculated in Question 2 by adding a new column, titled "Total Number of Accidental Deaths Due to All Causes," to the table in Question 1 and combining values from the table to complete the new column.

4. Use the new polynomial representing total number of accidental deaths per year to predict the number of accidental deaths in the years 2000 and 2002.

5. Create a bar graph that represents the data for total accidental deaths per year for the years 1990–1994 and your predictions for 2000 and 2002. Study your bar graph. Discuss what the graph implies about the future.

6. (Optional) Use a grapher to create the bar graph in Question 5. Compare it to your own graph. Can you draw the same conclusions? Explain.

See App. F for Group Activity answers and suggestions.

CHAPTER 4 HIGHLIGHTS

DEFINITIONS AND CONCEPTS	EXAMPLES

SECTION 4.1 EXPONENTS

a^n means the product of n factors, each of which is a.

$3^2 = 3 \cdot 3 = 9$

$(-5)^3 = (-5)(-5)(-5) = -125$

$\left(\frac{1}{2}\right)^4 = \frac{1}{2} \cdot \frac{1}{2} \cdot \frac{1}{2} \cdot \frac{1}{2} = \frac{1}{16}$

If m and n are integers and no denominators are 0,

Product Rule: $a^m \cdot a^n = a^{m+n}$

$x^2 \cdot x^7 = x^{2+7} = x^9$

Power of a Power Rule: $(a^m)^n = a^{mn}$

$(5^3)^8 = 5^{3 \cdot 8} = 5^{24}$

Power of a Product Rule: $(ab)^n = a^n b^n$

$(7y)^4 = 7^4 y^4$

Power of a Quotient Rule: $\left(\dfrac{a}{b}\right)^n = \dfrac{a^n}{b^n}$

$\left(\dfrac{x}{8}\right)^3 = \dfrac{x^3}{8^3}$

Quotient Rule: $\dfrac{a^m}{a^n} = a^{m-n}$

$\dfrac{x^9}{x^4} = x^{9-4} = x^5$

Zero Exponent: $a^0 = 1$, $a \neq 0$.

$5^0 = 1$, $x^0 = 1$, $x \neq 0$.

SECTION 4.2 ADDING AND SUBTRACTING POLYNOMIALS

A **term** is a number or the product of numbers and variables raised to powers.

Terms

$-5x$, $7a^2b$, $\frac{1}{4}y^4$, 0.2

The **numerical coefficient** or **coefficient** of a term is its numerical factor.

Term	Coefficient
$7x^2$	7
y	1
$-a^2b$	-1

A **polynomial** is a term or a finite sum of terms in which variables may appear in the numerator raised to whole number powers only.

Polynomials

$3x^2 - 2x + 1$ (Trinomial)

$-0.2a^2b - 5b^2$ (Binomial)

$\frac{5}{6}y^3$ (Monomial)

A **monomial** is a polynomial with exactly 1 term.

A **binomial** is a polynomial with exactly 2 terms.

A **trinomial** is a polynomial with exactly 3 terms.

The **degree of a term** is the sum of the exponents on the variables in the term.

Term	Degree
$-5x^3$	3
3(or $3x^0$)	0
$2a^2b^2c$	5

(continued)

DEFINITIONS AND CONCEPTS	EXAMPLES

SECTION 4.2 ADDING AND SUBTRACTING POLYNOMIALS

The **degree of a polynomial** is the greatest degree of any term of the polynomial.	Polynomial Degree $5x^2 - 3x + 2$ 2 $7y + 8y^2z^3 - 12$ $2 + 3 = 5$
To add polynomials, add like terms.	Add: $(7x^2 - 3x + 2) + (-5x - 6) = 7x^2 - 3x + 2 - 5x - 6$ $= 7x^2 - 8x - 4$
To subtract two polynomials, change the signs of the terms of the second polynomial, then add.	Subtract: $(17y^2 - 2y + 1) - (-3y^3 + 5y - 6)$ $= (17y^2 - 2y + 1) + (3y^3 - 5y + 6)$ $= 17y^2 - 2y + 1 + 3y^3 - 5y + 6$ $= 3y^3 + 17y^2 - 7y + 7$

SECTION 4.3 MULTIPLYING POLYNOMIALS

To multiply two polynomials, multiply each term of one polynomial by each term of the other polynomial, and then combine like terms.	Multiply: $(2x + 1)(5x^2 - 6x + 2)$ $= 2x(5x^2 - 6x + 2) + 1(5x^2 - 6x + 2)$ $= 10x^3 - 12x^2 + 4x + 5x^2 - 6x + 2$ $= 10x^3 - 7x^2 - 2x + 2$

SECTION 4.4 SPECIAL PRODUCTS

The **FOIL method** may be used when multiplying two binomials.	Multiply: $(5x - 3)(2x + 3)$ F O I L $(5x - 3)(2x + 3) = (5x)(2x) + (5x)(3) + (-3)(2x) + (-3)(3)$ $= 10x^2 + 15x - 6x - 9$ $= 10x^2 + 9x - 9$

Squaring a Binomial
 	
     $(a + b)^2 = a^2 + 2ab + b^2$
 	
     $(a - b)^2 = a^2 - 2ab + b^2$
	Square each binomial.
	     $(x + 5)^2 = x^2 + 2(x)(5) + 5^2$
	             $= x^2 + 10x + 25$
	     $(3x - 2y)^2 = (3x)^2 - 2(3x)(2y) + (2y)^2$
	             $= 9x^2 - 12xy + 4y^2$

Multiplying the Sum and Difference of Two Terms
     $(a + b)(a - b) = a^2 - b^2$
	Multiply.
	     $(6y + 5)(6y - 5) = (6y)^2 - 5^2$
	                 $= 36y^2 - 25$

(continued)

DEFINITIONS AND CONCEPTS	EXAMPLES
SECTION 4.5 NEGATIVE EXPONENTS AND SCIENTIFIC NOTATION	

DEFINITIONS AND CONCEPTS	EXAMPLES
If $a \neq 0$ and n is an integer, $$a^{-n} = \frac{1}{a^n}$$ Rules for exponents are true for positive and negative integers.	$$3^{-2} = \frac{1}{3^2} = \frac{1}{9}; \ 5x^{-2} = \frac{5}{x^2}$$ Simplify: $\left(\dfrac{x^{-2}y}{x^5}\right)^{-2} = \dfrac{x^4 y^{-2}}{x^{-10}}$ $$= x^{4-(-10)}y^{-2}$$ $$= \frac{x^{14}}{y^2}$$
A positive number is written in scientific notation if it is as the product of a number a, $1 \leq a < 10$, and an integer power r of 10. $$a \times 10^r$$	Numbers Written in Scientific Notation $$12{,}000 = 1.2 \times 10^4$$ $$0.00000568 = 5.68 \times 10^{-6}$$

SECTION 4.6 DIVISION OF POLYNOMIALS	

DEFINITIONS AND CONCEPTS	EXAMPLES
To divide a polynomial by a monomial: $$\frac{a+b}{c} = \frac{a}{c} + \frac{b}{c}$$ To divide a polynomial by a polynomial other than a monomial, use long division.	Divide: $$\frac{15x^5 - 10x^3 + 5x^2 - 2x}{5x^2} = \frac{15x^5}{5x^2} - \frac{10x^3}{5x^2} + \frac{5x^2}{5x^2} - \frac{2x}{5x^2}$$ $$= 3x^3 - 2x + 1 - \frac{2}{5x}$$ $$\begin{array}{r} 5x - 1 + \dfrac{-4}{2x+3} \\ 2x+3\overline{)10x^2 + 13x - 7} \\ \underline{10x^2 + 15x} \\ -2x - 7 \\ \underline{-2x - 3} \\ -4 \end{array}$$

CHAPTER 4 REVIEW

(4.1)   *State the base and the exponent for each expression.*

1. 3^2 base: 3; exponent: 2

2. $(-5)^4$ base: -5; exponent: 4

3. -5^4 base: 5; exponent: 4

Evaluate each expression.

4. 8^3 512

5. $(-6)^2$ 36

6. -6^2 -36

7. $-4^3 - 4^0$ -65

8. $(3b)^0$ 1

9. $\dfrac{8b}{8b}$ 1

Simplify each expression.

10. $5b^3 b^5 a^6$ $5a^6 b^8$

11. $2^3 \cdot x^0$ 8

12. $[(-3)^2]^3$ 729

13. $(2x^3)(-5x^2)$ $-10x^5$

14. $\left(\dfrac{mn}{q}\right)^2 \cdot \left(\dfrac{mn}{q}\right)$ $\dfrac{m^3 n^3}{q^3}$

15. $\left(\dfrac{3ab^2}{6ab}\right)^4$ $\dfrac{b^4}{16}$

16. $\dfrac{x^9}{x^4}$ x^5

17. $\dfrac{2x^7 y^8}{8xy^2}$ $\dfrac{x^6 y^6}{4}$

18. $\dfrac{12xy^6}{3x^4 y^{10}}$ $\dfrac{4}{x^3 y^4}$

19. $5a^7(2a^4)^3$ $40a^{19}$

20. $(2x)^2(9x)$ $36x^3$

21. $\dfrac{(-4)^2(3^3)}{(4^5)(3^2)}$ $\dfrac{3}{64}$

22. $\dfrac{(-7)^2(3^5)}{(-7)^3(3^4)}$ $-\dfrac{3}{7}$

23. $\dfrac{(2x)^0(-4)^2}{16x}$ $\dfrac{1}{x}$

24. $\dfrac{(8xy)(3xy)}{18x^2y^2}$ $\dfrac{4}{3}$

25. $m^0 + p^0 + 3q^0$ 5

26. $(-5a)^0 + 7^0 + 8^0$ 3

27. $(3xy^2 + 8x + 9)^0$ 1

28. $8x^0 + 9^0$ 9

29. $6(a^2b^3)^3$ $6a^6b^9$

30. $\dfrac{(x^3z)^a}{x^2z^2}$ $x^{3a-2}z^{a-2}$

(4.2)   *Find the degree of each term.*

31. $-5x^4y^3$ 7

32. $10x^3y^2z$ 6

33. $35a^5bc^2$ 8

34. $95xyz$ 3

Find the degree of each polynomial.

35. $y^5 + 7x - 8x^4$ 5

36. $9y^2 + 30y + 25$ 2

37. $-14x^2yb - 28x^2y^3b - 42x^2y^2$ 6

38. $6x^2y^2z^2 + 5x^2y^3 - 12xyz$ 6

39. The surface area of a box with a square base and a height of 5 units is given by the polynomial $2x^2 + 20x$. Fill in the table below by evaluating $2x^2 + 20x$ for the given values of x.

x	1	3	5.1	10
$2x^2 + 20x$	22	78	154.02	400

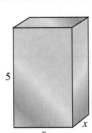

42. $-6a^2b - 3b^2 - q^2$
45. $2s^5 + 3s^4 + 4s^3 + s^2 - 7s - 6$
46. $-6m^7 - 3x^4 + 7m^6 - 4m^2$
51. $-12x^2a^2y^4$
56. $-7x^3 - 35x$
57. $-32y^3 + 48y$
58. $-2x^3 + 18x^2 - 2x$
59. $-3a^3b - 3a^2b - 3ab^2$
60. $-6a^4 + 8a^2 - 2a$
61. $42b^4 - 28b^2 + 14b$
62. $2x^2 - 12x - 14$

Combine like terms.

40. $6a^2b^2 + 4ab + 9a^2b^2$ $15a^2b^2 + 4ab$

41. $21x^2y^3 + 3xy + x^2y^3 + 6$ $22x^2y^3 + 3xy + 6$

42. $4a^2b - 3b^2 - 8q^2 - 10a^2b + 7q^2$

43. $2s^{14} + 3s^{13} + 12s^{12} - s^{10}$ cannot be combined

Add or subtract as indicated.

44. $(3k^2 + 2k + 6) + (5k^2 + k)$ $8k^2 + 3k + 6$

45. $(2s^5 + 3s^4 + 4s^3 + 5s^2) - (4s^2 + 7s + 6)$

46. $(2m^7 + 3x^4 + 7m^6) - (8m^7 + 4m^2 + 6x^4)$

47. Subtract $(4x^2 + 8x - 7)$ from the sum of $(x^2 + 7x + 9)$ and $(x^2 + 4)$. $-2x^2 - x + 20$

(4.3)   *Multiply each expression.*

48. $9x(x^2y)$ $9x^3y$

49. $-7(8xz^2)$ $-56xz^2$

50. $(6xa^2)(xya^3)$ $6x^2a^5y$

51. $(4xy)(-3xa^2y^3)$

52. $6(x + 5)$ $6x + 30$

53. $9(x - 7)$ $9x - 63$

54. $4(2a + 7)$ $8a + 28$

55. $9(6a - 3)$ $54a - 27$

56. $-7x(x^2 + 5)$

57. $-8y(4y^2 - 6)$

58. $-2(x^3 - 9x^2 + x)$

59. $-3a(a^2b + ab + b^2)$

60. $(3a^3 - 4a + 1)(-2a)$

61. $(6b^3 - 4b + 2)(7b)$

62. $(2x + 2)(x - 7)$

63. $(2x - 5)(3x + 2)$

64. $(4a - 1)(a + 7)$

65. $(6a - 1)(7a + 3)$

66. $(x + 7)(x^3 + 4x - 5)$

67. $(x + 2)(x^5 + x + 1)$

68. $(x^2 + 2x + 4)(x^2 + 2x - 4)$ $x^4 + 4x^3 + 4x^2 - 16$

69. $(x^3 + 4x + 4)(x^3 + 4x - 4)$ $x^6 + 8x^4 + 16x^2 - 16$

70. $(x + 7)^3$

71. $(2x - 5)^3$

(4.4)   *Use the special product rules to find each product.*

72. $(x + 7)^2$ $x^2 + 14x + 49$ **73.** $(x - 5)^2$

74. $(3x - 7)^2$

75. $(4x + 2)^2$

76. $(5x - 9)^2$

77. $(5x + 1)(5x - 1)$

78. $(7x + 4)(7x - 4)$

79. $(a + 2b)(a - 2b)$

80. $(2x - 6)(2x + 6)$

81. $(4a^2 - 2b)(4a^2 + 2b)$

(4.5)   *Simplify each expression.*

82. 7^{-2}

83. -7^{-2}

84. $2x^{-4}$

85. $(2x)^{-4}$

86. $\left(\dfrac{1}{5}\right)^{-3}$ 125

87. $\left(\dfrac{-2}{3}\right)^{-2}$ $\dfrac{9}{4}$

88. $2^0 + 2^{-4}$ $\dfrac{17}{16}$

89. $6^{-1} - 7^{-1}$ $\dfrac{1}{42}$

Simplify each expression. Assume that variables in an exponent represent positive integers only. Write each answer using positive exponents.

90. $\dfrac{1}{(2q)^{-3}}$ $8q^3$

91. $\dfrac{-1}{(qr)^{-3}}$ $-q^3r^3$

92. $\dfrac{r^{-3}}{s^{-4}}$ $\dfrac{s^4}{r^3}$

93. $\dfrac{rs^{-3}}{r^{-4}}$ $\dfrac{r^5}{s^3}$

94. $\dfrac{-6}{8x^{-3}r^4}$ $-\dfrac{3x^3}{4r^4}$

95. $\dfrac{-4s}{16s^{-3}}$ $-\dfrac{s^4}{4}$

96. $(2x^{-5})^{-3}$ $\dfrac{x^{15}}{8}$

97. $(3y^{-6})^{-1}$ $\dfrac{y^6}{3}$

63. $6x^2 - 11x - 10$ **64.** $4a^2 + 27a - 7$ **65.** $42a^2 + 11a - 3$ **66.** $x^4 + 7x^3 + 4x^2 + 23x - 35$
67. $x^6 + 2x^5 + x^2 + 3x + 2$ **70.** $x^3 + 21x^2 + 147x + 343$ **71.** $8x^3 - 60x^2 + 150x - 125$ **73.** $x^2 - 10x + 25$
74. $9x^2 - 42x + 49$ **75.** $16x^2 + 16x + 4$ **76.** $25x^2 - 90x + 81$ **77.** $25x^2 - 1$ **78.** $49x^2 - 16$
79. $a^2 - 4b^2$ **80.** $4x^2 - 36$ **81.** $16a^4 - 4b^2$

82. $\dfrac{1}{49}$ **83.** $-\dfrac{1}{49}$ **84.** $\dfrac{2}{x^4}$ **85.** $\dfrac{1}{16x^4}$ **98.** $\dfrac{a^2b^2c^4}{9}$

98. $(3a^{-1}b^{-1}c^{-2})^{-2}$

99. $(4x^{-2}y^{-3}z)^{-3}$ $\dfrac{x^6y^9}{64z^3}$

100. $\dfrac{5^{-2}x^8}{5^{-3}x^{11}}$ $\dfrac{5}{x^3}$

101. $\dfrac{7^5y^{-2}}{7^7y^{-10}}$ $\dfrac{y^8}{49}$

102. $\left(\dfrac{bc^{-2}}{bc^{-3}}\right)^4$ c^4

103. $\left(\dfrac{x^{-3}y^{-4}}{x^{-2}y^{-5}}\right)^{-3}$ $\dfrac{x^3}{y^3}$

104. $\dfrac{x^{-4}y^{-6}}{x^2y^7}$ $\dfrac{1}{x^6y^{13}}$

105. $\dfrac{a^5b^{-5}}{a^{-5}b^5}$ $\dfrac{a^{10}}{b^{10}}$

106. $-2^0 + 2^{-4}$ $-\dfrac{15}{16}$

107. $-3^{-2} - 3^{-3}$ $-\dfrac{4}{27}$

108. $a^{6m}a^{5m}$ a^{11m}

109. $\dfrac{(x^{5+h})^3}{x^5}$ x^{10+3h}

110. $(3xy^{2z})^3$ $27x^3y^{6z}$

111. $a^{m+2}a^{m+3}$ a^{2m+5}

Write each number in scientific notation.

112. 0.00027 2.7×10^{-4} **113.** 0.8868 8.868×10^{-1}

114. $80{,}800{,}000$ 8.08×10^7 **115.** $-868{,}000$ -8.68×10^5

116. The population of California is $29{,}760{,}000$. Write $29{,}760{,}000$ in scientific notation. 2.976×10^7

117. The radius of the earth is 4000 miles. Write 4000 in scientific notation. 4.0×10^3

Write each number in standard form.

118. 8.67×10^5 $867{,}000$ **119.** 3.86×10^{-3} 0.00386

120. 8.6×10^{-4} 0.00086 **121.** 8.936×10^5 $893{,}600$

122. The number of photons of light emitted by a 100-watt bulb every second is 1×10^{20}. Write 1×10^{20} in standard notation.

122. $100{,}000{,}000{,}000{,}000{,}000{,}000$

123. The real mass of all the galaxies in the constellation of Virgo is 3×10^{-25}. Write 3×10^{-25} in standard notation.

123. $0.0000000000000000000000003$

Simplify. Express each result in standard form.

124. $(8 \times 10^4)(2 \times 10^{-7})$ **125.** $\dfrac{8 \times 10^4}{2 \times 10^{-7}}$

124. 0.016 **125.** $400{,}000{,}000{,}000$

(4.6) *Perform each division.*

126. $\dfrac{4xy^2}{3xz^2y^3}$ $\dfrac{4}{3z^2y}$

127. $\dfrac{4xy^3}{32xy^2z}$ $\dfrac{y}{8z}$

128. $\dfrac{x^2 + 21x + 49}{7x^2}$

129. $\dfrac{5a^3b - 15ab^2 + 20ab}{-5ab}$

130. $(a^2 - a + 4) \div (a - 2)$ **129.** $-a^2 + 3b - 4$

131. $(4x^2 + 20x + 7) \div (x + 5)$

132. $\dfrac{a^3 + a^2 + 2a + 6}{a - 2}$ $a^2 + 3a + 8 + \dfrac{22}{a - 2}$

133. $\dfrac{9b^3 - 18b^2 + 8b - 1}{3b - 2}$ $3b^2 - 4b - \dfrac{1}{3b - 2}$

134. $\dfrac{4x^4 - 4x^3 + x^2 + 4x - 3}{2x - 1}$ $2x^3 - x^2 + 2 - \dfrac{1}{2x - 1}$

135. $\dfrac{-10x^2 - x^3 - 21x + 18}{x - 6}$ **128.** $\dfrac{1}{7} + \dfrac{3}{x} + \dfrac{7}{x^2}$

130. $a + 1 + \dfrac{6}{a - 2}$ **131.** $4x + \dfrac{7}{x + 5}$

135. $-x^2 - 16x - 117 - \dfrac{684}{x - 6}$

CHAPTER 4 TEST

Evaluate each expression.

1. 2^5 32 **2.** $(-3)^4$ 81 **3.** -3^4 -81
4. 4^{-3} $\dfrac{1}{64}$

Simplify each exponential expression.

5. $\left(\dfrac{5x^6y^3}{35x^7y}\right)^2$ $\dfrac{y^4}{49x^2}$ **6.** $\dfrac{7(xy)^4}{(xy)^2}$ $7x^2y^2$ **7.** $4(x^2y^3)^{-3}$

Simplify each expression. Write the result using only positive exponents.

8. $\left(\dfrac{x^2y^3}{x^3y^{-4}}\right)^{-2}$ $\dfrac{x^2}{y^{14}}$

9. $\dfrac{6^2x^{-4}y^{-1}}{6^3x^{-3}y^7}$ $\dfrac{1}{6xy^8}$

7. $\dfrac{4}{x^6y^9}$ **11.** 8.63×10^{-5}

Express each number in scientific notation.

10. $563{,}000$ 5.63×10^5 **11.** 0.0000863

Write each number in standard form.

12. 1.5×10^{-3} 0.0015 **13.** 6.23×10^4 $62{,}300$

14. Simplify. Write the answer in standard form.
$(1.2 \times 10^5)(3 \times 10^{-7})$ 0.036

15. Find the degree of the following polynomial.
$4xy^2 + 7xyz + 9x^3yz$ 5

16. Simplify by combining like terms.
$6xyz + 9x^2y - 3xyz + 9x^2y$ $3xyz + 18x^2y$

17. $16x^3 + 7x^2 - 3x - 13$ **20.** $3x^3 + 22x^2 + 41x + 14$

Perform the indicated operations.

17. $(8x^3 + 7x^2 + 4x - 7) + (8x^3 - 7x - 6)$

18.
$$\begin{array}{r} 5x^3 + x^2 + 5x - 2 \\ - (8x^3 - 4x^2 + x - 7) \\ \hline -3x^3 + 5x^2 + 4x + 5 \end{array}$$

21. $2x^5 - 5x^4 + 12x^3 - 8x^2 + 4x + 7$

25. $(8x + 3)^2$ **26.** $(x^2 - 9b)(x^2 + 9b)$

27. The height of the Bank of China in Hong Kong is 1001 feet. Neglecting air resistance, the height of an object dropped from this building at time t seconds is given by the polynomial $-16t^2 + 1001$. Find the height of the object at the given times below.

t	0 seconds	1 second	3 seconds	5 seconds
$-16t^2 + 1001$	1001 ft.	985 ft.	857 ft.	601 ft.

19. Subtract $(4x + 2)$ from the sum of $(8x^2 + 7x + 5)$ and $(x^3 - 8)$. $x^3 + 8x^2 + 3x - 5$

20. Multiply: $(3x + 7)(x^2 + 5x + 2)$.

21. Multiply $x^3 - x^2 + x + 1$ by $2x^2 - 3x + 7$ using the vertical format.

22. Use the FOIL method to multiply $(x + 7)(3x - 5)$.

Use special products to multiply each of the following.

23. $(3x - 7)(3x + 7)$ **24.** $(4x - 2)^2$

22. $3x^2 + 16x - 35$ **23.** $9x^2 - 49$

Divide.

28. $\dfrac{8xy^2}{4x^3y^3z} \quad \dfrac{2}{x^2yz}$

29. $\dfrac{4x^2 + 2xy - 7x}{8xy}$

30. $(x^2 + 7x + 10) \div (x + 5)$ $x + 2$

31. $\dfrac{27x^3 - 8}{3x + 2} \quad 9x^2 - 6x + 4 - \dfrac{16}{3x + 2}$

24. $16x^2 - 16x + 4$ **25.** $64x^2 + 48x + 9$

26. $x^4 - 81b^2$ **29.** $\dfrac{x}{2y} + \dfrac{1}{4} - \dfrac{7}{8y}$

CHAPTER 4   CUMULATIVE REVIEW

1. Translate each sentence into a mathematical statement.   *(Sec. 1.1, Ex. 3)*

a. Nine is less than or equal to eleven. $9 \le 11$

b. Eight is greater than one. $8 > 1$

c. Three is not equal to four. $3 \ne 4$

2. Write each of the following numbers as a product of primes.   *(Sec. 1.2, Ex. 1)*
a. 40 $2 \cdot 2 \cdot 2 \cdot 5$ **b.** 63 $3 \cdot 3 \cdot 7$

3. Add or subtract as indicated. Write each result in lowest terms.   *(Sec. 1.2, Ex. 5)*

a. $\dfrac{2}{7} + \dfrac{4}{7} \quad \dfrac{6}{7}$ **b.** $\dfrac{3}{10} + \dfrac{2}{10} \quad \dfrac{1}{2}$

c. $\dfrac{9}{7} - \dfrac{2}{7} \quad 1$ **d.** $\dfrac{5}{3} - \dfrac{1}{3} \quad \dfrac{4}{3}$

4. Decide whether 2 is a solution of $3x + 10 = 8x$.

5. Subtract 8 from -4.   *(Sec. 1.6, Ex. 2)* -12

6. If $x = -2$ and $y = -4$, evaluate each expression.

a. $5x - y \quad -6$ **b.** $x^3 - y^2 \quad -24$ **c.** $\dfrac{3x}{2y} \quad \dfrac{3}{4}$

7. Simplify each expression by combining like terms.

a. $2x + 3x + 5 + 2 \quad 5x + 7$

b. $-5a - 3 + a + 2 \quad -4a - 1$

c. $4y - 3y^2 \quad 4y - 3y^2$

d. $2.3x + 5x - 6 \quad 7.3x - 6$

8. Solve $-3x = 33$ for x.   *(Sec. 2.3, Ex. 2)* $x = -11$

9. Solve $3(x - 4) = 3x - 12$.   *(Sec. 2.4, Ex. 7)*

10. Solve $V = lwh$ for l.   *(Sec. 2.6, Ex. 4)* $l = \dfrac{V}{wh}$

11. The cost of a large hand-tossed pepperoni pizza at Domino's recently increased from $5.80 to $7.03. Find the percent increase.   *(Sec. 2.7, Ex. 7)*   21.2%

12. Rajiv Puri invested part of his $20,000 inheritance in a mutual funds account that pays 7% simple interest yearly and the rest in a certificate of deposit that pays 9% simple interest yearly. At the end of one year, Rajiv's investments earned $1550. Find the amount he invested at each rate.

13. Graph the linear equation $-5x + 3y = 15$.

14. Graph the line with x-intercept 1 and y-intercept -1.   *(Sec. 3.3, Ex. 2)*   See App. F.

4. *(Sec. 1.4, Ex. 3)*   solution   **6.** *(Sec. 1.7, Ex. 9)*   **7.** *(Sec. 2.1, Ex. 4)*   **9.** all real numbers
12. *(Sec. 2.8, Ex. 4)* $12,500 @ 7%; $7,500 @ 9%   **13.** *(Sec. 3.2, Ex. 3)*   See App. F.

15. Find the slope of the line through $(-1, -2)$ and $(2, 4)$. Graph the line.   *(Sec. 3.4, Ex. 2)*

16. Simplify each of the following expressions.

   a. $(x^2)^5$ x^{10} **b.** $(y^8)^2$ y^{16} **c.** $[(-5)^3]^4$ $(-5)^{12}$

17. Combine like terms.   *(Sec. 4.2, Ex. 5)*

   a. $-3x + 7x$ $4x$ **b.** $11x^2 + 5 + 2x^2 - 7$

18. Multiply: $(2x - y)^2$.   *(Sec. 4.3, Ex. 3)* $4x^2 - 4xy + y^2$

19. Use special products to square the following binomials.   *(Sec. 4.4, Ex. 5)*

   a. $(t + 2)^2$ $t^2 + 4t + 4$

   b. $(p - q)^2$ $p^2 - 2pq + q^2$

   c. $(2x + 3y)^2$ $4x^2 + 12xy + 9y^2$

   d. $(5r - 7s)^2$ $25r^2 - 70rs + 49s^2$

20. Simplify the following expressions. Write each result using positive exponents only.   *(Sec. 4.5, Ex. 4)*

   a. $(2x^3)(5x)^{-2}$ $\dfrac{2x}{25}$ **b.** $\left(\dfrac{3a^2}{b}\right)^{-3}$ $\dfrac{b^3}{27a^6}$

   c. $\dfrac{4^{-1}x^{-3}y}{4^{-3}x^2y^{-6}}$ $\dfrac{16y^7}{x^5}$ **d.** $(y^{-3}z^{-6})^{-6}$ $y^{18}z^{36}$

   e. $\left(\dfrac{-2x^3y}{xy^{-1}}\right)^3$ $-8x^6y^6$

21. Simplify: $\dfrac{8x^2y^2 - 16xy + 2x}{4xy}$.   *(Sec. 4.6, Ex. 3)*

15. $m = 2$; see App. F.   **16.** *(Sec. 4.1, Ex. 5)*   **17. b.** $13x^2 - 2$ **21.** $2xy - 4 + \dfrac{1}{2y}$

FACTORING POLYNOMIALS

CHOOSING AMONG BUILDING OPTIONS

Whether putting in a new floor, hanging new wallpaper, or retiling a bathroom, it is usually necessary to choose among several different building or decorating materials with different pricing schemes. Because it is often the case that only a fixed amount of money is available for such projects, it can be helpful to compare the choices by calculating how much area can be covered by a fixed dollar-value of the material.

IN THE CHAPTER GROUP ACTIVITY ON PAGE 333, YOU WILL HAVE THE OPPORTUNITY TO HELP SAM CHOOSE AMONG THREE DIFFERENT CHOICES OF MATERIALS FOR BUILDING A PATIO AROUND HIS NEW IN-GROUND SWIMMING POOL.

In Chapter 4, you learned how to multiply polynomials. This chapter deals with an operation that is the reverse process of multiplying, called factoring. Factoring is an important algebraic skill because this process allows us to write a sum as a product.

At the end of this chapter, we use factoring to help us solve equations other than linear equations, and in Chapter 6 we use factoring to simplify and perform arithmetic operations on rational expressions.

5.1 THE GREATEST COMMON FACTOR AND FACTORING BY GROUPING

OBJECTIVES

1. Find the greatest common factor of a list of integers.
2. Find the greatest common factor of a list of terms.
3. Factor out the greatest common factor from a polynomial.
4. Factor a polynomial by grouping.

When an integer is written as the product of two or more other integers, each of these integers is called a **factor** of the product. This is true for polynomials, also. When a polynomial is written as the product of two or more other polynomials, each of these polynomials is called a factor of the product. The process of writing a polynomial as a product is called **factoring.**

$$\underset{\text{factor}}{2} \cdot \underset{\text{factor}}{3} = \underset{\text{product}}{6} \qquad \underset{\text{factor}}{x^2} \cdot \underset{\text{factor}}{x^3} = \underset{\text{product}}{x^5} \qquad \underset{\text{factor}}{(x+2)}\underset{\text{factor}}{(x+3)} = \underset{\text{product}}{x^2 + 5x + 6}$$

Notice that factoring is the reverse process of multiplying.

$$x^2 + 5x + 6 = (x + 2)(x + 3)$$

multiplying

We begin our study of factoring by reviewing the **greatest common factor (GCF).** The GCF of a list of integers is the largest integer that is a factor of all the integers in the list. For example, the GCF of 12 and 20 is 4 because 4 is the largest integer that is a factor of both 12 and 20. With large integers, the GCF may not be easily found by inspection. When this happens, use the following steps.

TO FIND THE GCF OF A LIST OF INTEGERS

Step 1. Write each number as a product of prime numbers.

Step 2. Identify the common prime factors.

Step 3. The product of all common prime factors found in step 2 is the greatest common factor. If there are no common prime factors, the greatest common factor is 1.

Recall from Section 1.2 that a prime number is a whole number other than 1, whose only factors are 1 and itself.

EXAMPLE 1   Find the GCF of each list of numbers.

      a. 28 and 40   **b.** 55 and 21   **c.** 15, 18, and 66

Solution: **a.** Write each number as a product of primes.

$$28 = 2 \cdot 2 \cdot 7 = 2^2 \cdot 7$$
$$40 = 2 \cdot 2 \cdot 2 \cdot 5 = 2^3 \cdot 5$$

There are two common factors, each of which is 2, so the GCF is

$$\text{GCF} = 2 \cdot 2 = 4$$

    b. $55 = 5 \cdot 11$
        $21 = 3 \cdot 7$

There are no common prime factors; thus, the GCF is 1.

    c. $15 = 3 \cdot 5$
        $18 = 2 \cdot 3 \cdot 3 = 2 \cdot 3^2$
        $66 = 2 \cdot 3 \cdot 11$

The only prime factor common to all three numbers is 3, so the GCF is

$$\text{GCF} = 3$$

2   The greatest common factor of a list of variables raised to powers is found in a similar way. For example, the GCF of x^2, x^3, and x^5 is x^2 because each term contains a factor of x^2 and no higher power of x is a factor of each term.

$$x^2 = x^2$$
$$x^3 = x^2 \cdot x$$
$$x^5 = x^2 \cdot x^3$$

From this example, we see that **the GCF of a list of common variables raised to powers is the variable raised to the smallest exponent in the list.**

EXAMPLE 2   Find the GCF of each list of terms.

      a. $x^3, x^7,$ and x^3 **b.** $y, y^4,$ and y^7

Solution: **a.** The GCF is x^3, since 3 is the smallest exponent to which x is raised.

    b. The GCF is y^1 or y, since 1 is the smallest exponent on y.

In general, **the GCF of a list of terms is the product of all common factors.**

EXAMPLE 3 Find the greatest common factor of each list of terms.

     a. $6x^2$, $10x^3$, and $-8x$ **b.** $8y^2$, y^3, and y^5 **c.** a^3b^2, a^5b, and a^6b^2

Solution: **a.** The GCF of the numerical coefficients 6, 10, and -8 is 2.
       The GCF of variable factors x^2, x^3, and x is x.
       Thus, the GCF of the terms $6x^2$, $10x^3$, and $-8x$ is $2x$.

     b. The GCF of the numerical coefficients 8, 1, and 1 is 1.
       The GCF of variable factors y^2, y^3, and y^5 is y^2.
       Thus, the GCF of terms $8y^2$, y^3, and y^5 is $1y^2$ or y^2.

     c. The GCF of a^3, a^5, and a^6 is a^3.
       The GCF of b^2, b, and b^2 is b. Thus, the GCF of the terms is a^3b.

3   The first step in factoring a polynomial is to find the GCF of its terms. Once we do so, we can write the polynomial as a product by **factoring out** the GCF.

     The polynomial $8x + 14$, for example, contains two terms: $8x$ and 14. The GCF of these terms is 2. We factor out 2 from each term by writing each term as a product of 2 and the term's remaining factors.

$$8x + 14 = 2 \cdot 4x + 2 \cdot 7$$

Using the distributive property, we can write

$$8x + 14 = 2 \cdot 4x + 2 \cdot 7$$
$$= 2(4x + 7)$$

Thus, a factored form of $8x + 14$ is $2(4x + 7)$.

EXAMPLE 4 Factor each polynomial by factoring out the GCF.

     a. $6t + 18$ **b.** $y^5 - y^7$

Solution: **a.** The GCF of terms $6t$ and 18 is 6.

$$6t + 18 = 6 \cdot t + 6 \cdot 3$$
$$= 6(t + 3) \qquad \text{Apply the distributive property.}$$

      Our work can be checked by multiplying 6 and $(t + 3)$.

$$6(t + 3) = 6 \cdot t + 6 \cdot 3 = 6t + 18, \text{ the original polynomial.}$$

     b. The GCF of y^5 and y^7 is y^5. Thus,

$$y^5 - y^7 = (y^5)\,1 - (y^5)\,y^2$$
$$= y^5(1 - y^2)$$

EXAMPLE 5 Factor $-9a^5 + 18a^2 - 3a$.

Solution: $-9a^5 + 18a^2 - 3a = (3a)(-3a^4) + (3a)(6a) + (3a)(-1)$
                           $= 3a(-3a^4 + 6a - 1)$

In Example 5 we could have chosen to factor out a $-3a$ instead of $3a$. If we factor out a $-3a$, we have

$$-9a^5 + 18a^2 - 3a = (-3a)(3a^4) + (-3a)(-6a) + (-3a)(1)$$
$$= -3a(3a^4 - 6a + 1)$$

EXAMPLE 6   Factor $25x^4z + 15x^3z + 5x^2z$.

Solution: The greatest common factor is $5x^2z$.

$$25x^4z + 15x^3z + 5x^2z = 5x^2z(5x^2 + 3x + 1)$$

R E M I N D E R   Be careful when the GCF of the terms is the same as one of the terms in the polynomial. The greatest common factor of the terms of $8x^2 - 6x^3 + 2x$ is $2x$. When factoring out $2x$ from the terms of $8x^2 - 6x^3 + 2x$, don't forget a term of 1.

$$8x^2 - 6x^3 + 2x = 2x(4x) - 2x(3x^2) + 2x(1)$$
$$= 2x(4x - 3x^2 + 1)$$

Check by multiplying.

$$2x(4x - 3x^2 + 1) = 8x^2 - 6x^3 + 2x$$

EXAMPLE 7   Factor $5(x + 3) + y(x + 3)$.

Solution: The binomial $(x + 3)$ is the greatest common factor. Use the distributive property to factor out $(x + 3)$.

$$5(x + 3) + y(x + 3) = (x + 3)(5 + y)$$

EXAMPLE 8   Factor $3m^2n(a + b) - (a + b)$.

Solution: The greatest common factor is $(a + b)$.

$$3m^2n(a + b) - 1(a + b) = (a + b)(3m^2n - 1)$$

4   Once the GCF is factored out, we can often continue to factor the polynomial, using a variety of techniques. We discuss here a technique for factoring polynomials called **grouping.**

EXAMPLE 9   Factor $xy + 2x + 3y + 6$ by grouping. Check by multiplying.

Solution: The GCF of the first two terms is x, and the GCF of the last two terms is 3.

$$xy + 2x + 3y + 6 = x(y + 2) + 3(y + 2)$$

Notice that $x(y + 2) + 3(y + 2)$ is not a factored form of the original polynomial because it is a sum and not a product.

Next, factor out the common binomial factor of $(y + 2)$.

$$x(y + 2) + 3(y + 2) = (y + 2)(x + 3)$$

To check, multiply $(y + 2)$ by $(x + 3)$.

$$(y + 2)(x + 3) = xy + 2x + 3y + 6, \text{ the original polynomial.}$$

Thus, the factored form of $xy + 2x + 3y + 6$ is $(y + 2)(x + 3)$.

TO FACTOR A FOUR-TERM POLYNOMIAL BY GROUPING

Step 1.   Arrange the terms so that the first two terms have a common factor and the last two terms have a common factor.

Step 2.   For each pair of terms, use the distributive property to factor out the pair's greatest common factor.

Step 3.   If there is now a common binomial factor, factor it out.

Step 4.   If there is no common binomial factor in step 3, begin again, rearranging the terms differently. If no rearrangement leads to a common binomial factor, the polynomial cannot be factored.

EXAMPLE 10

Solution:

Factor $3x^2 + 4xy - 3x - 4y$ by grouping.

The first two terms have a common factor x. Factor -1 from the last two terms so that the common binomial factor of $(3x + 4y)$ appears.

$$3x^2 + 4xy - 3x - 4y = x(3x + 4y) - 1(3x + 4y)$$

Next, factor out the common factor $(3x + 4y)$.

$$= (3x + 4y)(x - 1)$$

R E M I N D E R   When **factoring** a polynomial, make sure the polynomial is written as a **product.** For example, it is true that

$$xy + 2x + 3y + 6 = x(y + 2) + 3(y + 2)$$

but $x(y + 2) + 3(y + 2)$ is not a **factored form** of the original polynomial since it is a **sum,** not a **product.** The factored form of $xy + 2x + 3y + 6$ is $(y + 2)(x + 3)$.

Factoring out a greatest common factor first makes factoring by any method easier, as we see in the next example.

EXAMPLE 11

Factor $4ax - 4ab - 2bx + 2b^2$.

Solution: First, factor out the common factor 2 from all four terms.

$$4ax - 4ab - 2bx + 2b^2$$

$= 2(2ax - 2ab - bx + b^2)$	Factor out 2 from all four terms.
$= 2[2a(x - b) - b(x - b)]$	Factor out common factors from each pair of terms.
$= 2(x - b)(2a - b)$	Factor out the common binomial.

Notice that we factored out $-b$ instead of b from the second pair of terms so that the binomial factor of each pair is the same.

MENTAL MATH

Find the prime factorization of the following integers mentally.

1. 14 $2 \cdot 7$

2. 15 $3 \cdot 5$

3. 10 $2 \cdot 5$

4. 70 $2 \cdot 5 \cdot 7$

Find the GCF of the following pairs of integers mentally.

5. 6, 15   3

6. 20, 15   5

7. 3, 18   3

8. 14, 35   7

EXERCISE SET 5.1

Find the GCF for each list. See Examples 1 through 3.

1. 32, 36   4

2. 36, 90   18

3. 12, 18, 36   6

4. 24, 14, 21   no GCF

5. y^2, y^4, y^7 y^2

6. x^3, x^2, x^3 x^2

7. $x^{10}y^2, xy^2, x^3y^3$ xy^2

8. p^7q, p^8q^2, p^9q^3 p^7q

9. $8x, 4$ 4

10. $9y, y$ y

11. $12y^4, 20y^3$ $4y^3$

12. $32x, 18x^2$ $2x$

13. $12x^3, 6x^4, 3x^5$ $3x^3$

14. $15y^2, 5y^7, 20y^3$ $5y^2$

15. $18x^2y, 9x^3y^3, 36x^3y$ $9x^2y$

16. $7x, 21x^2y^2, 14xy$ $7x$

Factor out the GCF from each polynomial. See Examples 4 through 6.

17. $3a + 6$ $3(a + 2)$

18. $18a - 12$ $6(3a - 2)$

19. $30x - 15$ $15(2x - 1)$

20. $42x - 7$ $7(6x - 1)$

21. $24cd^3 - 18c^2d$ $6cd(4d^2 - 3c)$

22. $25x^4y^3 - 15x^2y^2$ $5x^2y^2(5x^2y - 3)$

23. $-24a^4x + 18a^3x$ $-6a^3x(4a - 3)$

24. $-15a^2x + 9ax$ $3ax(-5a + 3)$

25. $12x^3 + 16x^2 - 8x$ $4x(3x^2 + 4x - 2)$

26. $6x^3 - 9x^2 + 12x$ $3x(2x^2 - 3x + 4)$

27. $5x^3y - 15x^2y + 10xy$ $5xy(x^2 - 3x + 2)$

28. $14x^3y + 7x^2y - 7xy$ $7xy(2x^2 + x - 1)$

29. Construct a binomial whose greatest common factor is $5a^3$.   answers may vary

30. Construct a trinomial whose greatest common factor is $2x^2$.   answers may vary

Factor out the GCF from each polynomial. See Examples 7 and 8.

31. $y(x + 2) + 3(x + 2)$ $(x + 2)(y + 3)$

32. $z(y + 4) + 3(y + 4)$ $(y + 4)(z + 3)$

33. $x(y - 3) - 4(y - 3)$ $(y - 3)(x - 4)$

34. $6(x + 2) - y(x + 2)$ $(x + 2)(6 - y)$

35. $2x(x + y) - (x + y)$ $(x + y)(2x - 1)$

36. $xy(y + 1) - (y + 1)$ $(y + 1)(xy - 1)$

Factor the following four-term polynomials by grouping. See Examples 9 through 11.

37. $5x + 15 + xy + 3y$ $(x + 3)(5 + y)$

38. $xy + y + 2x + 2$ $(x + 1)(y + 2)$

39. $2y - 8 + xy - 4x$ $(y - 4)(2 + x)$

40. $6x - 42 + xy - 7y$ $(x - 7)(6 + y)$

41. $3xy - 6x + 8y - 16$ $(y - 2)(3x + 8)$

42. $xy - 2yz + 5x - 10z$ $(x - 2z)(y + 5)$

43. $y^3 + 3y^2 + y + 3$ $(y + 3)(y^2 + 1)$

44. $x^3 + 4x + x^2 + 4$ $(x^2 + 4)(x + 1)$

Write an expression for the area of each shaded region. Then write the expression as a factored polynomial.

45.

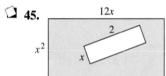

$12x^3 - 2x$; $2x(6x^2 - 1)$

46.

$32\pi x^2$

47.

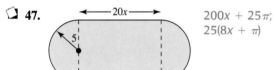

$200x + 25\pi$; $25(8x + \pi)$

48.

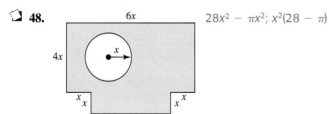

$28x^2 - \pi x^2$; $x^2(28 - \pi)$

Factor the following polynomials.

49. $3x - 6$ $3(x - 2)$ **50.** $4x - 16$ $4(x - 4)$

51. $-8x - 18$ $-2(4x + 9)$

52. $-6x - 40$ $-2(3x + 20)$

53. $32xy - 18x^2$ $2x(16y - 9x)$

54. $10xy - 15x^2$ $5x(2y - 3x)$

55. $4x - 8y + 4$ $4(x - 2y + 1)$

56. $7x + 21y - 7$ $7(x + 3y - 1)$

57. $8(x + 2) - y(x + 2)$ $(x + 2)(8 - y)$

58. $x(y^2 + 1) - 3(y^2 + 1)$ $(y^2 + 1)(x - 3)$

59. $-40x^8y^6 - 16x^9y^5$ $-8x^8y^5(5y + 2x)$

60. $-21x^3y - 49x^2y^2$ $-7x^2y(3x + 7y)$

61. $5x + 10$ $5(x + 2)$

62. $7x + 35$ $7(x + 5)$

63. $-3x + 12$ $-3(x - 4)$

64. $-10x + 20$ $-10(x - 2)$

65. $18x^3y^3 - 12x^3y^2 + 6x^5y^2$ $6x^3y^2(3y - 2 + x^2)$

66. $32x^3y^3 - 24x^2y^3 + 8x^2y^4$ $8x^2y^3(4x - 3 + y)$

67. $-2a^3 - 6a^2b$ $-2a^2(a + 3b)$

68. $-3b^3c - 9b^2$ $-3b^2(bc + 3)$

69. $y^2(x - 2) + (x - 2)$ $(x - 2)(y^2 + 1)$

70. $x(y + 4) + (y + 4)$ $(y + 4)(x + 1)$

71. $5xy + 15x + 6y + 18$ $(y + 3)(5x + 6)$

72. $2x^3 + x^2 + 8x + 4$ $(2x + 1)(x^2 + 4)$

73. $4x^2 - 8xy - 3x + 6y$ $(x - 2y)(4x - 3)$

74. $2x^3 - x^2 - 10x + 5$ $(2x - 1)(x^2 - 5)$

75. $126x^3yz + 210y^4z^3$ $42yz(3x^3 + 5y^3z^2)$

76. $231x^3y^2z - 143yz^2$ $11yz(21x^3y - 13z)$

77. $4y^2 - 12y + 4yz - 12z$ $4(y - 3)(y + z)$

78. $5x^2 - 20x^2y + 5z - 20zy$ $5(1 - 4y)(x^2 + z)$

79. $3y - 5x + 15 - xy$ $(3 - x)(5 + y)$

80. $2x - 9y + 18 - xy$ $(x + 9)(2 - y)$

81. $36x + 15y + 30 + 18xy$ $3(6x + 5)(2 + y)$

82. $21y - 15x + 15xy - 21$ $3(y - 1)(7 + 5x)$

83. $12x^2y - 42x^2 - 4y + 14$ $2(3x^2 - 1)(2y - 7)$

84. $90 + 15y^2 - 18x - 3xy^2$ $3(6 + y^2)(5 - x)$

85. Explain how you can tell whether a polynomial is written in factored form.   answers may vary

86. Construct a 4-term polynomial that can be factored by grouping.   answers may vary

Which of the following expressions is factored?

87. $(a + 6)(b - 2)$ factored

88. $(x + 5)(x + y)$ factored

89. $5(2y + z) - b(2y + z)$ not factored

90. $3x(a + 2b) + 2(a + 2b)$ not factored

Write an expression for the length of each rectangle.

91.

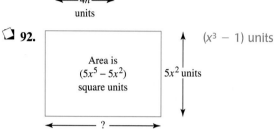

$(n^3 - 6)$ units

Area is
$(4n^4 - 24n)$
square
units

?

←4n→
units

92.

Area is
$(5x^5 - 5x^2)$
square units

$(x^3 - 1)$ units

$5x^2$ units

←——— ? ———→

Review Exercises

Multiply. See Section 4.4.

93. $(x + 2)(x + 5)$ $x^2 + 7x + 10$

94. $(y + 3)(y + 6)$ $y^2 + 9y + 18$

95. $(a - 7)(a - 8)$ $a^2 - 15a + 56$

96. $(z - 4)(z - 4)$ $z^2 - 8z + 16$

97. $(b + 1)(b - 4)$ $b^2 - 3b - 4$

98. $(x - 5)(x + 10)$ $x^2 + 5x - 50$

99. $(y - 9)(y + 2)$ $y^2 - 7y - 18$

100. $(a + 3)(a - 1)$ $a^2 + 2a - 3$

5.2 | Factoring Trinomials of the Form $x^2 + bx + c$

O B J E C T I V E S

TAPE BA 5.2

1 Factor trinomials of the form $x^2 + bx + c$.

2 Factor out the greatest common factor and then factor a trinomial of the form $x^2 + bx + c$.

1 In this section, we factor trinomials of the form $x^2 + bx + c$, where the numerical coefficient of the squared variable is 1. Recall that factoring a polynomial is the process of writing the polynomial as a product. For example, since $(x + 3)(x + 1) = x^2 + 4x + 3$, we say that the factored form of $x^2 + 4x + 3$ is

$$x^2 + 4x + 3 = (x + 3)(x + 1)$$

Notice that the product of the first terms of the binomials is $x \cdot x = x^2$, the first term of the trinomial. Also, the product of the last two terms of the binomials is $3 \cdot 1 = 3$, the third term of the trinomial. The sum of these same terms is $3 + 1 = 4$, the coefficient of the x term of the trinomial.

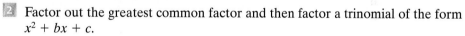

$$x^2 + 4x + 3 = (x + 3)(x + 1)$$

Many trinomials, such as the preceding, factor into two binomials. To factor $x^2 + 7x + 10$, assume that it factors into two binomials and begin by writing two pairs of parentheses. The first term of the trinomial is x^2, so we use x and x as the first terms of the binomial factors.

$$(x + \quad)(x + \quad)$$

To determine the last term of each binomial factor, we look for two integers whose product is 10 and whose sum is 7. Since our numbers must have a positive product and a positive sum, we list pairs of positive integer factors of 10 only.

POSITIVE FACTORS OF 10	SUM OF FACTORS
1, 10	$1 + 10 = 11$
2, 5	$2 + 5 = 7$

The correct pair of numbers is 2 and 5 because their product is 10 and their sum is 7. Now we can fill in the last terms of the binomial factors.

$$x^2 + 7x + 10 = (x + 2)(x + 5)$$

To see if we have factored correctly, multiply.

$$(x + 2)(x + 5) = x^2 + 2x + 5x + 10$$
$$= x^2 + 7x + 10 \qquad \text{Combine like terms.}$$

Since multiplication is commutative, the factored form of $x^2 + 7x + 10$ can also be written as $(x + 5)(x + 2)$.

TO FACTOR A TRINOMIAL OF THE FORM $x^2 + bx + c$

To factor a trinomial of the form $x^2 + bx + c$, look for two numbers whose product is c and whose sum is b. The factored form of $x^2 + bx + c$ is

$$(x + \text{one number})(x + \text{other number})$$

EXAMPLE 1 Factor $x^2 + 7x + 12$.

Solution: Begin by writing the first terms of the binomial factors.

$$(x + \quad)(x + \quad)$$

Next, look for two numbers whose product is 12 and whose sum is 7. Since our numbers must have a positive product and a positive sum, we look at positive pairs of factors of 12 only.

POSITIVE FACTORS OF 12	SUM OF FACTORS
1, 12	$1 + 12 = 13$
2, 6	$2 + 6 = 8$
3, 4	$3 + 4 = 7$

The correct pair of numbers is 3 and 4 because their product is 12 and their sum is 7. Use these numbers as the last terms of the binomial factors.

$$x^2 + 7x + 12 = (x + 3)(x + 4)$$

To check, multiply $(x + 3)$ by $(x + 4)$.

EXAMPLE 2   Factor $x^2 - 8x + 15$.

Solution: Begin by writing the first terms of the binomials.

$$(x + \quad)(x + \quad)$$

Now look for two numbers whose product is 15 and whose sum is -8. Since our numbers must have a positive product and a negative sum, we look at negative factors of 15 only.

Negative Factors of 15	Sum of Factors
$-1, -15$	$-1 + (-15) = -16$
$-3, -5$	$-3 + (-5) = -8$

The correct pair of numbers is -3 and -5 because their product is 15 and their sum is -8. Then

$$x^2 - 8x + 15 = (x - 3)(x - 5)$$

EXAMPLE 3   Factor $x^2 + 4x - 12$.

Solution: $x^2 + 4x - 12 = (x + \quad)(x + \quad)$

Look for two numbers whose product is -12 and whose sum is 4.

Factors of -12	Sum of Factors
$-1, 12$	$-1 + 12 = 11$
$1, -12$	$1 + (-12) = -11$
$-2, 6$	$-2 + 6 = 4$
$2, -6$	$2 + (-6) = -4$
$-3, 4$	$-3 + 4 = 1$
$3, -4$	$3 + (-4) = -1$

The correct pair of numbers is -2 and 6 since their product is -12 and their sum is 4. Hence

$$x^2 + 4x - 12 = (x - 2)(x + 6)$$

EXAMPLE 4   Factor $r^2 - r - 42$.

Solution: Because the variable in this trinomial is r, the first term of each binomial factor is r.

$$r^2 - r - 42 = (r + \ \ )(r + \ \ )$$

Find two numbers whose product is -42 and whose sum is -1, the numerical coefficient of r. The numbers are 6 and -7. Therefore,

$$r^2 - r - 42 = (r + 6)(r - 7)$$

EXAMPLE 5   Factor $a^2 + 2a + 10$.

Solution: Look for two numbers whose product is 10 and whose sum is 2. Neither 1 and 10 nor 2 and 5 give the required sum, 2. We conclude that $a^2 + 2a + 10$ is not factorable with integers. The polynomial $a^2 + 2a + 10$ is called a **prime polynomial.**

EXAMPLE 6   Factor $x^2 + 5yx + 6y^2$.

Solution: $x^2 + 5yx + 6y^2 = (x + \ \ )(x + \ \ )$

Look for two terms whose product is $6y^2$ and whose sum is $5y$, the coefficient of x in the middle term of the trinomial. The terms are $2y$ and $3y$ because $2y \cdot 3y = 6y^2$ and $2y + 3y = 5y$. Therefore,

$$x^2 + 5yx + 6y^2 = (x + 2y)(x + 3y)$$

2   Remember that the first step in factoring any polynomial is to factor out the greatest common factor (if there is one other than 1 or -1).

EXAMPLE 7   Factor $3m^2 - 24m - 60$.

Solution: First factor out the greatest common factor, 3, from each term.

$$3m^2 - 24m - 60 = 3(m^2 - 8m - 20)$$

Next, factor $m^2 - 8m - 20$ by looking for two factors of -20 whose sum is -8. The factors are -10 and 2.

$$3m^2 - 24m - 60 = 3(m + 2)(m - 10)$$

Remember to write the common factor 3 as part of the answer.

Check by multiplying.

$$3(m + 2)(m - 10) = 3(m^2 - 8m - 20)$$
$$= 3m^2 - 24m - 60$$

REMINDER   When factoring a polynomial, remember that factored out common factors are part of the final factored form. For example,

$$5x^2 - 15x - 50 = 5(x^2 - 3x - 10)$$
$$= 5(x + 2)(x - 5)$$

Thus, $5x^2 - 15x - 50$ **factored completely** is $5(x + 2)(x - 5)$.

MENTAL MATH

Complete the following.

1. $x^2 + 9x + 20 = (x + 4)(x + 5)$ **2.** $x^2 + 12x + 35 = (x + 5)(x + 7)$ **3.** $x^2 - 7x + 12 = (x - 4)(x - 3)$

4. $x^2 - 13x + 22 = (x - 2)(x - 11)$ **5.** $x^2 + 4x + 4 = (x + 2)(x + 2)$ **6.** $x^2 + 10x + 24 = (x + 6)(x + 4)$

EXERCISE SET 5.2

Factor each trinomial completely. See Examples 1 through 5.

1. $x^2 + 7x + 6$ $(x + 6)(x + 1)$

2. $x^2 + 6x + 8$ $(x + 4)(x + 2)$

3. $x^2 + 9x + 20$ $(x + 5)(x + 4)$

4. $x^2 + 13x + 30$ $(x + 3)(x + 10)$

5. $x^2 - 8x + 15$ $(x - 5)(x - 3)$

6. $x^2 - 9x + 14$ $(x - 7)(x - 2)$

7. $x^2 - 10x + 9$ $(x - 9)(x - 1)$

8. $x^2 - 6x + 9$ $(x - 3)(x - 3)$

9. $x^2 - 15x + 5$ prime

10. $x^2 - 13x + 30$ $(x - 3)(x - 10)$

11. $x^2 - 3x - 18$ $(x - 6)(x + 3)$

12. $x^2 - x - 30$ $(x - 6)(x + 5)$

13. $x^2 + 5x + 2$ prime

14. $x^2 - 7x + 5$ prime

Factor each trinomial completely. See Example 6.

15. $x^2 + 8xy + 15y^2$ $(x + 5y)(x + 3y)$

16. $x^2 + 6xy + 8y^2$ $(x + 4y)(x + 2y)$

17. $x^2 - 2xy + y^2$ $(x - y)(x - y)$

18. $x^2 - 11xy + 30y^2$ $(x - 6y)(x - 5y)$

19. $x^2 - 3xy - 4y^2$ $(x - 4y)(x + y)$

20. $x^2 - 4xy - 77y^2$ $(x - 11y)(x + 7y)$

Factor each trinomial completely. See Example 7.

21. $2z^2 + 20z + 32$ $2(z + 8)(z + 2)$

22. $3x^2 + 30x + 63$ $3(x + 7)(x + 3)$

23. $2x^3 - 18x^2 + 40x$ $2x(x - 5)(x - 4)$

24. $x^3 - x^2 - 56x$ $x(x - 8)(x + 7)$

25. $7x^2 + 14xy - 21y^2$ $7(x + 3y)(x - y)$

26. $6r^2 - 3rs - 3s^2$ $3(2r + s)(r - s)$

Factor each trinomial completely.

27. $x^2 + 15x + 36$ $(x + 12)(x + 3)$

28. $x^2 + 19x + 60$ $(x + 4)(x + 15)$

29. $x^2 - x - 2$ $(x - 2)(x + 1)$

30. $x^2 - 5x - 14$ $(x - 7)(x + 2)$

31. $r^2 - 16r + 48$ $(r - 12)(r - 4)$

32. $r^2 - 10r + 21$ $(r - 7)(r - 3)$

33. $x^2 - 4x - 21$ $(x - 7)(x + 3)$

34. $x^2 - 4x - 32$ $(x - 8)(x + 4)$

35. $x^2 + 7xy + 10y^2$ $(x + 5y)(x + 2y)$

36. $x^2 - 3xy - 4y^2$ $(x - 4y)(x + y)$

37. $r^2 - 3r + 6$ prime

38. $x^2 + 4x - 10$ prime

39. $2t^2 + 24t + 64$ $2(t + 8)(t + 4)$

40. $2t^2 + 20t + 50$ $2(t + 5)(t + 5)$

41. $x^3 - 2x^2 - 24x$ $x(x - 6)(x + 4)$

42. $x^3 - 3x^2 - 28x$ $x(x - 7)(x + 4)$

43. $x^2 - 16x + 63$ $(x - 9)(x - 7)$

44. $x^2 - 19x + 88$ $(x - 8)(x - 11)$

45. $x^2 + xy - 2y^2$ $(x + 2y)(x - y)$

46. $x^2 - xy - 6y^2$ $(x - 3y)(x + 2y)$

47. $3x^2 + 9x - 30$ $3(x + 5)(x - 2)$

48. $4x^2 - 4x - 48$ $4(x - 4)(x + 3)$

49. $3x^2 - 60x + 108$ $3(x - 18)(x - 2)$

50. $2x^2 - 24x + 70$ $2(x - 7)(x - 5)$

51. $x^2 - 18x - 144$ $(x - 24)(x + 6)$

52. $x^2 + x - 42$ $(x + 7)(x - 6)$

53. $6x^3 + 54x^2 + 120x$ $6x(x + 4)(x + 5)$

54. $3x^3 + 3x^2 - 126x$ $3x(x + 7)(x - 6)$

55. $2t^5 - 14t^4 + 24t^3$ $2t^3(t - 4)(t - 3)$

56. $3x^6 + 30x^5 + 72x^4$ $3x^4(x + 6)(x + 4)$

57. $5x^3y - 25x^2y^2 - 120xy^3$ $5xy(x - 8y)(x + 3y)$

58. $3x^2 - 6xy - 72y^2$ $3(x - 6y)(x + 4y)$

59. $4x^2y + 4xy - 12y$ $4y(x^2 + x - 3)$

60. $3x^2y - 9xy + 45y$ $3y(x^2 - 3x + 15)$

61. $2a^2b - 20ab^2 + 42b^3$ $2b(a - 7b)(a - 3b)$

62. $-1x^2z + 14xz^2 - 28z^3$ $-z(x^2 - 14xz + 28z^2)$

Find a positive value of b so that each trinomial is factorable.

63. $x^2 + bx + 15$ 8; 16   **64.** $y^2 + by + 20$ 9; 12; 21

65. $m^2 + bm - 27$ 6;26   **66.** $x^2 + bx - 14$ 5; 13

Find a positive value of c so that each trinomial is factorable.

67. $x^2 + 6x + c$ 5; 8; 9

68. $t^2 + 8t + c$ 7; 12; 15; 16

69. $y^2 - 4y + c$ 3; 4

70. $n^2 - 16n + c$ 15; 28; 39; 48; 55; 60; 63; 64

Complete the following sentences in your own words.

71. If $x^2 + bx + c$ is factorable and c is negative, then the signs of the last term factors of the binomial are opposite because . . . .   answers may vary

72. If $x^2 + bx + c$ is factorable and c is positive, then the signs of the last term factors of the binomials are the same because . . . .   answers may vary

Review Exercises

Multiply. See Section 4.4.

73. $(2x + 1)(x + 5)$ $2x^2 + 11x + 5$

74. $(3x + 2)(x + 4)$ $3x^2 + 14x + 8$

75. $(5y - 4)(3y - 1)$ $15y^2 - 17y + 4$

76. $(4z - 7)(7z - 1)$ $28z^2 - 53z + 7$

77. $(a + 3)(9a - 4)$ $9a^2 + 23a - 12$

78. $(y - 5)(6y + 5)$ $6y^2 - 25y - 25$

Graph each linear equation. See Section 3.2.

79. $y = -3x$ **80.** $y = 5x$

81. $y = 2x - 7$ **82.** $y = -x + 4$

graphical answers in App. F

A Look Ahead

EXAMPLE

Factor $2t^5y^2 - 22t^4y^2 + 56t^3y^2$.

Solution:

First, factor out the greatest common factor of $2t^3y^2$.

$$2t^5y^2 - 22t^4y^2 + 56t^3y^2 = 2t^3y^2(t^2 - 11t + 28)$$
$$= 2t^3y^2(t - 4)(t - 7)$$

Factor each trinomial completely.

❏ **83.** $2x^2y + 30xy + 100y$ $2y(x + 5)(x + 10)$

❏ **84.** $3x^2z^2 + 9xz^2 + 6z^2$ $3z^2(x + 2)(x + 1)$

❏ **85.** $-12x^2y^3 - 24xy^3 - 36y^3$ $-12y^3(x^2 + 2x + 3)$

❏ **86.** $-4x^2t^4 + 4xt^4 + 24t^4$ $-4t^4(x - 3)(x + 2)$

❏ **87.** $y^2(x + 1) - 2y(x + 1) - 15(x + 1)$

❏ **88.** $z^2(x + 1) - 3z(x + 1) - 70(x + 1)$

87. $(x + 1)(y - 5)(y + 3)$ **88.** $(x + 1)(z - 10)(z + 7)$

5.3 | FACTORING TRINOMIALS OF THE FORM $ax^2 + bx + c$

O B J E C T I V E S

1. Factor trinomials of the form $ax^2 + bx + c$.
2. Factor out a GCF before factoring a trinomial of the form $ax^2 + bx + c$.
3. Factor perfect square trinomials.
4. Factor trinomials of the form $ax^2 + bx + c$ by an alternate method.

TAPE
BA 5.3

1

In this section, we factor trinomials of the form $ax^2 + bx + c$, where the numerical coefficient of the squared variable is any integer, not just 1.

To begin, let's review the relationship between the numerical coefficients of the trinomial and the numerical coefficients of its factored form. For example, since $(2x + 1)(x + 6) = 2x^2 + 13x + 6$, the factored form of $2x^2 + 13x + 6$ is

$$2x^2 + 13x + 6 = (2x + 1)(x + 6)$$

Notice that $2x$ and x are factors of $2x^2$, the first term of the trinomial. Also, 6 and 1 are factors of 6, the last term of the trinomial, as shown:

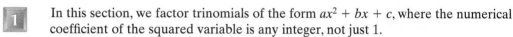

Also notice that $13x$ is the sum of the following products:

$$2x^2 + 13x + 6 = (2x + \underbrace{1)(x}_{1x} + 6)$$

$$\begin{array}{r} 1x \\ +12x \\ \hline 13x \end{array}$$

Let's use this information to factor $5x^2 + 7x + 2$. First, find factors of $5x^2$. Since all numerical coefficients in this trinomial are positive, use factors with positive numerical coefficients only.

Factors of $5x^2$ are $5x$ and x. Try these factors as first terms of the binomials. Thus far, we have

$$5x^2 + 7x + 2 = (5x \quad)(x \quad)$$

Next, find positive factors of 2. Positive factors of 2 are 1 and 2. Try possible combinations of these factors as second terms of the binomials until a middle term of $7x$ is obtained.

$$(5x + 1)(x + 2) = 5x^2 + 11x + 2$$

$$\underbrace{}_{1x}$$

$$\frac{+10x}{11x} \quad \textbf{Incorrect} \text{ middle term.}$$

Try switching factors 2 and 1.

$$(5x + 2)(x + 1) = 5x^2 + 7x + 2$$

$$\underbrace{}_{2x}$$

$$\frac{+5x}{7x} \quad \textbf{Correct} \text{ middle term.}$$

The factored form of $5x^2 + 7x + 2$ is

$$5x^2 + 7x + 2 = (5x + 2)(x + 1)$$

EXAMPLE 1 Factor $3x^2 + 11x + 6$.

Solution: Since all numerical coefficients are positive, use factors with positive numerical coefficients. First find factors of $3x^2$.

Factors of $3x^2$: $3x^2 = 3x \cdot x$

If factorable, the trinomial will be of the form:

$$3x^2 + 11x + 6 = (3x \quad)(x \quad)$$

Next, factor 6.

Factors of 6: $6 = 1 \cdot 6, \quad 6 = 2 \cdot 3$

Try combinations of factors of 6 until a middle term of $11x$ is obtained. First, try 1 and 6.

$$(3x + 1)(x + 6) = 3x^2 + 19x + 6$$

$$\underbrace{}_{1x}$$

$$\frac{+18x}{19x} \quad \textbf{Incorrect} \text{ middle term.}$$

Next, try 6 and 1.

$$(3x + 6)(x + 1)$$

Before multiplying, notice that the terms of the factor $3x + 6$ have a common factor of 3. The terms of the original trinomial $3x^2 + 11x + 6$ have no common factor other than 1, so the terms of the factored form of $3x^2 + 11x + 6$ contain no common factor other than 1, also. This means that $(3x + 6)(x + 1)$ is not the factored form.

Next, try 2 and 3 as last terms.

$$(3x + \underbrace{2)(x}_{2x} + 3) = 3x^2 + 11x + 6$$

$$\begin{array}{r} 2x \\ +9x \\ \hline 11x \end{array} \quad \textbf{Correct } \text{middle term.}$$

The factored form of $3x^2 + 11x + 6$ is

$$3x^2 + 11x + 6 = (3x + 2)(x + 3)$$

> REMINDER   If the terms of a trinomial have no common factor (other than 1), then the terms of neither of its binomial factors will contain a common factor (other than 1).

EXAMPLE 2   Factor $8x^2 - 22x + 5$.

Solution: Factors of $8x^2$: $8x^2 = 8x \cdot x$, $8x^2 = 4x \cdot 2x$

Try $8x$ and x.

$$8x^2 - 22x + 5 = (8x \qquad)(x \qquad)$$

Since the middle term $-22x$ has a negative numerical coefficient, factor 5 into negative factors.

Factors of 5: $5 = -1 \cdot -5$

Try -1 and -5.

$$(8x - \underbrace{1)(x}_{-1x} - 5) = 8x^2 - 41x + 5$$

$$\begin{array}{r} -1x \\ +(-40x) \\ \hline -41x \end{array} \quad \textbf{Incorrect } \text{middle term.}$$

Try -5 and -1.

$$(8x - \underbrace{5)(x}_{-5x} - 1) = 8x^2 - 13x + 5$$

$$\begin{array}{r} -5x \\ +(-8x) \\ \hline -13x \end{array} \quad \textbf{Incorrect } \text{middle term.}$$

Don't give up yet. Try other factors of $8x^2$. Try $4x$ and $2x$ with -1 and -5.

$$(4x - \underbrace{1)(2x}_{-2x} - 5) = 8x^2 - 22x + 5$$

$$\begin{array}{r} -2x \\ +(-20x) \\ \hline -22x \end{array} \quad \textbf{Correct } \text{middle term.}$$

The factored form of $8x^2 - 22x + 5$ is
$$8x^2 - 22x + 5 = (4x - 1)(2x - 5)$$

EXAMPLE 3 Factor $2x^2 + 13x - 7$.

Solution: Factors of $2x^2$: $2x^2 = 2x \cdot x$
Factors of -7: $-7 = -1 \cdot 7$, $-7 = 1 \cdot -7$

Try possible combinations. The combination that yields the middle term of $13x$ is
$$(2x - 1)(x + 7) = 2x^2 + 13x - 7$$
$$\underbrace{-1x}$$
$$\frac{+14x}{13x}$$ **Correct** middle term.

The factored form of $2x^2 + 13x - 7$ is
$$2x^2 + 13x - 7 = (2x - 1)(x + 7)$$

EXAMPLE 4 Factor $10x^2 - 13xy - 3y^2$.

Solution: Factors of $10x^2$: $10x^2 = 2x \cdot 5x$, $10x^2 = 10x \cdot x$
Factors of $-3y^2$: $-3y^2 = -3y \cdot y$, $-3y^2 = 3y \cdot -y$

Try possible combinations. The combination that yields the correct middle term is
$$(2x - 3y)(5x + y) = 10x^2 - 13xy - 3y^2$$
$$\underbrace{-15xy}$$
$$\frac{+ 2xy}{-13xy}$$ **Correct** middle term.

The factored form of $10x^2 - 13xy - 3y^2$ is
$$10x^2 - 13xy - 3y^2 = (2x - 3y)(5x + y)$$

2 Don't forget that the best first step in factoring any polynomial is to look for a common factor to factor out.

EXAMPLE 5 Factor $24x^4 + 40x^3 + 6x^2$.

Solution: Notice that all three terms have a common factor of $2x^2$. First, factor out $2x^2$.
$$24x^4 + 40x^3 + 6x^2 = 2x^2(12x^2 + 20x + 3)$$
Next, factor $12x^2 + 20x + 3$.
Factors of $12x^2$: $12x^2 = 6x \cdot 2x$, $12x^2 = 4x \cdot 3x$, $12x^2 = 12x \cdot x$

Since all terms in the trinomial have positive numerical coefficients, factor 3 using positive factors only.
Factors of 3: $3 = 1 \cdot 3$

The correct combination of factors is

$$2x^2(2x + 3)(6x + 1) = 2x^2(12x^2 + 20x + 3)$$

$$\underbrace{18x}$$

$$\underline{+\ 2x}$$

$$20x \qquad \textbf{Correct}\ \text{middle term.}$$

The factored form of $24x^4 + 40x^3 + 6x^2$ is

$$24x^4 + 40x^3 + 6x^2 = 2x^2(2x + 3)(6x + 1)$$

EXAMPLE 6   Factor $4x^2 - 12x + 9$.

Solution: Factors of $4x^2$: $4x^2 = 2x \cdot 2x, \qquad 4x^2 = 4x \cdot x$

Since the middle term $-12x$ has a negative numerical coefficient, factor 9 into negative factors only.

$$\text{Factors of 9:}\quad 9 = -3 \cdot -3, \quad 9 = -1 \cdot -9$$

The correct combination is

$$(2x - 3)(2x - 3) = 4x^2 - 12x + 9$$

$$\underbrace{-6x}$$

$$\underline{+(-6x)}$$

$$-12x \qquad \textbf{Correct}\ \text{middle term.}$$

Thus, $4x^2 - 12x + 9 = (2x - 3)(2x - 3)$, which can also be written as $(2x - 3)^2$.

3   Notice in Example 6 that $4x^2 - 12x + 9 = (2x - 3)^2$. The trinomial $4x^2 - 12x + 9$ is called a **perfect square trinomial** since it is the square of the binomial $2x - 3$.

In the last chapter, we learned a shortcut special product for squaring a binomial, recognizing that

$$(a + b)^2 = a^2 + 2ab + b^2$$

The trinomial $a^2 + 2ab + b^2$ is a perfect square trinomial, since it is the square of the binomial $a + b$. We can use this pattern to help us factor perfect square trinomials. To use this pattern, we must first be able to recognize a perfect square trinomial. A trinomial is a perfect square when its first term is the square of some expression a, its last term is the square of some expression b, and its middle term is twice the product of the expressions a and b. When a trinomial fits this description, its factored form is $(a + b)^2$.

PERFECT SQUARE TRINOMIALS

$$a^2 + 2ab + b^2 = (a + b)^2$$
$$a^2 - 2ab + b^2 = (a - b)^2$$

EXAMPLE 7 Factor $x^2 + 12x + 36$.

Solution: This trinomial is a perfect square trinomial since:

1. The first term is the square of x: $x^2 = (x)^2$.
2. The last term is the square of 6: $36 = (6)^2$.
3. The middle term is twice the product of x and 6: $12x = 2 \cdot x \cdot 6$.

Thus, $x^2 + 12x + 36 = (x + 6)^2$.

EXAMPLE 8 Factor $25x^2 + 25xy + 4y^2$.

Solution: Determine whether or not this trinomial is a perfect square by considering the same three questions.

1. Is the first term a square? Yes, $25x^2 = (5x)^2$.
2. Is the last term a square? Yes, $4y^2 = (2y)^2$.
3. Is the middle term twice the product of $5x$ and $2y$? **No.** $2(5x \cdot 2y) = 20xy$, not $25xy$.

Therefore, $25x^2 + 25xy + 4y^2$ is not a perfect square trinomial. It is factorable, though. Using earlier techniques, we find that $25x^2 + 25xy + 4y^2 = (5x + 4y)(5x + y)$.

> R E M I N D E R   A perfect square trinomial that is not recognized as such can be factored by other methods.

EXAMPLE 9 Factor $4m^2 - 4m + 1$.

Solution: This is a perfect square trinomial since $4m^2 = (2m)^2$, $1 = (1)^2$, and $4m = 2(2m \cdot 1)$.

$$4m^2 - 4m + 1 = (2m - 1)^2$$

4 Grouping can also be used to factor trinomials of the form $ax^2 + bx + c$. To use this method, write the trinomial as a four-term polynomial. For example, to factor $2x^2 + 11x + 12$ using grouping, find two numbers whose product is $2 \cdot 12 = 24$ and whose sum is 11. Since we want a positive product and a positive sum, we consider positive pairs of factors of 24 only.

FACTORS OF 24	SUM OF FACTORS
1, 24	1 + 24 = 25
2, 12	2 + 12 = 14
3, 8	3 + 8 = 11

The factors are 3 and 8. Use these factors to write the middle term $11x$ as $3x + 8x$. Replace $11x$ with $3x + 8x$ in the original trinomial and factor by grouping.

$$2x^2 + 11x + 12 = 2x^2 + 3x + 8x + 12$$
$$= (2x^2 + 3x) + (8x + 12)$$
$$= x(2x + 3) + 4(2x + 3)$$
$$= (2x + 3)(x + 4)$$

In general, we have the following:

TO FACTOR TRINOMIALS OF THE FORM $ax^2 + bx + c$ BY GROUPING

Step 1.   Find two numbers whose product is $a \cdot c$ and whose sum is b.

Step 2.   Write the middle term, bx, using the factors found in step 1.

Step 3.   Factor by grouping.

EXAMPLE 10   Factor $8x^2 - 14x + 5$ by grouping.

Solution: This trinomial is of the form $ax^2 + bx + c$ with $a = 8$, $b = -14$, and $c = 5$.

Step 1.   Find two numbers whose product is $a \cdot c$ or $8 \cdot 5 = 40$, and whose sum is b or -14. The numbers are -4 and -10.

Step 2.   Write $-14x$ as $-4x - 10x$ so that

$$8x^2 - 14x + 5 = 8x^2 - 4x - 10x + 5$$

Step 3.   Factor by grouping.

$$8x^2 - 4x - 10x + 5 = 4x(2x - 1) - 5(2x - 1)$$
$$= (2x - 1)(4x - 5)$$

EXAMPLE 11   Factor $3x^2 - x - 10$ by grouping.

Solution: In $3x^2 - x - 10$, $a = 3$, $b = -1$, and $c = -10$.

Step 1.   Find two numbers whose product is $a \cdot c$ or $3(-10) = -30$ and whose sum is b, -1. The numbers are -6 and 5.

Step 2. $3x^2 - x - 10 = 3x^2 - 6x + 5x - 10$

Step 3. $\qquad\qquad\;\; = 3x(x - 2) + 5(x - 2)$
$$= (x - 2)(3x + 5)$$

MENTAL MATH

State whether or not each trinomial is a perfect trinomial square.

1. $x^2 + 14x + 49$ yes

2. $9x^2 - 12x + 4$ yes

3. $y^2 + 2y + 4$ no

4. $x^2 - 4x + 2$ no

5. $9y^2 + 6y + 1$ yes

6. $y^2 - 16y + 64$ yes

EXERCISE SET 5.3

Factor completely. See Examples 1 through 5. (See Examples 10 and 11 for alternate method.)

1. $2x^2 + 13x + 15$ $(2x + 3)(x + 5)$

2. $3x^2 + 8x + 4$ $(3x + 2)(x + 2)$

3. $2x^2 - 9x - 5$ $(2x + 1)(x - 5)$

4. $3x^2 + 20x - 63$ $(3x - 7)(x + 9)$

5. $2y^2 - y - 6$ $(2y + 3)(y - 2)$

6. $8y^2 - 17y + 9$ $(y - 1)(8y - 9)$

7. $16a^2 - 24a + 9$ $(4a - 3)^2$

8. $25x^2 + 20x + 4$ $(5x + 2)^2$

9. $36r^2 - 5r - 24$ $(9r - 8)(4r + 3)$

10. $20r^2 + 27r - 8$ $(4r - 1)(5r + 8)$

11. $10x^2 + 17x + 3$ $(5x + 1)(2x + 3)$

12. $21x^2 - 41x + 10$ $(7x - 2)(3x - 5)$

13. $21x^2 - 48x - 45$ $3(7x + 5)(x - 3)$

14. $12x^2 - 14x - 10$ $2(3x - 5)(2x + 1)$

15. $12x^2 - 14x - 6$ $2(2x - 3)(3x + 1)$

16. $20x^2 - 2x + 6$ $2(10x^2 - x + 3)$

17. $4x^3 - 9x^2 - 9x$ $x(4x + 3)(x - 3)$

18. $6x^3 - 31x^2 + 5x$ $x(x - 5)(6x - 1)$

Factor the following perfect square trinomials. See Examples 6 through 9.

19. $x^2 + 22x + 121$ $(x + 11)^2$

20. $x^2 + 18x + 81$ $(x + 9)^2$

21. $x^2 - 16x + 64$ $(x - 8)^2$

22. $x^2 - 12x + 36$ $(x - 6)^2$

23. $16y^2 - 40y + 25$ $(4y - 5)^2$

24. $9y^2 + 48y + 64$ $(3y + 8)^2$

25. $x^2y^2 - 10xy + 25$ $(xy - 5)^2$

26. $4x^2y^2 - 28xy + 49$ $(2xy - 7)^2$

27. Describe a perfect square trinomial.

28. Write a perfect square trinomial that factors as $(x + 3y)^2$.   answers may vary

Factor the following completely.

29. $2x^2 - 7x - 99$ $(2x + 11)(x - 9)$

30. $2x^2 + 7x - 72$ $(2x - 9)(x + 8)$

31. $4x^2 - 8x - 21$ $(2x - 7)(2x + 3)$

32. $6x^2 - 11x - 10$ $(3x + 2)(2x - 5)$

33. $30x^2 - 53x + 21$ $(6x - 7)(5x - 3)$

34. $21x^3 - 6x - 30$ $3(7x^2 - 2x - 10)$

35. $24x^2 - 58x + 9$ $(4x - 9)(6x - 1)$

36. $36x^2 + 55x - 14$ $(4x + 7)(9x - 2)$

37. $9x^2 - 24xy + 16y^2$ $(3x - 4y)^2$

38. $25x^2 + 60xy + 36y^2$ $(5x + 6y)^2$

39. $x^2 - 14xy + 49y^2$ $(x - 7y)^2$

40. $x^2 + 10xy + 25y^2$ $(x + 5y)^2$

41. $2x^2 + 7x + 5$ $(2x + 5)(x + 1)$

42. $2x^2 + 7x + 3$ $(2x + 1)(x + 3)$

43. $3x^2 - 5x + 1$ prime

44. $3x^2 - 7x + 6$ prime

45. $-2y^2 + y + 10$ $(5 - 2y)(2 + y)$

46. $-4x^2 - 23x + 6$ $(-4x + 1)(x + 6)$

47. $16x^2 + 24xy + 9y^2$ $(4x + 3y)^2$

48. $4x^2 - 36xy + 81y^2$ $(2x - 9y)^2$

27. answers may vary

49. $8x^2y + 34xy - 84y$ $2y(4x - 7)(x + 6)$
50. $6x^2y^2 - 2xy^2 - 60y^2$ $2y^2(3x - 10)(x + 3)$
51. $3x^2 + x - 2$ $(3x - 2)(x + 1)$
52. $8y^2 + y - 9$ $(y - 1)(8y + 9)$
53. $x^2y^2 + 4xy + 4$ $(xy + 2)^2$
54. $x^2y^2 - 6xy + 9$ $(xy - 3)^2$
55. $49y^2 + 42xy + 9x^2$ $(7y + 3x)^2$
56. $16x^2 - 8xy + y^2$ $(4x - y)^2$
57. $3x^2 - 42x + 63$ $3(x^2 - 14x + 21)$
58. $5x^2 - 75x + 60$ $5(x^2 - 15x + 12)$
59. $42a^2 - 43a + 6$ $(7a - 6)(6a - 1)$
60. $54a^2 + 39ab - 8b^2$ $(9a + 8b)(6a - b)$
61. $18x^2 - 9x - 14$ $(6x - 7)(3x + 2)$
62. $8x^2 + 6x - 27$ $(4x + 9)(2x - 3)$
63. $25p^2 - 70pq + 49q^2$ $(5p - 7q)^2$
64. $36p^2 - 18pq + 9q^2$ $9(4p^2 - 2pq + q^2)$
65. $15x^2 - 16x - 15$ $(5x + 3)(3x - 5)$
66. $12x^2 + 7x - 12$ $(3x + 4)(4x - 3)$
67. $-27t + 7t^2 - 4$ $(7t + 1)(t - 4)$
68. $4t^2 - 7 - 3t$ $(4t - 7)(t + 1)$

Find a positive value of b so that each trinomial is factorable.

69. $3x^2 + bx - 5$ 2; 14   **70.** $2y^2 + by + 3$ 5; 7
71. $2z^2 + bz - 7$ 5; 13   **72.** $5x^2 + bx - 1$ 4

Find a positive value of c so that each trinomial is factorable.

73. $5x^2 + 7x + c$ 2   **74.** $7x^2 + 22x + c$ 3; 15; 16
75. $3x^2 - 8x + c$ 4; 5   **76.** $11y^2 - 40y + c$ 21; 29; 36

Review Exercises

Multiply the following. See Section 4.4.

77. $(x - 2)(x + 2)$ $x^2 - 4$
78. $(y^2 + 3)(y^2 - 3)$ $y^4 - 9$
79. $(a + 3)(a^2 - 3a + 9)$ $a^3 + 27$
80. $(z - 2)(z^2 + 2z + 4)$ $z^3 - 8$
81. $(y - 5)(y^2 + 5y + 25)$ $y^3 - 125$
82. $(m + 7)(m^2 - 7m + 49)$ $m^3 + 343$

As of June 1995, approximately 40% of U.S. households have a personal computer. The following graph shows the percent of households having a computer grouped according to household income. See Section 1.9.

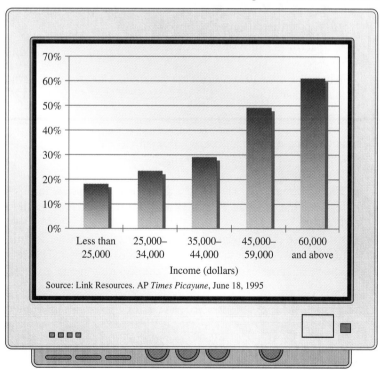

Source: Link Resources. AP *Times Picayune*, June 18, 1995

83. Which range of household income corresponds to the greatest percent of households having a personal computer? $60,000 and above

84. Which range of household income corresponds to the greatest increase in percent of households having a personal computer? $45,000–$59,000

85. Describe any trend you notice from this graph.

86. Why don't the percents shown in the graph add to 40%? answers may vary

A Look Ahead

EXAMPLE

Factor $10x^4y - 5x^3y^2 - 15x^2y^3$.

Solution:

Start by factoring out the greatest common factor of $5x^2y$.

$$10x^4y - 5x^3y^2 - 15x^2y^3 = 5x^2y(2x^2 - xy - 3y^2)$$

85. answers may vary

Next, factor $2x^2 - xy - 3y^2$. Try $2x$ and x as factors of $2x^2$ and $-3y$ and y as factors of $-3y^2$.

$$
\begin{array}{c}
(2x - 3y)(x + y) \\
\underbrace{\qquad}_{-3xy} \\
\underbrace{\qquad}_{+2xy} \\
\overline{\quad -xy \quad} \qquad \text{The middle term is correct.}
\end{array}
$$

Hence $10x^4y - 5x^3y^2 - 15x^2y^3 = 5x^2y(2x - 3y)(x + y)$.

Factor completely. See the example.

87. $-12x^3y^2 + 3x^2y^2 + 15xy^2$ $-3xy^2(4x - 5)(x + 1)$

88. $-12r^3x^2 + 38r^2x^2 + 14rx^2$ $-2rx^2(2r - 7)(3r + 1)$

89. $-30p^3q + 88p^2q^2 + 6pq^3$ $-2pq(p - 3q)(15p + q)$

90. $3x^3y^2 + 3x^2y^3 - 18xy^4$ $3xy^2(x + 3y)(x - 2y)$

91. $4x^2(y - 1)^2 + 10x(y - 1)^2 + 25(y - 1)^2$

92. $3x^2(a + 3)^3 - 28x(a + 3)^3 + 25(a + 3)^3$

91. $(y - 1)^2(4x^2 + 10x + 25)$

92. $(a + 3)^3(x - 1)(3x - 25)$

TAPE
BA 5.4

5.4   FACTORING BINOMIALS

O B J E C T I V E S

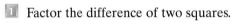

1   Factor the difference of two squares.

2   Factor the sum or difference of two cubes.

 When learning to multiply binomials in Chapter 4, we studied a special product, the product of the sum and difference of two terms, a and b

$$(a + b)(a - b) = a^2 - b^2$$

For example, the product of $x + 3$ and $x - 3$ is

$$(x + 3)(x - 3) = x^2 - 9$$

The binomial $x^2 - 9$ is called a **difference of squares.** In this section, we use the pattern for the product of a sum and difference to factor the binomial difference of squares.

To use this pattern to help us factor, we must be able to recognize a difference of squares. A binomial is a difference of squares when it is the difference of the square of some expression a and the square of some expression b.

> **DIFFERENCE OF TWO SQUARES**
> $$a^2 - b^2 = (a + b)(a - b)$$

EXAMPLE 1   Factor $4x^2 - 1$.

Solution: $4x^2 - 1$ is the difference of two squares since $4x^2 = (2x)^2$ and $1 = (1)^2$; therefore,

$$4x^2 - 1 = (2x)^2 - 1^2 = (2x + 1)(2x - 1)$$

Multiply to check.

EXAMPLE 2   Factor $25a^2 - 9b^2$.

Solution: $25a^2 - 9b^2 = (5a)^2 - (3b)^2 = (5a + 3b)(5a - 3b)$

EXAMPLE 3   Factor $9x^2 - 36$.

Solution: Remember when factoring always to check first for common factors. If there are common factors, factor out the GCF and then factor the resulting polynomial.

$$9x^2 - 36 = 9(x^2 - 4) \qquad \text{Factor out the GCF 9.}$$
$$= 9(x^2 - 2^2)$$
$$= 9(x + 2)(x - 2)$$

In this example, if we forget to factor out the GCF first, we still have the difference of two squares.

$$9x^2 - 36 = (3x)^2 - (6)^2 = (3x + 6)(3x - 6)$$

This binomial has not been factored completely since both terms of both binomial factors have a common factor of 3.

$$3x + 6 = 3(x + 2) \quad \text{and} \quad 3x - 6 = 3(x - 2)$$

Then

$$9x^2 - 36 = (3x + 6)(3x - 6) = 3(x + 2)3(x - 2) = 9(x + 2)(x - 2)$$

Factoring is easier if the GCF is factored out first before using other methods.

EXAMPLE 4   Factor $x^2 + 4$.

Solution: The binomial $x^2 + 4$ is the **sum** of squares since we can write $x^2 + 4$ as $x^2 + 2^2$. We might try to factor using $(x + 2)(x + 2)$ or $(x - 2)(x - 2)$. But when multiplying to check, neither factoring is correct.

$$(x + 2)(x + 2) = x^2 + 4x + 4$$
$$(x - 2)(x - 2) = x^2 - 4x + 4$$

In both cases, the product is a trinomial, not the required binomial. In fact, $x^2 + 4$ is a prime polynomial.

2   Although the sum of two squares usually does not factor, the sum or difference of two cubes can be factored and reveals factoring patterns. The pattern for the sum of cubes is illustrated by multiplying the binomial $x + y$ and the trinomial $x^2 - xy + y^2$.

$$
\begin{array}{r}
x^2 - xy + y^2 \\
x + y \\
\hline
x^2y - xy^2 + y^3 \\
x^3 - x^2y + xy^2 \\
\hline
x^3 \qquad\qquad + y^3
\end{array}
$$

$$(x + y)(x^2 - xy + y^2) = x^3 + y^3 \qquad \text{Sum of cubes.}$$

The pattern for the difference of two cubes is illustrated by multiplying the binomial $x - y$ by the trinomial $x^2 + xy + y^2$. The result is

$$(x - y)(x^2 + xy + y^2) = x^3 - y^3 \qquad \text{Difference of cubes.}$$

SUM OR DIFFERENCE OF TWO CUBES

$$a^3 + b^3 = (a + b)(a^2 - ab + b^2)$$
$$a^3 - b^3 = (a - b)(a^2 + ab + b^2)$$

EXAMPLE 5   Factor $x^3 + 8$.

Solution: First, write the binomial in the form $a^3 + b^3$.

$$x^3 + 8 = x^3 + 2^3 \qquad \text{Write in the form } a^3 + b^3.$$

If we replace a with x and b with 2 in the formula above, we have

$$x^3 + 2^3 = (x + 2)(x^2 - (x)(2) + 2^2)$$
$$= (x + 2)(x^2 - 2x + 4)$$

R E M I N D E R   When factoring sums or differences of cubes, notice the sign patterns.

same sign

$$x^3 + y^3 = (x + y)(x^2 - xy + y^2)$$

opposite sign   always positive

same sign

$$x^3 - y^3 = (x - y)(x^2 + xy + y^2)$$

opposite sign   always positive

EXAMPLE 6   Factor $y^3 - 27$.

Solution: $y^3 - 27 = y^3 - 3^3$ $\qquad$ Write in the form $a^3 - b^3$.

$\qquad\qquad\quad = (y - 3)[y^2 + (y)(3) + 3^2]$

$\qquad\qquad\quad = (y - 3)(y^2 + 3y + 9)$

EXAMPLE 7   Factor $64x^3 + 1$.

Solution: $64x^3 + 1 = (4x)^3 + 1^3$

$\qquad\qquad\quad = (4x + 1)[(4x)^2 - (4x)(1) + 1^2]$

$\qquad\qquad\quad = (4x + 1)(16x^2 - 4x + 1)$

EXAMPLE 8   Factor $54a^3 - 16b^3$.

Solution: Remember to factor out common factors first before using other factoring methods.

$54a^3 - 16b^3 = 2(27a^3 - 8b^3)$ $\qquad$ Factor out the GCF 2.

$\qquad\qquad\quad = 2[(3a)^3 - (2b)^3]$ $\qquad$ Difference of two cubes.

$\qquad\qquad\quad = 2(3a - 2b)[(3a)^2 + (3a)(2b) + (2b)^2]$

$\qquad\qquad\quad = 2(3a - 2b)(9a^2 + 6ab + 4b^2)$

MENTAL MATH

State each number as a square.

1. 1 1^2

2. 25 5^2

3. 81 9^2

4. 64 8^2

5. 9 3^2

6. 100 10^2

State each number as a cube.

7. 1 1^3

8. 64 4^3

9. 8 2^3

10. 27 3^3

EXERCISE SET 5.4

Factor the difference of two squares. See Examples 1 through 3.

1. $25y^2 - 9$ $(5y - 3)(5y + 3)$

2. $49a^2 - 16$ $(7a - 4)(7a + 4)$

3. $121 - 100x^2$ $(11 - 10x)(11 + 10x)$

4. $144 - 81x^2$ $9(4 - 3x)(4 + 3x)$

5. $12x^2 - 27$ $3(2x - 3)(2x + 3)$

6. $36x^2 - 64$ $4(3x - 4)(3x + 4)$

7. $169a^2 - 49b^2$ $(13a - 7b)(13a + 7b)$

8. $225a^2 - 81b^2$ $9(5a - 3b)(5a + 3b)$

9. $x^2y^2 - 1$ $(xy - 1)(xy + 1)$

10. $16 - a^2b^2$ $(4 - ab)(4 + ab)$

11. What binomial multiplied by $(x - 6)$ gives the difference of two squares? $(x + 6)$

⬭ **12.** What binomial multiplied by $(5 + y)$ gives the difference of two squares? $(5 - y)$

Factor the sum or difference of two cubes. See Examples 5 through 8.

13. $a^3 + 27$ $(a + 3)(a^2 - 3a + 9)$

14. $b^3 - 8$ $(b - 2)(b^2 + 2b + 4)$

15. $8a^3 + 1$ $(2a + 1)(4a^2 - 2a + 1)$

16. $64x^3 - 1$ $(4x - 1)(16x^2 + 4x + 1)$

17. $5k^3 + 40$ $5(k + 2)(k^2 - 2k + 4)$

18. $6r^3 - 162$ $6(r - 3)(r^2 + 3r + 9)$

19. $x^3y^3 - 64$ $(xy - 4)(x^2y^2 + 4xy + 16)$

20. $8x^3 - y^3$ $(2x - y)(4x^2 + 2xy + y^2)$

21. $x^3 + 125$ $(x + 5)(x^2 - 5x + 25)$

22. $a^3 - 216$ $(a - 6)(a^2 + 6a + 36)$

23. $24x^4 - 81xy^3$ $3x(2x - 3y)(4x^2 + 6xy + 9y^2)$

24. $375y^6 - 24y^3$ $3y^3(5y - 2)(25y^2 + 10y + 4)$

⬭ **25.** What binomial multiplied by $(4x^2 - 2xy + y^2)$ gives the sum or difference of two cubes? $(2x + y)$

⬭ **26.** What binomial multiplied by $(1 + 4y + 16y^2)$ gives the sum or difference of two cubes? $(1 - 4y)$

Factor the binomials completely.

27. $x^2 - 4$ $(x - 2)(x + 2)$

28. $x^2 - 36$ $(x + 6)(x - 6)$

29. $81 - p^2$ $(9 - p)(9 + p)$

30. $100 - t^2$ $(10 - t)(10 + t)$

31. $4r^2 - 1$ $(2r - 1)(2r + 1)$

32. $9t^2 - 1$ $(3t - 1)(3t + 1)$

33. $9x^2 - 16$ $(3x - 4)(3x + 4)$

34. $36y^2 - 25$ $(6y - 5)(6y + 5)$

35. $16r^2 + 1$ prime

36. $49y^2 + 1$ prime

37. $27 - t^3$ $(3 - t)(9 + 3t + t^2)$

38. $125 + r^3$ $(5 + r)(25 - 5r + r^2)$

39. $8r^3 - 64$ $8(r - 2)(r^2 + 2r + 4)$

40. $54r^3 + 2$ $2(3r + 1)(9r^2 - 3r + 1)$

41. $t^3 - 343$ $(t - 7)(t^2 + 7t + 49)$

42. $s^3 + 216$ $(s + 6)(s^2 - 6s + 36)$

43. $x^2 - 169y^2$ $(x - 13y)(x + 13y)$

44. $x^2 - 225y^2$ $(x - 15y)(x + 15y)$

45. $x^2y^2 - z^2$ $(xy - z)(xy + z)$

46. $x^3y^3 - z^3$ $(xy - z)(x^2y^2 + xyz + z^2)$

47. $x^3y^3 + 1$ $(xy + 1)(x^2y^2 - xy + 1)$

48. $x^2y^2 + z^2$ prime

49. $s^3 - 64t^3$ $(s - 4t)(s^2 + 4st + 16t^2)$

50. $8t^3 + s^3$ $(2t + s)(4t^2 - 2st + s^2)$

51. $18r^2 - 8$ $2(3r - 2)(3r + 2)$

52. $32t^2 - 50$ $2(4t - 5)(4t + 5)$

53. $9xy^2 - 4x$ $x(3y - 2)(3y + 2)$

54. $16xy^2 - 64x$ $16x(y - 2)(y + 2)$

55. $25y^4 - 100y^2$ $25y^2(y - 2)(y + 2)$

56. $xy^3 - 9xyz^2$ $xy(y - 3z)(y + 3z)$

57. $x^3y - 4xy^3$ $xy(x - 2y)(x + 2y)$

58. $12s^3t^3 + 192s^5t$ $12s^3t(t^2 + 16s^2)$

59. $8s^6t^3 + 100s^3t^6$ $4s^3t^3(2s^3 + 25t^3)$

60. $25x^5y + 121x^3y$ $x^3y(25x^2 + 121)$

61. $27x^2y^3 - xy^2$ $xy^2(27xy - 1)$

62. $8x^3y^3 + x^3y$ $x^3y(8y^2 + 1)$

Review Exercises

Divide the following. See Section 4.6.

63. $\dfrac{8x^4 + 4x^3 - 2x + 6}{2x}$ $\quad 4x^3 + 2x^2 - 1 + \dfrac{3}{x}$

64. $\dfrac{3y^4 + 9y^2 - 6y + 1}{3y^2}$ $\quad y^2 + 3 - \dfrac{2}{y} + \dfrac{1}{3y^2}$

Use long division to divide the following. See Section 4.6.

65. $\dfrac{2x^2 - 3x - 2}{x - 2}$ $\quad 2x + 1$

66. $\dfrac{4x^2 - 21x + 21}{x - 3}$ $\quad 4x - 9 - \dfrac{6}{x - 3}$

67. $\dfrac{3x^2 + 13x + 10}{x + 3}$ $\quad 3x + 4 - \dfrac{2}{x + 3}$

68. $\dfrac{5x^2 + 14x + 12}{x + 2}$ $\quad 5x + 4 + \dfrac{4}{x + 2}$

A Look Ahead

EXAMPLE:

Factor $(x + y)^2 - (x - y)^2$.

Solution:

Use the method for factoring the difference of squares.

$$(x + y)^2 - (x - y)^2 = [(x + y) + (x - y)]$$
$$[(x + y) - (x - y)]$$
$$= (x + y + x - y)(x + y - x + y)$$

$$= (2x)(2y)$$
$$= 4xy$$

Factor each difference of squares. See the example.

69. $x^4 - 16$ $(x - 2)(x + 2)(x^2 + 4)$

70. $x^6 - 1$ $(x - 1)(x + 1)(x^2 + x + 1)(x^2 - x + 1)$

71. $a^2 - (2 + b)^2$ $(a - 2 - b)(a + 2 + b)$

72. $(x + 3)^2 - y^2$ $(x - y + 3)(x + y + 3)$

73. $(x^2 - 4)^2 - (x - 2)^2$ $(x - 2)^2(x + 1)(x + 3)$

74. $(x^2 - 9) - (3 - x)$ $(x - 3)(x + 4)$

5.5 | CHOOSING A FACTORING STRATEGY

TAPE
BA 5.5

OBJECTIVE

 Factor polynomials completely.

A polynomial is factored completely when it is written as the product of prime polynomials. This section uses the various methods of factoring polynomials that have been discussed in earlier sections. Since these methods are applied throughout the remainder of this text, as well as in later courses, it is important to master the skills of factoring. The following is a set of guidelines for factoring polynomials.

TO FACTOR A POLYNOMIAL

Step 1. Are there any common factors? If so, factor out the GCF.

Step 2. How many terms are in the polynomial?

    a. If there are **two** terms, decide if one of the following can be applied.

        i. Difference of two squares: $a^2 - b^2 = (a - b)(a + b)$.

        ii. Difference of two cubes:
           $a^3 - b^3 = (a - b)(a^2 + ab + b^2)$.

        iii. Sum of two cubes: $a^3 + b^3 = (a + b)(a^2 - ab + b^2)$.

    b. If there are **three** terms, try one of the following.

        i. Perfect square trinomial: $a^2 + 2ab + b^2 = (a + b)^2$.

        ii. If not a perfect square trinomial, factor using the methods presented in Sections 5.2 and 5.3.

    c. If there are **four** or more terms, try factoring by grouping.

Step 3. See if any factors in the factored polynomial can be factored further.

Step 4. Check by multiplying.

EXAMPLE 1 Factor $10t^2 - 17t + 3$.

Solution: *Step 1.* The terms of this polynomial have no common factor (other than 1).

Step 2. There are three terms, so this polynomial is a trinomial. This trinomial is not a perfect square trinomial, so factor using methods from earlier sections.

$$\text{Factors of } 10t^2: \quad 10t^2 = 2t \cdot 5t, \quad 10t^2 = t \cdot 10t$$

Since the middle term, $-17t$, has a negative numerical coefficient, find negative factors of 3.

$$\text{Factors of 3:} \quad 3 = -1 \cdot -3$$

The correct combination is

$$(2t - 3)(5t - 1) = 10t^2 - 17t + 3$$

$$\underbrace{-15t}$$

$$\underbrace{-2t}$$

$$\overline{-17t} \qquad \text{Correct middle term.}$$

Step 3. No factor can be factored further, so we have factored completely.

Step 4. To check, multiply $2t - 3$ and $5t - 1$.

$$(2t - 3)(5t - 1) = 10t^2 - 2t - 15t + 3 = 10t^2 - 17t + 3$$

The factored form of $10t^2 - 17t + 3$ is $(2t - 3)(5t - 1)$.

EXAMPLE 2 Factor $2x^3 + 3x^2 - 2x - 3$.

Solution: *Step 1.* There are no factors common to all terms.

Step 2. Try factoring by grouping since this polynomial has four terms.

$$2x^3 + 3x^2 - 2x - 3 = x^2(2x + 3) - (2x + 3) \qquad \text{Factor out the greatest common factor for each pair of terms.}$$

$$= (2x + 3)(x^2 - 1) \qquad \text{Factor out } 2x + 3.$$

Step 3. The binomial $x^2 - 1$ can be factored further.

$$= (2x + 3)(x + 1)(x - 1) \qquad \text{Factor } x^2 - 1 \text{ as a difference of squares.}$$

Step 4. Check by finding the product of the three binomials.

The polynomial factored completely is $(2x + 3)(x + 1)(x - 1)$.

EXAMPLE 3   Factor $12m^2 - 3n^2$.

Solution: *Step 1.*   The terms of this binomial contain a greatest common factor of 3.

$$12m^2 - 3n^2 = 3(4m^2 - n^2) \quad \text{Factor out the greatest common factor.}$$

Step 2.   The binomial $4m^2 - n^2$ is a difference of squares.

$$= 3(2m + n)(2m - n) \quad \text{Factor the difference of squares.}$$

Step 3.   No factor can be factored further.

Step 4.   We check by multiplying.

$$3(2m + n)(2m - n) = 3(4m^2 - n^2) = 12m^2 - 3n^2$$

The factored form of $12m^2 - 3n^2$ is $3(2m + n)(2m - n)$.

EXAMPLE 4   Factor $x^3 + 27y^3$.

Solution: *Step 1.*   The terms of this binomial contain no common factor (other than 1).

Step 2.   This binomial is the sum of two cubes.

$$\begin{aligned} x^3 + 27y^3 &= (x)^3 + (3y)^3 \\ &= (x + 3y)[x^2 - 3xy + (3y)^2] \\ &= (x + 3y)(x^2 - 3xy + 9y^2) \end{aligned}$$

Step 3.   No factor can be factored further.

Step 4.   We check by multiplying.

$$\begin{aligned} (x + 3y)(x^2 - 3xy + 9y^2) &= x(x^2 - 3xy + 9y^2) \\ &\quad + 3y(x^2 - 3xy + 9y^2) \\ &= x^3 - 3x^2y + 9xy^2 + 3x^2y \\ &\quad - 9xy^2 + 27y^3 = x^3 + 27y^3 \end{aligned}$$

Thus, $x^3 + 27y^3$ factored completely is $(x + 3y)(x^2 - 3xy + 9y^2)$.

EXAMPLE 5   Factor $30a^2b^3 + 55a^2b^2 - 35a^2b$.

Solution: *Step 1.* $30a^2b^3 + 55a^2b^2 - 35a^2b = 5a^2b(6b^2 + 11b - 7)$ Factor out the GCF.

Step 2. $= 5a^2b(2b - 1)(3b + 7)$ Factor the resulting trinomial.

Step 3. No factor can be factored further.

Step 4. Check by multiplying.

The trinomial factored completely is $5a^2b(2b - 1)(3b + 7)$. ▬▬

EXERCISE SET 5.5

Factor the following completely. See Examples 1 through 5.

1. $a^2 + 2ab + b^2$ $(a + b)^2$

2. $a^2 - 2ab + b^2$ $(a - b)^2$

3. $a^2 + a - 12$ $(a - 3)(a + 4)$

4. $a^2 - 7a + 10$ $(a - 5)(a - 2)$

5. $a^2 - a - 6$ $(a + 2)(a - 3)$

6. $a^2 + 2a + 1$ $(a + 1)^2$

7. $x^2 + 2x + 1$ $(x + 1)^2$

8. $x^2 + x - 2$ $(x + 2)(x - 1)$

9. $x^2 + 4x + 3$ $(x + 1)(x + 3)$

10. $x^2 + x - 6$ $(x + 3)(x - 2)$

11. $x^2 + 7x + 12$ $(x + 3)(x + 4)$

12. $x^2 + x - 12$ $(x + 4)(x - 3)$

13. $x^2 + 3x - 4$ $(x + 4)(x - 1)$

14. $x^2 - 7x + 10$ $(x - 5)(x - 2)$

15. $x^2 + 2x - 15$ $(x + 5)(x - 3)$

16. $x^2 + 11x + 30$ $(x + 6)(x + 5)$

17. $x^2 - x - 30$ $(x - 6)(x + 5)$

18. $x^2 + 11x + 24$ $(x + 8)(x + 3)$

19. $2x^2 - 98$ $2(x - 7)(x + 7)$

20. $3x^2 - 75$ $3(x - 5)(x + 5)$

21. $x^2 + 3x + xy + 3y$ $(x + 3)(x + y)$

22. $3y - 21 + xy - 7x$ $(y - 7)(3 + x)$

23. $x^2 + 6x - 16$ $(x + 8)(x - 2)$

24. $x^2 - 3x - 28$ $(x - 7)(x + 4)$

25. $4x^3 + 20x^2 - 56x$ $4x(x + 7)(x - 2)$

26. $6x^3 - 6x^2 - 120x$ $6x(x - 5)(x + 4)$

27. $12x^2 + 34x + 24$ $2(3x + 4)(2x + 3)$

28. $8a^2 + 6ab - 5b^2$ $(2a - b)(4a + 5b)$

29. $4a^2 - b^2$ $(2a - b)(2a + b)$

30. $28 - 13x - 6x^2$ $(4 - 3x)(7 + 2x)$

31. $20 - 3x - 2x^2$ $(5 - 2x)(4 + x)$

32. $x^2 - 2x + 4$ prime

33. $a^2 + a - 3$ prime

34. $6y^2 + y - 15$ $(3y + 5)(2y - 3)$

35. $4x^2 - x - 5$ $(4x - 5)(x + 1)$

36. $x^2y - y^3$ $y(x - y)(x + y)$

37. $4t^2 + 36$ $4(t^2 + 9)$

38. $x^2 + x + xy + y$ $(x + 1)(x + y)$

39. $ax + 2x + a + 2$ $(x + 1)(a + 2)$

40. $18x^3 - 63x^2 + 9x$ $9x(2x^2 - 7x + 1)$

41. $12a^3 - 24a^2 + 4a$ $4a(3a^2 - 6a + 1)$

42. $x^2 + 14x - 32$ $(x + 16)(x - 2)$

43. $x^2 - 14x - 48$ prime

44. $16a^2 - 56ab + 49b^2$ $(4a - 7b)^2$

45. $25p^2 - 70pq + 49q^2$ $(5p - 7q)^2$

46. $7x^2 + 24xy + 9y^2$ $(7x + 3y)(x + 3y)$

47. $125 - 8y^3$ $(5 - 2y)(25 + 10y + 4y^2)$

48. $64x^3 + 27$ $(4x + 3)(16x^2 - 12x + 9)$

49. $-x^2 - x + 30$ $(5 - x)(6 + x)$

50. $-x^2 + 6x - 8$ $-(x - 2)(x - 4)$

51. $14 + 5x - x^2$ $(7 - x)(2 + x)$

52. $3 - 2x - x^2$ $(3 + x)(1 - x)$

53. $3x^4y + 6x^3y - 72x^2y$ $3x^2y(x + 6)(x - 4)$

54. $2x^3y + 8x^2y^2 - 10xy^3$ $2xy(x + 5y)(x - y)$

55. $5x^3y^2 - 40x^2y^3 + 35xy^4$ $5xy^2(x - 7y)(x - y)$

56. $4x^4y - 8x^3y - 60x^2y$ $4x^2y(x - 5)(x + 3)$

57. $12x^3y + 243xy$ $3xy(4x^2 + 81)$

58. $6x^3y^2 + 8xy^2$ $2xy^2(3x^2 + 4)$

59. $(x - y)^2 - z^2$ $(x - y - z)(x - y + z)$

60. $(x + 2y)^2 - 9$ $(x + 2y - 3)(x + 2y + 3)$

61. $3rs - s + 12r - 4$ $(s + 4)(3r - 1)$

62. $x^3 - 2x^2 + 3x - 6$ $(x - 2)(x^2 + 3)$

63. $4x^2 - 8xy - 3x + 6y$ $(4x - 3)(x - 2y)$

64. $4x^2 - 2xy - 7yz + 14xz$ $(2x - y)(2x + 7z)$

65. $6x^2 + 18xy + 12y^2$ $6(x + 2y)(x + y)$

66. $12x^2 + 46xy - 8y^2$ $2(x + 4y)(6x - y)$

67. $xy^2 - 4x + 3y^2 - 12$ $(x + 3)(y - 2)(y + 2)$

68. $x^2y^2 - 9x^2 + 3y^2 - 27$ $(y - 3)(y + 3)(x^2 + 3)$

69. $5(x + y) + x(x + y)$ $(5 + x)(x + y)$

70. $7(x - y) + y(x - y)$ $(x - y)(7 + y)$

71. $14t^2 - 9t + 1$ $(7t - 1)(2t - 1)$

72. $3t^2 - 5t + 1$ prime

73. $3x^2 + 2x - 5$ $(3x + 5)(x - 1)$

74. $7x^2 + 19x - 6$ $(7x - 2)(x + 3)$

75. $x^2 + 9xy - 36y^2$ $(x + 12y)(x - 3y)$

76. $3x^2 + 10xy - 8y^2$ $(3x - 2y)(x + 4y)$

77. $1 - 8ab - 20a^2b^2$ $(1 - 10ab)(1 + 2ab)$

78. $1 - 7ab - 60a^2b^2$ $(1 + 5ab)(1 - 12ab)$

79. $x^4 - 10x^2 + 9$ $(x - 3)(x + 3)(x - 1)(x + 1)$

80. $x^4 - 13x^2 + 36$ $(x - 3)(x + 3)(x - 2)(x + 2)$

81. $x^4 - 14x^2 - 32$ $(x - 4)(x + 4)(x^2 + 2)$

82. $x^4 - 22x^2 - 75$ $(x - 5)(x + 5)(x^2 + 3)$

83. $x^2 - 23x + 120$ $(x - 15)(x - 8)$

84. $y^2 + 22y + 96$ $(y + 16)(y + 6)$

85. $6x^3 - 28x^2 + 16x$ $2x(3x - 2)(x - 4)$

86. $6y^3 - 8y^2 - 30y$ $2y(3y + 5)(y - 3)$

87. $27x^3 - 125y^3$ $(3x - 5y)(9x^2 + 15xy + 25y^2)$

88. $216y^3 - z^3$ $(6y - z)(36y^2 + 6yz + z^2)$

89. $x^3y^3 + 8z^3$ $(xy + 2z)(x^2y^2 - 2xyz + 4z^2)$

90. $27a^3b^3 + 8$ $(3ab + 2)(9a^2b^2 - 6ab + 4)$

91. $2xy - 72x^3y$ $2xy(1 - 6x)(1 + 6x)$

92. $2x^3 - 18x$ $2x(x - 3)(x + 3)$

93. $x^3 + 6x^2 - 4x - 24$ $(x - 2)(x + 2)(x + 6)$

94. $x^3 - 2x^2 - 36x + 72$ $(x - 2)(x - 6)(x + 6)$

95. $6a^3 + 10a^2$ $2a^2(3a + 5)$

96. $4n^2 - 6n$ $2n(2n - 3)$

97. $a^2(a + 2) + 2(a + 2)$ $(a^2 + 2)(a + 2)$

98. $a - b + x(a - b)$ $(a - b)(1 + x)$

99. $x^3 - 28 + 7x^2 - 4x$ $(x - 2)(x + 2)(x + 7)$

100. $a^3 - 45 - 9a + 5a^2$ $(a - 3)(a + 3)(a + 5)$

101. Explain why it makes good sense to factor out the GCF first, before using other methods of factoring.   answers may vary

102. The sum of two squares usually does not factor. Is the sum of two squares $9x^2 + 81y^2$ factorable?
yes; $9(x^2 + 9y^2)$

Review Exercises

Solve each equation. See Section 2.4.

103. $x - 6 = 0$ $x = 6$ **104.** $y + 5 = 0$ $y = -5$

105. $2m + 4 = 0$ $m = -2$ **106.** $3x - 9 = 0$ $x = 3$

107. $5z - 1 = 0$ $z = \dfrac{1}{5}$ **108.** $4a + 2 = 0$ $a = -\dfrac{1}{2}$

Solve the following. See Section 2.6.

109. A suitcase has a volume of 960 cubic inches. Find x.
8 in.

10 inches

12 inches

x inches

110. The sail shown has an area of 25 square feet. Find its height, x.   5 ft.

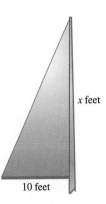

x feet

10 feet

List the x- and y-intercept points for each graph. See Section 3.3.

111.

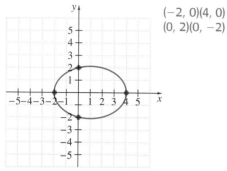

$(-2, 0)(4, 0)$
$(0, 2)(0, -2)$

112.

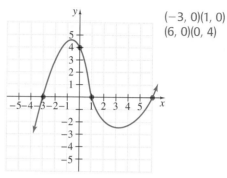

$(-3, 0)(1, 0)$
$(6, 0)(0, 4)$

113.

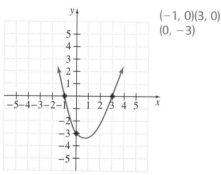

$(-1, 0)(3, 0)$
$(0, -3)$

114.

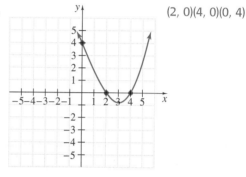

$(2, 0)(4, 0)(0, 4)$

5.6 SOLVING QUADRATIC EQUATIONS BY FACTORING

O B J E C T I V E S

1. Define quadratic equation.
2. Solve quadratic equations by factoring.
3. Solve equations with degree greater than 2 by factoring.

TAPE
BA 5.6

1. Linear equations, while versatile, are not versatile enough to model many real-life phenomena. For example, let's suppose an object is dropped from the top of a 256-foot cliff and we want to know how long before the object strikes the ground. The answer to this question is found by solving the equation $-16t^2 + 256 = 0$. (See Example 1 in the next section.) This equation is called a **quadratic equation** because it contains a variable with an exponent of 2 and no other variable in the equation contains an exponent greater than 2. In this section, we solve quadratic equations by factoring.

QUADRATIC EQUATION

A quadratic equation is one that can be written in the form $ax^2 + bx + c = 0$, where a, b, and c are real numbers, and $a \neq 0$.

Notice that the degree of the polynomial $ax^2 + bx + c$ is 2. Here are a few more examples of quadratic equations.

QUADRATIC EQUATIONS

$$3x^2 + 5x + 6 = 0 \qquad x^2 = 9 \qquad y^2 + y = 1$$

The form $ax^2 + bx + c = 0$ is called the **standard form** of a quadratic equation. The quadratic equations $3x^2 + 5x + 6 = 0$ and $-16t^2 + 256 = 0$ are in standard form. One side of the equation is 0 and the other side is a polynomial of degree 2 written in descending powers of the variable.

2   Some quadratic equations can be solved by making use of factoring and the **zero factor theorem.**

ZERO FACTOR THEOREM

If a and b are real numbers and if $ab = 0$, then $a = 0$ or $b = 0$.

This theorem states that if the product of two numbers is 0 then at least one of the numbers must be 0. If the equation

$$(x - 3)(x + 1) = 0$$

is a true statement, then either the factor $x - 3$ must be 0 or the factor $x + 1$ must be 0. In other words,

$$x - 3 = 0 \qquad \text{or} \qquad x + 1 = 0$$
$$x = 3 \qquad \text{or} \qquad x = -1$$

Thus, 3 and -1 are both solutions of the equation $(x - 3)(x + 1) = 0$. To check, replace x with 3 in the original equation. Then replace x with -1 in the original equation.

$$
\begin{aligned}
(x - 3)(x + 1) &= 0 \\
(3 - 3)(3 + 1) &= 0 \qquad &\text{Replace } x \text{ with 3.} \\
0(4) &= 0 \qquad &\text{True.} \\
(x - 3)(x + 1) &= 0 \\
(-1 - 3)(-1 + 1) &= 0 \qquad &\text{Replace } x \text{ with } -1. \\
(-4)(0) &= 0 \qquad &\text{True.}
\end{aligned}
$$

EXAMPLE 1   Solve $(x - 5)(2x + 7) = 0$.

Solution: Use the zero factor theorem; set each factor equal to 0 and solve the resulting linear equations.

$$
\begin{aligned}
(x - 5)(2x + 7) &= 0 \\
x - 5 = 0 \quad &\text{or} \quad 2x + 7 = 0 \\
x = 5 \quad &\text{or} \quad 2x = -7 \\
&\qquad\qquad x = -\frac{7}{2}
\end{aligned}
$$

If x is either 5 or $-\frac{7}{2}$, the product $(x - 5)(2x + 7)$ is 0. Check by replacing x with 5 in the original equation; then replace x with $-\frac{7}{2}$ in the original equation. Both 5 and $-\frac{7}{2}$ are solutions and the solution set is $\left\{5, -\frac{7}{2}\right\}$.

To use the zero factor theorem, one side of the quadratic equation must be 0 and the other side must be in factored form as in Example 1. If a quadratic equation is not in this form, first write the equation in standard form, then factor the polynomial.

EXAMPLE 2   Solve $x^2 - 9x = -20$.

Solution: First, write the equation in standard form; then factor.

$$x^2 - 9x = -20$$
$$x^2 - 9x + 20 = 0 \qquad \text{Write in standard form by adding 20 to both sides.}$$
$$(x - 4)(x - 5) = 0 \qquad \text{Factor.}$$

Next, use the zero factor theorem and set each factor equal to 0.

$$x - 4 = 0 \quad \text{or} \quad x - 5 = 0 \qquad \text{Set each factor equal to 0.}$$
$$x = 4 \quad \text{or} \qquad x = 5 \qquad \text{Solve.}$$

Check the solutions by replacing x with each value in the original equation. The solution set is $\{4, 5\}$.

The following steps may be used to solve a quadratic equation by factoring.

TO SOLVE QUADRATIC EQUATIONS BY FACTORING

Step 1.   Write the equation in standard form: $ax^2 + bx + c = 0$.
Step 2.   Factor the quadratic completely.
Step 3.   Set each factor containing a variable equal to 0.
Step 4.   Solve the resulting equations.
Step 5.   Check each solution in the original equation.

Since it is not always possible to factor a quadratic polynomial, not all quadratic equations can be solved by factoring. Other methods of solving quadratic equations are presented in Chapter 10.

EXAMPLE 3   Solve $x(2x - 7) = 4$.

Solution: First, write the equation in standard form; then factor.

$$x(2x - 7) = 4$$
$$2x^2 - 7x = 4 \qquad \text{Multiply.}$$
$$2x^2 - 7x - 4 = 0 \qquad \text{Write in standard form.}$$
$$(2x + 1)(x - 4) = 0 \qquad \text{Factor.}$$
$$2x + 1 = 0 \quad \text{or} \quad x - 4 = 0 \qquad \text{Set each factor equal to zero.}$$
$$2x = -1 \quad \text{or} \quad x = 4 \qquad \text{Solve.}$$
$$x = -\frac{1}{2}$$

Check both solutions $-\dfrac{1}{2}$ and 4. The solution set is $\left\{-\dfrac{1}{2}, 4\right\}$.

> **REMINDER**   To apply the zero factor theorem, one side of the equation must be 0 and the other side of the equation must be factored. To solve the equation $x(2x - 7) = 4$, for example, you may **not** set each factor equal to 4.

EXAMPLE 4   Solve $-2x^2 - 4x + 30 = 0$.

Solution: The equation is in standard form so we begin by factoring out a common factor of -2.

$$-2x^2 - 4x + 30 = 0$$
$$-2(x^2 + 2x - 15) = 0 \qquad \text{Factor out } -2.$$
$$-2(x + 5)(x - 3) = 0 \qquad \text{Factor the quadratic.}$$

Next, set each factor **containing a variable** equal to 0.

$$x + 5 = 0 \quad \text{or} \quad x - 3 = 0 \qquad \text{Set each factor containing a variable equal to 0.}$$
$$x = -5 \quad \text{or} \quad x = 3 \qquad \text{Solve.}$$

Note that the factor -2 is a constant term containing no variables and can never equal 0. The solution set is $\{-5, 3\}$.

3 Some equations involving polynomials of degree higher than 2 may also be solved by factoring and then applying the zero factor theorem.

EXAMPLE 5   Solve $3x^3 - 12x = 0$.

Solution: Factor the left side of the equation. Begin by factoring out the common factor of $3x$.

$$3x^3 - 12x = 0$$
$$3x(x^2 - 4) = 0 \qquad \text{Factor out the GCF } 3x.$$
$$3x(x + 2)(x - 2) = 0 \qquad \text{Factor } x^2 - 4, \text{ a difference of squares.}$$
$$3x = 0 \quad \text{or} \quad x + 2 = 0 \quad \text{or} \quad x - 2 = 0 \qquad \text{Set each factor equal to 0.}$$

$$x = 0 \quad \text{or} \quad x = -2 \text{ or} \quad x = 2 \quad \text{Solve.}$$

Thus, the equation $3x^3 - 12x = 0$ has three solutions: 0, -2, and 2. To check, replace x with each solution in the original equation.

Let $x = 0$	Let $x = -2$	Let $x = 2$
$3(0)^3 - 12(0) = 0$	$3(-2)^3 - 12(-2) = 0$	$3(2)^3 - 12(2) = 0$
$0 = 0$	$3(-8) + 24 = 0$	$3(8) - 24 = 0$
	$0 = 0$	$0 = 0$

Substituting 0, -2, or 2 into the original equation results each time in a true equation. The solution set is $\{-2, 0, 2\}$.

EXAMPLE 6   Solve $(5x - 1)(2x^2 + 15x + 18) = 0$.

Solution:

$$(5x - 1)(2x^2 + 15x + 18) = 0$$
$$(5x - 1)(2x + 3)(x + 6) = 0 \qquad \text{Factor the trinomial.}$$
$$5x - 1 = 0 \quad \text{or} \quad 2x + 3 = 0 \quad \text{or} \quad x + 6 = 0 \qquad \text{Set each factor equal to 0.}$$
$$5x = 1 \quad \text{or} \quad 2x = -3 \text{ or} \quad x = -6 \qquad \text{Solve.}$$
$$x = \frac{1}{5} \quad \text{or} \quad x = -\frac{3}{2}$$

The solutions are $\dfrac{1}{5}$, $-\dfrac{3}{2}$, and -6. Check by replacing x with each solution in the original equation. The solution set is $\left\{-6, -\dfrac{3}{2}, \dfrac{1}{5}\right\}$.

EXAMPLE 7   Solve $2x^3 - 4x^2 - 30x = 0$.

Solution: Begin by factoring out the GCF $2x$.

$$2x^3 - 4x^2 - 30x = 0$$
$$2x(x^2 - 2x - 15) = 0 \qquad \text{Factor out the GCF } 2x.$$
$$2x(x - 5)(x + 3) = 0 \qquad \text{Factor the quadratic.}$$
$$2x = 0 \quad \text{or} \quad x - 5 = 0 \quad \text{or} \quad x + 3 = 0 \qquad \text{Set each factor containing a variable equal to 0.}$$
$$x = 0 \quad \text{or} \quad x = 5 \quad \text{or} \quad x = -3 \qquad \text{Solve.}$$

Check by replacing x with each solution in the cubic equation. The solution set is $\{-3, 0, 5\}$.

In Chapter 3, we graphed linear equations in two variables, such as $y = 5x - 6$. Recall that to find the x-intercept of the graph of a linear equation, let $y = 0$ and solve for x. This is also how to find the x-intercepts of the graph of a **quadratic equation in two variables,** such as $y = x^2 - 5x + 4$.

EXAMPLE 8   Find the x-intercepts of the graph of $y = x^2 - 5x + 4$.

Solution: Let $y = 0$ and solve for x.

$$y = x^2 - 5x + 4$$
$$0 = x^2 - 5x + 4 \qquad \text{Let } y = 0$$
$$0 = (x - 1)(x - 4) \qquad \text{Factor.}$$
$$x - 1 = 0 \quad \text{or} \quad x - 4 = 0 \qquad \text{Set each factor equal to 0.}$$
$$x = 1 \quad \text{or} \qquad x = 4 \qquad \text{Solve.}$$

The x-intercepts of the graph of $y = x^2 - 5x + 4$ are 1 and also 4. The intercept points are $(1, 0)$ and $(4, 0)$.

The graph of $y = x^2 - 5x + 4$ is shown below.

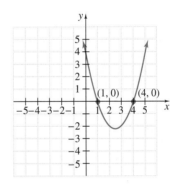

In general, a quadratic equation in two variables is one that can be written in the form $y = ax^2 + bx + c$ where $a \neq 0$. The graph of such an equation is called a **parabola** and will open up or down depending on the value of a.

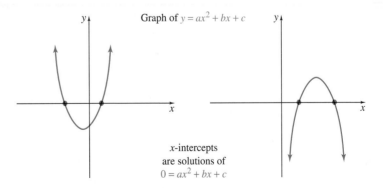

Graph of $y = ax^2 + bx + c$

x-intercepts
are solutions of
$0 = ax^2 + bx + c$

Notice that the x-intercepts of the graph of $y = ax^2 + bx + c$ are the real number solutions of $0 = ax^2 + bx + c$. Also, the real number solutions of $0 = ax^2 + bx + c$ are the x-intercepts of the graph of $y = ax^2 + bx + c$. We study more about graphs of quadratic equations in two variables in Chapter 10.

GRAPHING CALCULATOR EXPLORATIONS

A grapher may be used to find solutions of a quadratic equation whether the related quadratic polynomial is factorable or not. For example, let's use a grapher to approximate the solutions of $0 = x^2 + 4x - 3$. To do so, graph $y_1 = x^2 + 4x - 3$. Recall that the x-intercepts of this graph are the solutions of $0 = x^2 + 4x - 3$.

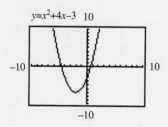

Notice that the graph appears to have an x-intercept between -5 and -4 and one between 0 and 1. Many graphers contain a TRACE feature. This feature activates a graph cursor that can be used to *trace* along a graph while the corresponding x- and y-coordinates are shown on the screen. Use the TRACE feature to confirm that x-intercepts lie between -5 and -4 and also 0 and 1. To approximate the x-intercepts to the nearest tenth, use a ROOT or a ZOOM feature on your grapher or redefine the viewing window. (A ROOT feature calculates the x-intercept. A ZOOM feature magnifies the viewing window around a specific location such as the graph cursor.) If we redefine the window to $[0, 1]$ on the x-axis and $[-1, 1]$ on the y-axis, the following graph is generated.

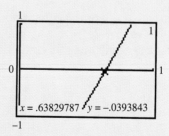

By using the TRACE feature, we can conclude that one x-intercept is approximately 0.6 to the nearest tenth. By repeating these steps for the other x-intercept, we find that it is approximately -4.6.

Use a grapher to approximate the real number solutions. If an equation has no real number solution, state so.

1. $3x^2 - 4x - 6 = 0$ $-0.9, 2.2$ 2. $x^2 - x - 9 = 0$ $-2.5, 3.5$
3. $2x^2 + x + 2 = 0$ No real solution   4. $-4x^2 - 5x - 4 = 0$ No real solution
5. $-x^2 + x + 5 = 0$ $-1.8, 2.8$ 6. $10x^2 + 6x - 3 = 0$ $-0.9, 0.3$

MENTAL MATH

Solve each equation by inspection.

1. $(a-3)(a-7)=0$ $a=3, a=7$

2. $(a-5)(a-2)=0$ $a=5, a=2$

3. $(x+8)(x+6)=0$ $x=-8, x=-6$

4. $(x+2)(x+3)=0$ $x=-2, x=-3$

5. $(x+1)(x-3)=0$ $x=-1, x=3$

6. $(x-1)(x+2)=0$ $x=1, x=-2$

EXERCISE SET 5.6

Solve each equation. See Example 1.

1. $(x-2)(x+1)=0$ $x=2, x=-1$

2. $(x+3)(x+2)=0$ $x=-3, x=-2$

3. $x(x+6)=0$ $x=0, x=-6$

4. $2x(x-7)=0$ $x=0, x=7$

5. $(2x+3)(4x-5)=0$ $x=-\frac{3}{2}, x=\frac{5}{4}$

6. $(3x-2)(5x+1)=0$ $x=\frac{2}{3}, x=-\frac{1}{5}$

7. $(2x-7)(7x+2)=0$ $x=\frac{7}{2}, x=-\frac{2}{7}$

8. $(9x+1)(4x-3)=0$ $x=-\frac{1}{9}, x=\frac{3}{4}$

9. Write a quadratic equation that has two solutions, 6 and -1. Leave the polynomial in the equation in factored form. $(x-6)(x+1)=0$

10. Write a quadratic equation that has two solutions, 0 and -2. Leave the polynomial in the equation in factored form. $x(x+2)=0$

Solve each equation. See Examples 2 through 4.

11. $x^2-13x+36=0$ $x=9, x=4$

12. $x^2+2x-63=0$ $x=-9, x=7$

13. $x^2+2x-8=0$ $x=-4, x=2$

14. $x^2-5x+6=0$ $x=3, x=2$

15. $x^2-4x=32$ $x=8, x=-4$

16. $x^2-5x=24$ $x=8, x=-3$

17. $x(3x-1)=14$ $x=\frac{7}{3}, x=-2$

18. $x(4x-11)=3$ $x=-\frac{1}{4}, x=3$

19. $3x^2+19x-72=0$ $x=\frac{8}{3}, x=-9$

20. $36x^2+x-21=0$ $x=\frac{3}{4}, x=-\frac{7}{9}$

21. Write a quadratic equation in standard form that has two solutions, 5 and 7. $x^2-12x+35=0$

22. Write an equation that has three solutions, 0, 1, and 2. $x^3-3x^2+2x=0$

Solve each equation. See Examples 5 through 7.

23. $x^3-12x^2+32x=0$ $x=0, x=8, x=4$

24. $x^3-14x^2+49x=0$ $x=0, x=7$

25. $(4x-3)(16x^2-24x+9)=0$ $x=\frac{3}{4}$

26. $(2x+5)(4x^2-10x+25)=0$ $x=-\frac{5}{2}$

27. $4x^3-x=0$ $x=0, x=\frac{1}{2}, x=-\frac{1}{2}$

28. $4y^3-36y=0$ $y=0, y=3, y=-3$

29. $32x^3-4x^2-6x=0$ $x=0, x=\frac{1}{2}, x=-\frac{3}{8}$

30. $15x^3+24x^2-63x=0$ $x=0, x=\frac{7}{5}, x=-3$

Find the x-intercepts of the graph of each equation. See Example 8.

31. $y=(3x+4)(x-1)$ $-\frac{4}{3}, 1$

32. $y=(5x-3)(x-4)$ $\frac{3}{5}, 4$

33. $y=x^2-3x-10$ $-2, 5$

34. $y=x^2+7x+6$ $-1, -6$

35. $y=2x^2+11x-6$ $-6, \frac{1}{2}$

36. $y=4x^2+11x+6$ $-2, -\frac{3}{4}$

For Exercises 37 through 42, match each equation with its graph.

A

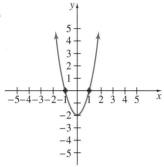

B

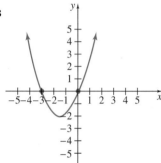

C

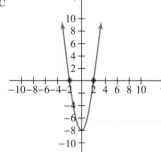

D

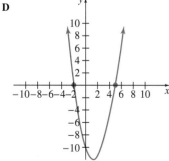

E

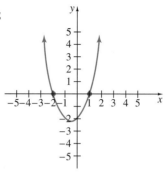

F

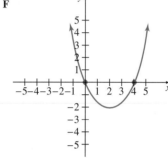

37. $y = (x + 2)(x - 1)$ E
38. $y = (x - 5)(x + 2)$ D
39. $y = x(x + 3)$ B
40. $y = x(x - 4)$ F
41. $y = 2x^2 - 8$ C
42. $y = 2x^2 - 2$ A

Solve each equation. Be careful. Some of the equations are quadratic and higher degree and some are linear.

43. $x(x + 7) = 0$ $x = 0, x = -7$
44. $y(6 - y) = 0$ $y = 0, y = 6$
45. $(x + 5)(x - 4) = 0$ $x = -5, x = 4$
46. $(x - 8)(x - 1) = 0$ $x = 8, x = 1$
47. $x^2 - x = 30$ $x = -5, x = 6$
48. $x^2 + 13x = -36$ $x = -9, x = -4$
49. $6y^2 - 22y - 40 = 0$ $y = -\frac{4}{3}, y = 5$
50. $3x^2 - 6x - 9 = 0$ $x = 3, x = -1$

51. $(2x + 3)(2x^2 - 5x - 3) = 0$ $x = -\frac{3}{2}, x = -\frac{1}{2}, x = 3$

52. $(2x - 9)(x^2 + 5x - 36) = 0$ $x = \frac{9}{2}, x = -9, x = 4$

53. $x^2 - 15 = -2x$ $x = -5, x = 3$

54. $x^2 - 26 = -11x$ $x = -13, x = 2$

55. $x^2 - 16x = 0$ $x = 0, x = 16$

56. $x^2 + 5x = 0$ $x = 0, x = -5$

57. $-18y^2 - 33y + 216 = 0$ $y = -\frac{9}{2}, y = \frac{8}{3}$

58. $-20y^2 + 145y - 35 = 0$ $y = \frac{1}{4}, y = 7$

59. $12x^2 - 59x + 55 = 0$ $x = \frac{5}{4}, x = \frac{11}{3}$

60. $30x^2 - 97x + 60 = 0$ $x = \frac{5}{6}, x = \frac{12}{5}$

61. $18x^2 + 9x - 2 = 0$ $x = -\frac{2}{3}, x = \frac{1}{6}$

62. $28x^2 - 27x - 10 = 0$ $x = \frac{5}{4}, x = -\frac{2}{7}$

63. $x(6x + 7) = 5$ $x = -\frac{5}{3}, x = \frac{1}{2}$

64. $4x(8x + 9) = 5$ $x = \frac{1}{8}, x = -\frac{5}{4}$

65. $4(x - 7) = 6$ $x = \frac{17}{2}$

66. $5(3 - 4x) = 9$ $x = \frac{3}{10}$

67. $5x^2 - 6x - 8 = 0$ $x = 2, x = -\frac{4}{5}$

68. $9x^2 + 6x + 2 = 0$ no real solution

69. $(y - 2)(y + 3) = 6$ $y = -4, y = 3$

70. $(y - 5)(y - 2) = 28$ $y = 9, y = -2$

71. $4y^2 - 1 = 0$ $y = \frac{1}{2}, y = -\frac{1}{2}$

72. $4y^2 - 81 = 0$ $y = \frac{9}{2}, y = -\frac{9}{2}$

73. $t^2 + 13t + 22 = 0$ $t = -2, t = -11$

74. $x^2 - 9x + 18 = 0$ $x = 6, x = 3$

75. $5t - 3 = 12$ $t = 3$

76. $9 - t = -1$ $t = 10$

77. $x^2 + 6x - 17 = -26$ $x = -3$

78. $x^2 - 8x - 4 = -20$ $x = 4$

79. $12x^2 + 7x - 12 = 0$ $x = \frac{3}{4}, x = -\frac{4}{3}$

80. $30x^2 - 11x - 30 = 0$ $x = -\frac{5}{6}, x = \frac{6}{5}$

81. $10t^3 - 25t - 15t^2 = 0$ $t = 0, t = \frac{5}{2}, t = -1$

82. $36t^3 - 48t - 12t^2 = 0$ $t = 0, t = \frac{4}{3}, t = -1$

83. A compass is accidentally thrown upward and out of an air balloon at a height of 300 feet. The height, y, of the compass at time x is given by the equation

$$y = -16x^2 + 20x + 300$$

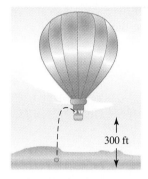

300 ft

a. Find the height of the compass at the given times by filling in the table below.

time x	0	1	2	3	4	5	6
height y	300	304	276	216	124	0	-156

b. Use the table to determine when the compass strikes the ground.   5 sec.

c. Use the table to approximate the maximum height of the compass.   304 ft.

d. Plot the points (x, y) on a rectangular coordinate system and connect them with a smooth curve. Explain your results.

84. A rocket is fired upward from the ground with an initial velocity of 100 feet per second. The height, y, of the rocket at any time x is given by the equation

$$y = -16x^2 + 100x$$

a. Find the height of the rocket at the given times by filling in the table below.

time x	0	1	2	3	4	5	6	7
height y	0	84	136	156	144	100	24	-84

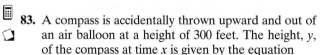
83. d. See App. F. Answers may vary.

b. Use the table to approximate when the rocket strikes the ground to the nearest tenth of a second.   6.3 sec.

c. Use the table to approximate the maximum height of the rocket.   156 ft.

d. Plot the points (x, y) on a rectangular coordinate system and connect them with a smooth curve. Explain your results.
See App. F. Answers may vary.

Review Exercises

Perform the following operations. Write all results in lowest terms. See Section 1.2.

85. $\dfrac{3}{5} + \dfrac{4}{9}$ $\dfrac{47}{45}$

86. $\dfrac{2}{3} + \dfrac{3}{7}$ $\dfrac{23}{21}$

87. $\dfrac{7}{10} - \dfrac{5}{12}$ $\dfrac{17}{60}$

88. $\dfrac{5}{9} - \dfrac{5}{12}$ $\dfrac{5}{36}$

89. $\dfrac{7}{8} \div \dfrac{7}{15}$ $\dfrac{15}{8}$

90. $\dfrac{5}{12} - \dfrac{3}{10}$ $\dfrac{7}{60}$

91. $\dfrac{4}{5} \cdot \dfrac{7}{8}$ $\dfrac{7}{10}$

92. $\dfrac{3}{7} \cdot \dfrac{12}{17}$ $\dfrac{36}{119}$

A Look Ahead

EXAMPLE   Solve $(x - 6)(2x - 3) = (x + 2)(x + 9)$.

Solution: $(x - 6)(2x - 3) = (x + 2)(x + 9)$
$$2x^2 - 15x + 18 = x^2 + 11x + 18$$
$$x^2 - 26x = 0$$
$$x(x - 26) = 0$$
$$x = 0 \quad \text{or} \quad x - 26 = 0$$
$$x = 26$$

Solve each equation. See the example.

93. $(x - 3)(3x + 4) = (x + 2)(x - 6)$ $x = 0, x = \dfrac{1}{2}$

94. $(2x - 3)(x + 6) = (x - 9)(x + 2)$ $x = 0, x = -16$

95. $(2x - 3)(x + 8) = (x - 6)(x + 4)$ $x = 0, x = -15$

96. $(x + 6)(x - 6) = (2x - 9)(x + 4)$ $x = 0, x = 1$

97. $(4x - 1)(x - 8) = (x + 2)(x + 4)$ $x = 0, x = 13$

98. $(5x - 2)(x + 3) = (2x - 3)(x + 2)$ $x = 0, x = -4$

5.7 | QUADRATIC EQUATIONS AND PROBLEM SOLVING

TAPE
BA 5.7

O B J E C T I V E

1 Solve problems that can be modeled by quadratic equations.

1 Some problems may be modeled by quadratic equations. To solve these problems, we use the same problem-solving steps that were introduced in Section 2.5. When solving these problems, keep in mind that a solution of an equation that models a problem may not be a solution to the problem. For example, a person's age or the length of a rectangle is always a positive number. Discard solutions that do not make sense as solutions of the problem.

EXAMPLE 1   For a TV commercial, a piece of luggage is dropped from a cliff 256 feet above the ground to show the durability of the luggage. Neglecting air resistance, the height h in feet of the luggage above the ground after t seconds is given by the quadratic equation

$$h = -16t^2 + 256$$

Find how long it takes for the luggage to hit the ground.

Solution: **1.** UNDERSTAND. Read and reread the problem. Then draw a picture of the problem.

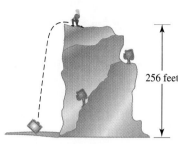

256 feet

The equation $h = -16t^2 + 256$ models the height of the falling luggage at time t. Familiarize yourself with this equation by finding the height of the luggage at $t = 1$ second and $t = 2$ seconds.

When $t = 1$ second, the height of the suitcase is $h = -16(1)^2 + 256 = 240$ feet

When $t = 2$ seconds, the height of the suitcase is $h = -16(2)^2 + 256 = 192$ feet

Since we have been given the needed equation, we proceed to step 4.

4. TRANSLATE. To find how long it takes the luggage to hit the ground, we want to know the value of t for which the height $h = 0$.

$$0 = -16t^2 + 256$$

5. COMPLETE. Solve the quadratic equation by factoring.

$$0 = -16t^2 + 256$$
$$0 = -16(t^2 - 16)$$
$$0 = -16(t - 4)(t + 4)$$

$$t - 4 = 0 \quad \text{or} \quad t + 4 = 0$$
$$t = 4 \quad \text{or} \quad t = -4$$

6. INTERPRET. Since the time t cannot be negative, the proposed solution is 4 seconds. To *check,* see that the height of the luggage when t is 4 seconds is 0.

When $t = 4$ seconds $h = -16(4)^2 + 256 = -256 + 256 = 0$ feet

State: The solution checks and the luggage hits the ground 4 seconds after it is dropped. ■

EXAMPLE 2   In May 1995, 24 inches of rain fell on Slidell, Louisiana, in just two days, causing approximately one-third of the houses in that community to flood. When a home floods, all flooring in the home must be removed and replaced. The Callacs' home flooded and the soiled rug in their den was removed and now needs to be replaced. The length of their den is 4 feet more than the width. If the area of the floor is 117 square feet, find its length and width.

Solution: **1.** UNDERSTAND. Read and reread the problem. Propose and check a solution.

2. ASSIGN. Let

$$x = \text{the width of the floor; then}$$
$$x + 4 = \text{the length of the floor since it is 4 feet longer.}$$

3. ILLUSTRATE. An illustration is shown to the left.

4. TRANSLATE. Here, we use the formula for the area of a rectangle.

In words: width $\cdot$ length $=$ area

Translate: $x \quad \cdot (x + 4) = \quad 117$

5. COMPLETE. $x(x + 4) = 117$

$$x^2 + 4x = 117 \qquad \text{Multiply.}$$
$$x^2 + 4x - 117 = 0 \qquad \text{Write in standard form.}$$
$$(x + 13)(x - 9) = 0 \qquad \text{Factor.}$$

Next, set each factor equal to 0.

$$x + 13 = 0 \qquad \text{or} \qquad x - 9 = 0$$
$$x = -13 \qquad \text{or} \qquad x = 9 \qquad \text{Solve.}$$

6. INTERPRET. The solutions are -13 and 9. Since x represents the width of the room, the solution -13 must be discarded. The proposed width is 9 feet and the proposed length is $x + 4$ or $9 + 4$ or 13 feet.

Check: The area of a 9-foot by 13-foot room is (9 feet)(13 feet) = 117 square feet. The proposed solution checks.

State: The floor is 9 feet by 13 feet.

EXAMPLE 3   The height of a triangular sail is 2 meters less than twice the length of the base. If the sail has an area of 30 square meters, find the length of its base and the height.

Solution: **1. UNDERSTAND.** Read and reread the problem. Propose and check a solution.

2. ASSIGN. Since we are finding the length of the base and the height, let

$$x = \text{the length of the base and since the height is 2 meters less than}$$
$$\text{twice the base,}$$
$$2x - 2 = \text{the height.}$$

3. ILLUSTRATE. An illustration is shown to the left.

4. TRANSLATE. We are given that the area of the triangle is 30 square meters, so we use the formula for area of a triangle.

In words: area of triangle $= \dfrac{1}{2} \cdot$ base $\cdot$ height

Translate: $30 \qquad = \dfrac{1}{2} \cdot \quad x \quad \cdot (2x - 2)$

5. COMPLETE. Here we solve the quadratic equation.

$$30 = \frac{1}{2}x(2x - 2)$$

$30 = x^2 - x$	Multiply.
$x^2 - x - 30 = 0$	Write in standard form.
$(x - 6)(x + 5) = 0$	Factor.
$x - 6 = 0 \quad \text{or} \quad x + 5 = 0$	Set each factor equal to 0.
$x = 6 \quad \text{or} \quad x = -5$	

6. INTERPRET. Since x represents the length of the base, discard the solution -5. The base of a triangle cannot be negative. The base is then 6 feet and the height is $2(6) - 2 = 10$ feet.

Check: To check this problem, recall that $\frac{1}{2}$ base · height = area, or

$$\frac{1}{2}(6)(10) = 30, \quad \text{the required area.}$$

State: The base of the triangular sail is 6 meters and the height is 10 meters.

The next example makes use of the **Pythagorean theorem** and consecutive integers. Before we review this theorem, recall that a **right triangle** is a triangle that contains a 90° or right angle. The **hypotenuse** of a right triangle is the side opposite the right angle and is the longest side of the triangle. The **legs** of a right triangle are the other sides of the triangle.

PYTHAGOREAN THEOREM

In a right triangle, the sum of the squares of the lengths of the two legs is equal to the square of the length of the hypotenuse.

$$(\text{leg})^2 + (\text{leg})^2 = (\text{hypotenuse})^2 \quad \text{or} \quad a^2 + b^2 = c^2$$

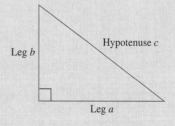

Study the following diagrams for a review of consecutive integers.

Consecutive integers:

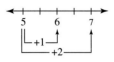

$$x \quad x+1 \quad x+2$$

Consecutive even integers:

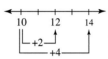

$$x \quad x+2 \quad x+4$$

Consecutive odd integers:

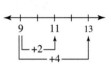

$$x \quad x+2 \quad x+4$$

EXAMPLE 4 Find the lengths of the sides of a right triangle if the lengths can be expressed by three consecutive even integers.

Solution: **1.** UNDERSTAND. Read and reread the problem. Let's propose and check a solution. If the length of one leg of the right triangle is 4 units, then the other leg is the next even integer, or 6 units, and the hypotenuse of the triangle is the next even integer, or 8 units. Remember that the hypotenuse is the longest side. Let's see if a triangle with sides of these lengths forms a right triangle. To do this, check to see whether the Pythagorean theorem holds true.

$$4^2 + 6^2 = 8^2$$
$$16 + 36 = 64$$
$$52 = 64 \qquad \text{False.}$$

4 units   8 units

6 units

Our guess does not check, but we now have a better understanding of the problem.

2. ASSIGN. Let x, $x + 2$, and $x + 4$ be three consecutive even integers. Since these integers represent lengths of the sides of a right triangle, we have

$$x = \text{one leg}$$
$$x + 2 = \text{other leg}$$
$$x + 4 = \text{hypotenuse (longest side)}$$

3. ILLUSTRATE.

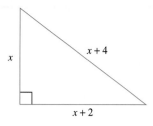

4. TRANSLATE. By the Pythagorean theorem, we have that

In words: (hypotenuse)^2 = (leg)^2 + (leg)^2

Translate: $(x + 4)^2 = (x)^2 + (x + 2)^2$

5. COMPLETE. Now solve the equation.

$$(x + 4)^2 = x^2 + (x + 2)^2$$
$$x^2 + 8x + 16 = x^2 + x^2 + 4x + 4 \qquad \text{Multiply.}$$
$$x^2 + 8x + 16 = 2x^2 + 4x + 4$$
$$x^2 - 4x - 12 = 0 \qquad\qquad \text{Write in standard form.}$$
$$(x - 6)(x + 2) = 0$$
$$x - 6 = 0 \quad \text{or} \quad x + 2 = 0$$
$$x = 6 \quad \text{or} \quad x = -2$$

6. INTERPRET. Discard $x = -2$ since length cannot be negative. If $x = 6$, then $x + 2 = 8$ and $x + 4 = 10$. To *check*, see that (hypotenuse)^2 = (leg)^2 + (leg)^2, or $10^2 = 6^2 + 8^2$, or $100 = 36 + 64$.

State: The sides of the right triangle have lengths 6 units, 8 units, and 10 units.

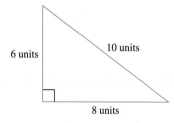

EXERCISE SET 5.7

Represent the given conditions using a single variable, x.

1. Two numbers whose sum is 36.   *x* and 36 − *x*

2. Two numbers whose sum is 10.   *x* and 10 − *x*

3. Two consecutive odd integers.

4. Two consecutive even integers.   *x* and *x* + 2 if *x* is an even integer

5. The length and width of a rectangle whose length is 4 centimeters less than three times the width.

6. The length and width of a rectangle whose length is twice its width.   width: *x*; length: 2*x*

7. A woman's age now and her age 10 years ago.

3. *x* and *x* + 2 if *x* is an odd integer   **5.** width: *x*; length: 3*x* − 4   **7.** age now: *x*; age 10 yrs ago: *x* − 10

8. The age of a man and the age of his son if the man is 5 years older than twice his son's age.

9. Three consecutive integers.

10. Three consecutive odd integers.

11. The three sides of a triangle if the first side is 2 inches less than twice the second side. The third side is 10 inches longer than the second side.

12. The three sides of a triangle if the first side is twice the second side and the third side is 5 more than three times the second side.

Solve. See Example 1.

13. An object is thrown upward from the top of an 80-foot building with an initial velocity of 64 feet per second. The height h of the object after t seconds is given by the quadratic

$$h = -16t^2 + 64t + 80$$

When will the object hit the ground?   5 sec.

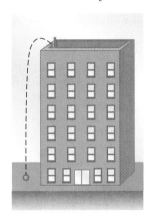

14. A hang-glider pilot accidentally drops her compass from the top of a 400-foot cliff. The height h of the compass after t seconds is given by the quadratic equation

$$h = -16t^2 + 400$$

When will the compass hit the ground?   5 sec.

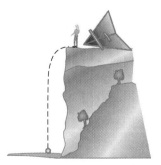

Solve the following. See Examples 2 and 3.

15. The length of a rectangle is 7 centimeters less than twice the width. Its area is 30 square centimeters. Find the dimensions of the rectangle.

16. The length of a rectangle is 9 inches more than its width. Its area is 112 square inches. Find the dimensions of the rectangle.

17. The altitude of a triangle is 8 centimeters more than twice the length of the base. If the triangle has an area of 96 square centimeters, find the length of its base and altitude.

18. The base of a triangle is 4 meters less than twice the length of the altitude. If the triangle has an area of 15 square meters, find the length of its base and altitude.   base: 6m; altitude: 5m

19. If the sides of a square are increased by 3 inches, the area becomes 64 square inches. Find the length of the sides of the original square.   5 in.

20. If the sides of a square are increased by 5 meters, the area becomes 100 square meters. Find the length of the sides of the original square.   5m

Solve. See Example 4.

21. Find the lengths of the sides of a right triangle if the hypotenuse is 10 centimeters longer than the short leg and 5 centimeters longer than the long leg.   hypotenuse: 25 cm; legs: 15 cm, 20 cm

22. Find the lengths of the sides of a right triangle if the length of the hypotenuse is 12 kilometers longer than the short leg and 6 kilometers longer than the long leg.   18 km; 24 km; 30 km

23. Find the length of the short leg of a right triangle if the long leg is 12 feet more than the short leg and the hypotenuse is 12 feet less than twice the short leg.   36 ft.

24. Find the length of the short leg of a right triangle if the long leg is 10 miles more than the short leg and the hypotenuse is 10 miles less than twice the short leg.   30 miles

Solve.

25. The sum of a number and its square is 132. Find the number.   −12 or 11

26. The sum of a number and its square is 182. Find the number.   −14 or 13

27. The sum of two numbers is 20, and the sum of their squares is 218. Find the numbers.   13 and 7

8. man's age: $2x + 5$; son's age: x **9.** $x, x + 1, x + 2$ if x is an integer **10.** x and $x + 2$ and $x + 4$ if x is an odd integer
11. second side: x; first side: $2x - 2$; third side: $x + 10$ **12.** second side: x; first side: $2x$; third side: $3x + 5$
15. width: 6 cm; length: 5 cm. **16.** width: 7 in.; length: 16 in. **17.** base: 8 cm; altitude: 24 cm.

28. The sum of two numbers is 25, and the sum of their squares is 325. Find the numbers.   10 and 15

29. If $D = \frac{n(n-3)}{2}$ is the formula for the number of diagonals D of a polygon with n sides, find the number of sides for a polygon with five diagonals.

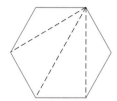

30. If $D = \frac{n(n-3)}{2}$ is the formula for the number of diagonals D of a polygon with n sides, find the number of sides for a polygon with 35 diagonals.

31. A rectangle has a perimeter of 42 miles and an area of 104 square miles. Find the dimensions of the rectangle.   length: 13 miles; width: 8 miles

32. A rectangle has a perimeter of 60 meters and an area of 209 square meters. Find the dimensions of the rectangle.   width: 11 m; length: 19 m

33. The sum of the squares of two consecutive integers is 9 greater than 8 times the smaller integer. Find the integers.   4 and 5 or −1 and 0

34. The square of the largest of three consecutive integers is equal to the sum of the squares of the other two. Find the integers.   3, 4, 5 or −1, 0, 1

35. A rectangular pool is surrounded by a walk 4 meters wide. The pool is 6 meters longer than its width. If the total area is 576 square meters more than the area of the pool, find the dimensions of the pool.   width: 29 m; length: 35 m

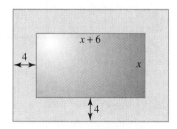

36. A rectangular garden is surrounded by a walk of uniform width. The area of the garden is 180 square yards. If the dimensions of the garden plus the walk are 16 yards by 24 yards, find the width of the walk. An illustration is at the top of the next column.   3 yd.

29. 5 sides   **30.** 10 sides

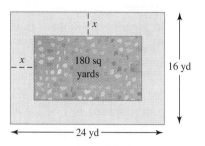

37. Find two consecutive even numbers whose product is 624.   −26 and −24 or 24 and 26

38. Find two consecutive odd numbers whose product is 399.   −19 and −21 or 19 and 21

39. One leg of a right triangle is 4 millimeters more than the smaller leg and the hypotenuse is 8 millimeters more than the smaller leg. Find the lengths of the sides of the triangle.

40. One leg of a right triangle is 9 centimeters longer than the other leg and the hypotenuse is 45 centimeters. Find the lengths of the legs of the triangle.   27 cm and 36 cm

41. The length of the base of a triangle is twice its altitude. If the area of the triangle is 100 square kilometers, find the altitude.   10 km

42. The altitude of a triangle is 2 millimeters less than the base. If the area is 60 square millimeters, find the base.   12 mm

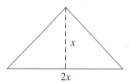

43. The sum of the squares of two consecutive negative integers is 221. Find the integers. −11 and −10

44. The sum of the squares of two consecutive even positive integers is 100. Find the integers.   6 and 8

45. Find the dimensions of a rectangle whose length is 2 yards more than twice its width and whose area is 60 square yards.   width: 5 yd.; length: 12 yd.

46. Find the dimensions of a rectangle whose length is 9 centimeters more than its width and whose area is 112 square centimeters.   width: 7 cm; length: 16 cm

39. 12 mm, 16 mm, and 20 mm

47. Find the dimensions of a rectangle whose width is 7 miles less than its length and whose area is 120 square miles.   length: 15 miles; width: 8 miles

48. Find the dimensions of a rectangle whose width is 2 inches less than half its length and whose area is 160 square inches.   width: 8 in.; length: 20 in.

49. At the end of 2 years, P dollars invested at an interest rate r compounded annually increases to an amount, A dollars, given by

$$A = P(1 + r)^2$$

Find the interest rate if $100 increased to $144 in 2 years.   20%

50. At the end of 2 years, P dollars invested at an interest rate r compounded annually increases to an amount, A dollars, given by

$$A = P(1 + r)^2$$

Find the interest rate if $2000 increased to $2420 in 2 years.   10%

51. If the cost, C, for manufacturing x units of a certain product is given by $C = x^2 - 15x + 50$, find the number of units manufactured at a cost of $9500.   105 units

52. If a switchboard handles n telephones, the number C of telephone connections it can make simultaneously is given by the equation $C = \frac{n(n - 1)}{2}$. Find how many telephones are handled by a switchboard making 120 telephone connections simultaneously.   16 telephones

53. Two boats travel at a right angle to each other after leaving the same dock at the same time. One hour later the boats are 17 miles apart. If one boat travels 7 miles per hour faster than the other boat, find the rate of each boat.   8 mph and 15 mph

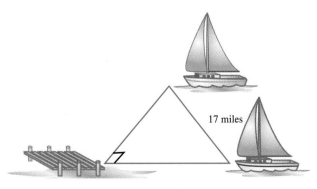

17 miles

54. The side of a square equals the width of a rectangle. The length of the rectangle is 6 meters longer than its width. The sum of the areas of the square and the rectangle is 176 square meters. Find the side of the square.   8 m

55. A rectangle has a perimeter of 42 yards and an area of 104 square yards. Find the dimensions of the rectangle.   length: 8 yd.; width: 13 yd.

56. Describe the kind of applied problem you find easiest to solve and why.   answers may vary

57. Describe the kind of applied problem you find most difficult to solve and why.   answers may vary

Review Exercises

The following double line graph shows a comparison of the number of farms in the United States and the size of the average farm. Use this graph for Exercises 58 through 63. See Sections 1.9 and 3.1.

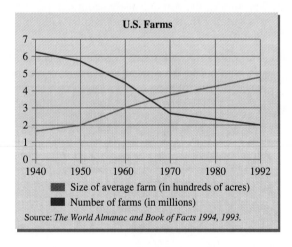

U.S. Farms

■ Size of average farm (in hundreds of acres)
■ Number of farms (in millions)

Source: *The World Almanac and Book of Facts 1994, 1993.*

58. Approximate the size of the average farm in 1940.

59. Approximate the size of the average farm in 1992.

60. Approximate the number of farms in 1940.

61. Approximate the number of farms in 1992.

58. 167 acres   **59.** 467 acres   **60.** 6.4 million   **61.** 2.1 million

62. Approximate the year that the broken lines in this graph intersect.   1966

⬡ **63.** In your own words, explain the meaning of the point of intersection in the graph.   answers may vary

⬡ **64.** Describe the trends shown in this graph and speculate as to why these trends have occurred.
answers may vary

Write each fraction in simplest form. See Section 1.2.

65. $\dfrac{20}{35}$ $\dfrac{4}{7}$ **66.** $\dfrac{24}{32}$ $\dfrac{3}{4}$ **67.** $\dfrac{27}{18}$ $\dfrac{3}{2}$

68. $\dfrac{15}{27}$ $\dfrac{5}{9}$ **69.** $\dfrac{14}{42}$ $\dfrac{1}{3}$ **70.** $\dfrac{45}{50}$ $\dfrac{9}{10}$

GROUP ACTIVITY

CHOOSING AMONG BUILDING OPTIONS

Sam has just had a 10-feet-by-15-feet, in-ground swimming pool installed in his backyard. He has $3000 left from the building project that he would like to spend on surrounding the pool with a patio, equally wide on all sides. He has talked to several local suppliers about options for building this patio and must choose among the following.

OPTION	MATERIAL	PRICE
A	Poured cement	$5 per square foot
B	Brick	$7.50 per square foot plus a $30 flat fee for delivering the bricks
C	Outdoor carpeting	$4.50 per square foot plus $10.86 per foot of the pool's perimeter to install an edging

1. Draw a diagram to represent the problem.

2. Write an algebraic expression that represents the total cost of the patio for each option.

3. If Sam plans to spend the entire $3000 he has saved for the patio, how wide would the patio be in each option? For each option, draw the final dimensions of the pool and patio to scale.

4. Which option should Sam choose? Why? Discuss the pros and cons of each option.

5. Summarize your findings and recommendations that could be used in a presentation to Sam.

See App. F. for Group Activity answers and suggestions.

Chapter 5   Highlights

Definitions and Concepts	Examples

Section 5.1   The Greatest Common Factor and Factoring by Grouping

Factoring is the process of writing an expression as a product.	Factor: $6 = 2 \cdot 3$ $x^2 + 5x + 6 = (x + 2)(x + 3)$

To find the GCF of a list of integers,

Step 1. Write each number as a product of primes.

Step 2. Identify the common prime factors.

Step 3. The product of all common factors is the greatest common factor. If there are no common prime factors, the GCF is 1.

Find the GCF of 12, 36, and 48.

$$12 = 2 \cdot 2 \cdot 3$$
$$36 = 2 \cdot 2 \cdot 3 \cdot 3$$
$$48 = 2 \cdot 2 \cdot 2 \cdot 2 \cdot 3$$

$GCF = 2 \cdot 2 \cdot 3 = 12$.

The GCF of a list of common variables raised to powers is the variable raised to the smallest exponent in the list.

The GCF of z^5, z^3, and z^{10} is z^3.

The GCF of a list of terms is the product of all common factors.

Find the GCF of $8x^2y$, $10x^3y^2$, and $25x^2y^3$.

The GCF of 8, 10, and 25 is 2.

The GCF of x^2, x^3, and x^2 is x^2.

The GCF of y, y^2, and y^3 is y.

The GCF of the terms is $2x^2y$.

To Factor by Grouping

Step 1. Arrange the terms so that the first two terms have a common factor and the last two have a common factor.

Step 2. For each pair of terms, factor out the pair's GCF.

Step 3. If there is now a common binomial factor, factor it out.

Step 4. If there is no common binomial factor, begin again, rearranging the terms differently. If no rearrangement leads to a common binomial factor, the polynomial cannot be factored.

Factor $10ax + 15a - 6xy - 9y$.

Step 1. $10ax + 15a - 6xy - 9y$

Step 2. $5a(2x + 3) - 3y(2x + 3)$

Step 3. $(2x + 3)(5a - 3y)$

Section 5.2   Factoring Trinomials of the Form $x^2 + bx + c$

To factor a trinomial of the form $x^2 + bx + c$, look for two numbers whose product is c and whose sum is b. The factored form is

$$(x + \text{one number})(x + \text{other number})$$

Factor: $x^2 + 7x + 12$

$3 + 4 = 7 \qquad 3 \cdot 4 = 12$

$(x + 3)(x + 4)$

(continued)

DEFINITIONS AND CONCEPTS	EXAMPLES

SECTION 5.3   FACTORING TRINOMIALS OF THE FORM $ax^2 + bx + c$

Method 1: To factor $ax^2 + bx + c$, try various combinations of factors of ax^2 and c until a middle term of bx is obtained when checking.	Factor: $3x^2 + 14x - 5$ Factors of $3x^2$: $3x, x$ Factors of -5: $-1, 5$ and $1, -5$. $\underbrace{(3x - 1)(x + 5)}$ $-1x$ $\underline{15x}$ $14x$ **correct** middle term

Method 2: Factor $ax^2 + bx + c$ by grouping.	Factor: $3x^2 + 14x - 5$
Step 1. Find two numbers whose product is $a \cdot c$ and whose sum is b.	*Step 1.* Find two numbers whose product is $3 \cdot (-5)$ or -15 and whose sum is 14. They are 15 and -1.
Step 2. Rewrite bx, using the factors found in step 1.	*Step 2* $3x^2 + 14x - 5$ $\quad = 3x^2 + 15x - 1x - 5$
Step 3. Factor by grouping.	*Step 3* $\quad = 3x(x + 5) - 1(x + 5)$ $\quad = (x + 5)(3x - 1)$

A **perfect square trinomial** is a trinomial that is the square of some binomial.	Perfect Square Trinomial = square of binomial $x^2 + 4x + 4 \qquad = (x + 2)^2$ $25x^2 - 10x + 1 \qquad = (5x - 1)^2$

Factoring perfect square trinomials: $a^2 + 2ab + b^2 = (a + b)^2$ $a^2 - 2ab + b^2 = (a - b)^2$	Factor: $x^2 + 6x + 9 = x^2 + 2(x \cdot 3) + 3^2 = (x + 3)^2$ $4x^2 - 12x + 9 = (2x)^2 - 2(2x \cdot 3) + 3^2 = (2x - 3)^2$

SECTION 5.4 FACTORING BINOMIALS

Difference of Squares $a^2 - b^2 = (a + b)(a - b)$ Sum or Difference of Cubes $a^3 + b^3 = (a + b)(a^2 - ab + b^2)$ $a^3 - b^3 = (a - b)(a^2 + ab + b^2)$	Factor: $x^2 - 9 = x^2 - 3^2 = (x + 3)(x - 3)$ $y^3 + 8 = y^3 + 2^3 = (y + 2)(y^2 - 2y + 4)$ $125z^3 - 1 = (5z)^3 - 1^3 = (5z - 1)(25z^2 + 5z + 1)$

SECTION 5.5 CHOOSING A FACTORING STRATEGY

To factor a polynomial,	Factor: $2x^4 - 6x^2 - 8$
Step 1. Factor out the GCF.	*Step 1.* $2x^4 - 6x^2 - 8 = 2(x^4 - 3x^2 - 4)$
Step 2. **a.** If two terms,	*Step 2.* **b. ii.** $\quad = 2(x^2 + 1)(x^2 - 4)$
$\quad$ **i.** $a^2 - b^2 = (a - b)(a + b)$	
$\quad$ **ii.** $a^3 - b^3 = (a - b)(a^2 + ab + b^2)$	
$\quad$ **iii.** $a^3 + b^3 = (a + b)(a^2 - ab + b^2)$	*(continued)*

DEFINITIONS AND CONCEPTS	**EXAMPLES**

SECTION 5.5 CHOOSING A FACTORING STRATEGY	

b. If three terms,

    i. $a^2 + 2ab + b^2 = (a + b)^2$

    ii. Methods in Sections 5.2 and 5.3

c. If four or more terms, try factoring by grouping.

Step 3.   See if any factors can be factored further.

Step 4.   Check by multiplying.

Step 3. $= 2(x^2 + 1)(x + 2)(x - 2)$

Step 4.   Check by multiplying.

$$2(x + 2)(x - 2)(x^2 + 1) = 2(x^2 - 4)(x^2 + 1)$$
$$= 2(x^4 - 3x^2 - 4)$$
$$= 2x^4 - 6x^2 - 8$$

SECTION 5.6 SOLVING QUADRATIC EQUATIONS BY FACTORING	

A **quadratic equation** is an equation that can be written in the form $ax^2 + bx + c = 0$ with a not 0. The form $ax^2 + bx + c = 0$ is called the **standard form** of a quadratic equation.

Quadratic Equation	Standard Form
$x^2 = 16$	$x^2 - 16 = 0$
$y = -2y^2 + 5$	$2y^2 + y - 5 = 0$

Zero Factor Theorem

If a and b are real numbers and if $ab = 0$, then $a = 0$ or $b = 0$.

If $(x + 3)(x - 1) = 0$, then $x + 3 = 0$ or $x - 1 = 0$

To solve quadratic equations by factoring,

Step 1.   Write the equation in standard form: $ax^2 + bx + c = 0$.

Step 2.   Factor the quadratic.

Step 3.   Set each factor containing a variable equal to 0.

Step 4.   Solve the equations.

Step 5.   Check in the original equation.

Solve: $3x^2 = 13x - 4$

Step 1. $3x^2 - 13x + 4 = 0$

Step 2. $(3x - 1)(x - 4) = 0$

Step 3. $3x - 1 = 0$ or $x - 4 = 0$

Step 4. $3x = 1$ or $x = 4$

              $x = \dfrac{1}{3}$

Step 5.   Check both $\frac{1}{3}$ and 4 in the original equation.

SECTION 5.7 QUADRATIC EQUATIONS AND PROBLEM SOLVING	

Problem-Solving Steps

A garden is in the shape of a rectangle whose length is two feet more than its width. If the area of the garden is 35 square feet, find its dimensions.

1. UNDERSTAND the problem.

1. Read and reread the problem. Guess a solution and check your guess.

2. ASSIGN a variable.

2. Let x be the width of the rectangular garden. Then $x + 2$ is the length.

(continued)

DEFINITIONS AND CONCEPTS	EXAMPLES

SECTION 5.7 QUADRATIC EQUATIONS AND PROBLEM SOLVING

3. ILLUSTRATE.	**3.** x ▭ $x + 2$
4. TRANSLATE.	**4.** In words: length · width = area Translate: $(x + 2)$ · x = 35
5. COMPLETE by solving.	**5.** $(x + 2)x = 35$ $x^2 + 2x - 35 = 0$ $(x - 5)(x + 7) = 0$ $x - 5 = 0$ or $x + 7 = 0$ $x = 5$ or $x = -7$
6. INTERPRET.	**6.** Discard the solution of -7 since x represents width. *Check:* If x is 5 feet then $x + 2 = 5 + 2 = 7$ feet. The area of a rectangle whose width is 5 feet and whose length is 7 feet is (5 feet)(7 feet) or 35 square feet. *State:* The garden is 5 feet by 7 feet.

CHAPTER 5 REVIEW

(5.1) *Complete the factoring.*

1. $6x^2 - 15x = 3x($ $)$ $2x - 5$

2. $2x^3y - 6x^2y^2 - 8xy^3 = 2xy($ $)$ $x^2 - 3xy - 4y^2$ or $(x - 4y)(x + y)$

Factor the GCF from each polynomial.

3. $20x^2 + 12x$ $4x(5x + 3)$

4. $6x^2y^2 - 3xy^3$ $3xy^2(2x - y)$

5. $-8x^3y + 6x^2y^2$ $-2x^2y(4x - 3y)$

6. $3x(2x + 3) - 5(2x + 3)$ $(2x + 3)(3x - 5)$

7. $5x(x + 1) - (x + 1)$ $(x + 1)(5x - 1)$

Factor.

8. $3x^2 - 3x + 2x - 2$ $(x - 1)(3x + 2)$

9. $6x^2 + 10x - 3x - 5$ $(2x - 1)(3x + 5)$

10. $3a^2 + 9ab + 3b^2 + ab$ $(a + 3b)(3a + b)$

(5.2) *Factor each trinomial.*

11. $x^2 + 6x + 8$ $(x + 4)(x + 2)$

12. $x^2 - 11x + 24$ $(x - 8)(x - 3)$

13. $x^2 + x + 2$ prime

14. $x^2 - 5x - 6$ $(x - 6)(x + 1)$

15. $x^2 + 2x - 8$ $(x + 4)(x - 2)$

16. $x^2 + 4xy - 12y^2$ $(x + 6y)(x - 2y)$

17. $x^2 + 8xy + 15y^2$ $(x + 5y)(x + 3y)$

18. $3x^2y + 6xy^2 + 3y^3$ $3y(x + y)^2$

19. $72 - 18x - 2x^2$ $2(3 - x)(12 + x)$

20. $32 + 12x - 4x^2$ $4(8 + 3x - x^2)$

(5.3) *Factor each trinomial.*

21. $2x^2 + 11x - 6$ $(2x - 1)(x + 6)$

22. $4x^2 - 7x + 4$ prime

23. $4x^2 + 4x - 3$ $(2x + 3)(2x - 1)$

24. $6x^2 + 5xy - 4y^2$ $(3x + 4y)(2x - y)$

25. $6x^2 - 25xy + 4y^2$ $(6x - y)(x - 4y)$

26. $18x^2 - 60x + 50$ $2(3x - 5)^2$

27. $2x^2 - 23xy - 39y^2$ $(2x + 3y)(x - 13y)$

28. $4x^2 - 28xy + 49y^2$ $(2x - 7y)^2$

29. $18x^2 - 9xy - 20y^2$ $(6x + 5y)(3x - 4y)$

30. $36x^3y + 24x^2y^2 - 45xy^3$ $3xy(2x + 3y)(6x - 5y)$

(5.4) *Factor each binomial.*

31. $4x^2 - 9$ $(2x - 3)(2x + 3)$

32. $9t^2 - 25s^2$ $(3t - 5s)(3t + 5s)$

33. $16x^2 + y^2$ prime

34. $x^3 - 8y^3$ $(x - 2y)(x^2 + 2xy + 4y^2)$

35. $8x^3 + 27$ $(2x + 3)(4x^2 - 6x + 9)$

36. $2x^3 + 8x$ $2x(x^2 + 4)$

37. $54 - 2x^3y^3$ $2(3 - xy)(9 + 3xy + x^2y^2)$

38. $9x^2 - 4y^2$ $(3x - 2y)(3x + 2y)$

39. $16x^4 - 1$ $(2x - 1)(2x + 1)(4x^2 + 1)$

40. $x^4 + 16$ prime

(5.5) *Factor.*

41. $2x^2 + 5x - 12$ $(2x - 3)(x + 4)$

42. $3x^2 - 12$ $3(x - 2)(x + 2)$

43. $x(x - 1) + 3(x - 1)$ $(x - 1)(x + 3)$

44. $x^2 + xy - 3x - 3y$ $(x + y)(x - 3)$

45. $4x^2y - 6xy^2$ $2xy(2x - 3y)$

46. $8x^2 - 15x - x^3$ $-x(x - 5)(x - 3)$

47. $125x^3 + 27$ $(5x + 3)(25x^2 - 15x + 9)$

48. $24x^2 - 3x - 18$ $3(8x^2 - x - 6)$

49. $(x + 7)^2 - y^2$ $(x + 7 - y)(x + 7 + y)$

50. $x^2(x + 3) - 4(x + 3)$ $(x + 3)(x - 2)(x + 2)$

(5.6) *Solve the following equations.*

51. $(x + 6)(x - 2) = 0$ $x = -6, x = 2$

52. $3x(x + 1)(7x - 2) = 0$ $x = 0, x = -1, x = \dfrac{2}{7}$

53. $4(5x + 1)(x + 3) = 0$ $x = -\dfrac{1}{5}, x = -3$

54. $x^2 + 8x + 7 = 0$ $x = -7, x = -1$

55. $x^2 - 2x - 24 = 0$ $x = -4, x = 6$

56. $x^2 + 10x = -25$ $x = -5$

57. $x(x - 10) = -16$ $x = 2, x = 8$

58. $(3x - 1)(9x^2 + 3x + 1) = 0$ $x = \dfrac{1}{3}$

59. $56x^2 - 5x - 6 = 0$ $x = -\dfrac{2}{7}, x = \dfrac{3}{8}$

60. $20x^2 - 7x - 6 = 0$ $x = \dfrac{3}{4}, x = -\dfrac{2}{5}$

61. $5(3x + 2) = 4$ $x = -\dfrac{2}{5}$

62. $6x^2 - 3x + 8 = 0$ no real solution

63. $12 - 5t = -3$ $t = 3$

64. $5x^3 + 20x^2 + 20x = 0$ $x = -2, x = 0$

65. $4t^3 - 5t^2 - 21t = 0$ $t = 0, t = -\dfrac{7}{4}, t = 3$

(5.7) *Solve the following problems.*

66. A flag for a local organization is in the shape of a rectangle whose length is 15 inches less than twice its width. If the area of the flag is 500 square inches, find its dimensions.   width: 20 in.; length: 25 in.

67. The base of a triangular sail is four times its height. If the area of the triangle is 162 square yards, find the base.   36 yd.

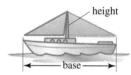

68. Find two consecutive positive integers whose product is 380.   19 and 20

69. A rocket is fired from the ground with an initial velocity of 440 feet per second. Its height h after t seconds is given by the equation

$$h = -16t^2 + 440t$$

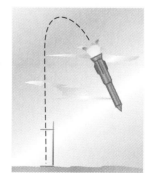

a. Find how many seconds pass before the rocket reaches a height of 2800 feet. Explain why two answers are obtained.

b. Find how many seconds pass before the rocket reaches the ground again.   27.5 sec.

70. An architect's squaring instrument is in the shape of a right triangle. Find the length of the long leg of the right triangle if the hypotenuse is 8 centime-ters longer than the long leg and the short leg is 8 centimeters shorter than the long leg.   32 cm

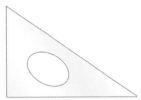

69a. 17.5 sec. and 10 sec. The rocket reaches a height of 2800 ft. on its way up and on its way back down.

CHAPTER 5 TEST

Factor each polynomial completely. If a polynomial cannot be factored, write "prime."

1. $9x^3 + 39x^2 + 12x$ $3x(3x + 1)(x + 4)$

2. $x^2 + x - 10$ prime

3. $x^2 + 4$ prime

4. $y^2 - 8y - 48$ $(y - 12)(y + 4)$

5. $3a^2 + 3ab - 7a - 7b$ $(3a - 7)(a + b)$

6. $3x^2 - 5x + 2$ $(3x - 2)(x - 1)$

7. $x^2 + 20x + 90$ prime

8. $x^2 + 14xy + 24y^2$ $(x + 12y)(x + 2y)$

9. $26x^6 - x^4$ $x^4(26x^2 - 1)$

10. $50x^3 + 10x^2 - 35x$ $5x(10x^2 + 2x - 7)$

11. $180 - 5x^2$ $5(6 - x)(6 + x)$

12. $64x^3 - 1$ $(4x - 1)(16x^2 + 4x + 1)$

13. $6t^2 - t - 5$ $(6t + 5)(t - 1)$

14. $xy^2 - 7y^2 - 4x + 28$ $(y - 2)(y + 2)(x - 7)$

15. $x - x^5$ $x(1 - x)(1 + x)(1 + x^2)$

16. $-xy^3 - x^3y$ $-xy(y^2 + x^2)$

Solve each equation.

17. $x^2 + 5x = 14$ $x = -7, x = 2$

18. $(x + 3)^2 = 16$ $x = -7, x = 1$

19. $3x(2x - 3)(3x + 4) = 0$ $x = 0, x = \frac{3}{2}, x = -\frac{4}{3}$

20. $5t^3 - 45t = 0$ $t = 0, t = 3, t = -3$

21. $3x^2 = -12x$ $x = 0, x = -4$

22. $t^2 - 2t - 15 = 0$ $t = -3, t = 5$

23. $7x^2 = 168 + 35x$ $x = -3, x = 8$

24. $6x^2 = 15x$ $x = 0, x = \frac{5}{2}$

Solve each problem.

25. Find the dimensions of a rectangular garden whose length is 5 feet longer than its width and whose area is 66 square feet.   width: 6 ft.; length: 11 ft.

26. A deck for a home is in the shape of a triangle. The length of the base of the triangle is 9 feet longer than its altitude. If the area of the triangle is 68 square feet, find the length of the base.   17 ft.

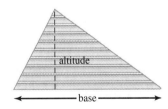

27. The sum of two numbers is 17, and the sum of their squares is 145. Find the numbers.   8 and 9

28. An object is dropped from the top of the Woolworth Building on Broadway in New York City. The height h of the object after t seconds is given by the equation

$$h = -16t^2 + 784$$

Find how many seconds pass before the object reaches the ground.   7 sec.

1. *(Sec. 1.1, Ex. 6)*   **2.** *(Sec. 1.4, Ex. 1)*   **7.** $x \geq 2$;   *(Sec. 2.9, Ex. 4)*   **9.** See App. F. *(Sec. 3.2, Ex. 5)*

CHAPTER 5   CUMULATIVE REVIEW

1. Find the absolute value of each number.
   a. $|4|$ 4   **b.** $|-5|$ 5   **c.** $|0|$ 0

2. Evaluate each expression if $x = 3$ and $y = 2$.
   a. $2x - y$ 4   **b.** $\dfrac{3x}{2y}$ $\dfrac{9}{4}$ **c.** $\dfrac{x}{y} + \dfrac{y}{2}$ $\dfrac{5}{2}$ **d.** $x^2 - y^2$ 5

3. The lowest point in North America is in Death Valley, at an elevation of 282 feet below sea level. Nearby, Mount Whitney reaches 14,494 feet, the highest point in the United States outside Alaska. How much of a variation in elevation is there between these two extremes?   14,776 ft. *(Sec. 1.6, Ex. 5)*

4. Solve $5t - 5 = 6t + 2$ for t. $t = -7$ *(Sec. 2.2, Ex. 3)*

5. Solve $4(2x - 3) + 7 = 3x + 5$. $x = 2$ *(Sec. 2.4, Ex. 1)*

6. A local cellular phone company charges Elaine Chapoton $50 per month and $0.36 per minute of phone use in her usage category. If Elaine was charged $99.68 for a month's cellular phone use, determine the number of whole minutes of phone use. 138 min. *(Sec. 2.5, Ex. 4)*

7. Solve $-2x \leq -4$, and graph the solution set.

8. Determine whether each ordered pair is a solution of the equation $x - 2y = 6$.   *(Sec. 3.1, Ex. 2)*

   a. $(6, 0)$ solution

   b. $(0, 3)$ not a solution

   c. $(2, -2)$ solution

9. Graph the linear equation $y = -\dfrac{1}{3}x$.

10. Identify the x- and y-intercepts and the intercept points.   *(Sec. 3.3, Ex. 1)*

B.

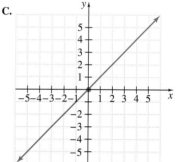

C.

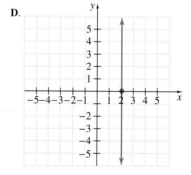

D.

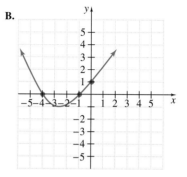

A.

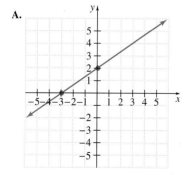

E.

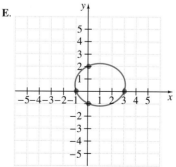

10. a. $x = -3$; $y = 2$; $(-3, 0)$; $(0, 2)$
   b. $x = -4$; $x = -1$; $y = 1$; $(-4, 0)$; $(-1, 0)$; $(0, 1)$
   c. $x = 0$; $y = 0$; $(0, 0)$

   d. $x = 2$; $(2, 0)$
   e. $x = -1$; $x = 3$; $y = -1$; $y = 2$; $(-1, 0)$; $(3, 0)$, $(0, -1)$; $(0, 2)$

11. Find the slope of the line whose equation is $-2x + 3y = 12$.

12. Graph $2x - y \geq 3$.   See App. F. *(Sec. 3.5, Ex. 3)*

13. Simplify each expression.   *(Sec. 4.1, Ex. 6)*

    a. $(st)^4$ s^4t^4

    b. $\left(\dfrac{m}{n}\right)^7$ $\dfrac{m^7}{n^7}$

    c. $(2a)^3$ $8a^3$

    d. $(-5x^2y^3z)^2$ $25x^4y^6z^2$

    e. $\left(\dfrac{2x^4}{3y^5}\right)^4$ $\dfrac{16x^{16}}{81y^{20}}$

14. Add: $(4x^3 - 6x^2 + 2x + 7) + (5x^2 - 2x)$.

15. Divide: $(x^2 + 7x + 12)$ by $(x + 3)$.

16. Factor each polynomial by factoring out the GCF.

    a. $6t + 18$ $6(t + 3)$ **b.** $y^5 - y^7$ $y^5(1 - y^2)$

17. Factor $x^2 - 8x + 15$. $(x - 5)(x - 3)$ *(Sec. 5.2, Ex. 2)*

18. Factor $3x^2 + 11x + 6$. $(3x + 2)(x + 3)$ *(Sec. 5.3, Ex. 1)*

19. Factor $9x^2 - 36$. $9(x + 2)(x - 2)$ *(Sec. 5.4, Ex. 3)*

20. Solve $(x - 5)(2x + 7) = 0$.

21. The height of a triangular sail is 2 meters less than twice the length of the base. If the sail has an area of 30 square meters, find the length of its base and the height.   base: 6 m; height: 10 m *(Sec. 5.7, Ex. 3)*

11. $\dfrac{2}{3}$ *(Sec. 3.4, Ex. 4)*   **14.** $4x^3 - x^2 + 7$ *(Sec. 4.2, Ex. 7)*   **15.** $x + 4$ *(Sec. 4.6, Ex. 5)*   **16.** *(Sec. 5.1, Ex. 4)*

20. $x = 5, x = -\dfrac{7}{2}$ *(Sec. 5.6, Ex. 1)*

RATIONAL EXPRESSIONS

COMPARING FORMULAS FOR DOSES OF MEDICATION

Depending on the medicine, too large a dose can be extremely dangerous and even fatal. Particularly for children, gauging the proper dose is critical. Mathematical models predicting the correct doses are available, but are, at best, approximations.

IN THE CHAPTER GROUP ACTIVITY ON PAGE 399, YOU WILL HAVE THE OPPORTUNITY TO INVESTIGATE TWO WELL-KNOWN FORMULAS FOR PREDICTING THE CORRECT DOSAGES FOR CHILDREN.

Ⅰn this chapter, we expand our knowledge of algebraic expressions to include another category called **rational expressions,** such as $\frac{x+1}{x}$. We explore the operations of addition, subtraction, multiplication, and division for these algebraic fractions, using principles similar to the principles for number fractions. Thus, the material in this chapter will make full use of your knowledge of number fractions.

6.1 SIMPLIFYING RATIONAL EXPRESSIONS

O B J E C T I V E S

1 Find the value of a rational expression given a replacement number.
2 Identify values for which a rational expression is undefined.
3 Write rational expressions in lowest terms.

TAPE
BA 6.1

1 As we reviewed in chapter 1, a rational number is a number that can be written as a quotient of integers. A **rational expression** is also a quotient; it is a quotient of polynomials.

RATIONAL EXPRESSION

A rational expression is an expression that can be written in the form $\frac{P}{Q}$, where P and Q are polynomials and Q does not equal 0.

RATIONAL EXPRESSIONS

$$\frac{3y^3}{8} \qquad \frac{-4p}{p^3 + 2p + 1} \qquad \frac{5x^2 - 3x + 2}{3x + 7}$$

Rational expressions have different values depending on what value replaces the variable. Next, we review the standard order of operations by finding values of rational expressions at given replacement values of the variable.

EXAMPLE 1 Find the value of $\dfrac{x+4}{2x-3}$ for the given replacement values.

a. $x = 5$ **b.** $x = -2$

Solution: **a.** Replace each x in the expression with 5 and then simplify.

$$\frac{x+4}{2x-3} = \frac{5+4}{2(5)-3} = \frac{9}{10-3} = \frac{9}{7}.$$

b. Replace each x in the expression with -2 and then simplify.

$$\frac{x+4}{2x-3} = \frac{-2+4}{2(-2)-3} = \frac{2}{-7} \quad \text{or} \quad -\frac{2}{7}.$$

For a negative fraction such as $\dfrac{2}{-7}$, recall from Chapter 1 that

$$\frac{2}{-7} = \frac{-2}{7} = -\frac{2}{7}$$

In general, for any fraction

$$\frac{-a}{b} = \frac{a}{-b} = -\frac{a}{b}, \qquad b \neq 0$$

This is also true for rational expressions. For example,

$$\underbrace{\frac{-(x+2)}{x}}_{\uparrow} = \frac{x+2}{-x} = -\frac{x+2}{x}$$

Notice the parentheses.

2 In the preceding box, notice that we wrote $b \neq 0$ for the denominator b. This is because the denominator of a rational expression must not equal 0 since division by 0 is not defined. This means we must be careful when replacing the variable in a rational expression by a number. For example, suppose we replace x with 5 in the rational expression $\dfrac{2+x}{x-5}$. The expression becomes

$$\frac{2+x}{x-5} = \frac{2+5}{5-5} = \frac{7}{0}$$

But division by 0 is undefined. Therefore, in this expression we can allow x to be any real number *except* 5. A rational expression is undefined for values that make the denominator 0.

EXAMPLE 2 Are there any values for x for which each expression is undefined?

a. $\dfrac{x}{x-3}$ **b.** $\dfrac{x^2+2}{x^2-3x+2}$ **c.** $\dfrac{x^3-6x^2-10x}{3}$ **d.** $\dfrac{2}{x^2+1}$

Solution: To find values for which a rational expression is undefined, find values that make the denominator 0.

a. The denominator of $\dfrac{x}{x-3}$ is 0 when $x - 3 = 0$ or when $x = 3$. Thus, when $x = 3$, the expression $\dfrac{x}{x-3}$ is undefined.

b. Set the denominator equal to zero.

$$x^2 - 3x + 2 = 0$$
$$(x-2)(x-1) = 0 \qquad \text{Factor.}$$
$$x - 2 = 0 \quad \text{or} \quad x - 1 = 0 \qquad \text{Set each factor equal to zero.}$$
$$x = 2 \quad \text{or} \qquad x = 1 \qquad \text{Solve.}$$

Thus, when $x = 2$ or $x = 1$, the denominator $x^2 - 3x + 2$ is 0. So the rational expression $\dfrac{x^2 + 2}{x^2 - 3x + 2}$ is undefined when $x = 2$ or when $x = 1$.

c. The denominator of $\dfrac{x^3 - 6x^2 - 10x}{3}$ is never zero, so there are no values of x for which this expression is undefined.

d. No matter which real number x is replaced by, the denominator $x^2 + 1$ does not equal 0, so there are no real numbers for which this expression is undefined.

[3] A fraction is said to be written in lowest terms or simplest form when the numerator and denominator have no common factors other than 1 (or -1). For example, the fraction $\frac{7}{10}$ is in lowest terms since the numerator and denominator have no common factors other than 1 (or -1).

The process of writing a rational expression in lowest terms or simplest form is called **simplifying** a rational expression. The following fundamental principle of rational expressions is used to simplify a rational expression.

FUNDAMENTAL PRINCIPLE OF RATIONAL EXPRESSIONS

If P, Q, and R are polynomials, and Q and R are not 0,

$$\frac{PR}{QR} = \frac{P}{Q}$$

Simplifying a rational expression is similar to simplifying a fraction. To simplify the fraction $\frac{15}{20}$, we factor the numerator and the denominator, look for common factors in both, and then use the fundamental principle.

$$\frac{15}{20} = \frac{3 \cdot 5}{2 \cdot 2 \cdot 5} = \frac{3}{2 \cdot 2} = \frac{3}{4}$$

To simplify the rational expression $\dfrac{x^2 - 9}{x^2 + x - 6}$, we also factor the numerator and denominator, look for common factors in both, and then use the fundamental principle of rational expressions.

$$\frac{x^2 - 9}{x^2 + x - 6} = \frac{(x - 3)(x + 3)}{(x - 2)(x + 3)} = \frac{x - 3}{x - 2}$$

This means that the rational expression $\dfrac{x^2 - 9}{x^2 + x - 6}$ has the same value as the rational expression $\dfrac{x - 3}{x - 2}$ for all values of x except 2 and -3. (Remember that when x is 2, the denominator of both rational expressions is 0 and when x is -3, the original rational expression has a denominator of 0.) As we simplify rational expressions, we will assume that the simplified rational expression is equal to the original rational expression for all real numbers except those for which either denominator is 0. The following steps may be used to simplify rational expressions.

TO SIMPLIFY A RATIONAL EXPRESSION

Step 1.   Completely factor the numerator and denominator.

Step 2.   Apply the fundamental principle of rational expressions to divide out common factors.

EXAMPLE 3   Write $\dfrac{21a^2b}{3a^5b}$ in simplest form.

Solution: Factor the numerator and denominator. Then apply the fundamental principle.

$$\frac{21a^2b}{3a^5b} = \frac{7 \cdot 3 \cdot a^2 \cdot b}{3 \cdot a^3 \cdot a^2 \cdot b} = \frac{7}{a^3}$$

EXAMPLE 4   Simplify: $\dfrac{5x - 5}{x^3 - x^2}$.

Solution: Factor the numerator and denominator if possible and then apply the fundamental principle.

$$\frac{5x - 5}{x^3 - x^2} = \frac{5(x - 1)}{x^2(x - 1)} = \frac{5}{x^2}$$

EXAMPLE 5   Write $\dfrac{x^2 + 8x + 7}{x^2 - 4x - 5}$ in lowest terms.

Solution: Factor the numerator and denominator and apply the fundamental principle.

$$\frac{x^2 + 8x + 7}{x^2 - 4x - 5} = \frac{(x + 7)(x + 1)}{(x - 5)(x + 1)} = \frac{x + 7}{x - 5}$$

EXAMPLE 6   Simplify: $\dfrac{x^2 + 4x + 4}{x^2 + 2x}$.

Solution: Factor the numerator and denominator and apply the fundamental principle.

$$\frac{x^2 + 4x + 4}{x^2 + 2x} = \frac{(x + 2)(x + 2)}{x(x + 2)} = \frac{x + 2}{x}$$

R E M I N D E R   When simplifying a rational expression, the fundamental principle applies to **common factors, not their common terms.** For example, $\frac{x + 2}{x}$ cannot be simplified any further because the numerator and denominator have no **common factors.**

EXAMPLE 7   Simplify each rational expression.

  a. $\dfrac{x + y}{y + x}$ **b.** $\dfrac{x - y}{y - x}$

Solution: **a.** The expression $\dfrac{x + y}{y + x}$ can be simplified by using the commutative property of addition to rewrite the denominator $y + x$ as $x + y$.

$$\frac{x + y}{y + x} = \frac{x + y}{x + y} = 1$$

b. The expression $\dfrac{x - y}{y - x}$ can be simplified by recognizing that $y - x$ and $x - y$ are opposites. In other words, $y - x = -1(x - y)$. Proceed as follows:

$$\frac{x - y}{y - x} = \frac{1 \cdot (x - y)}{(-1)(x - y)} = \frac{1}{-1} = -1$$

EXAMPLE 8   Simplify: $\dfrac{4 - x^2}{3x^2 - 5x - 2}$.

Solution: $\dfrac{4 - x^2}{3x^2 - 5x - 2} = \dfrac{(2 - x)(2 + x)}{(x - 2)(3x + 1)}$ Factor.

$$= \frac{(-1)(x - 2)(2 + x)}{(x - 2)(3x + 1)}$$ Write $2 - x$ as $-1(x - 2)$.

$$= \frac{(-1)(2 + x)}{3x + 1} \quad \text{or} \quad \frac{-2 - x}{3x + 1}$$ Simplify.

EXAMPLE 9   Simplify: $\dfrac{2x^2 - 2xy + 3x - 3y}{2x + 3}$.

Solution: First, factor the four-term numerator by grouping.

$$= \frac{2x^2 - 2xy + 3x - 3y}{2x + 3} = \frac{2x(x - y) + 3(x - y)}{2x + 3}$$

$$= \frac{(2x + 3)(x - y)}{2x + 3}$$ Factor.

$$= \frac{x - y}{1} \quad \text{or} \quad x - y$$ Simplify.

MENTAL MATH

Find any real numbers for which each equal rational expression is undefined. See Example 2.

1. $\dfrac{x + 5}{x}$ $x = 0$ **2.** $\dfrac{x^2 - 5x}{x - 3}$ $x = 3$ **3.** $\dfrac{x^2 + 4x - 2}{x(x - 1)}$ $x = 0, x = 1$ **4.** $\dfrac{x + 2}{(x - 5)(x - 6)}$ $x = 5, x = 6$

18. $x = -1, x = 0, x = 1$ **21.** answers may vary **22.** no; answers may vary **26.** $3(x + 7)$ **35.** answers may vary

EXERCISE SET 6.1

Find the value of the following expressions when $x = 2$, $y = -2$, and $z = -5$. See Example 1.

1. $\dfrac{x + 5}{x + 2}$ $\dfrac{7}{4}$

2. $\dfrac{x + 8}{2x + 5}$ $\dfrac{10}{9}$

3. $\dfrac{z - 8}{z + 2}$ $\dfrac{13}{3}$

4. $\dfrac{y - 2}{-5 + y}$ $\dfrac{4}{7}$

5. $\dfrac{x^2 + 8x + 2}{x^2 - x - 6}$ $-\dfrac{11}{2}$

6. $\dfrac{z^2 + 8}{z^3 - 25z}$ undefined

7. $\dfrac{x + 5}{x^2 + 4x - 8}$ $\dfrac{7}{4}$

8. $\dfrac{z^3 + 1}{z^2 + 1}$ $-\dfrac{62}{13}$

9. $\dfrac{y^3}{y^2 - 1}$ $-\dfrac{8}{3}$

10. $\dfrac{z}{z^2 - 5}$ $-\dfrac{1}{4}$

☐ 11. The total revenue R from the sale of a popular music compact disc is approximately given by the equation

$$R = \frac{150x^2}{x^2 + 3}$$

where x is the number of years since the CD has been released and revenue R is in millions of dollars.

a. Find the total revenue generated by the end of the first year. $37.5 million

b. Find the total revenue generated by the end of the second year. $85.7 million

c. Find the total revenue generated in the second year only. $48.2 million

☐ 12. For a certain model fax machine, the manufacturing cost C per machine is given by the equation

$$C = \frac{250x + 10{,}000}{x}$$

where x is the number of fax machines manufactured and cost C is in dollars per machine.

a. Find the cost per fax machine when manufacturing 100 fax machines. $350

b. Find the cost per fax machine when manufacturing 1000 fax machines $260

c. Does the cost per machine decrease or increase when more machines are manufactured? Explain why this is so.   yes; answers may vary

Find any real numbers for which each rational expression is undefined. See Example 2.

13. $\dfrac{x + 3}{x + 2}$ $x = -2$

14. $\dfrac{5x + 1}{x - 3}$ $x = 3$

15. $\dfrac{4x^2 + 9}{2x - 8}$ $x = 4$

16. $\dfrac{9x^3 + 4x}{15x + 45}$ $x = -3$

17. $\dfrac{9x^3 + 4}{15x + 30}$ $x = -2$

18. $\dfrac{19x^3 + 2}{x^3 - x}$

19. $\dfrac{x^2 - 5x - 2}{x^2 + 4}$ none

20. $\dfrac{9y^5 + y^3}{x^2 + 9}$ none

☐ 21. Explain why the denominator of a fraction or a rational expression must not equal zero.

☐ 22. Does $\dfrac{(x - 3)(x + 3)}{x - 3}$ have the same value as $x + 3$ for all real numbers? Explain why or why not.

Simplify each expression. See Examples 3 through 6.

23. $\dfrac{8x^5}{4x^9}$ $\dfrac{2}{x^4}$

24. $\dfrac{12y^7}{-2y^6}$ $-6y$

25. $\dfrac{5(x - 2)}{(x - 2)(x + 1)}$ $\dfrac{5}{x + 1}$

26. $\dfrac{9(x - 7)(x + 7)}{3(x - 7)}$

27. $\dfrac{-5a - 5b}{a + b}$ -5

28. $\dfrac{7x + 35}{x^2 + 5x}$ $\dfrac{7}{x}$

29. $\dfrac{x + 5}{x^2 - 4x - 45}$ $\dfrac{1}{x - 9}$

30. $\dfrac{x - 3}{x^2 - 6x + 9}$ $\dfrac{1}{x - 3}$

31. $\dfrac{5x^2 + 11x + 2}{x + 2}$ $5x + 1$

32. $\dfrac{12x^2 + 4x - 1}{2x + 1}$ $6x - 1$

33. $\dfrac{x^2 + x - 12}{2x^2 - 5x - 3}$ $\dfrac{x + 4}{2x + 1}$

34. $\dfrac{x^2 + 3x - 4}{x^2 - x - 20}$ $\dfrac{x - 1}{x - 5}$

☐ 35. Explain how to write a fraction in lowest terms.

☐ 36. Explain how to write a rational expression in lowest terms.   answers may vary

Simplify each expression. See Examples 7 through 9.

37. $\dfrac{x - 7}{7 - x}$ -1

38. $\dfrac{y - z}{z - y}$ -1

39. $\dfrac{y^2 - 2y}{4 - 2y}$ $-\dfrac{y}{2}$

40. $\dfrac{x^2 + 5x}{20 + 4x}$ $\dfrac{x}{4}$

41. $\dfrac{x^2 - 4x + 4}{4 - x^2}$ $\dfrac{2 - x}{x + 2}$

42. $\dfrac{x^2 + 10x + 21}{-2x - 14}$ $-\dfrac{x + 3}{2}$

43. $x + y$ **44.** $a + b$ **45.** $\dfrac{5 - y}{2}$ **46.** $\dfrac{x + 2}{y}$

43. $\dfrac{x^2 + xy + 2x + 2y}{x + 2}$ **44.** $\dfrac{ab + ac + b^2 + bc}{b + c}$

45. $\dfrac{5x + 15 - xy - 3y}{2x + 6}$ **46.** $\dfrac{xy - 6x + 2y - 12}{y^2 - 6y}$

Simplify each expression.

47. $\dfrac{15x^4 y^8}{-5x^8 y^3}$ $-\dfrac{3y^5}{x^4}$ **48.** $\dfrac{24a^3 b^3}{6a^2 b^4}$ $\dfrac{4a}{b}$

49. $\dfrac{(x - 2)(x + 3)}{5(x + 3)}$ $\dfrac{x - 2}{5}$ **50.** $\dfrac{-2(y - 9)}{(y - 9)^2}$ $-\dfrac{2}{y - 9}$

51. $\dfrac{-6a - 6b}{a + b}$ -6 **52.** $\dfrac{4a - 4y}{4y - 4a}$ -1

53. $\dfrac{2x^2 - 8}{4x - 8}$ $\dfrac{x + 2}{2}$ **54.** $\dfrac{5x^2 - 500}{35x + 350}$ $\dfrac{x - 10}{7}$

55. $\dfrac{11x^2 - 22x^3}{6x - 12x^2}$ $\dfrac{11x}{6}$ **56.** $\dfrac{16r^2 - 4s^2}{4r - 2s}$ $2(2r + s)$

57. $\dfrac{x + 7}{x^2 + 5x - 14}$ $\dfrac{1}{x - 2}$ **58.** $\dfrac{x - 10}{x^2 - 17x + 70}$ $\dfrac{1}{x - 7}$

59. $\dfrac{2x^2 + 3x - 2}{2x - 1}$ $x + 2$ **60.** $\dfrac{4x^2 + 24x}{x + 6}$ $4x$

61. $\dfrac{x^2 - 1}{x^2 - 2x + 1}$ $\dfrac{x + 1}{x - 1}$ **62.** $\dfrac{x^2 - 16}{x^2 - 8x + 16}$ $\dfrac{x + 4}{x - 4}$

63. $\dfrac{m^2 - 6m + 9}{m^2 - 9}$ $\dfrac{m - 3}{m + 3}$ **64.** $\dfrac{m^2 - 4m + 4}{m^2 + m - 6}$ $\dfrac{m - 2}{m + 3}$

65. $\dfrac{-2a^2 + 12a - 18}{9 - a^2}$ $\dfrac{2a - 6}{a + 3}$

66. $\dfrac{-4a^2 + 8a - 4}{2a^2 - 2}$ $\dfrac{-2(a - 1)}{a + 1}$

67. $\dfrac{2 - x}{x - 2}$ -1 **68.** $\dfrac{7 - y}{y - 7}$ -1

69. $\dfrac{x^2 - 1}{1 - x}$ $-x - 1$ **70.** $\dfrac{x^2 - xy}{2y - 2x}$ $-\dfrac{x}{2}$

71. $\dfrac{x^2 + 7x + 10}{x^2 - 3x - 10}$ $\dfrac{x + 5}{x - 5}$ **72.** $\dfrac{2x^2 + 7x - 4}{x^2 + 3x - 4}$ $\dfrac{2x - 1}{x - 1}$

73. $\dfrac{3x^2 + 7x + 2}{3x^2 + 13x + 4}$ $\dfrac{x + 2}{x + 4}$ **74.** $\dfrac{4x^2 - 4x + 1}{2x^2 + 9x - 5}$ $\dfrac{2x - 1}{x + 5}$

75. $\dfrac{x^2 + 3x - 2x - 6}{x^2 - 2x}$ $\dfrac{x + 3}{x}$

76. $\dfrac{ax + ay - bx - by}{x^2 - y^2}$ $\dfrac{a - b}{x - y}$

77. $\dfrac{x^3 + 8}{x + 2}$ $x^2 - 2x + 4$ **78.** $\dfrac{x^3 + 64}{x + 4}$ $x^2 - 4x + 16$

79. $\dfrac{x^2 + xy + 5x + 5y}{3x + 3y}$ $\dfrac{x + 5}{3}$

80. $\dfrac{x^2 + 2x + xy + 2y}{x^2 - 4}$ $\dfrac{x + y}{x - 2}$

81. $\dfrac{x^3 - 1}{1 - x}$ $-x^2 - x - 1$

82. $\dfrac{3 - x}{x^3 - 27}$ $\dfrac{-1}{x^2 + 3x + 9}$

How does the graph of $y = \frac{x^2 - 9}{x - 3}$ compare to the graph of $y = x + 3$?

Recall that $\frac{x^2 - 9}{x - 3} = \frac{(x + 3)(x - 3)}{x - 3} = x + 3$ as long as x is not 3.

This means that the graph of $y = \frac{x^2 - 9}{x - 3}$ is the same as the graph of $y = x + 3$ with $x \neq 3$. To graph $y = \frac{x^2 - 9}{x - 3}$, then, graph the linear equation $y = x + 3$ and place an open dot on the graph at 3. This open dot or interruption of the line at 3 means $x \neq 3$.

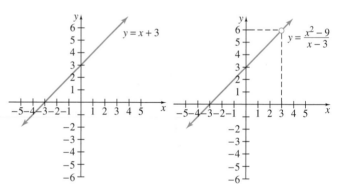

graphical answers in App. F

83. Graph $y = \dfrac{x^2 - 25}{x + 5}$.

84. Graph $y = \dfrac{x^2 - 16}{x - 4}$.

85. Graph $y = \dfrac{x^2 + x - 12}{x + 4}$.

86. Graph $y = \dfrac{x^2 - 6x + 8}{x - 2}$.

Review Exercises

Perform the indicated operations. See Section 1.2.

87. $\dfrac{1}{3} \cdot \dfrac{9}{11}$ $\dfrac{3}{11}$

88. $\dfrac{5}{27} \cdot \dfrac{2}{5}$ $\dfrac{2}{27}$

91. $\dfrac{5}{6} \cdot \dfrac{10}{11} \cdot \dfrac{2}{3}$ $\dfrac{50}{99}$

92. $\dfrac{4}{3} \cdot \dfrac{1}{7} \cdot \dfrac{10}{13}$ $\dfrac{40}{273}$

89. $\dfrac{1}{3} \div \dfrac{1}{4}$ $\dfrac{4}{3}$

90. $\dfrac{7}{8} \div \dfrac{1}{2}$ $\dfrac{7}{4}$

93. $\dfrac{13}{20} \div \dfrac{2}{9}$ $\dfrac{117}{40}$

94. $\dfrac{8}{15} \div \dfrac{5}{8}$ $\dfrac{64}{75}$

6.2 | MULTIPLYING AND DIVIDING RATIONAL EXPRESSIONS

OBJECTIVES

 Multiply rational expressions.

 Divide rational expressions.

 TAPE
BA 6.2

Just as simplifying rational expressions is similar to simplifying number fractions, multiplying and dividing rational expressions is similar to multiplying and dividing number fractions. To find the product of $\frac{3}{5}$ and $\frac{1}{4}$, multiply the numerators and then multiply the denominators of both fractions.

$$\frac{3}{5} \cdot \frac{1}{4} = \frac{3 \cdot 1}{5 \cdot 4} = \frac{3}{20}$$

Use this same procedure to multiply rational expressions.

MULTIPLYING RATIONAL EXPRESSIONS

Let P, Q, R, and S be polynomials. Then

$$\frac{P}{Q} \cdot \frac{R}{S} = \frac{PR}{QS}$$

as long as $Q \neq 0$ and $S \neq 0$.

EXAMPLE 1   Multiply.

a. $\dfrac{25x}{2} \cdot \dfrac{1}{y^3}$

b. $\dfrac{-7x^2}{5y} \cdot \dfrac{3y^5}{14x^2}$

Solution: To multiply rational expressions, multiply the numerators and then multiply the denominators of both expressions. Then write in lowest terms.

a. $\dfrac{25x}{2} \cdot \dfrac{1}{y^3} = \dfrac{25x \cdot 1}{2 \cdot y^3} = \dfrac{25x}{2y^3}$

The expression $\dfrac{25x}{2y^3}$ is in lowest terms.

b. $\dfrac{-7x^2}{5y} \cdot \dfrac{3y^5}{14x^2} = \dfrac{-7x^2 \cdot 3y^5}{5y \cdot 14x^2}$ Multiply.

The expression $\dfrac{-7x^2 \cdot 3y^5}{5y \cdot 14x^2}$ is not in lowest terms, so we factor the numerator and the denominator and apply the fundamental principle.

$$= \frac{-1 \cdot \boxed{7} \cdot 3 \cdot \boxed{x^2} \cdot \boxed{y} \cdot y^4}{5 \cdot 2 \cdot \boxed{7} \cdot \boxed{x^2} \cdot \boxed{y}}$$

$$= -\frac{3y^4}{10}$$

When multiplying rational expressions, it is usually best to factor each numerator and denominator. This will help us when we apply the fundamental principle to write the product in lowest terms.

EXAMPLE 2   Multiply: $\dfrac{x^2 + x}{3x} \cdot \dfrac{6}{5x + 5}$.

Solution: $\dfrac{x^2 + x}{3x} \cdot \dfrac{6}{5x + 5} = \dfrac{x(x + 1)}{3x} \cdot \dfrac{2 \cdot 3}{5(x + 1)}$ Factor numerators and denominators.

$$= \frac{x\,(x + 1) \cdot 2 \cdot \boxed{3}}{3x \cdot 5\,(x + 1)}$$ Multiply.

$$= \frac{2}{5}$$ Simplify.

The following steps may be used to multiply rational expressions.

TO MULTIPLY RATIONAL EXPRESSIONS

Step 1.   Completely factor numerators and denominators.

Step 2.   Multiply numerators and multiply denominators.

Step 3.   Simplify or write the product in lowest terms by applying the fundamental principle to all common factors.

EXAMPLE 3   Multiply $\dfrac{3x + 3}{5x - 5x^2} \cdot \dfrac{2x^2 + x - 3}{4x^2 - 9}$.

Solution: $\dfrac{3x + 3}{5x - 5x^2} \cdot \dfrac{2x^2 + x - 3}{4x^2 - 9} = \dfrac{3(x + 1)}{5x(1 - x)} \cdot \dfrac{(2x + 3)(x - 1)}{(2x - 3)(2x + 3)}$ Factor.

$$= \frac{3(x + 1)\,(2x + 3)(x - 1)}{5x(1 - x)(2x - 3)\,(2x + 3)}$$ Multiply.

$$= \frac{3(x + 1)(x - 1)}{5x(1 - x)(2x - 3)}$$ Apply the fundamental principle.

Next, recall that $x - 1$ and $1 - x$ are opposites so that $x - 1 = -1(1 - x)$.

$$= \frac{3(x + 1)(-1)\,(1 - x)}{5x\,(1 - x)\,(2x - 3)} \qquad \text{Write } x - 1 \text{ as } -1(1 - x).$$

$$= \frac{-3(x + 1)}{5x(2x - 3)} \qquad \text{Apply the fundamental principle.}$$

2   We can divide by a rational expression in the same way we divide by a fraction. To divide by a fraction, multiply by its reciprocal.

For example, to divide $\frac{3}{2}$ by $\frac{7}{8}$, multiply $\frac{3}{2}$ by $\frac{8}{7}$.

$$\frac{3}{2} \div \frac{7}{8} = \frac{3}{2} \cdot \frac{8}{7} = \frac{3 \cdot 4 \cdot 2}{2 \cdot 7} = \frac{12}{7}$$

> **DIVIDING RATIONAL EXPRESSIONS**
>
> Let P, Q, R, and S be polynomials. Then,
>
> $$\frac{P}{Q} \div \frac{R}{S} = \frac{P}{Q} \cdot \frac{S}{R} = \frac{PS}{QR}$$
>
> as long as $Q \neq 0$, $S \neq 0$, and $R \neq 0$.

EXAMPLE 4   Divide: $\dfrac{3x^3y^7}{40} \div \dfrac{4x^3}{y^2}$.

Solution: $\dfrac{3x^3y^7}{40} \div \dfrac{4x^3}{y^2} = \dfrac{3x^3y^7}{40} \cdot \dfrac{y^2}{4x^3}$ Multiply by the reciprocal of $\dfrac{4x^3}{y^2}$.

$$= \frac{3\,x^3y^9}{160\,x^3}$$

$$= \frac{3y^9}{160} \qquad \text{Simplify.}$$

EXAMPLE 5   Divide $\dfrac{(x - 1)(x + 2)}{10}$ by $\dfrac{2x + 4}{5}$.

Solution: $\dfrac{(x - 1)(x + 2)}{10} \div \dfrac{2x + 4}{5} = \dfrac{(x - 1)(x + 2)}{10} \cdot \dfrac{5}{2x + 4}$ Multiply by the reciprocal of $\dfrac{2x + 4}{5}$.

$$= \frac{(x - 1)\,(x + 2) \cdot 5}{5 \cdot 2 \cdot 2 \cdot (x + 2)} \qquad \text{Factor and multiply.}$$

$$= \frac{x - 1}{4} \qquad \text{Simplify.}$$

The following may be used to divide by a rational expression.

> **To Divide by a Rational Expression**
> Multiply by its reciprocal.

EXAMPLE 6   Divide: $\dfrac{6x+2}{x^2-1} \div \dfrac{3x^2+x}{x-1}$.

Solution: $\dfrac{6x+2}{x^2-1} \div \dfrac{3x^2+x}{x-1} = \dfrac{6x+2}{x^2-1} \cdot \dfrac{x-1}{3x^2+x}$ Multiply by the reciprocal.

$$= \dfrac{2\,(3x+1)\,(x-1)}{(x+1)\,(x-1)\cdot x\,(3x+1)}$$ Factor and multiply.

$$= \dfrac{2}{x(x+1)}$$ Simplify.

EXAMPLE 7   Divide: $\dfrac{2x^2-11x+5}{5x-25} \div \dfrac{4x-2}{10}$.

Solution: $\dfrac{2x^2-11x+5}{5x-25} \div \dfrac{4x-2}{10} = \dfrac{2x^2-11x+5}{5x-25} \cdot \dfrac{10}{4x-2}$ Multiply by the reciprocal.

$$= \dfrac{(2x-1)(x-5)\cdot 2\cdot 5}{5(x-5)\cdot \ 2(2x-1)}$$ Factor and multiply.

$$= \dfrac{1}{1} \quad \text{or} \quad 1$$ Simplify.

Mental Math

Find the following products. See Example 1.

1. $\dfrac{2}{y} \cdot \dfrac{x}{3}$ $\dfrac{2x}{3y}$ **2.** $\dfrac{3x}{4} \cdot \dfrac{1}{y}$ $\dfrac{3x}{4y}$ **3.** $\dfrac{5}{7} \cdot \dfrac{y^2}{x^2}$ $\dfrac{5y^2}{7x^2}$

4. $\dfrac{x^5}{11} \cdot \dfrac{4}{z^3}$ $\dfrac{4x^5}{11z^3}$ **5.** $\dfrac{9}{x} \cdot \dfrac{x}{5}$ $\dfrac{9}{5}$ **6.** $\dfrac{y}{7} \cdot \dfrac{3}{y}$ $\dfrac{3}{7}$

Exercise Set 6.2

Find each product and simplify if possible. See Examples 1 through 3.

1. $\dfrac{3x}{y^2} \cdot \dfrac{7y}{4x}$ $\dfrac{21}{4y}$ **2.** $\dfrac{9x^2}{y} \cdot \dfrac{4y}{3x^2}$ 12

3. $\dfrac{8x}{2} \cdot \dfrac{x^5}{4x^2}$ x^4 **4.** $\dfrac{6x^2}{10x^3} \cdot \dfrac{5x}{12}$ $\dfrac{1}{4}$

5. $-\dfrac{5a^2b}{30a^2b^2} \cdot b^3$ $-\dfrac{b^2}{6}$ **6.** $-\dfrac{9x^3y^2}{18xy^5} \cdot y^3$ $-\dfrac{x^2}{2}$

7. $\dfrac{x}{2x-14} \cdot \dfrac{x^2-7x}{5}$ $\dfrac{x^2}{10}$ **8.** $\dfrac{4x-24}{20x} \cdot \dfrac{5}{x-6}$ $\dfrac{1}{x}$

9. $\dfrac{6x+6}{5} \cdot \dfrac{10}{36x+36}$ $\dfrac{1}{3}$ **10.** $\dfrac{x^2+x}{8} \cdot \dfrac{16}{x+1}$ $2x$

12. $\dfrac{m-n}{m+n}$ **13.** $\dfrac{x+5}{x}$ **14.** $\dfrac{(a+3)^3}{(a-2)^2(a+2)}$ **16.** $\dfrac{4x^2}{(5x^2+3x)^2}$ or $\dfrac{4}{(5x+3)^2}$ sq. m

11. $\dfrac{m^2-n^2}{m+n}\cdot\dfrac{m}{m^2-mn}$ 1 **12.** $\dfrac{(m-n)^2}{m+n}\cdot\dfrac{m}{m^2-mn}$

13. $\dfrac{x^2-25}{x^2-3x-10}\cdot\dfrac{x+2}{x}$ **14.** $\dfrac{a^2+6a+9}{a^2-4}\cdot\dfrac{a+3}{a-2}$

❏ **15.** Find the area of the following rectangle.

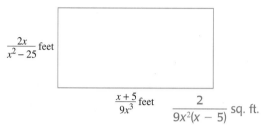

$\dfrac{2x}{x^2-25}$ feet

$\dfrac{x+5}{9x^3}$ feet $\dfrac{2}{9x^2(x-5)}$ sq. ft.

❏ **16.** Find the area of the following square.

$\dfrac{2x}{5x^2+3x}$ meters

Find each quotient and simplify. See Examples 4 through 7.

17. $\dfrac{5x^7}{2x^5}\div\dfrac{10x}{4x^3}$ x^4 **18.** $\dfrac{9y^4}{6y}\div\dfrac{y^2}{3}$ $\dfrac{9y}{2}$

19. $\dfrac{8x^2}{y^3}\div\dfrac{4x^2y^3}{6}$ $\dfrac{12}{y^6}$ **20.** $\dfrac{7a^2b}{3ab^2}\div\dfrac{21a^2b^2}{14ab}$ $\dfrac{14}{9b^2}$

21. $\dfrac{(x-6)(x+4)}{4x}\div\dfrac{2x-12}{8x^2}$ $x(x+4)$

22. $\dfrac{(x+3)^2}{5}\div\dfrac{5x+15}{25}$ **23.** $\dfrac{3x^2}{x^2-1}\div\dfrac{x^5}{(x+1)^2}$

24. $\dfrac{(x+1)}{(x+1)(2x+3)}\div\dfrac{20}{2x+3}$ $\dfrac{1}{20}$

25. $\dfrac{m^2-n^2}{m+n}\div\dfrac{m}{m^2+nm}$ **26.** $\dfrac{(m-n)^2}{m+n}\div\dfrac{m^2-mn}{m}$

27. $\dfrac{x+2}{7-x}\div\dfrac{x^2-5x+6}{x^2-9x+14}$ **28.** $(x-3)\div\dfrac{x^2+3x-18}{x}$

29. $\dfrac{x^2+7x+10}{1-x}\div\dfrac{x^2+2x-15}{x-1}$ $-\dfrac{x+2}{x-3}$

30. $\dfrac{a^2-b^2}{9}\cdot\dfrac{27x^2}{3b-3a}$ $-x^2(a+b)$

❏ **31.** Explain how to multiply rational expressions.

❏ **32.** Explain how to divide rational expressions.

31. answers may vary   **32.** answers may vary

Perform the indicated operations.

33. $\dfrac{5a^2b}{30a^2b^2}\cdot\dfrac{1}{b^3}$ $\dfrac{1}{6b^4}$ **34.** $\dfrac{9x^2y^2}{42xy^5}\cdot\dfrac{6}{x^5}$ $\dfrac{9}{7x^3y^3}$

35. $\dfrac{12x^3y}{8xy^7}\div\dfrac{7x^5y}{6x}$ $\dfrac{9}{7x^2y^7}$ **36.** $\dfrac{4y^2z}{3y^7z^7}\div\dfrac{12y}{6z}$ $\dfrac{2}{3y^6z^6}$

37. $\dfrac{5x-10}{12}\div\dfrac{4x-8}{8}$ $\dfrac{5}{6}$ **38.** $\dfrac{6x+6}{5}\div\dfrac{3x+3}{10}$ 4

39. $\dfrac{x^2+5x}{8}\cdot\dfrac{9}{3x+15}$ $\dfrac{3x}{8}$ **40.** $\dfrac{3x^2+12x}{6}\cdot\dfrac{9}{2x+8}$ $\dfrac{9x}{4}$

41. $\dfrac{7}{6p^2+q}\div\dfrac{14}{18p^2+3q}$ $\dfrac{3}{2}$ **42.** $\dfrac{5x-10}{12}\div\dfrac{4x-8}{8}$ $\dfrac{5}{6}$

43. $\dfrac{3x+4y}{x^2+4xy+4y^2}\cdot\dfrac{x+2y}{2}$ $\dfrac{3x+4y}{2(x+2y)}$

44. $\dfrac{2a+2b}{3}\div\dfrac{a^2-b^2}{a-b}$ $\dfrac{2}{3}$ **45.** $\dfrac{x^2-9}{x^2+8}\div\dfrac{3-x}{2x^2+16}$

46. $\dfrac{x^2-y^2}{3x^2+3xy}\cdot\dfrac{3x^2+6x}{3x^2-2xy-y^2}$ $\dfrac{x+2}{3x+y}$

47. $\dfrac{(x+2)^2}{x-2}\div\dfrac{x^2-4}{2x-4}$ **48.** $\dfrac{x^2-4}{2y}\div\dfrac{2-x}{6xy}$

49. $\dfrac{a^2+7a+12}{a^2+5a+6}\cdot\dfrac{a^2+8a+15}{a^2+5a+4}$ $\dfrac{(a+5)(a+3)}{(a+2)(a+1)}$

50. $\dfrac{b^2+2b-3}{b^2+b-2}\cdot\dfrac{b^2-4}{b^2+6b+8}$ $\dfrac{(b+3)(b-2)}{(b+2)(b+4)}$

51. $\dfrac{1}{-x-4}\div\dfrac{x^2-7x}{x^2-3x-28}$ $-\dfrac{1}{x}$

52. $\dfrac{x^2-10x+21}{7-x}\div(x+3)$ $-\dfrac{x-3}{x+3}$

53. $\dfrac{x^2-5x-24}{2x^2-2x-24}\cdot\dfrac{4x^2+4x-24}{x^2-10x+16}$ $\dfrac{2(x+3)}{x-4}$

54. $\dfrac{a^2-b^2}{a}\cdot\dfrac{a+b}{a^2+ab}$ **55.** $(x-5)\div\dfrac{5-x}{x^2+2}$

56. $\dfrac{2x^2+3xy+y^2}{x^2-y^2}\div\dfrac{1}{2x+2y}$ $\dfrac{2(x+y)(2x+y)}{x-y}$

57. $\dfrac{x^2-y^2}{x^2-2xy+y^2}\cdot\dfrac{y-x}{x+y}$ **58.** $\dfrac{x+3}{x^2-9}\cdot\dfrac{x^2-8x+15}{5x}$

59. $\dfrac{a^2+ac+ba+bc}{a-b}\div\dfrac{a+c}{a+b}$ $\dfrac{(a+b)^2}{a-b}$

22. $x+3$ **23.** $\dfrac{3(x+1)}{x^3(x-1)}$ **25.** m^2-n^2 **26.** $\dfrac{m-n}{m+n}$ **27.** $-\dfrac{x+2}{x-3}$ **28.** $\dfrac{x}{x+6}$ **45.** $-2(x+3)$ **47.** $\dfrac{2(x+2)}{x-2}$

48. $-3x(x+2)$ **54.** $\dfrac{(a-b)(a+b)}{a^2}$ **55.** $-(x^2+2)$ **57.** -1 **58.** $\dfrac{x-5}{5x}$

60. $\dfrac{x^2 + 2x - xy - 2y}{x^2 - y^2} \div \dfrac{2x + 4}{x + y}$ $\dfrac{1}{2}$

61. $\dfrac{3x^2 + 8x + 5}{x^2 + 8x + 7} \cdot \dfrac{x + 7}{x^2 + 4}$ $\dfrac{3x + 5}{x^2 + 4}$

62. $\dfrac{16x^2 + 2x}{16x^2 + 10x + 1} \cdot \dfrac{1}{4x^2 + 2x}$ $\dfrac{1}{(2x + 1)^2}$

63. Find the quotient of $\dfrac{x^2 - 9}{2x}$ and $\dfrac{x + 3}{8x^4}$. $4x^3(x - 3)$

64. Find the quotient of $\dfrac{4x^2 + 4x + 1}{4x + 2}$ and $\dfrac{4x + 2}{16}$. 4

65. $\dfrac{x^3 + 8}{x^2 - 2x + 4} \cdot \dfrac{4}{x^2 - 4}$ **66.** $\dfrac{9y}{3y - 3} \cdot \dfrac{y^3 - 1}{y^3 + y^2 + y}$ 3

67. $\dfrac{a^2 - ab}{6a^2 + 6ab} \div \dfrac{a^3 - b^3}{a^2 - b^2}$ **68.** $\dfrac{x^3 + 27y^3}{6x} \div \dfrac{x^2 - 9y^2}{x^2 - 3xy}$

65. $\dfrac{4}{x - 2}$ **67.** $\dfrac{a - b}{6(a^2 + ab + b^2)}$ **68.** $\dfrac{x^2 - 3xy + 9y^2}{6}$

Review Exercises

73. $-\dfrac{1}{5}$ **74.** $-\dfrac{5}{2}$

Perform each operation. See Section 1.2.

69. $\dfrac{1}{5} + \dfrac{4}{5}$ 1 **70.** $\dfrac{3}{15} + \dfrac{6}{15}$ $\dfrac{3}{5}$ **71.** $\dfrac{9}{9} - \dfrac{19}{9}$ $-\dfrac{10}{9}$

72. $\dfrac{4}{3} - \dfrac{8}{3}$ $-\dfrac{4}{3}$ **73.** $\dfrac{6}{5} + \left(\dfrac{1}{5} - \dfrac{8}{5}\right)$ **74.** $-\dfrac{3}{2} + \left(\dfrac{1}{2} - \dfrac{3}{2}\right)$

See Section 3.2.   graphical answers in App. F

75. Graph the linear equation $x - 2y = 6$.

76. Graph the linear equation $5x + y = 10$.

A Look Ahead

EXAMPLE   Perform the indicated operations.

$$\frac{15x^2 - x - 6}{12x^3} \cdot \frac{4x}{9 - 25x^2} \div \frac{x}{3x - 2}$$

Solution:

$$\frac{15x^2 - x - 6}{12x^3} \cdot \frac{4x}{9 - 25x^2} \div \frac{x}{3x - 2}$$

$$= \left(\frac{15x^2 - x - 6}{12x^3} \cdot \frac{4x}{9 - 25x^2}\right) \cdot \frac{3x - 2}{x}$$

$$= \frac{(3x - 2)(5x + 3) \cdot 4x(3x - 2)}{12x^3(3 - 5x)(3 + 5x) \cdot x}$$

$$= \frac{(3x - 2)^2}{3x^3(3 - 5x)}$$

Perform the following operations.

77. $\left(\dfrac{x^2 - y^2}{x^2 + y^2} \div \dfrac{x^2 - y^2}{3x}\right) \cdot \dfrac{x^2 + y^2}{6}$ $\dfrac{x}{2}$

78. $\left(\dfrac{x^2 - 9}{x^2 - 1} \cdot \dfrac{x^2 + 2x + 1}{2x^2 + 9x + 9}\right) \div \dfrac{2x + 3}{1 - x}$ $\dfrac{-(x - 3)(x + 1)}{(2x + 3)^2}$

79. $\left(\dfrac{2a + b}{b^2} \cdot \dfrac{3a^2 - 2ab}{ab + 2b^2}\right) \div \dfrac{a^2 - 3ab + 2b^2}{5ab - 10b^2}$

80. $\left(\dfrac{x^2y^2 - xy}{4x - 4y} \div \dfrac{3y - 3x}{8x - 8y}\right) \cdot \dfrac{y - x}{8}$ $\dfrac{xy(xy - 1)}{12}$

79. $\dfrac{5a(2a + b)(3a - 2b)}{b^2(a - b)(a + 2b)}$

6.3 | ## Adding and Subtracting Rational Expressions with Common Denominators and Least Common Denominator

TAPE
BA 6.3

O B J E C T I V E S

1. Add and subtract rational expressions with common denominators.
2. Find the least common denominator of a list of rational expressions.
3. Write a rational expression as an equivalent expression whose denominator is given.

1   Like multiplication and division, addition and subtraction of rational expressions is similar to addition and subtraction of rational numbers. For example, to add

fractions with common denominators such as $\frac{6}{5}$ and $\frac{2}{5}$, add the numerators and write the sum over the common denominator 5.

$$\frac{6}{5} + \frac{2}{5} = \frac{6+2}{5} = \frac{8}{5}$$

To subtract two fractions with common denominators, subtract the numerators and write the result over the common denominator.

$$\frac{6}{13} - \frac{5}{13} = \frac{6-5}{13} = \frac{1}{13}$$

Rational expressions with common denominators are added and subtracted in the same manner.

ADDING AND SUBTRACTING RATIONAL EXPRESSIONS WITH COMMON DENOMINATORS

If P, Q, and R are polynomials, then

$$\frac{P}{R} + \frac{Q}{R} = \frac{P+Q}{R} \qquad \text{and} \qquad \frac{P}{R} - \frac{Q}{R} = \frac{P-Q}{R}$$

as long as $R \neq 0$.

EXAMPLE 1   Add: $\dfrac{5m}{2n} + \dfrac{m}{2n}$.

Solution: $\dfrac{5m}{2n} + \dfrac{m}{2n} = \dfrac{5m+m}{2n}$ Add the numerators.

$= \dfrac{6m}{2n}$ Simplify the numerator by combining like terms.

$= \dfrac{3m}{n}$ Simplify the rational expression by applying the fundamental principle.

EXAMPLE 2   Subtract: $\dfrac{2y}{2y-7} - \dfrac{7}{2y-7}$.

Solution: $\dfrac{2y}{2y-7} - \dfrac{7}{2y-7} = \dfrac{2y-7}{2y-7}$ Subtract the numerators.

$= \dfrac{1}{1}$ or 1 Simplify.

EXAMPLE 3 Subtract: $\dfrac{3x^2 + 2x}{x - 1} - \dfrac{10x - 5}{x - 1}$.

Solution:

$$\frac{3x^2 + 2x}{x - 1} - \frac{10x - 5}{x - 1} = \frac{3x^2 + 2x - (10x - 5)}{x - 1} \qquad \text{Subtract the numerators.}$$
$$\text{Notice the parentheses.}$$

$$= \frac{3x^2 + 2x - 10x + 5}{x - 1} \qquad \text{Use the distributive property.}$$

$$= \frac{3x^2 - 8x + 5}{x - 1} \qquad \text{Combine like terms.}$$

$$= \frac{(x - 1)(3x - 5)}{x - 1} \qquad \text{Factor.}$$

$$= 3x - 5 \qquad \text{Simplify.}$$

R E M I N D E R   Notice how the numerator $10x - 5$ has been subtracted in Example 3.

| This $-$ sign applies to the entire numerator of $10x - 5$. | So parentheses are inserted here to indicate this. |

$$\frac{3x^2 + 2x}{x - 1} - \frac{10x - 5}{x - 1} = \frac{3x^2 + 2x - (10x - 5)}{x - 1}$$

To add and subtract fractions with **unlike** denominators, first find a least common denominator (LCD), and then write all fractions as equivalent fractions with the LCD.

For example, suppose we add $\frac{8}{3}$ and $\frac{2}{5}$. The LCD of denominators 3 and 5 is 15, since 15 is the smallest number that both 3 and 5 divide into evenly. Rewrite each fraction so that its denominator is 15. (Notice how we apply the fundamental principle.)

$$\frac{8}{3} + \frac{2}{5} = \frac{8(5)}{3(5)} + \frac{2(3)}{5(3)} = \frac{40}{15} + \frac{6}{15} = \frac{40 + 6}{15} = \frac{46}{15}$$

To add or subtract rational expressions with unlike denominators, we also first find an LCD and then write all rational expressions as equivalent expressions with the LCD. The **least common denominator LCD of a list of rational expressions** is a polynomial of least degree whose factors include all the factors of the denominators in the list.

TO FIND THE LEAST COMMON DENOMINATOR (LCD)

Step 1. Factor each denominator completely.

Step 2. The least common denominator LCD is the product of all unique factors found in step 1, each raised to a power equal to the greatest number of times that the factor appears in any one factored denominator.

EXAMPLE 4   Find the LCD for each pair.

a. $\dfrac{1}{8}, \dfrac{3}{22}$

b. $\dfrac{7}{5x}, \dfrac{6}{15x^2}$

Solution: **a.** Start by finding the prime factorization of each denominator.

$$8 = 2 \cdot 2 \cdot 2 = 2^3 \quad \text{and} \quad 22 = 2 \cdot 11$$

Next, write the product of all the unique factors, each raised to a power equal to the greatest number of times that the factor appears.

The greatest number of times that the factor 2 appears is 3.

The greatest number of times that the factor 11 appears is 1.

$$\text{LCD} = 2^3 \cdot 11^1 = 8 \cdot 11 = 88$$

b. Factor each denominator.

$$5x = 5 \cdot x \quad \text{and} \quad 15x^2 = 3 \cdot 5 \cdot x^2$$

The greatest number of times that the factor 5 appears is 1.

The greatest number of times that the factor 3 appears is 1.

The greatest number of times that the factor x appears is 2.

$$\text{LCD} = 3^1 \cdot 5^1 \cdot x^2 = 15x^2$$

EXAMPLE 5   Find the LCD of $\dfrac{7x}{x+2}$ and $\dfrac{5x^2}{x-2}$.

Solution: The denominators $x + 2$ and $x - 2$ are completely factored already. The factor $x + 2$ appears once and the factor $x - 2$ appears once.

$$\text{LCD} = (x + 2)(x - 2)$$

EXAMPLE 6   Find the LCD of $\dfrac{6m^2}{3m+15}$ and $\dfrac{2}{(m+5)^2}$.

Solution: Factor each denominator.

$$3m + 15 = 3(m + 5)$$
$(m + 5)^2$ is already factored.

The greatest number of times that the factor 3 appears is 1.

The greatest number of times that the factor $m + 5$ appears *in any one denominator* is 2.

$$\text{LCD} = 3(m + 5)^2$$

EXAMPLE 7 Find the LCD of $\dfrac{t - 10}{t^2 - t - 6}$ and $\dfrac{t + 5}{t^2 + 3t + 2}$.

Solution: Start by factoring each denominator.

$$t^2 - t - 6 = (t - 3)(t + 2)$$
$$t^2 + 3t + 2 = (t + 1)(t + 2)$$
$$\text{LCD} = (t - 3)(t + 2)(t + 1)$$

EXAMPLE 8 Find the LCD of $\dfrac{2}{x - 2}$ and $\dfrac{10}{2 - x}$.

Solution: The denominators $x - 2$ and $2 - x$ are opposites. That is, $2 - x = -1(x - 2)$. Use $x - 2$ or $2 - x$ as the LCD.

$$\text{LCD} = x - 2 \qquad \text{or} \qquad \text{LCD} = 2 - x$$

3 Next we practice writing a rational expression as an equivalent rational expression with a given denominator. To do this, we apply the fundamental principle, which says that $\frac{PR}{QR} = \frac{P}{Q}$, or equivalently that $\frac{P}{Q} = \frac{PR}{QR}$. This can be seen by recalling that multiplying an expression by 1 produces an equivalent expression. In other words,

$$\frac{P}{Q} = \frac{P}{Q} \cdot 1 = \frac{P}{Q} \cdot \frac{R}{R} = \frac{PR}{QR}.$$

EXAMPLE 9 Write $\dfrac{4b}{9a}$ as an equivalent fraction with the given denominator.

$$\frac{4b}{9a} = \frac{}{27a^2b}$$

Solution: Ask yourself: "What do we multiply $9a$ by to get $27a^2b$?" The answer is $3ab$, since $9a(3ab) = 27a^2b$. Multiply the numerator and denominator by $3ab$.

$$\frac{4b}{9a} = \frac{4b(3ab)}{9a(3ab)} = \frac{12ab^2}{27a^2b}$$

EXAMPLE 10 Write the rational expression as an equivalent rational expression with the given denominator.

$$\frac{5}{x^2 - 4} = \frac{}{(x - 2)(x + 2)(x - 4)}$$

Solution: First, factor the denominator $x^2 - 4$.

$$\frac{5}{x^2 - 4} = \frac{5}{(x + 2)(x - 2)}$$

If we multiply the original denominator $(x + 2)(x - 2)$ by $x - 4$, the result is the new denominator $(x + 2)(x - 2)(x - 4)$. Thus, multiply the numerator and the denominator by $x - 4$.

$$\frac{5}{x^2 - 4} = \frac{5}{(x - 2)(x + 2)} = \frac{5(x - 4)}{(x - 2)(x + 2)(x - 4)}$$
$$= \frac{5x - 20}{(x - 2)(x + 2)(x - 4)}$$

MENTAL MATH

Perform the indicated operation.

1. $\dfrac{2}{3} + \dfrac{1}{3}$ 1

2. $\dfrac{5}{11} + \dfrac{1}{11}$ $\dfrac{6}{11}$

3. $\dfrac{3x}{9} + \dfrac{4x}{9}$ $\dfrac{7x}{9}$

4. $\dfrac{3y}{8} + \dfrac{2y}{8}$ $\dfrac{5y}{8}$

5. $\dfrac{8}{9} - \dfrac{7}{9}$ $\dfrac{1}{9}$

6. $-\dfrac{4}{12} - \dfrac{3}{12}$ $-\dfrac{7}{12}$

7. $\dfrac{7}{5} - \dfrac{10y}{5}$ $\dfrac{7 - 10y}{5}$

8. $\dfrac{12x}{7} - \dfrac{4x}{7}$ $\dfrac{8x}{7}$

EXERCISE SET 6.3

Add or subtract as indicated. Simplify the result if possible. See examples 1 through 3.

1. $\dfrac{a}{13} + \dfrac{9}{13}$ $\dfrac{a + 9}{13}$

2. $\dfrac{x + 1}{7} + \dfrac{6}{7}$ $\dfrac{x + 7}{7}$

3. $\dfrac{9}{3 + y} + \dfrac{y + 1}{3 + y}$ $\dfrac{y + 10}{3 + y}$

4. $\dfrac{9}{y + 9} + \dfrac{y}{y + 9}$ 1

5. $\dfrac{4m}{3n} + \dfrac{5m}{3n}$ $\dfrac{3m}{n}$

6. $\dfrac{3p}{2} + \dfrac{11p}{2}$ $7p$

7. $\dfrac{2x + 1}{x - 3} + \dfrac{3x + 6}{x - 3}$

8. $\dfrac{4p - 3}{2p + 7} + \dfrac{3p + 8}{2p + 7}$

9. $\dfrac{7}{8} - \dfrac{3}{8}$ $\dfrac{1}{2}$

10. $\dfrac{4}{5} - \dfrac{13}{5}$ $-\dfrac{9}{5}$

11. $\dfrac{4m}{m - 6} - \dfrac{24}{m - 6}$ 4

12. $\dfrac{8y}{y - 2} - \dfrac{16}{y - 2}$ 8

13. $\dfrac{2x^2}{x - 5} - \dfrac{25 + x^2}{x - 5}$ $x + 5$

14. $\dfrac{6x^2}{2x - 5} - \dfrac{25 + 2x^2}{2x - 5}$

15. $\dfrac{-3x^2 - 4}{x - 4} - \dfrac{12 - 4x^2}{x - 4}$

16. $\dfrac{7x^2 - 9}{2x - 5} - \dfrac{16 + 3x^2}{2x - 5}$

17. $\dfrac{2x + 3}{x + 1} - \dfrac{x + 2}{x + 1}$ 1

18. $\dfrac{1}{x^2 - 2x - 15} - \dfrac{4 - x}{x^2 - 2x - 15}$ $\dfrac{x - 3}{x^2 - 2x - 15}$

19. $\dfrac{3}{x^3} + \dfrac{9}{x^3}$ $\dfrac{12}{x^3}$

20. $\dfrac{5}{xy} + \dfrac{8}{xy}$ $\dfrac{13}{xy}$

21. $\dfrac{5}{x + 4} - \dfrac{10}{x + 4}$ $-\dfrac{5}{x + 4}$

22. $\dfrac{4}{2x + 1} - \dfrac{8}{2x + 1}$

23. $\dfrac{x}{x + y} - \dfrac{2}{x + y}$ $\dfrac{x - 2}{x + y}$

24. $\dfrac{y + 1}{y + 2} - \dfrac{3}{y + 2}$ $\dfrac{y - 2}{y + 2}$

25. $\dfrac{8x}{2x + 5} + \dfrac{20}{2x + 5}$ 4

26. $\dfrac{12y - 5}{3y - 1} + \dfrac{1}{3y - 1}$ 4

7. $\dfrac{5x + 7}{x - 3}$ **8.** $\dfrac{7p + 5}{2p + 7}$ **14.** $2x + 5$ **15.** $x + 4$ **16.** $2x + 5$ **22.** $-\dfrac{4}{2x + 1}$

28. $x - 2$ **42.** $12y(y + 3)$ **43.** $6(x + 1)^2$ **45.** $8 - x$ or $x - 8$ **46.** $7 - 3x$ or $3x - 7$ **47.** $40x^3(x - 1)^2$
48. $36x(x + 2)^2$ **49.** $(2x + 1)(2x - 1)$ **50.** $2(2x - 1)(2x + 1)$ **54.** answers may vary

27. $\dfrac{5x + 4}{x - 1} - \dfrac{2x + 7}{x - 1}$ $\quad 3 \quad$ **28.** $\dfrac{x^2 + 9x}{x + 7} - \dfrac{4x + 14}{x + 7}$

29. $\dfrac{a}{a^2 + 2a - 15} - \dfrac{3}{a^2 + 2a - 15}$ $\quad \dfrac{1}{a + 5}$

30. $\dfrac{3y}{y^2 + 3y - 10} - \dfrac{6}{y^2 + 3y - 10}$ $\quad \dfrac{3}{y + 5}$

31. $\dfrac{2x + 3}{x^2 - x - 30} - \dfrac{x - 2}{x^2 - x - 30}$ $\quad \dfrac{1}{x - 6}$

32. $\dfrac{3x - 1}{x^2 + 5x - 6} - \dfrac{2x - 7}{x^2 + 5x - 6}$ $\quad \dfrac{1}{x - 1}$

❑ **33.** A square-shaped pasture has a side of length $\dfrac{5}{x - 2}$ meters. Express its perimeter as a rational expression.

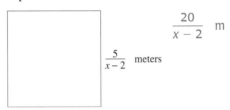
$$\dfrac{20}{x - 2} \text{ m}$$
$\dfrac{5}{x - 2}$ meters

❑ **34.** The following trapezoid has sides of indicated length. Find its perimeter.

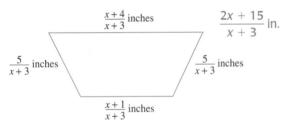
$\dfrac{x + 4}{x + 3}$ inches $\qquad \dfrac{2x + 15}{x + 3}$ in.

$\dfrac{5}{x + 3}$ inches $\qquad \dfrac{5}{x + 3}$ inches

$\dfrac{x + 1}{x + 3}$ inches

❑ **35.** Describe the process for adding and subtracting two rational expressions with the same denominators. answers may vary

❑ **36.** Explain the similarities between subtracting $\frac{3}{8}$ from $\frac{7}{8}$ and subtracting $\frac{6}{x + 3}$ from $\frac{9}{x + 3}$. answers may vary

Find the LCD for the following lists of rational expressions. See examples 4 through 8.

37. $\dfrac{2}{3}, \dfrac{4}{33}$ $\quad 33$ **38.** $\dfrac{8}{20}, \dfrac{4}{15}$ $\quad 60$ **39.** $\dfrac{19}{2x}, \dfrac{5}{4x^3}$ $\quad 4x^3$

40. $\dfrac{17x}{4y^5}, \dfrac{2}{8y}$ $\quad 8y^5$ **41.** $\dfrac{9}{8x}, \dfrac{3}{2x + 4}$ $\quad 8x(x + 2)$

42. $\dfrac{1}{6y}, \dfrac{3x}{4y + 12}$ **43.** $\dfrac{1}{3x + 3}, \dfrac{8}{2x^2 + 4x + 2}$

58. $\dfrac{136axyz}{32y^3x^2z}$ **59.** $\dfrac{18}{2(x + 3)}$ **60.** $\dfrac{4xy + y}{3y(x + 2)}$ **61.** $\dfrac{9ab + 2b}{5b(a + 2)}$ **62.** $\dfrac{2(5 + y)}{4(x^2 + 5)}$ **63.** $\dfrac{x(x + 1)}{x(x + 4)(x + 2)(x + 1)}$

64. $\dfrac{5x(x - 5)}{(x - 1)(x - 5)(x + 3)}$ **65.** $\dfrac{15x(x - 7)}{3x(2x + 1)(x - 7)(x - 5)}$ **66.** $\dfrac{x(x - 9)(x + 5)}{x(x + 3)(x + 5)(3x + 1)}$ **68.** $\dfrac{6x + 12}{(x + 3)(x - 3)(x + 2)}$

44. $\dfrac{19x + 5}{4x - 12}, \dfrac{3}{2x^2 - 12x + 18}$ $\quad 4(x - 3)^2$

45. $\dfrac{5}{x - 8}, \dfrac{3}{8 - x}$ **46.** $\dfrac{2x + 5}{3x - 7}, \dfrac{5}{7 - 3x}$

47. $\dfrac{4 + x}{8x^2(x - 1)^2}, \dfrac{17}{10x^3(x - 1)}$ **48.** $\dfrac{2x + 3}{9x(x + 2)}, \dfrac{9x + 5}{12(x + 2)^2}$

49. $\dfrac{9x + 1}{2x + 1}, \dfrac{3x - 5}{2x - 1}$ **50.** $\dfrac{5}{4x - 2}, \dfrac{7}{4x + 2}$

51. $\dfrac{5x + 1}{2x^2 + 7x - 4}, \dfrac{3x}{2x^2 + 5x - 3}$ $\quad (2x - 1)(x + 4)(x + 3)$

52. $\dfrac{4}{x^2 + 4x + 3}, \dfrac{4x - 2}{x^2 + 10x + 21}$ $\quad (x + 3)(x + 1)(x + 7)$

❑ **53.** Write some instructions to help a friend who is having difficulty finding the LCD of two rational expressions. answers may vary

❑ **54.** Explain why the LCD of rational expressions $\dfrac{7}{x + 1}$ and $\dfrac{9x}{(x + 1)^2}$ is $(x + 1)^2$ and not $(x + 1)^3$.

Rewrite each rational expression as an equivalent rational expression whose denominator is the given polynomial. See Examples 9 and 10.

55. $\dfrac{3}{2x}; 4x^2$ $\quad \dfrac{6x}{4x^2}$ **56.** $\dfrac{3}{9y^5}; 72y^9$ $\quad \dfrac{24y^4}{72y^9}$

57. $\dfrac{6}{3a}; 12ab^2$ $\quad \dfrac{24b^2}{12ab^2}$ **58.** $\dfrac{17a}{4y^2x}; 32y^3x^2z$

59. $\dfrac{9}{x + 3}; 2(x + 3)$ **60.** $\dfrac{4x + 1}{3x + 6}; 3y(x + 2)$

61. $\dfrac{9a + 2}{5a + 10}; 5b(a + 2)$ **62.** $\dfrac{5 + y}{2x^2 + 10}; 4(x^2 + 5)$

63. $\dfrac{x}{x^3 + 6x^2 + 8x}; x(x + 4)(x + 2)(x + 1)$

64. $\dfrac{5x}{x^2 + 2x - 3}; (x - 1)(x - 5)(x + 3)$

65. $\dfrac{5}{2x^2 - 9x - 5}; 3x(2x + 1)(x - 7)(x - 5)$

66. $\dfrac{x - 9}{3x^2 + 10x + 3}; x(x + 3)(x + 5)(3x + 1)$

67. $\dfrac{9y - 1}{15x^2 - 30}; 30x^2 - 60$ $\quad \dfrac{18y - 2}{30x^2 - 60}$

68. $\dfrac{6}{x^2 - 9}; (x + 3)(x - 3)(x + 2)$

69. $\dfrac{1}{x^2 - 16}$; $x(x-4)^2(x+4)$ $\dfrac{x(x-4)}{x(x-4)^2(x+4)}$

70. $\dfrac{-3}{x^2 - 9}$; $(x-3)(x+3)^2$ $\dfrac{-3x-9}{(x-3)(x+3)^2}$

Write each rational expression as an equivalent expression with a denominator of $x - 2$.

71. $\dfrac{5}{2-x}$ $-\dfrac{5}{x-2}$

72. $\dfrac{8y}{2-x}$ $-\dfrac{8y}{x-2}$

73. $-\dfrac{7+x}{2-x}$ $\dfrac{7+x}{x-2}$

74. $\dfrac{x-3}{-(x-2)}$ $\dfrac{3-x}{x-2}$

75. $x = 0, x = 3$ **76.** $x = 0, x = -5$ **77.** $x = -5, x = -1$ **78.** $x = 1, x = 5$

Review Exercises

Solve the following quadratic equations by factoring. See Section 5.6.

75. $x(x-3) = 0$

76. $2x(x+5) = 0$

77. $x^2 + 6x + 5 = 0$

78. $x^2 - 6x + 5 = 0$

Perform each operation. See Section 1.2.

79. $\dfrac{2}{3} + \dfrac{5}{7}$ $\dfrac{29}{21}$

80. $\dfrac{9}{10} - \dfrac{3}{5}$ $\dfrac{3}{10}$

81. $\dfrac{2}{6} - \dfrac{3}{4}$ $-\dfrac{5}{12}$

82. $\dfrac{11}{15} + \dfrac{5}{9}$ $\dfrac{58}{45}$

6.4 | ADDING AND SUBTRACTING RATIONAL EXPRESSIONS WITH UNLIKE DENOMINATORS

TAPE
BA 6.4

O B J E C T I V E

1 Add and subtract rational expressions with unlike denominators.

1 In the previous section, we practiced all the skills we need to add and subtract rational expressions with unlike denominators. The steps are as follows:

TO ADD OR SUBTRACT RATIONAL EXPRESSIONS WITH UNLIKE DENOMINATORS

Step 1. Find the LCD of the rational expressions.

Step 2. Rewrite each rational expression as an equivalent expression whose denominator is the LCD found in *step 1*.

Step 3. Add or subtract numerators and write the sum or difference over the common denominator.

Step 4. Write the rational expression in lowest terms.

EXAMPLE 1   Perform the indicated operation.

a. $\dfrac{a}{4} - \dfrac{2a}{8}$

b. $\dfrac{3}{10x^2} + \dfrac{7}{25x}$

Solution: **a.** First, find the LCD. Since $4 = 2^2$ and $8 = 2^3$, the LCD $= 2^3 = 8$. Write each fraction as an equivalent fraction with the denominator 8; then subtract.

$$\frac{a}{4} - \frac{2a}{8} = \frac{a(2)}{4(2)} - \frac{2a}{8} = \frac{2a}{8} - \frac{2a}{8} = \frac{2a - 2a}{8} = \frac{0}{8} = 0$$

b. Since $10x^2 = 2 \cdot 5 \cdot x \cdot x$ and $25x = 5 \cdot 5 \cdot x$, the LCD $= 2 \cdot 5^2 \cdot x^2 = 50x^2$. Write each fraction as an equivalent fraction with a denominator of $50x^2$.

$$\frac{3}{10x^2} + \frac{7}{25x} = \frac{3(5)}{10x^2(5)} + \frac{7(2x)}{25x(2x)}$$

$$= \frac{15}{50x^2} + \frac{14x}{50x^2}$$

$$= \frac{15 + 14x}{50x^2} \qquad \text{Add numerators, and use the common denominator.} \quad \blacksquare$$

EXAMPLE 2   Subtract: $\dfrac{6x}{x^2 - 4} - \dfrac{3}{x + 2}$.

Solution: Since $x^2 - 4 = (x + 2)(x - 2)$,

$$\text{LCD} = (x - 2)(x + 2)$$

Write equivalent expressions with the LCD as denominators.

$$\frac{6x}{x^2 - 4} - \frac{3}{x + 2} = \frac{6x}{(x - 2)(x + 2)} - \frac{3(x - 2)}{(x + 2)(x - 2)}$$

$$= \frac{6x - 3(x - 2)}{(x + 2)(x - 2)} \qquad \text{Subtract numerators, and use the common denominator.}$$

$$= \frac{6x - 3x + 6}{(x + 2)(x - 2)} \qquad \text{Apply the distributive property in the numerator.}$$

$$= \frac{3x + 6}{(x + 2)(x - 2)} \qquad \text{Combine like terms in the numerator.}$$

Next, factor the numerator to see if this rational expression can be simplified.

$$= \frac{3x + 6}{(x + 2)(x - 2)}$$

$$= \frac{3\,(x + 2)}{(x + 2)\,(x - 2)} \qquad \text{Factor.}$$

$$= \frac{3}{x - 2} \qquad \text{Apply the fundamental principle to simplify.} \quad \blacksquare$$

EXAMPLE 3   Add: $\dfrac{2}{3t} + \dfrac{5}{t + 1}$.

Solution: The LCD is $3t(t + 1)$. Write each rational expression as an equivalent rational expression with a denominator of $3t(t + 1)$.

$$\frac{2}{3t} + \frac{5}{t + 1} = \frac{2(t + 1)}{3t(t + 1)} + \frac{5(3t)}{(t + 1)(3t)}$$

$$= \frac{2(t + 1) + 5(3t)}{3t(t + 1)}$$

Add numerators, and use the common denominator.

$$= \frac{2t + 2 + 15t}{3t(t + 1)}$$

Apply the distributive property in the numerator.

$$= \frac{17t + 2}{3t(t + 1)}$$

Combine like terms in the numerator.

EXAMPLE 4   Subtract: $\dfrac{7}{x - 3} - \dfrac{9}{3 - x}$.

Solution: To find a common denominator, notice that $x - 3$ and $3 - x$ are opposites. That is, $3 - x = -(x - 3)$. Write the denominator $3 - x$ as $-(x - 3)$ and simplify.

$$\frac{7}{x - 3} - \frac{9}{3 - x} = \frac{7}{x - 3} - \frac{9}{-(x - 3)}$$

$$= \frac{7}{x - 3} - \frac{-9}{x - 3}$$

Apply $\dfrac{a}{-b} = \dfrac{-a}{b}$.

$$= \frac{7 - (-9)}{x - 3}$$

Subtract numerators, and use the common denominator.

$$= \frac{16}{x - 3}$$

EXAMPLE 5   Add: $1 + \dfrac{m}{m + 1}$.

Solution: Recall that 1 is the same as $\dfrac{1}{1}$. The LCD of $\dfrac{1}{1}$ and $\dfrac{m}{m + 1}$ is $m + 1$.

$$1 + \frac{m}{m + 1} = \frac{1}{1} + \frac{m}{m + 1}$$

Write 1 as $\frac{1}{1}$.

$$= \frac{1(m + 1)}{1(m + 1)} + \frac{m}{m + 1}$$

Multiply both the numerator and the denominator of $\frac{1}{1}$ by $m + 1$.

$$= \frac{m + 1 + m}{m + 1}$$

Add numerators, and use the common denominator.

$$= \frac{2m + 1}{m + 1}$$

Combine like terms in the numerator.

EXAMPLE 6   Subtract: $\dfrac{3}{2x^2 + x} - \dfrac{2x}{6x + 3}$.

Solution: First, factor the denominators.

$$\frac{3}{2x^2 + x} - \frac{2x}{6x + 3} = \frac{3}{x(2x + 1)} - \frac{2x}{3(2x + 1)}$$

The LCD is $3x(2x + 1)$. Write equivalent expressions with denominators of $3x(2x + 1)$.

$$= \frac{3(3)}{x(2x + 1)(3)} - \frac{2x(x)}{3(2x + 1)(x)}$$

$$= \frac{9 - 2x^2}{3x(2x + 1)} \qquad \text{Subtract numerators, and use the com-mon denominator.}$$

EXAMPLE 7   Add: $\dfrac{2x}{x^2 + 2x + 1} + \dfrac{x}{x^2 - 1}$.

Solution: First, factor the denominators.

$$\frac{2x}{x^2 + 2x + 1} + \frac{x}{x^2 - 1} = \frac{2x}{(x + 1)(x + 1)} + \frac{x}{(x + 1)(x - 1)}$$

Write the rational expressions as equivalent expressions with denominators of $(x + 1)(x + 1)(x - 1)$, the LCD.

$$= \frac{2x(x - 1)}{(x + 1)(x + 1)(x - 1)} + \frac{x(x + 1)}{(x + 1)(x - 1)(x + 1)}$$

$$= \frac{2x(x - 1) + x(x + 1)}{(x + 1)^2(x - 1)} \qquad \text{Add numerators, and use the common denominator.}$$

$$= \frac{2x^2 - 2x + x^2 + x}{(x + 1)^2(x - 1)} \qquad \text{Apply the distributive property in the numerator.}$$

$$= \frac{3x^2 - x}{(x + 1)^2(x - 1)} \qquad \text{or} \qquad \frac{x(3x - 1)}{(x + 1)^2(x - 1)}$$

The numerator was factored as a last step to see if the rational expression could be simplified further.

EXERCISE SET 6.4

Perform the indicated operations. See Example 1.

1. $\dfrac{4}{2x} + \dfrac{9}{3x} \quad \dfrac{5}{x}$

2. $\dfrac{15}{7a} + \dfrac{8}{6a} \quad \dfrac{73}{21a}$

3. $\dfrac{15a}{b} + \dfrac{6b}{5} \quad \dfrac{75a + 6b^2}{5b}$

4. $\dfrac{4c}{d} - \dfrac{8x}{5} \quad \dfrac{20c - 8dx}{5d}$

5. $\dfrac{3}{x} + \dfrac{5}{2x^2} \quad \dfrac{6x + 5}{2x^2}$

6. $\dfrac{14}{3x^2} + \dfrac{6}{x} \quad \dfrac{14 + 18x}{3x^2}$

8. $\dfrac{7}{x + 4}$ **9.** $\dfrac{17x + 30}{2(x - 2)(x + 2)}$ **10.** $\dfrac{3x - 7}{(x - 2)(x + 2)}$ **12.** $\dfrac{x^2 + 2x + 3}{(x + 1)(x - 1)}$ **13.** $\dfrac{5 + 10y - y^3}{y^2(2y + 1)}$ **14.** $\dfrac{2x - 3}{8x - 6}$

Perform the indicated operations. See Examples 2 and 3.

7. $\dfrac{6}{x + 1} + \dfrac{9}{2x + 2} \quad \dfrac{21}{2(x + 1)}$

8. $\dfrac{8}{x + 4} - \dfrac{3}{3x + 12}$

9. $\dfrac{15}{2x - 4} + \dfrac{x}{x^2 - 4}$

10. $\dfrac{3}{x + 2} - \dfrac{1}{x^2 - 4}$

11. $\dfrac{3}{4x} + \dfrac{8}{x - 2} \quad \dfrac{35x - 6}{4x(x - 2)}$

12. $\dfrac{x}{x + 1} + \dfrac{3}{x - 1}$

15. answers may vary   **16.** answers may vary

13. $\dfrac{5}{y^2} - \dfrac{y}{2y+1}$

14. $\dfrac{x}{4x-3} - \dfrac{3}{8x-6}$

15. In your own words, explain how to add two rational expressions with different denominators.

16. In your own words, explain how to subtract two rational expressions with different denominators.

19. $-\dfrac{1}{x^2-1}$ **20.** $-\dfrac{16}{25x^2-1}$ **22.** $\dfrac{4}{x-3}$

Add or subtract as indicated. See Example 4.

17. $\dfrac{6}{x-3} + \dfrac{8}{3-x}$ $-\dfrac{2}{x-3}$ **18.** $\dfrac{9}{x-3} + \dfrac{9}{3-x}$ 0

19. $\dfrac{-8}{x^2-1} - \dfrac{7}{1-x^2}$ **20.** $\dfrac{-9}{25x^2-1} + \dfrac{7}{1-25x^2}$

21. $\dfrac{x}{x^2-4} - \dfrac{2}{4-x^2}$ $\dfrac{1}{x-2}$ **22.** $\dfrac{5}{2x-6} - \dfrac{3}{6-2x}$

Add or subtract as indicated. See Example 5.

23. $\dfrac{5}{x} + 2$ $\dfrac{5+2x}{x}$ **24.** $\dfrac{7}{x^2} - 5x$ $\dfrac{7-5x^3}{x^2}$

25. $\dfrac{5}{x-2} + 6$ $\dfrac{6x-7}{x-2}$ **26.** $\dfrac{6y}{y+5} + 1$ $\dfrac{7y+5}{y+5}$

27. $\dfrac{y+2}{y+3} - 2$ $\dfrac{-y+4}{y+3}$ **28.** $\dfrac{7}{2x-3} - 3$ $\dfrac{16-6x}{2x-3}$

29. Two angles are said to be complementary if their sum is 90°. If one angle measures $\frac{40}{x}$ degrees, find the measure of its complement.

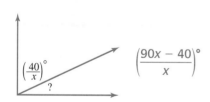

30. Two angles are said to be supplementary if their sum is 180°. If one angle measures $\frac{x+2}{x}$ degrees, find the measure of its supplement.

Perform the indicated operations. See Examples 6 and 7.

31. $\dfrac{5x}{x+2} - \dfrac{3x-4}{x+2}$ 2 **32.** $\dfrac{7x}{x-3} - \dfrac{4x+9}{x-3}$ 3

33. $\dfrac{3x^4}{x} - \dfrac{4x^2}{x^2}$ $3x^3 - 4$ **34.** $\dfrac{5x}{6} + \dfrac{15x^2}{2}$ $\dfrac{5x+45x^2}{6}$

35. $\dfrac{1}{x+3} - \dfrac{1}{(x+3)^2}$ **36.** $\dfrac{5x}{(x-2)^2} - \dfrac{3}{x-2}$

37. $\dfrac{4}{5b} + \dfrac{1}{b-1}$ $\dfrac{9b-4}{5b(b-1)}$ **38.** $\dfrac{1}{y+5} + \dfrac{2}{3y}$

39. $\dfrac{2}{m} + 1$ $\dfrac{2+m}{m}$ **40.** $\dfrac{6}{x} - 1$ $\dfrac{6-x}{x}$

41. $\dfrac{6}{1-2x} - \dfrac{4}{2x-1}$ **42.** $\dfrac{10}{3n-4} - \dfrac{5}{4-3n}$

43. $\dfrac{7}{(x+1)(x-1)} + \dfrac{8}{(x+1)^2}$ $\dfrac{15x-1}{(x+1)^2(x-1)}$

44. $\dfrac{5x+2}{(x+1)(x+5)} - \dfrac{2}{x+5}$ $\dfrac{3x}{(x+1)(x+5)}$

45. $\dfrac{x}{x^2-1} - \dfrac{2}{x^2-2x+1}$ $\dfrac{x^2-3x-2}{(x-1)^2(x+1)}$

46. $\dfrac{x}{x^2-4} - \dfrac{5}{x^2-4x+4}$ $\dfrac{x^2-7x-10}{(x+2)(x-2)^2}$

47. $\dfrac{3a}{2a+6} - \dfrac{a-1}{a+3}$ **48.** $\dfrac{1}{x+y} - \dfrac{y}{x^2-y^2}$

49. $\dfrac{5}{2-x} + \dfrac{x}{2x-4}$ $\dfrac{x-10}{2(x-2)}$ **50.** $\dfrac{-1}{a-2} + \dfrac{4}{4-2a}$

51. $\dfrac{-7}{y^2-3y+2} - \dfrac{2}{y-1}$ **52.** $\dfrac{2}{x^2+4x+4} + \dfrac{1}{x+2}$

53. $\dfrac{13}{x^2-5x+6} - \dfrac{5}{x-3}$ **54.** $\dfrac{27}{y^2-81} + \dfrac{3}{2(y+9)}$

55. $\dfrac{8}{(x+2)(x-2)} + \dfrac{4}{(x+2)(x-3)}$

56. $\dfrac{5}{6x^2(x+2)} + \dfrac{4x}{x(x+2)^2}$ $\dfrac{24x^2+5x+10}{6x^2(x+2)^2}$

57. $\dfrac{5}{9x^2-4} + \dfrac{2}{3x-2}$ $\dfrac{6x+9}{(3x-2)(3x+2)}$

58. $\dfrac{4}{x^2-x-6} + \dfrac{x}{x^2+5x+6}$ $\dfrac{x^2+x+12}{(x-3)(x+3)(x+2)}$

59. $\dfrac{x+8}{x^2-5x-6} + \dfrac{x+1}{x^2-4x-5}$ $\dfrac{2x^2-2x-46}{(x+1)(x-6)(x-5)}$

35. $\dfrac{x+2}{(x+3)^2}$ **36.** $\dfrac{2x+6}{(x-2)^2}$ **38.** $\dfrac{5y+10}{3y(y+5)}$ **41.** $\dfrac{10}{1-2x}$ **42.** $\dfrac{15}{3n-4}$ **47.** $\dfrac{a+2}{2(a+3)}$ **48.** $\dfrac{x-2y}{(x+y)(x-y)}$

50. $\dfrac{3}{a-2}$ **51.** $\dfrac{-2y-3}{(y-1)(y-2)}$ **52.** $\dfrac{x+4}{(x+2)^2}$ **53.** $\dfrac{-5x+23}{(x-3)(x-2)}$ **54.** $\dfrac{3}{2(y-9)}$ **55.** $\dfrac{12x-32}{(x+2)(x-2)(x-3)}$

60. $\dfrac{x^2 - 3x - 2}{(x + 10)(x + 2)(x - 2)}$ **64.** $\dfrac{z}{3(9z - 5)}$ **67.** $\dfrac{25a}{9(a - 2)}$ **68.** $\dfrac{9}{4(x - 1)}$ **73.** $(x - 1)(x^2 + x + 1)$

60. $\dfrac{x}{x^2 + 12x + 20} - \dfrac{1}{x^2 + 8x - 20}$

61. A board of length $\frac{3}{x + 4}$ inches was cut into two pieces. If one piece is $\frac{1}{x - 4}$ inches, express the length of the other board as a rational expression.

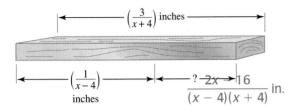

$\dfrac{2x + 16}{(x - 4)(x + 4)}$ in.

62. The length of the rectangle is $\frac{3}{y - 5}$ feet, while its width is $\frac{2}{y}$ feet. Find its perimeter and then find its area.

$\dfrac{3}{y-5}$ feet $\dfrac{10y - 20}{y(y - 5)}$ ft.; $\dfrac{6}{y^2 - 5y}$ sq. ft.

$\dfrac{2}{y}$ feet

Perform the indicated operations. Addition, subtraction, multiplication, and division of rational expressions are included here.

63. $\dfrac{15x}{x + 8} \cdot \dfrac{2x + 16}{3x}$ 10

64. $\dfrac{9z + 5}{15} \cdot \dfrac{5z}{81z^2 - 25}$

65. $\dfrac{8x + 7}{3x + 5} - \dfrac{2x - 3}{3x + 5}$ 2

66. $\dfrac{2z^2}{4z - 1} - \dfrac{z - 2z^2}{4z - 1}$ z

67. $\dfrac{5a + 10}{18} \div \dfrac{a^2 - 4}{10a}$

68. $\dfrac{9}{x^2 - 1} \div \dfrac{12}{3x + 3}$

69. $\dfrac{5}{x^2 - 3x + 2} + \dfrac{1}{x - 2}$ $\dfrac{x + 4}{(x - 2)(x - 1)}$

70. $\dfrac{4}{2x^2 + 5x - 3} + \dfrac{2}{x + 3}$ $\dfrac{4x + 2}{(2x - 1)(x + 3)}$

71. Explain when the LCD is the product of the denominators. answers may vary

72. Explain when the LCD is the same as one of the denominators of a rational expression to be added or subtracted. answers may vary

Review Exercises

Factor the following. See Sections 5.1 and 5.4.

73. $x^3 - 1$ **74.** $8y^3 + 1$

75. $125z^3 + 8$ **76.** $a^3 - 27$

77. $xy + 2x + 3y + 6$ **78.** $x^2 - x + xy - y$

Find the slope of each line. See Section 3.4.

79.
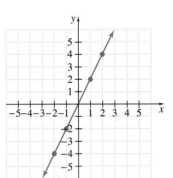
$m = 2$

80.
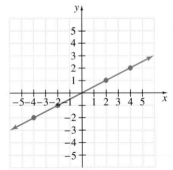
$m = \dfrac{1}{2}$

81.
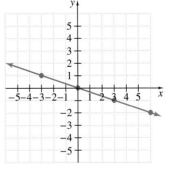
$m = -\dfrac{1}{3}$

74. $(2y + 1)(4y^2 - 2y + 1)$ **75.** $(5z + 2)(25z^2 - 10z + 4)$ **76.** $(a - 3)(a^2 + 3a + 9)$ **77.** $(x + 3)(y + 2)$

78. $(x - 1)(x + y)$

82.

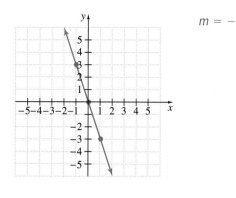

$m = -3$

Write each expression with an LCD of $(x - 4)(x + 4)(x - 5)$.

$$= \frac{3(x - 5)}{(x - 4)(x + 4)(x - 5)} + \frac{2(x + 4)}{(x - 4)(x + 4)(x - 5)}$$

$$- \frac{4(x - 4)}{(x - 4)(x + 4)(x - 5)}$$

$$= \frac{3(x - 5) + 2(x + 4) - 4(x - 4)}{(x - 4)(x + 4)(x - 5)}$$

$$= \frac{3x - 15 + 2x + 8 - 4x + 16}{(x - 4)(x + 4)(x - 5)}$$

$$= \frac{x + 9}{(x - 4)(x + 4)(x - 5)}$$

A Look Ahead

EXAMPLE

Perform the indicated operations:

$$\frac{3}{x^2 - 16} + \frac{2}{x^2 - 9x + 20} - \frac{4}{x^2 - x - 20}$$

Solution:

Factor the denominators.

$$= \frac{3}{(x + 4)(x - 4)} + \frac{2}{(x - 4)(x - 5)} - \frac{4}{(x - 5)(x + 4)}$$

Add or subtract as indicated.

83. $\dfrac{5}{x^2 - 4} + \dfrac{2}{x^2 - 4x + 4} - \dfrac{3}{x^2 - x - 6}$

84. $\dfrac{8}{x^2 + 6x + 5} - \dfrac{3x}{x^2 + 4x - 5} + \dfrac{2}{x^2 - 1}$

85. $\dfrac{9}{x^2 + 9x + 14} - \dfrac{3x}{x^2 + 10x + 21} + \dfrac{4}{x^2 + 5x + 6}$

86. $\dfrac{10}{x^2 - 3x - 4} - \dfrac{8}{x^2 + 6x + 5} - \dfrac{9}{x^2 + x - 20}$

87. $\dfrac{5 + x}{x^3 - 27} + \dfrac{x}{x^3 + 3x^2 + 9x}$ $\dfrac{2(x + 1)}{(x - 3)(x^2 + 3x + 9)}$

88. $\dfrac{x + 5}{x^3 + 1} - \dfrac{3}{2x^2 - 2x + 2}$ $\dfrac{-x + 7}{2(x + 1)(x^2 - x + 1)}$

83. $\dfrac{4x^2 - 15x + 6}{(x - 2)^2(x + 2)(x - 3)}$ **84.** $\dfrac{-3x^2 + 7x + 2}{(x + 5)(x + 1)(x - 1)}$

85. $\dfrac{-3x^2 + 7x + 55}{(x + 2)(x + 7)(x + 3)}$ **86.** $\dfrac{-7x + 73}{(x - 4)(x + 1)(x + 5)}$

6.5 | SIMPLIFYING COMPLEX FRACTIONS

O B J E C T I V E S

1. Define complex fractions.
2. Simplify complex fractions using method 1.
3. Simplify complex fractions using method 2.

TAPE
BA 6.5

A rational expression whose numerator or denominator or both numerator and denominator contain fractions is called a **complex rational expression** or a **complex fraction**. Our goal in this section is to write complex fractions such as the ones shown next in simplest form. A complex fraction is in simplest form when it is in the form P/Q, where P and Q are polynomials that have no common factors.

EXAMPLES OF COMPLEX FRACTIONS

$$\dfrac{4}{2 - \dfrac{1}{2}} \qquad \dfrac{\dfrac{3}{2}}{\dfrac{4}{7} - x} \qquad \dfrac{\dfrac{1}{x+2}}{x+2 - \dfrac{1}{x}}$$

The parts of a complex fraction are

$$\dfrac{\dfrac{x}{y+2}}{7 + \dfrac{1}{y}} \left. \begin{array}{l} \\ \end{array} \right\} \leftarrow \text{Numerator of complex fraction.}$$

$\leftarrow$ Main fraction bar.

$\leftarrow$ Denominator of complex fraction.

2 In this section, two methods of simplifying complex fractions are represented. The first method presented uses the fact that the main fraction bar indicates division.

METHOD 1: TO SIMPLIFY A COMPLEX FRACTION

Step 1. Add or subtract fractions in the numerator or denominator so that the numerator is a single fraction and the denominator is a single fraction.

Step 2. Perform the indicated division by multiplying the numerator of the complex fraction by the reciprocal of the denominator of the complex fraction.

Step 3. Write the rational expression in lowest terms.

EXAMPLE 1 Simplify the complex fraction $\dfrac{\dfrac{5}{8}}{\dfrac{2}{3}}$.

Solution: Since the numerator and denominator of the complex fraction are already single fractions, proceed to step 2: perform the indicated division by multiplying the numerator $\dfrac{5}{8}$ by the reciprocal of the denominator $\dfrac{2}{3}$.

$$\dfrac{\dfrac{5}{8}}{\dfrac{2}{3}} = \dfrac{5}{8} \cdot \dfrac{3}{2} \qquad \text{The reciprocal of } \dfrac{2}{3} \text{ is } \dfrac{3}{2}.$$

$$= \dfrac{5 \cdot 3}{8 \cdot 2} = \dfrac{15}{16}$$

EXAMPLE 2   Simplify: $\dfrac{\dfrac{2}{3} + \dfrac{1}{5}}{\dfrac{2}{3} - \dfrac{2}{9}}.$

Solution: Simplify above and below the main fraction bar separately. First, add $\dfrac{2}{3}$ and $\dfrac{1}{5}$ to obtain a single fraction in the numerator; then subtract $\dfrac{2}{9}$ from $\dfrac{2}{3}$ to obtain a single fraction in the denominator.

$$\dfrac{\dfrac{2}{3} + \dfrac{1}{5}}{\dfrac{2}{3} - \dfrac{2}{9}} = \dfrac{\dfrac{2(5)}{3(5)} + \dfrac{1(3)}{5(3)}}{\dfrac{2(3)}{3(3)} - \dfrac{2}{9}}$$ The LCD of the numerator's fractions is 15.

The LCD of the denominator's fractions is 9.

$$= \dfrac{\dfrac{10}{15} + \dfrac{3}{15}}{\dfrac{6}{9} - \dfrac{2}{9}}$$ Simplify.

$$= \dfrac{\dfrac{13}{15}}{\dfrac{4}{9}}$$ Add the numerator's fractions.

Subtract the denominator's fractions.

Next, perform the indicated division by multiplying the numerator of the complex fraction by the reciprocal of the denominator of the complex fraction.

$$\dfrac{\dfrac{13}{15}}{\dfrac{4}{9}} = \dfrac{13}{15} \cdot \dfrac{9}{4}$$ The reciprocal of $\dfrac{4}{9}$ is $\dfrac{9}{4}$.

$$= \dfrac{13 \cdot 3 \cdot \boxed{3}}{\boxed{3} \cdot 5 \cdot 4} = \dfrac{39}{20}$$

EXAMPLE 3   Simplify: $\dfrac{\dfrac{x + 1}{y}}{\dfrac{x}{y} + 2}.$

Solution: The numerator of this complex fraction is already a single fraction. For the denominator, write 2 as $\dfrac{2}{1}$ and add $\dfrac{x}{y}$ and $\dfrac{2}{1}$ to get a single fraction.

$$\dfrac{\dfrac{x + 1}{y}}{\dfrac{x}{y} + \dfrac{2}{1}} = \dfrac{\dfrac{x + 1}{y}}{\dfrac{x}{y} + \dfrac{2(y)}{1(y)}}$$ The LCD of $\dfrac{x}{y}$ and $\dfrac{2}{1}$ is y.

$$= \frac{\dfrac{x+1}{y}}{\dfrac{x+2y}{y}}$$ Add the fractions in the denominator.

$$= \frac{x+1}{y} \cdot \frac{y}{x+2y}$$ Multiply by the reciprocal of $\dfrac{x+2y}{y}$.

$$= \frac{y(x+1)}{y(x+2y)}$$

$$= \frac{x+1}{x+2y}$$ Apply the fundamental principle to write in lowest terms.

EXAMPLE 4 Simplify: $\dfrac{\dfrac{1}{z} - \dfrac{1}{2}}{\dfrac{1}{3} - \dfrac{z}{6}}$.

Solution: Subtract to get a single fraction in the numerator and a single fraction in the denominator of the complex fraction.

$$\frac{\dfrac{1}{z} - \dfrac{1}{2}}{\dfrac{1}{3} - \dfrac{z}{6}} = \frac{\dfrac{2}{2z} - \dfrac{z}{2z}}{\dfrac{2}{6} - \dfrac{z}{6}}$$ The LCD of the numerator's fractions is $2z$.

The LCD of the denominator's fractions is 6.

$$= \frac{\dfrac{2-z}{2z}}{\dfrac{2-z}{6}}$$

$$= \frac{2-z}{2z} \cdot \frac{6}{2-z}$$ Multiply by the reciprocal of $\dfrac{2-z}{6}$.

$$= \frac{2 \cdot 3 \cdot (2-z)}{2 \cdot z \cdot (2-z)}$$ Factor.

$$= \frac{3}{z}$$ Write in lowest terms.

Next we study a second method for simplifying complex fractions. In this method, we multiply the numerator and the denominator of the complex fraction by the LCD of all fractions in the complex fraction.

Method 2: To Simplify a Complex Fraction

Step 1. Find the LCD of all the fractions in the complex fraction.

Step 2. Multiply both the numerator and the denominator of the complex fraction by the LCD from *step 1*.

Step 3. Perform the indicated operations and write in lowest terms.

We use method 2 to rework Example 2.

EXAMPLE 5   Simplify: $\dfrac{\dfrac{2}{3} + \dfrac{1}{5}}{\dfrac{2}{3} - \dfrac{2}{9}}$.

Solution: The LCD of $\dfrac{2}{3}, \dfrac{1}{5}, \dfrac{2}{3}$, and $\dfrac{2}{9}$ is 45. Multiply the numerator and the denominator of the complex fraction by 45, perform the indicated operations, and write in lowest terms.

$$\frac{\dfrac{2}{3} + \dfrac{1}{5}}{\dfrac{2}{3} - \dfrac{2}{9}} = \frac{45\left(\dfrac{2}{3} + \dfrac{1}{5}\right)}{45\left(\dfrac{2}{3} - \dfrac{2}{9}\right)}$$

$$= \frac{45\left(\dfrac{2}{3}\right) + 45\left(\dfrac{1}{5}\right)}{45\left(\dfrac{2}{3}\right) - 45\left(\dfrac{2}{9}\right)}$$ Apply the distributive property.

$$= \frac{30 + 9}{30 - 10} = \frac{39}{20}$$ Simplify.

EXAMPLE 6   Simplify $\dfrac{\dfrac{x}{y} + \dfrac{3}{2x}}{\dfrac{x}{2} + y}$ using method 2.

Solution: The LCD of $\dfrac{x}{y}, \dfrac{3}{2x}, \dfrac{x}{2}$, and $\dfrac{y}{1}$ is $2xy$. Multiply both the numerator and the denominator of the complex fraction by $2xy$.

$$\frac{\dfrac{x}{y} + \dfrac{3}{2x}}{\dfrac{x}{2} + y} = \frac{2xy\left(\dfrac{x}{y} + \dfrac{3}{2x}\right)}{2xy\left(\dfrac{x}{2} + y\right)}$$

$$= \frac{2xy\left(\dfrac{x}{y}\right) + 2xy\left(\dfrac{3}{2x}\right)}{2xy\left(\dfrac{x}{2}\right) + 2xy(y)}$$ Apply the distributive property.

$$= \frac{2x^2 + 3y}{x^2y + 2xy^2}$$

EXERCISE SET 6.5

Simplify each complex fraction. See Example 1.

1. $\dfrac{\frac{1}{2}}{\frac{3}{4}}$ $\quad \frac{2}{3}$

2. $\dfrac{\frac{1}{8}}{-\frac{5}{12}}$ $\quad -\frac{3}{10}$

3. $\dfrac{-\frac{4x}{9}}{-\frac{2x}{3}}$ $\quad \frac{2}{3}$

4. $\dfrac{-\frac{6y}{11}}{\frac{4y}{9}}$ $\quad -\frac{27}{22}$

5. $\dfrac{\frac{1+x}{6}}{\frac{1+x}{3}}$ $\quad \frac{1}{2}$

6. $\dfrac{\frac{6x-3}{5x^2}}{\frac{2x-1}{10x}}$ $\quad \frac{6}{x}$

7. $\dfrac{\frac{(y+1)(y-1)}{6}}{\frac{(y+1)(y+2)}{8}}$ $\quad \frac{4(y-1)}{3(y+2)}$

8. $\dfrac{\frac{(y+4)(2y-1)}{9}}{\frac{(y+3)(2y-1)}{18}}$

9. If the distance formula $d = r \cdot t$ is solved for t, then $t = \frac{d}{r}$. Use this formula to find t if distance d is $\frac{20x}{3}$ miles and rate r is $\frac{5x}{9}$ miles per hour. Write t in simplified form.   12 hr

10. If the formula for area of a rectangle, $A = l \cdot w$, is solved for w, then $w = \frac{A}{l}$. Use this formula to find w if area A is $\frac{4x-2}{3}$ square meters and length l is $\frac{6x-3}{5}$ meters. Write w in simplified form. $\frac{10}{9}$ m

Simplify each complex fraction. See Examples 2 through 6 and use either method.

11. $\dfrac{\frac{1}{2}+\frac{2}{3}}{\frac{5}{9}-\frac{5}{6}}$ $\quad -\frac{21}{5}$

12. $\dfrac{\frac{3}{4}-\frac{1}{2}}{\frac{3}{8}+\frac{1}{6}}$ $\quad \frac{6}{13}$

13. $\dfrac{2+\frac{7}{10}}{1+\frac{3}{5}}$ $\quad \frac{27}{16}$

14. $\dfrac{4-\frac{11}{12}}{5+\frac{1}{4}}$ $\quad \frac{37}{63}$

15. $\dfrac{\frac{1}{3}}{\frac{1}{2}-\frac{1}{4}}$ $\quad \frac{4}{3}$

16. $\dfrac{\frac{7}{10}-\frac{3}{5}}{\frac{1}{2}}$ $\quad \frac{1}{5}$

17. $\dfrac{-\frac{2}{9}}{-\frac{14}{3}}$ $\quad \frac{1}{21}$

18. $\dfrac{\frac{3}{8}}{\frac{4}{15}}$ $\quad \frac{45}{32}$

19. $\dfrac{-\frac{5}{12x^2}}{\frac{25}{16x^3}}$ $\quad -\frac{4x}{15}$

20. $\dfrac{-\frac{7}{8y}}{\frac{21}{4y}}$ $\quad -\frac{1}{6}$

21. $\dfrac{\frac{m}{n}-1}{\frac{m}{n}+1}$ $\quad \frac{m-n}{m+n}$

22. $\dfrac{\frac{x}{2}+2}{\frac{x}{2}-2}$ $\quad \frac{x+4}{x-4}$

23. $\dfrac{\frac{1}{5}-\frac{1}{x}}{\frac{7}{10}+\frac{1}{x^2}}$ $\quad \frac{2x(x-5)}{7x^2+10}$

24. $\dfrac{\frac{1}{y^2}+\frac{2}{3}}{\frac{1}{y}-\frac{5}{6}}$ $\quad \frac{6+4y^2}{6y-5y^2}$

25. $\dfrac{1+\frac{1}{y-2}}{y+\frac{1}{y-2}}$ $\quad \frac{1}{y-1}$

26. $\dfrac{x-\frac{1}{2x+1}}{1-\frac{x}{2x+1}}$

27. $\dfrac{\frac{4y-8}{16}}{\frac{6y-12}{4}}$ $\quad \frac{1}{6}$

28. $\dfrac{\frac{7y+21}{3}}{\frac{3y+9}{8}}$ $\quad \frac{56}{9}$

29. $\dfrac{\frac{x}{y}+1}{\frac{x}{y}-1}$ $\quad \frac{x+y}{x-y}$

30. $\dfrac{\frac{3}{5y}+8}{\frac{3}{5y}-8}$ $\quad \frac{3+40y}{3-40y}$

31. $\dfrac{1}{2+\frac{1}{3}}$ $\quad \frac{3}{7}$

32. $\dfrac{3}{1-\frac{4}{3}}$ $\quad -9$

33. $\dfrac{\frac{ax+ab}{x^2-b^2}}{\frac{x+b}{x-b}}$ $\quad \frac{a}{x+b}$

34. $\dfrac{\frac{m+2}{m-2}}{\frac{2m+4}{m^2-4}}$ $\quad \frac{m+2}{2}$

35. $\dfrac{\frac{-3+y}{4}}{\frac{8+y}{28}}$ $\quad \frac{7(y-3)}{8+y}$

36. $\dfrac{\frac{-x+2}{18}}{\frac{8}{9}}$ $\quad \frac{-x+2}{16}$

8. $\dfrac{2y+8}{y+3}$ 26. $\dfrac{2x^2+x-1}{x+1}$

37. $\dfrac{3 + \dfrac{12}{x}}{1 - \dfrac{16}{x^2}}$ $\dfrac{3x}{x-4}$

38. $\dfrac{\dfrac{5}{6} + \dfrac{1}{2}}{\dfrac{1}{5} - 4}$ $-\dfrac{20}{57}$

39. $\dfrac{2 + \dfrac{6}{x}}{1 - \dfrac{9}{x^2}}$ $\dfrac{2x}{x-3}$

40. $\dfrac{\dfrac{3}{8} + 2}{\dfrac{5}{4} - \dfrac{1}{3}}$ $\dfrac{57}{22}$

41. $\dfrac{\dfrac{8}{x+4} + 2}{\dfrac{12}{x+4} - 2}$ $-\dfrac{x+8}{x-2}$

42. $\dfrac{\dfrac{25}{x+5} + 5}{\dfrac{3}{x+5} - 5}$ $-\dfrac{5x+50}{5x+22}$

43. $\dfrac{\dfrac{s}{r} + \dfrac{r}{s}}{\dfrac{s}{r} - \dfrac{r}{s}}$ $\dfrac{s^2 + r^2}{s^2 - r^2}$

44. $\dfrac{\dfrac{2}{x} + \dfrac{x}{2}}{\dfrac{2}{x} - \dfrac{x}{2}}$ $\dfrac{4 + x^2}{4 - x^2}$

45. Explain how to simplify a complex fraction using method 1.   answers may vary

46. Explain how to simplify a complex fraction using method 2.   answers may vary

To find the average of two numbers, we find their sum and divide by 2. For example, the average of 65 and 81 is found by simplifying $\frac{65 + 81}{2}$. This simplifies to $\frac{146}{2} = 73$.

47. Find the average of $\frac{1}{3}$ and $\frac{3}{4}$.

48. Write the average of $\frac{3}{n}$ and $\frac{5}{n^2}$ as a simplified rational expression.

49. In electronics, when two resistors R_1 (read R sub 1) and R_2 (read R sub 2) are connected in parallel, the total resistance is given by the complex fraction
$\dfrac{1}{\dfrac{1}{R_1} + \dfrac{1}{R_2}}$. Simplify this expression.

Resistance R_1 R_2

50. Astronomers occasionally need to know the day of the week a particular date fell on. The complex fraction

$$\dfrac{J + \dfrac{3}{2}}{7},$$

where J is the *Julian day number,* is used to make this calculation. Simplify this expression.

Review Exercises

Solve the following linear and quadratic equations. See Sections 2.4 and 5.6.

51. $3x + 5 = 7$ $x = \dfrac{2}{3}$

52. $5x - 1 = 8$ $x = \dfrac{9}{5}$

53. $2x^2 - x - 1 = 0$

54. $4x^2 - 9 = 0$

Simplify the following rational expressions. See Section 6.1.

55. $\dfrac{2 + x}{x + 2}$ 1

56. $\dfrac{y^2 + y}{y + y^2}$ 1

57. $\dfrac{2 - x}{x - 2}$ -1

58. $\dfrac{z - 4}{4 - z}$ -1

A Look Ahead

EXAMPLE

Simplify $\dfrac{1 + x^{-1}}{x^{-1}}$.

Solution:

$$\dfrac{1 + x^{-1}}{x^{-1}} = \dfrac{1 + \dfrac{1}{x}}{\dfrac{1}{x}}$$

$$= \dfrac{x\left(1 + \dfrac{1}{x}\right)}{x\left(\dfrac{1}{x}\right)}$$

$$= \dfrac{x + 1}{1} \text{ or } x + 1$$

Simplify the following.

59. $\dfrac{x^{-1} + 2^{-1}}{x^{-2} - 4^{-1}}$ $\dfrac{2x}{2 - x}$

60. $\dfrac{3^{-1} - x^{-1}}{9^{-1} - x^{-2}}$ $\dfrac{3x}{x - 3}$

61. $\dfrac{x + y^{-1}}{\dfrac{x}{y}}$ $\dfrac{xy + 1}{x}$

62. $\dfrac{x - xy^{-1}}{\dfrac{1 + x}{y}}$ $\dfrac{xy - x}{1 + x}$

63. $\dfrac{y^{-2}}{1 - y^{-2}}$ $\dfrac{1}{y^2 - 1}$

64. $\dfrac{4 + x^{-1}}{3 + x^{-1}}$ $\dfrac{4x + 1}{3x + 1}$

47. $\frac{13}{24}$ **48.** $\frac{3n + 5}{2n^2}$ **49.** $\frac{R_1 R_2}{R_2 + R_1}$ **50.** $\dfrac{2J + 3}{14}$ **53.** $x = -\dfrac{1}{2}, x = 1$ **54.** $x = \dfrac{3}{2}, x = -\dfrac{3}{2}$

6.6 | SOLVING EQUATIONS CONTAINING RATIONAL EXPRESSIONS

TAPE
BA 6.6

OBJECTIVES

 Solve equations containing rational expressions.

2 Understand the difference between solving an equation and performing arithmetic operations on rational expressions.

3 Solve equations containing rational expressions for a specified variable.

1 In Chapter 2, we solved equations containing fractions. In this section, we continue the work we began in Chapter 2 by solving equations containing rational expressions.

EXAMPLES OF EQUATIONS CONTAINING RATIONAL EXPRESSIONS

$$\frac{x}{5} + \frac{x + 2}{9} = 8 \quad \text{and} \quad \frac{x + 1}{9x - 5} = \frac{2}{3x}$$

To solve, use the multiplication property of equality to clear the equation of fractions by multiplying both sides of the equation by the LCD.

TO SOLVE AN EQUATION CONTAINING RATIONAL EXPRESSIONS

Step 1. Multiply both sides of the equation by the LCD of all rational expressions in the equation.

Step 2. Remove any grouping symbols and solve the resulting equation.

Step 3. Check the solution in the original equation.

EXAMPLE 1 Solve: $\dfrac{t - 4}{2} - \dfrac{t - 3}{9} = \dfrac{5}{18}$.

Solution: The LCD of denominators 2, 9, and 18 is 18. Multiply both sides of the equation by 18.

$$18\left(\frac{t - 4}{2} - \frac{t - 3}{9}\right) = 18\left(\frac{5}{18}\right)$$

$$18\left(\frac{t - 4}{2}\right) - 18\left(\frac{t - 3}{9}\right) = 18\left(\frac{5}{18}\right) \qquad \text{Apply the distributive property.}$$

$$9(t - 4) - 2(t - 3) = 5 \qquad \text{Simplify.}$$

$$9t - 36 - 2t + 6 = 5 \qquad \text{Use the distributive property.}$$

$$7t - 30 = 5 \qquad \text{Combine like terms.}$$

$$7t = 35 \quad \text{or} \quad t = 5 \qquad \text{Solve for } t.$$

To check, substitute 5 for t in the original equation.

$$\frac{t-4}{2} - \frac{t-3}{9} = \frac{5}{18}$$

$$\frac{5-4}{2} - \frac{5-3}{9} = \frac{5}{18} \qquad \text{Replace } t \text{ with 5.}$$

$$\frac{1}{2} - \frac{2}{9} = \frac{5}{18} \qquad \text{Simplify.}$$

Next, subtract the fractions on the left by writing each with a denominator of 18.

$$\frac{1(9)}{2(9)} - \frac{2(2)}{9(2)} = \frac{5}{18}$$

$$\frac{9-4}{18} = \frac{5}{18}$$

$$\frac{5}{18} = \frac{5}{18} \qquad \text{True.}$$

Since the statement is true, 5 is the solution and the solution set is {5}.   ▬▬▬

Recall from Section 6.1 that a rational expression is defined for all real numbers except those that make the denominator of the expression 0. This means that if an equation contains rational expressions with variables in the denominator, we must be certain that the proposed solution does not make the denominator 0. If replacing the variable with the proposed solution makes the denominator 0, the rational expression is undefined and this proposed solution must be rejected. It is called an **extraneous solution.**

EXAMPLE 2   Solve the equation $3 - \dfrac{6}{x} = x + 8$.

Solution: In this equation, 0 cannot be a solution because if x is 0, the rational expression $\dfrac{6}{x}$ is undefined. The LCD is x. Multiply both sides of the equation by x.

$$x\left(3 - \frac{6}{x}\right) = x(x+8)$$

$$x(3) - x\left(\frac{6}{x}\right) = x \cdot x + x \cdot 8 \qquad \text{Apply the distributive property.}$$

$$3x - 6 = x^2 + 8x \qquad \text{Simplify.}$$

Write the quadratic equation in standard form and solve for x.

$$0 = x^2 + 5x + 6$$

$$0 = (x+3)(x+2) \qquad \qquad \text{Factor.}$$

$$x + 3 = 0 \quad \text{or} \quad x + 2 = 0 \qquad \text{Set each factor equal to 0 and solve.}$$

$$x = -3 \quad \text{or} \qquad x = -2$$

Notice that neither -3 nor -2 makes the denominator in the original equation equal to 0. To check these solutions, replace x in the original equation by -3, and then by -2. This will verify that both -3 and -2 are solutions. The solution set is $\{-3, -2\}$.

EXAMPLE 3 Solve the equation $\dfrac{x+1}{x+3} = \dfrac{1}{2x}$.

Solution: Neither -3 nor 0 can be a solution of this equation since each makes a denominator in this equation 0.

Multiply both sides of the equation by the LCD $2x(x+3)$.

$$2x(x+3)\left(\frac{x+1}{x+3}\right) = 2x(x+3)\left(\frac{1}{2x}\right)$$

$$2x(x+1) = (x+3)(1) \qquad \text{Simplify.}$$

$$2x^2 + 2x = x + 3 \qquad \text{Apply the distributive property.}$$

$$2x^2 + x - 3 = 0. \qquad \text{Write the quadratic equation in the standard form.}$$

$$(2x+3)(x-1) = 0 \qquad \text{Factor.}$$

$$2x+3 = 0 \quad \text{or} \quad x-1 = 0 \qquad \text{Set each factor equal to zero.}$$

$$x = -\frac{3}{2} \quad \text{or} \qquad x = 1 \qquad \text{Solve.}$$

Check to see that $-\dfrac{3}{2}$ and 1 are both solutions. The solution set is $\left\{-\dfrac{3}{2}, 1\right\}$.

EXAMPLE 4 Solve the equation $x + \dfrac{14}{x-2} = \dfrac{7x}{x-2} + 1$.

Solution: In this equation, 2 can't be a solution. The LCD is $x - 2$. Multiply both sides of the equation by $x - 2$.

$$(x-2)\left(x + \frac{14}{x-2}\right) = (x-2)\left(\frac{7x}{x-2} + 1\right)$$

$$(x-2)(x) + (x-2)\left(\frac{14}{x-2}\right) = (x-2)\left(\frac{7x}{x-2}\right) + (x-2)(1)$$

$$x^2 - 2x + 14 = 7x + x - 2 \qquad \text{Simplify.}$$

$$x^2 - 2x + 14 = 8x - 2 \qquad \text{Combine like terms.}$$

$$x^2 - 10x + 16 = 0 \qquad \text{Write the quadratic equation in standard form.}$$

$$(x-8)(x-2) = 0 \qquad \text{Factor.}$$

$$x-8 = 0 \quad \text{or} \quad x-2 = 0 \qquad \text{Set each factor equal to 0.}$$

$$x = 8 \quad \text{or} \qquad x = 2 \qquad \text{Solve.}$$

As we stated earlier, 2 can't be a solution of the original equation. Replacing x with 8 in the original equation, we find that 8 is a solution. The only solution is 8 and the solution set is $\{8\}$.

If an equation contains rational expressions with variables in the denominator, make sure that you identify the values for which these expressions are not defined. These values cannot be solutions of the equation.

EXAMPLE 5   Solve the equation $\dfrac{3a}{3a-2} - \dfrac{5}{3a^2+7a-6} = 1$.

Solution: Since $3a^2 + 7a - 6 = (3a - 2)(a + 3)$, both $\dfrac{2}{3}$ and -3 cannot be solutions. The LCD is $(3a - 2)(a + 3)$, so we multiply both sides of the equation by the LCD.

$$(3a-2)(a+3)\left(\dfrac{3a}{3a-2} - \dfrac{5}{(3a-2)(a+3)}\right) = (3a-2)(a+3)(1)$$

$$(3a-2)(a+3)\left(\dfrac{3a}{3a-2}\right) - (3a-2)(a+3)\left(\dfrac{5}{(3a-2)(a+3)}\right)$$
$$= (3a-2)(a+3)(1)$$

$$3a(a+3) - 5 = (3a-2)(a+3) \qquad \text{Simplify.}$$
$$3a^2 + 9a - 5 = 3a^2 + 7a - 6 \qquad \text{Apply the distributive property.}$$
$$9a - 5 = 7a - 6 \qquad \text{Subtract } 3a^2 \text{ from both sides.}$$
$$2a = -1 \qquad \begin{array}{l}\text{Subtract } 7a \text{ from both sides and}\\ \text{add 5 to both sides.}\end{array}$$

$$a = -\dfrac{1}{2} \qquad \text{Divide both sides by 2.}$$

Check that $-\dfrac{1}{2}$ is the solution. The solution set is $\left\{-\dfrac{1}{2}\right\}$.

2   At this point, let's make sure you understand the difference between solving an equation containing rational expressions and performing operations on rational expressions.

EXAMPLE 6   **a.** Solve for x: $\dfrac{x}{4} + 2x = 9$.   **b.** Add: $\dfrac{x}{4} + 2x$.

Solution: **a.** This is an equation to solve for x. Begin by multiplying both sides by the LCD, 4.

$$4\left(\dfrac{x}{4} + 2x\right) = 4(9)$$

$$4\left(\dfrac{x}{4}\right) + 4(2x) = 4(9) \qquad \text{Apply the distributive property.}$$

$$x + 8x = 36 \qquad \text{Simplify.}$$
$$9x = 36 \qquad \text{Combine like terms.}$$
$$x = 4 \qquad \text{Solve.}$$

Check to see that 4 is the solution and the solution set is $\{4\}$.

b. This example is **not an equation** to solve; it is an addition to perform. To add these rational expressions, find the LCD and write each rational expression as an equivalent expression whose denominator is the LCD. The LCD is 4.

$$\frac{x}{4} + 2x = \frac{x}{4} + \frac{2x(4)}{4}$$

$$= \frac{x + 8x}{4} \qquad \text{Add.}$$

$$= \frac{9x}{4} \qquad \text{Combine like terms in the numerator.}$$

3   The last example in this section is an equation containing several variables. We are directed to solve for one of them. The steps used in the preceeding examples can be applied to solve equations for a specified variable as well.

EXAMPLE 7   Solve for x: $\dfrac{2x}{a} - 5 = \dfrac{3x}{b} + a$.

Solution: The LCD is ab. Multiply both sides of the equation by ab.

$$ab\left(\frac{2x}{a} - 5\right) = ab\left(\frac{3x}{b} + a\right)$$

$$ab\left(\frac{2x}{a}\right) - ab(5) = ab\left(\frac{3x}{b}\right) + ab(a) \qquad \text{Apply the distributive property.}$$

$$2xb - 5ab = 3xa + a^2b \qquad \text{Simplify.}$$

Next, write the equation so that all terms containing the variable x appear on one side of the equation. To do this, subtract $3xa$ from both sides and add $5ab$ to both sides.

$$2xb - 3xa = a^2b + 5ab$$

$$x(2b - 3a) = a^2b + 5ab \qquad \text{Factor out } x \text{ from each term on the left side.}$$

$$\frac{x(2b - 3a)}{2b - 3a} = \frac{a^2b + 5ab}{2b - 3a} \qquad \text{Divide both sides by } 2b - 3a.$$

$$x = \frac{a^2b + 5ab}{2b - 3a} \qquad \text{Simplify.}$$

MENTAL MATH

Solve each equation for the variable.

1. $\dfrac{x}{5} = 2$ $x = 10$ **2.** $\dfrac{x}{8} = 4$ $x = 32$ **3.** $\dfrac{z}{6} = 6$ $z = 36$ **4.** $\dfrac{y}{7} = 8$ $y = 56$

20. $y = -1, y = 12$ **21.** $y = 5$ **22.** $a = -1$ **23.** $x = -\dfrac{10}{9}$

EXERCISE SET 6.6

24. $x = 2$ **29.** expression; $\dfrac{3 + 2x}{3x}$

Solve each equation. See Examples 1 and 2.

1. $\dfrac{x}{5} + 3 = 9$ $x = 30$ **2.** $\dfrac{x}{5} - 2 = 9$ $x = 55$ **3.** $\dfrac{x}{2} + \dfrac{5x}{4} = \dfrac{x}{12}$ $x = 0$ **4.** $\dfrac{x}{6} + \dfrac{4x}{3} = \dfrac{x}{18}$ $x = 0$

30. expression; $\dfrac{18 + 5a}{6a}$ **31.** equation; $x = 3$ **32.** equation; $a = 18$ **33.** expression; $\dfrac{x - 1}{x(x + 1)}$

5. $x = -5, x = 2$ 6. $y = -1, y = 7$ 9. $x = 3$ 10. $a = -6$ 17. extraneous solution 18. $y = 7$ 19. $x = 6, x = -4$

5. $2 + \dfrac{10}{x} = x + 5$

6. $6 + \dfrac{5}{y} = y - \dfrac{2}{y}$

7. $\dfrac{a}{5} = \dfrac{a - 3}{2}$ $a = 5$

8. $\dfrac{2b}{5} = \dfrac{b + 2}{6}$ $b = \dfrac{10}{7}$

9. $\dfrac{x - 3}{5} + \dfrac{x - 2}{2} = \dfrac{1}{2}$

10. $\dfrac{a + 5}{4} + \dfrac{a + 5}{2} = \dfrac{a}{8}$

Recall that two angles are supplementary if the sum of their measures is 180°. Find the measures of the following supplementary angles.

11. 100°, 80°

12. 30°, 150°

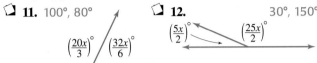

Recall that two angles are complementary if the sum of their measures is 90°. Find the measures of the following complementary angles.

13. 22.5°, 67.5°

14. 40°, 50°

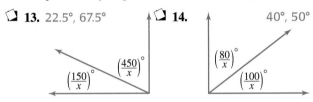

Solve each equation. See Examples 3 through 5.

15. $\dfrac{9}{2a - 5} = -2$ $a = \dfrac{1}{4}$

16. $\dfrac{6}{4 - 3x} = 3$ $x = \dfrac{2}{3}$

17. $\dfrac{y}{y + 4} + \dfrac{4}{y + 4} = 3$

18. $\dfrac{5y}{y + 1} - \dfrac{3}{y + 1} = 4$

19. $\dfrac{2x}{x + 2} - 2 = \dfrac{x - 8}{x - 2}$

20. $\dfrac{4y}{y - 3} - 3 = \dfrac{3y - 1}{y + 3}$

21. $\dfrac{4y}{y - 4} + 5 = \dfrac{5y}{y - 4}$

22. $\dfrac{2a}{a + 2} - 5 = \dfrac{7a}{a + 2}$

23. $\dfrac{7}{x - 2} + 1 = \dfrac{x}{x + 2}$

24. $1 + \dfrac{3}{x + 1} = \dfrac{x}{x - 1}$

25. $\dfrac{x + 1}{x + 3} = \dfrac{2x^2 - 15x}{x^2 + x - 6} - \dfrac{x - 3}{x - 2}$ $x = \dfrac{11}{14}$

26. $\dfrac{3}{x + 3} = \dfrac{12x + 19}{x^2 + 7x + 12} - \dfrac{5}{x + 4}$ $x = 2$

27. $\dfrac{y}{2y + 2} + \dfrac{2y - 16}{4y + 4} = \dfrac{2y - 3}{y + 1}$ extraneous solution

28. $\dfrac{1}{x + 2} = \dfrac{4}{x^2 - 4} - \dfrac{1}{x - 2}$ extraneous solution

Determine whether each of the following is an equation or an expression. If it is an equation, then solve it for its variable. If it is an expression, perform the indicated operation. See Example 6.

29. $\dfrac{1}{x} + \dfrac{2}{3}$

30. $\dfrac{3}{a} + \dfrac{5}{6}$

31. $\dfrac{1}{x} + \dfrac{2}{3} = \dfrac{3}{x}$

32. $\dfrac{3}{a} + \dfrac{5}{6} = 1$

33. $\dfrac{2}{x + 1} - \dfrac{1}{x}$

34. $\dfrac{4}{x - 3} - \dfrac{1}{x}$

35. $\dfrac{2}{x + 1} - \dfrac{1}{x} = 1$

36. $\dfrac{4}{x - 3} - \dfrac{1}{x} = \dfrac{6}{x(x - 3)}$

37. Explain the difference between solving an equation such as $\dfrac{x}{2} + \dfrac{3}{4} = \dfrac{x}{4}$ for x and performing an operation such as adding: $\dfrac{x}{2} + \dfrac{3}{4}$.

38. When solving an equation such as $\dfrac{y}{4} = \dfrac{y}{2} - \dfrac{1}{4}$, we may multiply all terms by 4. When subtracting two rational expressions such as $\dfrac{y}{2} - \dfrac{1}{4}$, we may not. Explain why.

Solve each equation. **40.** $x = -\dfrac{20}{9}$

39. $\dfrac{2x}{7} - 5x = 9$ $x = -\dfrac{21}{11}$

40. $\dfrac{4x}{8} - 5x = 10$

41. $\dfrac{2}{y} + \dfrac{1}{2} = \dfrac{5}{2y}$ $y = 1$

42. $\dfrac{6}{3y} + \dfrac{3}{y} = 1$ $y = 5$

43. $\dfrac{4x + 10}{7} = \dfrac{8}{2}$ $x = \dfrac{9}{2}$

44. $\dfrac{1}{2} = \dfrac{x + 1}{8}$ $x = 3$

45. $2 + \dfrac{3}{a - 3} = \dfrac{a}{a - 3}$ extraneous solution

46. $\dfrac{2y}{y - 2} - \dfrac{4}{y - 2} = 4$

47. $\dfrac{5}{x} + \dfrac{2}{3} = \dfrac{7}{2x}$ $x = -\dfrac{9}{4}$

48. $\dfrac{5}{3} - \dfrac{3}{2x} = \dfrac{5}{4}$ $x = \dfrac{18}{5}$

49. $\dfrac{2a}{a + 4} = \dfrac{3}{a - 1}$

50. $\dfrac{5}{3x - 8} = \dfrac{x}{x - 2}$

51. $\dfrac{x + 1}{3} - \dfrac{x - 1}{6} = \dfrac{1}{6}$

52. $\dfrac{3x}{5} - \dfrac{x - 6}{3} = \dfrac{1}{5}$ $x = -\dfrac{27}{4}$

51. $x = -2$

53. $\dfrac{4r - 1}{r^2 + 5r - 14} + \dfrac{2}{r + 7} = \dfrac{1}{r - 2}$ $r = \dfrac{12}{5}$

54. $\dfrac{2t + 3}{t - 1} - \dfrac{2}{t + 3} = \dfrac{5 - 6t}{t^2 + 2t - 3}$ $t = -\dfrac{1}{2}, t = -6$

55. $\dfrac{t}{t - 4} = \dfrac{t + 4}{6}$

56. $\dfrac{15}{x + 4} = \dfrac{x - 4}{x}$

34. expression; $\dfrac{3x + 3}{x(x - 3)}$ **35.** equation; no solution **36.** equation; $x = 1$ **37.** answers may vary

38. answers may vary **46.** extraneous solution **49.** $a = -\dfrac{3}{2}, a = 4$ **50.** $x = 1, x = \dfrac{10}{3}$ **55.** $t = -2; t = 8$

56. $x = 16, x = -1$

57. $\dfrac{x}{2x + 6} + \dfrac{x + 1}{3x + 9} = \dfrac{2}{4x + 12}$ $x = \dfrac{1}{5}$

58. $\dfrac{a}{5a - 5} - \dfrac{a - 2}{2a - 2} = \dfrac{5}{4a - 4}$ $a = -\dfrac{5}{6}$

Solve each equation for the indicated variable. See Example 7.

59. $\dfrac{D}{R} = T$; for R $R = \dfrac{D}{T}$ **60.** $\dfrac{A}{W} = L$; for W

61. $\dfrac{3}{x} = \dfrac{5y}{x + 2}$; for y **62.** $\dfrac{7x - 1}{2x} = \dfrac{5}{y}$; for y

63. $\dfrac{3a + 2}{3b - 2} = -\dfrac{4}{2a}$; for b $b = -\dfrac{3a^2 + 2a - 4}{6}$

64. $\dfrac{6x + y}{7x} = \dfrac{3x}{h}$; for h $h = \dfrac{21x^2}{6x + y}$

65. $\dfrac{A}{BH} = \dfrac{1}{2}$; for B **66.** $\dfrac{V}{\pi r^2 h} = 1$; for h

67. $\dfrac{C}{\pi r} = 2$; for r $r = \dfrac{C}{2\pi}$ **68.** $\dfrac{3V}{A} = H$; for V $V = \dfrac{AH}{3}$

69. $\dfrac{1}{a} = \dfrac{1}{b} + \dfrac{1}{c}$; for a **70.** $\dfrac{1}{2} - \dfrac{1}{x} = \dfrac{1}{y}$; for x

71. $\dfrac{m^2}{6} - \dfrac{n}{3} = \dfrac{p}{2}$; for n **72.** $\dfrac{x^2}{r} + \dfrac{y^2}{t} = 1$; for r

Solve each equation.

73. $\dfrac{5}{a^2 + 4a + 3} + \dfrac{2}{a^2 + a - 6} - \dfrac{3}{a^2 - a - 2} = 0$

74. $-\dfrac{2}{a^2 + 2a - 8} + \dfrac{1}{a^2 + 9a + 20} = \dfrac{-4}{a^2 + 3a - 10}$

graphical answers in App. F

Review Exercises

Graph the line passing through the given point with the given slope. See Section 3.4.

75. $(4, 3), m = -3$ **76.** $(0, -1), m = \dfrac{1}{2}$

77. $(-5, 2), m = \dfrac{3}{2}$ **78.** $(-1, -4), m = -\dfrac{1}{5}$

Identify the x- and y-intercepts. Also, write the intercepts as ordered pairs of numbers. See Section 3.3.

79.

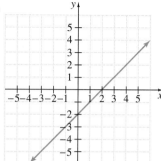

$x = 2, y = -2,$
$(2, 0), (0, -2)$

80.

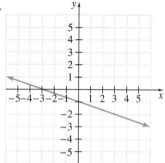

$x = -3, y = -1,$
$(-3, 0), (0, -1)$

81.

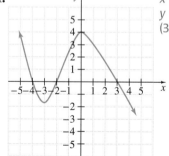

$x = -4, x = -2, x = 3,$
$y = 4, (-4, 0), (-2, 0),$
$(3, 0), (0, 4)$

82.

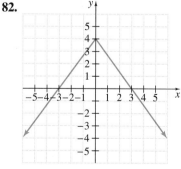

$x = -3, x = 3,$
$y = 4, (-3, 0),$
$(3, 0), (0, 4)$

60. $W = \dfrac{A}{L}$ **61.** $y = \dfrac{3x + 6}{5x}$ **62.** $y = \dfrac{10x}{7x - 1}$ **65.** $B = \dfrac{2A}{H}$ **66.** $h = \dfrac{V}{\pi r^2}$ **69.** $a = \dfrac{bc}{c + b}$ **70.** $x = \dfrac{2y}{y - 2}$

71. $n = \dfrac{m^2 - 3p}{2}$ **72.** $r = \dfrac{x^2 t}{t - y^2}$ **73.** $a = \dfrac{17}{4}$ **74.** $a = -\dfrac{4}{3}$

6.7 | RATIO AND PROPORTION

O B J E C T I V E S

TAPE
BA 6.7

1 Use fractional notation to express ratios.

2 Identify and solve proportions.

3 Use proportions to solve problems.

4 Determine unit pricing.

1 A **ratio** is the quotient of two numbers or two quantities.

> ### RATIO
>
> If a and b are two numbers and $b \neq 0$, the **ratio of a to b** is the quotient of a and b. The ratio of a to b can also be written as
>
> $$a{:}b \text{ or as } \frac{a}{b}$$

EXAMPLE 1   Write a ratio for each phrase. Use fractional notation.

a. The ratio of 2 parts salt to 5 parts water

b. The ratio of 18 inches to 2 feet

Solution: **a.** The ratio of 2 parts salt to 5 parts water is $\frac{2}{5}$.

b. When comparing measurements, use the same unit of measurement in the numerator as in the denominator. Here, we write 2 feet as $2 \cdot 12$ inches, or 24 inches. The ratio of 18 inches to 2 feet is then $\frac{18}{24} = \frac{3}{4}$ in lowest terms. ■

2 If two ratios are equal, we say the ratios are **in proportion** to each other. A **proportion** is a mathematical statement that two ratios are equal.

For example, the equation $\frac{1}{2} = \frac{4}{8}$ is a proportion, as is $\frac{x}{5} = \frac{8}{10}$, because both sides of the equations are ratios. When we want to emphasize the equation as a proportion, we

read the proportion $\dfrac{1}{2} = \dfrac{4}{8}$ as "one is to two as four is to eight"

In the proportion $\frac{1}{2} = \frac{4}{8}$, the 1 and the 8 are called the **extremes**; the 2 and the 4 are called the **means**.

For any true proportion, the product of the means equals the product of the extremes.

For example, since the proportion

$$\frac{1}{2} = \frac{4}{8}$$

is true, then it must also be true that

$$1 \cdot 8 = 2 \cdot 4$$

extremes   means

or 8 = 8

To see why this is so, multiply both sides of the proportion $\frac{a}{b} = \frac{c}{d}$ by the LCD bd.

$$\frac{a}{b} = \frac{c}{d}$$

$$bd\left(\frac{a}{b}\right) = bd\left(\frac{c}{d}\right) \qquad \text{Multiply both sides by } bd.$$

$$ad = bc \qquad \text{Simplify.}$$

product   product
of extremes   of means

The products ad and bc are also called **cross products.** Equating these cross products is called **cross multiplication.**

$$\frac{a}{b} = \frac{c}{d} \rightarrow ad = bc$$

CROSS MULTIPLICATION

For any ratios $\frac{a}{b}$ and $\frac{c}{d}$, if

$$\frac{a}{b} = \frac{c}{d}, \text{ then } ad = bc.$$

Cross multiplication can be used to solve a proportion for an unknown.

EXAMPLE 2   Solve for x: $\dfrac{45}{x} = \dfrac{5}{7}$.

Solution: To solve, cross multiply.

$$\frac{45}{x} = \frac{5}{7}$$

$$45 \cdot 7 = x \cdot 5 \qquad \text{Cross multiply.}$$

$$\frac{315}{5} = \frac{5x}{5} \qquad \text{Divide both sides by 5.}$$

$$63 = x \qquad \text{Simplify.}$$

To check, substitute 63 for x in the original proportion. The solution set is $\{63\}$.

EXAMPLE 3   Solve for x: $\dfrac{x}{x - 2} = \dfrac{x - 3}{x + 1}$.

Solution: To solve, cross multiply.

$$\frac{x}{x - 2} = \frac{x - 3}{x + 1}$$

$$x(x + 1) = (x - 2)(x - 3) \qquad \text{Cross multiply.}$$

$$x^2 + x = x^2 - 5x + 6 \qquad \text{Multiply.}$$

$$x = -5x + 6 \qquad \text{Subtract } x^2 \text{ from both sides.}$$

$$6x = 6 \qquad \text{Add } 5x \text{ to both sides.}$$

$$x = 1 \qquad \text{Divide both sides by 6.}$$

To check, substitute 1 for x in the original proportion. The solution set is $\{1\}$.

3   Proportions can be used to model and solve many real-life problems. When using proportions in this way, it is important to judge whether the solution is reasonable. Doing so helps us to decide if the proportion has been formed correctly. We use the same problem-solving steps that were introduced in Section 2.5.

EXAMPLE 4   Three boxes of 3.5-inch high-density diskettes cost \$37.47. How much should 5 boxes cost?

Solution: **1.** UNDERSTAND the problem. To do so, read and reread the problem. Guess the solution and check the guess. We know that the cost of 5 boxes is more than the cost of 3 boxes, or \$37.47, and less than the cost of 6 boxes, which is double the cost of 3 boxes, or 2(\$37.47) = \$74.94. Let's guess that 5 boxes cost \$60.00. To check this guess, see if 3 boxes is to 5 boxes as the *price* of 3 boxes is to the *price* of 5 boxes. In other words, see if

$$\frac{3 \text{ boxes}}{5 \text{ boxes}} = \frac{\text{price of 3 boxes}}{\text{price of 5 boxes}}$$

or

$$\frac{3}{5} = \frac{37.47}{60.00}$$

To see if the proportion is true, we cross multiply.

$$3(60.00) = 5(37.47)$$

or

$$180.00 = 187.35 \qquad \textbf{Not a true statement.}$$

Thus, $60 is not correct but we now have a better understanding of the problem.

2. ASSIGN a variable. Let x = price of 5 boxes of diskettes.
3. ILLUSTRATE the problem. No illustration is needed.
4. TRANSLATE the problem.

$$\text{In words: } \frac{3 \text{ boxes}}{5 \text{ boxes}} = \frac{\text{price of 3 boxes}}{\text{price of 5 boxes}}$$

$$\text{Translate: } \frac{3}{5} = \frac{37.47}{x}$$

5. COMPLETE the work. Here, we solve the proportion.

$$\frac{3}{5} = \frac{37.47}{x}$$

$$3x = 5(37.47) \qquad \text{Cross multiply.}$$

$$3x = 187.35$$

$$x = 62.45 \qquad \text{Divide both sides by 3.}$$

6. INTERPRET the results. First *check* the solution. To do so, see that 3 boxes is to 5 boxes as $37.47 is to $62.45. Also, notice that our solution is a reasonable one as discussed in step 1.

Next *state* the conclusions. Five boxes of high-density diskettes cost $62.45. ▬▬

The proportion $\dfrac{5 \text{ boxes}}{3 \text{ boxes}} = \dfrac{\text{price of 5 boxes}}{\text{price of 3 boxes}}$ could also have been used to solve the problem above.

EXAMPLE 5 To estimate the number of people in Jackson, population 50,000, who have no health insurance, 250 people were polled, and 39 of those polled had no insurance. How many people in the city might we expect to be uninsured?

Solution:
1. UNDERSTAND. Read and reread the problem. Guess the solution and check your guess.
2. ASSIGN. Let x = number of people in the city with no health insurance.
3. ILLUSTRATE. No diagram or chart is needed.
4. TRANSLATE.

In words: $\dfrac{\text{total number polled}}{\text{number polled with no insurance}} = \dfrac{\text{total city population}}{\text{number in city with no insurance}}$

Translate: $\dfrac{250}{39} = \dfrac{50{,}000}{x}$

5. COMPLETE. Solve the proportion for x.

$$\dfrac{250}{39} = \dfrac{50{,}000}{x}$$

$250x = 39(50{,}000)$ Cross multiply.

$250x = 1{,}950{,}000$

$x = 7800$ Divide both sides by 250.

6. INTERPRET. *Check* the solution and *state* the conclusion. We expect the city of Jackson to have approximately 7800 citizens with no health insurance. ▬▬▬

4

When shopping for an item offered in many different sizes, it is important to be able to determine the best buy, or the best price per unit. To find the unit price of an item, divide the total price of the item by the total number of units.

$$\text{unit price} = \dfrac{\text{total price}}{\text{number of units}}$$

For example, if a 16-ounce can of green beans is priced at $0.88, its unit price is

$$\text{unit price} = \dfrac{\$0.88}{16} = \$0.055$$

▬▬▬

EXAMPLE 6

A supermarket offers a 14-ounce box of cereal for $3.79 and an 18-ounce box of the same brand of cereal for $4.99. Which is the better buy?

Solution:

To find the better buy, we compare unit prices. The following unit prices were rounded to three decimal places.

SIZE	PRICE	UNIT PRICE
14-ounce	$3.79	$\dfrac{\$3.79}{14} \approx \$0.271$
18-ounce	$4.99	$\dfrac{\$4.99}{18} \approx \$0.277$

The 14-ounce box of cereal has the lower unit price so it is the better buy.

▬▬▬

EXERCISE SET 6.7

Write each ratio in fractional notation in lowest terms. See Example 1.

1. 2 megabytes to 15 megabytes $\frac{2}{15}$

2. 18 disks to 41 disks $\frac{18}{41}$

3. 10 inches to 12 inches $\frac{5}{6}$

4. 15 miles to 40 miles $\frac{3}{8}$

5. 5 quarts to 3 gallons $\frac{5}{12}$

6. 8 inches to 3 feet $\frac{2}{9}$

7. 4 nickels to 2 dollars $\frac{1}{10}$

8. 12 quarters to 2 dollars $\frac{3}{2}$

9. 175 centimeters to 5 meters $\frac{7}{20}$

10. 90 centimeters to 4 meters $\frac{9}{40}$

11. 190 minutes to 3 hours $\frac{19}{18}$

12. 60 hours to 2 days $\frac{5}{4}$

13. Suppose someone tells you that the ratio of 11 inches to 2 feet is $\frac{11}{2}$. How do you correct that person and explain the error?   answers may vary

14. Write a ratio that can be written in fractional notation as $\frac{3}{2}$.   answers may vary

Solve each proportion. See Examples 2 and 3.

15. $\frac{2}{3} = \frac{x}{6}$ $x = 4$

16. $\frac{x}{2} = \frac{16}{6}$ $x = \frac{16}{3}$

17. $\frac{x}{10} = \frac{5}{9}$ $x = \frac{50}{9}$

18. $\frac{9}{4x} = \frac{6}{2}$ $x = \frac{3}{4}$

19. $\frac{4x}{6} = \frac{7}{2}$ $x = \frac{21}{4}$

20. $\frac{a}{5} = \frac{3}{2}$ $a = \frac{15}{2}$

21. $\frac{a}{25} = \frac{12}{10}$ $a = 30$

22. $\frac{n}{10} = 9$ $n = 90$

23. $\frac{x-3}{x} = \frac{4}{7}$ $x = 7$

24. $\frac{y}{y-16} = \frac{5}{3}$ $y = 40$

25. $\frac{5x+1}{x} = \frac{6}{3}$ $x = -\frac{1}{3}$

26. $\frac{3x-2}{5} = \frac{4x}{1}$ $x = -\frac{2}{17}$

27. $\frac{x+1}{2x+3} = \frac{2}{3}$ $x = -3$

28. $\frac{x+1}{x+2} = \frac{5}{3}$ $x = -\frac{7}{2}$

29. $\frac{9}{5} = \frac{12}{3x+2}$ $x = \frac{14}{9}$

30. $\frac{6}{11} = \frac{27}{3x-2}$ $x = \frac{103}{6}$

31. $\frac{3}{x+1} = \frac{5}{2x}$ $x = 5$

32. $\frac{7}{x-3} = \frac{8}{2x}$ $x = -4$

33. $x = -\frac{2}{3}$ **34.** $x = -\frac{9}{7}$ **43.** 73 brown M&M's

33. $\frac{x+1}{x} = \frac{x+2}{x-2}$

34. $\frac{x-1}{x+3} = \frac{x+3}{x}$

35. If x is 10, is $\frac{2}{x}$ in proportion to $\frac{x}{50}$? Explain why or why not.   yes; answers may vary

36. For what value of x is $\frac{x}{x-1}$ in proportion to $\frac{x+1}{x}$? Explain your result.   none; answers may vary

Given the following prices charged for various sizes of an item, find the best buy. See Example 6.

37. Laundry detergent
110 ounces for $5.79
240 ounces for $13.99   110 oz. for $5.79

38. Jelly
10 ounces for $1.14
15 ounces for $1.69   15 oz. for $1.69

39. Tuna (in cans)
6 ounces for $0.69
8 ounces for $0.90
16 ounces for $1.89   8 oz. for $0.90

40. Picante sauce
10 ounces for $0.99
16 ounces for $1.69
30 ounces for $3.29   10 oz. for $0.99

Solve. See Examples 4 and 5.

41. The ratio of the weight of an object on Earth to the weight of the same object on Pluto is 100 to 3. If an elephant weighs 4100 pounds on Earth, find the elephant's weight on Pluto.   123 lb.

42. If a 170-pound person weighs approximately 65 pounds on Mars, how much does a 9000-pound satellite weigh?   3441 lb.

43. In a bag of M&M's, 28 out of 80 M&M's were found to be the color brown. How many M&M's would you expect to be brown from a bag containing 208 M&M's? (Round to the nearest whole.)

44. On an architect's blueprint, 1 inch corresponds to 4 feet. Find the length of a wall represented by a line that is $3\frac{7}{8}$ inches long on the blueprint. $15\frac{1}{2}$ ft.

45. There are 110 calories per 28.4 grams of Crispy Rice cereal. Find how many calories are in 42.6 grams of this cereal.   165 calories

51. 360 sq. ft.

46. A box of flea and tick powder instructs the user to mix 4 tablespoons of powder with 1 gallon of water. Find how much powder should be mixed with 5 gallons of water.   20 tablespoons

47. Miss Babola's new Mazda gets 35 miles per gallon. Find how far she can drive if the tank contains 13.5 gallons of gas.   472.5 miles

48. In a week of city driving, Miss Babola noticed that she was able to drive 418.5 miles on a tank of gas (13.5 gallons). Find how many miles per gallon Miss Babola got in city traffic.   31 mpg

49. Ken Hall, a tailback, holds the high school sports record for total yards rushed in a season. In 1953, he rushed for 4045 total yards in 12 games. Find his average rushing yards per game.   337 yds/game

50. A recent headline read, "Women earn bigger check in 1 of every 6 couples." If there are 23,000 couples in a nearby metropolitan area, how many women would you expect to earn bigger paychecks?   3833 women

51. A human factors expert recommends that there be at least 9 square feet of floor space in a college classroom for every student in the class. Find the minimum floor space that 40 students need.

52. There are 1280 calories in a 14-ounce portion of Eagle Brand Milk. Find how many calories are in 2 ounces of Eagle Brand Milk. $182\frac{6}{7}$ calories

53. Due to space problems at a local university, a 20-foot by 12-foot conference room is converted into a classroom. Find the maximum number of students the room can accommodate. (See Exercise 51.)   26 students

54. The manufacturers of cans of salted mixed nuts state that the ratio of peanuts to other nuts is 3 to 2. If 324 peanuts are in a can, find how many other nuts should also be in the can.   216 nuts

55. If Sam Abney can travel 343 miles in 7 hours, find how far he can travel if he maintains the same speed for 5 hours.   245 miles

56. The instructions on a bottle of plant food read as follows: "Use four tablespoons plant food per 3 gallons of water." Find how many tablespoons of plant food should be mixed into 6 gallons of water.   8 tablespoons

57. To mix weed killer with water correctly, it is necessary to mix 8 teaspoons of weed killer with 2 gallons of water. Find how many gallons of water are needed to mix with the entire box if it contains 36 teaspoons of weed killer.   9 gal.

58. There are 290 milligrams of sodium per 1-ounce serving of Rice Crispies. Find how many milligrams of sodium are in three 1-ounce servings of the cereal.   870 mg

59. Mr. Lin's contract states that he will be paid $153 per 8-hour day to teach mathematics. Find how much he earns per hour rounded to the nearest cent. $19.13 per hr

60. Mr. Gonzales, a pool contractor, bases the cost of labor on the volume of the pool to be constructed. The cost of labor on a wading pool of 803 cubic feet is $750.00. If the customer decided to cut the

66. answers may vary   **68.** $m = -\dfrac{4}{3}$; downward   **69.** $m = 3$; upward   **70.** $m = \dfrac{11}{4}$; upward

volume by a third, find the cost of labor for the smaller pool. $500

61. An accountant finds that Country Collections earned $35,063 during its first 6 months. Find how much the business earned **each week** on average rounded to the nearest cent. $1348.58

The following graph shows the capacity of the world to generate electricity from the wind.

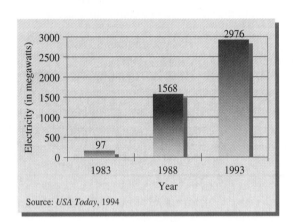

Source: *USA Today*, 1994

62. Find the increase in megawatt capacity during the 5-year period from 1983 to 1988.   1471 megawatts

71. $m = -\dfrac{9}{5}$; downward   **72.** undefined slope; vertical   **73.** $m = 0$; horizontal

63. Find the increase in megawatt capacity during the 5-year period from 1988 to 1993.   1408 megawatts

64. If the trend shown on this graph continues, approximate the number of megawatts available from the wind in 1998.   4400 megawatts

In general, 1000 megawatts will serve the average electricity needs of 560,000 people. Use this fact and the preceding graph to answer the following.

65. In 1993, the number of megawatts that can be generated from wind will serve the electricity needs of how many people?   1,666,560 people

66. How many megawatts of electricity are needed to serve the city or town in which you live?

67. If $x = ad$, $y = bc$, and $\frac{a}{b} = \frac{c}{d}$, what true statement can you make about the value of x as compared to the value of y?   Their ratio is 1.

Review Exercises

Find the slope of the line through each pair of points. Use the slope to determine whether the line is vertical, horizontal, or moves upward or downward from left to right. See Section 3.4.

68. $(-2, 5), (4, -3)$ **69.** $(0, 4)\,(2, 10)$

70. $(-3, -6)\,(1, 5)$ **71.** $(-2, 7)\,(3, -2)$

72. $(3, 7)\,(3, -2)$ **73.** $(0, -4)\,(2, -4)$

6.8   RATIONAL EQUATIONS AND PROBLEM SOLVING

O B J E C T I V E S

1. Translate sentences to equations containing rational expressions.
2. Solve problems involving work.
3. Solve problems involving distance.
4. Solve problems involving similar triangles.

TAPE
BA 6.8

1   In this section, we solve problems that can be modeled by equations containing rational expressions. To solve these problems, we use the same problem-solving steps that were first introduced in Section 2.5. In our first example, our goal is to find an unknown number.

EXAMPLE 1   The quotient of a number and 6 minus $\dfrac{5}{3}$ is the quotient of the number and 2. Find the number.

Solution: **1.** UNDERSTAND. Read and reread the problem, guess a solution, and check your guess. For example, if the unknown number is 2, then we see if the quotient of 2 and 6, or $\dfrac{2}{6} - \dfrac{5}{3}$ is equal to the quotient of 2 and 2, or $\dfrac{2}{2}$. $\dfrac{2}{6} - \dfrac{5}{3} = \dfrac{1}{3} - \dfrac{5}{3} = -\dfrac{4}{3}$, and not $\dfrac{2}{2}$. Don't forget that the purpose of a guess is to better understand the problem.

2. ASSIGN. Let x = the unknown number.

3. ILLUSTRATE. No illustration is needed.

4. TRANSLATE:

In words:	The quotient of x and 6	minus	$\dfrac{5}{3}$	is	the quotient of x and 2.
Translate:	$\dfrac{x}{6}$	$-$	$\dfrac{5}{3}$	$=$	$\dfrac{x}{2}$

5. COMPLETE. Here, we solve the equation $\dfrac{x}{6} - \dfrac{5}{3} = \dfrac{x}{2}$. Begin solving this equation by multiplying both sides of the equation by the LCD 6.

$$6\left(\dfrac{x}{6} - \dfrac{5}{3}\right) = 6\left(\dfrac{x}{2}\right)$$

$$6\left(\dfrac{x}{6}\right) - 6\left(\dfrac{5}{3}\right) = 6\left(\dfrac{x}{2}\right) \qquad \text{Apply the distributive property.}$$

$$x - 10 = 3x \qquad \text{Simplify.}$$

$$-10 = 2x \qquad \text{Subtract } x \text{ from both sides.}$$

$$-\dfrac{10}{2} = \dfrac{2x}{2} \qquad \text{Divide both sides by 2.}$$

$$-5 = x \qquad \text{Simplify.}$$

6. INTERPRET. *Check:* To check, verify that "the quotient of -5 and 6 minus $\dfrac{5}{3}$ is the quotient of -5 and 2, or $-\dfrac{5}{6} - \dfrac{5}{3} = -\dfrac{5}{2}$. *State:* The unknown number is -5.

The next example is often called a work problem. Work problems usually involve people or machines doing a certain task.

EXAMPLE 2   Sam Waterton and Frank Schaffer work in a plant that manufactures automobiles. Sam can complete a quality control tour of the plant in 3 hours while his assistant, Frank, needs 7 hours to complete the same job. The regional manager is coming to inspect the plant facilities, so both Sam and Frank are directed to complete a quality control tour together. How long will this take?

Solution: **1.** UNDERSTAND. Read and reread the problem. The key idea here is the relationship between the **time** (hours) it takes to complete the job and the **part of the job** completed in 1 unit of time (hour). For example, if the **time** it takes Sam to complete the job is 3 hours, the **part of the job** he can complete in 1 hour is $\frac{1}{3}$. Similarly, Frank can complete $\frac{1}{7}$ of the job in 1 hour.

2. ASSIGN. Let x represent the **time** in hours it takes Sam and Frank to complete the job together. Then $\frac{1}{x}$ represents the **part of the job** they complete in 1 hour.

3. ILLUSTRATE. Here we summarize the information discussed above in a chart.

	HOURS TO COMPLETE TOTAL JOB	**PART OF JOB COMPLETED IN 1 HOUR**
Sam	3	$\frac{1}{3}$
Frank	7	$\frac{1}{7}$
Together	x	$\frac{1}{x}$

4. TRANSLATE.

In words:	part of job Sam completed in 1 hour	added to	part of job Frank completed in 1 hour	is equal to	part of job they completed together in 1 hour
Translate:	$\frac{1}{3}$	$+$	$\frac{1}{7}$	$=$	$\frac{1}{x}$

5. COMPLETE. Here, we solve the equation $\frac{1}{3} + \frac{1}{7} = \frac{1}{x}$. Begin solving the equation by multiplying both sides of the equation by the LCD $21x$.

$$21x\left(\frac{1}{3}\right) + 21x\left(\frac{1}{7}\right) = 21x\left(\frac{1}{x}\right)$$

$$7x + 3x = 21 \qquad \text{Simplify.}$$

$$10x = 21$$

$$x = \frac{21}{10} \quad \text{or} \quad 2\frac{1}{10} \text{ hours}$$

6. INTERPRET. *Check:* Our proposed solution is $2\frac{1}{10}$ hours. This proposed solution is reasonable since $2\frac{1}{10}$ hours is more than half of Sam's time and less than half of Frank's time. Check this solution in the originally stated problem. *State:* Sam and Frank can complete the quality control tour in $2\frac{1}{10}$ hours. ▬▬▬

3 Next we look at a problem solved by the distance formula.

▬▬▬▬▬

EXAMPLE 3   A car travels 180 miles in the same time that a semi truck travels 120 miles. If the car's speed is 20 miles per hour faster than the truck's, find the car's speed and the truck's speed.

Solution: **1. UNDERSTAND.** Read and reread the problem. Next, guess a solution. Suppose that the truck's speed is 45 miles per hour. Then the car's speed is 20 miles per hour more, or 65 miles per hour.

We are given that the car travels 180 miles in the same time that the truck travels 120 miles. To find the time it takes that car to travel 180 miles, remember that since $d = rt$, we know that $\dfrac{d}{r} = t$.

CAR'S TIME

TRUCK'S TIME

$$t = \frac{d}{r} = \frac{180}{65} = 2\frac{50}{65} = 2\frac{10}{13} \text{ hours} \qquad t = \frac{d}{r} = \frac{120}{45} = 2\frac{30}{45} = 2\frac{2}{3} \text{ hours}$$

Since the times are not the same, we have not guessed the speeds correctly.

2. ASSIGN. Let

$$x = \text{the speed of the truck.}$$

Since the car's speed is 20 miles per hour faster than the truck's, then

$$x + 20 = \text{the speed of the car}$$

3. ILLUSTRATE. Use the formula $d = r \cdot t$ or **d**istance = **r**ate (speed) $\cdot$ **t**ime. Prepare a chart to organize the information in the problem. Recall that if $d = r \cdot t$, then $t = \dfrac{d}{r}$.

	distance	=	rate	$\cdot$	time
Truck	120		x		$\dfrac{120}{x} \left(\dfrac{\text{distance}}{\text{rate}}\right)$
Car	180		$x + 20$		$\dfrac{180}{x + 20} \left(\dfrac{\text{distance}}{\text{rate}}\right)$

4. TRANSLATE. Since the car and the truck traveled the same amount of time, we have that

In words: car's time   =   truck's time

Translate: $\dfrac{180}{x + 20} = \dfrac{120}{x}$

5. COMPLETE. Begin solving the equation by cross multiplying.

$$\dfrac{180}{x + 20} = \dfrac{120}{x}$$

$180x = 120(x + 20)$ Cross multiply.

$180x = 120x + 2400$ Use the distributive property.

$60x = 2400$ Subtract $120x$ from both sides.

$x = 40$ Divide both sides by 60.

6. INTERPRET. The speed of the truck is 40 miles per hour. The speed of the car must then be $x + 20$ or 60 miles per hour. To *check*, find the time it takes the car to travel 180 miles and the time it takes the truck to travel 120 miles.

<table>
<tr><td align="center">CAR'S TIME</td><td align="center">TRUCK'S TIME</td></tr>
<tr><td align="center">$t = \dfrac{d}{r} = \dfrac{180}{60} = 3$ hours</td><td align="center">$t = \dfrac{d}{r} = \dfrac{120}{40} = 3$ hours</td></tr>
</table>

Since both travel the same amount of time, the proposed solution is correct. *State:* The car's speed is 60 miles per hour and the truck's speed is 40 miles per hour.

4

Similar triangles have the same shape but not necessarily the same size. In similar triangles, the measures of corresponding angles are equal, and corresponding sides are in proportion.

If triangle ABC and triangle XYZ below are similar, then we know that the measure of angle A = the measure of angle X, the measure of angle B = the measure of angle Y, and the measure of angle C = the measure of angle Z, and we also know that corresponding sides are in proportion: $\dfrac{a}{x} = \dfrac{b}{y} = \dfrac{c}{z}$.

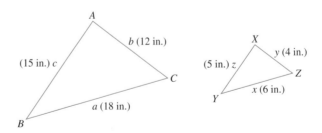

In this section, we will position similar triangles so that they have the same orientation.

To show that corresponding sides are in proportion for the triangles above, write the ratios of the corresponding sides.

$$\frac{a}{x} = \frac{18}{6} = 3 \qquad \frac{b}{y} = \frac{12}{4} = 3 \qquad \frac{c}{z} = \frac{15}{5} = 3$$

EXAMPLE 4   If the following two triangles are similar, find the missing length x.

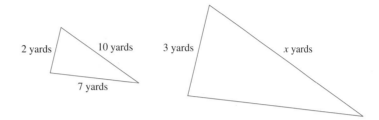

Solution: Since the triangles are similar, their corresponding sides are in proportion and we have

$$\frac{2}{3} = \frac{10}{x}$$

To solve, cross multiply.

$$2x = 30 \qquad \text{Cross multiply.}$$
$$x = 15 \qquad \text{Divide both sides by 2.}$$

The missing length is 15 yards.

EXERCISE SET 6.8

Solve the following. See Example 1.

1. Three times the reciprocal of a number equals 9 times the reciprocal of 6. Find the number.   2

2. Twelve divided by the sum of x and 2 equals the quotient of 4 and the difference of x and 2. Find x.   4

3. If twice a number added to 3 is divided by the number plus 1, the result is three halves. Find the number.   −3

4. A number added to the product of 6 and the reciprocal of the number equals −5. Find the number.

See Example 2.

5. Smith Engineering found that an experienced surveyor surveys a roadbed in 4 hours. An apprentice surveyor needs 5 hours to survey the same stretch

of road. If the two work together, find how long it takes them to complete the job. $2\frac{2}{9}$ hr.

6. An experienced bricklayer constructs a small wall in 3 hours. The apprentice completes the job in 6 hours. Find how long it takes if they work together.

7. In 2 minutes, a conveyor belt moves 300 pounds of recyclable aluminum from the delivery truck to a storage area. A smaller belt moves the same quantity of cans the same distance in 6 minutes. If both belts are used, find how long it takes to move the cans to the storage area. $1\frac{1}{2}$ min.

8. Find how long it takes the conveyor belts described in Exercise 7 to move 1200 pounds of cans. (*Hint:* Think of 1200 pounds as four 300-pound jobs.)

4. −3 or −2   **6.** 2 hr.   **8.** 6 min.

See Example 3.

9. A jogger begins her workout by jogging to the park, a distance of 12 miles. She then jogs home at the same speed but along a different route. This return trip is 18 miles and her time is one hour longer. Find her jogging speed. Complete the accompanying chart and use it to find her jogging speed.

	distance	=	rate	·	time
Trip to park	12				x
Return trip	18				$x + 1$

10. A boat can travel 9 miles upstream in the same amount of time it takes to travel 11 miles downstream. If the current of the river is 3 miles per hour, complete the chart below and use it to find the speed of the boat in still water.

	distance	=	rate	·	time
Upstream	9		$r - 3$		
Downstream	11		$r + 3$		

11. A cyclist rode the first 20-mile portion of his workout at a constant speed. For the 16-mile cooldown portion of his workout, he reduced his speed by 2 miles per hour. Each portion of the workout took the same time. Find the cyclist's speed during the first portion and find his speed during the cooldown portion.   1st portion: 10 mph; cooldown: 8 mph

12. A semi truck travels 300 miles through the flatland in the same amount of time that it travels 180 miles through mountains. The rate of the truck is 20 miles per hour slower in the mountains than in the flatland. Find both the flatland rate and mountain rate.   flatland: 50 mph; mountains: 30 mph

Given that the following pairs of triangles are similar, find the missing lengths. See Example 4.

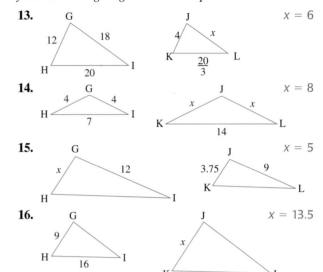

13. $x = 6$

14. $x = 8$

15. $x = 5$

16. $x = 13.5$

Solve the following.

17. One-fourth equals the quotient of a number and 8. Find the number.   2

18. Four times a number added to 5 is divided by 6. The result is $\frac{7}{2}$. Find the number.   4

19. Marcus and Tony work for Lombardo's Pipe and Concrete. Mr. Lombardo is preparing an estimate for a customer. He knows that Marcus lays a slab of concrete in 6 hours. Tony lays the same size slab in 4 hours. If both work on the job and the cost of labor is $45.00 per hour, decide what the labor estimate should be. $108.00

20. Mr. Dodson can paint his house by himself in 4 days. His son needs an additional day to complete the job if he works by himself. If they work together, find how long it takes to paint the house. $2\frac{2}{9}$ days

21. While road testing a new make of car, the editor of a consumer magazine finds that he can go 10 miles into a 3-mile-per-hour wind in the same amount of time he can go 11 miles with a 3-mile-per-hour wind behind him. Find the speed of the car in still air.   63 mph

22. A fisherman on Pearl River rows 9 miles downstream in the same amount of time he rows 3 miles upstream. If the current is 6 miles per hour, find how long it takes him to cover the 12 miles.   1 hr.

9. Trip to park rate: r To park time: $\frac{12}{r}$
    Return trip rate: r Return time: $\frac{18}{r}$
    $r = 6$ mph

10. Upstream time: $\frac{9}{r-3}$; Downstream time: $\frac{11}{r+3}$; $r = 30$ mph

28. car: 4 hr; jet: 3 hr   **31.** 3 hr

Find the unknown length in the following pairs of similar triangles.

23.

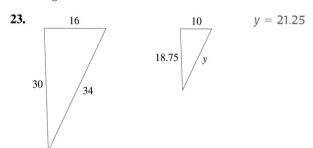

$y = 21.25$

24.

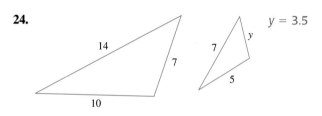

$y = 3.5$

Solve the following.

25. Two divided by the difference of a number and 3 minus 4 divided by a number plus 3, equals 8 times the reciprocal of the difference of the number squared and 9. What is the number?   5

26. If 15 times the reciprocal of a number is added to the ratio of 9 times a number minus 7 and the number plus 2, the result is 9. What is the number?   3

27. A pilot flies 630 miles with a tail wind of 35 miles per hour. Against the wind, he flies only 455 miles. Find the rate of the plane in still air.   217 mph

28. A marketing manager travels 1080 miles in a corporate jet and then an additional 240 miles by car. If the car ride takes one hour longer then the jet ride takes, and if the rate of the jet is 6 times the rate of the car, find the time the manager travels by jet and find the time the manager travels by car.

29. A cyclist rides 16 miles per hour on level ground on a still day. He finds that he rides 48 miles with the wind behind him in the same amount of time that he rides 16 miles into the wind. Find the rate of the wind.   8 mph

30. The current on a portion of the Mississippi River is 3 miles per hour. A barge can go 6 miles upstream in the same amount of time it takes to go 10 miles downstream. Find the speed of the boat in still water.   12 mph

31. One custodian cleans a suite of offices in 3 hours. When a second worker is asked to join the regular custodian, the job takes only $1\frac{1}{2}$ hours. How long does it take the second worker to do the same job alone?

32. One person proofreads copy for a small newspaper in 4 hours. If a second proofreader is also employed, the job can be done in $2\frac{1}{2}$ hours. How long does it take for the second proofreader to do the same job alone? $6\frac{2}{3}$ hr

33. One pipe fills a storage pool in 20 hours. A second pipe fills the same pool in 15 hours. When a third pipe is added and all three are used to fill the pond, it takes only 6 hours. Find how long it takes the third pipe to do the job.   20 hr

34. Mr. Jamison can do an audit in 400 hours. Mr. Ling can do the same audit in 300 hours with the help of a new computer program. How long will it take if both are assigned to the audit? $171\frac{3}{7}$ hr

35. A toy maker wishes to make a triangular mainsail for a toy sailboat that will be the same shape as a regular-size sailboat's mainsail. Use the following diagram to find the missing dimensions.
$x = 4.4$ ft; $y = 5.6$ ft

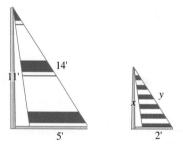

36. A seamstress wishes to make a doll's triangular diaper that will have the same shape as a full-size diaper. Use the following diagram to find the missing dimensions of the doll's diaper.
$x = 9.6$ in; $y = 8$ in

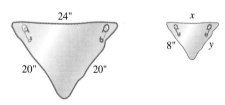

37. In 6 hours, an experienced cook prepares enough pies to supply a local restaurant's daily order. Another cook prepares the same number of pies in 7 hours. Together with a third cook, they prepare the pies in 2 hours. Find the work rate of the third cook. $5\frac{1}{4}$ hr

38. Mrs. Smith balances the company books in 8 hours. It takes her assistant half again as long to do the same job. If they work together, find how long it takes them to balance the books. $4\frac{4}{5}$ hr

39. One pump fills a tank 3 times as fast as another pump. If the pumps work together, they fill the tank in 21 minutes. How long does it take for each pump to fill the tank?
first pump: 28 min; second pump: 84 min

One of the great algebraists of ancient times was a man named Diophantus. Little is known of his life other than that he lived and worked in Alexandria. Some historians believe he lived during the first century of the Christian era, about the time of Nero. The only clue to his personal life is the following epigram found in a collection called the Palatine Anthology.

God granted him youth for a sixth of his life and added a twelfth part to this. He clothed his cheeks in down. He lit him the light of wedlock after a seventh part and five years after his marriage, He granted him a son. Alas, lateborn wretched child. After attaining the measure of half his father's life, cruel fate overtook him, thus leaving Diophantus during the last four years of his life only such consolation as the science of numbers. How old was Diophantus at his death?*

*From *The Nature and Growth of Modern Mathematics,* Edna Kramer, 1970, Fawcett Premier Books, Vol. 1, pages 107–108.

We are looking for Diophantus' age when he died, so let x represent that age. If we sum the parts of his life, we should get the total age.

Parts of his life $\begin{cases} \dfrac{1}{6} \cdot x + \dfrac{1}{12} \cdot x \text{ is the time of his youth.} \\[2mm] \dfrac{1}{7} \cdot x \text{ is the time between his youth and when he married.} \\[2mm] 5 \text{ years is the time between his marriage and the birth of his son.} \\[2mm] \dfrac{1}{2} \cdot x \text{ is the time Diophantus had with his son.} \\[2mm] 4 \text{ years is the time between his son's death and his own.} \end{cases}$

The sum of these parts should equal Diophantus' age when he died.

$$\frac{1}{6} \cdot x + \frac{1}{12} \cdot x + \frac{1}{7} \cdot x + 5 + \frac{1}{2} \cdot x + 4 = x$$

40. Solve the epigram. 84 yr

41. How old was Diophantus when his son was born? How old was the son when he died? 35 yr; 42 yr

42. Solve the following epigram:

I was four when my mother packed my lunch and sent me off to school. Half my life was spent in school and another sixth was spent on a farm. Alas, hard times befell me. My crops and cattle fared poorly and my land was sold. I returned to school for 3 years and have spent one tenth of my life teaching. How old am I? 30 yr

43. Write an epigram describing your life. Be sure that none of the time periods in your epigram overlap.
answers may vary

Review Exercises

Graph each linear equation by finding intercepts. See Section 3.3. graphical answers in App. F

44. $5x + y = 10$ **45.** $-x + 3y = 6$

46. $x = -3y$ **47.** $y = 2x$

48. $x - y = -2$ **49.** $y - x = -5$

GROUP ACTIVITY

COMPARING FORMULAS FOR DOSES OF MEDICATION

MATERIALS:
• Calculator

Two dose formulas for prescribing medicines to children are well known among doctors. Unlike formulas for, say, area or distance, these dose formulas describe only an approximate relationship. Young's Rule and Cowling's Rule both relate a child's age A in years and an adult dose D of medication to the proper child's dose C. The formulas are most accurate when applied to children between the ages of 2 and 13.

$$\text{Young's Rule:} \quad C = \frac{DA}{A + 12}$$

$$\text{Cowling's Rule:} \quad C = \frac{D(A + 1)}{24}$$

1. Let the adult dose $D = 1000$ mg. Create a table comparing the doses predicted by both formulas for ages $A = 2, 3, \ldots, 13$.

2. Use the data from your table in Question 1 to form sets of ordered pairs for each formula. In your ordered pairs let x represent the age of the child. Graph the ordered pairs for each formula on the same graph. Describe the shapes of the graphed data.

3. Use your table, graph, or both to decide whether either formula will consistently predict a larger dose than the other. If so, which one? If not, is there an age at which the doses predicted by one becomes greater than the doses predicted by the other? If so, estimate that age.

4. Use your graph to estimate for what age the difference in the two predicted doses is greatest. Verify your answer by adding another column or row to your table giving the absolute value of the differences between Young's doses and Cowling's doses.

5. Does Cowling's Rule ever predict exactly the adult dose? If so, at what age? Explain. Does Young's Rule ever predict exactly the adult dose? If so, at what age? Explain.

6. Many doctors prefer to use formulas that relate doses to factors other than a child's age. Why is age not necessarily the most important factor when predicting a child's dose? What other factors might be used?

See App. F for Group Activity answers and suggestions.

CHAPTER 6   HIGHLIGHTS

DEFINITIONS AND CONCEPTS	EXAMPLES

SECTION 6.1   SIMPLIFYING RATIONAL EXPRESSIONS

A **rational expression** is an expression that can be written in the form $\frac{P}{Q}$, where P and Q are polynomials and Q does not equal 0.

Rational Expressions

$$\frac{7y^3}{4}, \quad \frac{x^2 + 6x + 1}{x - 3}, \quad \frac{-5}{s^3 + 8}$$

To find values for which a rational expression is undefined, find values for which the denominator is 0.

Find any values for which the expression $\frac{5y}{y^2 - 4y + 3}$ is undefined.

$$y^2 - 4y + 3 = 0 \qquad \text{Set denominator equal to 0.}$$

$$(y - 3)(y - 1) = 0 \qquad \text{Factor.}$$

$$y - 3 = 0 \text{ or } y - 1 = 0 \qquad \text{Set each factor equal to 0.}$$

$$y = 3 \text{ or } y = 1 \qquad \text{Solve.}$$

The expression is undefined when y is 3 and when y is 1.

Fundamental Principle of Rational Expressions

If P and Q are polynomials, and Q and R are not 0, then

$$\frac{PR}{QR} = \frac{P}{Q}$$

By the fundamental principle,

$$\frac{(x - 3)(x + 1)}{x(x + 1)} = \frac{x - 3}{x}$$

as long as $x \neq 0$ and $x \neq -1$.

To simplify a rational expression,

Step 1. Factor the numerator and denominator.

Step 2. Apply the fundamental principle to divide out common factors.

Simplify $\frac{4x + 20}{x^2 - 25}$.

$$\frac{4x + 20}{x^2 - 25} = \frac{4(x + 5)}{(x + 5)(x - 5)} = \frac{4}{x - 5}$$

SECTION 6.2   MULTIPLYING AND DIVIDING RATIONAL EXPRESSIONS

To multiply rational expressions,

Step 1. Factor numerators and denominators.

Step 2. Multiply numerators and multiply denominators.

Step 3. Write the product in lowest terms.

$$\frac{P}{Q} \cdot \frac{R}{S} = \frac{PR}{QS}$$

Multiply: $\frac{4x + 4}{2x - 3} \cdot \frac{2x^2 + x - 6}{x^2 - 1}$

$$\frac{4x + 4}{2x - 3} \cdot \frac{2x^2 + x - 6}{x^2 - 1} = \frac{4(x + 1)}{2x - 3} \cdot \frac{(2x - 3)(x + 2)}{(x + 1)(x - 1)}$$

$$= \frac{4(x + 1)(2x - 3)(x + 2)}{(2x - 3)(x + 1)(x - 1)}$$

$$= \frac{4(x + 2)}{(x - 1)}$$

(continued)

DEFINITIONS AND CONCEPTS	EXAMPLES

SECTION 6.2   MULTIPLYING AND DIVIDING RATIONAL EXPRESSIONS

To divide by a rational expression, multiply by the reciprocal.

$$\frac{P}{Q} \div \frac{R}{S} = \frac{P}{Q} \cdot \frac{S}{R} = \frac{PS}{QR}$$

Divide: $\dfrac{15x + 5}{3x^2 - 14x - 5} \div \dfrac{15}{3x - 12}$

$$\frac{15x + 5}{3x^2 - 14x - 5} \div \frac{15}{3x - 12} = \frac{5\,(3x + 1)}{(3x + 1)(x - 5)} \cdot \frac{3\,(x - 4)}{3 \cdot 5}$$

$$= \frac{x - 4}{x - 5}$$

SECTION 6.3   ADDING AND SUBTRACTING RATIONAL EXPRESSIONS WITH COMMON DENOMINATORS AND LEAST COMMON DENOMINATOR

To add or subtract rational expressions with the same denominator, add or subtract numerators, and place the sum or difference over a common denominator.

$$\frac{P}{R} + \frac{Q}{R} = \frac{P + Q}{R}$$

$$\frac{P}{R} - \frac{Q}{R} = \frac{P - Q}{R}$$

Perform indicated operations.

$$\frac{5}{x + 1} + \frac{x}{x + 1} = \frac{5 + x}{x + 1}$$

$$\frac{2y + 7}{y^2 - 9} - \frac{y + 4}{y^2 - 9} = \frac{(2y + 7) - (y + 4)}{y^2 - 9}$$

$$= \frac{2y + 7 - y - 4}{y^2 - 9}$$

$$= \frac{y + 3}{(y + 3)(y - 3)}$$

$$= \frac{1}{y - 3}$$

To find the least common denominator (LCD),

Step 1.   Factor the denominators.

Step 2.   The LCD is the product of all unique factors, each raised to a power equal to the greatest number of times that it appears in any one factored denominator.

Find the LCD for

$$\frac{7x}{x^2 + 10x + 25} \text{ and } \frac{11}{3x^2 + 15x}$$

$$x^2 + 10x + 25 = (x + 5)(x + 5)$$

$$3x^2 + 15x = 3x(x + 5)$$

$$\text{LCD} = 3x(x + 5)(x + 5) \text{ or } 3x(x + 5)^2$$

SECTION 6.4   ADDING AND SUBTRACTING RATIONAL EXPRESSIONS WITH UNLIKE DENOMINATORS

To add or subtract rational expressions with unlike denominators.

Step 1.   Find the LCD.

Step 2.   Rewrite each rational expression as an equivalent expression whose denominator is the LCD.

Step 3.   Add or subtract numerators and place the sum or difference over the common denominator.

Perform the indicated operation.

$$\frac{9x + 3}{x^2 - 9} - \frac{5}{x - 3}$$

$$= \frac{9x + 3}{(x + 3)(x - 3)} - \frac{5}{x - 3}$$

LCD is $(x + 3)(x - 3)$.

(continued)

DEFINITIONS AND CONCEPTS	**EXAMPLES**

SECTION 6.4   ADDING AND SUBTRACTING RATIONAL EXPRESSIONS WITH UNLIKE DENOMINATORS

Step 4.   Write the result in lowest terms.

$$= \frac{9x + 3}{(x + 3)(x - 3)} - \frac{5(x + 3)}{(x - 3)(x + 3)}$$

$$= \frac{9x + 3 - 5(x + 3)}{(x + 3)(x - 3)}$$

$$= \frac{9x + 3 - 5x - 15}{(x + 3)(x - 3)}$$

$$= \frac{4x - 12}{(x + 3)(x - 3)}$$

$$= \frac{4\,(x - 3)}{(x + 3)\,(x - 3)} = \frac{4}{x + 3}$$

SECTION 6.5   COMPLEX FRACTIONS

Method 1:   To Simplify a Complex Fraction

Step 1.   Add or subtract fractions in the numerator and the denominator of the complex fraction.

Step 2.   Perform the indicated division.

Step 3.   Write the result in lowest terms.

Simplify.

$$\frac{\dfrac{1}{x} + 2}{\dfrac{1}{x} - \dfrac{1}{y}} = \frac{\dfrac{1}{x} + \dfrac{2x}{x}}{\dfrac{y}{xy} - \dfrac{x}{xy}}$$

$$= \frac{\dfrac{1 + 2x}{x}}{\dfrac{y - x}{xy}}$$

$$= \frac{1 + 2x}{x} \cdot \frac{x\,y}{y - x}$$

$$= \frac{y(1 + 2x)}{y - x}$$

Method 2:   To Simplify a Complex Fraction

Step 1.   Find the LCD of all fractions in the complex fraction.

Step 2.   Multiply the numerator and the denominator of the complex fraction by the LCD.

Step 3.   Perform indicated operations and write in lowest terms.

$$\frac{\dfrac{1}{x} + 2}{\dfrac{1}{x} - \dfrac{1}{y}} = \frac{xy\left(\dfrac{1}{x} + 2\right)}{xy\left(\dfrac{1}{x} - \dfrac{1}{y}\right)}$$

$$= \frac{xy\left(\dfrac{1}{x}\right) + xy(2)}{xy\left(\dfrac{1}{x}\right) - xy\left(\dfrac{1}{y}\right)}$$

$$= \frac{y + 2xy}{y - x} \quad \text{or} \quad \frac{y(1 + 2x)}{y - x}$$

DEFINITIONS AND CONCEPTS	EXAMPLES

SECTION 6.6 SOLVING EQUATIONS CONTAINING RATIONAL EXPRESSIONS

To solve an equation containing rational expressions,

Step 1. Multiply both sides of the equation by the LCD of all rational expressions in the equation.

Step 2. Remove any grouping symbols and solve the resulting equation.

Step 3. Check the solution in the original equation.

Solve: $\dfrac{5x}{x+2} + 3 = \dfrac{4x-6}{x+2}$

$$(x+2)\left(\dfrac{5x}{x+2} + 3\right) = (x+2)\left(\dfrac{4x-6}{x+2}\right)$$

$$(x+2)\left(\dfrac{5x}{x+2}\right) + (x+2)(3) = (x+2)\left(\dfrac{4x-6}{x+2}\right)$$

$$5x + 3x + 6 = 4x - 6$$
$$4x = -12$$
$$x = -3$$

The solution checks and the solution set is $\{-3\}$.

SECTION 6.7 RATIO AND PROPORTION

A **ratio** is the quotient of two numbers or two quantities.

The ratio of *a* to *b* can be written as

Fractional Notation $\qquad$ Colon Notation

$\dfrac{a}{b}$ $\qquad\qquad$ $a{:}b$

Write the ratio of 5 hours to 1 day using fractional notation.

$$\dfrac{5 \text{ hours}}{1 \text{ day}} = \dfrac{5 \text{ hours}}{24 \text{ hours}} = \dfrac{5}{24}$$

A **proportion** is a mathematical statement that two ratios are equal.

Proportions

$$\dfrac{2}{3} = \dfrac{8}{12} \qquad\qquad \dfrac{x}{7} = \dfrac{15}{35}$$

In the proportion $\frac{2}{3} = \frac{8}{12}$, the 2 and the 12 are called the **extremes;** the 3 and the 8 are called the **means.**

extremes $\rightarrow 2 8$
means $\rightarrow \dfrac{2}{3} = \dfrac{8}{12}$

In the proportion $\frac{a}{b} = \frac{c}{d}$, the products *ad* and *bc* are called **cross products.**

Cross Products

$$\dfrac{2}{3} = \dfrac{8}{12} \quad\rightarrow\quad 3 \cdot 8 \text{ or } 24$$
$$\phantom{\dfrac{2}{3} = \dfrac{8}{12}} \quad\rightarrow\quad 2 \cdot 12 \text{ or } 24$$

In a true proportion, the product of the means equals the product of the extremes.

Cross multiplication:

If $\dfrac{a}{b} = \dfrac{c}{d}$, then $ad = bc$.

Solve: $\dfrac{3}{4} = \dfrac{x}{x-1}$

$$\dfrac{3}{4} = \dfrac{x}{x-1}$$
$$3(x-1) = 4x \qquad \text{Cross multiply.}$$
$$3x - 3 = 4x$$
$$-3 = x$$

(continued)

DEFINITIONS AND CONCEPTS	EXAMPLES
SECTION 6.7 RATIO AND PROPORTION	

Problem-Solving Steps	A sample of 200 size C batteries contained 3 defective batteries. How many defective batteries might we expect in a shipment of 25,000 size C batteries?
1. UNDERSTAND. Read and reread the problem.	
2. ASSIGN.	Let x = number of defective batteries in the shipment.
3. ILLUSTRATE.	In words: $\dfrac{\text{total number in sample}}{\text{defective number in sample}} = \dfrac{\text{total number in shipment}}{\text{defective number in shipment}}$
4. TRANSLATE.	Translate: $\dfrac{200}{3} = \dfrac{25{,}000}{x}$
5. COMPLETE.	Solve: $\dfrac{200}{3} = \dfrac{25{,}000}{x}$ $200x = 3(25{,}000)$ Cross multiply. $200x = 75{,}000$ $x = 375$
6. INTERPRET.	**Check** the solution and **state** the conclusion. We expect the shipment of 25,000 batteries to contain 375 defective batteries.

SECTION 6.8 RATIONAL EQUATIONS AND PROBLEM SOLVING	

Problem-Solving Steps	A small plane and a car leave Kansas City, Missouri, and head for Minneapolis, Minnesota, a distance of 450 miles. The speed of the plane is 3 times the speed of the car, and the plane arrives 6 hours ahead of the car. Find the speed of the car.
1. UNDERSTAND. Read and reread the problem.	
2. ASSIGN.	Let x = the speed of the car. Then $3x$ = the speed of the plane.
3. ILLUSTRATE.	

	DISTANCE = RATE	•	TIME
Car	450	x	$\dfrac{450}{x}\left(\dfrac{\text{distance}}{\text{rate}}\right)$
Plane	450	$3x$	$\dfrac{450}{3x}\left(\dfrac{\text{distance}}{\text{rate}}\right)$

(continued)

DEFINITIONS AND CONCEPTS	EXAMPLES
SECTION 6.8 RATIONAL EQUATIONS AND PROBLEM SOLVING	

4. TRANSLATE.

In words:

Plane's time	+	6 hours	=	car's time

Translate:

$$\frac{450}{3x} \quad + \quad 6 \quad = \quad \frac{450}{x}$$

5. COMPLETE.

Solve:
$$\frac{450}{3x} + 6 = \frac{450}{x}$$

$$3x\left(\frac{450}{3x}\right) + 3x(6) = 3x\left(\frac{450}{x}\right)$$

$$450 + 18x = 1350$$

$$18x = 900$$

$$x = 50$$

6. INTERPRET.

Check the solution and **state** the conclusion. The speed of the car is 50 miles per hour.

---

2. $x = \dfrac{5}{2}, x = -\dfrac{3}{2}$ **12.** $\dfrac{x + a}{x - c}$ **13.** $\dfrac{x + 5}{x - 3}$ **19.** $-\dfrac{2x(2x + 5)}{(x - 6)^2}$ **22.** $(x - 6)(x - 3)$

CHAPTER 6 REVIEW

(6.1) *Find any real number for which each rational expression is undefined.*

1. $\dfrac{x + 5}{x^2 - 4}$ $x = 2, x = -2$ **2.** $\dfrac{5x + 9}{4x^2 - 4x - 15}$

Find the value of each rational expression when $x = 5$, $y = 7$, and $z = -2$.

3. $\dfrac{z^2 - z}{z + xy}$ $\dfrac{2}{11}$ **4.** $\dfrac{x^2 + xy - z^2}{x + y + z}$ $\dfrac{28}{5}$

Simplify each rational expression.

5. $\dfrac{x + 2}{x^2 - 3x - 10}$ $\dfrac{1}{x - 5}$ **6.** $\dfrac{x + 4}{x^2 + 5x + 4}$ $\dfrac{1}{x + 1}$

7. $\dfrac{x^3 - 4x}{x^2 + 3x + 2}$ $\dfrac{x(x - 2)}{x + 1}$ **8.** $\dfrac{5x^2 - 125}{x^2 + 2x - 15}$ $\dfrac{5(x - 5)}{x - 3}$

9. $\dfrac{x^2 - x - 6}{x^2 - 3x - 10}$ $\dfrac{x - 3}{x - 5}$ **10.** $\dfrac{x^2 - 2x}{x^2 + 2x - 8}$ $\dfrac{x}{x + 4}$

11. $\dfrac{x^2 + 6x + 5}{2x^2 + 11x + 5}$ $\dfrac{x + 1}{2x + 1}$ **12.** $\dfrac{x^2 + xa + xb + ab}{x^2 - xc + bx - bc}$

13. $\dfrac{x^2 + 5x - 2x - 10}{x^2 - 3x - 2x + 6}$ **14.** $\dfrac{x^2 - 9}{9 - x^2}$ -1

15. $\dfrac{4 - x}{x^3 - 64}$ $-\dfrac{1}{x^2 + 4x + 16}$

(6.2) *Perform the indicated operations and simplify.*

16. $\dfrac{15x^3 y^2}{z} \cdot \dfrac{z}{5xy^3}$ $\dfrac{3x^2}{y}$ **17.** $\dfrac{-y^3}{8} \cdot \dfrac{9x^2}{y^3}$ $-\dfrac{9x^2}{8}$

18. $\dfrac{x^2 - 9}{x^2 - 4} \cdot \dfrac{x - 2}{x + 3}$ $\dfrac{x - 3}{x + 2}$ **19.** $\dfrac{2x + 5}{x - 6} \cdot \dfrac{2x}{-x + 6}$

20. $\dfrac{x^2 - 5x - 24}{x^2 - x - 12} \div \dfrac{x^2 - 10x + 16}{x^2 + x - 6}$ $\dfrac{x + 3}{x - 4}$

21. $\dfrac{4x + 4y}{xy^2} \div \dfrac{3x + 3y}{x^2 y}$ $\dfrac{4x}{3y}$

22. $\dfrac{x^2 + x - 42}{x - 3} \cdot \dfrac{(x - 3)^2}{x + 7}$ **23.** $\dfrac{2a + 2b}{3} \cdot \dfrac{a - b}{a^2 - b^2}$ $\dfrac{2}{3}$

24. $\dfrac{x^2 - 9x + 14}{x^2 - 5x + 6} \cdot \dfrac{x + 2}{x^2 - 5x - 14}$ $\dfrac{1}{x - 3}$

31. $\dfrac{5x + 2}{3x - 1}$ **36.** $\dfrac{x^2 - 3x - 10}{(x + 2)(x - 5)(x + 9)}$ **39.** $\dfrac{-2x + 10}{(x - 3)(x - 1)}$ **40.** $\dfrac{14x + 58}{(x + 3)(x + 7)}$ **44.** $-\dfrac{x}{x - 1}$

25. $(x - 3) \cdot \dfrac{x}{x^2 + 3x - 18}$ $\quad \dfrac{x}{x + 6}$

26. $\dfrac{2x^2 - 9x + 9}{8x - 12} \div \dfrac{x^2 - 3x}{2x}$ $\quad \dfrac{1}{2}$

27. $\dfrac{x^2 - y^2}{x^2 + xy} \div \dfrac{3x^2 - 2xy - y^2}{3x^2 + 6x}$ $\quad \dfrac{3(x + 2)}{3x + y}$

28. $\dfrac{x^2 - y^2}{8x^2 - 16xy + 8y^2} \div \dfrac{x + y}{4x - y}$ $\quad \dfrac{4x - y}{8(x - y)}$

29. $\dfrac{x - y}{4} \div \dfrac{y^2 - 2y - xy + 2x}{16x + 24}$ $\quad -\dfrac{2(2x + 3)}{y - 2}$

30. $\dfrac{y - 3}{4x + 3} \div \dfrac{9 - y^2}{4x^2 - x - 3}$ $\quad -\dfrac{x - 1}{y + 3}$

(6.3) *Perform the indicated operations and simplify.*

31. $\dfrac{5x - 4}{3x - 1} + \dfrac{6}{3x - 1}$ **32.** $\dfrac{4x - 5}{3x^2} - \dfrac{2x + 5}{3x^2}$

33. $\dfrac{9x + 7}{6x^2} - \dfrac{3x + 4}{6x^2}$ $\quad \dfrac{2x + 1}{2x^2}$ **32.** $\dfrac{2x - 10}{3x^2}$

Find the LCD of each pair of rational expressions.

34. $\dfrac{x + 4}{2x}, \dfrac{3}{7x}$ $\quad 14x$

35. $\dfrac{x - 2}{x^2 - 5x - 24}, \dfrac{3}{x^2 + 11x + 24}$ $\quad (x - 8)(x + 8)(x + 3)$

Rewrite the following rational expressions as equivalent expressions whose denominator is the given polynomial.

36. $\dfrac{x + 2}{x^2 + 11x + 18}, (x + 2)(x - 5)(x + 9)$

37. $\dfrac{3x - 5}{x^2 + 4x + 4}, (x + 2)^2 \cdot (x + 3)$ $\quad \dfrac{3x^2 + 4x - 15}{(x + 2)^2(x + 3)}$

(6.4) *Perform the indicated operations and simplify.*

38. $\dfrac{4}{5x^2} - \dfrac{6}{y}$ $\quad \dfrac{4y - 30x^2}{5x^2y}$ **39.** $\dfrac{2}{x - 3} - \dfrac{4}{x - 1}$

40. $\dfrac{x + 7}{x + 3} - \dfrac{x - 3}{x + 7}$ **41.** $\dfrac{4}{x + 3} - 2$ $\quad \dfrac{-2x - 2}{x + 3}$

42. $\dfrac{3}{x^2 + 2x - 8} + \dfrac{2}{x^2 - 3x + 2}$ $\quad \dfrac{5x + 5}{(x + 4)(x - 2)(x - 1)}$

43. $\dfrac{2x - 5}{6x + 9} - \dfrac{4}{2x^2 + 3x}$ $\quad \dfrac{x - 4}{3x}$

45. $\dfrac{x^2 + 2x - 3}{(x + 2)^2}$ **46.** $\dfrac{x^2 + 2x + 4}{4x}; \dfrac{x + 2}{32}$ **47.** $\dfrac{29x}{12(x - 1)}; \dfrac{3xy}{5(x - 1)}$

44. $\dfrac{x - 1}{x^2 - 2x + 1} - \dfrac{x + 1}{x - 1}$ **45.** $\dfrac{x - 1}{x^2 + 4x + 4} + \dfrac{x - 1}{x + 2}$

Find the perimeter and the area of each figure.

46.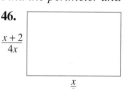

47.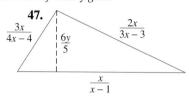

(6.5) *Simplify each complex fraction.*

48. $\dfrac{\dfrac{5x}{27}}{-\dfrac{10xy}{21}}$ $\quad -\dfrac{7}{18y}$ **49.** $\dfrac{\dfrac{8x}{x^2 - 9}}{\dfrac{4}{x + 3}}$ $\quad \dfrac{2x}{x - 3}$

50. $\dfrac{\dfrac{3}{5} + \dfrac{2}{7}}{\dfrac{1}{5} + \dfrac{5}{6}}$ $\quad \dfrac{6}{7}$ **51.** $\dfrac{\dfrac{2}{a} + \dfrac{1}{2a}}{a + \dfrac{a}{2}}$ $\quad \dfrac{5}{3a^2}$

52. $\dfrac{3 - \dfrac{1}{y}}{2 - \dfrac{1}{y}}$ $\quad \dfrac{3y - 1}{2y - 1}$ **53.** $\dfrac{2 + \dfrac{1}{x^2}}{\dfrac{1}{x} + \dfrac{2}{x^2}}$ $\quad \dfrac{2x^2 + 1}{x + 2}$

54. $\dfrac{\dfrac{1}{a} + \dfrac{1}{b}}{\dfrac{1}{ab}}$ $\quad b + a$ **55.** $\dfrac{\dfrac{6}{x + 2} + 4}{\dfrac{8}{x + 2} - 4}$ $\quad -\dfrac{7 + 2x}{2x}$

(6.6) *Solve each equation for the variable or perform the indicated operation.*

56. $\dfrac{x + 4}{9} = \dfrac{5}{9}$ $\quad x = 1$ **57.** $\dfrac{n}{10} = 9 - \dfrac{n}{5}$ $\quad n = 30$

58. $\dfrac{5y - 3}{7} = \dfrac{15y - 2}{28}$ $\quad y = 2$

59. $\dfrac{2}{x + 1} - \dfrac{1}{x - 2} = -\dfrac{1}{2}$ $\quad x = 3, x = -4$

60. $\dfrac{1}{a + 3} + \dfrac{1}{a - 3} = -\dfrac{5}{a^2 - 9}$ $\quad a = -\dfrac{5}{2}$

61. $\dfrac{y}{2y + 2} + \dfrac{2y - 16}{4y + 4} = \dfrac{y - 3}{y + 1}$ $\quad$ extraneous solution

62. $\dfrac{4}{x + 3} + \dfrac{8}{x^2 - 9} = 0$ $\quad x = 1$

63. $\dfrac{2}{x-3} - \dfrac{4}{x+3} = \dfrac{8}{x^2-9}$ $x = 5$

64. $\dfrac{x-3}{x+1} - \dfrac{x-6}{x+5} = 0$ $x = \dfrac{9}{7}$

65. $x + 5 = \dfrac{6}{x}$ $x = -6, x = 1$

Solve the equation for the indicated variable.

66. $\dfrac{4A}{5b} = x^2$, for b $b = \dfrac{4A}{5x^2}$ **67.** $\dfrac{x}{7} + \dfrac{y}{8} = 10$, for y

(6.7)   *Write each phrase as a ratio in fractional notation.*

68. 20 cents to 1 dollar $\dfrac{1}{5}$

69. four parts red to six parts white $\dfrac{2}{3}$

Solve each proportion.

70. $\dfrac{x}{2} = \dfrac{12}{4}$ $x = 6$ **71.** $\dfrac{20}{1} = \dfrac{x}{25}$ $x = 500$

72. $\dfrac{32}{100} = \dfrac{100}{x}$ $x = 312.5$ **73.** $\dfrac{20}{2} = \dfrac{c}{5}$ $c = 50$

74. $\dfrac{2}{x-1} = \dfrac{3}{x+3}$ $x = 9$ **75.** $\dfrac{4}{y-3} = \dfrac{2}{y-3}$

76. $\dfrac{y+2}{y} = \dfrac{5}{3}$ $y = 3$ **77.** $\dfrac{x-3}{3x+2} = \dfrac{2}{6}$

Given the following prices charged for various sizes of an item, find the best buy.

▦ **78.** Shampoo

      10 ounces for $1.29

      16 ounces for $2.15   10 oz. for $1.29

▦ **79.** Frozen green beans

      8 ounces for $0.89

      15 ounces for $1.63

      20 ounces for $2.36   15 oz. for $1.63

Solve.

80. A machine can process 300 parts in 20 minutes. Find how many parts can be processed in 45 minutes.

81. As his consulting fee, Mr. Visconti charges $90.00 per day. Find how much he charges for 3 hours of consulting. Assume an 8-hour work day. $33.75

82. One fund raiser can address 100 letters in 35 minutes. Find how many he can address in 55 minutes.

(6.8)   *Solve each problem.*

83. Five times the reciprocal of a number equals the sum of $\frac{3}{2}$ times the reciprocal of the number and $\frac{7}{6}$. What is the number?   3

84. The reciprocal of a number equals the reciprocal of the difference of 4 and the number. Find the number.   2

85. A car travels 90 miles in the same time that a car traveling 10 miles per hour slower travels 60 miles. Find the speed of each car.   30 mph; 20 mph

86. The speed of a bayou near Lafayette, Louisiana, is 4 miles per hour. A paddle boat travels 48 miles upstream in the same amount of time it takes to travel 72 miles downstream. Find the speed of the boat in still water.   20 mph

87. When Mark and Maria manicure Mr. Stergeon's lawn, it takes them 5 hours. If Mark works alone, it takes 7 hours. Find how long it takes Maria alone. $17\frac{1}{2}$ hr

88. It takes pipe A 20 days to fill a fish pond. Pipe B takes 15 days. Find how long it takes both pipes together to fill the pond. $8\frac{4}{7}$ days

67. $y = \dfrac{560 - 8x}{7}$ **75.** extraneous solution

77. extraneous solution   **80.** 675 parts

82. $157\dfrac{1}{7}$ or 157 letters

CHAPTER 6 TEST

1. Find any real numbers for which the following expression is undefined.

$$\dfrac{x+5}{x^2+4x+3}$$ $x = -1, x = -3$

2. For a certain computer desk, the manufacturing cost C per desk (in dollars) is

$$C = \dfrac{100x + 3000}{x}$$

where x is the number of desks manufactured.

a. Find the average cost per desk when manufacturing 200 computer desks. $115

b. Find the average cost per desk when manufacturing 1000 computer desks. $103

Simplify each rational expression.

3. $\dfrac{3x - 6}{5x - 10}$ $\dfrac{3}{5}$

4. $\dfrac{x + 10}{x^2 - 100}$ $\dfrac{1}{x - 10}$

5. $\dfrac{x + 6}{x^2 + 12x + 36}$ $\dfrac{1}{x + 6}$

6. $\dfrac{x + 3}{x^3 + 27}$ $\dfrac{1}{x^2 - 3x + 9}$

7. $\dfrac{2m^3 - 2m^2 - 12m}{m^2 - 5m + 6}$

8. $\dfrac{ay + 3a + 2y + 6}{ay + 3a + 5y + 15}$

9. $\dfrac{y - x}{x^2 - y^2}$ $-\dfrac{1}{x + y}$

Perform the indicated operation and simplify if possible.

10. $\dfrac{x^2 - 13x + 42}{x^2 + 10x + 21} \div \dfrac{x^2 - 4}{x^2 + x - 6}$ $\dfrac{(x - 6)(x - 7)}{(x + 7)(x + 2)}$

11. $\dfrac{3}{x - 1} \cdot (5x - 5)$ 15

12. $\dfrac{y^2 - 5y + 6}{2y + 4} \cdot \dfrac{y + 2}{2y - 6}$

13. $\dfrac{5}{2x + 5} - \dfrac{6}{2x + 5}$

14. $\dfrac{5a}{a^2 - a - 6} - \dfrac{2}{a - 3}$

15. $\dfrac{6}{x^2 - 1} + \dfrac{3}{x + 1}$ $\dfrac{3}{x - 1}$

16. $\dfrac{x^2 - 9}{x^2 - 3x} \div \dfrac{xy + 5x + 3y + 15}{2x + 10}$ $\dfrac{2(x + 5)}{x(y + 5)}$

17. $\dfrac{x + 2}{x^2 + 11x + 18} + \dfrac{5}{x^2 - 3x - 10}$ $\dfrac{x^2 + 2x + 35}{(x + 9)(x + 2)(x - 5)}$

18. $\dfrac{4y}{y^2 + 6y + 5} - \dfrac{3}{y^2 + 5y + 4}$ $\dfrac{4y^2 + 13y - 15}{(y + 4)(y + 5)(y + 1)}$

Solve each equation.

19. $\dfrac{4}{y} - \dfrac{5}{3} = \dfrac{-1}{5}$ $y = \dfrac{30}{11}$

20. $\dfrac{5}{y + 1} = \dfrac{4}{y + 2}$

21. $\dfrac{a}{a - 3} = \dfrac{3}{a - 3} - \dfrac{3}{2}$ extraneous solution

22. $\dfrac{10}{x^2 - 25} = \dfrac{3}{x + 5} + \dfrac{1}{x - 5}$ extraneous solution

Simplify each complex fraction.

23. $\dfrac{\dfrac{5x^2}{yz^2}}{\dfrac{10x}{z^3}}$ $\dfrac{xz}{2y}$

24. $\dfrac{\dfrac{b}{a} - \dfrac{a}{b}}{\dfrac{b}{a} + \dfrac{b}{a}}$ $\dfrac{b^2 - a^2}{2b^2}$

25. $\dfrac{5 - \dfrac{1}{y^2}}{\dfrac{1}{y} + \dfrac{2}{y^2}}$ $\dfrac{5y^2 - 1}{y + 2}$

26. In a sample of 85 fluorescent bulbs, 3 were found to be defective. At this rate, how many defective bulbs should be found in 510 bulbs? 18 bulbs

27. One number plus five times its reciprocal is equal to six. Find the number. 5 or 1

28. A pleasure boat traveling down the Red River takes the same time to go 14 miles upstream as it takes to go 16 miles downstream. If the current of the river is 2 miles per hour, find the speed of the boat in still water. 30 mph

29. An inlet pipe can fill a tank in 12 hours. A second pipe can fill the tank in 15 hours. If both pipes are used, find how long it takes to fill the tank. $6\dfrac{2}{3}$ hr

30. Decide which is the best buy in crackers.

    6 ounces for $1.19

    10 ounces for $2.15   6 oz. for $1.19

    16 ounces for $3.25

7. $\dfrac{2m(m + 2)}{m - 2}$ **8.** $\dfrac{a + 2}{a + 5}$ **12.** $\dfrac{y - 2}{4}$ **13.** $-\dfrac{1}{2x + 5}$

14. $\dfrac{3a - 4}{(a - 3)(a + 2)}$ **20.** $y = -6$

CHAPTER 6 CUMULATIVE REVIEW

1. Write each sentence as an equation. Let x represent the unknown number. *(Sec. 1.4, Ex. 4)*

  a. The quotient of 15 and a number is 4. $\dfrac{15}{x} = 4$

  b. Three subtracted from 12 is a number.

1b. $12 - 3 = x$ **1c.** $4x + 17 = 21$

  c. Four times a number added to 17 is 21.

2. Find each sum. *(Sec. 1.5, Ex. 1)*

  a. $-3 + (-7)$ -10 **b.** $5 + (+12)$ 17

  c. $(-1) + (-20)$ -21 **d.** $-2 + (-10)$ -12

3. Name the property illustrated by each true statement. *(Sec. 1.8, Ex. 7)*

   a. $2 \cdot 3 = 3 \cdot 2$ **b.** $3(x + 5) = 3x + 15$

   c. $2 + (4 + 8) = (2 + 4) + 8$

4. Solve $3 - x = 7$ for x. $x = -4$ *(Sec. 2.2, Ex. 6)*

5. Twice a number added to seven is the same as three subtracted from the number. Find the number. -10 *(Sec. 2.4, Ex. 8)*

6. Solve $P = 2l + 2w$ for w.

7. Solve $x + 4 \le -6$ for x. Graph the solution set.

8. Graph the linear equation $y = 3x$.   See App. F

9. Simplify each quotient. *(Sec. 4.1, Ex. 7)*

   a. $\dfrac{x^5}{x^2}$ x^3 **b.** $\dfrac{4^7}{4^3}$ 256

   c. $\dfrac{(-3)^5}{(-3)^2}$ -27 **d.** $\dfrac{2x^5y^2}{xy}$ $2x^4y$

10. Use the distributive property to find each product.

   a. $5x(2x^3 + 6)$ $10x^4 + 30x$

   b. $-3x^2(5x^2 + 6x - 1)$ $-15x^4 - 18x^3 + 3x^2$

   c. $(3n^2 - 5n + 4)(2n)$ $6n^3 - 10n^2 + 8n$

11. Simplify each expression. Write answers with positive exponents. *(Sec. 4.5, Ex. 3)*

   a. $\dfrac{y}{y^{-2}}$ y^3 **b.** $\dfrac{p^{-4}}{q^{-9}}$ $\dfrac{q^9}{p^4}$ **c.** $\dfrac{x^{-5}}{x^7}$ $\dfrac{1}{x^{12}}$

12. Divide: $\dfrac{2x^4 - x^3 + 3x^2 + x - 1}{x^2 + 1}$. *(Sec. 4.6, Ex. 8)*

13. Factor $x^2 + 7x + 12$.   **14.** Factor $x^3 + 8$.

15. Solve $2x^3 - 4x^2 - 30x = 0$. *(Sec. 5.6, Ex. 7)*

16. Multiply: $\dfrac{x^2 + x}{3x} \cdot \dfrac{6}{5x + 5}$. $\dfrac{2}{5}$ *(Sec. 6.2, Ex. 2)*

17. Subtract: $\dfrac{3x^2 + 2x}{x - 1} - \dfrac{10x - 5}{x - 1}$. $3x - 5$

18. Subtract: $\dfrac{6x}{x^2 - 4} - \dfrac{3}{x + 2}$. $\dfrac{3}{x - 2}$ *(Sec. 6.4, Ex. 2)*

19. Simplify $\dfrac{\dfrac{1}{z} - \dfrac{1}{2}}{\dfrac{1}{3} - \dfrac{z}{6}} \cdot \dfrac{3}{z}$ *(Sec. 6.5, Ex. 4)*

20. Solve $\dfrac{t - 4}{2} - \dfrac{t - 3}{9} = \dfrac{5}{18}$. $t = 5$ *(Sec. 6.6, Ex. 1)*

21. Sam Waterton and Frank Schaffer work in a plant that manufactures automobiles. Sam can complete a quality control tour of the plant in 3 hours while his assistant, Frank, needs 7 hours to complete the same job. The regional manager is coming to inspect the plant facilities, so both Sam and Frank are directed to complete a quality control tour together. How long will this take?

   $2\dfrac{1}{10}$ hr *(Sec. 6.8, Ex. 2)*

3a. commutative property of multiplication   **3b.** distributive property

3c. associative property of addition   **6.** $w = \dfrac{P - 2l}{2}$ *(Sec. 2.6, Ex. 6)*   **7.** $x \le -10$; (number line: $-13\ -12\ -11\ -10\ -9$) *(Sec. 2.9, Ex. 3)*

8. *(Sec. 3.2, Ex. 4)*   **10.** *(Sec. 4.3, Ex. 1)*   **12.** $2x^2 - x + 1 + \dfrac{2x - 2}{x^2 + 1}$ **13.** $(x + 3)(x + 4)$ *(Sec. 5.2, Ex. 1)*

14. $(x + 2)(x^2 - 2x + 4)$ *(Sec. 5.4, Ex. 5)*   **15.** $x = 0, x = -3, x = 5$ **17.** *(Sec. 6.3, Ex. 3)*

FURTHER GRAPHING

MATCHING DESCRIPTIONS OF LINEAR DATA TO THEIR EQUATIONS AND GRAPHS

From the way sales of record albums have changed to the trend in U.S. production of potatoes to the decrease in the numbers of U.S. Air Force personnel on active duty, many situations can be described using a linear equation. With the data's annual rate of change or two points on the line, it is possible to find an equation that summarizes the situation.

IN THE CHAPTER GROUP ACTIVITY ON PAGE 444, WE WILL HAVE THE OPPORTUNITY TO MATCH A DESCRIPTION OF A SITUATION TO ITS LINEAR EQUATION AND ITS GRAPH.

In Chapter 3, linear equations and inequalities in two variables were graphed on a rectangular coordinate system. We found that the graph of a linear equation in two variables is simply a picture of its ordered pair solutions and this picture each time is a line. Chapter 3 also included a study of the tilt of a line, called its slope. Equations in two variables and the slope of their graphs have many applications in the fields of science and business, just to name a few. For example, every business would like a graph of its profits to show a positive slope over time.

In this chapter, our study of slope and two variable equations is continued. Two-variable equations lead directly to the concept of function, perhaps the most important concept in all mathematics. Functions are introduced in the concluding section of this chapter.

7.1 | THE SLOPE-INTERCEPT FORM

O B J E C T I V E S

1. Use the slope-intercept form to find the slope and the y-intercept of a line.
2. Use the slope-intercept form to determine whether two lines are parallel, perpendicular, or neither.
3. Use the slope-intercept form to write an equation of a line.
4. Use the slope-intercept form to graph a linear equation.

TAPE
BA 7.1

In Section 3.4, we learned that the slant of a line is called its slope. Recall that the slope of a line m is the ratio of vertical change (change in y) to the corresponding horizontal change (change in x) as we move along the line.

EXAMPLE 1   Find the slope of the line whose equation is $y = \frac{3}{4}x + 6$.

Solution: To find the slope of this line, find any two ordered pair solutions of the equation. Let's find and use intercept points as our two ordered pair solutions. Recall from Section 3.2 that the graph of an equation of the form $y = mx + b$ has y-intercept b. This means that the graph of $y = \frac{3}{4}x + 6$ has a y-intercept of 6 as shown below.

To find the y-intercept, let $x = 0$.

Then $y = \frac{3}{4}x + 6$ becomes

$$y = \frac{3}{4} \cdot 0 + 6 \text{ or}$$

$$y = 6$$

The y-intercept point is $(0, 6)$, as expected.

To find the x-intercept, let $y = 0$.

Then $y = \frac{3}{4}x + 6$ becomes

$$0 = \frac{3}{4}x + 6 \quad \text{or}$$

$$0 = 3x + 24 \quad \text{Multiply both sides by 4.}$$
$$-24 = 3x \qquad \text{Subtract 24 from both sides.}$$
$$-8 = x \qquad \text{Divide both sides by 3.}$$

The x-intercept point is $(-8, 0)$.

Use the points $(0, 6)$ and $(-8, 0)$ to find the slope. Let (x_1, y_1) be $(0, 6)$ and (x_2, y_2) be $(-8, 0)$. Then

$$m = \frac{\text{change in } y}{\text{change in } x} = \frac{y_2 - y_1}{x_2 - x_1} = \frac{0 - 6}{-8 - 0} = \frac{-6}{-8} = \frac{3}{4}$$

The slope of the line is $\frac{3}{4}$.

Analyzing the results of Example 1, you may notice a striking pattern:

The slope of $y = \frac{3}{4}x + 6$ is $\frac{3}{4}$, the same as the coefficient of x.

Also as mentioned earlier, the y-intercept is 6, the same as the constant term.

When a linear equation is written in the form $y = mx + b$, not only is b the y-intercept of the line, but m is its slope. The form $y = mx + b$ is appropriately called the **slope-intercept form.**

SLOPE-INTERCEPT FORM

When a linear equation in two variables is written in slope-intercept form,

$$y = mx + b$$

then m is the slope of the line and b is the y-intercept of the line.

EXAMPLE 2   Find the slope and the y-intercept of the line whose equation is $5x + y = 2$.

Solution: Write the equation in slope-intercept form by solving the equation for y.

$$5x + y = 2$$
$$y = -5x + 2 \qquad \text{Subtract } 5x \text{ from both sides.}$$

The coefficient of x, -5, is the slope and the constant term, 2, is the y-intercept.

EXAMPLE 3   Find the slope and the y-intercept of the line whose equation is $3x - 4y = 4$.

Solution: Write the equation in slope-intercept form by solving for y.

$$3x - 4y = 4$$
$$-4y = -3x + 4 \qquad\qquad \text{Subtract } 3x \text{ from both sides.}$$
$$\frac{-4y}{-4} = \frac{-3x}{-4} + \frac{4}{-4} \qquad\qquad \text{Divide both sides by } -4.$$
$$y = \frac{3}{4}x - 1 \qquad\qquad \text{Simplify.}$$

The coefficient of x, $\frac{3}{4}$, is the slope, and the constant term -1 is the y-intercept.

2   The slope-intercept form can be used to determine whether two lines are parallel or perpendicular. Recall that nonvertical parallel lines have the same slope and different y-intercepts. Also, nonvertical perpendicular lines have slopes whose product is -1.

EXAMPLE 4   Determine whether the graphs of $y = -\dfrac{1}{5}x + 1$ and $2x + 10y = 3$ are parallel lines, perpendicular lines, or neither.

Solution: The graph of $y = -\dfrac{1}{5}x + 1$ is a line with slope $-\dfrac{1}{5}$ and with y-intercept 1. To find the slope and the y-intercept of the graph of $2x + 10y = 3$, write this equation in the slope-intercept form. To do this, solve the equation for y.

$$2x + 10y = 3$$

$$10y = -2x + 3 \qquad \text{Subtract } 2x \text{ from both sides.}$$

$$y = -\frac{1}{5}x + \frac{3}{10} \qquad \text{Divide both sides by 10.}$$

The graph of this equation is a line with slope $-\dfrac{1}{5}$ and y-intercept $\dfrac{3}{10}$. Because both lines have the same slopes but different y-intercepts, these lines are parallel.

3   The slope-intercept form can also be used to write the equation of a line given its slope and y-intercept.

EXAMPLE 5   Find an equation of the line with y-intercept -3 and the slope of $\dfrac{1}{4}$.

Solution: We are given the slope and the y-intercept. Let $m = \dfrac{1}{4}$ and $b = -3$, and write the equation in slope-intercept form, $y = mx + b$.

$$y = mx + b$$

$$y = \frac{1}{4}x + (-3) \qquad \text{Let } m = \frac{1}{4} \text{ and } b = -3.$$

$$y = \frac{1}{4}x - 3 \qquad \text{Simplify.}$$

4   We now use the slope-intercept form of the equation of a line to graph a linear equation.

EXAMPLE 6   Graph the equation $y = \dfrac{3}{5}x - 2$.

Solution: Since the equation $y = \dfrac{3}{5}x - 2$ is written in slope-intercept form $y = mx + b$, the slope of its graph is $\dfrac{3}{5}$ and the y-intercept is -2. To graph, begin by plotting the intercept point $(0, -2)$. From this point, find another point of the graph by using the slope $\dfrac{3}{5}$ and recalling that slope is $\dfrac{\text{rise}}{\text{run}}$. Start at the intercept point and move 3 units up since the numerator of the slope is 3; then move 5 units to the right since the denominator of the slope is 5. Stop at the point $(5, 1)$. The line through $(0, -2)$ and $(5, 1)$ is the graph of $y = \dfrac{3}{5}x - 2$.

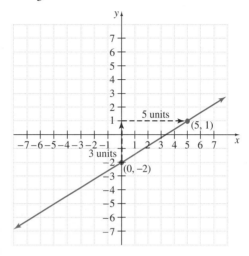

EXAMPLE 7   Use the slope-intercept form to graph the equation $4x + y = 1$.

Solution: First, write the given equation in slope-intercept form.

$$4x + y = 1$$
$$y = -4x + 1$$

The graph of this equation will have slope -4 and y-intercept 1. To graph this line, first plot the intercept point $(0, 1)$. To find another point of the graph, use the slope -4, which can be written as $\dfrac{-4}{1}$. Start at the point $(0, 1)$ and move 4 units down (since the numerator of the slope is -4), then move 1 unit to the right (since the denominator of the slope is 1). We arrive at the point $(1, -3)$. The line through $(0, 1)$ and $(1, -3)$ is the graph of $4x + y = 1$.

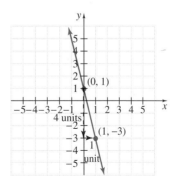

GRAPHING CALCULATOR EXPLORATIONS

A grapher is a very useful tool for discovering patterns. To discover the change in the graph of a linear equation caused by a change in slope, try the following. Use a standard window and graph a linear equation in the form $y = mx + b$. Recall that the graph of such an equation will have slope m and y-intercept b.

First graph $y = x + 3$. To do so, press the $\boxed{Y =}$ key and enter $Y_1 = x + 3$. Notice that this graph has slope 1 and that the y-intercept is 3. Next, on the same set of axes, graph $y = 2x + 3$ and $y = 3x + 3$ by pressing $\boxed{Y =}$ and entering $Y_2 = 2x + 3$ and $Y_3 = 3x + 3$.

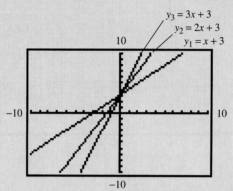

Notice the difference in the graph of each equation as the slope changes from 1 to 2 to 3. How would the graph of $y = 5x + 3$ appear? To see the change in the graph caused by a change in negative slope, try graphing $y = -x + 3$, $y = -2x + 3$, and $y = -3x + 3$ on the same set of axes.

Use a grapher to graph the following equations. For each exercise, graph the first equation and use its graph to predict the appearance of the other equations. Then graph the other equations on the same set of axes and check your prediction. graphical answers in App. F.

1. $y = x$; $y = 6x$, $y = -6x$

2. $y = -x$; $y = -5x$, $y = -10x$

3. $y = \frac{1}{2}x + 2$; $y = \frac{3}{4}x + 2$, $y = x + 2$

4. $y = x + 1$; $y = \frac{5}{4}x + 1$, $y = \frac{5}{2}x + 1$

5. $y = -7x + 5$; $y = 7x + 5$

6. $y = 3x - 1$; $y = -3x - 1$

MENTAL MATH

Identify the slope and the y-intercept of the graph of each equation.

1. $y = 2x - 1$ $\quad m = 2$; $(0, -1)$

2. $y = -7x + 3$ $\quad m = -7$; $(0, 3)$

3. $y = x + \frac{1}{3}$ $\quad m = 1$; $\left(0, \frac{1}{3}\right)$

4. $y = -x - \frac{2}{9}$ $\quad m = -1$, $\left(0, -\frac{2}{9}\right)$

5. $y = \frac{5}{7}x - 4$ $\quad m = \frac{5}{7}$; $(0, -4)$

6. $y = -\frac{1}{4}x + \frac{3}{5}$ $\quad m = -\frac{1}{4}$; $\left(0, \frac{3}{5}\right)$

1. $m = -2$; $(0, 4)$ **3.** $m = -\dfrac{1}{9}$; $\left(0, \dfrac{1}{9}\right)$ **4.** $m = -\dfrac{1}{7}$; $\left(0, \dfrac{6}{7}\right)$ **5.** $m = \dfrac{4}{3}$; $(0, -4)$ **6.** $m = \dfrac{1}{3}$; $(0, -2)$

EXERCISE SET 7.1

7. $m = -1$; $(0, 0)$ **9.** $m = 0$; $(0, -3)$ **11.** $m = \dfrac{1}{5}$; $(0, 4)$

Determine the slope and the y-intercept of the graph of each equation. See Examples 1–3.

D.

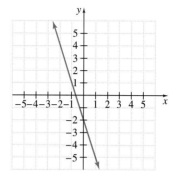

1. $2x + y = 4$

2. $-3x + y = -5$ $m = 3$; $(0, -5)$

3. $x + 9y = 1$

4. $x + 7y = 6$

5. $4x - 3y = 12$

6. $2x - 6y = 12$

7. $x + y = 0$

8. $x - y = 0$ $m = 1$; $(0, 0)$

9. $y = -3$

10. $y + 2 = 0$ $m = 0$; $(0, -2)$

11. $-x + 5y = 20$

12. $-8x + 3y = 11$ $m = \dfrac{8}{3}$; $\left(0, \dfrac{11}{3}\right)$

Match each linear equation with its graph.

13. $y = 2x + 1$ B

14. $y = -x + 1$ C

15. $y = -3x - 2$ D

16. $y = \dfrac{5}{3}x - 2$ A

A.

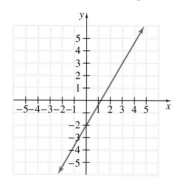

B.

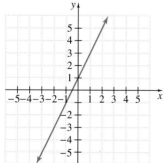

C.

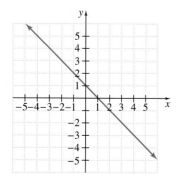

18. perpendicular   **21.** perpendicular   **22.** parallel

Determine whether the lines are parallel, perpendicular, or neither. See Example 4.

17. $x - 3y = -6$
$3x - y = 0$ neither

18. $-5x + y = -6$
$x + 5y = 5$

19. $2x - 7y = 1$
$2y = 7x - 2$ neither

20. $y = 4x - 2$
$4x + y = 5$ neither

21. $10 + 3x = 5y$
$5x + 3y = 1$

22. $-x + 2y = -2$
$2x = 4y + 3$

23. $6x = 5y + 1$
$-12x + 10y = 1$

24. $11x + 10y = 3$
$11y = 10x + 5$

 25. Explain how to write the equation of a line if you are given the slope of the line and its *y*-intercept.

26. Explain why the graphs of $y = x$ and $y = -x$ are perpendicular lines.   answers may vary

23. parallel   **24.** perpendicular

Use the slope-intercept form of the linear equation to write an equation of each line with given slope and y-intercept. See Example 5.

27. Slope -1; *y*-intercept 1 $y = -x + 1$

28. Slope $\dfrac{1}{2}$; *y*-intercept -6 $y = \dfrac{1}{2}x - 6$

29. Slope 2; *y*-intercept $\dfrac{3}{4}$ $y = 2x + \dfrac{3}{4}$

30. Slope -3; *y*-intercept $-\dfrac{1}{5}$ $y = -3x - \dfrac{1}{5}$

31. Slope $\dfrac{2}{7}$; *y*-intercept 0 $y = \dfrac{2}{7}x$

32. Slope $-\dfrac{4}{5}$; *y*-intercept 0 $y = -\dfrac{4}{5}x$

25. answers may vary

*Use the slope-intercept form to graph each equation.
See Examples 6 and 7.* graphical answers in App. F

33. $y = \dfrac{2}{3}x + 5$ **34.** $y = \dfrac{1}{4}x - 3$

35. $y = -\dfrac{3}{5}x - 2$ **36.** $y = -\dfrac{5}{7}x + 5$

37. $y = 2x + 1$ **38.** $y = -4x - 1$

39. $y = -5x$ **40.** $y = 6x$

41. $4x + y = 6$ **42.** $-3x + y = 2$

43. $x - y = -2$ **44.** $x + y = 3$

45. $3x + 5y = 10$ **46.** $2x + 3y = 5$

47. $4x - 7y = -14$ **48.** $3x - 4y = 4$

49. Show that the slope of a line whose equation is $y = mx + b$ is m. To do so, find two ordered pair solutions of this equation and use the slope formula. For example, if $x = 0$, then $y = 0 \cdot x + b$ or $y = b$. Thus, one ordered pair solution is $(0, b)$. Find one more ordered pair solution and find the slope. answers may vary

50. The equation relating Celsius temperature and Fahrenheit temperature is linear and its graph is shown to the right. This line contains the point $(0, 32)$, which means that $0°$ Celsius is equivalent to $32°$ Fahrenheit. This line also passes through the point $(100, 212)$.

a. Use this graph and approximate the Fahrenheit temperature equivalent to $20°$ Celsius. $68°$ F

b. Use the graph and approximate the Celsius temperature equivalent to $80°$ Fahrenheit. $27°$ C

c. Use the ordered pairs $(0, 32)$ and $(100, 212)$ and the slope-intercept form and write an equation of the line relating Celsius temperatures and Fahrenheit temperatures.

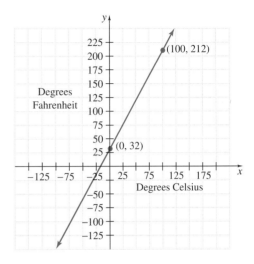

d. Use the linear equation found in part (c) and check your approximations from (a) and (b).

51. The equation relating temperatures on the Kelvin scale and the Celsius scale is linear. If $0°$ Celsius is equivalent to 273 Kelvin and $100°$ Celsius is equivalent to 373 Kelvin, find the linear equation that models the relationship between these two temperature scales. (*Hint:* See Exercise 50 and use the ordered pairs $(0, 273)$ and $(100, 373)$.) K = C + 273

Review Exercises graphical answers in App. F

Graph each equation. See Section 3.3.

52. $y = 5$ **53.** $y = -1$

54. $x = -3$ **55.** $x = 2$

*Write each equation in standard form: $Ax + By = C$.
See Section 3.2.*

56. $y - 5 = 2(x - 3)$ **57.** $y - 2 = 3(x - 6)$

58. $y + 2 = -1(x + 5)$ **59.** $y + 4 = -6(x + 1)$

7.2 | THE POINT-SLOPE FORM

O B J E C T I V E S

1 Use the point-slope form to find an equation of a line given its slope and a point of the line.

2 Use the point-slope form to find an equation of a line given two points of the line.

3 Find equations of vertical and horizontal lines.

4 Use the point-slope form to solve problems.

TAPE
BA 7.2

50. c. $F = \dfrac{9}{5}C + 32$ **50. d.** $68°F; 27°C$ **56.** $2x - y = 1$ **57.** $3x - y = 16$ **58.** $x + y = -7$

59. $6x + y = -10$

From the last section, if we know that if the y-intercept of a line and its slope are given, then the line can be graphed and an equation of this line can be found. Our goal in this section is to answer the following: If any two points of a line are given, can an equation of the line be found? To answer this question, first let's see if we can find an equation of a line given one point (not necessarily the y-intercept) and the slope of the line.

Suppose that the slope of a line is -3 and the line contains the point $(1, 4)$. For any other point with ordered pair (x, y) to be on the line, its coordinates must satisfy the slope equation

$$\frac{y - 4}{x - 1} = -3$$

Now multiply both sides of this equation by $x - 1$.

$$y - 4 = -3(x - 1)$$

This equation is a linear equation whose graph is a line that contains the point $(1, 4)$ and has slope -3. This form of a linear equation is called the **point-slope form.**

In general, when the slope of a line and any point on the line are known, the equation of the line can be found. To do this, use the slope formula to write the slope of a line that passes through points (x, y) and (x_1, y_1). We have

$$\frac{y - y_1}{x - x_1} = m$$

Multiply both sides of this equation by $x - x_1$ to obtain

$$y - y_1 = m(x - x_1)$$

POINT-SLOPE FORM OF THE EQUATION OF A LINE

The point-slope form of the equation of a line is $y - y_1 = m(x - x_1)$, where m is the slope of the line and (x_1, y_1) is a point on the line.

EXAMPLE 1   Find an equation of the line passing through $(-1, 5)$ with slope -2. Write the equation in standard form: $Ax + By = C$.

Solution: Since the slope and a point on the line are given, use point-slope form $y - y_1 = m(x - x_1)$ to write the equation. Let $m = -2$ and $(-1, 5) = (x_1, y_1)$.

$$
\begin{aligned}
y - y_1 &= m(x - x_1) \\
y - 5 &= -2[x - (-1)] \qquad &&\text{Let } m = -2 \text{ and } (x_1, y_1) = (-1, 5). \\
y - 5 &= -2(x + 1) \qquad &&\text{Simplify.} \\
y - 5 &= -2x - 2 \qquad &&\text{Use the distributive property.}
\end{aligned}
$$

$$y = -2x + 3 \qquad \text{Add 5 to both sides.}$$
$$2x + y = 3 \qquad \text{Add } 2x \text{ to both sides.}$$

In standard form, the equation is $2x + y = 3$.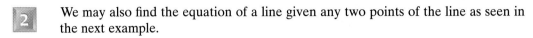

2 We may also find the equation of a line given any two points of the line as seen in the next example.

EXAMPLE 2 Find an equation of the line through $(2, 5)$ and $(-3, 4)$. Write the equation in standard form.

Solution: First, use the two given points to find the slope of the line. Let (x_1, y_1) be $(2, 5)$ and (x_2, y_2) be $(-3, 4)$.

$$m = \frac{y_2 - y_1}{x_2 - x_1} = \frac{4 - 5}{-3 - 2} = \frac{-1}{-5} = \frac{1}{5}$$

Next, use the slope $\frac{1}{5}$ and either one of the given points to write the equation in point-slope form. We use $(2, 5)$. Let $x_1 = 2$, $y_1 = 5$, and $m = \frac{1}{5}$.

$$y - y_1 = m(x - x_1) \qquad \text{Use point-slope form.}$$

$$y - 5 = \frac{1}{5}(x - 2) \qquad \text{Let } x_1 = 2, y_1 = 5, \text{ and } m = \frac{1}{5}.$$

$$5(y - 5) = 5 \cdot \frac{1}{5}(x - 2) \qquad \text{Multiply both sides by 5 to clear fractions.}$$

$$5y - 25 = x - 2 \qquad \text{Use the distributive property and simplify.}$$

$$-x + 5y - 25 = -2 \qquad \text{Subtract } x \text{ from both sides.}$$

$$-x + 5y = 23 \qquad \text{Add 25 to both sides.}$$

> **REMINDER** Multiply both sides of the equation $-x + 5y = 23$ by -1, and it becomes $x - 5y = -23$. Both $-x + 5y = 23$ and $x - 5y = -23$ are in standard form, and they are equations of the same line.

3 Recall from Section 3.3 that the graph of $x = c$, where c is a real number, is a vertical line with x-intercept c and the graph of $y = c$ is a horizontal line with y-intercept c.

EXAMPLE 3 Find an equation of the vertical line through $(-1, 5)$.

Solution: The equation of a vertical line can be written in the form $x = c$, so an equation for a vertical line passing through $(-1, 5)$ is $x = -1$.

EXAMPLE 4   Find an equation of the line parallel to the line $y = 7$ and passing through $(-2, -3)$.

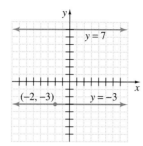

Solution: Since the graph of $y = 7$ is a horizontal line, any line parallel to it is also horizontal. The equation of a horizontal line can be written in the form $y = c$. An equation for the horizontal line passing through $(-2, -3)$ is $y = -3$.

    Problems occurring in many fields can be modeled by linear equations in two variables. The next example is from the field of marketing and shows how consumer demand of a product depends on the price of the product.

EXAMPLE 5   The Whammo Company has learned that by pricing a newly released Frisbee at $6, sales will reach 2000 Frisbees per day. Raising the price to $8 will cause the sales to fall to 1500 Frisbees per day.

a. Assume that the relationship between sales price and number of Frisbees sold is linear and write an equation describing this relationship. Write the equation in slope-intercept form.

b. Predict the daily sales of Frisbees if the price is $7.50.

Solution: **a.** First, use the given information and write two ordered pairs. Ordered pairs will be in the form (sales price, number sold) so that our ordered pairs are (6, 2000) and (8, 1500). Use the point-slope form to write an equation. To do so, we find the slope of the line that contains these points.

$$m = \frac{2000 - 1500}{6 - 8} = \frac{500}{-2} = -250$$

Next, use the slope and either one of the points to write the equation in point-slope form. We use (6, 2000).

$$y - y_1 = m(x - x_1)$$ Use point-slope form.
$$y - 2000 = -250(x - 6)$$ Let $x_1 = 6$ and $y_1 = 2000$.
$$y - 2000 = -250x + 1500$$ Use the distributive property.
$$y = -250x + 3500$$ Write in slope-intercept form.

b. To predict the sales if the price is $7.50, we find y when $x = 7.50$.

$$y = -250x + 3500$$
$$y = -250(7.50) + 3500$$ Let $x = 7.50$.

$$y = -1875 + 3500$$
$$y = 1625$$

If the price is \$7.50, sales will reach 1625 Frisbees per day. ▬▬▬

The preceding example may also be solved by using ordered pairs of the form (number sold, sales price).

FORMS OF LINEAR EQUATIONS

$Ax + By = C$	**Standard form** of a linear equation. A and B are not both 0.
$y = mx + b$	**Slope-intercept form** of a linear equation. The slope is m and the y-intercept is b.
$y - y_1 = m(x - x_1)$	**Point-slope form** of a linear equation. The slope is m and (x_1, y_1) is a point on the line.
$y = c$	**Horizontal line** The slope is 0 and the y-intercept is c.
$x = c$	**Vertical line** The slope is undefined and the x-intercept is c.

PARALLEL AND PERPENDICULAR LINES

Nonvertical parallel lines have the same slope.

The product of the slopes of two nonvertical pependicular lines is -1.

MENTAL MATH

The graph of each equation below is a line. Use the equation to identify the slope and a point of the line.

1. $y - 8 = 3(x - 4)$ $m = 3$; answers vary, Ex. (4, 8)

2. $y - 1 = 5(x - 2)$ $m = 5$; answers vary, Ex. (2, 1)

3. $y + 3 = -2(x - 10)$ $m = -2$; answers vary, Ex. (10, -3)

4. $y + 6 = -7(x - 2)$ $m = -7$; answers vary, Ex. (2, -6)

5. $y = \frac{2}{5}(x + 1)$ $m = \frac{2}{5}$; answers vary, Ex. (-1, 0)

6. $y = \frac{3}{7}(x + 4)$ $m = \frac{3}{7}$; answers vary, Ex. (-4, 0)

EXERCISE SET 7.2

Use the point-slope form of the linear equation to find an equation of each line with the given slope and passing through the given point. Then write the equation in standard form. See Example 1.

1. Slope 6; through $(2, 2)$ **2.** Slope 4; through $(1, 3)$

3. Slope -8; through $(-1, -5)$ $8x + y = -13$

4. Slope -2; through $(-11, -12)$ $2x + y = -34$

5. Slope $\frac{1}{2}$; through $(5, -6)$ $x - 2y = 17$

6. Slope $\frac{2}{3}$; through $(-8, 9)$ $2x - 3y = -43$

Find an equation of the line through the given points. Write the equation in standard form. See Example 2.

7. Through $(3, 2)$ and $(5, 6)$ $2x - y = 4$

1. $6x - y = 10$ **2.** $4x - y = 1$

8. Through $(6, 2)$ and $(8, 8)$ $3x - y = 16$
9. Through $(-1, 3)$ and $(-2, -5)$ $8x - y = -11$
10. Through $(-4, 0)$ and $(6, -1)$ $x + 10y = -4$
11. Through $(2, 3)$ and $(-1, -1)$ $4x - 3y = -1$
12. Through $(0, 0)$ and $\left(\dfrac{1}{2}, \dfrac{1}{3}\right)$ $2x - 3y = 0$

Find an equation of each line. See Example 3.
13. Vertical line through $(0, 2)$ $x = 0$
14. Horizontal line through $(1, 4)$ $y = 4$
15. Horizontal line through $(-1, 3)$ $y = 3$
16. Vertical line through $(-1, 3)$ $x = -1$
17. Vertical line through $\left(\dfrac{-7}{3}, \dfrac{-2}{5}\right)$ $x = -\dfrac{7}{3}$
18. Horizontal line through $\left(\dfrac{2}{7}, 0\right)$ $y = 0$

Find an equation of each line. See Example 4.
19. Parallel to $y = 5$, through $(1, 2)$ $y = 2$
20. Perpendicular to $y = 5$, through $(1, 2)$ $x = 1$
21. Perpendicular to $x = -3$, through $(-2, 5)$ $y = 5$
22. Parallel to $y = -4$, through $(0, -3)$ $y = -3$
23. Parallel to $x = 0$, through $(6, -8)$ $x = 6$
24. Perpendicular to $x = 7$, through $(-5, 0)$ $y = 0$

Solve. See Example 5.
25. A rock is dropped from the top of a 400-foot cliff. After 1 second, the rock is traveling 32 feet per second. After 3 seconds, the rock is traveling 96 feet per second.

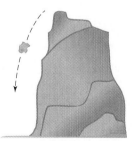

a. Assume that the relationship between time and speed is linear and write an equation describing this relationship. Use ordered pairs of the form (time, speed). $s = 32t$
b. Use this equation to determine the speed of the rock 4 seconds after it was dropped.   128 ft/sec

26. The value of a building bought in 1980 may be depreciated (or decreased) as time passes for income tax purposes. Seven years after the building was bought, this value was \$165,000 and 12 years after it was bought, this value was \$140,000.

a. If the relationship between number of years past 1980 and the depreciated value of the building is linear, write an equation describing this relationship. Use ordered pairs of the form (years past 1980, value of building). $V = -5000t + 200,000$
b. Use this equation to estimate the depreciated value of the building in the year 2000.   \$100,000

Find an equation of each line described. Write each equation in standard form.

27. With slope $-\dfrac{1}{2}$ through $\left(0, \dfrac{5}{3}\right)$ $3x + 6y = 10$

28. With slope $\dfrac{5}{7}$, through $(0, -3)$ $5x - 7y = 21$

29. Slope 1, through $(-7, 9)$ $x - y = -16$
30. Slope 5, through $(6, -8)$ $5x - y = 38$
31. Through $(10, 7)$ and $(7, 10)$ $x + y = 17$
32. Through $(5, -6)$ and $(-6, 5)$ $x + y = -1$
33. Through $(6, 7)$, parallel to the x-axis $y = 7$
34. Through $(0, -5)$, parallel to the y-axis $x = 0$

35. Slope $-\dfrac{4}{7}$, through $(-1, -2)$ $4x + 7y = -18$

36. Slope $-\dfrac{3}{5}$, through $(4, 4)$ $3x + 5y = 32$

37. Through $(-8, 1)$ and $(0, 0)$ $x + 8y = 0$
38. Through $(2, 3)$ and $(0, 0)$ $3x - 2y = 0$
39. Through $(0, 0)$ with slope 3 $3x - y = 0$
40. Through $(0, -2)$ with slope -1 $x + y = -2$
41. Through $(-6, -6)$ and $(0, 0)$ $x - y = 0$
42. Through $(0, 0)$ and $(4, 4)$ $x - y = 0$
43. Slope -5, y-intercept 7 $5x + y = 7$

44. Slope -2; y-intercept -4 $2x + y = -4$

45. Through $(-1, 5)$ and $(0, -6)$ $11x + y = -6$

46. Through $(4, 0)$ and $(0, -5)$ $5x - 4y = 20$

47. With undefined slope, through $\left(-\dfrac{3}{4}, 1\right)$ $x = -\frac{3}{4}$

48. With slope 0, through $(6.7, 12.1)$ $y = 12.1$

49. Through $(-2, -3)$, perpendicular to the y-axis

50. Through $(0, 12)$, perpendicular to the x-axis

51. With slope 7, through $(1, 3)$ $7x - y = 4$

52. With slope -10, through $(5, -1)$ $10x + y = 49$

53. A linear equation can be written that relates the radius of a circle to its circumference. A circle with a radius of 10 centimeters has a circumference of approximately 63 centimeters. A circle with a radius of 15 centimeters has a circumference of approximately 94 centimeters. Use the ordered pairs $(10, 63)$ and $(15, 94)$ to write a linear equation that approximates the relationship between the radius of the circle and its circumference. $31x - 5y = -5$

54. Del Monte Fruit Company is studying the sales of a pineapple sauce to see if this product is to be continued. At the end of its first year, profits on this product amounted to $30,000. At the end of the fourth year, profits are $66,000.

   a. Assume that the relationship between years on the market and profit is linear and write an equation describing this relationship. Use ordered pairs of the form (years on the market, profit). $p = 12{,}000t + 18{,}000$

   b. Use this equation to predict the profit at the end of 7 years. $102,000

55. The value of a computer bought in 1990 depreciates or decreases as time passes. Two years after the computer was bought, it was worth $2600 and 5 years after it was bought, it was worth $2000.

   a. If the relationship between number of years past 1990 and value of computer is linear, write an equation describing this relationship. Use ordered pairs of the form (years past 1990, value of computer). $V = -200t + 3000$

   b. Use this equation to estimate the value of the computer in the year 2000. $1000

56. The Pool Fun Company has learned that, by pricing a newly released Fun Noodle at $3, sales will reach 10,000 Fun Noodles per day during the sum-

mer. Raising the price to $5 will cause the sales to fall to 8000 Fun Noodles per day.

   a. Assume that the relationship between sales price and number of Fun Noodles sold is linear and write an equation describing this relationship. Use ordered pairs of the form (sales price, number sold). $S = -1000p + 13{,}000$

   b. Predict the daily sales of Fun Noodles if the price is $3.50.   9500 Fun Noodles

57. Given the equation of a nonvertical line, explain how to find the slope without finding two points on the line.   answers may vary

58. Given two points on a nonvertical line, explain how to use the point-slope form to find the equation of the line.   answers may vary

59. Write an equation in standard form of the line that contains the point $(-1, 2)$ and is

   a. parallel to the line $y = 3x - 1$. $3x - y = -5$

   b. perpendicular to the line $y = 3x - 1$.

60. Write an equation in standard form of the line that contains the point $(4, 0)$ and is

   a. parallel to the line $y = -2x + 3$. $2x + y = 8$

   b. perpendicular to the line $y = -2x + 3$.

61. Write an equation in standard form of the line that contains the point $(3, -5)$ and is

   a. parallel to the line $3x + 2y = 7$. $3x + 2y = -1$

   b. perpendicular to the line $3x + 2y = 7$.

62. Write an equation in standard form of the line that contains the point $(-2, 4)$ and is

   a. parallel to the line $x + 3y = 6$. $x + 3y = 10$

   b. perpendicular to the line $x + 3y = 6$.

Review Exercises   graphical answers in App. F

Graph each linear equation. See Section 3.2.

63. $y = 2x - 6$ **64.** $y = -5x$

65. $x + 3y = 5$ **66.** $5x - y = 4$

67. $y = -2$ **68.** $x = 4$

49. $y = -3$ **50.** $x = 0$ **59. b.** $x + 3y = 5$ **60. b.** $x - 2y = 4$ **61. b.** $2x - 3y = 21$ **62. b.** $3x - y = -10$

7.3 | GRAPHING NONLINEAR EQUATIONS

TAPE
BA 7.3

O B J E C T I V E S

1. Identify and graph nonlinear equations.
2. Write ordered pairs from the graphs of nonlinear equations.

Recall from Section 3.2 that a linear equation in two variables is an equation that can be written in the form $Ax + By = C$ where A and B are not both 0. We also know that the graph of a linear equation is a line. Linear equations are important in their own right, but they are not sufficient for describing all relationships between two quantities. For example, the area, y, of a square is related to the length of its side x, by the nonlinear equation $y = x^2$.

$$y = x^2$$

$\uparrow$
x
$\downarrow$

1 In this section, we practice identifying and graphing nonlinear equations.

EXAMPLE 1   Graph the equation $y = x^2$.

Solution: This equation is not linear and its graph is not a line. Recall that the graph of an equation is a picture of its ordered pair solutions. To graph this equation, begin by finding ordered pair solutions. Because this graph is solved for y, we choose x-values and find corresponding y-values.

x	y
−3	9
−2	4
−1	1
0	0
1	1
2	4
3	9

If $x = -3$, then $y = (-3)^2$, or 9
If $x = -2$, then $y = (-2)^2$, or 4
If $x = -1$, then $y = (-1)^2$, or 1
If $x = 0$, then $y = 0^2$, or 0
If $x = 1$, then $y = 1^2$, or 1
If $x = 2$, then $y = 2^2$, or 4
If $x = 3$, then $y = 3^2$, or 9

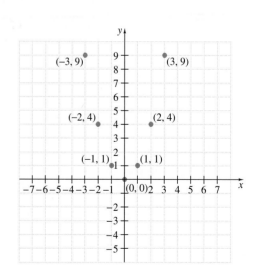

To complete the graph, connect these plotted points with a smooth curve as shown below.

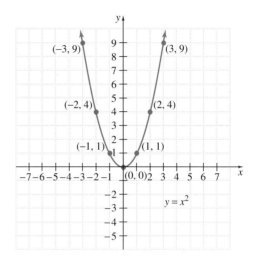

This curve is called a **parabola.** The equation $y = x^2$ is called a **quadratic equation** and the graph of a quadratic equation is a parabola.
We will study more about parabolas in Chapter 10.

QUADRATIC EQUATION IN TWO VARIABLES

A quadratic equation in two variables is an equation that can be written in the form

$$y = ax^2 + bx + c$$

where a, b, and c are real numbers and a is not 0. The graph of a quadratic equation is a parabola.

EXAMPLE 2   Graph the equation $y = |x|$.

Solution: This is not a linear equation and its graph is not a line. Because we do not know the shape of the graph of this equation, we find many ordered pair solutions. Choose x-values and substitute to find corresponding y-values.

x	y
-4	4
-2	2
0	0
$\dfrac{1}{2}$	$\dfrac{1}{2}$
1	1
3	3

If $x = -4$, then $y = |-4|$, or 4

If $x = -2$, then $y = |-2|$, or 2

If $x = 0$, then $y = |0|$, or 0

If $x = \dfrac{1}{2}$, then $y = \left|\dfrac{1}{2}\right|$, or $\dfrac{1}{2}$

If $x = 1$, then $y = |1|$, or 1

If $x = 3$, then $y = |3|$, or 3

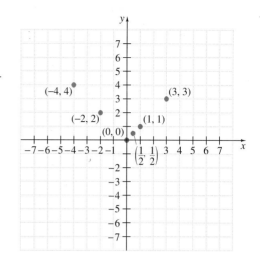

From the plotted ordered pairs, we see that the graph of this absolute value equation is V-shaped. The completed graph is shown below.

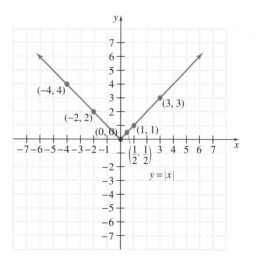

EXAMPLE 3   Graph the equation $y = |x| - 3$.

Solution: The graph of $y = |x| - 3$ is simply the graph of $y = |x|$ moved down 3 units. To see this, choose x-values and substitute to find corresponding y-values.

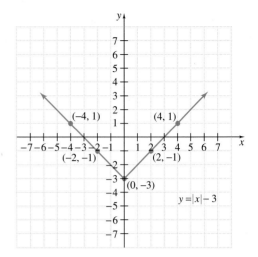

If $x = -4$, then $y = |-4| - 3$, or 1

If $x = -2$, then $y = |-2| - 3$, or -1

If $x = 0$, then $y = |0| - 3$, or -3

If $x = 2$, then $y = |2| - 3$, or -1

If $x = 4$, then $y = |4| - 3$, or 1

x	y
-4	1
-2	-1
0	-3
2	-1
4	1

Notice that the graph of $y = |x| - 3$ is the graph of $y = |x|$ moved down 3 units.

In general, the shape of the graph of $y = |x| + k$ where k is a constant is V-shaped.

2 The concept of a relation and a function is introduced in the next section. To prepare for this section, we review reading ordered pairs from a graph.

EXAMPLE 4 Find the missing coordinate so that each ordered pair is a solution of the equation graphed below.

a. $(-1, \quad)$ **b.** $(-2, \quad)$ **c.** $(\quad, -1)$

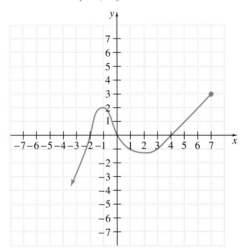

Solution: **a.** To find the y-coordinate that corresponds to an x-value of -1, find -1 on the x-axis and move vertically until the graph is reached. From the point on the graph, move horizontally until the y-axis is reached and read the y-value. The y-value is 2, so the ordered pair is $(-1, 2)$.

   b. The completed ordered pair is $(-2, 0)$.

c. The *y*-value of −1 corresponds to two *x*-values as shown below. The completed ordered pairs are (1, −1) and (3, −1).

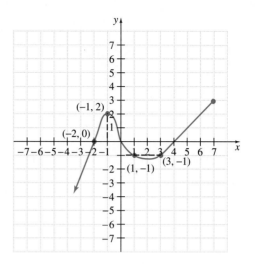

EXAMPLE 5   Use the graph of Example 4 to
   a. find the coordinates of the point on the graph with the greatest *y*-value;
   b. find the coordinates of the point on the graph with the least *y*-value;
   c. list the *x*-intercept points and the *y*-intercept points.

Solution: **a.** The point on the graph with the greatest *y*-value is the "highest" point. This point has coordinates (7, 3).

   b. The point on the graph with the least *y*-value is the "lowest" point. The arrow shown on the graph means that the graph continues in the same manner. This means that this graph contains no lowest point, that is, no point with the least *y*-value.

   c. The *x*-intercept points are (−2, 0), (0, 0), and (4, 0). The *y*-intercept point is (0, 0).

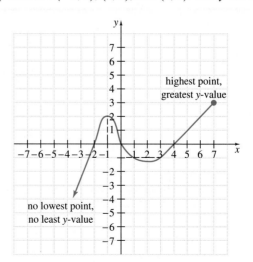

GRAPHING CALCULATOR EXPLORATIONS

We can use a grapher to approximate the x-intercepts of the graph of an equation such as the quadratic equation $y = x^2 - 2x - 4$. When we use a standard window, the graph of this equation looks like this:

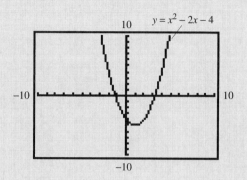

The graph appears to have one x-intercept between -2 and -1 and one between 3 and 4. To find the x-intercept between 3 and 4 to the nearest hundredth, we can use a ROOT feature, a ZOOM feature, which magnifies a portion of the graph around the cursor, or we can redefine our window. If we redefine our window to

$$\text{Xmin} = 2 \qquad \text{Ymin} = -1$$
$$\text{Xmax} = 5 \qquad \text{Ymax} = 1$$
$$\text{Xscl} = 1 \qquad \text{Yscl} = 1$$

and trace along the graph, the resulting screen is

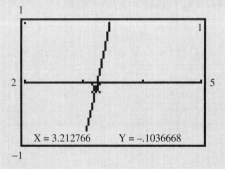

By using the TRACE feature, we can now see that one of the intercepts is between 3.21 and 3.25. To approximate to the nearest hundredth, ZOOM again or redefine the window to *(continued)*

Xmin = 3.2   Ymin = −0.1

Xmax = 3.3   Ymax = 0.1

Xscl = 1   Yscl = 1

If we use the TRACE feature again, we see that, to the nearest hundredth, the x-intercept is 3.23. By repeating this process, we can approximate the other x-intercept to be −1.23.

To check, find y when $x = 3.23$ and when $x = -1.23$. Both of these y-values should be close to 0. (They will not be exactly 0 since we approximated these solutions.)

3. (−0.87, 0), (2.78, 0)   **4.** (−30.66, 0), (−0.34, 0)

Approximate to two decimals places the x-intercepts of the graph of each equation.

1. $y = x^2 + 3x - 2$ (0.56, 0), (−3.56, 0)   **2.** $y = 5x^2 - 7x + 1$ (0.16, 0), (1.24, 0)

3. $y = 2.3x^2 - 4.4x - 5.6$ **4.** $y = 0.2x^2 + 6.2x + 2.1$

5. $y = 2|x| - 1.3$ (−0.65, 0), (0.65, 0)   **6.** $y = |-3x| + 2.7$ none

7. $y = 1.7x^3 + 5.9$ (−1.51, 0)   **8.** $y = -3.6x^3 - 1.3$ (−0.71, 0)

EXERCISE SET 7.3

Graph each nonlinear equation. See Examples 1 through 3.   graphical answers in App. F

1. $y = x^2 + 2$ **2.** $y = x^2 - 2$ **3.** $y = |x| - 1$

4. $y = |x| + 3$ **5.** $y = -x^2$ **6.** $y = -|x|$

7. $y = |x + 5|$ **8.** $y = (x - 5)^2$

C.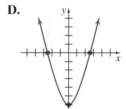

D.

Match each equation with its graph.

9. $y = |x| + 1$ C   **10.** $y = |x| - 4$ F

11. $y = x^2 - 6$ D   **12.** $y = x^2 + 2$ B

13. $y = 3x - 2$ A   **14.** $y = x + 1$ E

A.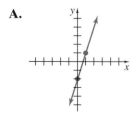

B.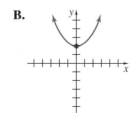

E.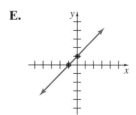

F.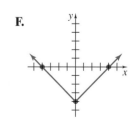

Use the graph to complete each ordered pair. See Example 4.

15.

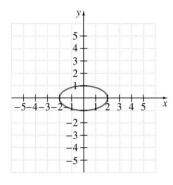

(0,)
(,0)

(0, 1), (0, −1),
(−2, 0), (2, 0)

16.

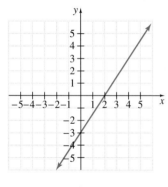

(3,)
(, −2)

(3, 3)
(1, −2),
(−1, −2)

17.

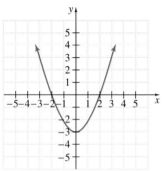

(2,)
(, −2)

(2, 0),
$\left(\frac{2}{3}, -2\right)$

18.

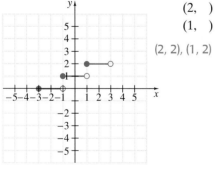

(2,)
(1,)

(2, 2), (1, 2)

19.

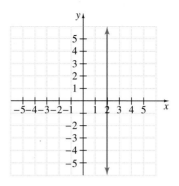

(2,)
(2, any real number)

20.

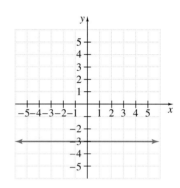

(, −3)
(any real number, −3)

Answer the following questions about the given graph. See Example 5.

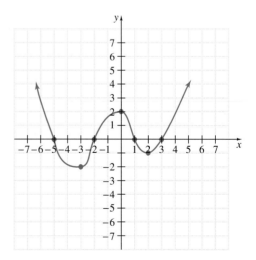

21. Find the coordinates of the point with the greatest *y*-value.   There is no such point.

22. Find the coordinates of the point with the least *y*-value.   (−3, −2)

23. Between $x = 1$ and $x = 3$, find the coordinates of the point with the least y-value. $(2, -1)$

24. List the x- and y-intercept points for the graph.
$(-5, 0), (-2, 0), (1, 0), (3, 0), (0, 2)$

Determine whether each equation is linear or not. Then graph each equation. graphical answers in App. F

25. $x + y = 3$

26. $y - x = 8$

27. $y = 4x$

28. $y = 6x$

29. $y = 4x - 2$

30. $y = 6x - 5$

31. $y = |x| + 3$

32. $y = |x| + 2$

33. $2x - y = 5$

34. $4x - y = 7$

35. $y = 2x^2$

36. $y = 3x^2$

37. $y = (x - 3)^2$

38. $y = (x + 3)^2$

39. $y = -2x$

40. $y = -3x$

41. $y = -2x + 3$

42. $y = -3x + 2$

43. $y = x^2 + 4x$

44. $y = x^2 - 2x$

45. Graph the equation $y = x^3$ by completing the table of values, plotting the ordered pair solutions on a rectangular coordinate system, and then connecting the points with a smooth curve.

x	-3	-2	-1	0	1	2	3
y	-27	-8	-1	0	1	8	27

See App. F.

46. Graph the equation $y = x^4$ by completing the table of values, plotting the ordered pair solutions on a

rectangular coordinate system, and then connecting the points with a smooth curve.

x	-3	-2	-1	0	1	2	3
y	81	16	1	0	1	16	81

See App. F.

Review Exercises

Find the value of $x^2 - 3x + 1$ for each given value of x. See Section 2.1.

47. 2 -1

48. 5   11

49. -1 5

50. -3 19

For each graph, determine whether any x-values correspond to two or more y-values. See Section 7.3.

51.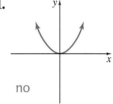

no

52.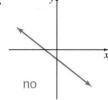

no

53.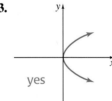

yes

54.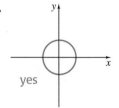

yes

7.4   FUNCTIONS

O B J E C T I V E S

1 Identify relations, domains, and ranges.

2 Identify functions.

3 Use the vertical line test.

4 Use function notation.

TAPE
BA 7.4

In previous sections, we have discussed the relationships between two quantities. For example, the relationship between the length of the side of a square x and its area y is described by the equation $y = x^2$. These variables x and y are related in the following way: for any given value of x, we can find the corresponding value of y by squaring the x-value. Ordered pairs can be used to write down solutions of this equation. For example, (2, 4) is a solution of $y = x^2$, and this notation tells us that the x-value 2 is related to the y-value 4 for this equation. In other words, when the length of the side of a square is 2 units, its area is 4 units square.

A set of ordered pairs is called a **relation.** The set of all x-coordinates is called the **domain** of a relation, and the set of all y-coordinates is called the **range** of a relation. Equations such as $y = x^2$ are also called relations since equations in two variables define a set of ordered pair solutions.

EXAMPLE 1   Find the domain and the range of the relation $\{(0, 2), (3, 3), (-1, 0), (3, -2)\}$.

Solution: The domain is the set of all x-values or $\{-1, 0, 3\}$, and the range is the set of all y-values, or $\{-2, 0, 2, 3\}$.

2   Some relations are also functions.

FUNCTION

A function is a set of ordered pairs that assigns to each x-value exactly one y-value.

EXAMPLE 2   Which of the following relations are also functions?

a. $\{(-1, 1), (2, 3), (7, 3), (8, 6)\}$ **b.** $\{(0, -2), (1, 5), (0, 3), (7, 7)\}$

Solution: **a.** Although the ordered pairs (2, 3) and (7, 3) have the same y-value, each x-value is assigned to only one y-value so this set of ordered pairs is a function.

b. The x-value 0 is assigned to two y-values, -2 and 3, so this set of ordered pairs is not a function.

EXAMPLE 3   Is the relation $y = x^2$ a function?

Solution: The relation $y = x^2$ is a function if each x-value corresponds to a single y-value. For each value substituted in the equation $y = x^2$, the squaring of the x-value gives a single result so that only one y-value is produced from each x-value. Thus $y = x^2$ defines a function.

EXAMPLE 4   Is the relation $x = y^2$ a function?

Solution: In $x = y^2$, if $y = 3$, then $x = 9$. Also, if $y = -3$, then $x = 9$. In other words, the x-value 9 corresponds to two y-values, 3 and -3. Thus $x = y^2$ is not a function.

3   As we have seen so far, not all relations are functions. Consider the graphs of $x = y^2$ and $y = x^2$ shown next. For the graph of $x = y^2$, the x-value 9, for example, corresponds to two y-values, 3 and -3, as shown by the vertical line. Recall from Example 4 that $x = y^2$ is not a function. For the graph of $y = x^2$, each x-value corresponds to only one y-value. Notice that a vertical line drawn will intersect the graph at no more than one point. Recall from Example 3 that $y = x^2$ is a function.

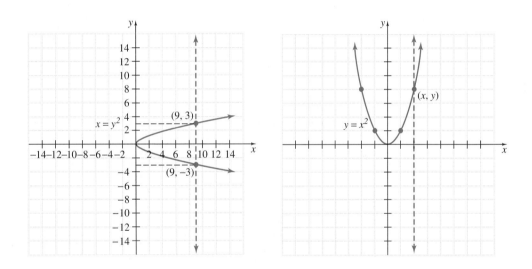

Graphs can be used to help determine whether a relation is also a function by the following vertical line test.

> **VERTICAL LINE TEST**
>
> If a vertical line can be drawn so that it intersects a graph more than once, the graph is not the graph of a function.

EXAMPLE 5   Which of the following graphs are graphs of functions?

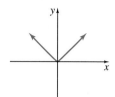   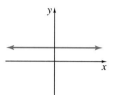   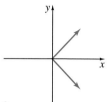

a.   **b.**   **c.**

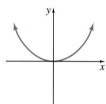   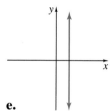   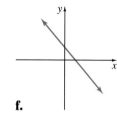

d.   **e.**   **f.**

Solution: **a.** This graph is the graph of a function since no vertical line will intersect this graph more than once.

b. This graph is also the graph of a function.

c. This graph is not the graph of a function. Note that vertical lines can be drawn that intersect the graph in two points.

d. This graph is the graph of a function.

e. This graph is not the graph of a function. A vertical line can be drawn that intersects this line at every point.

f. The graph of this linear equation is the graph of a function.

Recall that the graph of a linear equation is a line, and a line that is not vertical will pass the vertical line test. **Thus, all linear equations are functions except those of the form $x = c$, which are vertical lines.**

EXAMPLE 6   Which of the following linear equations are functions?

     a. $y = x$

     b. $y = 2x + 1$

     c. $y = 5$

     d. $x = -1$

Solution: (a), (b), and (c) are functions because their graphs are nonvertical lines.

(d) This is not a function because its graph is a vertical line. ▄▄▄▄▄

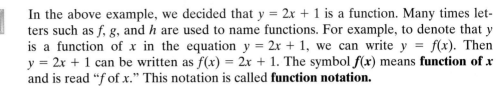

In the above example, we decided that $y = 2x + 1$ is a function. Many times letters such as f, g, and h are used to name functions. For example, to denote that y is a function of x in the equation $y = 2x + 1$, we can write $y = f(x)$. Then $y = 2x + 1$ can be written as $f(x) = 2x + 1$. The symbol $f(x)$ means **function of x** and is read "f of x." This notation is called **function notation.**

The notation $f(1)$ means to replace x with 1 and find the resulting y or function value. Since

$$f(x) = 2x + 1$$

then

$$f(1) = 2(1) + 1 = 3$$

This means that, when $x = 1$, y or $f(x) = 3$. Now find $f(2)$, $f(0)$, and $f(-1)$.

$$f(x) = 2x + 1 \qquad f(x) = 2x + 1 \qquad f(x) = 2x + 1$$
$$f(2) = 2(2) + 1 \qquad f(0) = 2(0) + 1 \qquad f(-1) = 2(-1) + 1$$
$$= 4 + 1 \qquad\qquad = 0 + 1 \qquad\qquad = -2 + 1$$
$$= 5 \qquad\qquad\quad = 1 \qquad\qquad\quad = -1$$

R E M I N D E R   Note that $f(x)$ is a special symbol in mathematics used to denote a function. The symbol $f(x)$ is read "f of x." It does **not** mean $f \cdot x$ (f times x).

EXAMPLE 7   Given $g(x) = x^2 - 3$, find the following.

a. $g(2)$ **b.** $g(-2)$ **c.** $g(0)$

Solution:
a. $g(x) = x^2 - 3$ **b.** $g(x) = x^2 - 3$ **c.** $g(x) = x^2 - 3$
$\quad\; g(2) = 2^2 - 3 \qquad\quad g(-2) = (-2)^2 - 3 \qquad\;\; g(0) = 0^2 - 3$
$\qquad\quad = 4 - 3 \qquad\qquad\qquad = 4 - 3 \qquad\qquad\qquad = 0 - 3$
$\qquad\quad = 1 \qquad\qquad\qquad\quad = 1 \qquad\qquad\qquad\quad = -3$ ▄▄▄▄▄

Many of the equations that were graphed in the last section were functions.

EXAMPLE 8   Graph the function $f(x) = 2 - x^2$.

Solution: Recall from the previous section that $f(x) = 2 - x^2$ or $y = 2 - x^2$ is a quadratic equation whose graph is a parabola. To graph this parabola, find ordered pair solutions of the form $(x, f(x))$ or (x, y), plot them, and connect them with a smooth curve.

x	f(x)
-3	-7
-2	-2
-1	1
0	2
1	1
2	-2
3	-7

$f(-3) = 2 - (-3)^2 = 2 - 9 = -7$

$f(-2) = 2 - (-2)^2 = 2 - 4 = -2$

$f(-1) = 2 - (-1)^2 = 2 - 1 = 1$

$f(0) = 2 - 0^2 = 2 - 0 = 2$

$f(1) = 2 - 1^2 = 2 - 1 = 1$

$f(2) = 2 - 2^2 = 2 - 4 = -2$

$f(3) = 2 - 3^2 = 2 - 9 = -7$

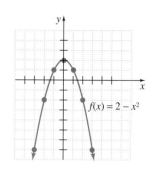

Notice that the graph of this function passes the vertical line test.

Functions can often be found in magazines, newspapers, books, and other printed material in the form of tables or graphs as shown below.

EXAMPLE 9 The graph below shows the sunrise time for Indianapolis, Indiana, for the year. Use this graph to answer the questions below.

a. Approximate the time of sunrise on February 1.

b. Approximate the date when the sun rises at 5 A.M.

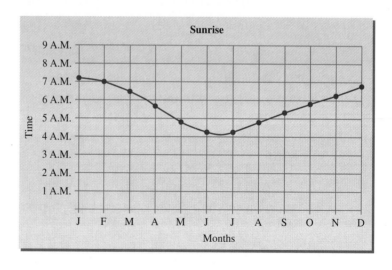

c. Is this the graph of a function?

Solution: **a.** To approximate the time of sunrise on February 1, find the mark on the horizontal axis that corresponds to February 1. From this mark, move vertically up ward until the graph is reached. From the point on the graph, move horizontally to the left until the vertical axis is reached. The vertical axis reads 7 A.M.

b. To approximate the date when the sun rises at 5 A.M., find 5 A.M. on the time axis and move horizontally to the right. Notice that you will reach the graph twice corresponding to two dates for which the sun rises at 5 A.M. Follow both points on the graph vertically downward until the horizontal axis is reached. The sun rises at 5 A.M. at approximately the end of the month of April and the middle of the month of August.

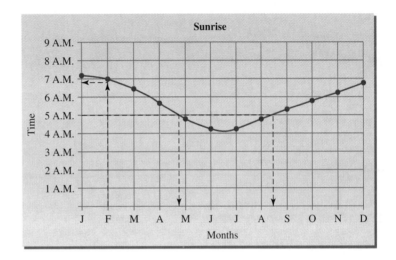

c. The graph is the graph of a function. In other words, for every day of the year in Indianapolis, there is exactly one sunrise time.

We now practice finding the domain and the range of a function. The domain of our functions will be the set of all possible real numbers that x can be replaced by. The range is the set of corresponding y-values.

EXAMPLE 10   Find the domain of each function.

a. $g(x) = \dfrac{1}{x}$ **b.** $f(x) = 2x + 1$

Solution: **a.** Recall that we cannot divide by 0 so that the domain of $g(x)$ is the set of all real numbers except 0.

b. In this function, x can be any real number. The domain of $f(x)$ is the set of all real numbers.

EXAMPLE 11 Find the domain and the range of each function graphed.

a.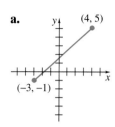

b.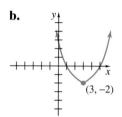

Solution: **a.**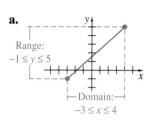

b.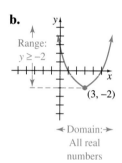

EXERCISE SET 7.4

Find the domain and the range of each relation. See Example 1.

1. {(2, 4), (0, 0), (−7, 10), (10, −7)}

2. {(3, −6), (1, 4), (−2, −2)}

3. {(0, −2), (1, −2) (5, −2)}

4. {(5, 0), (5, −3), (5, 4), (5, 3)}

Determine which relations are also functions. See Example 2.

5. {(1, 1), (2, 2), (−3, −3), (0, 0)}   yes

6. {(1, 2), (3, 2), (4, 2)}   yes

7. {(−1, 0), (−1, 6), (−1, 8)}   no

8. {(11, 6), (−1, −2), (0, 0), (3, −2)}   yes

Decide whether the equation describes a function. See Examples 3, 4, and 6.

9. $y = x + 1$ yes

10. $y = x − 1$ yes

11. $x = 2y^2$ no

12. $y = x^2$ yes

13. $y − x = 7$ yes

14. $2x − 3y = 9$ yes

15. $y = \dfrac{1}{x}$ yes

16. $y = \dfrac{1}{x − 3}$ yes

17. $x = 5$ no

18. $y = 7$ yes

19. $y = x^3$ yes

20. $x = y^3$ yes

21. $y < 2x + 1$ no

22. $y \leq x − 4$ no

23. $y = x + 3$ yes

24. $y = 7x + 2$ yes

Use the vertical line test to determine whether each graph is the graph of a function. See Example 5.

25. 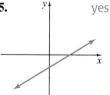   yes

26. 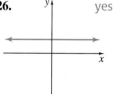   yes

27. 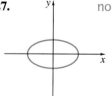   no

28. 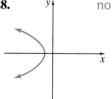   no

1. domain: {−7, 0, 2, 10}; range {−7, 0, 4, 10}   **2.** domain: {−2, 1, 3}; range {−6, −2, 4}
3. domain: {0, 1, 5}; range: {−2}   **4.** domain: {5}; range: {−3, 0, 3, 4}

29. 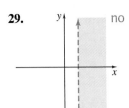 no   **30.**  no

31. 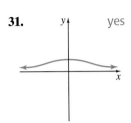 yes   **32.** 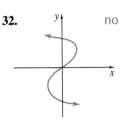 no

Given the following functions, find f(−2), f(0), and f(3). See Example 7.

33. $f(x) = 2x - 5$ **34.** $f(x) = 3 - 7x$
35. $f(x) = x^2 + 2$ 6, 2, 11   **36.** $f(x) = x^2 - 4$ 0, −4, 5
37. $f(x) = x^3$ −8, 0, 27   **38.** $f(x) = -x^3$ 8, 0, −27
39. $f(x) = |x|$ 2, 0, 3   **40.** $f(x) = |2 - x|$ 4, 2, 1

Given the following functions, find h(−1), h(0), and h(4). See Example 7.

41. $h(x) = 5x$ −5, 0, 20   **42.** $h(x) = -3x$ 3, 0, −12
43. $h(x) = 2x^2 + 3$ 5, 3, 35   **44.** $h(x) = -x^2$ −1, 0, −16
45. $h(x) = -x^2 - 2x + 3$ **46.** $h(x) = -x^2 + 4x - 3$
47. $h(x) = 6$ 6, 6, 6   **48.** $h(x) = -12$

Graph each function. See Example 8. graphical answers
                                        in App. F.
49. $f(x) = x + 5$ **50.** $g(x) = x - 2$
51. $h(x) = -2x - 1$ **52.** $H(x) = -5x + 2$
53. $g(x) = -|x|$ **54.** $f(x) = -2|x|$
55. $F(x) = 4 - x^2$ **56.** $G(x) = 3 - x^2$
57. $h(x) = |x - 2|$ **58.** $g(x) = |x + 1|$

Use the graph in Example 9 to answer the questions below.

59. Approximate the time of sunrise on September 1 in Indianapolis.   5:20 A.M.

60. Approximate the date(s) when the sun rises in Indianapolis at 6 A.M.

61. Describe the change in sunrise over the year for Indianapolis.   answers may vary

62. When, in Indianapolis, is the earliest sunrise? What point on the graph does this correspond to?

Find the domain of each function. See Example 10.

63. $f(x) = 3x - 7$ **64.** $g(x) = 5 - 2x$
65. $h(x) = \dfrac{1}{x + 5}$ **66.** $f(x) = \dfrac{1}{x - 6}$
67. $g(x) = |x + 1|$ **68.** $h(x) = |2x|$

Find the domain and the range of each function graphed. See Example 11.

69. 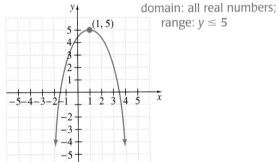   domain: all real numbers; range: $y \geq -4$

70. 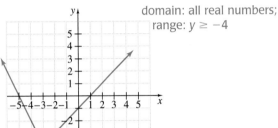   domain: all real numbers; range: $y \leq 5$

71. 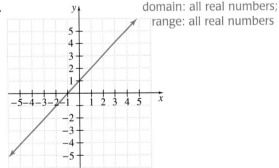   domain: all real numbers; range: all real numbers

---

33. −9, −5, 1   **34.** 17, 3, −18   **45.** 4, 3, −21   **46.** −8, −3, −3   **48.** −12, −12, −12
60. at the end of the month of March and at the middle of the month of October
62. 4:20 A.M. in the middle of the month of June   **63.** all real numbers   **64.** all real numbers   **65.** all real numbers except −5   **66.** all real numbers except 6   **67.** all real numbers   **68.** all real numbers

72.

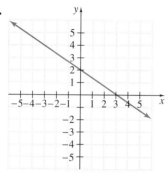

domain: all real numbers;
range: all real numbers

73.

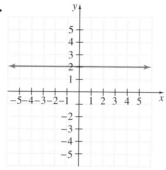

domain: all real numbers;
range: $y = 2$

74.

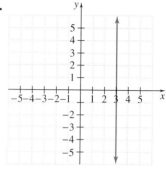

domain: $x = 3$;
range: all real numbers

The graph below shows the sunset times for Seward, Alaska. Use this graph to answer Exercises 75 through 79.

75. Approximate the time of sunset on June 1.   9 P.M.

76. Approximate the time of sunset on November 1. 4 P.M.

77. Approximate the date(s) when the sunset is 3 P.M.

78. Approximate the date(s) when the sunset is 9 P.M.

79. Is this graph the graph of a function? Why or why not?   yes; it passes the vertical line test

80. In your own words define **(a)** function; **(b)** domain; **(c)** range.   answers may vary

81. Explain the vertical line test and how it is used.

82. Since $y = x + 7$ is a function, rewrite the equation using function notation. $f(x) = x + 7$

83. Forensic scientists use the function

$$H(x) = 2.59x + 47.24$$

to estimate the height of a woman given the length x of her femur bone.

a. Estimate the height of a woman whose femur measures 46 centimeters.   166.38 cm

b. Estimate the height of a woman whose femur measures 39 centimeters.   148.25 cm

84. The dosage in milligrams D of Ivermectin, a heartworm preventive for a dog who weighs x pounds, is given by the function

$$D(x) = \frac{136}{25}x$$

a. Find the proper dosage for a dog that weighs 35 pounds.   190.4 mg

b. Find the proper dosage for a dog that weighs 70 pounds.   380.8 mg

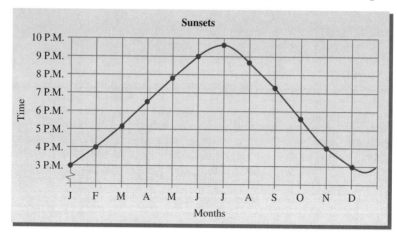

77. January 1, December 1   **78.** June 1, end of July   **81.** answers may vary

Review Exercises

Find the perimeter of the following figures. See Section 6.4.

85.

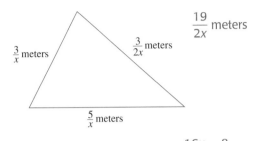

$\frac{3}{x}$ meters $\frac{3}{2x}$ meters $\frac{19}{2x}$ meters

$\frac{5}{x}$ meters

86.

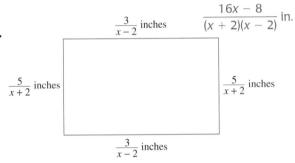

$\frac{3}{x-2}$ inches $\frac{16x-8}{(x+2)(x-2)}$ in.

$\frac{5}{x+2}$ inches $\frac{5}{x+2}$ inches

$\frac{3}{x-2}$ inches

Approximate the coordinates of the point of intersection. See Section 3.1.

87.

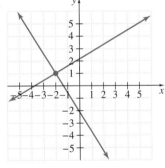

$(-2, 1)$

88.

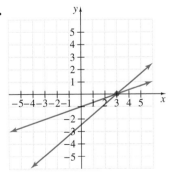

$(3, 0)$

89.

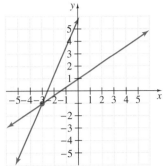

$(-3, -1)$

90.

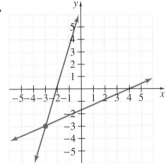

$(-3, -3)$

A Look Ahead

EXAMPLE If $f(x) = x^2 + 2x + 1$, find the following:

   a. $f(\pi)$ **b.** $f(c + d)$

Solution:

   a. $f(x) = x^2 + 2x + 1$
            $f(\pi) = \pi^2 + 2\pi + 1$

   b. $f(x) = x^2 + 2x + 1$
         $f(c + d) = (c + d)^2 + 2(c + d) + 1$
                  $= c^2 + 2cd + d^2 + 2c + 2d + 1$

Given the following functions, find the indicated values.

91. $f(x) = 2x + 7$;   **a.** $f(2)$ 11
    b. $f(a)$ $2a + 7$ **c.** $f(a + 2)$ $2a + 11$

92. $g(x) = -3x + 12$;   **a.** $g(s)$ $-3s + 12$
    b. $g(r)$ $-3r + 12$ **c.** $g(r + s)$ $-3r - 3s + 12$

93. $h(x) = x^2 + 7$;   **a.** $h(3)$ 16
    b. $h(a)$ $a^2 + 7$ **c.** $h(a - 3)$ $a^2 - 6a + 16$

94. $f(x) = x^2 - 12$;   **a.** $f(12)$ 132
    b. $f(12 + a)$ **c.** $f(12 - a)$132 $- 24a + a^2$

94. b. 132 $+ 24a + a^2$

Group Activity

Matching Descriptions of Linear Data to Their Equations and Graphs

atch each given equation to its corresponding graph. Then for each of the following descriptions of situations, match the description to its paired equation and graph that model the situation. Assume that the relationship described in each situation is linear, and assume that $x = 0$ represents 1990.

Equations

a. $y = -20.7x + 63.5$

b. $y = 27x + 535$

c. $y = 428$

d. $y = 26.5x - 422$

e. $y = 0.73x + 19.35$

f. $y = -20.7x + 270.5$

g. $y = -300$

h. $y = -27x + 535$

i. $y = -0.73x + 19.35$

j. $y = 26.5x + 422$

Graphs

(i)

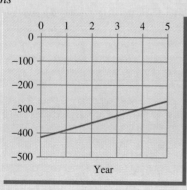

(ii)

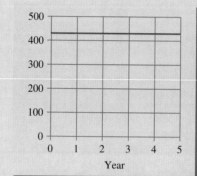

(iii)

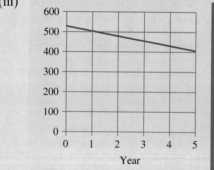

(iv)

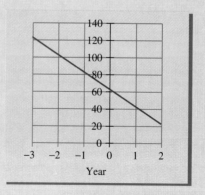

See App. F. for Group Activity answers and suggestions.

(v)

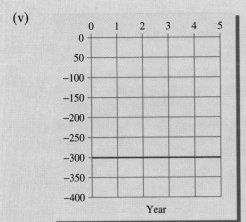

(viii)

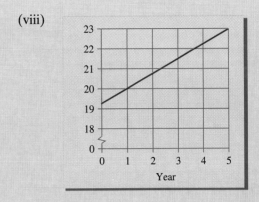

(vi)

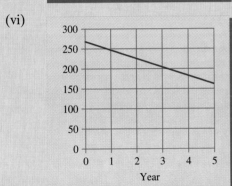

(ix)

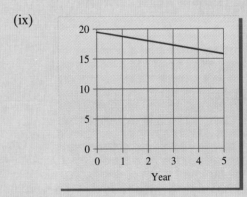

(vii)

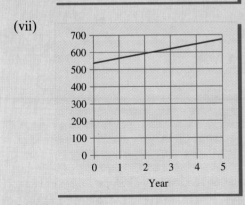

(x)

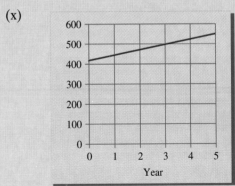

Situations

A. The average SAT score in 1988 was 428, and in 1995 it was 428. *Source:* College Entrance Examination Board

B. In 1985, sales of long-playing record albums were 167.0 million and have since declined at a rate of 20.7 million per year. *Source:* Recording Industry Association of America

C. The number of U.S. airline passengers in 1992 was 475 million and in 1994 was 528 million. *Source:* Air Transport Association of America

(continued)

D. The number of U.S. Air Force personnel on active duty was 535 thousand in 1990 and 400 thousand in 1995. *Source:* U.S. Department of Defense

E. U.S. production of potatoes was 23.0 million tons in 1995 and had been increasing at a rate of 0.73 million tons per year. *Source:* National Agricultural Statistics Service

CHAPTER 7   HIGHLIGHTS

DEFINITIONS AND CONCEPTS	EXAMPLES

SECTION 7.1   THE SLOPE-INTERCEPT FORM

Slope-Intercept Form

$$y = mx + b$$

m is the slope of the line.

b is the y-intercept.

Find the slope and the y-intercept of the line whose equation is $2x + 3y = 6$.

Solve for y: $2x + 3y = 6$

$$3y = -2x + 6 \quad \text{Subtract } 2x.$$

$$y = -\frac{2}{3}x + 2 \quad \text{Divide by 3.}$$

The slope of the line is $-\frac{2}{3}$ and the y-intercept is 2.

Find the equation of the line with slope 3 and y-intercept -1.

The equation is $y = 3x - 1$.

SECTION 7.2   THE POINT-SLOPE FORM

Point-Slope Form

$$y - y_1 = m(x - x_1)$$

m is the slope.

(x_1, y_1) is a point of the line.

Find an equation of the line with slope $\frac{3}{4}$ that contains the point $(-1, 5)$.

$$y - 5 = \frac{3}{4}(x - (-1))$$

$$4(y - 5) = 3(x + 1) \quad \text{Multiply by 4.}$$

$$4y - 20 = 3x + 3 \quad \text{Distribute.}$$

$$-3x + 4y = 23 \quad \text{Subtract } 3x \text{ and add 20.}$$

DEFINITIONS AND CONCEPTS	EXAMPLES

SECTION 7.3   GRAPHING NONLINEAR EQUATIONS

A **quadratic equation in two variables** is an equation that can be written in the form $y = ax^2 + bx + c$ where $a, b,$ and c are real numbers and a is not 0.

The graph of a quadratic equation is a parabola.

Quadratic Equations

$y = x^2 + 3, \quad y = 7x^2 - 2x + 3$

Graph $y = x^2 + 2x$.

x	y
-3	3
-2	0
-1	-1
0	0
1	3

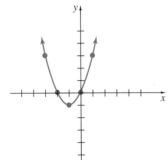

SECTION 7.4   FUNCTIONS

A set of ordered pairs is a **relation.** The set of all x-coordinates is called the **domain** of the relation and the set of all y-coordinates is called the **range** of the relation.

A **function** is a set of ordered pairs that assigns to each x-value exactly one y-value.

Vertical Line Test

If a vertical line can be drawn so that it intersects a graph more than once, the graph is not the graph of a function.

The domain of the relation

$\{(0, 5), (2, 5), (4, 5), (5, -2)\}$ is $\{0, 2, 4, 5\}$. The range is $\{-2, 5\}$.

Which are graphs of functions?

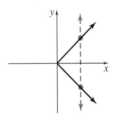

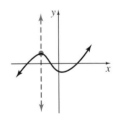

This graph is not the graph of a function.

This graph is the graph of a function.

The symbol $f(x)$ means **function of x.** This notation is called **function** notation.

If $f(x) = 2x^2 + 6x - 1$, find $f(3)$.

$$f(3) = 2(3)^2 + 6 \cdot 3 - 1$$
$$= 2 \cdot 9 + 18 - 1$$
$$= 18 + 18 - 1$$
$$= 35$$

CHAPTER 7 REVIEW

(7.1) *Determine the slope and the y-intercept of the graph of each equation.*

1. $3x + y = 7$

2. $x - 6y = -1$

3. $y = 2$ $m = 0; (0, 2)$

4. $x = -5$

Determine whether the lines are parallel, perpendicular, or neither.

5. $x - y = -6$
   $x + y = 3$ perpendicular

6. $3x + y = 7$ parallel
   $-3x - y = 10$

7. $y = 4x + \dfrac{1}{2}$
   $4x + 2y = 1$ neither

Write an equation of each line in the slope-intercept form.

8. slope -5; y-intercept $\dfrac{1}{2}$ **9.** slope $\dfrac{2}{3}$; y-intercept 6
   graphical answers in App. F

Use the slope-intercept form to graph each equation.

10. $y = -3x$

11. $y = 3x - 1$

12. $-x + 2y = 8$

13. $5x - 3y = 15$

Match each equation with its graph.

14. $y = 2x + 1$ D

15. $y = -4x$ C

16. $y = 2x$ A

17. $y = 2x - 1$ B

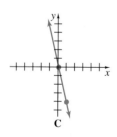

A

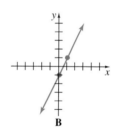

B

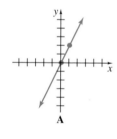

C

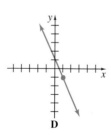

D

(7.2) *Write an equation of each line in standard form.*

18. With slope 4, through $(2, 0)$ $4x - y = 8$

19. With slope -3, through $(0, -5)$ $3x + y = -5$

20. With slope $\dfrac{1}{2}$, through $\left(0, -\dfrac{7}{2}\right)$ $x - 2y = 7$

21. With slope 0, through $(-2, -3)$ $y = -3$

22. With 0 slope, through the origin $y = 0$

23. With slope -6, through $(2, -1)$ $6x + y = 11$

24. With slope 12, through $\left(\dfrac{1}{2}, 5\right)$ $12x - y = 1$

25. Through $(0, 6)$ and $(6, 0)$ $x + y = 6$

26. Through $(0, -4)$ and $(-8, 0)$ $x + 2y = -8$

27. Vertical line, through $(5, 7)$ $x = 5$

28. Horizontal line, through $(-6, 8)$ $y = 8$

29. Through $(6, 0)$, perpendicular to $y = 8$. $x = 6$

30. Through $(10, 12)$, perpendicular to $x = -2$ $y = 12$

31. Write an equation in standard form of the line that contains $(5, 0)$ and is   **b.** $x - 3y = 5$

   a. parallel to the line $y = -3x + 7$. $3x + y = 15$

   b. perpendicular to the line $y = -3x + 7$.

(7.3) *Graph each equation.* graphical answers in App. F

32. $y = x + 5$

33. $y = x^2 + 5$

34. $y = |x| + 5$

35. $y = -x^2 + 1$

36. $y = |5x|$

37. $y = x^2 + 4x$

Use the given graph to answer the following exercises.

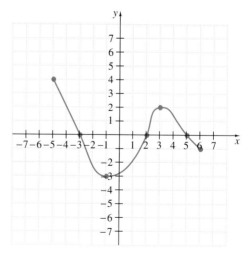

1. $m = -3; (0, 7)$ **2.** $m = \dfrac{1}{6}; \left(0, \dfrac{1}{6}\right)$ **4.** undefined slope; no y-intercept   **8.** $y = -5x + \dfrac{1}{2}$ **9.** $y = \dfrac{2}{3}x + 6$

38. Find the coordinates of the point with the greatest y-value. $(-5, 4)$

39. Find the coordinates of the point with the least y-value. $(-1, -3)$

40. Find the coordinates of the point with the least x-value. $(-5, 4)$

41. Find the coordinates of the point with the greatest x-value. $(6, -1)$

(7.4) *Determine which of the following are functions.*

42. $\{(7, 1), (7, 5), (2, 6)\}$ no

43. $\{(0, -1), (5, -1), (2, 2)\}$ yes

44. $7x - 6y = 1$ yes

45. $y = 7$ yes

46. $x = 2$ no   **47.** $y = x^3$ yes

48. $x + y < 6$ no

49. no

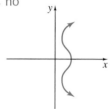

50. yes

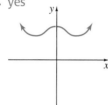

Given the following functions, find the indicated function value.

51. Given $f(x) = -2x + 6$, find

   a. $f(0)$ 6   **b.** $f(-2)$ 10   **c.** $f\left(\dfrac{1}{2}\right)$ 5

52. Given $h(x) = -5 - 3x$, find

   a. $h(2)$ -11 **b.** $h(-3)$ 4   **c.** $h(0)$ -5

53. Given $g(x) = x^2 + 12x$, find

   a. $g(3)$ 45   **b.** $g(-5)$ -35 **c.** $g(0)$ 0

54. Given $h(x) = 6 - |x|$, find

   a. $h(-1)$ 5   **b.** $h(1)$ 5   **c.** $h(-4)$ 2

Find the domain of each function.

55. $f(x) = 2x + 7$

all real numbers

56. $g(x) = \dfrac{7}{x - 2}$

all real numbers except 2

Find the domain and the range of each function graphed.

57.

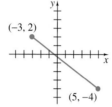

58.

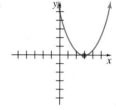

59.

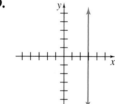

60.

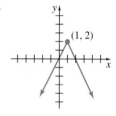

CHAPTER 7 TEST

1. $m = \dfrac{7}{3}; \left(0, -\dfrac{2}{3}\right)$

1. Determine the slope and the y-intercept of the graph of $7x - 3y = 2$.

2. Determine whether the graphs of $y = 2x - 6$ and $-4x = 2y$ are parallel lines, perpendicular lines, or neither.   neither

Find equations of the following lines. Write the equation in standard form.

3. With slope of $-\dfrac{1}{4}$, through $(2, 2)$ $x + 4y = 10$

4. Through the origin and $(6, -7)$ $7x + 6y = 0$

5. Through $(2, -5)$ and $(1, 3)$ $8x + y = 11$

6. Through $(-5, -1)$ and parallel to $x = 7$ $x = -5$

7. With slope $\dfrac{1}{8}$ and y-intercept 12 $x - 8y = -96$

Graph each equation.   graphical answers in App. F

8. $x - 4y = 4$ **9.** $y = -3x$

10. $y = |x + 1|$ **11.** $y = x^2 + 1$

Which of the following are functions?

12. $7x - y = 12$ yes   **13.** $y = \dfrac{1}{x + 1}$ yes

57. domain: $-3 \le x \le 5$; range: $-4 \le y \le 2$ **58.** domain: all real numbers; range: $y \ge 0$ **59.** domain: $x = 3$; range: all real numbers   **60.** domain: all real numbers; range: $y \le 2$

14. yes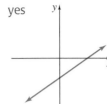

15. no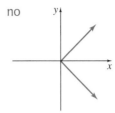

16. no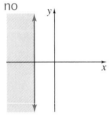

20. Find the domain of $y = \dfrac{1}{x + 1}$. all real numbers except -1

Find the domain and the range of each function graphed.

21.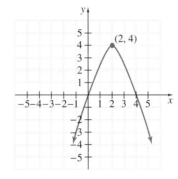

domain: all real numbers; range: $y \leq 4$

22.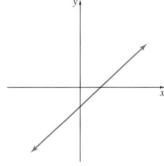

domain: all real numbers; range: all real numbers

Given the following functions, find the indicated function values.

17. $f(x) = 2x - 4$

   a. $f(-2)$ -8 **b.** $f(0.2)$ -3.6 **c.** $f(0)$. -4

18. $h(x) = x^3 - x$

   a. $h(-1)$ 0 **b.** $h(0)$ 0 **c.** $h(4)$. 60

19. $g(x) = 6$

   a. $g(0)$ 6 **b.** $g(a)$ 6 **c.** $g(242)$. 6

CHAPTER 7 CUMULATIVE REVIEW

1. Find the sums.   *(Sec. 1.5, Ex. 3)*

   a. $3 + (-7) + (-8)$ -12

   b. $[7 + (-10)] + [-2 + (-4)]$ -9

2. The bar graph on the next page shows Disney's top six animated films before 1995 and the amount of money they generated at theatres.

   a. Find the film shown which generated the most income for Disney and approximate the income.   *The Lion King; $315 million*

   b. How much more money did the film *Aladdin* make than the film *Beauty and the Beast*? $70 million   *(Sec. 1.9, Ex. 2)*

3. Write the following phrases as algebraic expressions and simplify if possible. Let x represent the unknown number.   *(Sec. 2.1, Ex. 8)*

   a. Twice a number, added to 6. $2x + 6$

   b. The difference of a number and 4, divided by 7.

   c. Five added to 3 times the sum of a number and 1.

4. Solve for x: $\dfrac{5}{2}x = 15$. $x = 6$ *(Sec. 2.3, Ex. 1)*

5. Solve $2x < -4$, and graph the solution set. $x < -2$;   *(Sec. 2.9, Ex. 5)*

6. Graph the linear equation $y = 3x + 6$.   See App. F. *(Sec. 3.2, Ex. 6)*

3. b. $\dfrac{x - 4}{7}$ **c.** $3x + 8$

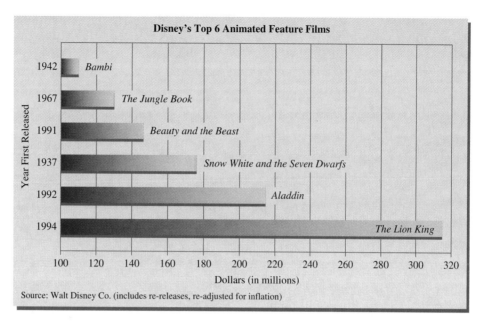

Disney's Top 6 Animated Feature Films

1942 Bambi

1967 The Jungle Book

1991 Beauty and the Beast

1937 Snow White and the Seven Dwarfs

1992 Aladdin

1994 The Lion King

Year First Released

Dollars (in millions)

Source: Walt Disney Co. (includes re-releases, re-adjusted for inflation)

(Sec. 3.4, Ex. 6)

7. Find the slope of the line $x = 5$. undefined slope

8. Find the degree of each polynomial and tell whether the polynomial is a monomial, binomial, trinomial, or none of these. *(Sec. 4.2, Ex. 2)*

 a. $-2t^2 + 3t + 6$ 2; trinomial

 b. $15x - 10$ 1; binomial

 c. $7x + 3x^3 + 2x^2 - 1$ 3; none of these

 d. $7x^2y - 6xy$ 3; binomial

9. Factor $x^2 + 4x - 12$. $(x + 6)(x - 2)$ *(Sec. 5.2, Ex. 3)*

10. Solve $x^2 - 9x = -20$. $x = 4, x = 5$ *(Sec. 5.6, Ex. 2)*

11. Divide: $\dfrac{2x^2 - 11x + 5}{5x - 25} \div \dfrac{4x - 2}{10}$. 1 *(Sec. 6.2, Ex. 7)*

12. Write $\dfrac{4b}{9a}$ as an equivalent fraction with a denominator of $27a^2b$. $\dfrac{12ab^2}{27a^2b}$ *(Sec. 6.3, Ex. 9)*

13. Add: $1 + \dfrac{m}{m + 1}$. $\dfrac{2m + 1}{m + 1}$ *(Sec. 6.4, Ex. 5)*

14. Simplify: $\dfrac{\dfrac{x + 1}{y}}{\dfrac{x}{y} + 2}$. $\dfrac{x + 1}{x + 2y}$ *(Sec. 6.5, Ex. 3)*

15. Solve the equation $3 - \dfrac{6}{x} = x + 8$.

 (Sec. 6.6, Ex. 2) $x = -3, x = -2$

(Sec. 7.1, Ex. 2)

16. Find the slope and the y-intercept of the line whose equation is $5x + y = 2$. $m = -5$; (0, 2)

17. Find the equation of the vertical line through $(-1, 5)$. $x = -1$ *(Sec. 7.2, Ex. 3)*

18. Graph the equation $y = x^2$. See App. F. *(Sec. 7.3, Ex. 1)*

19. Find the missing coordinate so that each ordered pair is a solution of the equation graphed below.

 a. $(-1, \quad)$ **b.** $(-2, \quad)$ **c.** $(\quad, -1)$

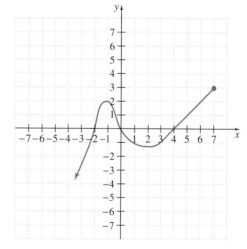

20. Given $g(x) = x^2 - 3$, find the following.

 a. $g(2)$ 1 **b.** $g(-2)$ 1 **c.** $g(0)$ -3 *(Sec. 7.4, Ex. 7)*

19. a. $(-1, \ 2)$ **b.** $(-2, 0)$ **c.** $\left(-2\dfrac{2}{3}, -1\right)$ $(1, -1), (3, -1)$ *(Sec. 7.3, Ex. 4)*

CHAPTER

8

SOLVING SYSTEMS OF LINEAR EQUATIONS

ANALYZING THE COURSES OF SHIPS

From overhead photographs or satellite imagery of ships on the ocean, defense analysts can tell a lot about a ship's immediate course by looking at its wake. Assuming that two ships will maintain their present courses, it is possible to extend their paths, based on their wakes visible in the photograph, and find possible points of collision.

IN THE CHAPTER GROUP ACTIVITY ON PAGE 487, YOU WILL HAVE THE OPPORTUNITY TO ANALYZE THE COURSES OF TWO SHIPS TO IDENTIFY A POSSIBLE POINT OF COLLISION.

In Chapters 3 and 7, we graphed equations containing two variables. Equations like these are often needed to represent relationships between two different values. For example, an economist attempts to predict what effects a price change will have on the sales prospects of calculators. There are many opportunities to compare and contrast two such equations, called a **system of equations.** This chapter presents **linear systems** and ways we solve these systems and apply them to real-life situations.

8.1 SOLVING SYSTEMS OF LINEAR EQUATIONS BY GRAPHING

TAPE
BA 8.1

OBJECTIVES

1. Determine if an ordered pair is a solution of a system of equations in two variables.
2. Solve a system of linear equations by graphing.
3. Identify a consistent system of equations, an inconsistent system of equations, and dependent equations.
4. Without graphing, determine the number of solutions of a system.

A **system of linear equations** consists of two or more linear equations. In this section, we focus on solving systems of linear equations containing two equations in two variables. Examples of such linear systems are

$$\begin{cases} 3x - 3y = 0 \\ x = 2y \end{cases} \qquad \begin{cases} x - y = 0 \\ 2x + y = 10 \end{cases} \qquad \begin{cases} y = 7x - 1 \\ y = 4 \end{cases}$$

A **solution** of a system of two equations in two variables is an ordered pair of numbers that is a solution of both equations in the system.

EXAMPLE 1 Which of the following ordered pairs is a solution of the given system?

$$\begin{cases} 2x - 3y = 6 \qquad \text{First equation.} \\ \quad\; x = 2y \qquad \text{Second equation.} \end{cases}$$

a. $(12, 6)$ **b.** $(0, -2)$

Solution: If an ordered pair is a solution of both equations, it is a solution of the system.

a. Replace x with 12 and y with 6 in both equations.

$2x - 3y = 6$	First equation.	$x = 2y$	Second equation.
$2(12) - 3(6) = 6$	Let $x = 12$ and $y = 6$.	$12 = 2(6)$	Let $x = 12$ and $y = 6$.
$24 - 18 = 6$	Simplify.	$12 = 12$	True.
$6 = 6$	True.		

Since $(12, 6)$ is a solution of both equations, it is a solution of the system.

b. Start by replacing x with 0 and y with -2 in both equations.

$2x - 3y = 6$	First equation.		$x = 2y$	Second equation.
$2(0) - 3(-2) = 6$	Let $x = 0$ and $y = -2$.		$0 = 2(-2)$	Let $x = 0$ and $y = -2$.
$0 + 6 = 6$	Simplify.		$0 = -4$	False.
$6 = 6$	True.			

While $(0, -2)$ is a solution of the first equation, it is not a solution of the second equation, so it is **not** a solution of the system.

2   Since a solution of a system of two equations in two variables is a solution common to both equations, it is also a point common to the graphs of both equations. Let's practice finding solutions of both equations in a system, that is, solutions of a system, by graphing and identifying points of intersection.

EXAMPLE 2   Solve the system of equations by graphing.

$$\begin{cases} -x + 3y = 10 \\ x + y = 2 \end{cases}$$

Solution: On a single set of axes, graph each linear equation. The two lines appear to intersect at the point $(-1, 3)$. Verify this point as a solution of the system by replacing x with -1 and y with 3 in both equations and seeing that true statements result. The solution is $(-1, 3)$.

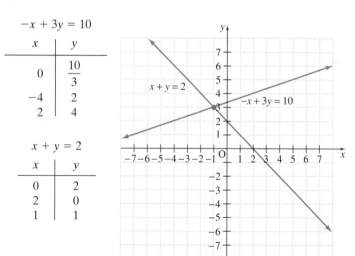

$-x + 3y = 10$

x	y
0	$\dfrac{10}{3}$
-4	2
2	4

$x + y = 2$

x	y
0	2
2	0
1	1

REMINDER   Neatly drawn graphs can help when "guessing" the solution of a system of linear equations by graphing.

A system of equations that has at least one solution as in Example 2 is said to be a **consistent system.** A system that has no solution is said to be an **inconsistent system.**

EXAMPLE 3 Solve the following system of equations by graphing.

$$\begin{cases} 2x + y = 7 \\ 2y = -4x \end{cases}$$

Solution: Graph each of the two lines in the system.

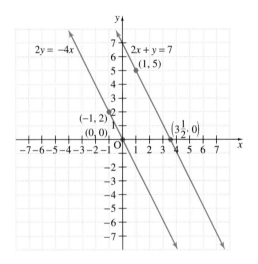

The lines **appear** to be parallel. To confirm this, write both equations in slope-intercept form by solving each equation for y.

$2x + y = 7$	First equation.	$2y = -4x$	Second equation.
$y = -2x + 7$	Subtract $2x$ from both sides.	$\dfrac{2y}{2} = \dfrac{-4x}{2}$	Divide both sides by 2.
		$y = -2x$	

Recall that when an equation is written in slope-intercept form, the coefficient of x is the slope. Since both equations have the same slope, -2, but different y-intercepts, the lines are parallel and have no points in common. Thus, there is no solution of the system and the system is inconsistent.

In Examples 2 and 3, the graphs of the two linear equations of each system are different. When this happens, we call these equations **independent equations.** If the graphs of the two equations in a system are identical, we call the equations **dependent equations.**

EXAMPLE 4   Solve the system of equations by graphing.

$$\begin{cases} x - y = 3 \\ -x + y = -3 \end{cases}$$

Solution: Graph each line.

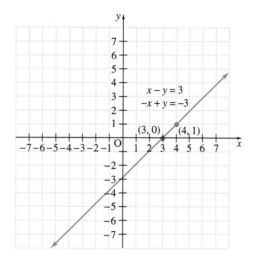

These graphs **appear** to be identical. To confirm this, write each equation in slope-intercept form.

$x - y = 3$	First equation.	$-x + y = -3$	Second equation.
$-y = -x + 3$	Subtract x from both sides.	$\boldsymbol{y = x - 3}$	Add x to both sides.
$\dfrac{-y}{-1} = \dfrac{-x}{-1} + \dfrac{3}{-1}$	Divide both sides by -1.		
$\boldsymbol{y = x - 3}$			

The equations are identical and so must be their graphs. The lines have an infinite number of points in common. Thus, there is an infinite number of solutions of the system and this is a consistent system. The equations are dependent equations.

As we have seen, three different situations can occur when graphing the two lines associated with the equations in a linear system:

One point of intersection: one solution

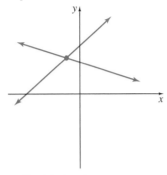

Consistent system
(at least one solution)
Independent equations
(graphs of equations differ)

Parallel lines: no solution

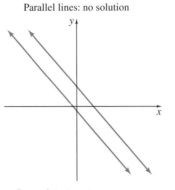

Inconsistent system
(no solution)
Independent equations
(graphs of equations differ)

Same line: infinite number of solutions

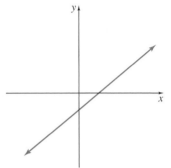

Consistent system
(at least one solution)
Dependent equations
(graphs of equations identical)

 You may have suspected by now that graphing alone is not an accurate way to solve a system of linear equations. For example, a solution of $\left(\frac{1}{2}, \frac{2}{9}\right)$ is unlikely to be read correctly from a graph. The next two sections present two accurate methods of solving these systems. In the meantime, we can decide how many solutions a system has by writing each equation in the slope-intercept form.

EXAMPLE 5   Without graphing, determine the number of solutions of the system.

$$\begin{cases} \dfrac{1}{2}x - y = 2 \\ x = 2y + 5 \end{cases}$$

Solution: First write each equation in slope-intercept form.

$\dfrac{1}{2}x - y = 2$	First equation.	$x = 2y + 5$	Second equation.
$\dfrac{1}{2}x = y + 2$	Add y to both sides.	$x - 5 = 2y$	Subtract 5 from both sides.
		$\dfrac{x}{2} - \dfrac{5}{2} = \dfrac{2y}{2}$	Divide both sides by 2.
$\dfrac{1}{2}x - 2 = y$	Subtract 2 from both sides.	$\dfrac{1}{2}x - \dfrac{5}{2} = y$	Simplify.

The slope of each line is $\frac{1}{2}$, but they have different y-intercepts. This tells us that the lines representing these equations are parallel. Since the lines are parallel, the system has no solution and is inconsistent.

EXAMPLE 6   Determine the number of solutions of the system.

$$\begin{cases} 3x - y = 4 \\ x + 2y = 8 \end{cases}$$

Solution: Once again, the slope-intercept form helps determine how many solutions this system has.

$3x - y = 4$	First equation.	$x + 2y = 8$	Second equation.
$3x = y + 4$	Add y to both sides.	$x = -2y + 8$	Add $-2y$ to both sides.
$3x - 4 = y$	Subtract 4 from both sides.	$x - 8 = -2y$	Subtract 8 from both sides.
		$\dfrac{x}{-2} - \dfrac{8}{-2} = \dfrac{-2y}{-2}$	Divide both sides by -2.
		$-\dfrac{1}{2}x + 4 = y$	Simplify.

The slope of the second line is $-\frac{1}{2}$, whereas the slope of the first line is 3. Since the slopes are not equal, the two lines are neither parallel nor identical and must intersect. Therefore, this system has one solution and is consistent.

GRAPHING CALCULATOR EXPLORATIONS

A grapher may be used to approximate solutions of systems of equations by graphing each equation on the same set of axes and approximating any points of intersection.

Solve each system of equations. Approximate the solutions to two decimal places.

1. $\begin{cases} y = -2.68x + 1.21 \\ y = 5.22x - 1.68 \end{cases}$ (0.37, 0.23)

2. $\begin{cases} y = 4.25x + 3.89 \\ y = -1.88x + 3.21 \end{cases}$ (−0.11, 3.42)

3. $\begin{cases} 4.3x - 2.9y = 5.6 \\ 8.1x + 7.6y = -14.1 \end{cases}$ (0.03, −1.89)

4. $\begin{cases} -3.6x - 8.6y = 10 \\ -4.5x + 9.6y = -7.7 \end{cases}$ (−0.41, −0.99)

MENTAL MATH

Identify each system of linear equations as consistent or inconsistent. Also state whether the equations of the lines are independent or dependent.

1.

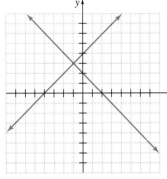

consistent, independent

2.

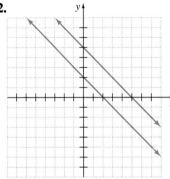

inconsistent, independent

3.

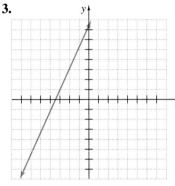

consistent, dependent

4.

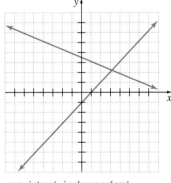

consistent, independent

5.

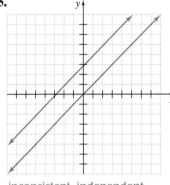

inconsistent, independent

6.

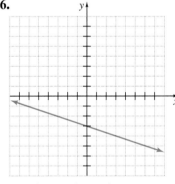

consistent, dependent

7.

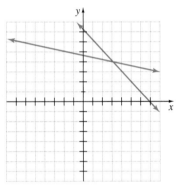

consistent, independent

8.

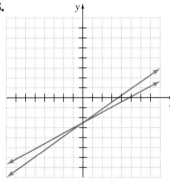

consistent, independent

EXERCISE SET 8.1

Determine whether any of the following ordered pairs satisfy the system of the linear equations. See Example 1.

1. $\begin{cases} x + y = 8 \\ 3x + 2y = 21 \end{cases}$

   a. $(2, 4)$ no
   b. $(5, 3)$ yes
   c. $(1, 9)$ no

2. $\begin{cases} 2x + y = 5 \\ x + 3y = 5 \end{cases}$

   a. $(5, 0)$ no
   b. $(1, 2)$ no
   c. $(2, 1)$ yes

3. $\begin{cases} 3x - y = 5 \\ x + 2y = 11 \end{cases}$

   a. $(2, -1)$ no
   b. $(3, 4)$ yes
   c. $(0, -5)$ no

4. $\begin{cases} 2x - 3y = 8 \\ x - 2y = 6 \end{cases}$

   a. $(4, 0)$ no
   b. $(-2, -4)$ yes
   c. $(7, 2)$ no

5. $\begin{cases} 2y = 4x \\ 2x - y = 0 \end{cases}$

   a. $(-3, -6)$ yes
   b. $(0, 0)$ yes
   c. $(1, 2)$ yes

6. $\begin{cases} 4x = 1 - y \\ x - 3y = -8 \end{cases}$

   a. $(0, 1)$ no
   b. $(1, -3)$ no
   c. $(-2, 2)$ no

7. Construct a system of two linear equations that has $(2, 5)$ as a solution.   answers may vary

8. Construct a system of two linear equations that has $(0, 1)$ as a solution.   answers may vary

9. $(2, 3)$; consistent; independent
10. $(3, 5)$; consistent; independent

Solve each system of equations by graphing the equations on the same set of axes. Tell whether the system is consistent or inconsistent and whether the equations are dependent or independent. See Examples 2 through 4.
Graphical answers in App. F

9. $\begin{cases} y = x + 1 \\ y = 2x - 1 \end{cases}$

10. $\begin{cases} y = 3x - 4 \\ y = x + 2 \end{cases}$

11. $\begin{cases} 2x + y = 0 \\ 3x + y = 1 \end{cases}$

12. $\begin{cases} 2x + y = 1 \\ 3x + y = 0 \end{cases}$

13. $\begin{cases} y = -x - 1 \\ y = 2x + 5 \end{cases}$

14. $\begin{cases} y = x - 1 \\ y = -3x - 5 \end{cases}$

15. $\begin{cases} 2x - y = 6 \\ y = 2 \end{cases}$

16. $\begin{cases} x + y = 5 \\ x = 4 \end{cases}$

17. $\begin{cases} x + y = 5 \\ x + y = 6 \end{cases}$

18. $\begin{cases} 2x + y = 4 \\ x + y = 2 \end{cases}$

19. $\begin{cases} y - 3x = -2 \\ 6x - 2y = 4 \end{cases}$

20. $\begin{cases} y + 2x = 3 \\ 4x = 2 - 2y \end{cases}$

21. $\begin{cases} x - 2y = 2 \\ 3x + 2y = -2 \end{cases}$

22. $\begin{cases} x + 3y = 7 \\ 2x - 3y = -4 \end{cases}$

32. intersecting, one solution   **33.** parallel, no solution
34. identical lines, infinite number of solutions

23. $\begin{cases} \dfrac{1}{2}x + y = -1 \\ x = 4 \end{cases}$

24. $\begin{cases} x + \dfrac{3}{4}y = 2 \\ x = -1 \end{cases}$

25. $\begin{cases} y = x - 2 \\ y = 2x + 3 \end{cases}$

26. $\begin{cases} y = x + 5 \\ y = -2x - 4 \end{cases}$

27. $\begin{cases} x + y = 7 \\ x - y = 3 \end{cases}$

28. $\begin{cases} x + y = -4 \\ x - y = 2 \end{cases}$

29. Explain how to use a graph to determine the number of solutions of a system.   answers may vary

30. The ordered pair $(-2, 3)$ is a solution of all three independent equations:

$$x + y = 1$$
$$2x - y = -7$$
$$x + 3y = 7$$

Describe the graph of all three equations on the same axes.   answers may vary

23. $(4, -3)$; consistent; independent
24. $(-1, 4)$; consistent; independent

Without graphing, decide:

   a. Are the graphs of the equations identical lines, parallel lines, or lines intersecting at a single point?

   b. How many solutions does the system have? See Examples 5 and 6.

31. $\begin{cases} 4x + y = 24 \\ x + 2y = 2 \end{cases}$

32. $\begin{cases} 3x + y = 1 \\ 3x + 2y = 6 \end{cases}$

33. $\begin{cases} 2x + y = 0 \\ 2y = 6 - 4x \end{cases}$

34. $\begin{cases} 3x + y = 0 \\ 2y = -6x \end{cases}$

35. $\begin{cases} 6x - y = 4 \\ \dfrac{1}{2}y = -2 + 3x \end{cases}$

36. $\begin{cases} 3x - y = 2 \\ \dfrac{1}{3}y = -2 + 3x \end{cases}$

37. $\begin{cases} x = 5 \\ y = -2 \end{cases}$

38. $\begin{cases} y = 3 \\ x = -4 \end{cases}$

39. $\begin{cases} 3y - 2x = 3 \\ x + 2y = 9 \end{cases}$

40. $\begin{cases} 2y = x + 2 \\ y - 2x = 3 \end{cases}$

41. $\begin{cases} 6y + 4x = 6 \\ 3y - 3 = -2x \end{cases}$

42. $\begin{cases} 8y + 6x = 4 \\ 4y - 2 = 3x \end{cases}$

43. $\begin{cases} x + y = 4 \\ x + y = 3 \end{cases}$

44. $\begin{cases} 2x + y = 0 \\ y = -2x + 1 \end{cases}$

11. $(1, -2)$; consistent; independent   **12.** $(-1, 3)$; consistent; independent   **13.** $(-2, 1)$; consistent; independent
14. $(-1, -2)$; consistent; independent   **15.** $(4, 2)$; consistent; independent   **16.** $(4, 1)$; consistent; independent
17. no solution; inconsistent; independent   **18.** $(2, 0)$; consistent; independent   **19.** infinite number of solutions; consistent; dependent   **20.** no solution; inconsistent; independent   **21.** $(0, -1)$; consistent; independent
22. $(1, 2)$; consistent; independent   **25.** $(-5, -7)$; consistent; independent   **26.** $(-3, 2)$; consistent; independent
27. $(5, 2)$; consistent; independent   **28.** $(-1, -3)$; consistent; independent   **31.** intersecting, one solution

45. Explain how writing each equation in a linear system in the point-slope form helps determine the number of solutions of a system. answers may vary

46. Is it possible for a system of two linear equations in two variables to be inconsistent, but with dependent equations? Why or why not?
answers may vary

The double line graph below shows the number of pounds of fishery products from U.S. domestic catch and from imports.

Fishery Products: Domestic Catch and Imports

Source: U.S. Bureau of the Census, *Statistical Abstract of the United States: 1994*, 113th ed., Washington, DC, 1994.

b. graph several ordered pairs from each table and sketch the two lines.   See App. F

Does your graph confirm the solution from part **(a)**?   yes

x	y		x	y
1	3		1	6
2	5		2	7
3	7		3	8
4	9		4	9
5	11		5	10

35. identical lines, infinite number of solutions
36. intersecting, one solution
37. intersecting, one solution
38. intersecting, one solution
39. intersecting, one solution
40. intersecting, one solution
41. identical lines, infinite number of solutions
42. intersecting, one solution
43. parallel, no solutions
44. parallel, no solutions

47. In what year(s) is the number of pounds of fishery products imported equal to the number of pounds of domestic catch?   1984, 1988

48. In what year(s) is the number of pounds of fishery products imported greater than the number of pounds of domestic catch?   1984 through 1988

49. In the next column are tables of values for two linear equations. Using the tables,   **a.** (4, 9)

   a. find a solution of the corresponding system.

Review Exercises

Solve each equation. See Section 2.4.

50. $5(x - 3) + 3x = 1$ **51.** $-2x + 3(x + 6) = 17$

52. $4\left(\frac{y + 1}{2}\right) + 3y = 0$ **53.** $-y + 12\left(\frac{y - 1}{4}\right) = 3$

54. $8a - 2(3a - 1) = 6$ **55.** $3z - (4z - 2) = 9$

50. $x = 2$ **51.** $x = -1$ **52.** $y = -\frac{2}{5}$ **53.** $y = 3$
54. $a = 2$ **55.** $z = -7$

8.2   SOLVING SYSTEMS OF LINEAR EQUATIONS BY SUBSTITUTION

OBJECTIVE

   1 Use the substitution method to solve a system of linear equations.

TAPE
BA 8.2

As we mentioned in the previous section, graphing is not an accurate method for solving systems of equations. In this section, we discuss a second, more accurate method for solving systems of equations. This method is called the **substitution method** and is introduced in the next example.

EXAMPLE 1   Solve the system

$$\begin{cases} 2x + y = 10 & \text{First equation.} \\ x = y + 2 & \text{Second equation.} \end{cases}$$

Solution: The second equation in this system is $x = y + 2$. Since this tells us that x and $y + 2$ have the same value, we may substitute $y + 2$ for x in the first equation. Doing so results in an equation in just one variable, y, for which we can then solve.

$$2x + y = 10 \qquad \text{First equation.}$$

$$2(y + 2) + y = 10 \qquad \text{Substitute } y + 2 \text{ for } x \text{ since } x = y + 2.$$
$$2y + 4 + y = 10 \qquad \text{Use the distributive property.}$$
$$3y + 4 = 10$$
$$3y = 6$$
$$y = 2$$

The y-value of the ordered pair solution of the system is 2. To find the corresponding x-value, replace y with 2 in the equation $x = y + 2$ and solve for x.

$$x = y + 2$$
$$x = 2 + 2 \qquad \text{Let } y = 2.$$
$$x = 4$$

The solution of the system is the ordered pair $(4, 2)$. Since an ordered pair solution must satisfy both linear equations in the system, we could have chosen the equation $2x + y = 10$ to find the corresponding x-value. The resulting x-value is the same.

To check, see that $(4, 2)$ satisfies both equations of the original system.

First equation	Second equation	
$2x + y = 10$	$x = y + 2$	
$2(4) + 2 = 10$	$4 = 2 + 2$	Let $x = 4$ and $y = 2$.
$10 = 10$ True.	$4 = 4$	True.

The solution of the system is $(4, 2)$.

A graph of the two equations shows two lines intersecting at the point $(4, 2)$.

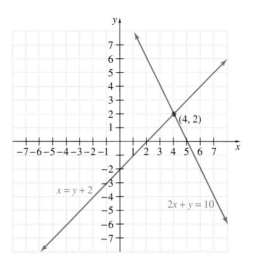

To solve a system of equations by substitution, we first need an equation solved for one of its variables.

EXAMPLE 2   Solve the system $\begin{cases} 2x + 2y = 7 \\ 2x + 2y = 13 \end{cases}$

Solution: Choose an equation and solve for x or y. We will solve the first equation for x by subtracting $2y$ from both sides.

$$x + 2y = 7 \qquad \text{First equation.}$$
$$x = 7 - 2y$$

Since $x = 7 - 2y$, substitute $7 - 2y$ for x in the second equation and solve for y.

$$2x + 2y = 13 \qquad \text{Second equation}$$
$$2(7 - 2y) + 2y = 13 \qquad \text{Let } x = 7 - 2y.$$
$$14 - 4y + 2y = 13 \qquad \text{Distributive property.}$$
$$14 - 2y = 13 \qquad \text{Simplify.}$$
$$-2y = -1 \qquad \text{Subtract 14 from both sides.}$$
$$y = \frac{1}{2} \qquad \text{Divide both sides by } -2.$$

To find x, let $y = \frac{1}{2}$ in the equation $x = 7 - 2y$.

$$x = 7 - 2y$$
$$x = 7 - 2\left(\frac{1}{2}\right) \qquad \text{Let } y = \frac{1}{2}$$
$$x = 7 - 1$$
$$x = 6$$

The solution is $\left(6, \frac{1}{2}\right)$. Check the solution in both equations of the original system.

The following steps may be used to solve a system of equations by the substitution method.

To Solve a System of Linear Equations by the Substitution Method

Step 1. Solve one of the equations for one of its variables.

Step 2. Substitute the expression for the variable found in step 1 into the other equation.

Step 3. Solve the equation from step 2 to find the value of one variable.

Step 4. Substitute the value found in step 3 in any equation containing both variables to find the value of the other variable.

Step 5. Check the proposed solution in the original system.

EXAMPLE 3   Solve the system $\begin{cases} 7x - 3y = -14 \\ -3x + y = 6 \end{cases}$

Solution: To avoid introducing fractions, we will solve the second equation for y.

$$-3x + y = 6 \qquad \text{Second equation.}$$
$$y = 3x + 6$$

Next, substitute $3x + 6$ for y in the first equation.

$$7x - 3y = -14 \qquad \text{First equation.}$$
$$7x - 3(3x + 6) = -14$$
$$7x - 9x - 18 = -14$$
$$-2x - 18 = -14$$
$$-2x = 4$$
$$\frac{-2x}{-2} = \frac{4}{-2}$$
$$x = -2$$

To find the corresponding y-value, substitute -2 for x in the equation $y = 3x + 6$. Then $y = 3(-2) + 6$ or $y = 0$. The solution of the system is $(-2, 0)$. Check this solution in both equations of the system.

> R E M I N D E R   When solving a system of equations by the substitution
> method, begin by solving an equation for one of its variables. If possible, solve for
> a variable that has a coefficient of 1 or -1. This way, we avoid working with time-
> consuming fractions.

EXAMPLE 4   Solve the system $\begin{cases} \frac{1}{2}x - y = 3 \\ x = 6 + 2y \end{cases}$ by substitution.

Solution: The second equation is already solved for x in terms of y. Thus we substitute
$6 + 2y$ for x in the first equation and solve for y.

$$\frac{1}{2}x - y = 3 \qquad \text{First equation.}$$

$$\frac{1}{2}(6 + 2y) - y = 3 \qquad \text{Let } x = 6 + 2y.$$

$$3 + y - y = 3$$

$$3 = 3$$

Arriving at a true statement such as $3 = 3$ indicates that the two linear equations in the original system are equivalent. This means that their graphs are identical and there is an infinite number of solutions to the system. Any solution of one
equation is also a solution of the other. The equations are dependent.

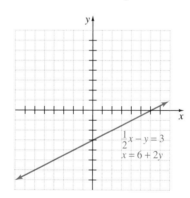

EXAMPLE 5   Use substitution to solve the system.

$$\begin{cases} 6x + 12y = 5 \\ -4x - 8y = 0 \end{cases}$$

Solution: Choose the second equation and solve for y.

$$-4x - 8y = 0 \qquad \text{Second equation.}$$

$$-8y = 4x \qquad \text{Add } 4x \text{ to both sides.}$$

$$\frac{-8y}{-8} = \frac{4x}{-8} \qquad \text{Divide both sides by } -8.$$

$$y = -\frac{1}{2}x \qquad \text{Simplify.}$$

Now replace y with $-\frac{1}{2}x$ in the first equation.

$$6x + 12y = 5 \qquad \text{First equation.}$$

$$6x + 12\left(-\frac{1}{2}x\right) = 5 \qquad \text{Let } y = -\frac{1}{2}x.$$

$$6x + (-6x) = 5 \qquad \text{Simplify.}$$

$$0 = 5 \qquad \text{Combine like terms.}$$

The false statement $0 = 5$ indicates that this system has no solution and is inconsistent. The graph of the linear equations in the system is a pair of parallel lines.

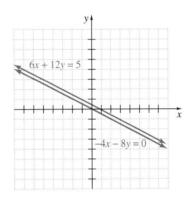

EXERCISE SET 8.2

Solve each system of equations by the substitution method. See Example 1.

1. $\begin{cases} x + y = 3 \\ x = 2y \end{cases}$ (2, 1)

2. $\begin{cases} x + y = 20 \\ x = 3y \end{cases}$ (15, 5)

3. $\begin{cases} x + y = 6 \\ y = -3x \end{cases}$ (-3, 9)

4. $\begin{cases} x + y = 6 \\ y = -4x \end{cases}$ (-2, 8)

5. $\begin{cases} 3x + 2y = 16 \\ x = 3y - 2 \end{cases}$ (4, 2)

6. $\begin{cases} 2x + 3y = 18 \\ x = 2y - 5 \end{cases}$ (3, 4)

Solve each system of equations by the substitution method. See Examples 2 through 5.

7. $\begin{cases} x + 2y = 6 \\ 2x + 3y = 8 \end{cases}$ (-2, 4)

8. $\begin{cases} x + 3y = -5 \\ 2x + 2y = 6 \end{cases}$ (7, -4)

9. $\begin{cases} 2x - 5y = 1 \\ 3x + y = -7 \end{cases}$ (-2, -1)

10. $\begin{cases} 4x + 2y = 5 \\ 2x + y = -4 \end{cases}$

11. $\begin{cases} 2y = x + 2 \\ 6x - 12y = 0 \end{cases}$

12. $\begin{cases} 3y = x + 6 \\ 4x + 12y = 0 \end{cases}$ (-3, 1)

13. $\begin{cases} \frac{1}{3}x - y = 2 \\ x - 3y = 6 \end{cases}$

14. $\begin{cases} \frac{1}{4}x - 2y = 1 \\ x - 8y = 4 \end{cases}$

15. Explain how to identify an inconsistent system when using the substitution method.

16. Occasionally, when using the substitution method, the equation $0 = 0$ is obtained. Explain how this result indicates that the equations are dependent.

10. no solution   **11.** no solution   **13.** infinite number of solutions   **14.** infinite number of solutions
15. answers may vary   **16.** answers may vary

Solve each system of equations by the substitution method.

17. $\begin{cases} 3x - 4y = 10 \\ \quad\quad x = 2y \end{cases}$ (10, 5)

18. $\begin{cases} 3x - 4y = 10 \\ \quad\quad y = 2x \end{cases}$

19. $\begin{cases} \quad y = 3x + 1 \\ 4y - 8x = 12 \end{cases}$ (2, 7)

20. $\begin{cases} \quad y = 2x + 3 \\ 5y - 7x = 18 \end{cases}$ (1, 5)

21. $\begin{cases} 4x + \quad y = 11 \\ 2x + 5y = 1 \end{cases}$ (3, −1)

22. $\begin{cases} 3x + \quad y = -14 \\ 4x + 3y = -22 \end{cases}$

23. $\begin{cases} 2x - 3y = -9 \\ 3x = y + 4 \end{cases}$ (3, 5)

24. $\begin{cases} 8x - 3y = -4 \\ 7x = y + 3 \end{cases}$ (1, 4)

25. $\begin{cases} 6x - 3y = 5 \\ \quad x + 2y = 0 \end{cases}$ $\left(\frac{2}{3}, -\frac{1}{3}\right)$

26. $\begin{cases} 10x - 5y = -21 \\ \quad x + 3y = 0 \end{cases}$

27. $\begin{cases} 3x - \quad y = 1 \\ 2x - 3y = 10 \end{cases}$

28. $\begin{cases} 2x - \quad y = -7 \\ 4x - 3y = -11 \end{cases}$

29. $\begin{cases} \quad -x + 2y = 10 \\ -2x + 3y = 18 \end{cases}$

30. $\begin{cases} \quad -x + 3y = 18 \\ -3x + 2y = 19 \end{cases}$

31. $\begin{cases} 5x + 10y = 20 \\ 2x + 6y = 10 \end{cases}$ (2, 1)

32. $\begin{cases} 2x + \quad 4y = 6 \\ 5x + 10y = 15 \end{cases}$

33. $\begin{cases} 3x + 6y = 9 \\ 4x + 8y = 16 \end{cases}$

34. $\begin{cases} 6x + 3y = 12 \\ 9x + 6y = 15 \end{cases}$

35. $\begin{cases} y = 2x + 9 \\ y = 7x + 10 \end{cases}$

36. $\begin{cases} y = 5x - 3 \\ y = 8x + 4 \end{cases}$

Solve each system by the substitution method. First simplify each equation by combining like terms.

37. $-5y + 6y = 3x + 2(x - 5) - 3x + 5$
$4(x + y) - x + y = -12$ (1, −3)

38. $5x + 2y - 4x - 2y = 2(2y + 6) - 7$
$3(2x - y) - 4x = 1 + 9$ (5, 0)

Use a graphing calculator to solve each system.

39. $\begin{cases} y = 5.1x + 14.56 \\ y = -2x - 3.9 \end{cases}$

40. $\begin{cases} y = 3.1x - 16.35 \\ y = -9.7x + 28.45 \end{cases}$

41. $\begin{cases} 3x + 2y = 14.05 \\ 5x + \quad y = 18.5 \end{cases}$

42. $\begin{cases} x + \quad y = -15.2 \\ -2x + 5y = -19.3 \end{cases}$

39. (−2.6, 1.3)   **40.** (3.5, −5.5)   **41.** (3.28, 2.11)
42. (−8.1, −7.1)

Review Exercises

Write equivalent equations by multiplying both sides of the given equation by the given nonzero number. See Section 2.4.

43. $3x + 2y = 6; -2$
$-6x - 4y = -12$

44. $-x + y = 10; 5$
$-5x + 5y = 50$

45. $-4x + y = 3; 3$
$-12x + 3y = 9$

46. $5a - 7b = -4; -4$
$-20a + 28b = 16$

Add the binomials. See Section 4.2.

47. $3n + 6m$
$\underline{2n - 6m}$
$5n$

48. $-2x + 5y$
$\underline{2x + 11y}$
$16y$

49. $-5a - 7b$
$\underline{5a - 8b}$
$-15b$

50. $9q + p$
$\underline{-9q - p}$
0

18. (−2, −4)   **22.** (−4, −2)   **26.** $\left(-\frac{9}{5}, \frac{3}{5}\right)$
27. (−1, −4)   **28.** (−5, −3)   **29.** (−6, 2)
30. (−3, 5)   **32.** infinite number of solutions
33. no solution   **34.** (3, −2)   **35.** $\left(-\frac{1}{5}, \frac{43}{5}\right)$
36. $\left(-\frac{7}{3}, -\frac{44}{3}\right)$

8.3 | SOLVING SYSTEMS OF LINEAR EQUATIONS BY ADDITION

O B J E C T I V E

1 Use the addition method to solve a system of linear equations.

TAPE
BA 8.3

1   We have seen that substitution is an accurate way to solve a linear system. Another method for solving a system of equations accurately is the **addition** or **elimination method.** The addition method is based on the addition property of equality: adding equal quantities to both sides of an equation does not change the solution of the equation. In symbols,

$$\text{if } A = B \text{ and } C = D, \text{ then } A + C = B + D.$$

EXAMPLE 1   Solve the system $\begin{cases} x + y = 7 \\ x - y = 5 \end{cases}$ by the addition method.

Solution: Since the left side of each equation is equal to the right side, we add equal quantities by adding the left sides of the equations together and the right sides of the equations together. If we choose wisely, this adding gives us an equation in one variable, x, which we can solve for x.

$x + y = 7$	First equation.
$\underline{x - y = 5}$	Second equation.
$2x \quad\;\; = 12$	Add the equations.
$x = 6$	Divide both sides by 2.

The x-value of the solution is 6. To find the corresponding y-value, let $x = 6$ in either equation of the system. We will use the first equation.

$x + y = 7$	First equation.
$6 + y = 7$	Let $x = 6$.
$y = 7 - 6$	Solve for y.
$y = 1$	Simplify.

The solution is $(6, 1)$. Check this in both equations.
Check:

FIRST EQUATION	SECOND EQUATION	
$x + y = 7$	$x - y = 5$	
$6 + 1 = 7$	$6 - 1 = 5$	Let $x = 6$ and $y = 1$.
$7 = 7$ True.	$5 = 5$ True.	

Thus, the solution of the system is $(6, 1)$ and the graphs of the two equations intersect at the point $(6, 1)$ as shown.

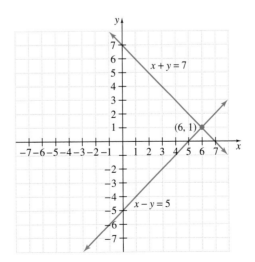

EXAMPLE 2 Use the addition method to solve $\begin{cases} -2x + y = 2 \\ -x + 3y = -4 \end{cases}$

Solution: If we simply add the two equations, the result is still an equation in two variables. However, our goal is to eliminate one of the variables. Notice what happens if we multiply *both sides* of the first equation by -3, which we are allowed to do by the multiplication property of equality. The system

$$\begin{cases} -3(-2x + y) = -3(2) \\ -x + 3y = -4 \end{cases} \quad \text{simplifies to} \quad \begin{cases} 6x - 3y = -6 \\ -x + 3y = -4 \end{cases}$$

Now add the resulting equations and the y variable is eliminated.

$$\begin{aligned} 6x - 3y &= -6 \\ \underline{-x + 3y} &= \underline{-4} \\ 5x &= -10 \qquad \text{Add.} \\ x &= -2 \qquad \text{Divide both sides by 5.} \end{aligned}$$

To find the corresponding y-value, let $x = -2$ in any of the preceding equations containing both variables. We use the first equation of the original system.

$$\begin{aligned} -2x + y &= 2 \qquad \text{First equation.} \\ -2(-2) + y &= 2 \qquad \text{Let } x = -2. \\ 4 + y &= 2 \\ y &= -2 \end{aligned}$$

The solution is $(-2, -2)$. Check this ordered pair in both equations of the original system.

In Example 2, the decision to multiply the first equation by -3 was no accident. To eliminate a variable when adding two equations, the coefficient of the variable in one equation must be the opposite of its coefficient in the other equation.

> **REMINDER** Be sure to multiply *both sides* of an equation by a chosen number when solving by the addition method. A common mistake is to multiply only the side containing the variables.

EXAMPLE 3 Use the addition method to solve $\begin{cases} 2x - y = 7 \\ 8x - 4y = 1 \end{cases}$

Solution: Multiply both sides of the first equation by -4 and the resulting coefficient of x is -8, the opposite of 8, the coefficient of x in the second equation. The system

$$\begin{cases} -4(2x - y) = -4(7) \\ 8x - 4y = 1 \end{cases} \quad \text{simplifies to} \quad \begin{cases} -8x + 4y = -28 \\ 8x - 4y = 1 \end{cases}$$

Now add the resulting equations.

$$-8x + 4y = -28$$
$$\underline{8x - 4y = 1}$$
$$0 = -27 \quad \text{False.}$$

Since $0 = -27$ is a false statement, the system has no solution. This system is inconsistent. The equations, if graphed, are parallel lines.

As we discovered with the substitution method, if our attempts to solve a system yield a true statement like $0 = 0$, the system has an infinite number of solutions. The equations, if graphed, are identical lines and dependent.

EXAMPLE 4   Solve the system $\begin{cases} 3x + 4y = 13 \\ 5x - 9y = 6 \end{cases}$ by the addition method.

Solution: We can eliminate the variable y by multiplying the first equation by 9 and the second equation by 4.

$$\begin{cases} 9(3x + 4y) = 9(13) \\ 4(5x - 9y) = 4(6) \end{cases} \quad \text{simplifies to} \quad \begin{cases} 27x + 36y = 117 \\ 20x - 36y = 24 \end{cases}$$

$$47x \qquad = 141 \quad \text{Add the equations.}$$
$$x = 3$$

To find the corresponding y-value, let $x = 3$ in any equation containing 2 variables. Doing so in any of these equations will give $y = 1$. The solution to this system is $(3, 1)$. check to see that $(3, 1)$ satisfies each equation in the original system.

If we had decided to eliminate x instead of y in Example 4, the first equation could have been multiplied by 5 and the second by -3. Try solving the original system this way to check that the solution is $(3, 1)$.

TO SOLVE A SYSTEM OF TWO LINEAR EQUATIONS BY THE ADDITION METHOD

Step 1.   Rewrite each equation in standard form $Ax + By = C$.

Step 2.   If necessary, multiply one or both equations by a nonzero number so that the coefficients of a chosen variable in the system are opposites.

Step 3.   Add the equations.

Step 4.   Find the value of one variable by solving the resulting equation from step 3.

Step 5.   Find the value of the second variable by substituting the value found in step 4 into either of the original equations.

Step 6.   Check the proposed solution in the original system.

EXAMPLE 5 Use the addition method to solve $\begin{cases} -x - \dfrac{y}{2} = \dfrac{5}{2} \\ -\dfrac{x}{2} + \dfrac{y}{4} = 0 \end{cases}$

Solution: Begin by clearing each equation of fractions. Multiply both sides of the first equation by the LCD 2 and multiply both sides of the second equation by the LCD 4. Then

$$\begin{cases} 2\left(-x - \dfrac{y}{2}\right) = 2\left(\dfrac{5}{2}\right) \\ 4\left(-\dfrac{x}{2} + \dfrac{y}{4}\right) = 4(0) \end{cases} \quad \text{simplifies to} \quad \begin{cases} -2x - y = 5 \\ -2x + y = 0 \end{cases}$$

Now add the resulting equations in the simplified system.

$$\begin{array}{r} -2x - y = 5 \\ -2x + y = 0 \\ \hline -4x = 5 \end{array} \quad \text{Add.}$$

$$x = -\dfrac{5}{4}$$

To find y, we could replace x with $-\dfrac{5}{4}$ in an equation with two variables.

Instead, let's go back to the simplified system and multiply by appropriate factors to eliminate the variable x and solve for y. To do this, multiply the first equation in the simplified system by -1. Then

$$\begin{cases} -1(-2x - y) = -1(5) \\ -2x + y = 0 \end{cases} \quad \text{simplifies to} \quad \begin{cases} 2x + y = -5 \\ -2x + y = 0 \end{cases}$$

$$\begin{array}{r} 2y = -5 \quad \text{Add.} \\ y = -\dfrac{5}{2} \end{array}$$

The solution is $\left(-\dfrac{5}{4}, -\dfrac{5}{2}\right)$.

EXERCISE SET 8.3

Solve each system of equations by the addition method. See Example 1.

1. $\begin{cases} 3x + y = 5 \\ 6x - y = 4 \end{cases}$ (1, 2)

2. $\begin{cases} 4x + y = 13 \\ 2x - y = 5 \end{cases}$ (3, 1)

3. $\begin{cases} x - 2y = 8 \\ -x + 5y = -17 \end{cases}$ (2, −3)

4. $\begin{cases} x - 2y = -11 \\ -x + 5y = 23 \end{cases}$ (−3, 4)

5. $\begin{cases} 3x + 2y = 11 \\ 5x - 2y = 29 \end{cases}$ (5, −2)

6. $\begin{cases} 4x + 2y = 2 \\ 3x - 2y = 12 \end{cases}$ (2, −3)

Solve each system of equations by the addition method. See Examples 2 through 4.

7. $\begin{cases} 3x + y = -11 \\ 6x - 2y = -2 \end{cases}$ (−2, −5)

8. $\begin{cases} 4x + y = -13 \\ 6x - 3y = -15 \end{cases}$ (−3, −1)

9. $\begin{cases} x + 5y = 18 \\ 3x + 2y = -11 \end{cases}$ $(-7, 5)$ **10.** $\begin{cases} x + 4y = 14 \\ 5x + 3y = 2 \end{cases}$ $(-2, 4)$

11. $\begin{cases} 2x - 5y = 4 \\ 3x - 2y = 4 \end{cases}$ $\left(\dfrac{12}{11}, -\dfrac{4}{11}\right)$ **12.** $\begin{cases} 6x - 5y = 7 \\ 4x - 6y = 7 \end{cases}$ $\left(\dfrac{7}{16}, -\dfrac{7}{8}\right)$

13. $\begin{cases} 2x + 3y = 0 \\ 4x + 6y = 3 \end{cases}$ no solution **14.** $\begin{cases} -x + 5y = -1 \\ 3x - 15y = 3 \end{cases}$ infinite number of solutions

Solve each system of equations by the addition method. See Example 5.

15. $\begin{cases} \dfrac{x}{3} + \dfrac{y}{6} = 1 \\ \dfrac{x}{2} - \dfrac{y}{4} = 0 \end{cases}$ $\left(\dfrac{3}{2}, 3\right)$ **16.** $\begin{cases} \dfrac{x}{2} + \dfrac{y}{8} = 3 \\ x - \dfrac{y}{4} = 0 \end{cases}$ $(3, 12)$

17. $\begin{cases} x - \dfrac{y}{3} = -1 \\ -\dfrac{x}{2} + \dfrac{y}{8} = \dfrac{1}{4} \end{cases}$ $(1, 6)$ **18.** $\begin{cases} 2x - \dfrac{3y}{4} = -3 \\ x + \dfrac{y}{9} = \dfrac{13}{9} \end{cases}$ $(3, 12)$

19. When solving a system of equations by the addition method, how do we know when the system has no solution? answers may vary

20. To solve the system $\begin{cases} 2x - 3y = 5 \\ 5x + 2y = 6 \end{cases}$, explain why the addition method might be preferred rather than the substitution method. answers may vary

Solve each system by the addition method.

21. $\begin{cases} x + y = 6 \\ x - y = 6 \end{cases}$ $(6, 0)$ **22.** $\begin{cases} x - y = 1 \\ -x + 2y = 0 \end{cases}$ $(2, 1)$

23. $\begin{cases} 3x + y = 4 \\ 9x + 3y = 6 \end{cases}$ no solution **24.** $\begin{cases} 2x + y = 6 \\ 4x + 2y = 12 \end{cases}$

25. $\begin{cases} 3x - 2y = 7 \\ 5x + 4y = 8 \end{cases}$ $\left(2, -\dfrac{1}{2}\right)$ **26.** $\begin{cases} 6x - 5y = 25 \\ 4x + 15y = 13 \end{cases}$

27. $\begin{cases} \dfrac{2}{3}x + 4y = -4 \\ 5x + 6y = 18 \end{cases}$ $(6, -2)$ **28.** $\begin{cases} \dfrac{3}{2}x + 4y = 1 \\ 9x + 24y = 5 \end{cases}$

29. $\begin{cases} 4x - 6y = 8 \\ 6x - 9y = 12 \end{cases}$ **30.** $\begin{cases} 9x - 3y = 12 \\ 12x - 4y = 18 \end{cases}$

31. $\begin{cases} \dfrac{x}{3} - y = 2 \\ -\dfrac{x}{2} + \dfrac{3y}{2} = -3 \end{cases}$ **32.** $\begin{cases} \dfrac{x}{2} + \dfrac{y}{4} = 1 \\ -\dfrac{x}{4} - \dfrac{y}{8} = 1 \end{cases}$

33. $\begin{cases} 8x = -11y - 16 \\ 2x + 3y = -4 \end{cases}$ $(-2, 0)$ **34.** $\begin{cases} 10x + 3y = -12 \\ 5x = -4y - 16 \end{cases}$

35. Use the system of linear equations below to answer the questions.

$\begin{cases} x + y = 5 \\ 3x + 3y = b \end{cases}$ **35b.** any real number except 15

a. Find the value of b so that the equations are dependent and the system has an infinite number of solutions. $b = 15$

b. Find a value of b so that the system is inconsistent and there are no solutions to the system.

36. Use the system of linear equations below to answer the questions.

$\begin{cases} x + y = 4 \\ 2x + by = 8 \end{cases}$

a. Find the value of b so that the equations are dependent and the system has an infinite number of solutions. $b = 2$

b. Find a value of b so that the system is consistent and the system has a single solution. any real number except 2

Solve each system by either the addition method or the substitution method.

37. $\begin{cases} 2x - 3y = -11 \\ y = 4x - 3 \end{cases}$ $(2, 5)$ **38.** $\begin{cases} 4x - 5y = 6 \\ y = 3x - 10 \end{cases}$ $(4, 2)$

39. $\begin{cases} x + 2y = 1 \\ 3x + 4y = -1 \end{cases}$ $(-3, 2)$ **40.** $\begin{cases} x + 3y = 5 \\ 5x + 6y = -2 \end{cases}$ $(-4, 3)$

41. $\begin{cases} 2y = x + 6 \\ 3x - 2y = -6 \end{cases}$ $(0, 3)$ **42.** $\begin{cases} 3y = x + 14 \\ 2x - 3y = -16 \end{cases}$

43. $\begin{cases} y = 2x - 3 \\ y = 5x - 18 \end{cases}$ $(5, 7)$ **44.** $\begin{cases} y = 6x - 5 \\ y = 4x - 11 \end{cases}$

45. $\begin{cases} x + \dfrac{1}{6}y = \dfrac{1}{2} \\ 3x + 2y = 3 \end{cases}$ $\left(\dfrac{1}{3}, 1\right)$ **46.** $\begin{cases} x + \dfrac{1}{3}y = \dfrac{5}{12} \\ 8x + 3y = 4 \end{cases}$

47. $\begin{cases} \dfrac{x + 2}{2} = \dfrac{y + 11}{3} \\ \dfrac{x}{2} = \dfrac{2y + 16}{6} \end{cases}$ **48.** $\begin{cases} \dfrac{x + 5}{2} = \dfrac{y + 14}{4} \\ \dfrac{x}{3} = \dfrac{2y + 2}{6} \end{cases}$

infinite number of solutions   **48.** $(3, 2)$

Solve each system by the addition method.

49. $\begin{cases} 2x + 3y = 14 \\ 3x - 4y = -69.1 \end{cases}$ $(-8.9, 10.6)$

50. $\begin{cases} 5x - 2y = -19.8 \\ -3x + 5y = -3.7 \end{cases}$ $(-5.6, -4.1)$

24. infinite number of solutions **26.** $\left(4, -\dfrac{1}{5}\right)$ **28.** no solution **29.** infinite number of solutions **30.** no solution

31. infinite number of solutions **32.** no solution **34.** $(0, -4)$ **42.** $(-2, 4)$ **44.** $(-3, -23)$ **46.** $\left(-\dfrac{1}{4}, 2\right)$

Review Exercises

Rewrite the following sentences using mathematical symbols. Do not solve the equations. See Sections 2.4 and 2.5.

51. Twice a number added to 6 is 3 less than the number.

52. The sum of three consecutive integers is 66.

51. $2x + 6 = x - 3$ **52.** $n + (n + 1) + (n + 2) = 66$ **53.** $20 - 3x = 2$

53. Three times a number subtracted from 20 is 2.

54. Twice the sum of 8 and a number is the difference of the number and 20. $2(8 + x) = x - 20$

55. The product of 4 and the sum of a number and 6 is twice a number. $4(n + 6) = 2n$

56. The quotient of twice a number and 7 is subtracted from the reciprocal of the number. $\frac{1}{x} - \frac{2x}{7}$

8.4 | SYSTEMS OF LINEAR EQUATIONS AND PROBLEM SOLVING

O B J E C T I V E

 Use a system of equations to solve problems.

TAPE
BA 8.4

1 Many of the word problems solved earlier using one-variable equations can also be solved using two equations in **two** variables. We use the same problem-solving steps that have been used throughout this text. The only difference is that two variables are assigned to represent the two unknown quantities and that the problem is translated into **two** equations.

PROBLEM-SOLVING STEPS

1. UNDERSTAND the problem. During this step don't work with variables (except for known formulas), but simply become comfortable with the problem. Some ways of accomplishing this are listed below.

- Read and reread the problem.

- Construct a drawing.

- Look up an unknown formula.

- Propose a solution and check. Pay careful attention to how to check your proposed solution. This will help later when writing an equation to model the problem.

2. ASSIGN two variables to two unknowns in the problem. Use these variables to represent any other unknown quantities.

3. ILLUSTRATE the problem. A diagram or chart using the assigned variables can often help you to visualize the known facts.

4. TRANSLATE the problem into two equations.

5. COMPLETE the work. This often means to solve the system of two equations.

6. INTERPRET the results: *Check* the proposed solution in the stated problem and *state* your conclusion.

Review the preceding steps while we solve the following problem about numbers.

EXAMPLE 1   Find two numbers whose sum is 37 and whose difference is 21.

Solution: **1.** UNDERSTAND. Read and reread the problem and guess a solution. If one number is 20 and their sum is 37, this means that the other number is 17 because $20 + 17 = 37$. Is their difference 21? No, since $20 - 17 = 3$. Our guess is incorrect, but we now have a better understanding of the problem.

2. ASSIGN. Since we are looking for two numbers, we let

x = first number

y = second number

3. ILLUSTRATE. No illustration is needed.

4. TRANSLATE. Since we have assigned two variables to this problem, we translate our problem into two equations.

In words:	two numbers whose sum	is	37
Translate:	$x + y$	=	37

In words:	two numbers whose difference	is	21
Translate:	$x - y$	=	21

5. COMPLETE. Here we solve the system

$$\begin{cases} x + y = 37 \\ x - y = 21 \end{cases}$$

Notice that the coefficients of the variable y are opposites. Let's then solve by the addition method and begin by adding the equations.

$$\begin{array}{l} x + y = 37 \\ \underline{x - y = 21} \\ 2x \quad = 58 \qquad \text{Add the equations.} \\ x \quad = \dfrac{58}{2} = 29 \qquad \text{Divide both sides by 2.} \end{array}$$

Let $x = 29$ in the first equation to find y.

$$\begin{array}{ll} x + y = 37 & \text{First equation.} \\ 29 + y = 37 & \\ \quad y = 37 - 29 = 8 & \end{array}$$

6. INTERPRET. The solution of the system is (29,8), corresponding to the numbers 29 and 8. To *check,* notice that the sum of 29 and 8 is $29 + 8 = 37$, the required sum, and their difference is $29 - 8 = 21$, the required difference.

State: The numbers are 29 and 8.

EXAMPLE 2 The Barnum and Bailey Circus is in town. Admission for 4 adults and 2 children is $22, while admission for 2 adults and 3 children is $16.

    a. What is the price of an adult's ticket?

    b. What is the price of a child's ticket?

    c. A special rate of $60 is charged for advance sales to groups of 20 persons. Should a group of 4 adults and 16 children use the group rate? Why or why not?

Solution: **1.** UNDERSTAND. Read and reread the problem and guess a solution. Let's suppose that the price of an adult's ticket is $5 and the price of a child's ticket is $4. To check our proposed solution, let's see if admission for 4 adults and 2 children is $22. Admission for 4 adults is 4($5) or $20 and admission for 2 children is 2($4) or $8. This gives a total admission of $20 + $8 = $28, not the required $22. Again though, we have accomplished the purpose of this process. We have a better understanding of the problem.

    2. ASSIGN. Let

        A = the price of an adult's ticket

        C = the price of a child's ticket

    3. ILLUSTRATE. No illustration is needed.

    4. TRANSLATE. Translate into two equations.

In words:	admission for 4 adults	and	admission for 2 children	is	$22	or
Translate:	$4A$	$+$	$2C$	$=$	22	

In words:	admission for 2 adults	and	admission for 3 children	is	$16	or
Translate:	$2A$	$+$	$3C$	$=$	16	

    5. COMPLETE. Solve the system:

$$\begin{cases} 4A + 2C = 22 \\ 2A + 3C = 16 \end{cases}$$

Since both equations are written in the form $Ax + By = C$, we solve by the addition method.

Multiply the second equation by -2 to eliminate the variable A. Then

$$\begin{cases} 4A + 2C = 22 \\ -2(2A + 3C) = -2(16) \end{cases} \quad \text{simplifies to} \quad \begin{cases} 4A + 2C = 22 \\ -4A - 6C = -32 \end{cases}$$

$$\overline{ -4C = -10} \quad \text{Add the equations.}$$

$$-4C = -10$$

$$C = \frac{5}{2} = 2.5 \text{ or } \$2.50, \text{ the children's ticket price.}$$

To find A, replace C with 2.5 in the first equation.

$$4A + 2C = 22 \qquad \text{First equation.}$$
$$4A + 2(2.5) = 22 \qquad \text{Let } C = 2.5.$$
$$4A + 5 = 22$$
$$4A = 17$$
$$A = \frac{17}{4} = 4.25 \text{ or } \$4.25, \text{ the adult's ticket price.}$$

6. **INTERPRET.** To *check* these solutions, notice that 4 adults and 2 children will pay $4(\$4.25) + 2(\$2.50) = \$17 + \$5 = \$22$, the required amount. Also, the price for 2 adults and 3 children is $2(\$4.25) + 3(\$2.50) = \$8.50 + \$7.50 = \$16$, the required amount.

State the solution and answer the three questions.

a. Since $A = 4.25$, the price of an adult's ticket is $4.25.

b. Since $C = 2.5$, the price of a child's ticket is $2.50.

c. The regular admission price for 4 adults and 16 children is

$$4(\$4.25) + 16(\$2.50) = \$17.00 + \$40.00$$
$$= \$57.00$$

This is $3 less than the special group rate of $60, so they should **not** request the group rate.

EXAMPLE 3

Albert and Louis are 15 miles away from each other when they start walking toward one another. After 2 hours they meet. If Louis walks one mile per hour faster than Albert, find both walking speeds.

Solution:

1. **UNDERSTAND.** Read and reread the problem. Let's guess a solution and use the formula $d = r \cdot t$ to check. Suppose that Louis's rate is 4 miles per hour. Then since Louis's rate is 1 mile per hour faster, this means that Albert's rate is 3 miles per hour. To check, see if they can walk a total of 15 miles in 2 hours. Louis's distance is rate $\cdot$ time $= 4(2) = 8$ miles and Albert's distance is rate $\cdot$ time $= 3(2) = 6$ miles. Their total distance is 8 miles $+$ 6 miles $=$ 14 miles, not the required 15 miles. Now that we have a better understanding of the problem, let's model it with a system of equations.

2. **ASSIGN.** Let

$$x = \text{Albert's rate in miles per hour}$$
$$y = \text{Louis's rate in miles per hour}$$

3. ILLUSTRATE. Use the facts stated in the problem and the formula $d = rt$ to fill in the following chart.

$$r \cdot t = d$$

	r	t	d
Albert	x	2	$2x$
Louis	y	2	$2y$

4. TRANSLATE. Translate the problem into two equations using both variables.

In words: Albert's distance $+$ Louis's distance $=$ 15

Translate: $2x$ $+$ $2y$ $=$ 15

In words: Louis's rate is 1 mile per hour faster than Albert's

Translate: y $=$ $x + 1$

5. COMPLETE. The system of equations we are solving is

$$\begin{cases} y = x + 1 \\ 2x + 2y = 15 \end{cases}$$

Use substitution to solve the system since the first equation is solved for y.

$$2x + 2y = 15 \qquad \text{Second equation}$$

$$2x + 2(x + 1) = 15 \qquad \text{Replace } y \text{ with } x + 1.$$
$$2x + 2x + 2 = 15$$
$$4x = 13$$
$$x = \frac{13}{4} = 3.25$$
$$y = x + 1 = 3.25 + 1 = 4.25$$

6. INTERPRET. Albert's proposed rate is 3.25 miles per hour and Louis' proposed rate is 4.25 miles per hour.

To *check*, use the formula $d = rt$, and find that in 2 hours Albert's distance is $(3.25)(2)$ miles or 6.5 miles. In 2 hours, Louis's distance is $(4.25)(2)$ miles or 8.5 miles. The total distance walked is 6.5 miles + 8.5 miles or 15 miles, the given distance.

State: Albert walks at a rate of 3.25 miles per hour and Louis walks at a rate of 4.25 miles per hour.

EXAMPLE 4   Guy, a chemistry teaching assistant, needs 10 liters of a 20% saline solution (salt water) for his 2 P.M. laboratory class. Unfortunately, the only mixtures on hand are a 5% saline solution and a 25% saline solution. How much of each solution should he mix to produce the 20% solution?

Solution: **1. UNDERSTAND.** Read and reread the problem. Next let's guess the solution. Suppose that we need 4 liters of the 5% solution. Then we need $10 - 4 = 6$ liters of the 25% solution. To see if this gives us 10 liters of a 20% saline solution, let's find the amount of pure salt in each solution.

	concentration rate	$\times$	amount of solution	$=$	amount of pure salt
5% solution:	0.05	$\times$	4 liters	$=$	0.2 liters
25% solution:	0.25	$\times$	6 liters	$=$	1.5 liters
20% solution:	0.20	$\times$	10 liters	$=$	2 liters

Since 0.2 liters + 1.5 liters = 1.7 liters, not 2 liters, our guess is incorrect, but we have gained some insight into how to model and check this problem.

2. ASSIGN. Let

x = number of liters of 5% solution

y = number of liters of 25% solution

3. ILLUSTRATE. Use a table to organize the given data.

	Concentration Rate	Liters of Solution	Liters of Pure Salt
First Solution	5%	x	$0.05x$
Second Solution	25%	y	$0.25y$
Mixture Needed	20%	10	$(0.20)(10)$

4. TRANSLATE. We translate into two equations.

In words: liters of 5% solution $+$ liters of 25% solution $=$ 10

Translate: x $+$ y $=$ 10

In words: salt in 5% solution $+$ salt in 25% solution $=$ salt in mixture

Translate: $0.05x$ $+$ $0.25y$ $=$ $(0.20)(10)$

5. COMPLETE. Here we solve the system

$$\begin{cases} x + y = 10 \\ 0.05x + 0.25y = 2 \end{cases}$$

To solve, by the addition method multiply the first equation by -25 and the second equation by 100. Then

$$\begin{cases} -25(x + y) = -25(10) \\ 100(0.05x + 0.25y) = 100(2) \end{cases} \quad \text{simplifies to} \quad \begin{array}{r} -25x - 25y = -250 \\ 5x + 25y = 200 \\ \hline -20x = -50 \quad \text{Add.} \\ x = 2.5 \end{array}$$

To find y, let $x = 2.5$ in the first equation of the original system.

$$x + y = 10$$
$$2.5 + y = 10 \qquad \text{Let } x = 2.5.$$
$$y = 7.5$$

6. **INTERPRET.** Thus, we propose that Guy needs to mix 2.5 liters of 5% saline solution with 7.5 liters of 25% saline solution. This *checks* since $2.5 + 7.5 = 10$, the required number of liters. Also, the sum of the liters of salt in the two solutions equals the liters of salt in the required mixture:

$$0.05(2.5) + 0.25(7.5) = 0.20(10)$$
$$0.125 + 1.875 = 2$$

State: Guy needs 2.5 liters of a 5% saline solution and 7.5 liters of a 25% solution.

EXERCISE SET 8.4

Write a system of equations describing the known facts. Do not solve the system.

1. Two numbers add up to 15 and have a difference of 7. $x + y = 15 \quad x - y = 7$

2. The total of two numbers is 16. The first number plus 2 more than 3 times the second equals 18.

3. Keiko has a total of $6500, which she has invested in two accounts. The larger account is $800 greater than the smaller account.

4. Fran has four times as much invested in his savings account as in his checking account. The total amount invested is $2300. $x + y = 2300 \quad y = 4x$

Assuming the facts given, choose the correct solution from among the options given.

5. The length of a rectangle is 3 feet longer than the width. The perimeter is 30 feet. Find the dimensions of the rectangle. (c)

a. Length = 8 feet; width = 5 feet
b. Length = 8 feet; width = 7 feet
c. Length = 9 feet; width = 6 feet

6. An isosceles triangle, a triangle with two sides of equal length, has a perimeter of 20 inches. Each of the equal sides is one inch longer than the unequal side. Find the lengths of the three sides. (b)

a. 6 inches, 6 inches, and 7 inches
b. 7 inches, 7 inches, and 6 inches
c. 6 inches, 7 inches, and 8 inches

7. A small plane takes 4 hours to travel 400 miles when flying against the wind. Only 2 hours are needed to complete the trip when the plane flies with the wind. Find the wind speed and the speed of the plane in still air. (a)

a. Wind = 50 mph; plane = 150 mph
b. Wind = 150 mph; plane = 50 mph
c. Wind = 100 mph; plane = 100 mph

2. $x + y = 16$
$x + 3y + 2 = 18$

3. larger: $x \qquad x + y = 6500$
smaller: $y \qquad x = y + 800$

❏ **8.** Two CDs and 4 cassette tapes cost a total of $40. However, 3 CDs and 5 cassette tapes cost $55. Find the price of each type of recording.   (c)

   a. CDs = $12; cassettes = $4
   b. CDs = $15; cassettes = $2
   c. CDs = $10; cassettes = $5

❏ **9.** Lynn has a total of 100 coins, all of which are either dimes or quarters. The total value of the coins is $13.00. Find the number of each type of coin.   (a)

   a. 80 dimes; 20 quarters
   b. 20 dimes; 44 quarters
   c. 60 dimes; 40 quarters

❏ **10.** Yolanda has 28 gallons of saline solution available in the pharmacy in two large containers. One container holds three times as much as the other container. Find the capacity of each container.   (c)

   a. 15 gallons; 5 gallons
   b. 20 gallons; 8 gallons
   c. 21 gallons; 7 gallons

Solve. See Example 1.

11. Two numbers total 83 and have a difference of 17. Find the two numbers.   33 and 50

12. The sum of two numbers is 76 and their difference is 52. Find the two numbers.   64 and 12

Solve. See Example 2.

13. Jane has been pricing Amtrak train fares for a group trip to New York. Three adults and 4 children must pay $159. Two adults and 3 children must pay $112. Find the price of an adult's ticket, and find the price of a child's ticket.

14. Last month Kathryn purchased 5 cassettes and 2 compact disks at Wall-to-Wall Sound for $65. This month she bought 3 cassettes and 4 compact disks for $81. Find the price of each cassette, and find the price of each compact disk.

15. Taylor has 80 coins in a jar, all of which are either quarters or nickels. The total value is $14.60. How many of each type of coin does he have?

16. Rita purchased 40 stamps, a mixture of 32¢ and 19¢ stamps. Find the number of each type of stamp if she spent $12.15.   32¢ stamps: 35; 19¢ stamps: 5

Solve. See Example 3.

17. Alan can row 18 miles down the Delaware River in 2 hours, but the return trip took him $4\frac{1}{2}$ hours. Find the rate Alan could row in still water, and find the rate of the current.

	D	=	*R*	·	*T*
DOWNSTREAM	18		$x + y$		2
UPSTREAM	18		$x - y$		$4\frac{1}{2}$

18. The Schultz family took a canoe 10 miles down the Allegheny River in 1 hour and 15 minutes. After lunch it took them 4 hours to return. Find the rate of the current.   2.75 mph

	D	=	*R*	·	*T*
DOWNSTREAM	10		$x + y$		$1\frac{1}{4}$
UPSTREAM	10		$x - y$		4

19. Rich and Pat are frequent flyers with Delta Airlines. They often fly from Philadelphia to Chicago, a distance of 780 miles. On one particular trip they fly into the wind, and the flight takes 2 hours. The return trip, with the wind behind them, only takes $1\frac{1}{2}$ hours. Find the speed of the wind and find the speed of the plane in still air.

20. With a strong wind behind it, a United Airlines jet flies 2400 miles from Los Angeles to Orlando in 4 hours and 45 minutes. The return trip takes 6 hours, as the plane flies into the wind. Find the speed of the plane in still air, and find the wind speed to the nearest tenth mile per hour.

Solve. See Example 4.

21. Kay is a chemist with Gemco Pharmaceutical. She needs to prepare 12 ounces of a 9% hydrochloric acid solution. To get this solution, find the amount of 4% and the amount of 12% solution she should mix. $4\frac{1}{2}$ oz. of 4% solution; $7\frac{1}{2}$ oz. of 12% solution

22. Hope is preparing 15 liters of a 25% saline solution. Hope has two other saline solutions with strengths of 40% and 10%. Find the amount of 40% solution and the amount of 10% solution she should mix to get 15 liters of a 25% solution.

13. adult's: $29; child's: $18   **14.** cassettes: $7; CDs: $15   **15.** 27 nickels, 53 quarters
17. rowing: $6\frac{1}{2}$ mph; current: $2\frac{1}{2}$ mph   **19.** plane: 455 mph; wind: 65 mph
20. plane: 452.6 mph; wind: 52.6 mph   **22.** 7.5 liters of 40% solution; 7.5 liters of 10% solution

23. Guillermo blends coffee for Maxwell House. He needs to prepare 200 pounds of blended coffee beans selling for $3.95 per pound. He intends to do this by blending together a high-quality bean costing $4.95 per pound and a cheaper bean costing $2.65 per pound. To the nearest pound, find how much high-quality coffee and how much cheaper coffee bean he should blend.

24. Macadamia nuts cost an astounding $16.50 per pound, but research by Planter's Peanuts says that mixed nuts sell better if macadamias are included. The standard mix costs $9.25 per pound. Find how many pounds of macadamias and how many pounds of the standard mix should be combined to produce 40 pounds that will cost $10 per pound. Find the amounts to the nearest tenth of a pound. 4.1 lb macadamia nuts; 35.9 lb standard mix

Solve.

25. A first number plus twice a second number is 8. Twice the first number plus the second totals 25. Find the numbers. 14 and −3

26. One number is 4 more than twice the second number. Their total is 25. Find the numbers. 18 and 7

27. Carrie and Raymond had a pottery stand at the annual Skippack Craft Fair. They sold some of their pottery at the original price of $9.50 each, but later dropped the price to $7.50 each. If they sold all 90 pieces and took in $721, find how many they sold at the original price. 23

28. Trinity Church held its annual supper and fed a total of 387 people. They charged $6.75 for adults and $3.50 for children. If they took in $2433.50, find how many adults and how many children attended the dinner. 332 adults; 55 children

29. Joe has decided to fence off a garden plot behind his house, using his house as the "fence" along one side of the garden. The length (which runs parallel to the house) is 3 feet less than twice the width. Find the dimensions if 33 feet of fencing is used along the three sides requiring it. $x = 9$ ft.; $y = 15$ ft

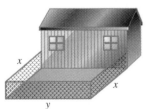

23. 113 lb. high quality; 87 lb. cheaper quality
30. $x = 40$ft.; $y = 72$ ft.

30. Anne plans to erect 152 feet of fencing around her rectangular horse pasture. A river bank serves as one side of the rectangle. If each width is 4 feet longer than half the length, find the dimensions.

31. Jim began a 186-mile bicycle trip to build up stamina for a triathlete competition. Unfortunately, his bicycle chain broke, so he finished the trip walking. The whole trip took 6 hours. If Jim walks at a rate of 4 miles per hour and rides at 40 mph, find the amount of time he spent on the bicycle. $4\frac{1}{2}$ hr.

32. In Canada, eastbound and westbound trains travel along the same track, with sidings to pull onto to avoid accidents. Two trains are now 150 miles apart, with the westbound train traveling twice as fast as the eastbound train. A warning must be issued to pull one train onto a siding or else the trains will crash in $1\frac{1}{4}$ hours. Find the speed of the eastbound train and the speed of the westbound train. westbound: 80 mph; eastbound: 40 mph

33. Joan rented a car from Hertz, which rents its cars for a daily fee plus an additional charge per mile driven. Joan recalls that a car rented for 5 days and driven for 300 miles cost her $178, while a car rented for 4 days and driven for 500 miles costs $197. Find the daily fee, and find the mileage charge. $23 daily fee; $0.21 per mile

34. Cyril and Anoa Nantambu operate a small construction and supply company. In July they charged the Shaffers $1702.50 for 65 hours of labor and 3 tons of material. In August the Shaffers paid $1349 for 49 hours of labor and $2\frac{1}{2}$ tons of material. Find the cost per hour of labor and the cost per ton of material. labor: $13.50 per hr.; material: $275 per ton

Review Exercises   graphical answers in App. F

Graph each linear inequality. See Section 6.5.

35. $y < 3 - x$

36. $y \geq 4 - 2x$

37. $2x - y \geq 6$

38. $3x + 5y < 15$

8.5 | Systems of Linear Inequalities

O B J E C T I V E

 Solve a system of linear inequalities.

Earlier we solved linear inequalities in two variables. Just as two linear equations make a system of linear equations, two linear inequalities make a **system of linear inequalities.** Systems of inequalities are very important in a process called linear programming. Many businesses use linear programming to find the most profitable way to use limited resources such as employees, machines, or buildings.

A **solution of a system of linear inequalities** is an ordered pair that satisfies each inequality in the system. The set of all such ordered pairs is the solution set of the system. Graphing this set gives us a picture of the solution set. We can graph a system of inequalities by graphing each inequality in the system and identifying the region of overlap.

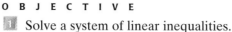

EXAMPLE 1 Graph the solution of the system $\begin{cases} 3x \geq y \\ x + 2y \leq 8 \end{cases}$

Solution: We begin by graphing each inequality on the same set of axes. The graph of the solution of the system is the region contained in the graphs of both inequalities. It is their intersection.

First, graph $3x \geq y$. The boundary line is the graph of $3x = y$. Sketch a solid boundary line since the inequality $3x \geq y$ means $3x > y$ or $3x = y$. The test point $(1, 0)$ satisfies the inequality, so shade the half-plane that includes $(1, 0)$.

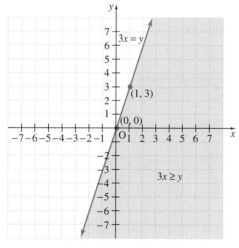

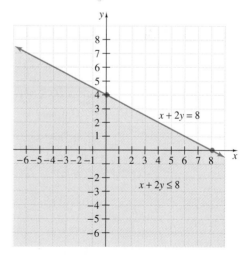

Next, sketch a solid boundary line $x + 2y = 8$ on the same set of axes. The test point $(0, 0)$ satisfies the inequality $x + 2y \leq 8$, so shade the half-plane that includes $(0, 0)$. (For clarity, the graph of $x + 2y \leq 8$ is shown on a separate set of axes.)

An ordered pair solution of the system must satisfy both inequalities. These solutions are points that lie in both shaded regions. The solution of the system is the darkest shaded region. This solution includes parts of both boundary lines.

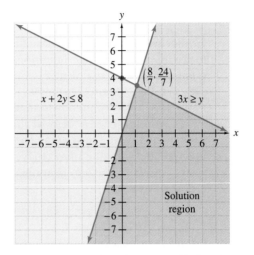

In linear programming, it is sometimes necessary to find the coordinates of the **corner point:** the point at which the two boundary lines intersect. To find the point of intersection, solve the related linear system

$$\begin{cases} 3x = y \\ x + 2y = 8 \end{cases}$$

by the substitution method or the addition method. The lines intersect at $\left(\frac{8}{7}, \frac{24}{7}\right)$, the corner point of the graph.

TO GRAPH THE SOLUTIONS OF A SYSTEM OF LINEAR INEQUALITIES

Step 1.   Graph each inequality in the system on the same set of axes.

Step 2.   The solutions of the system are the points common to the graphs of all the inequalities in the system.

EXAMPLE 2   Graph the solution of the system $\begin{cases} x - y < 2 \\ x + 2y > -1 \end{cases}$

Solution: Graph both inequalities on the same set of axes. Both boundary lines are dashed lines since the inequality symbols are $<$ and $>$. The solution of the system is the region shown by the darkest shading. In this example, the boundary lines are not a part of the solution.

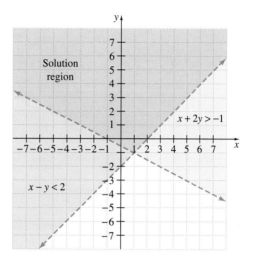

EXAMPLE 3 Graph the solution of the system $\begin{cases} -3x + 4y < 12 \\ x \geq 2 \end{cases}$

Solution: Graph both inequalities on the same set of axes.

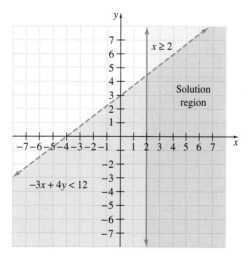

The solution of the system is the darkest shaded region, including a portion of the line $x = 2$.

EXERCISE SET 8.5   graphical answers in App. F

Graph the solution of each system of linear inequalities. See Examples 1 through 3.

1. $\begin{cases} y \geq x + 1 \\ y \geq 3 - x \end{cases}$

2. $\begin{cases} y \geq x - 3 \\ y \geq -1 - x \end{cases}$

3. $\begin{cases} y < 3x - 4 \\ y \leq x + 2 \end{cases}$

4. $\begin{cases} y \leq 2x + 1 \\ y > x + 2 \end{cases}$

5. $\begin{cases} y \leq -2x - 2 \\ y \geq x + 4 \end{cases}$

6. $\begin{cases} y \leq 2x + 4 \\ y \geq -x - 5 \end{cases}$

7. $\begin{cases} y \geq -x + 2 \\ y \leq 2x + 5 \end{cases}$

8. $\begin{cases} y \geq x - 5 \\ y \leq -3x + 3 \end{cases}$

9. $\begin{cases} x \geq 3y \\ x + 3y \leq 6 \end{cases}$

10. $\begin{cases} -2x < y \\ x + 2y < 3 \end{cases}$

11. $\begin{cases} y + 2x \geq 0 \\ 5x - 3y \leq 12 \end{cases}$

12. $\begin{cases} y + 2x \leq 0 \\ 5x + 3y \geq -2 \end{cases}$

13. $\begin{cases} 3x - 4y \geq -6 \\ 2x + y \leq 7 \end{cases}$

14. $\begin{cases} 4x - y \geq -2 \\ 2x + 3y \leq -8 \end{cases}$

15. $\begin{cases} x \leq 2 \\ y \geq -3 \end{cases}$

16. $\begin{cases} x \geq -3 \\ y \geq -2 \end{cases}$

17. $\begin{cases} y \geq 1 \\ x < -3 \end{cases}$

18. $\begin{cases} y > 2 \\ x \geq -1 \end{cases}$

19. $\begin{cases} 2x + 3y < -8 \\ x \geq -4 \end{cases}$

20. $\begin{cases} 3x + 2y \leq 6 \\ x < 2 \end{cases}$

21. $\begin{cases} 2x - 5y \leq 9 \\ y \leq -3 \end{cases}$

22. $\begin{cases} 2x + 5y \leq -10 \\ y \geq 1 \end{cases}$

23. $\begin{cases} y \geq \dfrac{1}{2}x + 2 \\ y \leq \dfrac{1}{2}x - 3 \end{cases}$

24. $\begin{cases} y \geq \dfrac{-3}{2}x + 3 \\ y < \dfrac{-3}{2}x + 6 \end{cases}$

For each system of inequalities, choose the corresponding graph.

25. $\begin{cases} y < 5 \\ x > 3 \end{cases}$ C

26. $\begin{cases} y > 5 \\ x < 3 \end{cases}$ A

27. $\begin{cases} y \leq 5 \\ x < 3 \end{cases}$ D

28. $\begin{cases} y > 5 \\ x \geq 3 \end{cases}$ B

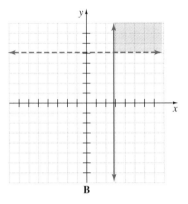

B

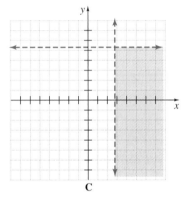

C

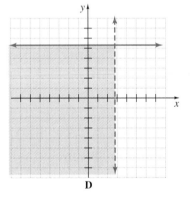

D

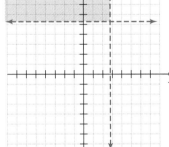

A

29. Tony Noellert budgets his time at work today. Part of the day he can write bills; the rest of the day he can use to write purchase orders. The total time available is at most 8 hours. Less than 3 hours is to be spent writing bills.

  a. Write a system of inequalities to describe the situation. (Let x = hours available for writing

bills and y = hours available for writing pur-
chase orders.) $x + y \leq 8$ $x < 3$ $x \geq 0$ $y \geq 0$

b. Graph the solutions of the system.   See App. F

30. Marty Thieme plans to invest money in two different accounts: a standard savings account and a
riskier money market fund. The total of the two
investments can be no more than $1000. At least
$300 is to be put into the money market fund.

a. Write a system of inequalities to describe the
situation. Let x = amount invested in the savings account and y = amount invested in the
money market fund. $x + y \leq 1000$ $y \geq 300$ $x \geq 0$

b. Graph the solutions to the system in part **(a)**.

31. Explain how to decide which region to shade to
show the solution region of the following system.
$\begin{cases} x \geq 3 \\ y \geq -2 \end{cases}$ answers may vary

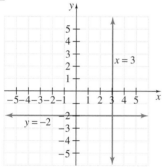

32. Describe the location of the solution region of the
system $\begin{cases} x > 0 \\ y > 0. \end{cases}$ answers may vary

Review Exercises

Evaluate each expression. See Section 3.1.

33. $(-3)^2$ 9

34. $(-5)^3$ -125

35. $\left(\dfrac{2}{3}\right)^2$ $\dfrac{4}{9}$

36. $\left(\dfrac{3}{4}\right)^3$ $\dfrac{27}{64}$

Perform the indicated operations.

37. $-2 - (-3)$ 1

38. $5 - 11$ -6

39. $8 + (-13)$ -5

40. $-12 + (-1)$ -13

30. b. See App. F

GROUP ACTIVITY

ANALYZING THE COURSES OF SHIPS

MATERIALS:
• Ruler   • Grapher (optional)

Investigate the courses and possibility of collision of the two ships shown in the figure on
the next page. Assume that the ships will maintain their present courses.

(continued)

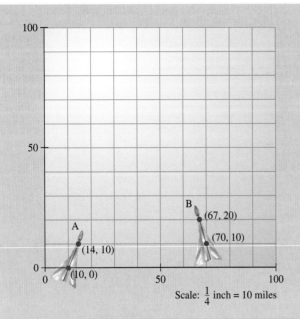

Scale: $\frac{1}{4}$ inch = 10 miles

1. Using each ship's wake as a guide, extend the paths of the ships on the figure. Estimate the coordinates of the point of intersection of the ships' courses from the grid. If the ships continue in these courses, it is possible that they will collide at the point of intersection of their paths.

2. Using the coordinates labeled on each ship's wake, find a linear equation that describes each path.

3. (Optional) Use a grapher to graph both equations in the same window. Use the Intersect or Trace feature to estimate the point of intersection of the two paths. Compare this estimate to your estimate in Question 1.

4. Solve the system of two linear equations by hand to find the point of intersection of the two paths. Compare your answer to your estimates from Questions 1 and 3.

5. Plot the point of intersection you found in Question 4 on the figure. Use the figure's scale to find each ship's distance from this point of collision by measuring from the bow (tip) of each ship with a ruler. Given the present positions and courses of the two ships find a relationship between their speeds that would ensure their collision.

See App. F for Group Activity answers and suggestions.

CHAPTER 8   HIGHLIGHTS

DEFINITIONS AND CONCEPTS	EXAMPLES
SECTION 8.1 SOLVING SYSTEMS OF LINEAR EQUATIONS BY GRAPHING	
A **system of linear equations** consists of two or more linear equations.	Systems of Linear Equations $$\begin{cases} 2x + y = 6 \\ x = -3y \end{cases} \qquad \begin{cases} -3x + 5y = 10 \\ x - 4y = -2 \end{cases}$$

DEFINITIONS AND CONCEPTS	EXAMPLES

SECTION 8.1   SOLVING SYSTEMS OF LINEAR EQUATIONS BY GRAPHING

A **solution** of a system of two equations in two variables is an ordered pair of numbers that is a solution of both equations in the system.	Determine whether $(-1, 3)$ is a solution of the system: $$\begin{cases} 2x - y = -5 \\ x = 3y - 10 \end{cases}$$ Replace x with -1 and y with 3 in both equations. $$\begin{array}{ll} 2x - y = -5 & x = 3y - 10 \\ 2(-1) - 3 = -5 & -1 = 3 \cdot 3 - 10 \\ \qquad -5 = -5 \quad \text{True} & -1 = -1 \quad \text{True} \end{array}$$ $(-1, 3)$ is a solution of the system.
Graphically, a solution of a system is a point common to the graphs of both equations.	Solve by graphing: $\begin{cases} 3x - 2y = -3 \\ x + y = 4 \end{cases}$ 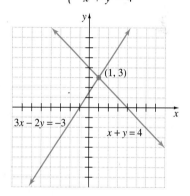
A system of equations with at least one solution is a **consistent system.** A system that has no solution is an **inconsistent system.** If the graphs of two linear equations are identical, the equations are **dependent.** If their graphs are different, the equations are **independent.**	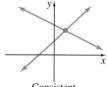 Consistent and independent 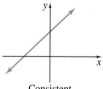 Consistent and dependent 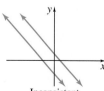 Inconsistent and independent

(continued)

DEFINITIONS AND CONCEPTS	**EXAMPLES**

SECTION 8.2 SOLVING SYSTEMS OF LINEAR EQUATIONS BY SUBSTITUTION

To solve a system of linear equations by the substitution method	Solve by substitution.
Step 1 Solve one equation for a variable.	$$\begin{cases} 3x + 2y = 1 \\ x = y - 3 \end{cases}$$
Step 2 Substitute the expression for the variable into the other equation.	Substitute $y - 3$ for x in the first equation.
Step 3 Solve the equation from step 2 to find the value of one variable.	$$3x + 2y = 1$$ $$3(y - 3) + 2y = 1$$
Step 4 Substitute the value from step 3 in either original equation to find the value of the other variable.	$$3y - 9 + 2y = 1$$ $$5y = 10$$ $$y = 2 \quad \text{Divide by 5.}$$
Step 5 Check the solution in both equations.	To find x, substitute 2 for y in $x = y - 3$ so that $x = 2 - 3$ or -1. The solution $(-1, 2)$ checks.

SECTION 8.3 SOLVING SYSTEMS OF LINEAR EQUATIONS BY ADDITION

To solve a system of linear equations by the addition method	Solve by addition.
Step 1 Rewrite each equation in standard form $Ax + By = C$.	$$\begin{cases} x - 2y = 8 \\ 3x + y = -4 \end{cases}$$
Step 2 Multiply one or both equations by a nonzero number so that the coefficient of a variable are opposites.	Multiply both sides of the first equation by -3. $$\begin{cases} -3x + 6y = -24 \\ \underline{3x + y = -4} \end{cases}$$
Step 3 Add the equations.	$$7y = -28 \quad \text{Add}$$
Step 4 Find the value of one variable by solving the resulting equation.	$$y = -4 \quad \text{Divide by 7.}$$
Step 5 Substitute the value from step 4 into either original equation to find the value of the other variable.	To find x, let $y = -4$ in an original equation. $$x - 2(-4) = 8 \quad \text{First equation.}$$ $$x + 8 = 8$$ $$x = 0$$
Step 6 Check the solution in both equations.	The solution $(0, -4)$ checks.
If solving a system of linear equations by substitution or addition yields a true statement such as $-2 = -2$, then the graphs of the equations in the system are identical and there is an infinite number of solutions of the system.	Solve: $\begin{cases} 2x - 6y = -2 \\ x = 3y - 1 \end{cases}$ Substitute $3y - 1$ for x in the first equation. $$2(3y - 1) - 6y = -2$$ $$6y - 2 - 6y = -2$$ $$-2 = -2 \quad \text{True.}$$ The system has an infinite number of solutions.

DEFINITIONS AND CONCEPTS	EXAMPLES

SECTION 8.3 SOLVING SYSTEMS OF LINEAR EQUATIONS BY ADDITION

If solving a system of linear equations yields a false statement such as $0 = 3$, the graphs of the equations in the system are parallel lines and the system has no solution.	Solve: $\begin{cases} 5x - 2y = 6 \\ -5x + 2y = -3 \end{cases}$ $0 = 3$ False. The system has no solution.

SECTION 8.4 SYSTEMS OF LINEAR EQUATIONS AND PROBLEM SOLVING

Problem-Solving Steps

1. UNDERSTAND. Read and reread the problem.

Two angles are supplementary if their sum is 180°. The larger of two supplementary angles is three times the smaller, decreased by twelve. Find the measure of each angle.

2. ASSIGN.

Let x = measure of smaller angle, then y = measure of larger angle.

3. ILLUSTRATE.

4. TRANSLATE.

In words: the sum of supplementary angles   is   180°

Translate: $x + y = 180$

In words: larger angle = 3 times smaller − 12

Translate: $y = 3x - 12$

Solve the system

$$\begin{cases} x + y = 180 \\ y = 3x - 12 \end{cases}$$

5. COMPLETE.

Use the substitution method and replace y with $3x - 12$ in the first equation.

$$\begin{aligned} x + y &= 180 \\ x + (3x - 12) &= 180 \\ 4x &= 192 \\ x &= 48 \end{aligned}$$

6. INTERPRET.

Since $y = 3x - 12$, then $y = 3 \cdot 48 - 12$ or 132.

The solution checks. The smaller angle measures 48° and the larger angle measures 132°.

(continued)

Definitions and Concepts	**Examples**

Section 8.5 Systems of Linear Inequalities

A system of linear inequalities consists of two or more linear inequalities.

To graph a system of inequalities, graph each inequality in the system. The overlapping region is the solution of the system.

System of Linear Inequalities

$$\begin{cases} x - y \geq 3 \\ y \leq -2x \end{cases}$$

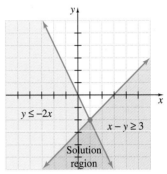

---

5. $(3, -1)$ **6.** $(-1, 1)$ **7.** $(-3, -2)$ **8.** no solution

9. $\left(\dfrac{1}{2}, \dfrac{1}{2}\right)$ **10.** $(4, -1)$ graphical answers in App. F

CHAPTER 8 REVIEW

(8.1) *Determine whether any of the following ordered pairs satisfy the system of linear equations.*

1. $\begin{cases} 2x - 3y = 12 \\ 3x + 4y = 1 \end{cases}$

   a. $(12, 4)$ no

   b. $(3, -2)$ yes

   c. $(-3, 6)$ no

2. $\begin{cases} 4x + y = 0 \\ -8x - 5y = 9 \end{cases}$

   a. $\left(\dfrac{3}{4}, -3\right)$ yes

   b. $(-2, 8)$ no

   c. $\left(\dfrac{1}{2}, -2\right)$ no

3. $\begin{cases} 5x - 6y = 18 \\ 2y - x = -4 \end{cases}$

   a. $(-6, -8)$ no

   b. $\left(3, \dfrac{5}{2}\right)$ no

   c. $\left(3, -\dfrac{1}{2}\right)$ yes

4. $\begin{cases} 2x + 3y = 1 \\ 3y - x = 4 \end{cases}$

   a. $(2, 2)$ no

   b. $(-1, 1)$ yes

   c. $(2, -1)$ no

Solve each system of equations by graphing.

5. $\begin{cases} 2x + y = 5 \\ 3y = -x \end{cases}$

6. $\begin{cases} 3x + y = -2 \\ 2x - y = -3 \end{cases}$

7. $\begin{cases} y - 2x = 4 \\ x + y = -5 \end{cases}$

8. $\begin{cases} y - 3x = 0 \\ 2y - 3 = 6x \end{cases}$

9. $\begin{cases} 3x + y = 2 \\ 3x - 6 = -9y \end{cases}$

10. $\begin{cases} 2y + x = 2 \\ x - y = 5 \end{cases}$

Without graphing, **(a)** *decide whether the graphs of the system are identical lines, parallel lines, or lines intersecting at a single point and* **(b)** *determine the number of solutions for each system.*

11. $\begin{cases} 2x - y = 3 \\ y = 3x + 1 \end{cases}$

12. $\begin{cases} 3x + y = 4 \\ y = -3x + 1 \end{cases}$

13. $\begin{cases} \dfrac{2}{3}x + \dfrac{1}{6}y = 0 \\ \phantom{\dfrac{2}{3}x} y = -4x \end{cases}$

14. $\begin{cases} \dfrac{1}{4}x + \dfrac{1}{8}y = 0 \\ \phantom{\dfrac{1}{4}x} y = -6x \end{cases}$

11. intersecting, one solution   **12.** parallel, no solutions
13. identical, infinite number of solutions
14. intersecting, one solution

(8.2) *Solve the following systems of equations by the substitution method. If there is a single solution, give the ordered pair. If not, state whether the system is inconsistent or whether the equations are dependent.*

15. $\begin{cases} y = 2x + 6 \\ 3x - 2y = -11 \end{cases}$ $(-1, 4)$ **16.** $\begin{cases} y = 3x - 7 \\ 2x - 3y = 7 \end{cases}$ $(2, -1)$

17. $\begin{cases} x + 3y = -3 \\ 2x + y = 4 \end{cases}$ $(3, -2)$ **18.** $\begin{cases} 3x + y = 11 \\ x + 2y = 12 \end{cases}$ $(2, 5)$

19. $\begin{cases} 4y = 2x - 3 \\ x - 2y = 4 \end{cases}$ **20.** $\begin{cases} 2x = 3y - 18 \\ x + 4y = 2 \end{cases}$ $(-6, 2)$

21. $\begin{cases} 2(3x - y) = 7x - 5 \\ 3(x - y) = 4x - 6 \end{cases}$ **22.** $\begin{cases} 4(x - 3y) = 3x - 1 \\ 3(4y - 3x) = 1 - 8x \end{cases}$

23. $\begin{cases} \dfrac{3}{4}x + \dfrac{2}{3}y = 2 \\ 3x + y = 18 \end{cases}$ $(8, -6)$ **24.** $\begin{cases} \dfrac{2}{5}x + \dfrac{3}{4}y = 1 \\ x + 3y = -2 \end{cases}$ $(10, -4)$

(8.3) *Solve the following systems of equations by the addition method. If there is a single solution, give the ordered pair. If not, state whether the system is inconsistent or whether the equations are dependent.*

25. $\begin{cases} 2x + 3y = -6 \\ x - 3y = -12 \end{cases}$ $(-6, 2)$ **26.** $\begin{cases} 4x + y = 15 \\ -4x + 3y = -19 \end{cases}$ $(4, -1)$

27. $\begin{cases} 2x - 3y = -15 \\ x + 4y = 31 \end{cases}$ $(3, 7)$ **28.** $\begin{cases} x - 5y = -22 \\ 4x + 3y = 4 \end{cases}$ $(-2, 4)$

29. $\begin{cases} 2x = 6y - 1 \\ \dfrac{1}{3}x - y = \dfrac{-1}{6} \end{cases}$ **30.** $\begin{cases} 8x = 3y - 2 \\ \dfrac{4}{7}x - y = \dfrac{-5}{2} \end{cases}$ $\left(\dfrac{7}{8}, 3\right)$

31. $\begin{cases} 5x = 6y + 25 \\ -2y = 7x - 9 \end{cases}$ $\left(2, -2\dfrac{1}{2}\right)$ **32.** $\begin{cases} -4x = 8 + 6y \\ -3y = 2x - 3 \end{cases}$

33. $\begin{cases} 3(x - 4) = -2y \\ 2x = 3(y - 19) \end{cases}$ $(-6, 15)$ **34.** $\begin{cases} 4(x + 5) = -3y \\ 3x - 2(y + 18) \end{cases}$

35. $\begin{cases} \dfrac{2x + 9}{3} = \dfrac{y + 1}{2} \\ \dfrac{x}{3} = \dfrac{y - 7}{6} \end{cases}$ $(-3, 1)$ **36.** $\begin{cases} \dfrac{2 - 5x}{4} = \dfrac{2y - 4}{2} \\ \dfrac{x + 5}{3} = \dfrac{y}{5} \end{cases}$ $(-2, 5)$

(8.4) *Solve by writing and solving a system of linear equations.*

37. The sum of two numbers is 16. Three times the larger number decreased by the smaller number is 72. Find the two numbers. -6 and 22

38. The Forrest Theater can seat a total of 360 people. They take in $15,150 when every seat is sold. If orchestra section tickets cost $45 and balcony tickets cost $35, find the number of people that can be seated in the orchestra section. 255 people

39. A riverboat can head 340 miles upriver in 19 hours, but the return trip takes only 14 hours. Find the current of the river and find the speed of the ship in still water to the nearest tenth of a mile.

	D	=	R	•	T
UPRIVER	340		$x + y$		19
DOWNRIVER	340		$x - y$		14

40. Sam Abney invested $9000 one year ago. Part of the money was invested at 6%, the rest at 10%. If the total interest earned in one year was $652.80, find how much was invested at each rate.

41. Ancient Greeks thought that the most pleasing dimensions for a picture are those where the length is approximately 1.6 times longer than the width. This ratio is known as the Golden Ratio. If Sandreka Walker has 6 feet of framing material, find the dimensions of the largest frame she can make that satisfies the Golden Ratio. Find the dimensions to the nearest hundredth of a foot.

42. Find the amount of 6% acid solution and the amount of 14% acid solution Pat should combine to prepare 50 cc's (cubic centimeters) of a 12% solution. 12.5 cc of 6% solution; 37.5 cc of 14% solution

43. The Deli charges $3.80 for a breakfast of 3 eggs and 4 strips of bacon. The charge is $2.75 for 2 eggs and 3 strips of bacon. Find the cost of each egg and the cost of each strip of bacon.

44. An exercise enthusiast alternates between jogging and walking. He traveled 15 miles during the past 3 hours. He jogs at a rate of 7.5 miles per hour and walks at a rate of 4 miles per hour. Find how much time, to the nearest hundredth of an hour, he actually spent jogging. 0.86 hr.

(8.5) *Graph the solution of the following systems of linear inequalities.* graphical answers in App. F

45. $\begin{cases} y \geq 2x - 3 \\ y \leq -2x + 1 \end{cases}$ **46.** $\begin{cases} y \leq -3x - 3 \\ y \leq 2x + 7 \end{cases}$

47. $\begin{cases} x + 2y > 0 \\ x - y \leq 6 \end{cases}$ **48.** $\begin{cases} x - 2y \geq 7 \\ x + y \leq -5 \end{cases}$

49. $\begin{cases} 3x - 2y \leq 4 \\ 2x + y \geq 5 \end{cases}$ **50.** $\begin{cases} 4x - y \leq 0 \\ 3x - 2y \geq -5 \end{cases}$

51. $\begin{cases} -3x + 2y > -1 \\ y < -2 \end{cases}$ **52.** $\begin{cases} -2x + 3y > -7 \\ x \geq -2 \end{cases}$

19. no solutions, inconsistent **21.** $(3, 1)$ **22.** infinite number of solutions, dependent
29. infinite number of solutions, dependent **32.** no solutions, inconsistent **34.** $(4, -12)$
39. ship: 21.1 mph; current: 3.2 mph **40.** $6180 at 6%; $2820 at 10% **41.** width: 1.15 ft. length: 1.85 ft.
43. one egg: $0.40; one strip of bacon: $0.65

CHAPTER 8 TEST

Is the ordered pair a solution of the given linear system?

1. $\begin{cases} 2x - 3y = 5 \\ 6x + y = 1 \end{cases}$; $(1, -1)$ no

2. $\begin{cases} 4x - 3y = 24 \\ 4x + 5y = -8 \end{cases}$; $(3, -4)$ yes

3. Use graphing to find the solutions of the system

$\begin{cases} y - x = 6 \\ y + 2x = -6 \end{cases}$ $(-4, 2)$, See App. F

4. Use the substitution method to solve the system

$\begin{cases} 3x - 2y = -14 \\ x + 3y = -1 \end{cases}$ $(-4, 1)$

5. Use the substitution method to solve the system

$\begin{cases} \dfrac{1}{2}x + 2y = -\dfrac{15}{4} \\ 4x = -y \end{cases}$ $\left(\dfrac{1}{2}, -2\right)$

6. Use the addition method to solve the system

$\begin{cases} 3x + 5y = 2 \\ 2x - 3y = 14 \end{cases}$ $(4, -2)$

7. Use the addition method to solve the system

$\begin{cases} 5x - 6y = 7 \\ 7x - 4y = 12 \end{cases}$ $\left(2, \dfrac{1}{2}\right)$

Solve each system using the substitution method or the addition method.

8. $\begin{cases} 3x + y = 7 \\ 4x + 3y = 1 \end{cases}$ $(4, -5)$

9. $\begin{cases} 3(2x + y) = 4x + 20 \\ x - 2y = 3 \end{cases}$ $(7, 2)$

10. $\begin{cases} \dfrac{x - 3}{2} = \dfrac{2 - y}{4} \\ \dfrac{7 - 2x}{3} = \dfrac{y}{2} \end{cases}$ $(5, -2)$

11. 20 \$1.00 bills; 42 \$5.00 bills

11. Lisa has a bundle of money consisting of \$1 bills and \$5 bills. There are 62 bills in the bundle. The total value of the bundle is \$230. Find the number of \$1 bills and the number of \$5 bills.

12. Don has invested \$4000, part at 5% simple annual interest and the rest at 9%. Find how much he invested at each rate if the total interest after 1 year is \$311.   \$1225 at 5%; \$2775 at 9%

Graph the solutions of the following systems of linear inequalities.

13. $\begin{cases} y + 2x \le 4 \\ y \ge 2 \end{cases}$ See App. F

14. $\begin{cases} 2y - x \ge 1 \\ x + y \ge -4 \end{cases}$ See App. F

CHAPTER 8   CUMULATIVE REVIEW

1. Find each quotient. Write all answers in lowest terms. *(Sec. 1.2, Ex. 4)*

   a. $\dfrac{4}{5} \div \dfrac{5}{16}$ $\dfrac{64}{25}$ **b.** $\dfrac{7}{10} \div 14$ $\dfrac{1}{20}$ **c.** $\dfrac{3}{8} \div \dfrac{3}{10}$ $\dfrac{5}{4}$

2. During three days, a share of Lamplighter's International stock recorded the following gains and losses: *(Sec. 1.5, Ex. 4)*

Monday	Tuesday	Wednesday
a gain of \$2	a loss of \$1	a loss of \$3

Find the overall gain or loss for the stock for the three days.   \$2 loss

(Sec. 1.7, Ex. 4)

3. Find the multiplicative inverse of each number.

   a. 22 $\dfrac{1}{22}$ **b.** $\dfrac{3}{16}$ $\dfrac{16}{3}$ **c.** -10 $-\dfrac{1}{10}$ **d.** $-\dfrac{9}{13}$ $-\dfrac{13}{9}$

4. a. The sum of two numbers is 8. If one number is 3, find the other number.   5

   b. The sum of two numbers is 8. If one number is x, write an expression representing the other number. $8 - x$ *(Sec. 2.2, Ex. 7)*

5. Solve $-2(x - 5) + 10 = -3(x + 2) + x$.

6. Write each number as a percent. *(Sec. 2.7, Ex. 2)*

   a. 0.73 73%   **b.** 1.39 139%   **c.** $\dfrac{1}{4}$ 25%

5. *(Sec. 2.4, Ex. 6)* no solution

7. Solve $2x + 7 \le x - 11$, and graph the solution set.

8. Graph $x - 3y = 6$ by finding and plotting intercept points. *(Sec. 3.3, Ex. 3)* See App. F

9. Graph $x > 2y$. *(Sec. 3.5, Ex. 4)* See App. F

10. Simplify the following.

 a. $\left(\dfrac{-5x^2}{y^3}\right)^2$ **b.** $\dfrac{(x^3)^4 x}{x^7}$ **c.** $\dfrac{(2x)^5}{x^3}$ **d.** $\dfrac{(a^2 b)^3}{a^3 b^2}$

11. Add $(-2x^2 + 5x - 1)$ and $(-2x^2 + x + 3)$.

12. Divide: $\dfrac{4x^2 + 7 + 8x^3}{2x + 3}$.

13. Solve $x(2x - 7) = 4$. *(Sec. 5.6, Ex. 3)* $x = -\dfrac{1}{2}, x = 4$

14. Find the lengths of the sides of a right triangle if the lengths can be expressed by three consecutive even integers. *(Sec. 5.7, Ex. 4)* 6, 8, 10

15. Subtract: $\dfrac{2y}{2y - 7} - \dfrac{7}{2y - 7}$. *(Sec. 6.3, Ex. 2)* 1

16. Find the slope of the line whose equation is $y = \dfrac{3}{4}x + 6$. *(Sec. 7.1, Ex. 1)* $m = \dfrac{3}{4}$

17. Which of the following ordered pairs is a solution of the given system?

$$\begin{cases} 2x - 3y = 6 & \text{First equation.} \quad \text{\textit{(Sec. 8.1, Ex. 1)}} \\ \quad\ x = 2y & \text{Second equation.} \end{cases}$$

 a. $(12, 6)$ yes **b.** $(0, -2)$ no

18. Without graphing, determine the number of solutions of the system.

$$\begin{cases} \dfrac{1}{2}x - y = 2 \\ \quad\ x = 2y + 5 \end{cases} \quad \text{\textit{(Sec. 8.1, Ex. 5)} no solution}$$

19. Solve the system $\begin{cases} x + 2y = 7 \\ 2x + 2y = 13 \end{cases}$.

20. Use the addition method to solve $\begin{cases} -x - \dfrac{y}{2} = \dfrac{5}{2} \\ -\dfrac{x}{2} + \dfrac{y}{4} = 0 \end{cases}$.

21. Find two numbers whose sum is 37 and whose difference is 21. *(Sec. 8.4, Ex. 1)* 29 and 8

22. Graph the solution of the system $\begin{cases} -3x + 4y < 12 \\ \quad\ x \ge 2 \end{cases}$.

---

7. *(Sec. 2.9, Ex. 7)* $x \le -18$; $\underset{-21\,-20\,-19\,-18\,-17}{\longleftarrow\!\!+\!\!+\!\!+\!\!\bullet\!\!+\!\!\longrightarrow}$ **10.** *(Sec. 4.1, Ex. 9)* **a.** $\dfrac{25x^4}{y^6}$ **10b.** x^6 **10c.** $32x^2$ **10d.** $a^3 b$

11. *(Sec. 4.2, Ex. 6)* $-4x^2 + 6x + 2$ **12.** *(Sec. 4.6, Ex. 7)* $4x^2 - 4x + 6 + \dfrac{-11}{2x + 3}$ **19.** *(Sec. 8.2, Ex. 2)* $\left(6, \dfrac{1}{2}\right)$

20. *(Sec. 8.3, Ex. 5)* $\left(-\dfrac{5}{4}, -\dfrac{5}{2}\right)$ **22.** *(Sec. 8.5, Ex. 3)* See App. F.

ROOTS AND RADICALS

INVESTIGATING THE DIMENSIONS OF CYLINDERS

It is often difficult to estimate visually the location of a diameter of a cylinder, such as a tin can or a water pipe, and, thus, to measure the length of its radius. By measuring the volume and height of a cylinder, it is possible to calculate the length of its radius using square roots.

IN THE CHAPTER GROUP ACTIVITY ON PAGE 537, YOU WILL HAVE THE OPPORTUNITY TO MEASURE THE VOLUMES AND HEIGHTS OF SEVERAL DIFFERENT-SIZED CANS TO CALCULATE THE LENGTHS OF THEIR RADII.

Having spent the last chapter studying equations, we return now to algebraic expressions. We expand on your skills of operating on expressions—adding, subtracting, multiplying, dividing, and raising to powers—to include finding roots. Just as subtraction is defined by addition and division by multiplication, finding roots is defined by raising to powers. This chapter also includes working with equations that contain roots and solving problems that can be modeled by such equations.

9.1   INTRODUCTION TO RADICALS

O B J E C T I V E S

  Find square roots of perfect squares.

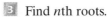 Find cube roots of perfect cubes.

3 Find *n*th roots.

4 Identify rational and irrational numbers.

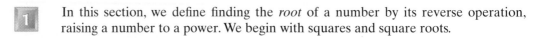

TAPE BA 9.1

1 In this section, we define finding the *root* of a number by its reverse operation, raising a number to a power. We begin with squares and square roots.

The square of 5 is $5^2 = 25$.
The square of -5 is $(-5)^2 = 25$.
The square of $\frac{1}{2}$ is $\left(\frac{1}{2}\right)^2 = \frac{1}{4}$.

The reverse operation of squaring a number is finding the *square root* of a number. For example,

A square root of 25 is 5, because $5^2 = 25$.
A square root of 25 is also -5, because $(-5)^2 = 25$.
A square root of $\frac{1}{4}$ is $\frac{1}{2}$, because $\left(\frac{1}{2}\right)^2 = \frac{1}{4}$.

In general, a number b is a square root of a number a if $b^2 = a$.

Notice that both 5 and -5 are square roots of 25. The symbol $\sqrt{}$ is used to denote the **positive** or **principal square root** of a number. For example,

$$\sqrt{25} = 5 \text{ since } 5^2 = 25 \text{ and } 5 \text{ is positive.}$$

The symbol $-\sqrt{}$ is used to denote the **negative square root.** For example,

$$-\sqrt{25} = -5$$

> **SQUARE ROOT**
>
> The positive or principal square root of a positive number a is written as $\sqrt{a}$. The negative square root of a is written as $-\sqrt{a}$.
>
> $$\sqrt{a} = b \qquad \text{only if } b^2 = a \text{ and } b > 0$$
>
> Also, the square root of 0, written as $\sqrt{0}$, is 0.

The symbol $\sqrt{}$ is called a **radical** or **radical sign.** The expression within or under a radical sign is called the **radicand.** An expression containing a radical is called a **radical expression.**

$$\sqrt{a} \overset{\longleftarrow}{} \text{radical sign}$$
$$\overset{\longleftarrow}{} \text{radicand}$$

EXAMPLE 1   Find each square root.

   a. $\sqrt{36}$ **b.** $\sqrt{64}$ **c.** $-\sqrt{25}$ **d.** $\sqrt{\dfrac{9}{100}}$ **e.** $\sqrt{0}$

Solution: **a.** $\sqrt{36} = 6$, because $6^2 = 36$ and 6 is positive.

   b. $\sqrt{64} = 8$, because $8^2 = 64$ and 8 is positive.

   c. $-\sqrt{25} = -5$. The negative sign in front of the radical indicates the negative square root of 25.

   d. $\sqrt{\dfrac{9}{100}} = \dfrac{3}{10}$ because $\left(\dfrac{3}{10}\right)^2 = \dfrac{9}{100}$ and $\dfrac{3}{10}$ is positive.

   e. $\sqrt{0} = 0$ because $0^2 = 0$.

Is the square root of a negative number a real number? For example, is $\sqrt{-4}$ a real number? To answer this question, we ask ourselves, is there a real number whose square is -4? Since there is no real number whose square is -4, we say that $\sqrt{-4}$ is not a real number. In general,

A square root of a negative number is not a real number.

2   Finding roots can be extended to other roots such as cube roots. For example, since $2^3 = 8$, we call 2 the **cube root** of 8. In symbols, we write

$$\sqrt[3]{8} = 2$$

> **CUBE ROOT**
>
> The **cube root** of a real number a is written as $\sqrt[3]{a}$, and
>
> $$\sqrt[3]{a} = b \qquad \text{only if } b^3 = a$$

From the above definition, we have

$$\sqrt[3]{27} = 3, \qquad \text{since } 3^3 = 27$$
$$\sqrt[3]{-64} = -4, \qquad \text{since } (-4)^3 = -64$$

Notice that unlike the square root of a negative number, the cube root of a negative number is a real number. This is so because while we cannot find a real number whose **square** is negative, we **can** find a real number whose **cube** is negative. In fact, the cube of a negative number is a negative number. Therefore, the cube root of a negative number is a negative number.

EXAMPLE 2   Find the cube roots.

   a. $\sqrt[3]{1}$ **b.** $\sqrt[3]{-27}$ **c.** $\sqrt[3]{\dfrac{1}{125}}$

Solution: **a.** $\sqrt[3]{1} = 1$ because $1^3 = 1$.
   b. $\sqrt[3]{-27} = -3$ because $(-3)^3 = -27$
   c. $\sqrt[3]{\dfrac{1}{125}} = \dfrac{1}{5}$ because $\left(\dfrac{1}{5}\right)^3 = \dfrac{1}{125}$.

3   Just as we can raise a real number to powers other than 2 or 3, we can find roots other than square roots and cube roots. In fact, we can take the nth root of a number where n is any natural number. An **nth root** of a number a is a number whose nth power is a. The natural number n is called the **index**.
   In symbols, the nth root of a is written as $\sqrt[n]{a}$. The index 2 is usually omitted for square roots.

> R E M I N D E R   If the index is even, such as $\sqrt{}, \sqrt[4]{}, \sqrt[6]{}$, and so on, the radicand must be nonnegative for the root to be a real number. For example,
>
> $$\sqrt[4]{16} = 2 \text{ but } \sqrt[4]{-16} \text{ is not a real number}$$
> $$\sqrt[6]{64} = 2 \text{ but } \sqrt[6]{-64} \text{ is not a real number}$$

EXAMPLE 3   Simplify the following expressions.

   a. $\sqrt[4]{16}$ **b.** $\sqrt[5]{-32}$ **c.** $-\sqrt[3]{8}$ **d.** $\sqrt[4]{-81}$

Solution: **a.** $\sqrt[4]{16} = 2$ because $2^4 = 16$ and 2 is positive.

b. $\sqrt[5]{-32} = -2$ because $(-2)^5 = -32$.

c. $-\sqrt[3]{8} = -2$ since $\sqrt[3]{8} = 2$.

d. $\sqrt[4]{-81}$ is not a real number since the index 4 is even and the radicand -81 is negative.

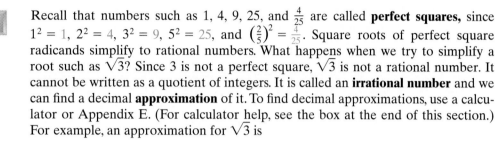

Recall that numbers such as 1, 4, 9, 25, and $\frac{4}{25}$ are called **perfect squares,** since $1^2 = 1$, $2^2 = 4$, $3^2 = 9$, $5^2 = 25$, and $\left(\frac{2}{5}\right)^2 = \frac{4}{25}$. Square roots of perfect square radicands simplify to rational numbers. What happens when we try to simplify a root such as $\sqrt{3}$? Since 3 is not a perfect square, $\sqrt{3}$ is not a rational number. It cannot be written as a quotient of integers. It is called an **irrational number** and we can find a decimal **approximation** of it. To find decimal approximations, use a calculator or Appendix E. (For calculator help, see the box at the end of this section.) For example, an approximation for $\sqrt{3}$ is

$$\sqrt{3} \approx 1.732$$
$$\uparrow$$

approximation symbol

Radicands can also contain variables. Since the square root of a negative number is not a real number, we want to make sure variables in the radicand do not have replacement values that would make the radicand negative. To avoid negative radicands, assume for the rest of this chapter that **if a variable appears in the radicand of a radical expression, it represents positive numbers only.** Then

$$\sqrt{y^2} = y \qquad \text{because } (y)^2 = y^2$$

Also,

$$\sqrt{x^8} = x^4 \qquad \text{because } (x^4)^2 = x^8$$

Also,

$$\sqrt{9x^2} = 3x \qquad \text{because } (3x)^2 = 9x^2$$

EXAMPLE 4   Simplify the following expressions. Assume that each variable represents a positive number.

a. $\sqrt{x^2}$ **b.** $\sqrt{x^6}$ **c.** $\sqrt[3]{27y^6}$ **d.** $\sqrt{16x^{16}}$

Solution: **a.** $\sqrt{x^2} = x$ because x times itself equals x^2.

b. $\sqrt{x^6} = x^3$ because $(x^3)^2 = x^6$.

c. $\sqrt[3]{27y^6} = 3y^2$ because $(3y^2)^3 = 27y^6$.

d. $\sqrt{16x^{16}} = 4x^8$ because $(4x^8)^2 = 16x^{16}$.

SCIENTIFIC CALCULATOR EXPLORATIONS

To simplify or approximate square roots using a calculator, locate the key marked $\boxed{\sqrt{}}$.
To simplify $\sqrt{25}$, press $\boxed{25}$ $\boxed{\sqrt{}}$. The display should read $\boxed{5}$.
To approximate $\sqrt{30}$, press $\boxed{30}$ $\boxed{\sqrt{}}$. The display should read $\boxed{5.4772256}$. This is
an approximation for $\sqrt{30}$. Then a three-decimal-place approximation is

$$\sqrt{30} \approx 5.477$$

Is this answer reasonable? Since 30 is between perfect squares 25 and 36, $\sqrt{30}$ is
between $\sqrt{25} = 5$ and $\sqrt{36} = 6$. The calculator result is then reasonable since
5.4772256 is between 5 and 6.
Use a calculator to approximate to three decimal places.

1. $\sqrt{7}$ 2.646
2. $\sqrt{14}$ 3.742
3. $\sqrt{10}$ 3.162
4. $\sqrt{200}$ 14.142
5. $\sqrt{82}$ 9.055
6. $\sqrt{46}$ 6.782

EXERCISE SET 9.1

Find each root. See Examples 1 through 3.

1. $\sqrt{16}$ 4
2. $\sqrt{9}$ 3
3. $\sqrt{81}$ 9
4. $\sqrt{49}$ 7
5. $\sqrt{\dfrac{1}{25}}$ $\dfrac{1}{5}$
6. $\sqrt{\dfrac{1}{64}}$ $\dfrac{1}{8}$
7. $-\sqrt{100}$ -10
8. $-\sqrt{36}$ -6
9. $\sqrt[3]{64}$ 4
10. $\sqrt[3]{-1}$ -1
11. $-\sqrt[3]{27}$ -3
12. $-\sqrt[3]{8}$ -2
13. $\sqrt[3]{\dfrac{1}{8}}$ $\dfrac{1}{2}$
14. $\sqrt[3]{\dfrac{1}{64}}$ $\dfrac{1}{4}$
15. $\sqrt[3]{-125}$ -5
16. $\sqrt[3]{-27}$ -3
17. $\sqrt[5]{32}$ 2
18. $\sqrt[4]{-1}$
19. $\sqrt[4]{81}$ 3
20. $\sqrt{121}$ 11
21. $\sqrt{-4}$
22. $\sqrt[5]{\dfrac{1}{32}}$ $\dfrac{1}{2}$
23. $\sqrt[3]{\dfrac{1}{27}}$ $\dfrac{1}{3}$
24. $\sqrt[4]{256}$ 4
25. $\sqrt{\dfrac{9}{25}}$ $\dfrac{3}{5}$
26. $\sqrt[3]{\dfrac{8}{27}}$ $\dfrac{2}{3}$
27. $-\sqrt{49}$ -7
28. $-\sqrt[4]{625}$ -5

29. Explain why the square root of a negative number is not a real number.   answers may vary

30. Explain why the cube root of a negative number is a real number.   answers may vary

Find each root. Assume that each variable represents a nonnegative real number. See Example 4.

31. $\sqrt{z^2}$ z
32. $\sqrt{y^{10}}$ y^5
33. $\sqrt{x^4}$ x^2
34. $\sqrt{z^6}$ z^3
35. $\sqrt{9x^8}$ $3x^4$
36. $\sqrt{36x^{12}}$ $6x^6$
37. $\sqrt{x^2y^6}$ xy^3
38. $\sqrt{y^4z^{18}}$ y^2z^9
39. $\sqrt[3]{x^{15}}$ x^5
40. $\sqrt[3]{y^{12}}$ y^4
41. $\sqrt{x^{12}}$ x^6
42. $\sqrt{z^{16}}$ z^8
43. $\sqrt{81x^2}$ $9x$
44. $\sqrt{100z^4}$ $10z^2$
45. $-\sqrt{144y^{14}}$
46. $-\sqrt{121z^{22}}$
47. $\sqrt{x^2y^2}$ xy
48. $\sqrt{y^{20}z^{30}}$
49. $\sqrt{16x^{16}}$ $4x^8$
50. $\sqrt{36y^{36}}$ $6y^{18}$

Find each root that is a real number.

51. $\sqrt{0}$ 0
52. $\sqrt[3]{0}$ 0
53. $-\sqrt[5]{\dfrac{1}{32}}$ $-\dfrac{1}{2}$
54. $-\sqrt[3]{\dfrac{27}{125}}$ $-\dfrac{3}{5}$
55. $\sqrt{-64}$
56. $\sqrt[3]{-64}$ -4
57. $-\sqrt{64}$ -8
58. $\sqrt[6]{64}$ 2
59. $-\sqrt{169}$

18. not a real number   21. not a real number   45. $-12y^7$ 46. $-11z^{11}$ 48. $y^{10}z^{15}$ 55. not a real number
59. -13

60. $\sqrt[4]{-16}$ **61.** $\sqrt{1}$ 1 **62.** $\sqrt[3]{1}$ 1

63. $\sqrt{\dfrac{25}{64}}$ $\dfrac{5}{8}$ **64.** $\sqrt{\dfrac{1}{100}}$ $\dfrac{1}{10}$ **65.** $-\sqrt[3]{-8}$ 2

66. $-\sqrt[3]{-27}$ 3

67. Simplify $\sqrt{\sqrt{81}}$. 3 **68.** Simplify $\sqrt[3]{\sqrt[3]{1}}$. 1

*Determine whether each square root is rational or irrational. If it is rational, find its **exact value.** If it is irrational, use a calculator and write a three-decimal-place* ***approximation.***

69. $\sqrt{9}$ **70.** $\sqrt{8}$ **71.** $\sqrt{37}$

72. $\sqrt{36}$ **73.** $\sqrt{169}$ **74.** $\sqrt{160}$

75. $\sqrt{4}$ **76.** $\sqrt{27}$ irrational, 5.196

77. A fence is to be erected around a square garden with an area of 324 square feet. Each side of this garden has a length of $\sqrt{324}$ feet. Write a one-decimal-point approximation of this length. 18.0 ft.

78. The roof of the warehouse shown needs to be shingled. The total area of the roof is exactly $240\sqrt{41}$ square feet. Approximate this area to the nearest whole number. 1537 sq. ft.

79. A standard baseball diamond is a square with 90-foot sides connecting the bases. The distance from first base to third base is $90 \cdot \sqrt{2}$ feet. Approximate $\sqrt{2}$ accurate to two decimal places and use it to approximate the distance $90 \cdot \sqrt{2}$ feet. $\sqrt{2} \approx 1.41$; 126.90 ft.

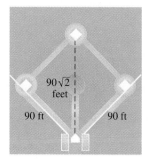

80. Graph $y = \sqrt{x}$. (*Hint:* Complete the table below, plot the ordered pair solutions, and draw a smooth curve through the points. Remember that since the radicand cannot be negative, this particular graph begins at the point with coordinates (0,0).)

x	y
0	0
1	1
3	1.7 (approximate)
4	2
9	3 See App. F

81. Graph $y = \sqrt[3]{x}$ (Complete the table below, plot the ordered pair solutions, and draw a smooth curve through the points.)

x	y
-8	-2
-2	-1.3 (approximate)
-1	-1
0	0
1	1
2	1.3 (approximate)
8	2 See App. F

graphical answers in App. F

Use a grapher and graph each function. Observe the graph from left to right and give the ordered pair that corresponds to the "beginning" of the graph. Then tell why the graph starts at that point.

82. $y = \sqrt{x - 2}$ (2, 0) **83.** $y = \sqrt{x + 3}$ $(-3, 0)$

84. $y = \sqrt{x + 4}$ $(-4, 0)$ **85.** $y = \sqrt{x - 5}$ (5, 0)

Review Exercises

Write each integer as a product of two integers such that one of the factors is a perfect square. For example, in $18 = 9 \cdot 2$, 9 is a perfect square.

86. 50 $25 \cdot 2$ **87.** 8 $4 \cdot 2$ **88.** 32

89. 75 $25 \cdot 3$ **90.** 28 $4 \cdot 7$ **91.** 44 $4 \cdot 11$

92. 27 $9 \cdot 3$ **93.** 90 $9 \cdot 10$

60. not a real number **69.** rational, 3 **70.** irrational, 2.828 **71.** irrational, 6.083 **72.** rational, 6
73. rational, 13 **74.** irrational, 12.649 **75.** rational, 2 **88.** $16 \cdot 2$ or $4 \cdot 8$

9.2 | SIMPLIFYING RADICALS

OBJECTIVES

 Use the product rule to simplify radicals.

2 Use the quotient rule to simplify radicals.

TAPE BA 9.2

1 Much of our work with expressions in this book has involved finding ways to write expressions in their simplest form. Writing radicals in simplest form requires recognizing several patterns, or rules, which we present here. Notice that

$$\sqrt{9 \cdot 16} = \sqrt{144} = 12$$

Also,

$$\sqrt{9} \cdot \sqrt{16} = 3 \cdot 4 = 12$$

Since both expressions simplify to 12, we can write

$$\sqrt{9 \cdot 16} = \sqrt{9} \cdot \sqrt{16}$$

This suggests the following product rule for square roots.

PRODUCT RULE FOR SQUARE ROOTS

If $\sqrt{a}$ and $\sqrt{b}$ are real numbers, then

$$\sqrt{a \cdot b} = \sqrt{a} \cdot \sqrt{b}$$

The product rule states that the square root of a product is equal to the product of the square roots. We use this rule to write radical expressions in **simplest form.** A radical is written in simplest form when the radicand has no perfect square factors other than 1. To simplify $\sqrt{20}$, for example, factor 20 so that one of its factors is a perfect square factor.

$$\sqrt{20} = \sqrt{4 \cdot 5} \qquad \text{Factor 20.}$$
$$= \sqrt{4} \cdot \sqrt{5} \qquad \text{Apply the product rule.}$$
$$= 2\sqrt{5} \qquad \text{Write } \sqrt{4} \text{ as 2.}$$

The notation $2\sqrt{5}$ means $2 \cdot \sqrt{5}$. Since the radicand 5 has no perfect square factor other than 1, $2\sqrt{5}$ is in simplest form.

When factoring a radicand, look for at least one factor that is a perfect square. Review the table of perfect squares in Appendix E to help locate perfect square factors more quickly.

When simplifying a radical, realize that a radical expression in simplest form does *not mean* a decimal approximation. The simplest form of a radical expression is an exact form and may still contain a radical.

> REMINDER When simplifying a radical, use **factors** of the radicand; **do not** write the radicand as a sum. For example, **do not** write $\sqrt{20}$ as $\sqrt{4 + 16}$ because $\sqrt{4 + 16} \neq \sqrt{4} + \sqrt{16}$. Correctly simplified, $\sqrt{20} = \sqrt{4 \cdot 5} = \sqrt{4} \cdot \sqrt{5} = 2\sqrt{5}$.

EXAMPLE 1   Simplify each expression.

a. $\sqrt{54}$ **b.** $\sqrt{12}$ **c.** $\sqrt{200}$ **d.** $\sqrt{35}$

Solution: **a.** Try to factor 54 so that at least one of the factors is a perfect square. Since 9 is a perfect square and $54 = 9 \cdot 6$,

$$\begin{aligned} \sqrt{54} &= \sqrt{9 \cdot 6} && \text{Factor 54.} \\ &= \sqrt{9} \cdot \sqrt{6} && \text{Apply the product rule.} \\ &= 3\sqrt{6} && \text{Write } \sqrt{9} \text{ as 3.} \end{aligned}$$

b.
$$\begin{aligned} \sqrt{12} &= \sqrt{4 \cdot 3} && \text{Factor 12.} \\ &= \sqrt{4} \cdot \sqrt{3} && \text{Apply the product rule.} \\ &= 2\sqrt{3} && \text{Write } \sqrt{4} \text{ as 2.} \end{aligned}$$

c. The largest perfect square factor of 200 is 100.

$$\begin{aligned} \sqrt{200} &= \sqrt{100 \cdot 2} && \text{Factor 200.} \\ &= \sqrt{100} \cdot \sqrt{2} && \text{Apply the product rule.} \\ &= 10\sqrt{2} && \text{Write } \sqrt{100} \text{ as 10.} \end{aligned}$$

d. The radicand 35 contains no perfect square factors other than 1. Thus $\sqrt{35}$ is in simplest form.

In Example 1, part **(c)**, what happens if we don't use the largest perfect square factor of 200? Although using the largest perfect square factor saves time, the result is the same no matter what perfect square factor is used. For example, it is also true that $200 = 4 \cdot 50$. Then

$$\begin{aligned} \sqrt{200} &= \sqrt{4} \cdot \sqrt{50} \\ &= 2 \cdot \sqrt{50} \end{aligned}$$

Since $\sqrt{50}$ is not in simplest form, we continue.

$$\begin{aligned} \sqrt{200} &= 2 \cdot \sqrt{50} \\ &= 2 \cdot \sqrt{25} \cdot \sqrt{2} \\ &= 2 \cdot 5 \cdot \sqrt{2} \\ &= 10\sqrt{2} \end{aligned}$$

2

Next, let's examine the square root of a quotient.

$$\sqrt{\frac{16}{4}} = \sqrt{4} = 2$$

Also,

$$\frac{\sqrt{16}}{\sqrt{4}} = \frac{4}{2} = 2$$

Since both expressions equal 2, we can write

$$\sqrt{\frac{16}{4}} = \frac{\sqrt{16}}{\sqrt{4}}$$

This suggests the following quotient rule.

QUOTIENT RULE FOR SQUARE ROOTS

If $\sqrt{a}$ and $\sqrt{b}$ are real numbers and $b \neq 0$, then

$$\sqrt{\frac{a}{b}} = \frac{\sqrt{a}}{\sqrt{b}}$$

The quotient rule states that the square root of a quotient is equal to the quotient of the square roots.

EXAMPLE 2 Simplify the following.

a. $\sqrt{\dfrac{25}{36}}$ b. $\sqrt{\dfrac{3}{64}}$ c. $\sqrt{\dfrac{4}{81}}$

Solution: Use the quotient rule.

a. $\sqrt{\dfrac{25}{36}} = \dfrac{\sqrt{25}}{\sqrt{36}} = \dfrac{5}{6}$ b. $\sqrt{\dfrac{3}{64}} = \dfrac{\sqrt{3}}{\sqrt{64}} = \dfrac{\sqrt{3}}{8}$

c. $\sqrt{\dfrac{40}{81}} = \dfrac{\sqrt{40}}{\sqrt{81}}$

                 Use the quotient rule.

    $= \dfrac{\sqrt{4} \cdot \sqrt{10}}{9}$ Apply the product rule and write $\sqrt{81}$ as 9.

    $= \dfrac{2\sqrt{10}}{9}$ Write $\sqrt{4}$ as 2.

EXAMPLE 3 Simplify each expression. Assume that variables represent positive numbers only.

a. $\sqrt{x^5}$ **b.** $\sqrt{8y^2}$ **c.** $\sqrt{\dfrac{45}{x^6}}$

Solution: **a.** $\sqrt{x^5} = \sqrt{x^4 \cdot x} = \sqrt{x^4} \cdot \sqrt{x} = x^2\sqrt{x}$

b. $\sqrt{8y^2} = \sqrt{4 \cdot 2 \cdot y^2} = \sqrt{4y^2 \cdot 2} = \sqrt{4y^2} \cdot \sqrt{2} = 2y\sqrt{2}$

c. $\sqrt{\dfrac{45}{x^6}} = \dfrac{\sqrt{45}}{\sqrt{x^6}} = \dfrac{\sqrt{9 \cdot 5}}{x^3} = \dfrac{\sqrt{9} \cdot \sqrt{5}}{x^3} = \dfrac{3\sqrt{5}}{x^3}$

The product and quotient rules also apply to roots other than square roots. In general, we have the following product and quotient rules for radicals:

PRODUCT RULE FOR RADICALS

If $\sqrt[n]{a}$ and $\sqrt[n]{b}$ are real numbers, then

$$\sqrt[n]{a \cdot b} = \sqrt[n]{a} \cdot \sqrt[n]{b}$$

QUOTIENT RULE FOR RADICALS

If $\sqrt[n]{a}$ and $\sqrt[n]{b}$ are real numbers and $b \neq 0$, then

$$\sqrt[n]{\dfrac{a}{b}} = \dfrac{\sqrt[n]{a}}{\sqrt[n]{b}}$$

For example, to simplify cube roots, look for perfect cube factors of the radicand. For example, 8 is a perfect cube, since $2^3 = 8$.

To simplify $\sqrt[3]{48}$, factor 48 as $8 \cdot 6$.

$$\sqrt[3]{48} = \sqrt[3]{8 \cdot 6} \qquad \text{Factor 48.}$$
$$= \sqrt[3]{8} \cdot \sqrt[3]{6} \qquad \text{Apply the product rule.}$$
$$= 2\sqrt[3]{6} \qquad \text{Write } \sqrt[3]{8} \text{ as 2.}$$

$2\sqrt[3]{6}$ is in simplest form since the radicand 6 contains no perfect cube factors other than 1.

EXAMPLE 4 Simplify each expression.

a. $\sqrt[3]{54}$ **b.** $\sqrt[3]{18}$ **c.** $\sqrt[3]{\dfrac{7}{8}}$ **d.** $\sqrt[3]{\dfrac{40}{27}}$

Solution: **a.** $\sqrt[3]{54} = \sqrt[3]{27 \cdot 2} = \sqrt[3]{27} \cdot \sqrt[3]{2} = 3\sqrt[3]{2}$

b. The number 18 contains no perfect cube factors, so $\sqrt[3]{18}$ cannot be simplified further.

c. $\sqrt[3]{\dfrac{7}{8}} = \dfrac{\sqrt[3]{7}}{\sqrt[3]{8}} = \dfrac{\sqrt[3]{7}}{2}$

d. $\sqrt[3]{\dfrac{40}{27}} = \dfrac{\sqrt[3]{40}}{\sqrt[3]{27}} = \dfrac{\sqrt[3]{8 \cdot 5}}{3} = \dfrac{\sqrt[3]{8} \cdot \sqrt[3]{5}}{3} = \dfrac{2\sqrt[3]{5}}{3}$

EXAMPLE 5 Simplify each expression. Assume variables represent positive numbers only.

  a. $\sqrt[3]{x^5}$ **b.** $\sqrt[3]{40y^7}$ **c.** $\sqrt[3]{\dfrac{16}{x^6}}$

Solution: **a.** $\sqrt[3]{x^5} = \sqrt[3]{x^3 \cdot x^2} = \sqrt[3]{x^3} \cdot \sqrt[3]{x^2} = x\sqrt[3]{x^2}$

  b. $\sqrt[3]{40y^7} = \sqrt[3]{8 \cdot 5 \cdot y^6 \cdot y} = \sqrt[3]{8y^6 \cdot 5y} = \sqrt[3]{8y^6} \cdot \sqrt[3]{5y} = 2y^2\sqrt[3]{5y}$

  c. $\sqrt[3]{\dfrac{16}{x^6}} = \dfrac{\sqrt[3]{16}}{\sqrt[3]{x^6}} = \dfrac{\sqrt[3]{8 \cdot 2}}{x^2} = \dfrac{\sqrt[3]{8} \cdot \sqrt[3]{2}}{x^2} = \dfrac{2\sqrt[3]{2}}{x^2}$

MENTAL MATH

Simplify each expression. Assume that all variables represent nonnegative real numbers.

1. $\sqrt{4 \cdot 9}$ 6   **2.** $\sqrt{9 \cdot 36}$ 18   **3.** $\sqrt{x^2}$ x   **4.** $\sqrt{y^4}$ y^2

5. $\sqrt{0}$ 0   **6.** $\sqrt{1}$ 1   **7.** $\sqrt{25x^4}$ $5x^2$ **8.** $\sqrt{49x^2}$ $7x$

EXERCISE SET 9.2

Use the product rule to simplify each expression. See Examples 1 and 4.

1. $\sqrt{20}$ $2\sqrt{5}$ **2.** $\sqrt{44}$ $2\sqrt{11}$ **3.** $\sqrt{18}$ $3\sqrt{2}$

4. $\sqrt{45}$ $3\sqrt{5}$ **5.** $\sqrt{50}$ $5\sqrt{2}$ **6.** $\sqrt{28}$ $2\sqrt{7}$

7. $\sqrt{33}$ $\sqrt{33}$ **8.** $\sqrt{98}$ $7\sqrt{2}$ **9.** $\sqrt[3]{24}$ $2\sqrt[3]{3}$

10. $\sqrt[3]{81}$ $3\sqrt[3]{3}$ **11.** $\sqrt[3]{250}$ $5\sqrt[3]{2}$ **12.** $\sqrt[3]{40}$ $2\sqrt[3]{5}$

13. By using replacement values for a and b, show that $\sqrt{a^2 + b^2}$ does not equal $a + b$.

14. By using replacement values for a and b, show that $\sqrt{a + b}$ does not equal $\sqrt{a} + \sqrt{b}$.

13. answers may vary   **14.** answers may vary

Use the quotient rule and the product rule. Simplify each expression. See Examples 2 and 4.

15. $\sqrt{\dfrac{8}{25}}$ $\dfrac{2\sqrt{2}}{5}$ **16.** $\sqrt{\dfrac{63}{16}}$ $\dfrac{3\sqrt{7}}{4}$

17. $\sqrt{\dfrac{27}{121}}$ $\dfrac{3\sqrt{3}}{11}$ **18.** $\sqrt{\dfrac{24}{169}}$ $\dfrac{2\sqrt{6}}{13}$

19. $\sqrt{\dfrac{9}{4}}$ $\dfrac{3}{2}$ **20.** $\sqrt{\dfrac{100}{49}}$ $\dfrac{10}{7}$

21. $\sqrt{\dfrac{125}{9}}$ $\dfrac{5\sqrt{5}}{3}$ **22.** $\sqrt{\dfrac{27}{100}}$ $\dfrac{\sqrt[3]{3}}{10}$

23. $\sqrt[3]{\dfrac{5}{64}}$ $\dfrac{\sqrt[3]{5}}{4}$

24. $\sqrt[3]{\dfrac{32}{125}}$ $\dfrac{2\sqrt[3]{4}}{5}$

25. $\sqrt[3]{\dfrac{7}{8}}$ $\dfrac{\sqrt[3]{7}}{2}$

26. $\sqrt[3]{\dfrac{10}{27}}$ $\dfrac{\sqrt[3]{10}}{3}$

Simplify each expression. Assume that all variables represent positive numbers only. See Examples 3 and 5.

27. $\sqrt{x^7}$ $x^3\sqrt{x}$

28. $\sqrt{y^3}$ $y\sqrt{y}$

29. $\sqrt{\dfrac{88}{x^4}}$ $\dfrac{2\sqrt{22}}{x^2}$

30. $\sqrt{\dfrac{x^{11}}{81}}$ $\dfrac{x^5\sqrt{x}}{9}$

31. $\sqrt[3]{x^{16}}$ $x^5\sqrt[3]{x}$

32. $\sqrt[3]{y^{20}}$ $y^6\sqrt[3]{y^2}$

33. $\sqrt[3]{\dfrac{2}{x^9}}$ $\dfrac{\sqrt[3]{2}}{x^3}$

34. $\sqrt[3]{\dfrac{48}{x^{12}}}$ $\dfrac{2\sqrt[3]{6}}{x^4}$

Simplify each expression. Assume that all variables represent positive numbers only.

35. $\sqrt{60}$ $2\sqrt{15}$

36. $\sqrt{90}$ $3\sqrt{10}$

37. $\sqrt{180}$ $6\sqrt{5}$

38. $\sqrt{150}$ $5\sqrt{6}$

39. $\sqrt{52}$ $2\sqrt{13}$

40. $\sqrt{75}$ $5\sqrt{3}$

41. $\sqrt{\dfrac{11}{36}}$ $\dfrac{\sqrt{11}}{6}$

42. $\sqrt{\dfrac{30}{49}}$ $\dfrac{\sqrt{30}}{7}$

43. $-\sqrt{\dfrac{27}{144}}$ $-\dfrac{\sqrt{3}}{4}$

44. $-\sqrt{\dfrac{84}{121}}$ $-\dfrac{2\sqrt{21}}{11}$

45. $\sqrt[3]{\dfrac{15}{64}}$ $\dfrac{\sqrt[3]{15}}{4}$

46. $\sqrt[3]{\dfrac{4}{27}}$ $\dfrac{\sqrt[3]{4}}{3}$

47. $\sqrt[3]{80}$ $2\sqrt[3]{10}$

48. $\sqrt[3]{108}$ $3\sqrt[3]{4}$

49. $\sqrt[4]{48}$ $2\sqrt[4]{3}$

50. $\sqrt[4]{162}$ $3\sqrt[4]{2}$

51. $\sqrt{x^{13}}$ $x^6\sqrt{x}$

52. $\sqrt{y^{17}}$ $y^8\sqrt{y}$

53. $\sqrt{75x^2}$ $5x\sqrt{3}$

54. $\sqrt{72y^2}$ $6y\sqrt{2}$

55. $\sqrt{96x^4y^2}$ $4x^2y\sqrt{6}$

56. $\sqrt{40x^8y^{10}}$ $2x^4y^5\sqrt{10}$

57. $\sqrt{\dfrac{12}{y^2}}$ $\dfrac{2\sqrt{3}}{y}$

58. $\sqrt{\dfrac{63}{x^4}}$ $\dfrac{3\sqrt{7}}{x^2}$

59. $\sqrt{\dfrac{9x}{y^2}}$ $\dfrac{3\sqrt{x}}{y}$

60. $\sqrt{\dfrac{6y^2}{x^4}}$ $\dfrac{y\sqrt{6}}{x^2}$

61. $\sqrt[3]{-8x^6}$ $-2x^2$

62. $\sqrt[3]{-54y^6}$ $-3y^2\sqrt[3]{2}$

63. If a cube is to have a volume of 80 cubic inches, then each side must be $\sqrt[3]{80}$ inches long. Simplify the radical representing the side length. $2\sqrt[3]{10}$ in.

64. Jeannie is swimming across a 40-foot-wide river, trying to head straight across to the opposite shore. However, the current is strong enough to move her downstream 100 feet by the time she reaches land. (See the figure.) Because of the current, the actual distance she swam is $\sqrt{11,600}$ feet. Simplify this radical. $20\sqrt{29}$ ft.

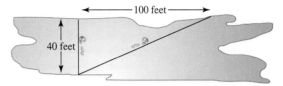

The cost C in dollars per day to operate a small delivery service is given by $C = 100\sqrt[3]{n} + 700$ where n is the number of deliveries per day.   **65.** $1700

65. Find the cost if the number of deliveries is 1000.

66. Approximate the cost if the number of deliveries is 500. $1493.70

Review Exercises

Perform the following operations. See Sections 4.2 and 4.3.   **69.** $2x^2 - 7x - 15$ **70.** $3x - 2$

67. $6x + 8x$ $14x$

68. $(6x)(8x)$ $48x^2$

69. $(2x + 3)(x - 5)$

70. $(2x + 3) + (x - 5)$

71. $9y^2 - 9y^2$ 0

72. $(9y^2)(-8y^2)$ $-72y^4$

The following pairs of triangles are similar. Find the unknown lengths. See Section 6.7.

73.

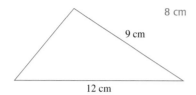

8 cm

9 cm

6 cm

x

12 cm

74.

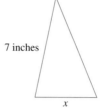

$\frac{21}{5}$ in.

7 inches

5 inches

3 inches

x

TAPE BA 9.3

9.3 | ADDING AND SUBTRACTING RADICALS

O B J E C T I V E S

1. Add or subtract like radicals.
2. Simplify radical expressions, and then add or subtract any like radicals.

To combine like terms, we use the distributive property.

$$5x + 3x = (5 + 3)x = 8x$$

The distributive property can also be applied to expressions containing radicals. For example,

$$5\sqrt{2} + 3\sqrt{2} = (5 + 3)\sqrt{2} = 8\sqrt{2}$$

Also,

$$9\sqrt{5} - 6\sqrt{5} = (9 - 6)\sqrt{5} = 3\sqrt{5}$$

Radical terms $5\sqrt{2}$ and $3\sqrt{2}$ are **like radicals,** as are $9\sqrt{5}$ and $6\sqrt{5}$.

> **LIKE RADICALS**
>
> **Like radicals** are radical expressions that have the same index and the same radicand.

From the examples above, we can see that **only like radicals can be combined** in this way. For example, the expression $2\sqrt{3} + 3\sqrt{2}$ cannot be further simplified since the radicals are not like radicals. Also, the expression $4\sqrt{7} + 4\sqrt[3]{7}$ cannot be further simplified because the radicals are not like radicals since the indices are different.

EXAMPLE 1   Simplify by combining like radical terms.

a. $4\sqrt{5} + 3\sqrt{5}$ **b.** $\sqrt{10} - 6\sqrt{10}$ **c.** $2\sqrt[3]{7} - 5\sqrt[3]{7} - 3\sqrt[3]{7}$ **d.** $2\sqrt{6} + 2\sqrt[3]{6}$

Solution: **a.** $4\sqrt{5} + 3\sqrt{5} = (4 + 3)\sqrt{5} = 7\sqrt{5}$
b. $\sqrt{10} - 6\sqrt{10} = 1\sqrt{10} - 6\sqrt{10} = (1 - 6)\sqrt{10} = -5\sqrt{10}$
c. $2\sqrt[3]{7} - 5\sqrt[3]{7} - 3\sqrt[3]{7} = (2 - 5 - 3)\sqrt[3]{7} = -6\sqrt[3]{7}$
d. $2\sqrt{6} + 2\sqrt[3]{6}$ cannot be simplified further since the indices are not the same.

2   At first glance, it appears that the expression $\sqrt{50} + \sqrt{8}$ cannot be simplified further because the radicands are different. However, the product rule can be used to simplify each radical, and then further simplification might be possible.

EXAMPLE 2   Add or subtract by first simplifying each radical.

a. $\sqrt{50} + \sqrt{8}$
b. $7\sqrt{12} - \sqrt{75}$
c. $\sqrt{25} - \sqrt{27} - 2\sqrt{18} - \sqrt{16}$
d. $2\sqrt[3]{27} - \sqrt[3]{54}$

Solution: **a.** First simplify each radical.

$$\sqrt{50} + \sqrt{8} = \sqrt{25 \cdot 2} + \sqrt{4 \cdot 2}$$ Factor radicands.
$$= \sqrt{25} \cdot \sqrt{2} + \sqrt{4} \cdot \sqrt{2}$$ Apply the product rule.
$$= 5\sqrt{2} + 2\sqrt{2}$$ Simplify $\sqrt{25}$ and $\sqrt{4}$.
$$= 7\sqrt{2}$$ Add like radicals.

b. $7\sqrt{12} - \sqrt{75} = 7\sqrt{4 \cdot 3} - \sqrt{25 \cdot 3}$ Factor radicands.
$$= 7\sqrt{4} \cdot \sqrt{3} - \sqrt{25} \cdot \sqrt{3}$$ Apply the product rule.
$$= 7 \cdot 2\sqrt{3} - 5\sqrt{3}$$ Simplify $\sqrt{4}$ and $\sqrt{25}$.
$$= 14\sqrt{3} - 5\sqrt{3}$$ Multiply.
$$= 9\sqrt{3}$$ Subtract like radicals.

c. $\sqrt{25} - \sqrt{27} - 2\sqrt{18} - \sqrt{16}$
$$= 5 - \sqrt{9 \cdot 3} - 2\sqrt{9 \cdot 2} - 4$$ Factor radicands.
$$= 5 - \sqrt{9} \cdot \sqrt{3} - 2\sqrt{9} \cdot \sqrt{2} - 4$$ Apply the product rule.
$$= 5 - 3\sqrt{3} - 2 \cdot 3\sqrt{2} - 4$$ Simplify.
$$= 1 - 3\sqrt{3} - 6\sqrt{2}$$ Write $5 - 4$ as 1 and $2 \cdot 3$ as 6.

d. $2\sqrt[3]{27} - \sqrt[3]{54} = 2 \cdot 3 - \sqrt[3]{27 \cdot 2}$
$$= 6 - 3\sqrt[3]{2}$$

No further simplification is possible.

If radical expressions contain variables, we proceed in a similar way. Simplify radicals using the product and quotient rules. Then add or subtract any like radicals.

EXAMPLE 3 Simplify each radical expression. Assume variables represent positive numbers.

    a. $2\sqrt{x^2} - \sqrt{25x} + \sqrt{x}$ **b.** $3\sqrt[3]{54x^4} + 5x\sqrt[3]{16x}$

Solution: **a.** $2\sqrt{x^2} - \sqrt{25x} + \sqrt{x}$

$$=2x - \sqrt{25} \cdot \sqrt{x} + \sqrt{x}$$ Write $\sqrt{x^2}$ as x and apply the product rule.

$$= 2x - 5\sqrt{x} + 1\sqrt{x}$$ Simplify.

$$= 2x - 4\sqrt{x}$$ Add like radicals.

    b. $3\sqrt[3]{54x^4} + 5x\sqrt[3]{16x}$

$$=3\sqrt[3]{27x^3 \cdot 2x} + 5x\sqrt[3]{8 \cdot 2x}$$ Factor.

$$= 3\sqrt[3]{27x^3} \cdot \sqrt[3]{2x} + 5x\sqrt[3]{8} \cdot \sqrt[3]{2x}$$ Apply the product rule.

$$= 3 \cdot 3x \cdot \sqrt[3]{2x} + 5x \cdot 2 \cdot \sqrt[3]{2x}$$ Simplify.

$$=9x\sqrt[3]{2x} + 10x\sqrt[3]{2x}$$ Multiply.

$$= 19x\sqrt[3]{2x}$$ Add like radicals.

MENTAL MATH

Simplify each expression by combining like radicals.

1. $3\sqrt{2} + 5\sqrt{2}$ $8\sqrt{2}$ **2.** $2\sqrt{3} + 7\sqrt{3}$ $9\sqrt{3}$ **3.** $5\sqrt{x} + 2\sqrt{x}$ $7\sqrt{x}$

4. $8\sqrt{x} + 3\sqrt{x}$ $11\sqrt{x}$ **5.** $5\sqrt{7} - 2\sqrt{7}$ $3\sqrt{7}$ **6.** $8\sqrt{6} - 5\sqrt{6}$ $3\sqrt{6}$

EXERCISE SET 9.3

Simplify each expression by combining like radicals where possible. See Example 1.

1. $4\sqrt{3} - 8\sqrt{3}$ $-4\sqrt{3}$ **2.** $\sqrt{5} - 9\sqrt{5}$ $-8\sqrt{5}$

3. $3\sqrt{6} + 8\sqrt{6} - 2\sqrt{6} - 5$ $9\sqrt{6} - 5$

4. $12\sqrt{2} - 3\sqrt{2} + 8\sqrt{2} + 10$ $17\sqrt{2} + 10$

5. $6\sqrt{5} - 5\sqrt{5} + \sqrt{2}$ **6.** $4\sqrt{3} + \sqrt{5} - 3\sqrt{3}$

7. $2\sqrt[3]{3} + 5\sqrt[3]{3} - \sqrt{3}$ **8.** $8\sqrt[3]{4} + 2\sqrt[3]{4} + 4$

9. $2\sqrt[3]{2} - 7\sqrt[3]{2} - 6$ **10.** $5\sqrt[3]{9} + 2 - 11\sqrt[3]{9}$

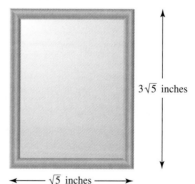

11. Find the perimeter of the rectangular picture frame. $8\sqrt{5}$ in.

$3\sqrt{5}$ inches

$\longleftarrow \sqrt{5}$ inches $\longrightarrow$

5. $\sqrt{5} + \sqrt{2}$ **6.** $\sqrt{3} + \sqrt{5}$ **7.** $7\sqrt[3]{3} - \sqrt{3}$ **8.** $10\sqrt[3]{4} + 4$ **9.** $-5\sqrt[3]{2} - 6$ **10.** $2 - 6\sqrt[3]{9}$

12. Find the perimeter of the plot of land. $80\sqrt{6}$ ft.

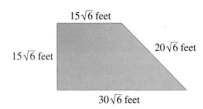

15√6 feet

15√6 feet

20√6 feet

30√6 feet

13. In your own words, describe like radicals.

14. In the expression $\sqrt{5} + 2 - 3\sqrt{5}$, explain why 2 and -3 cannot be combined.   answers may vary

Add or subtract by first simplifying each radical and then combining any like radical terms. Assume that all variables represent positive real numbers. See Examples 2 and 3.

15. $\sqrt{12} + \sqrt{27}$ $5\sqrt{3}$ **16.** $\sqrt{50} + \sqrt{18}$ $8\sqrt{2}$

17. $\sqrt{45} + 3\sqrt{20}$ $9\sqrt{5}$

18. $2\sqrt{54} - \sqrt{20} + \sqrt{45} - \sqrt{24}$ $4\sqrt{6} + \sqrt{5}$

19. $2\sqrt{8} - \sqrt{128} + \sqrt{48} + \sqrt{18}$ $-\sqrt{2} + 4\sqrt{3}$

20. $\sqrt[3]{81} + \sqrt[3]{24}$ $5\sqrt[3]{3}$ **21.** $\sqrt[3]{32} - \sqrt[3]{4}$ $\sqrt[3]{4}$

22. $4\sqrt[3]{9} - \sqrt[3]{243}$ $\sqrt[3]{9}$ **23.** $4x - 3\sqrt{x^2} + \sqrt{x}$

24. $x - 6\sqrt{x^2} + 2\sqrt{x}$ $-5x + 2\sqrt{x}$

25. $\sqrt{25x} + \sqrt{36x} - 11\sqrt{x}$ 0

26. $3\sqrt{x^3} - x\sqrt{4x}$ $x\sqrt{x}$ **27.** $\sqrt{16x} - \sqrt{x^3}$

28. $\sqrt[3]{8x^3} + x\sqrt[3]{27}$ $5x$

Simplify the following. Assume that all variables represent nonnegative real numbers.

29. $12\sqrt{5} - \sqrt{5} - 4\sqrt{5}$ **30.** $\sqrt{5} + \sqrt[3]{5}$ $\sqrt{5} + \sqrt[3]{5}$

31. $\sqrt{5} + \sqrt{5}$ $2\sqrt{5}$ **32.** $4 + 8\sqrt{2} - 9$

33. $6 - 2\sqrt{3} - \sqrt{3}$ $6 - 3\sqrt{3}$ **34.** $8 - \sqrt{2} - 5\sqrt{2}$

35. $\sqrt{75} + \sqrt{48}$ $9\sqrt{3}$ **36.** $5\sqrt{32} - \sqrt{72}$

37. $2\sqrt{80} - \sqrt{45}$ $5\sqrt{5}$

38. $\sqrt{8} + \sqrt{9} + \sqrt{18} + \sqrt{81}$ $5\sqrt{2} + 12$

39. $\sqrt{6} + \sqrt{16} + \sqrt{24} + \sqrt{25}$ $3\sqrt{6} + 9$

40. $\sqrt{\dfrac{5}{9}} + \sqrt{\dfrac{5}{81}}$ $\dfrac{4\sqrt{5}}{9}$ **41.** $\sqrt{\dfrac{3}{64}} + \sqrt{\dfrac{3}{16}}$ $\dfrac{3\sqrt{3}}{8}$

42. $\sqrt{\dfrac{3}{4}} - \sqrt{\dfrac{3}{64}}$ $\dfrac{3\sqrt{3}}{8}$ **43.** $2\sqrt[3]{8} + 2\sqrt[3]{16}$

44. $3\sqrt[3]{27} + 3\sqrt[3]{81}$ **45.** $\sqrt[3]{8} + \sqrt[3]{54} - 5$

46. $2\sqrt{45} - 2\sqrt{20}$ $2\sqrt{5}$ **47.** $5\sqrt{18} + 2\sqrt{32}$ $23\sqrt{2}$

48. $\sqrt{35} - \sqrt{140}$ $-\sqrt{35}$ **49.** $5\sqrt{2xz^2} + z\sqrt{98x}$

50. $3\sqrt{9x} + 2\sqrt{x}$ $11\sqrt{x}$ **51.** $5\sqrt{x} + 4\sqrt{4x}$ $13\sqrt{x}$

52. $\sqrt{9x} + \sqrt{81x} - 11\sqrt{x}$ **53.** $\sqrt{3x^3} + 3x\sqrt{x}$

54. $x\sqrt{4x} + \sqrt{9x^3}$ $5x\sqrt{x}$

55. $\sqrt{32x^2} + \sqrt[3]{32x^2} + \sqrt{4x^2}$ $4x\sqrt{2} + 2\sqrt[3]{4x^2} + 2x$

56. $\sqrt{18x^2} + \sqrt[3]{24x^3} + \sqrt{2x^2}$ $4x\sqrt{2} + 2x\sqrt[3]{3}$

57. $\sqrt{40x} + \sqrt[3]{40x^4} - 2\sqrt{10x} - \sqrt[3]{5x^4}$ $x\sqrt[3]{5x}$

58. $\sqrt{72x^2} + \sqrt[3]{54x} - x\sqrt{50} - 3\sqrt[3]{2x}$ $x\sqrt{2}$

59. A water trough is to be made of wood. Each of the two triangular end pieces has an area of $\dfrac{3\sqrt{27}}{4}$ square feet. The two side panels are both rectangular. In simplest radical form, find the total area of the wood needed.

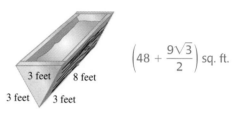

3 feet   8 feet

3 feet   3 feet

$\left(48 + \dfrac{9\sqrt{3}}{2}\right)$ sq. ft.

60. Eight wooden braces are to be attached along the diagonals of the vertical sides of a storage bin. Each of four of these diagonals has a length of $\sqrt{52}$ feet, while each of the other four has a length of $\sqrt{80}$ feet. In simplest radical form, find the total length of the wood needed for these braces.

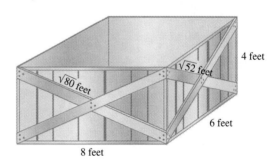

4 feet

$\sqrt{52}$ feet

$\sqrt{80}$ feet

6 feet

8 feet

Review Exercises

Square each binomial. See Section 4.4.

61. $(x + 6)^2$ **62.** $(3x + 2)^2$

63. $(2x - 1)^2$ **64.** $(x - 5)^2$

Solve each system of linear equations. See Section 8.2.

65. $\begin{cases} x = 2y \\ x + 5y = 14 \end{cases}$ (4, 2)   **66.** $\begin{cases} y = -5x \\ x + y = 16 \end{cases}$ (-4, 20)

13. answers may vary   **23.** $x + \sqrt{x}$ **27.** $4\sqrt{x} - x\sqrt{x}$ **29.** $7\sqrt{5}$ **32.** $-5 + 8\sqrt{2}$ **34.** $8 - 6\sqrt{2}$ **36.** $14\sqrt{2}$

43. $4 + 4\sqrt[3]{2}$ **44.** $9 + 9\sqrt[3]{3}$ **45.** $-3 + 3\sqrt[3]{2}$ **49.** $12z\sqrt{2x}$ **52.** $\sqrt{x}$ **53.** $x\sqrt{3x} + 3x\sqrt{x}$

60. $(8\sqrt{13} + 16\sqrt{5})$ ft.   **61.** $x^2 + 12x + 36$ **62.** $9x^2 + 12x + 4$ **63.** $4x^2 - 4x + 1$ **64.** $x^2 - 10x + 25$

9.4 | MULTIPLYING AND DIVIDING RADICALS

O B J E C T I V E S

TAPE BA 9.4

1 Multiply radicals.

2 Divide radicals.

3 Rationalize denominators.

4 Rationalize using conjugates.

 In Section 9.2 we used the product and quotient rules for radicals to help us simplify radicals. In this section, we use these rules to simplify products and quotients of radicals.

> **PRODUCT RULE FOR RADICALS**
> If $\sqrt[n]{a}$ and $\sqrt[n]{b}$ are real numbers, then
> $$\sqrt[n]{a} \cdot \sqrt[n]{b} = \sqrt[n]{a \cdot b}$$

This property says that the product of the nth roots of two numbers is the nth root of the product of the two numbers. For example,

$$\sqrt{3} \cdot \sqrt{2} = \sqrt{3 \cdot 2} = \sqrt{6}$$

Also,

$$\sqrt[3]{5} \cdot \sqrt[3]{7} = \sqrt[3]{5 \cdot 7} = \sqrt[3]{35}$$

EXAMPLE 1 Find the following products.

a. $\sqrt{7} \cdot \sqrt{3}$

b. $\sqrt{3} \cdot \sqrt{15}$

c. $\sqrt[3]{4} \cdot \sqrt[3]{18}$

d. $2\sqrt{6} \cdot 5\sqrt{2}$

e. $(3\sqrt{2})^2$

Solution: a. $\sqrt{7} \cdot \sqrt{3} = \sqrt{7 \cdot 3} = \sqrt{21}$

b. $\sqrt{3} \cdot \sqrt{15} = \sqrt{45}$. Next, simplify $\sqrt{45}$.

$$\sqrt{45} = \sqrt{9 \cdot 5} = \sqrt{9} \cdot \sqrt{5} = 3\sqrt{5}$$

c. $\sqrt[3]{4} \cdot \sqrt[3]{18} = \sqrt[3]{4 \cdot 18} = \sqrt[3]{4 \cdot 2 \cdot 9} = \sqrt[3]{8 \cdot 9} = \sqrt[3]{8} \cdot \sqrt[3]{9} = 2\sqrt[3]{9}$

d. $2\sqrt{6} \cdot 5\sqrt{2} = 2 \cdot 5\sqrt{6 \cdot 2} = 10\sqrt{12}$. Next, simplify $\sqrt{12}$.

$$10\sqrt{12} = 10\sqrt{4 \cdot 3} = 10\sqrt{4} \cdot \sqrt{3} = 10 \cdot 2 \cdot \sqrt{3} = 20\sqrt{3}$$

e. $(3\sqrt{2})^2 = 3^2 \cdot (\sqrt{2})^2 = 9 \cdot 2 = 18$

When multiplying radical expressions containing more than one term, use the same techniques we use to multiply other algebraic expressions with more than one term.

EXAMPLE 2   Find the product and simplify.

a. $\sqrt{5}(\sqrt{5} - \sqrt{2})$ b. $(\sqrt{x} + \sqrt{2})(\sqrt{3} - \sqrt{2})$

Solution: a. Using the distributive property, we have

$$\sqrt{5}(\sqrt{5} - \sqrt{2}) = \sqrt{5} \cdot \sqrt{5} - \sqrt{5} \cdot \sqrt{2}$$
$$= 5 - \sqrt{10}$$

b. Use the FOIL method of multiplication.

$$\overset{\text{F}\qquad\qquad\text{O}\qquad\qquad\text{I}\qquad\qquad\text{L}}{(\sqrt{x} + \sqrt{2})(\sqrt{3} - \sqrt{2}) = \sqrt{x} \cdot \sqrt{3} - \sqrt{x} \cdot \sqrt{2} + \sqrt{2} \cdot \sqrt{3} - \sqrt{2} \cdot \sqrt{2}}$$
$$= \sqrt{3x} - \sqrt{2x} + \sqrt{6} - \sqrt{4} \qquad \text{Apply the product rule.}$$
$$= \sqrt{3x} - \sqrt{2x} + \sqrt{6} - 2 \qquad \text{Simplify.}$$

Special products can be used to multiply expressions containing radicals.

EXAMPLE 3   Find the product and simplify.

a. $(\sqrt{5} - 7)(\sqrt{5} + 7)$ b. $(\sqrt{7x} + 2)^2$

Solution: a. Recall from Chapter 4 that $(a - b)(a + b) = a^2 - b^2$. Then
$$(\sqrt{5} - 7)(\sqrt{5} + 7) = (\sqrt{5})^2 - 7^2$$
$$= 5 - 49$$
$$= -44$$

b. Recall that $(a + b)^2 = a^2 + 2ab + b^2$. Then
$$(\sqrt{7x} + 2)^2 = (\sqrt{7x})^2 + 2(\sqrt{7x})(2) + (2)^2$$
$$= 7x + 4\sqrt{7x} + 4$$

2   To simplify quotients of rational expressions, we use the quotient rule.

QUOTIENT RULE FOR RADICALS

If $\sqrt[n]{a}$ *and* $\sqrt[n]{b}$ are real numbers and $b \neq 0$, then

$$\frac{\sqrt[n]{a}}{\sqrt[n]{b}} = \sqrt[n]{\frac{a}{b}}, \text{ providing } b \neq 0$$

EXAMPLE 4 Find the quotient and simplify.

a. $\dfrac{\sqrt{14}}{\sqrt{2}}$ b. $\dfrac{\sqrt{100}}{\sqrt{5}}$ c. $\dfrac{\sqrt[3]{32}}{\sqrt[3]{4}}$ d. $\dfrac{\sqrt{12x^3}}{\sqrt{3x}}$

Solution: Use the quotient rule and then simplify the resulting radicand.

a. $\dfrac{\sqrt{14}}{\sqrt{2}} = \sqrt{\dfrac{14}{2}} = \sqrt{7}$

b. $\dfrac{\sqrt{100}}{\sqrt{5}} = \sqrt{\dfrac{100}{5}} = \sqrt{20} = \sqrt{4 \cdot 5} = \sqrt{4} \cdot \sqrt{5} = 2\sqrt{5}$

c. $\dfrac{\sqrt[3]{32}}{\sqrt[3]{4}} = \sqrt[3]{\dfrac{32}{4}} = \sqrt[3]{8} = 2$

d. $\dfrac{\sqrt{12x^3}}{\sqrt{3x}} = \sqrt{\dfrac{12x^3}{3x}} = \sqrt{4x^2} = 2x$

⬛ It is sometimes easier to work with radical expressions if the denominator does not contain a radical. To eliminate the radical in the denominator of a radical expression, we use the fact that we can multiply the numerator and the denominator of a fraction by the same nonzero number. This is equivalent to multiplying the fraction by 1. To eliminate the radical in the denominator of $\dfrac{\sqrt{5}}{\sqrt{2}}$, multiply the numerator and the denominator by $\sqrt{2}$. Then

$$\frac{\sqrt{5}}{\sqrt{2}} = \frac{\sqrt{5} \cdot \sqrt{2}}{\sqrt{2} \cdot \sqrt{2}} = \frac{\sqrt{10}}{2}$$

This process is called **rationalizing** the denominator.

EXAMPLE 5 Rationalize each denominator.

a. $\dfrac{2}{\sqrt{7}}$ b. $\dfrac{\sqrt{5}}{\sqrt{12}}$ c. $\sqrt{\dfrac{1}{18x}}$

Solution: **a.** To eliminate the radical in the denominator of $\dfrac{2}{\sqrt{7}}$, multiply the numerator and the denominator by $\sqrt{7}$.

$$\frac{2}{\sqrt{7}} = \frac{2 \cdot \sqrt{7}}{\sqrt{7} \cdot \sqrt{7}} = \frac{2\sqrt{7}}{7}$$

b. We can multiply the numerator and denominator by $\sqrt{12}$, but see what happens if we simplify first.

$$\frac{\sqrt{5}}{\sqrt{12}} = \frac{\sqrt{5}}{\sqrt{4 \cdot 3}} = \frac{\sqrt{5}}{2\sqrt{3}}$$

To rationalize the denominator now, multiply the numerator and the denominator by $\sqrt{3}$.

$$\frac{\sqrt{5}}{2\sqrt{3}} = \frac{\sqrt{5} \cdot \sqrt{3}}{2\sqrt{3} \cdot \sqrt{3}} = \frac{\sqrt{15}}{2 \cdot 3} = \frac{\sqrt{15}}{6}$$

c. $\sqrt{\dfrac{1}{18x}} = \dfrac{\sqrt{1}}{\sqrt{18x}} = \dfrac{1}{\sqrt{9} \cdot \sqrt{2x}} = \dfrac{1}{3\sqrt{2x}}$

To rationalize the denominator, multiply the numerator and denominator by $\sqrt{2x}$.

$$\frac{1}{3\sqrt{2x}} = \frac{1 \cdot \sqrt{2x}}{3\sqrt{2x} \cdot \sqrt{2x}} = \frac{\sqrt{2x}}{3 \cdot 2x} = \frac{\sqrt{2x}}{6x}$$

As a general rule, simplify a radical expression first and then rationalize the denominator.

EXAMPLE 6   Rationalize each denominator.

a. $\dfrac{5}{\sqrt[3]{4}}$ **b.** $\dfrac{\sqrt[3]{7}}{\sqrt[3]{3}}$

Solution: **a.** Since the denominator contains a cube root, we multiply the numerator and the denominator by a factor that gives the **cube root of a perfect cube** in the denominator. Recall that $\sqrt[3]{8} = 2$ and that the denominator $\sqrt[3]{4}$ multiplied by $\sqrt[3]{2}$ is $\sqrt[3]{4 \cdot 2}$ or $\sqrt[3]{8}$.

$$\frac{5}{\sqrt[3]{4}} = \frac{5 \cdot \sqrt[3]{2}}{\sqrt[3]{4} \cdot \sqrt[3]{2}} = \frac{5\sqrt[3]{2}}{\sqrt[3]{8}} = \frac{5\sqrt[3]{2}}{2}$$

b. Recall that $\sqrt[3]{27} = 3$. Multiply the denominator $\sqrt[3]{3}$ by $\sqrt[3]{9}$ and the result is $\sqrt[3]{3 \cdot 9}$ or $\sqrt[3]{27}$.

$$\frac{\sqrt[3]{7}}{\sqrt[3]{3}} = \frac{\sqrt[3]{7} \cdot \sqrt[3]{9}}{\sqrt[3]{3} \cdot \sqrt[3]{9}} = \frac{\sqrt[3]{63}}{\sqrt[3]{27}} = \frac{\sqrt[3]{63}}{3}$$

4   To rationalize a denominator that is a sum, such as the denominator in

$$\frac{2}{4 + \sqrt{3}}$$

we multiply the numerator and the denominator by $4 - \sqrt{3}$. The expressions $4 + \sqrt{3}$ and $4 - \sqrt{3}$ are called **conjugates** of each other. When a radical expression such as $4 + \sqrt{3}$ is multiplied by its conjugate $4 - \sqrt{3}$, the product simplifies to an expression that contains no radicals.

$$(a + b)(a - b) = a^2 - b^2$$
$$(4 + \sqrt{3})(4 - \sqrt{3}) = 4^2 - (\sqrt{3})^2 = 16 - 3 = 13$$

Then

$$\frac{2}{4 + \sqrt{3}} = \frac{2(4 - \sqrt{3})}{(4 + \sqrt{3})(4 - \sqrt{3})} = \frac{2(4 - \sqrt{3})}{13}$$

EXAMPLE 7 Rationalize each denominator and simplify.

a. $\dfrac{2}{1 + \sqrt{3}}$ **b.** $\dfrac{\sqrt{5} + 4}{\sqrt{5} - 1}$

Solution: **a.** Multiply the numerator and the denominator of this fraction by the conjugate of $1 + \sqrt{3}$, that is, by $1 - \sqrt{3}$.

$$\frac{2}{1 + \sqrt{3}} = \frac{2(1 - \sqrt{3})}{(1 + \sqrt{3})(1 - \sqrt{3})}$$

$$= \frac{2(1 - \sqrt{3})}{1^2 - (\sqrt{3})^2}$$

$$= \frac{2(1 - \sqrt{3})}{1 - 3}$$

$$= \frac{2(1 - \sqrt{3})}{-2}$$

$$= -\frac{\boxed{2}(1 - \sqrt{3})}{\boxed{2}} \qquad \frac{a}{-b} = -\frac{a}{b}$$

$$= -1(1 - \sqrt{3}) \qquad \text{Simplify.}$$

$$= -1 + \sqrt{3}$$

b. $\dfrac{\sqrt{5} + 4}{\sqrt{5} - 1} = \dfrac{(\sqrt{5} + 4)(\sqrt{5} + 1)}{(\sqrt{5} - 1)(\sqrt{5} + 1)}$ Multiply the numerator and denominator by $\sqrt{5} + 1$, the conjugate of $\sqrt{5} - 1$.

$$= \frac{5 + \sqrt{5} + 4\sqrt{5} + 4}{5 - 1} \qquad \text{Multiply.}$$

$$= \frac{9 + 5\sqrt{5}}{4} \qquad \text{Simplify.}$$

EXAMPLE 8 Simplify $\dfrac{12 - \sqrt{18}}{9}$.

Solution: First simplify $\sqrt{18}$.

$$\frac{12 - \sqrt{18}}{9} = \frac{12 - \sqrt{9 \cdot 2}}{9} = \frac{12 - 3\sqrt{2}}{9}$$

Next, factor out a common factor of 3 from the terms in the numerator and the denominator and simplify.

$$\frac{12 - 3\sqrt{2}}{9} = \frac{\boxed{3}(4 - \sqrt{2})}{\boxed{3} \cdot 3} = \frac{4 - \sqrt{2}}{3}$$

MENTAL MATH

Find each product. Assume that variables represent nonnegative real numbers.

1. $\sqrt{2} \cdot \sqrt{3}$ $\sqrt{6}$

2. $\sqrt{5} \cdot \sqrt{7}$ $\sqrt{35}$

3. $\sqrt{1} \cdot \sqrt{6}$ $\sqrt{6}$

4. $\sqrt{7} \cdot \sqrt{x}$ $\sqrt{7x}$

5. $\sqrt{10} \cdot \sqrt{y}$ $\sqrt{10y}$

6. $\sqrt{x} \cdot \sqrt{y}$ $\sqrt{xy}$

EXERCISE SET 9.4

Find each product and simplify. See Examples 1 through 3.

1. $\sqrt{8} \cdot \sqrt{2}$ 4

2. $\sqrt{3} \cdot \sqrt{12}$ 6

3. $\sqrt{10} \cdot \sqrt{5}$ $5\sqrt{2}$

4. $3\sqrt{2} \cdot 5\sqrt{14}$ $30\sqrt{7}$

5. $\sqrt[3]{12} \cdot \sqrt[3]{4}$ $2\sqrt[3]{6}$

6. $\sqrt[3]{9} \cdot \sqrt[3]{6}$ $3\sqrt[3]{2}$

7. $\sqrt{10}(\sqrt{2} + \sqrt{5})$

8. $\sqrt{6}(\sqrt{3} + \sqrt{2})$

9. $(3\sqrt{5} - \sqrt{10})(\sqrt{5} - 4\sqrt{3})$

10. $(2\sqrt{3} - 6)(\sqrt{3} - 4\sqrt{2})$ $6 - 8\sqrt{6} - 6\sqrt{3} + 24\sqrt{2}$

11. $(\sqrt{x} + 6)(\sqrt{x} - 6)$

12. $(2\sqrt{5} + 1)(2\sqrt{5} - 1)$

13. $(\sqrt{3} + 8)^2$ $67 + 16\sqrt{3}$

14. $(\sqrt{x} - 7)^2$

15. Find the area of a rectangular room whose length is $13\sqrt{2}$ meters and width is $5\sqrt{6}$ meters.

16. Find the volume of a microwave oven whose length is $\sqrt{3}$ feet, width is $\sqrt{2}$ feet, and height is $\sqrt{2}$ feet.

$\sqrt{2}$ feet
$\sqrt{3}$ feet
$\sqrt{2}$ feet $2\sqrt{3}$ cubic ft.

Find each quotient and simplify. See Example 4.

17. $\dfrac{\sqrt{32}}{\sqrt{2}}$ 4

18. $\dfrac{\sqrt{40}}{\sqrt{10}}$ 2

19. $\dfrac{\sqrt{90}}{\sqrt{5}}$ $3\sqrt{2}$

20. $\dfrac{\sqrt{96}}{\sqrt{8}}$ $2\sqrt{3}$

21. $\dfrac{\sqrt{75y^5}}{\sqrt{3y}}$ $5y^2$

22. $\dfrac{\sqrt{24x^7}}{\sqrt{6x}}$ $2x^3$

Rationalize each denominator and simplify. See Example 5.

23. $\sqrt{\dfrac{3}{5}}$ $\dfrac{\sqrt{15}}{5}$

24. $\sqrt{\dfrac{2}{3}}$ $\dfrac{\sqrt{6}}{3}$

25. $\dfrac{1}{\sqrt{6y}}$ $\dfrac{\sqrt{6y}}{6y}$

26. $\dfrac{1}{\sqrt{10z}}$ $\dfrac{\sqrt{10z}}{10z}$

27. $\sqrt{\dfrac{5}{18}}$ $\dfrac{\sqrt{10}}{6}$

28. $\sqrt{\dfrac{7}{12}}$ $\dfrac{\sqrt{21}}{6}$

Rationalize each denominator. See Example 6.

29. $\dfrac{6}{\sqrt[3]{2}}$ $3\sqrt[3]{4}$

30. $\dfrac{3}{\sqrt[3]{5}}$ $\dfrac{3\sqrt[3]{25}}{5}$

31. $\sqrt[3]{\dfrac{1}{9}}$ $\dfrac{\sqrt[3]{3}}{3}$

32. $\sqrt[3]{\dfrac{8}{11}}$ $\dfrac{2\sqrt[3]{121}}{11}$

33. $\sqrt[3]{\dfrac{2}{9x^2}}$ $\dfrac{\sqrt[3]{6x}}{3x}$

34. $\sqrt[3]{\dfrac{3}{4y^2}}$ $\dfrac{\sqrt[3]{6y}}{2y}$

35. If a circle has area A, then the formula for the

radius r of the circle is

$$r = \sqrt{\dfrac{A}{\pi}} \quad \dfrac{\sqrt{A\pi}}{\pi}$$

Simplify this expression by rationalizing the denominator.

36. If a round ball has volume V, then the formula for the radius r of the ball is

$$r = \sqrt[3]{\dfrac{3V}{4\pi}} \quad \dfrac{\sqrt[3]{6V\pi^2}}{2\pi}$$

Simplify this expression by rationalizing the denominator.

7. $2\sqrt{5} + 5\sqrt{2}$ **8.** $3\sqrt{2} + 2\sqrt{3}$ **9.** $15 - 12\sqrt{15} - 5\sqrt{2} + 4\sqrt{30}$ **11.** $x - 36$ **12.** 19 **14.** $x + 49 - 14\sqrt{x}$

15. $130\sqrt{3}$ sq. meters

43. $\sqrt{30} + 5 + \sqrt{6} + \sqrt{5}$ **44.** $3 + \sqrt{6} + \sqrt{3} + \sqrt{2}$ **57.** $\sqrt{30} + \sqrt{42}$ **58.** $\sqrt{30} - \sqrt{70}$ **59.** $4x\sqrt{5} - 60\sqrt{x}$

37. When rationalizing the denominator of $\dfrac{\sqrt{2}}{\sqrt{3}}$, explain why both the numerator and the denominator must be multiplied by $\sqrt{3}$.   answers may vary

38. When rationalizing the denominator of $\dfrac{\sqrt[3]{2}}{\sqrt[3]{3}}$, explain why both the numerator and the denominator must be multiplied by $\sqrt[3]{9}$.   answers may vary

74. $\dfrac{\sqrt{54x^3}}{\sqrt{2x}}$ $3x\sqrt{3}$ **75.** $\dfrac{\sqrt{24x^3y^4}}{\sqrt{2xy}}$ **76.** $\dfrac{\sqrt{96x^5y^3}}{\sqrt{3x^2y}}$

77. $2\sqrt[3]{5} \cdot 6\sqrt[3]{2}$ $12\sqrt[3]{10}$ **78.** $8\sqrt[3]{4} \cdot 7\sqrt[3]{7}$ $56\sqrt[3]{28}$

79. $\sqrt[3]{15} \cdot \sqrt[3]{25}$ $5\sqrt[3]{3}$ **80.** $\sqrt[3]{4} \cdot \sqrt[3]{4}$ $2\sqrt[3]{2}$

81. $\dfrac{\sqrt[3]{54x^2}}{\sqrt[3]{2}}$ $3\sqrt[3]{x^2}$ **82.** $\dfrac{\sqrt[3]{80y^2}}{\sqrt[3]{10}}$ $2\sqrt[3]{y^2}$

Rationalize each denominator and simplify. See Example 7.

39. $\dfrac{3}{\sqrt{2} + 1}$ $3\sqrt{2} - 3$ **40.** $\dfrac{6}{\sqrt{5} + 2}$ $6\sqrt{5} - 12$

41. $\dfrac{2}{\sqrt{10} - 3}$ $2\sqrt{10} + 6$ **42.** $\dfrac{4}{2 - \sqrt{3}}$ $8 + 4\sqrt{3}$

43. $\dfrac{\sqrt{5} + 1}{\sqrt{6} - \sqrt{5}}$ **44.** $\dfrac{\sqrt{3} + 1}{\sqrt{3} - \sqrt{2}}$

Simplify the following. See Example 8.

45. $\dfrac{6 + 2\sqrt{3}}{2}$ $3 + \sqrt{3}$ **46.** $\dfrac{9 + 6\sqrt{2}}{3}$ $3 + 2\sqrt{2}$

47. $\dfrac{18 - 12\sqrt{5}}{6}$ $3 - 2\sqrt{5}$ **48.** $\dfrac{8 - 20\sqrt{3}}{4}$ $2 - 5\sqrt{3}$

49. $\dfrac{15\sqrt{3} + 5}{5}$ $3\sqrt{3} + 1$ **50.** $\dfrac{8 + 16\sqrt{2}}{8}$ $1 + 2\sqrt{2}$

Multiply or divide as indicated and simplify.

51. $2\sqrt{3} \cdot 4\sqrt{15}$ $24\sqrt{5}$ **52.** $3\sqrt{14} \cdot 4\sqrt{2}$ $24\sqrt{7}$

53. $(2\sqrt{5})^2$ 20 **54.** $(3\sqrt{10})^2$ 90

55. $(6\sqrt{x})^2$ $36x$ **56.** $(8\sqrt{y})^2$ $64y$

57. $\sqrt{6}(\sqrt{5} + \sqrt{7})$ **58.** $\sqrt{10}(\sqrt{3} - \sqrt{7})$

59. $4\sqrt{5x}(\sqrt{x} - 3\sqrt{5})$ **60.** $3\sqrt{7y}(\sqrt{y} - 2\sqrt{7})$

61. $(\sqrt{3} + \sqrt{5})(\sqrt{2} - \sqrt{5})$ $\sqrt{6} - \sqrt{15} + \sqrt{10} - 5$

62. $(\sqrt{6} + \sqrt{3})(\sqrt{6} - \sqrt{3})$ 3

63. $(\sqrt{7} - 2\sqrt{3})(\sqrt{7} + 2\sqrt{3})$ -5

64. $(\sqrt{2} - 4\sqrt{5})(\sqrt{2} + 4\sqrt{5})$ -78

65. $(\sqrt{x} - 3)(\sqrt{x} + 3)$ **66.** $(2\sqrt{y} + 5)(2\sqrt{y} - 5)$

67. $(\sqrt{6} + 3)^2$ $15 + 6\sqrt{6}$ **68.** $(2 + \sqrt{7})^2$ $11 + 4\sqrt{7}$

69. $(3\sqrt{x} - 5)^2$ **70.** $(2\sqrt{x} - 7)^2$

71. $\dfrac{\sqrt{150}}{\sqrt{2}}$ $5\sqrt{3}$ **72.** $\dfrac{\sqrt{120}}{\sqrt{3}}$ $2\sqrt{10}$ **73.** $\dfrac{\sqrt{72y^5}}{\sqrt{3y^3}}$ $2y\sqrt{6}$

Rationalize each denominator and simplify.

83. $\sqrt{\dfrac{2}{15}}$ $\dfrac{\sqrt{30}}{15}$ **84.** $\sqrt{\dfrac{11}{14}}$ $\dfrac{\sqrt{154}}{14}$ **85.** $\sqrt{\dfrac{3}{20}}$ $\dfrac{\sqrt{15}}{10}$

86. $\sqrt{\dfrac{3}{50}}$ $\dfrac{\sqrt{6}}{10}$ **87.** $\dfrac{3x}{\sqrt{2x}}$ $\dfrac{3\sqrt{2x}}{2}$ **88.** $\dfrac{5y}{\sqrt{3y}}$ $\dfrac{5\sqrt{3y}}{3}$

89. $\sqrt[3]{\dfrac{5}{4}}$ $\dfrac{\sqrt[3]{10}}{2}$ **90.** $\sqrt[3]{\dfrac{7}{9}}$ $\dfrac{\sqrt[3]{21}}{3}$ **91.** $\dfrac{4}{2 - \sqrt{5}}$

92. $\dfrac{2}{1 - \sqrt{2}}$ **93.** $\dfrac{5}{3 + \sqrt{10}}$ **94.** $\dfrac{5}{\sqrt{6} + 2}$

95. $\dfrac{2\sqrt{3}}{\sqrt{15} + 2}$ **96.** $\dfrac{3\sqrt{2}}{\sqrt{10} + 2}$ **97.** $\dfrac{\sqrt{3} + 1}{\sqrt{2} - 1}$

98. $\dfrac{\sqrt{2} - 2}{2 - \sqrt{3}}$ $2\sqrt{2} + \sqrt{6} - 2\sqrt{3} - 4$

It is often more convenient to work with a radical expression whose numerator is rationalized. Rationalize the numerator of each expression by multiplying numerator and denominator by the conjugate of the numerator.

99. $\dfrac{\sqrt{3} + 1}{\sqrt{2} - 1}$ **100.** $\dfrac{\sqrt{2} - 2}{2 - \sqrt{3}}$

Review Exercises

Simplify the following expressions. See Section 6.1.

101. $\dfrac{3x + 12}{3}$ $x + 4$ **102.** $\dfrac{12x + 8}{4}$ $3x + 2$

103. $\dfrac{6x^2 - 3x}{3x}$ $2x - 1$ **104.** $\dfrac{8y^2 - 2y}{2y}$ $4y - 1$

Solve each equation. See Sections 2.4 and 4.3.

105. $x + 5 = 7^2$ $x = 44$ **106.** $2y - 1 = 3^2$ $y = 5$

107. $4z^2 + 6z - 12 = (2z)^2$ $z = 2$

108. $9x^2 + 5x + 4 = (3x + 1)^2$ $x = 3$

60. $3y\sqrt{7} - 42\sqrt{y}$ **65.** $x - 9$ **66.** $4y - 25$ **69.** $9x - 30\sqrt{x} + 25$ **70.** $4x - 28\sqrt{x} + 49$ **75.** $2xy\sqrt{3y}$

76. $4xy\sqrt{2x}$ **91.** $-8 - 4\sqrt{5}$ **92.** $-2 - 2\sqrt{2}$ **93.** $5\sqrt{10} - 15$ **94.** $\dfrac{5\sqrt{6}}{2} - 5$ **95.** $\dfrac{6\sqrt{5} - 4\sqrt{3}}{11}$

96. $\sqrt{5} - \sqrt{2}$ **97.** $\sqrt{6} + \sqrt{3} + \sqrt{2} + 1$ **99.** $\dfrac{2}{\sqrt{6} - \sqrt{2} - \sqrt{3} + 1}$ **100.** $\dfrac{2}{-2\sqrt{2} - 4 + \sqrt{6} + 2\sqrt{3}}$

9.5 | SOLVING EQUATIONS CONTAINING RADICALS

TAPE BA 9.5

O B J E C T I V E

 Solve equations containing square roots of variables.

In this section, we solve **radical equations** such as

$$\sqrt{x + 3} = 5$$

To do so, we rely on the following squaring property.

SQUARING PROPERTY OF EQUALITY

If $a = b$, then $a^2 = b^2$

Unfortunately, this squaring property does not guarantee that all solutions of the new equation are solutions of the original equation. For example, if we square both sides of the equation

$$x = 2$$

we have

$$x^2 = 4$$

This new equation has two solutions, 2 and -2, while the original equation $x = 2$ has only one solution. For this reason:

Always check proposed solutions of radical equations in the original equation.

EXAMPLE 1   Solve for x: $\sqrt{x + 3} = 5$.

Solution: To solve for x, use the squaring property of equality and square both sides of the equation.

$$\sqrt{x + 3} = 5$$
$$(\sqrt{x + 3})^2 = 5^2 \qquad \text{Square both sides.}$$
$$x + 3 = 25 \qquad \text{Simplify.}$$
$$x = 22 \qquad \text{Subtract 3 from both sides.}$$

To check this proposed solution, replace x with 22 in the original equation.

$$\sqrt{x + 3} = 5 \qquad \text{Original equation.}$$
$$\sqrt{22 + 3} = 5 \qquad \text{Let } x = 22.$$
$$\sqrt{25} = 5$$
$$5 = 5 \qquad \text{True.}$$

Since a true statement results, 22 is the solution, and the solution set is {22}.

EXAMPLE 2   Solve the equation $\sqrt{y + 2} + 9 = 16$.

Solution: First, write the equation so that the radical is by itself on one side of the equation. Then square both sides.

$$\sqrt{y + 2} + 9 = 16$$
$$\sqrt{y + 2} = 7 \qquad \text{Subtract 9 from both sides.}$$
$$(\sqrt{y + 2})^2 = 7^2 \qquad \text{Square both sides.}$$
$$y + 2 = 49 \qquad \text{Simplify.}$$
$$y = 47 \qquad \text{Subtract 2 from both sides.}$$

Check this solution by replacing y with 47 in the original equation.

$$\sqrt{y + 2} + 9 = 16$$
$$\sqrt{47 + 2} + 9 = 16 \qquad \text{Replace } y \text{ with 47.}$$
$$\sqrt{49} + 9 = 16$$
$$7 + 9 = 16$$
$$16 = 16 \qquad \text{True.}$$

The solution set is {47}.

EXAMPLE 3   Solve for x: $\sqrt{x} + 6 = 4$.

Solution: First isolate the radical. Then square both sides.

$$\sqrt{x} + 6 = 4$$
$$\sqrt{x} = -2 \qquad \text{Subtract 6 from both sides to isolate the radical.}$$

Recall that $\sqrt{x}$ is the principal or nonnegative square root of x so that $\sqrt{x}$ cannot equal -2 and thus this equation has no solution. We arrive at the same conclusion if we continue by applying the squaring property.

$$\sqrt{x} = -2$$
$$(\sqrt{x})^2 = (-2)^2 \qquad \text{Square both sides.}$$
$$x = 4 \qquad \text{Simplify.}$$

To check, replace x with 4 in the original equation.

$$\sqrt{x} + 6 = 4 \qquad \text{Original equation.}$$
$$\sqrt{4} + 6 = 4 \qquad \text{Let } x = 4.$$
$$2 + 6 = 4 \qquad \textbf{False.}$$

Since 4 **does not** satisfy the original equation this equation has no solution. In set notation, we can write { }.

Example 3 makes it very clear that we **must** check proposed solutions in the original equation to determine if they are truly solutions. If a proposed solution does not work, we say that the value is an **extraneous solution.**

The following set of guidelines will help you solve radical equations containing square roots.

TO SOLVE A RADICAL EQUATION CONTAINING SQUARE ROOTS

Step 1. Arrange terms so that one radical is by itself on one side of the equation. That is, isolate a radical.

Step 2. Square both sides of the equation.

Step 3. Simplify both sides of the equation.

Step 4. If the equation still contains a radical term, repeat *steps 1 through 3.*

Step 5. Solve the equation.

Step 6. Check all solutions in the original equation for extraneous solutions.

EXAMPLE 4 Solve the equation $\sqrt{x} = \sqrt{5x - 2}$.

Solution: Each of the radicals is already isolated, since each is by itself on one side of the equation. Begin solving by squaring both sides.

$$\sqrt{x} = \sqrt{5x - 2} \qquad \text{Original equation.}$$
$$(\sqrt{x})^2 = (\sqrt{5x - 2})^2 \qquad \text{Square both sides.}$$
$$x = 5x - 2 \qquad \text{Simplify.}$$
$$-4x = -2 \qquad \text{Subtract } 5x \text{ from both sides.}$$
$$x = \frac{-2}{-4} = \frac{1}{2} \qquad \text{Divide both sides by } -4 \text{ and simplify.}$$

The proposed solution is $\frac{1}{2}$. To check, replace x with $\frac{1}{2}$ in the original equation.

$$\sqrt{x} = \sqrt{5x - 2} \qquad \text{Original equation.}$$
$$\sqrt{\frac{1}{2}} = \sqrt{5 \cdot \frac{1}{2} - 2} \qquad \text{Let } x = \frac{1}{2}.$$
$$\sqrt{\frac{1}{2}} = \sqrt{\frac{5}{2} - 2} \qquad \text{Multiply.}$$
$$\sqrt{\frac{1}{2}} = \sqrt{\frac{5}{2} - \frac{4}{2}} \qquad \text{Write 2 as } \frac{4}{2}.$$
$$\sqrt{\frac{1}{2}} = \sqrt{\frac{1}{2}} \qquad \text{True.}$$

This statement is true, so the solution is $\frac{1}{2}$ and the solution set is $\left\{\frac{1}{2}\right\}$.

It is helpful to recall the following special product before proceeding to the next example.

REMINDER

$$(a + b)^2 = a^2 + 2ab + b^2$$

For example,

$$(x - 3)^2 = x^2 - 6x + 9$$

EXAMPLE 5   Solve the equation $\sqrt{x + 3} - x = -3$.

Solution: First, isolate the radical by adding x to both sides. Then square both sides.

$$\sqrt{x + 3} - x = -3$$

$\sqrt{x + 3} = x - 3$	Add x to both sides.
$(\sqrt{x + 3})^2 = (x - 3)^2$	Square both sides.
$x + 3 = x^2 - 6x + 9$	Write $(x - 3)^2$ as $x^2 - 6x + 9$.

To solve this resulting quadratic equation, write the equation in standard form by subtracting x and 3 from both sides.

$3 = x^2 - 7x + 9$	Subtract x from both sides.
$0 = x^2 - 7x + 6$	Subtract 3 from both sides.
$0 = (x - 6)(x - 1)$	Factor.
$0 = x - 6$ or $0 = x - 1$	Set each factor equal to zero.
$6 = x$ or $1 = x$	Solve for x.

Both of these proposed solutions are checked below.

Let $x = 6$	Let $x = 1$
$\sqrt{x + 3} - x = -3$	$\sqrt{x + 3} - x = -3$
$\sqrt{6 + 3} - 6 = -3$	$\sqrt{1 + 3} - 1 = -3$
$\sqrt{9} - 6 = -3$	$\sqrt{4} - 1 = -3$
$3 - 6 = -3$	$2 - 1 = -3$
$-3 = -3$ True.	$1 = -3$ False.

Since replacing x with 1 resulted in a false statement, 1 is not a solution. The only solution is 6, and the solution set is {6}.

EXAMPLE 6   Solve for y: $\sqrt{4y^2 + 5y - 15} = 2y$.

Solution: The radical is already isolated, so start by squaring both sides.

$$\sqrt{4y^2 + 5y - 15} = 2y$$
$$(\sqrt{4y^2 + 5y - 15})^2 = (2y)^2$$

$4y^2 + 5y - 15 = 4y^2$	Simplify.
$5y - 15 = 0$	Subtract $4y^2$ from both sides.
$5y = 15$	Add 15 to both sides.
$y = 3$	Divide both sides by 5.

Replace y with 3 in the original equation to check this proposed solution.

$\sqrt{4y^2 + 5y - 15} = 2y$	Original equation.
$\sqrt{4 \cdot 3^2 + 5 \cdot 3 - 15} = 2 \cdot 3$	Let $y = 3$.
$\sqrt{4 \cdot 9 + 15 - 15} = 6$	Simplify.
$\sqrt{36} = 6$	
$6 = 6$	True.

This statement is true, so the solution set is {3}.

---

7. $x = 2$ **8.** $x = 3$ **9.** $x = -3$ **10.** $x = 4$ **11.** $x = 2$ **12.** $x = 3$
13. no solution   **14.** no solution   **19.** no solution   **21.** $x = -2$ **22.** $x = 4$
23. $x = \dfrac{3}{2}$ **24.** no solution   **25.** $x = 4$ **26.** $x = 1$ **27.** $x = 3$ **28.** $x = 2$

30. no solution   **31.** $x = 12$ **32.** $x = 10$ **33.** $x = 3, x = 1$ **34.** no solution
35. $x = -1$ **36.** $x = -2, x = -1$

EXERCISE SET 9.5

Solve each equation. See Examples 1 through 3.

1. $\sqrt{x} = 9$ $x = 81$
2. $\sqrt{x} = 4$ $x = 16$
3. $\sqrt{x + 5} = 2$ $x = -1$
4. $\sqrt{x + 12} = 3$ $x = -3$
5. $\sqrt{x} + 3 = 7$ $x = 16$
6. $\sqrt{x} + 5 = 10$ $x = 25$

Solve each equation. Be sure to check for extraneous solutions. See Examples 4 through 6.

7. $\sqrt{4x - 3} = \sqrt{x + 3}$
8. $\sqrt{5x - 4} = \sqrt{x + 8}$
9. $\sqrt{x + 7} = x + 5$
10. $\sqrt{x + 5} = x - 1$
11. $\sqrt{9x^2 + 2x - 4} = 3x$
12. $\sqrt{4x^2 + 3x - 9} = 2x$

Solve each equation. Be sure to check for extraneous solutions.

13. $\sqrt{x + 6} + 5 = 3$
14. $\sqrt{2x - 1} + 7 = 1$
15. $\sqrt{2x + 6} = 4$ $x = 5$
16. $\sqrt{3x + 7} = 5$ $x = 6$
17. $\sqrt{x} - 2 = 5$ $x = 49$
18. $4\sqrt{x} - 7 = 5$ $x = 9$
19. $3\sqrt{x} + 5 = 2$
20. $3\sqrt{x} + 5 = 8$ $x = 1$
21. $\sqrt{x + 6} + 1 = 3$
22. $\sqrt{x + 5} + 2 = 5$
23. $\sqrt{2x + 1} + 3 = 5$
24. $\sqrt{3x - 1} + 4 = 1$

25. $\sqrt{x} = \sqrt{3x - 8}$
26. $\sqrt{x} = \sqrt{4x - 3}$
27. $\sqrt{4x} - \sqrt{2x + 6} = 0$
28. $\sqrt{5x + 6} - \sqrt{8x} = 0$
29. $\sqrt{x} = x - 6$ $x = 9$
30. $\sqrt{x} = x + 6$
31. $\sqrt{2x + 1} = x - 7$
32. $\sqrt{2x + 5} = x - 5$
33. $x = \sqrt{2x - 2} + 1$
34. $\sqrt{1 - 8x} + 2 = x$
35. $\sqrt{1 - 8x} - x = 4$
36. $\sqrt{3x + 7} - x = 3$
37. $\sqrt{2x + 5} - 1 = x$
38. $x = \sqrt{4x - 7} + 1$
39. $\sqrt{16x^2 - 3x + 6} = 4x$
40. $\sqrt{9x^2 - 2x + 8} = 3x$
41. $\sqrt{16x^2 + 2x + 2} = 4x$
42. $\sqrt{4x^2 + 3x - 2} = 2x$
43. $\sqrt{2x^2 + 6x + 9} = 3$
44. $\sqrt{3x^2 + 6x + 4} = 2$

45. A number is 6 more than its principal square root. Find the number.   9

46. A number is 4 more than the principal square root of twice the number. Find the number.   8

47. The formula $b = \sqrt{\dfrac{V}{2}}$ can be used to determine the length b of a side of the base of a square-based pyramid with height 6 units and volume V cubic units.

37. $x = 2$ **38.** $x = 4, x = 2$ **39.** $x = 2$ **40.** $x = 4$ **41.** no solution   **42.** $x = \dfrac{2}{3}$ **43.** $x = 0, x = -3$
44. $x = 0, x = -2$

a. Find the length of the side of the base that produces a pyramid with each volume. (Round to the nearest tenth of a unit.)

V	20	200	2000
b	3.2	10	31.6

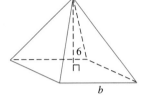

b. Notice in the table above that volume V has been increased by a factor of 10 each time. Does the corresponding radius increase by a factor of 10 each time also? no

48. The formula $r = \sqrt{\dfrac{V}{2\pi}}$ can be used to determine the radius r of a cylinder with height 2 units and volume V cubic units.

a. Find the radius needed to manufacture a cylinder with each volume. (Round to the nearest tenth of a unit.)

V	10	100	1000
r	1.3	4.0	12.6

2 units

b. Notice in the table above that volume V has been increased by a factor of 10 each time. Does the corresponding radius increase by a factor of 10 each time also? no

49. Explain why proposed solutions of radical equations must be checked in the original equation.
answers may vary

50 7.30 **52.** 0.76 **53** 0.48

Graphers can be used to solve equations. To solve $\sqrt{x-2} = x-5$, for example, graph $y_1 = \sqrt{x-2}$ and $y_2 = x - 5$ on the same set of axes. Use Trace and Zoom features or an Intersect feature to find the point of intersection of the graphs. The x-value of the point is the solution of the equation. Use a grapher to solve the equations below. Approximate solutions to the nearest hundredth.

50. $\sqrt{x-2} = x - 5$ **51.** $\sqrt{x+1} = 2x - 3$ 2.43

52. $-\sqrt{x+4} = 5x - 6$ **53.** $-\sqrt{x+5} = -7x + 1$

Review Exercises

Translate each sentence into an equation, and then solve. See Section 2.5.

54. If 8 is subtracted from the product of 3 and x, the result is 19. Find x. $3x - 8 = 19$; $x = 9$

55. If 3 more than x is subtracted from twice x, the result is 11. Find x. $2x - (x + 3) = 11$; $x = 14$

56. The length of a rectangle is twice the width. The perimeter is 24 inches. Find the length.

57. The length of a rectangle is 2 inches longer than the width. The perimeter is 24 inches. Find the length. $2x + 2(x + 2) = 24$. length: 7 in.

Translate each sentence into two equations with two unknowns and solve the resulting systems of equations. See Section 8.4.

58. The sum of two numbers is 20 and their difference is 8. Find the numbers.

59. The difference of two numbers is 58. If one number is three times the other number, find the numbers.

56. $2x + 2(2x) = 24$; length: 8 in.
58. $x + y = 20$ **59.** $x - y = 58$
$x - y = 8$; $y = 3x$;
6 and 14   29 and 87

9.6 RADICAL EQUATIONS AND PROBLEM SOLVING

O B J E C T I V E S

1 Use the Pythagorean formula to solve problems.

2 Use the distance formula to solve problems.

3 Solve problems using formulas containing radicals.

1 Applications of radicals can be found in geometry, finance, science, and other areas of technology. Our first application involves the Pythagorean theorem, giving a formula that relates the lengths of the three sides of a right triangle. We first studied the Pythagorean theorem in Chapter 5 and we review it here.

THE PYTHAGOREAN THEOREM

If a and b are the lengths of the legs of a right triangle and c is the length of the hypotenuse, then $a^2 + b^2 = c^2$.

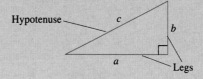

EXAMPLE 1 Find the length of the hypotenuse c of a right triangle whose legs are $a = 6$ inches and $b = 8$ inches long.

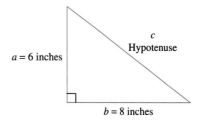

Solution: Because this is a right triangle, we use the Pythagorean theorem.

$$a^2 + b^2 = c^2 \qquad \text{Pythagorean formula.}$$
$$6^2 + 8^2 = c^2 \qquad \text{Substitute the lengths of the legs.}$$
$$36 + 64 = c^2 \qquad \text{Simplify.}$$
$$100 = c^2$$

Since c represents a length, we know that c is positive and is the principal square root of 100.

$$100 = c^2$$
$$\sqrt{100} = c \qquad \text{Apply the definition of principal square root.}$$
$$10 = c \qquad \text{Simplify.}$$

The hypothenuse has a length of 10 inches.

EXAMPLE 2   A surveyor must determine the distance across a lake. He first inserts poles at points P and Q on opposite sides of the lake, as pictured. Perpendicular to line PQ, he finds another point R. If the length of $\overline{PR}$ is 320 feet and the length of $\overline{QR}$ is 240 feet, what is the distance across the lake?

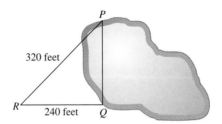

Solution: **1.** UNDERSTAND. Read and reread the problem. Guess a solution and check using the Pythagorean theorem.

2. ASSIGN. By creating a line perpendicular to line PQ, the surveyor deliberately constructed a right triangle. The hypotenuse, $\overline{PR}$, has a length of 320 feet, so we let $c = 320$ in the Pythagorean theorem. The side $\overline{QR}$ is one of the legs, so let $a = 240$, and $b = $ the unknown length.

3. ILLUSTRATE.

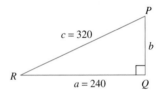

4. TRANSLATE.

$$a^2 + b^2 = c^2 \qquad \text{Use the Pythagorean theorem.}$$
$$240^2 + b^2 = 320^2 \qquad \text{Let } c = 320 \text{ and } a = 240.$$

5. COMPLETE.

$$57{,}600 + b^2 = 102{,}400$$
$$b^2 = 44{,}800 \qquad \text{Subtract 57,600 from both sides.}$$
$$b = \sqrt{44{,}800} \qquad \text{Apply the definition of principal square root.}$$

6. INTERPRET.

Check: See that $240^2 + (\sqrt{44{,}800})^2 = 320^2$.

State: The distance across the lake is **exactly** $\sqrt{44{,}800}$ feet. The surveyor can now use a calculator to find that $\sqrt{44{,}800}$ feet is **approximately** 211.6601 feet, so the distance across the lake is roughly 212 feet.

2   A second important application of radicals is in finding the distance between two points in the plane. By using the Pythagorean theorem, the following formula can be derived.

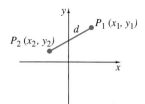

DISTANCE FORMULA

The distance d between two points with coordinates (x_1, y_1) and (x_2, y_2) is given by

$$d = \sqrt{(x_2 - x_1)^2 + (y_2 - y_1)^2}$$

EXAMPLE 3   Find the distance between $(-1, 9)$ and $(-3, -5)$.

Solution: Use the distance formula with $(x_1, y_1) = (-1, 9)$ and $(x_2, y_2) = (-3, -5)$.

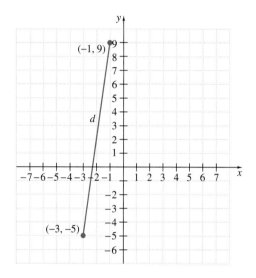

$$d = \sqrt{(x_2 - x_1)^2 + (y_2 - y_1)^2} \qquad \text{The distance formula.}$$
$$= \sqrt{[-3 - (-1)]^2 + (-5 - 9)^2} \qquad \text{Substitute known values.}$$
$$= \sqrt{(-2)^2 + (-14)^2} \qquad \text{Simplify.}$$
$$= \sqrt{4 + 196}$$
$$= \sqrt{200} = 10\sqrt{2} \qquad \text{Simplify the radical.}$$

The distance is **exactly** $10\sqrt{2}$ units or **approximately** 14.1 units.

3   The Pythagorean theorem and the distance formula are both extremely important results in mathematics and should be memorized. But there are other applications involving formulas containing radicals that are not quite as well known. For example:

EXAMPLE 4   A formula used to determine the velocity v, in feet per second, of an object (neglecting air resistance) after it has fallen a certain height is $v = \sqrt{2gh}$, where g is the acceleration due to gravity, and h is the height the object has fallen. On Earth, the acceleration g due to gravity is approximately 32 feet per second per second. Find the velocity of a watermelon after it has fallen 5 feet.

Solution: We are told that g = 32 feet per second per second. To find the velocity v when h = 5 feet, use the velocity formula.

$$v = \sqrt{2gh} \qquad \text{The velocity formula.}$$
$$= \sqrt{2 \cdot 32 \cdot 5} \qquad \text{Substitute known values.}$$
$$= \sqrt{320}$$
$$= 8\sqrt{5} \qquad \text{Simplify the radicand.}$$

The velocity of the watermelon after it falls 5 feet is **exactly** $8\sqrt{5}$ feet per second, or **approximately** 17.9 feet per second.

EXERCISE SET 9.6

Use the Pythagorean theorem to find the unknown side of each right triangle. See Example 1.

1.
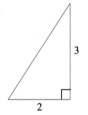
$\sqrt{13}$

2.
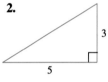
$\sqrt{34}$

3.
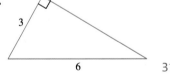
$3\sqrt{3}$

4.
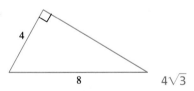
$4\sqrt{3}$

5.
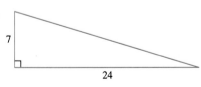
25

6.
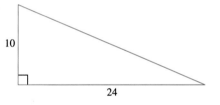

7.
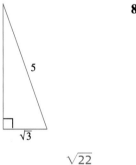
$\sqrt{22}$

8.
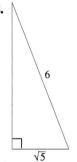
$\sqrt{5}$ $\sqrt{31}$

9.
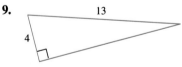
$3\sqrt{17}$

10.
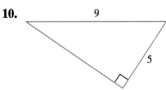
$2\sqrt{14}$

Find the length of the unknown side of each right triangle with sides a, b, and c, where c is the hypotenuse. See Example 1.

11. $a = 4, b = 5$ $\sqrt{41}$ **12.** $a = 2, b = 7$ $\sqrt{53}$

13. $b = 2, c = 6$ $4\sqrt{2}$ **14.** $b = 1, c = 5$ $2\sqrt{6}$

15. $a = \sqrt{10}, c = 10$ $3\sqrt{10}$ **16.** $a = \sqrt{7}, c = \sqrt{35}$ $2\sqrt{7}$

Find the length of x.

17.

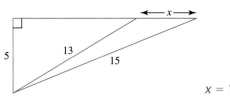

$x = -4 + 2\sqrt{10}$

18.

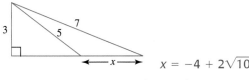

$x = 10\sqrt{2} - 12$

Use the distance formula to find the distance between the points given. See Example 3.

19. $(3, 6,) (5, 11)$ $\sqrt{29}$ **20.** $(2, 3), (9, 7)$ $\sqrt{65}$

21. $(-3, 1), (5, -2)$ $\sqrt{73}$ **22.** $(-2, 6), (3, -2)$ $\sqrt{89}$

23. $(3, -2), (1, -8)$ $2\sqrt{10}$ **24.** $(-5, 8), (-2, 2)$ $3\sqrt{5}$

25. $\left(\frac{1}{2}, 2\right), (2, -1)$ $\frac{3\sqrt{5}}{2}$ **26.** $\left(\frac{1}{3}, 1\right), (1, -1)$ $\frac{2\sqrt{10}}{3}$

27. $(3, -2), (5, 7)$ $\sqrt{85}$ **28.** $(-2, -3), (-1, 4)$ $5\sqrt{2}$

Solve the following. See Examples 2 and 4.

29. A wire is used to anchor a 20-foot-high pole. One end of the wire is attached to the top of the pole. The other end is fastened to a stake five feet away from the bottom of the pole. Find the length of the wire, to the nearest tenth of a foot.   20.6 ft.

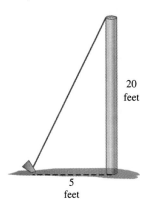

30. David and Stephanie leave the seashore at the same time. David drives northward at a rate of 30 miles per hour, while Stephanie drives west at 60 mph. Find how far apart they are after 3 hours to the nearest mile.   201 miles

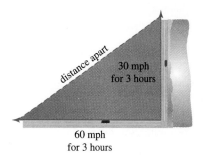

31. For a square-based pyramid, the formula $b = \sqrt{\frac{3V}{H}}$ describes the relationship between the length b of one side of the base, the volume V, and the height H. Find the volume if each side of the base is 6 feet long, and the pyramid is 2 feet high.   24 cubic ft.

32. The formula $t = \frac{\sqrt{d}}{4}$ relates the distance d, in feet, that an object falls in t seconds, assuming that air resistance does not slow down the object. Find how long, to the nearest hundredth of a second, it takes an object to reach the ground from the top of the Sears Tower in Chicago, a distance of 1454 feet.   9.53 sec.

33. Beverly wants to determine the distance at certain points across a pond on her property. She is able to measure the distances shown on the following diagram. Find how wide the lake is to the nearest tenth of a foot.   51.2 ft.

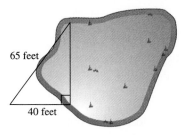

34. Carl needs to connect two underground pipelines, which are offset by 3 feet, as pictured in the diagram. Neglecting the joints needed to join the

pipes, find the length of the shortest possible connecting pipe rounded to the nearest hundredth of a foot.   4.24 ft.

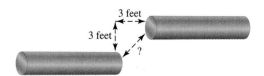

 35. Bruce needs to attach a diagonal brace to a rectangular frame in order to make it structurally sound. If the framework is 6 feet by 10 feet, find how long the brace needs to be to the nearest tenth of a foot.   11.7 ft.

36. Connie is flying a kite. She let out 80 feet of string and attached the string to a stake in the ground. The kite is now directly above her brother Mike, who is 32 feet away from Connie. Find the height of the kite to the nearest foot.   73 ft.

37. Using the formula $b = \sqrt{\frac{3V}{H}}$, find the exact length of the base b, if the volume V is 18 cubic inches and the height H is 1 foot. $\frac{3\sqrt{2}}{2}$ in.

38. Police use the formula $s = \sqrt{30fd}$ to estimate the speed s of a car in miles per hour. In this formula, d represents the distance the car skidded in feet and f represents the coefficient of friction. The value of f depends on the type of road surface, and for wet concrete f is 0.35. Find how fast a car was moving if it skidded 280 feet on wet concrete, to the nearest mile per hour.   54 mph

39. The coefficient of friction of a certain dry road is 0.95. Use the formula in Exercise 38 to find how far a car will skid on this dry road if it is traveling at a rate of 60 mph. Round the length to the nearest foot.   126 ft.

40. The formula $v = \sqrt{2.5r}$ can be used to estimate the maximum safe velocity, v, in miles per hour, at which a car can travel if it is driven along a curved road with a **radius of curvature,** r, in feet. To the nearest whole number, find the maximum safe speed if a cloverleaf exit on an expressway has a radius of curvature of 300 feet.   27 mph

41. Use the formula from Exercise 40 to find the radius of curvature if the safe velocity is 30 mph.   360 ft.

42. The maximum distance d in kilometers that you can see from a height of h meters is given by $d = 3.5\sqrt{h}$. Find how far you can see from the top of the Texas Commerce Tower in Houston, a height of 3051 meters. Round to the nearest tenth of a kilometer.   193.3 km

43. Use the formula from Exercise 42 to determine how high above the ground you need to be to see 40 kilometers. Round to the nearest tenth of a meter.   130.6 m

44. Railroad tracks are invariably made up of relatively short sections of rail connected by expansion joints. To see why this construction is necessary, consider a single rail 100 feet long (or 1200 inches). On an extremely hot day, suppose it expands 1 inch in the hot sun to a new length of 1201 inches. Theoretically, the track would bow upward as pictured.

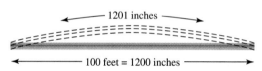

Let us approximate the bulge in the railroad this way.

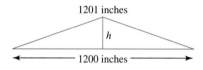

Calculate the height h of the bulge to the nearest tenth of an inch.   24.5 in.

45. Based on the results of Exercise 44, explain why railroads use short sections of rail connected by expansion joints.   answers may vary

Review Exercises

Simplify using rules for exponents. See Sections 4.1 and 4.2.

46. 2^5 32

47. $(-3)^3$ -27

48. $\left(-\frac{1}{5}\right)^2$ $\frac{1}{25}$

49. $\left(\frac{2}{7}\right)^3$ $\frac{8}{343}$

50. $x^2 \cdot x^3$ x^5

51. $x^4 \cdot x^2$ x^6

52. $y^3 \cdot y$ y^4

53. $x \cdot x^7$ x^8

TAPE BA 9.7

9.7 | RATIONAL EXPONENTS

O B J E C T I V E S

1 Evaluate exponential expressions of the form $a^{1/n}$.

2 Evaluate exponential expressions of the form $a^{m/n}$.

3 Evaluate exponential expressions of the form $a^{-m/n}$.

4 Use rules for exponents to simplify expressions containing fractional exponents.

1

Radical notation is widely used, as we've seen. In this section, we study an alternate notation, one that proves to be more efficient and compact. This alternate notation makes use of expressions containing an exponent that is a rational number but not necessarily an integer, for example,

$$3^{1/2}, \quad 2^{-3/4}, \quad \text{and} \quad y^{5/6}$$

In giving meaning to rational exponents, keep in mind that we want the rules for operating with them to be the same as the rules for operating with integer exponents. These rules are repeated here for review.

SUMMARY OF EXPONENT RULES

If m and n are integers and a, b, and c are real numbers, then

Product Rule for Exponents:	$a^m \cdot a^n = a^{m+n}$
Power of a Power Rule for Exponents:	$(a^m)^n = a^{m \cdot n}$
Power of a Product or Quotient:	$(ab)^n = a^n b^n$ and
	$\left(\dfrac{a}{c}\right)^n = \dfrac{a^n}{c^n}, c \neq 0$
Quotient Rule for Exponents:	$\dfrac{a^m}{a^n} = a^{m-n}, a \neq 0$
Zero Exponent:	$a^0 = 1, a \neq 0$
Negative Exponent:	$a^{-n} = \dfrac{1}{a^n}, a \neq 0$

If the rule $(a^m)^n = a^{m \cdot n}$ is to hold for fractional exponents, it should be true that

$$(3^{1/2})^2 = 3^{1/2 \cdot 2} = 3^1 = 3$$

Also, we know that

$$(\sqrt{3})^2 = 3$$

Since the square of both $3^{1/2}$ and $\sqrt{3}$ is 3, it would be reasonable to say that

$$3^{1/2} \text{ means } \sqrt{3}$$

In general, we have the following:

DEFINITION OF $a^{1/n}$

If n is a positive integer and $\sqrt[n]{a}$ is a real number, then

$$a^{1/n} = \sqrt[n]{a}$$

Notice that the denominator of the rational exponent is the same as the index of the corresponding radical.

EXAMPLE 1   Write in radical notation. Then simplify.

   a. $25^{1/2}$ **b.** $8^{1/3}$ **c.** $-16^{1/4}$ **d.** $(-27)^{1/3}$ **e.** $\left(\dfrac{1}{9}\right)^{1/2}$

Solution: **a.** $25^{1/2} = \sqrt{25} = 5$.
   b. $8^{1/3} = \sqrt[3]{8} = 2$.
   c. In $-16^{1/4}$, the base of the exponent is 16. Thus **the negative sign is not affected by the exponent;** so $-16^{1/4} = -\sqrt[4]{16} = -2$.
   d. The parentheses show that -27 is the base. $(-27)^{1/3} = \sqrt[3]{-27} = -3$.
   e. $\left(\dfrac{1}{9}\right)^{1/2} = \sqrt{\dfrac{1}{9}} = \dfrac{1}{3}$.

   ————

2   In Example 1, each rational exponent has a numerator of 1. What happens if the numerator is some other positive integer? Consider $8^{2/3}$. Since $\frac{2}{3}$ is the same $\frac{1}{3} \cdot 2$, we reason that

$$8^{2/3} = 8^{1/3 \cdot 2} = (8^{1/3})^2 = (\sqrt[3]{8})^2 = 2^2 = 4.$$

The denominator 3 of the rational exponent is the same as the index of the radical. The numerator 2 of the fractional exponent indicates that the radical base is to be squared.

DEFINITION OF $a^{m/n}$

If m and n are integers with $n > 0$ and if a is a positive number, then

$$a^{m/n} = (a^{1/n})^m = (\sqrt[n]{a})^m$$

Also,

$$a^{m/n} = (a^m)^{1/n} = \sqrt[n]{a^m}$$

EXAMPLE 2   Simplify each expression.

     a. $4^{3/2}$ **b.** $27^{2/3}$ **c.** $-16^{3/4}$

Solution: **a.** $4^{3/2} = (4^{1/2})^3 = (\sqrt{4})^3 = 2^3 = 8.$

           b. $27^{2/3} = (27^{1/3})^2 = (\sqrt[3]{27})^2 = 3^2 = 9.$

           c. The negative sign is **not** affected by the exponent since the base of the exponent is 16. $-16^{3/4} = -(16^{1/4})^3 = -(\sqrt[4]{16})^3 = -2^3 = -8.$

R E M I N D E R   Recall that

$$-3^2 = -(3 \cdot 3) = -9$$

and

$$(-3)^2 = (-3)(-3) = 9$$

In other words, without parentheses the exponent 2 applies to the base 3, **not** -3. The same is true of rational exponents. For example,

$$-16^{1/2} = -\sqrt{16} = -4$$

and

$$(-27)^{1/3} = \sqrt[3]{-27} = -3$$

---

3   If the exponent is a negative rational number, use the following definition.

> **DEFINITION OF $a^{-m/n}$**
>
> If $a^{m/n}$ is a nonzero real number, then
>
> $$a^{-m/n} = \frac{1}{a^{m/n}}$$

EXAMPLE 3   Write each expression with a positive exponent and then simplify.

     a. $36^{-1/2}$ **b.** $16^{-3/4}$ **c.** $-9^{1/2}$ **d.** $32^{-4/5}$

Solution: **a.** $36^{-1/2} = \dfrac{1}{36^{1/2}} = \dfrac{1}{\sqrt{36}} = \dfrac{1}{6}.$

           b. $16^{-3/4} = \dfrac{1}{16^{3/4}} = \dfrac{1}{(\sqrt[4]{16})^3} = \dfrac{1}{2^3} = \dfrac{1}{8}.$

           c. $-9^{1/2} = -\sqrt{9} = -3.$

           d. $32^{-4/5} = \dfrac{1}{32^{4/5}} = \dfrac{1}{(\sqrt[5]{32})^4} = \dfrac{1}{2^4} = \dfrac{1}{16}.$

It can be shown that the properties of integer exponents hold for rational exponents. By using these properties and definitions, we can now simplify products and quotients of expressions containing rational exponents.

EXAMPLE 4 Simplify each expression. Write results with positive exponents only. Assume that all variables represent positive numbers.

a. $3^{1/2} \cdot 3^{3/2}$ **b.** $\dfrac{5^{1/3}}{5^{2/3}}$ **c.** $(x^{1/4})^{12}$ **d.** $\dfrac{x^{1/5}}{x^{-4/5}}$ **e.** $\left(\dfrac{y^{3/5}}{z^{1/4}}\right)^2$

Solution: **a.** $3^{1/2} \cdot 3^{3/2} = 3^{1/2+3/2} = 3^{4/2} = 3^2 = 9.$

b. $\dfrac{5^{1/3}}{5^{2/3}} = 5^{(1/3)-(2/3)} = 5^{-1/3} = \dfrac{1}{5^{1/3}}.$

c. $(x^{1/4})^{12} = x^{1/4 \cdot 12} = x^3.$

d. $\dfrac{x^{1/5}}{x^{-4/5}} = x^{1/5-(-4/5)} = x^{5/5} = x^1$ or $x.$

e. $\left(\dfrac{y^{3/5}}{z^{1/4}}\right)^2 = \dfrac{y^{3/5 \cdot 2}}{z^{1/4 \cdot 2}} = \dfrac{y^{6/5}}{z^{1/2}}.$

EXERCISE SET 9.7

Simplify each expression. See Examples 1 and 2.

1. $8^{1/3}$ 2 **2.** $16^{1/4}$ 2 **3.** $9^{1/2}$ 3

4. $16^{1/2}$ 4 **5.** $16^{3/4}$ 8 **6.** $27^{2/3}$ 9

7. $32^{2/5}$ 4 **8.** $64^{5/6}$ 32

Simplify each expression. See Example 3.

9. $-16^{-1/4}$ $-\dfrac{1}{2}$ **10.** $-8^{-1/3}$ $-\dfrac{1}{2}$ **11.** $16^{-3/2}$ $\dfrac{1}{64}$

12. $27^{-4/3}$ $\dfrac{1}{81}$ **13.** $81^{-3/2}$ $\dfrac{1}{729}$ **14.** $32^{-2/5}$ $\dfrac{1}{4}$

15. $\left(\dfrac{4}{25}\right)^{-1/2}$ $\dfrac{5}{2}$ **16.** $\left(\dfrac{8}{27}\right)^{-1/3}$ $\dfrac{3}{2}$

17. Explain the meaning of the numbers 2, 3, and 4 in the exponential expression $4^{3/2}$.

18. Explain why $-4^{1/2}$ is a real number but $(-4)^{1/2}$ is not.   answers may vary

Simplify each expression. Write each answer with positive exponents. Assume that all variables represent positive numbers. See Example 4.

19. $2^{1/3} \cdot 2^{2/3}$ 2 **20.** $4^{2/5} \cdot 4^{3/5}$ 4 **21.** $\dfrac{4^{3/4}}{4^{1/4}}$ 2

17. answers may vary

22. $\dfrac{9^{7/2}}{9^{3/2}}$ 81 **23.** $\dfrac{x^{1/6}}{x^{5/6}}$ $\dfrac{1}{x^{2/3}}$ **24.** $\dfrac{x^{1/4}}{x^{3/4}}$ $\dfrac{1}{x^{1/2}}$

25. $(x^{1/2})^6$ x^3 **26.** $(x^{1/3})^6$ x^2

27. Explain how simplifying $x^{1/2} \cdot x^{1/3}$ is similar to simplifying $x^2 \cdot x^3$.   answers may vary

28. Explain how simplifying $(x^{1/2})^{1/3}$ is similar to simplifying $(x^2)^3$.   answers may vary

Simplify each expression.

29. $81^{1/2}$ 9 **30.** $(-27)^{1/3}$ -3 **31.** $(-8)^{1/3}$ -2

32. $36^{1/2}$ 6 **33.** $-81^{1/4}$ -3 **34.** $-64^{1/3}$ -4

35. $\left(\dfrac{1}{81}\right)^{1/2}$ $\dfrac{1}{9}$ **36.** $\left(\dfrac{9}{16}\right)^{1/2}$ $\dfrac{3}{4}$ **37.** $\left(\dfrac{27}{64}\right)^{1/3}$ $\dfrac{3}{4}$

38. $\left(\dfrac{16}{81}\right)^{1/4}$ $\dfrac{2}{3}$ **39.** $9^{3/2}$ 27 **40.** $16^{3/2}$ 64

41. $64^{3/2}$ 512 **42.** $64^{2/3}$ 16 **43.** $-8^{2/3}$ -4

44. $(8)^{2/3}$ 4 **45.** $4^{5/2}$ 32 **46.** $9^{4/2}$ 81

47. $\left(\dfrac{4}{9}\right)^{3/2}$ $\dfrac{8}{27}$ **48.** $\left(\dfrac{8}{27}\right)^{2/3}$ $\dfrac{4}{9}$ **49.** $\left(\dfrac{1}{81}\right)^{3/4}$ $\dfrac{1}{27}$

50. $\left(\dfrac{1}{32}\right)^{3/5}$ $\dfrac{1}{8}$ **51.** $4^{-1/2}$ $\dfrac{1}{2}$ **52.** $9^{-1/2}$ $\dfrac{1}{3}$

53. $125^{-1/3}$ $\dfrac{1}{5}$ **54.** $216^{-1/3}$ $\dfrac{1}{6}$ **55.** $625^{-3/4}$ $\dfrac{1}{125}$

56. $256^{-5/8}$ $\dfrac{1}{32}$

Simplify each expression. Write each answer with positive exponents. Assume that all variables represent positive numbers.

57. $3^{4/3} \cdot 3^{2/3}$ 9 **58.** $2^{5/4} \cdot 2^{3/4}$ 4 **59.** $\dfrac{6^{2/3}}{6^{1/3}}$ $6^{1/3}$

60. $\dfrac{3^{3/5}}{3^{1/5}}$ $3^{2/5}$ **61.** $(x^{2/3})^9$ x^6 **62.** $(x^6)^{3/4}$ $x^{9/2}$

63. $\dfrac{6^{1/3}}{6^{-5/3}}$ 36 **64.** $\dfrac{2^{-3/4}}{2^{5/4}}$ $\dfrac{1}{4}$ **65.** $\dfrac{3^{-3/5}}{3^{2/5}}$ $\dfrac{1}{3}$

66. $\dfrac{5^{1/4}}{5^{-3/4}}$ 5 **67.** $\left(\dfrac{x^{1/3}}{y^{3/4}}\right)^2$ $\dfrac{x^{2/3}}{y^{3/2}}$ **68.** $\left(\dfrac{x^{1/2}}{y^{2/3}}\right)^6$ $\dfrac{x^3}{y^4}$

69. $\left(\dfrac{x^{2/5}}{y^{3/4}}\right)^8$ $\dfrac{x^{16/5}}{y^6}$ **70.** $\left(\dfrac{x^{3/4}}{y^{1/6}}\right)^3$ $\dfrac{x^{9/4}}{y^{1/2}}$

 71. If a population grows at a rate of 8% annually, the formula $P = P_O(1.08)^N$ can be used to estimate the total population P after N years have passed, assuming the original population is P_O. Find the population after $1\frac{1}{2}$ years if the original population of 10,000 people is growing at a rate of 8% annually.

72. Money grows in a certain savings account at a rate of 4% compounded annually. The amount of money A in the account at time t is given by the formula

$$A = P(1.04)^t$$

where P is the original amount deposited in the account. Find the amount of money in the account after $3\frac{3}{4}$ years if \$200 was initially deposited.

Use a calculator and approximate each to three decimal places.

73. $5^{3/4}$ 3.344 **74.** $20^{1/8}$ 1.454

75. $18^{3/5}$ 5.665 **76.** $42^{3/10}$ 3.069

Review Exercises

Solve each system of linear inequalities by graphing on a single coordinate system. See Section 8.5.

77. $\begin{cases} x + y < 6 \\ y \geq 2x \end{cases}$ See App. F   **78.** $\begin{cases} 2x - y \geq 3 \\ x < 5 \end{cases}$

Solve each quadratic equation. See Section 5.6.

79. $x^2 - 4 = 3x$ **80.** $x^2 + 2x = 8$

81. $2x^2 - 5x - 3 = 0$ **82.** $3x^2 + x - 2 = 0$

71. 11,224 people   **72.** \$231.69   **78.** See App. F

79. $x = -1, x = 4$ **80.** $x = -4, x = 2$

81. $x = -\dfrac{1}{2}, x = 3$ **82.** $x = \dfrac{2}{3}, x = -1$

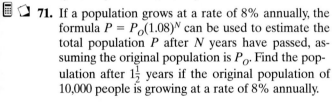

GROUP ACTIVITY

INVESTIGATING THE DIMENSIONS OF CYLINDERS

MATERIALS:
• Calculator
• Several empty cans of different sizes
• 2-cup (16-fluid-ounce) transparent measuring cup with metric markings (in milliliters)
• Metric ruler
• Water

(continued)

$\mathbf{T}$he volume V (in cubic units) of a cylinder is given by the formula $V = \pi r^2 h$, where r is the radius of the cylinder and h is its height. You will investigate the radii of several cylindrical cans by completing the following table.

CAN	VOLUME (ML)	HEIGHT (CM)	CALCULATED RADIUS (CM)	MEASURED RADIUS (CM)	DIFFERENCE BETWEEN CALCULATED AND MEASURED RADII
A					
B					
C					
D					

1. For each can, measure its volume by filling it with water and pouring the water into the measuring cup. Find the volume of the water in milliliters (ml). Record the volumes of the cans in the table. (Remember that $1 \text{ ml} = 1 \text{ cm}^3$).

2. For each can, use a ruler to measure its height in centimeters (cm). Record the heights in the table.

3. Use the formula $V = \pi r^2 h$ to calculate an estimate of each can's radius. Record these calculated radii in the table.

4. Try to measure the radius of each can and record these measured radii in the table. (Remember that radius $= \frac{1}{2}$ diameter.)

5. Find the difference between the calculated radius and the measured radius for each can. Are the differences very large? Discuss the factors that could account for the differences.

See App. F for Group Activity answers and suggestions.

CHAPTER 9 HIGHLIGHTS

DEFINITIONS AND CONCEPTS	EXAMPLES
SECTION 9.1 INTRODUCTION TO RADICALS	
The **positive or principal square root** of a positive number a is written as $\sqrt{a}$. The **negative square root** of a is written as $-\sqrt{a}$. $\sqrt{a} = b$ only if $b^2 = a$ and $b > 0$. A square root of a negative number is not a real number.	$\sqrt{25} = 5 \qquad \sqrt{100} = 10$ $-\sqrt{9} = -3 \qquad \sqrt{\dfrac{4}{49}} = \dfrac{2}{7}$ $\sqrt{-4}$ is not a real number.

DEFINITIONS AND CONCEPTS	EXAMPLES

SECTION 9.1 INTRODUCTION TO RADICALS

The **cube root** of a real number a is written as $\sqrt[3]{a}$ and $\sqrt[3]{a} = b$ only if $b^3 = a$.

$$\sqrt[3]{64} = 4 \qquad \sqrt[3]{-8} = -2$$

The **n^{th} root** of a number a is written as $\sqrt[n]{a}$ and $\sqrt[n]{a} = b$ only if $b^n = a$.

$$\sqrt[4]{81} = 3$$
$$\sqrt[5]{-32} = -2$$

The natural number n is called the **index,** the symbol $\sqrt{}$ is called a **radical,** and the expression within the radical is called the **radicand.**

(Note: If the index is even, the radicand must be non-negative for the root to be a real number.)

SECTION 9.2 SIMPLIFYING RADICALS

Product rule for radicals

If $\sqrt[n]{a}$ and $\sqrt[n]{b}$ are real numbers, then
$$\sqrt[n]{a} \cdot \sqrt[n]{b} = \sqrt[n]{a \cdot b}$$

$$\sqrt{2} \cdot \sqrt{3} = \sqrt{6}$$
$$\sqrt[3]{7} \cdot \sqrt[3]{2} = \sqrt[3]{14}$$

A square root is in **simplified form** if the radicand contains no perfect square factors other than 1. To simplify a square root, factor the radicand so that one of its factors is a perfect square factor.

$$\sqrt{45} = \sqrt{9 \cdot 5}$$
$$= \sqrt{9} \cdot \sqrt{5}$$
$$= 3\sqrt{5}$$
in simplest form.

To simplify cube roots, factor the radicand so that one of its factors is a perfect cube.

$$\sqrt[3]{48} = \sqrt[3]{8 \cdot 6}$$
$$= \sqrt[3]{8} \cdot \sqrt[3]{6}$$
$$= 2\sqrt[3]{6}$$

Quotient rule for radicals

If $\sqrt[n]{a}$ and $\sqrt[n]{b}$ are real numbers and $b \neq 0$, then
$$\sqrt[n]{\frac{a}{b}} = \frac{\sqrt[n]{a}}{\sqrt[n]{b}}$$

$$\sqrt{\frac{18}{x^6}} = \frac{\sqrt{9 \cdot 2}}{\sqrt{x^6}} = \frac{\sqrt{9} \cdot \sqrt{2}}{x^3} = \frac{3\sqrt{2}}{x^3}$$
$$\sqrt[3]{\frac{18}{x^6}} = \frac{\sqrt[3]{18}}{\sqrt[3]{x^6}} = \frac{\sqrt[3]{18}}{x^2}$$

SECTION 9.3 ADDING AND SUBTRACTING RADICALS

Like radicals are radical expressions that have the same index and the same radicand.

Like Radicals
$$5\sqrt{2}, \ -7\sqrt{2}, \ \sqrt{2}$$
$$-\sqrt[3]{11}, \ 3\sqrt[3]{11}$$

To combine like radicals, use the distributive property.

$$2\sqrt{7} - 13\sqrt{7} = (2 - 13)\sqrt{7} = -11\sqrt{7}$$

$$\sqrt[3]{24} + \sqrt[3]{8} + \sqrt[3]{81}$$
$$= \sqrt[3]{8 \cdot 3} + 2 + \sqrt[3]{27 \cdot 3}$$
$$= \sqrt[3]{8} \cdot \sqrt[3]{3} + 2 + \sqrt[3]{27} \cdot \sqrt[3]{3}$$
$$= 2\sqrt[3]{3} + 2 + 3\sqrt[3]{3}$$
$$= (2 + 3)\sqrt[3]{3} + 2$$
$$= 5\sqrt[3]{3} + 2$$

(continued)

DEFINITIONS AND CONCEPTS	EXAMPLES

SECTION 9.4 MULTIPLYING AND DIVIDING RADICALS

The product and quotient rules for radicals may be used to simplify products and quotients of radicals.

Perform indicated operations and simplify.
$$(2\sqrt{5})^2 = 2^2 \cdot (\sqrt{5})^2 = 4 \cdot 5 = 20$$
Multiply.
$$(\sqrt{3x} + 1)(\sqrt{5} - \sqrt{3})$$
$$= \sqrt{15x} - \sqrt{9x} + \sqrt{5} - \sqrt{3}$$
$$= \sqrt{15x} - 3\sqrt{x} + \sqrt{5} - \sqrt{3}$$
$$\frac{\sqrt[3]{56x^4}}{\sqrt[3]{7x}} = \sqrt[3]{\frac{56x^4}{7x}} = \sqrt[3]{8x^3} = 2x$$

The process of eliminating the radical in the denominator of a radical expression is called **rationalizing the denominator.**

The **conjugate** of $a + b$ is $a - b$.

To rationalize a denominator that is a sum or difference of radicals, multiply the numerator and the denominator by the conjugate of the denominator.

Rationalize the denominator.
$$\frac{5}{\sqrt{11}} = \frac{5 \cdot \sqrt{11}}{\sqrt{11} \cdot \sqrt{11}} = \frac{5\sqrt{11}}{11}$$
The conjugate of $2 + \sqrt{3}$ is $2 - \sqrt{3}$.

Rationalize the denominator.
$$\frac{5}{6 - \sqrt{5}} = \frac{5(6 + \sqrt{5})}{(6 - \sqrt{5})(6 + \sqrt{5})}$$
$$= \frac{5(6 + \sqrt{5})}{36 + 6\sqrt{5} - 6\sqrt{5} - 5}$$
$$= \frac{5(6 + \sqrt{5})}{31}$$

SECTION 9.5 SOLVING EQUATIONS CONTAINING RADICALS

To solve a radical equation containing square roots.

Step 1. Get one radical by itself on one side of the equation.

Step 2. Square both sides of the equation.

Step 3. Simplify both sides of the equation.

Step 4. If the equation still contains a radical term, repeat *steps 1* through *3*.

Step 5. Solve the equation.

Step 6. Check solutions in the original equation.

Solve:
$$\sqrt{2x - 1} - x = -2$$
$$\sqrt{2x - 1} = x - 2$$
$$(\sqrt{2x - 1})^2 = (x - 2)^2 \qquad \text{Square both sides.}$$
$$2x - 1 = x^2 - 4x + 4$$
$$0 = x^2 - 6x + 5$$
$$0 = (x - 1)(x - 5) \qquad \text{Factor.}$$
$$x - 1 = 0 \quad \text{or} \quad x - 5 = 0$$
$$x = 1 \quad \text{or} \qquad x = 5 \qquad \text{Solve.}$$

Check both proposed solutions in the original equation. 5 checks but 1 does not. The only solution is 5.

SECTION 9.6 RADICAL EQUATIONS AND PROBLEM SOLVING

1. UNDERSTAND. Read and reread the problem.

A gutter is mounted on the eaves of a house 15 feet above the ground. A garden is adjacent to the house so that the closest a ladder can be placed to the house is 6 feet. How long a ladder is needed for installing the gutter?

DEFINITIONS AND CONCEPTS	EXAMPLES

SECTION 9.6   RADICAL EQUATIONS AND PROBLEM SOLVING

2. ASSIGN.

3. ILLUSTRATE.

Let x = the length of the ladder.

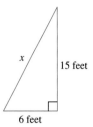

4. TRANSLATE.

Here, we use the Pythagorean theorem. The unknown length x is the hypotenuse.

In words: $(\text{leg})^2 + (\text{leg})^2 = (\text{hypotenuse})^2$

Translate: $6^2 + 15^2 = x^2$

5. COMPLETE.

$$36 + 225 = x^2$$
$$261 = x^2$$
$$\sqrt{261} = x \quad \text{or} \quad x = 3\sqrt{29}$$

6. INTERPRET.

Check and state. The ladder needs to be $3\sqrt{29}$ feet or approximately 16.2 feet long.

SECTION 9.7   RATIONAL EXPONENTS

If n is a positive integer and $\sqrt[n]{a}$ is a real number, then $a^{1/n} = \sqrt[n]{a}$.

$$9^{1/2} = \sqrt{9} = 3$$
$$(-8)^{1/3} = \sqrt[3]{-8} = -2$$

If m and n are integers with $n > 0$ and a is positive, then

$a^{m/n} = (a^{1/n})^m = (\sqrt[n]{a})^m$. Also

$a^{m/n} = (a^m)^{1/n} = \sqrt[n]{a^m}$.

$$25^{3/2} = (25^{1/2})^3 = (\sqrt{25})^3 = 5^3 \text{ or } 125$$

If $a^{m/n}$ is a nonzero real number, then

$$a^{-m/n} = \frac{1}{a^{m/n}}$$

$$81^{-3/4} = \frac{1}{81^{3/4}} = \frac{1}{(\sqrt[4]{81})^3}$$
$$= \frac{1}{3^3} = \frac{1}{27}$$

Properties for integer exponents hold for rational exponents also.

$$a^m \cdot a^n = a^{m+n}$$
$$(a^m)^n = a^{mn}$$
$$\frac{a^m}{a^n} = a^{m-n}$$

$$x^{1/2} \cdot x^{1/4} = x^{1/2 + 1/4} = x^{2/4 + 1/4} = x^{3/4}$$
$$(x^{2/3})^{1/5} = x^{2/3 \cdot 1/5} = x^{2/15}$$
$$\frac{x^{5/6}}{x^{1/6}} = x^{5/6 - 1/6} = x^{4/6} = x^{2/3}$$

CHAPTER 9 REVIEW

(9.1) *Find the root. Indicate if the expression is not a real number.*

1. $\sqrt{81}$ 9

2. $-\sqrt{49}$ -7

3. $\sqrt[3]{27}$ 3

4. $\sqrt[4]{16}$ 2

5. $-\sqrt{\dfrac{9}{64}}$ $-\dfrac{3}{8}$

6. $\sqrt{\dfrac{36}{81}}$ $\dfrac{2}{3}$

7. $\sqrt[4]{-\dfrac{16}{81}}$

8. $\sqrt[3]{-\dfrac{27}{64}}$ $-\dfrac{3}{4}$

Determine whether each of the following is rational or irrational. If rational, find the exact value. If irrational, use a calculator to find an approximation accurate to three decimal places.

9. $\sqrt{76}$ irrational, 8.718

10. $\sqrt{576}$ rational, 24

Find the following roots. Assume that variables represent positive numbers only.

11. $\sqrt{x^{12}}$ x^6

12. $\sqrt{x^8}$ x^4

13. $\sqrt{9x^6y^2}$ $3x^3y$

14. $\sqrt{25x^4y^{10}}$ $5x^2y^5$

15. $-\sqrt[3]{8x^6}$ $-2x^2$

16. $-\sqrt[4]{16x^8}$ $-2x^2$

(9.2) *Simplify each expression using the product rule. Assume that variables represent nonnegative real numbers.*

17. $\sqrt{54}$ $3\sqrt{6}$

18. $\sqrt{88}$ $2\sqrt{22}$

19. $\sqrt{150x^3y^6}$ $5xy^3\sqrt{6x}$

20. $\sqrt{92x^8y^5}$ $2x^4y^2\sqrt{23y}$

21. $\sqrt[3]{54}$ $3\sqrt[3]{2}$

22. $\sqrt[3]{88}$ $2\sqrt[3]{11}$

23. $\sqrt[4]{48x^3y^6}$ $2y\sqrt[4]{3x^3y^2}$

24. $\sqrt[4]{162x^8y^5}$ $3x^2y\sqrt[4]{2y}$

Simplify each expression using the quotient rule. Assume that variables represent positive real numbers.

25. $\sqrt{\dfrac{18}{25}}$ $\dfrac{3\sqrt{2}}{5}$

26. $\sqrt{\dfrac{75}{64}}$ $\dfrac{5\sqrt{3}}{8}$

27. $\sqrt{\dfrac{45x^2y^2}{4x^6}}$ $\dfrac{3y\sqrt{5}}{2x^2}$

28. $\sqrt{\dfrac{20x^5}{9x^2}}$ $\dfrac{2x\sqrt{5x}}{3}$

29. $\sqrt[4]{\dfrac{9}{16}}$ $\dfrac{\sqrt[4]{9}}{2}$

30. $\sqrt[3]{\dfrac{40}{27}}$ $\dfrac{2\sqrt[3]{5}}{3}$

31. $\sqrt[3]{\dfrac{3y^6}{8x^3}}$ $\dfrac{y^2\sqrt[3]{3}}{2x}$

32. $\sqrt[4]{\dfrac{5x^2}{81x^8}}$ $\dfrac{\sqrt[4]{5x^2}}{3x^2}$

(9.3) *Add or subtract by combining like radicals.*

33. $3\sqrt[3]{2} + 2\sqrt[3]{3} - 4\sqrt[3]{2}$

34. $5\sqrt{2} + 2\sqrt[3]{2} - 8\sqrt{2}$

35. $\sqrt{6} + 2\sqrt[3]{6} - 4\sqrt[3]{6} + 5\sqrt{6}$ $6\sqrt{6} - 2\sqrt[3]{6}$

36. $3\sqrt{5} - \sqrt[3]{5} - 2\sqrt{5} + 3\sqrt[3]{5}$ $\sqrt{5} + 2\sqrt[3]{5}$

Add or subtract by simplifying each radical and then combining like terms. Assume that variables represent nonnegative real numbers.

37. $\sqrt{28} + \sqrt{63} + \sqrt[3]{56}$

38. $\sqrt{75} + \sqrt{48} - \sqrt[4]{16}$

39. $\sqrt{\dfrac{5}{9}} - \sqrt{\dfrac{5}{36}}$ $\dfrac{\sqrt{5}}{6}$

40. $\sqrt{\dfrac{11}{25}} + \sqrt{\dfrac{11}{16}}$ $\dfrac{9\sqrt{11}}{20}$

41. $2\sqrt[3]{125x^3} - 5x\sqrt[3]{8}$ 0

42. $3\sqrt[3]{16x^4} - 2x\sqrt[3]{2x}$

(9.4) *Find the product and simplify if possible.*

43. $3\sqrt{10} \cdot 2\sqrt{5}$ $30\sqrt{2}$

44. $2\sqrt[3]{4} \cdot 5\sqrt[3]{6}$ $20\sqrt[3]{3}$

45. $\sqrt{3}(2\sqrt{6} - 3\sqrt{12})$

46. $4\sqrt{5}(2\sqrt{10} - 5\sqrt{5})$

47. $(\sqrt{3} + 2)(\sqrt{6} - 5)$

48. $(2\sqrt{5} + 1)(4\sqrt{5} - 3)$

Find the quotient and simplify if possible. Assume that variables represent positive real numbers.

49. $\dfrac{\sqrt{96}}{\sqrt{3}}$ $4\sqrt{2}$

50. $\dfrac{\sqrt{160}}{\sqrt{8}}$ $2\sqrt{5}$

51. $\dfrac{\sqrt{15x^6y}}{\sqrt{12x^3y^9}}$ $\dfrac{x\sqrt{5x}}{2y^4}$

52. $\dfrac{\sqrt{50xy^8}}{\sqrt{72x^7y^3}}$ $\dfrac{5y^2\sqrt{y}}{6x^3}$

Rationalize each denominator and simplify.

53. $\sqrt{\dfrac{5}{6}}$ $\dfrac{\sqrt{30}}{6}$

54. $\sqrt{\dfrac{7}{10}}$ $\dfrac{\sqrt{70}}{10}$

55. $\sqrt{\dfrac{3}{2x}}$ $\dfrac{\sqrt{6x}}{2x}$

56. $\sqrt{\dfrac{6}{5y}}$ $\dfrac{\sqrt{30y}}{5y}$

57. $\sqrt{\dfrac{7}{20y^2}}$ $\dfrac{\sqrt{35}}{10y}$

58. $\sqrt{\dfrac{5z}{12x^2}}$ $\dfrac{\sqrt{15z}}{6x}$

59. $\sqrt[3]{\dfrac{7}{9}}$ $\dfrac{\sqrt[3]{21}}{3}$

60. $\sqrt[3]{\dfrac{3}{4}}$ $\dfrac{\sqrt[3]{6}}{2}$

61. $\sqrt[3]{\dfrac{3}{2x^2}}$ $\dfrac{\sqrt[3]{12x}}{2x}$

62. $\sqrt[3]{\dfrac{5x}{4y}}$ $\dfrac{\sqrt[3]{10xy^2}}{2y}$

63. $\dfrac{3}{\sqrt{5} - 2}$ $3\sqrt{5} + 6$

64. $\dfrac{8}{\sqrt{10} - 3}$ $8\sqrt{10} + 24$

65. $\dfrac{8}{\sqrt{6} + 2}$ $4\sqrt{6} - 8$

66. $\dfrac{12}{\sqrt{15} - 3}$ $2\sqrt{15} + 6$

67. $\dfrac{\sqrt{2}}{4 + \sqrt{2}}$ $\dfrac{2\sqrt{2} - 1}{7}$

68. $\dfrac{\sqrt{3}}{5 + \sqrt{3}}$ $\dfrac{5\sqrt{3} - 3}{22}$

7. not a real number   **33.** $2\sqrt[3]{3} - \sqrt[3]{2}$ **34.** $-3\sqrt{2} + 2\sqrt[3]{2}$ **37.** $5\sqrt{7} + 2\sqrt[3]{7}$ **38.** $9\sqrt{3} - 2$ **42.** $4x\sqrt[3]{2x}$
45. $6\sqrt{2} - 18$ **46.** $40\sqrt{2} - 100$ **47.** $3\sqrt{2} - 5\sqrt{3} + 2\sqrt{6} - 10$ **48.** $37 - 2\sqrt{5}$

69. $\dfrac{2\sqrt{3}}{\sqrt{3}-5}$ $\quad -\dfrac{3+5\sqrt{3}}{11}$ **70.** $\dfrac{7\sqrt{2}}{\sqrt{2}-4}$ $\quad -1-2\sqrt{2}$

(9.5) *Solve the following radical equations.*

71. $\sqrt{2x}=6$ $\quad x=18$ **72.** $\sqrt{x+3}=4$ $\ x=13$

73. $\sqrt{x}+3=8$ $\ x=25$ **74.** $\sqrt{x}+8=3$

75. $\sqrt{2x+1}=x-7$ $\ x=12$ **76.** $\sqrt{3x+1}=x-1$

77. $\sqrt{x+3}+x=9$ $\ x=6$ **78.** $\sqrt{2x}+x=4$ $\ x=2$

(9.6) *Use the Pythagorean theorem to find the length of the unknown side.*

79.
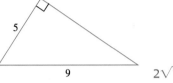
$2\sqrt{14}$

80.
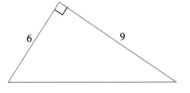
$\sqrt{117}$

81. Romeo is standing 20 feet away from the wall below Juliet's balcony during a school play. Juliet is on the balcony, 12 feet above the ground. Find how far apart Romeo and Juliet are. $\ 4\sqrt{34}$ ft.

82. The diagonal of a rectangle is 10 inches long. If the width of the rectangle is 5 inches, find the length of the rectangle. $\ 5\sqrt{3}$ in.

Use the distance formula to find the distance between the points.

83. $(6,-2)$ and $(-3,5)$ **84.** $(2,8)$ and $(-6,10)$

74. no solution **76.** $x=5$ **83.** $\sqrt{130}$ **84.** $2\sqrt{17}$

Use the formula $r=\sqrt{\dfrac{S}{4\pi}}$, *where* r = *the radius of a sphere and* S = *the surface area of the sphere, for Exercises 85 and 86.*

85. Find the radius of a sphere to the nearest tenth of an inch if the area is 72 square inches. 2.4 in.

86. Find the exact surface area of a sphere if its radius is 6 inches. (Do not approximate π.) 144π sq. in.

(9.7) *Write each of the following with fractional exponents and simplify if possible. Assume that variables represent nonnegative real numbers.*

87. $\sqrt{a^5}$ $\ a^{5/2}$ **88.** $\sqrt[5]{a^3}$ $\ a^{3/5}$

89. $\sqrt[6]{x^{15}}$ $\ x^{5/2}$ **90.** $\sqrt[4]{x^{12}}$ $\ x^3$

Simplify each of the following expressions.

91. $16^{1/2}$ $\ 4$ **92.** $36^{1/2}$ $\ 6$

93. $(-8)^{1/3}$ $\ -2$ **94.** $(-32)^{1/5}$ $\ -2$

95. $-64^{3/2}$ $\ -512$ **96.** $-8^{2/3}$ $\ -4$

97. $\left(\dfrac{16}{81}\right)^{3/4}$ $\ \dfrac{8}{27}$ **98.** $\left(\dfrac{9}{25}\right)^{3/2}$ $\ \dfrac{27}{125}$

99. $25^{-1/2}$ $\ \dfrac{1}{5}$ **100.** $64^{-2/3}$ $\ \dfrac{1}{16}$

Simplify each expression using positive exponents only. Assume that variables represent positive real numbers.

101. $8^{1/3}\cdot 8^{4/3}$ $\ 32$ **102.** $4^{3/2}\cdot 4^{1/2}$ $\ 16$

103. $\dfrac{3^{1/6}}{3^{5/6}}$ $\ \dfrac{1}{3^{2/3}}$ **104.** $\dfrac{2^{1/4}}{2^{-3/5}}$ $\ 2^{17/20}$

105. $(x^{-1/3})^6$ $\ \dfrac{1}{x^2}$ **106.** $\left(\dfrac{x^{1/2}}{y^{1/3}}\right)^2$ $\ \dfrac{x}{y^{2/3}}$

Simplify the following. Indicate if the expression is not a real number.

1. $\sqrt{16}$ $\ 4$ **2.** $\sqrt[3]{125}$ $\ 5$

3. $16^{3/4}$ $\ 8$ **4.** $\left(\dfrac{9}{16}\right)^{1/2}$ $\ \dfrac{3}{4}$

5. $\sqrt[4]{-81}$ not a real number **6.** $27^{-2/3}$ $\ \dfrac{1}{9}$

Simplify each radical expression. Assume that variables represent positive numbers only.

7. $\sqrt{54}$ $\ 3\sqrt{6}$ **8.** $\sqrt{92}$ $\ 2\sqrt{23}$

9. $\sqrt{3x^6}$ $\ x^3\sqrt{3}$ **10.** $\sqrt{8x^4y^7}$ $\ 2x^2y^3\sqrt{2y}$

11. $\sqrt{9x^9}$ $\ 3x^4\sqrt{x}$ **12.** $\sqrt[3]{40}$ $\ 2\sqrt[3]{5}$

13. $\sqrt[3]{8x^6y^{10}}$ $\ 2x^2y^3\sqrt[3]{y}$ **14.** $\sqrt{12}-2\sqrt{75}$ $\ -8\sqrt{3}$

15. $\sqrt{2x^2} + \sqrt[3]{54} - x\sqrt{18}$ **16.** $\sqrt{\dfrac{5}{16}} \quad \dfrac{\sqrt{5}}{4}$

17. $\sqrt[3]{\dfrac{2x^3}{27}} \quad \dfrac{x\sqrt[3]{2}}{3}$ **18.** $3\sqrt{8x} \quad 6\sqrt{2x}$

Rationalize the denominator.

19. $\sqrt{\dfrac{2}{3}} \quad \dfrac{\sqrt{6}}{3}$ **20.** $\sqrt[3]{\dfrac{5}{9}} \quad \dfrac{\sqrt[3]{15}}{3}$

21. $\sqrt[3]{\dfrac{3}{4x^2}} \quad \dfrac{\sqrt[3]{6x}}{2x}$ **22.** $\dfrac{8}{\sqrt{6}+2} \quad 4\sqrt{6}-8$

15. $3\sqrt[3]{2} - 2x\sqrt{2}$ **25.** $x = 5$

CHAPTER 9 CUMULATIVE REVIEW

1. Simplify $3[4(5 + 2) - 10]$.   54 *(Sec. 1.3, Ex. 4)*

2. Solve $2x + 3x - 5 + 7 = 10x + 3 - 6x - 4$ for x.

3. The circle graph below shows how much money homeowners in the United States spend annually on maintaining their homes. Use this graph to answer the questions below.   *(Sec. 2.7, Ex. 4)*

Yearly Home Maintenance in the U.S.

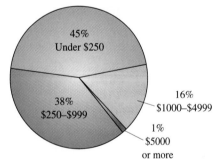

a. What percent of homeowners spend under $250 on yearly home maintenance?   45%

b. What percent of homeowners spend less than $1000 per year on maintenance?   83%

c. How many of the 22,000 homeowners in a town called Fairview might we expect to spend under $250 a year on home maintenance?

d. Find the number of degrees in the 16% sector.

23. $\dfrac{2\sqrt{3}}{\sqrt{3}-3} \quad -1 - \sqrt{3}$

Solve each of the following radical equations.

24. $\sqrt{x} + 8 = 11$ $x = 9$ **25.** $\sqrt{3x - 6} = \sqrt{x + 4}$

26. $\sqrt{2x - 2} = x - 5$ $x = 9$

27. Find the length of the unknown leg of a right triangle if the other leg is 8 inches long and the hypotenuse is 12 inches long. $4\sqrt{5}$ in.

28. Find the distance between $(-3, 6)$ and $(-2, 8)$. $\sqrt{5}$

Simplify each expression using positive exponents only.

29. $16^{-3/4} \cdot 16^{-1/4} \quad \dfrac{1}{16}$ **30.** $(x^{2/3})^5 \quad x^{10/3}$

2. $x = -3$ *(Sec. 2.2, Ex. 4)*
3c. 9900 homeowners   **3d.** 57.6°.

4. $x > \dfrac{13}{7}$;   *(Sec. 2.9, Ex. 8)*

4. Solve $-5x + 7 < 2(x - 3)$, and graph the solution set.

5. Graph $y = -3$.   See App. F *(Sec. 3.3, Ex. 7)*

6. Graph $5x + 4y \le 20$.   See App. F *(Sec. 3.5, Ex. 5)*

7. Find the product: $(3x + 2)(2x - 5)$.

8. Write the following numbers in standard notation, without exponents.   *(Sec. 4.5, Ex. 6)*

  a. 1.02×10^5 102,000 **b.** 7.358×10^{-3} 0.007358

  c. 8.4×10^7 **d.** 3.007×10^{-5}

9. Factor $-9a^5 + 18a^2 - 3a$.

10. Are there any values for x for which each expression is undefined?   *(Sec. 6.1, Ex. 2)*

  a. $\dfrac{x}{x - 3}$ $x = 3$ **b.** $\dfrac{x^2 + 2}{x^2 - 3x + 2}$

  c. $\dfrac{x^3 - 6x^2 - 10x}{3}$ none **d.** $\dfrac{2}{x^2 + 1}$ none

11. Simplify $\dfrac{4 - x^2}{3x^2 - 5x - 2}$. $-\dfrac{2 + x}{3x + 1}$ *(Sec. 6.1, Ex. 8)*

12. Perform the indicated operation.   *(Sec. 6.4, Ex. 1)*

  a. $\dfrac{a}{4} - \dfrac{2a}{8}$ 0   **b.** $\dfrac{3}{10x^2} + \dfrac{7}{25x} \quad \dfrac{15 + 14x}{50x^2}$

13. Solve the equation $\dfrac{x + 1}{x + 3} = \dfrac{1}{2x}$.

7. $6x^2 - 11x - 10$ *(Sec. 4.3, Ex. 2)*   **8c.** 84,000,000   **8d.** 0.00003007   **9.** $-3a(3a^4 - 6a + 1)$ *(Sec. 5.1, Ex.5)*
10b. $x = 1, x = 2$ **13.** $x = -\dfrac{3}{2}, x = 1$ *(Sec. 6.6, Ex. 3)*

14. $2x + y = 3$ *(Sec. 7.2, Ex. 1)*

14. Find an equation of the line passing through $(-1, 5)$ with slope -2. Write the equation in standard form: $Ax + By = C$.

15. Solve the system $\begin{cases} 3x + 4y = 13 \\ 5x - 9y = 6 \end{cases}$ by the addition method. (3, 1) *(Sec. 8.3, Ex. 4)*

16. Graph the solution of the system $\begin{cases} x - y < 2 \\ x + 2y > -1 \end{cases}$.

17. Find the cube roots. *(Sec. 9.1, Ex. 2)*

    a. $\sqrt[3]{1}$ 1   **b.** $\sqrt[3]{-27}$ -3 **c.** $\sqrt[3]{\dfrac{1}{125}}$ $\dfrac{1}{5}$

18. Simplify each expression. *(Sec. 9.2, Ex. 1)*

    a. $\sqrt{54}$ **b.** $\sqrt{12}$ **c.** $\sqrt{200}$ **d.** $\sqrt{35}$

19. Simplify each radical expression. Assume variables represent positive numbers. *(Sec. 9.3, Ex. 3)*

    a. $2\sqrt{x^2} - \sqrt{25x} + \sqrt{x}$ $2x - 4\sqrt{x}$

    b. $3\sqrt[3]{54x^4} + 5x\sqrt[3]{16x}$ $19x\sqrt[3]{2x}$

20. Rationalize each denominator. *(Sec. 9.4, Ex. 5)*

    a. $\dfrac{2}{\sqrt{7}}$ $\dfrac{2\sqrt{7}}{7}$ **b.** $\dfrac{\sqrt{5}}{\sqrt{12}}$ $\dfrac{\sqrt{15}}{6}$ **c.** $\sqrt{\dfrac{1}{18x}}$ $\dfrac{\sqrt{2x}}{6x}$

21. Solve the equation $\sqrt{x} = \sqrt{5x - 2}$.

22. A formula used to determine the velocity v, in feet per second, of an object (neglecting air resistance) after it has fallen a certain height is $v = \sqrt{2gh}$, where g is the acceleration due to gravity, and h is the height the object has fallen. On Earth, the acceleration g due to gravity is approximately 32 feet per second per second. Find the velocity of a watermelon after it has fallen 5 feet.

    $8\sqrt{5}$ ft. per sec. *(Sec. 9.6, Ex. 4)*

23. Write in radical notation. Then simplify.

    a. $25^{1/2}$ 5   **b.** $8^{1/3}$ 2   **c.** $-16^{1/4}$ -2

    d. $(-27)^{1/3}$ -3 **e.** $\left(\dfrac{1}{9}\right)^{1/2}$ $\dfrac{1}{3}$

16. See App. F *(Sec. 8.5, Ex. 2)* **18a.** $3\sqrt{6}$ **18b.** $2\sqrt{3}$ **18c.** $10\sqrt{2}$ **18d.** $\sqrt{35}$ **21.** $x = \dfrac{1}{2}$ *(Sec. 9.5, Ex. 4)*

23. *(Sec. 9.7, Ex. 1)*

SOLVING QUADRATIC EQUATIONS

MODELING A PHYSICAL SITUATION

When water comes out of a water fountain, it initially heads upward, but then the effects of gravity cause the water to fall. The curve formed by the stream of water can be modeled using a quadratic equation (parabola).

An engineer preparing a design for a new water fountain must consider many factors, such as the initial velocity at which the water leaves the fountain. Then he or she may use such a mathematical model to be sure that the water will land within the planned catchment area, if someone doesn't drink it.

IN THE CHAPTER GROUP ACTIVITY ON PAGE 584, YOU WILL HAVE THE OPPORTUNITY TO MODEL THE PARABOLIC PATH OF WATER AS IT LEAVES A DRINKING FOUNTAIN.

An important part of the study of algebra is learning to use methods for solving equations. In Chapter 2, we presented techniques for solving linear equations in one variable. In Chapter 5, we solved quadratic equations in one variable by factoring the quadratic expressions. We now present other methods for solving quadratic equations in one variable.

10.1 SOLVING QUADRATIC EQUATIONS BY THE SQUARE ROOT METHOD

O B J E C T I V E S

TAPE
BA 10.1

1 Review factoring to solve quadratic equations.

2 Use the square root property to solve quadratic equations.

1 Recall that a quadratic equation is an equation that can be written in the form

$$ax^2 + bx + c = 0$$

where a, b, and c are real numbers and $a \neq 0$.

To solve quadratic equations by factoring, use the **zero factor theorem:** If the product of two numbers is zero, then at least one of the two numbers is zero. Example 1 reviews the process of solving quadratic equations by factoring.

EXAMPLE 1   Solve $x^2 - 4 = 0$.

Solution:

$$x^2 - 4 = 0$$

$$(x + 2)(x - 2) = 0 \qquad \text{Factor.}$$

$$x + 2 = 0 \quad \text{or} \quad x - 2 = 0 \qquad \text{Apply the zero factor theorem.}$$

$$x = -2 \quad \text{or} \quad x = 2 \qquad \text{Solve each equation.}$$

The solution set is $\{-2, 2\}$.

2 Consider solving $x^2 - 4 = 0$ another way. First, add 4 to both sides of the equation.

$$x^2 - 4 = 0$$

$$x^2 = 4 \qquad \text{Add 4 to both sides.}$$

The variable x must be a number whose square is 4. Therefore, $x = 2$ or $x = -2$, written compactly as $x = \pm 2$. This reasoning is an example of the square root property.

SQUARE ROOT PROPERTY

If $X^2 = a$ for $a \geq 0$, then $X = \pm\sqrt{a}$.

EXAMPLE 2   Use the square root property to solve $x^2 - 9 = 0$.

Solution: $x^2 - 9 = 0$

$x^2 = 9$ Add 9 to both sides.

Next, apply the square root property.

$x = \pm\sqrt{9} = \pm 3$

Check to verify these solutions in the original equation. The solution set is $\{-3, 3\}$.

EXAMPLE 3   Use the square root property to solve $2x^2 = 7$.

Solution: First, solve for x^2 by dividing both sides by 2. Then apply the square root property.

$2x^2 = 7$

$x^2 = \dfrac{7}{2}$ Divide both sides by 2.

$x = \pm\sqrt{\dfrac{7}{2}}$ Use the square root property.

Next, rationalize the denominator of each solution by multiplying numerator and denominator by $\sqrt{2}$.

$x = \pm\sqrt{\dfrac{7}{2}}$

$x = \pm\dfrac{\sqrt{7}}{\sqrt{2}}$ Apply the quotient rule.

$x = \pm\dfrac{\sqrt{7} \cdot \sqrt{2}}{\sqrt{2} \cdot \sqrt{2}} = \pm\dfrac{\sqrt{14}}{2}$

Remember to check both solutions. The solution set is $\left\{-\dfrac{\sqrt{14}}{2}, \dfrac{\sqrt{14}}{2}\right\}$.

EXAMPLE 4   Use the square root property to solve $(x - 3)^2 = 16$.

Solution: To use the square root property, consider X to be $x - 3$ and a to be 16.

$(x - 3)^2 = 16$

$(x - 3) = \pm\sqrt{16}$ Use the square root property.

$x - 3 = \pm 4$ Simplify.

Now solve for x by adding 3 to both sides.

$$x = 3 \pm 4$$
$$x = 3 + 4 \quad \text{or} \quad x = 3 - 4$$
$$x = 7 \quad \text{or} \quad x = -1$$

The solution set is $\{-1, 7\}$.

EXAMPLE 5 Use the square root property to solve $(x + 1)^2 = 8$.

Solution:
$$(x + 1)^2 = 8$$
$$x + 1 = \pm\sqrt{8} \qquad \text{Apply the square root property.}$$
$$x + 1 = \pm 2\sqrt{2} \qquad \text{Simplify the radical.}$$
$$x = -1 \pm 2\sqrt{2} \qquad \text{Solve for } x.$$

The solutions are both $-1 + 2\sqrt{2}$ and $-1 - 2\sqrt{2}$. The solution set is $\{-1 + 2\sqrt{2}, -1 - 2\sqrt{2}\}$.

EXAMPLE 6 Use the square root property to solve $(x - 1)^2 = -2$.

Solution: This equation has no real solution because there is no real number whose square is -2.

EXAMPLE 7 Use the square root property to solve $(5x - 2)^2 = 10$.

Solution:
$$(5x - 2)^2 = 10$$
$$5x - 2 = \pm\sqrt{10} \qquad \text{Apply the square root property.}$$
$$5x = 2 \pm\sqrt{10} \qquad \text{Add 2.}$$
$$x = \frac{2 \pm \sqrt{10}}{5} \qquad \text{Divide by 5.}$$

The solution set is $\left\{ \dfrac{2 + \sqrt{10}}{5}, \dfrac{2 - \sqrt{10}}{5} \right\}$.

EXERCISE SET 10.1

Solve each equation by factoring. See Example 1.

1. $k^2 - 9 = 0$ $k = \pm 3$ **2.** $k^2 - 49 = 0$ $k = \pm 7$

3. $m^2 + 2m = 15$ $m = -5, 3$ **4.** $m^2 + 6m = 7$ $m = -7, 1$

5. $x = \pm\dfrac{9\sqrt{2}}{2}$

5. $2x^2 - 81 = 0$ **6.** $3p^4 - 9p^3 = 0$ $p = 0, 3$

7. $4a^2 - 36 = 0$ $a = \pm 3$ **8.** $7a^2 - 175 = 0$ $a = \pm 5$

Use the square root property to solve each quadratic equation. See Examples 2 and 3.

9. $x^2 = 64$ $x = \pm 8$ **10.** $x^2 = 121$ $x = \pm 11$

11. $p^2 = \dfrac{1}{49}$ $p = \pm\dfrac{1}{7}$ **12.** $p^2 = \dfrac{1}{16}$ $p = \pm\dfrac{1}{4}$

13. $y^2 = -36$ **14.** $y^2 = -25$

15. $2x^2 = 50$ $x = \pm 5$ **16.** $5x^2 = 20$ $x = \pm 2$

17. $3x^2 = 4$ $x = \pm\dfrac{2\sqrt{3}}{3}$ **18.** $5x^2 = 9$ $x = \pm\dfrac{3\sqrt{5}}{5}$

19. The area of a square room is 225 square feet. Find the dimensions of the room.   15 ft. by 15 ft.

13. no real solution
14. no real solution
21. $x = -2, 12$ **23.** $x = -2 \pm\sqrt{7}$
24. $x = 7 \pm\sqrt{2}$ **25.** $m = 0, 1$
26. $m = -\dfrac{2}{3}, 0$ **27.** $p = -2 \pm\sqrt{10}$

20. Explain why the equation $x^2 = -9$ has no real solution.   answers may vary

Use the square root property to solve each quadratic equation. See Examples 4 through 7.

21. $(x - 5)^2 = 49$ **22.** $(x + 2)^2 = 25$ $x = -7, 3$

23. $(x + 2)^2 = 7$ **24.** $(x - 7)^2 = 2$

25. $\left(m - \dfrac{1}{2}\right)^2 = \dfrac{1}{4}$ **26.** $\left(m + \dfrac{1}{3}\right)^2 = \dfrac{1}{9}$

27. $(p + 2)^2 = 10$ **28.** $(p - 7)^2 = 13$

29. $(3y + 2)^2 = 100$ **30.** $(4y - 3)^2 = 81$

31. $(z - 4)^2 = -9$ **32.** $(z + 7)^2 = -20$

33. $(2x - 11)^2 = 50$ **34.** $(3x - 17)^2 = 28$

35. An isosceles right triangle has legs of equal length. If the hypotenuse is 20 centimeters long, use the Pythagorean theorem to find the length of the legs.

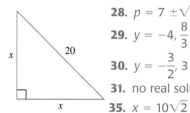

28. $p = 7 \pm\sqrt{13}$
29. $y = -4, \dfrac{8}{3}$
30. $y = -\dfrac{3}{2}, 3$
31. no real solution
35. $x = 10\sqrt{2}$ cm

36. A 27-inch-square TV is advertised in the local paper. If 27 inches is the measure of the diagonal

of the picture tube, use the Pythagorean theorem to find the measure of the side of the picture tube.

32. no real solution

33. $x = \dfrac{11 \pm 5\sqrt{2}}{2}$

34. $x = \dfrac{17 \pm 2\sqrt{7}}{3}$

39. $x = 9, 17$

$x = \dfrac{27\sqrt{2}}{2}$ in.

Solve each quadratic equation by using the square root property.

37. $q^2 = 100$ $q = \pm 10$ **38.** $q^2 = 25$ $q = \pm 5$

39. $(x - 13)^2 = 16$ **40.** $(x + 10)^2 = 25$

41. $z^2 = 12$ $z = \pm 2\sqrt{3}$ **42.** $z^2 = 18$ $z = \pm 3\sqrt{2}$

43. $(x + 5)^2 = 10$ **44.** $(x - 2)^2 = 13$

45. $m^2 = -10$ **46.** $m^2 = -25$

47. $2y^2 = 11$ **48.** $3y^2 = 13$

49. $(2p - 5)^2 = 121$ **50.** $(3p - 1)^2 = 4$

51. $(3x - 1)^2 = 7$ **52.** $(5x + 3)^2 = 3$

53. $(3x - 7)^2 = 32$ **54.** $(5x - 11)^2 = 54$

40. $x = -15, -5$ **43.** $x = -5 \pm \sqrt{10}$ **44.** $x = 2 \pm \sqrt{13}$

Solve each quadratic equation by using the square root property. Use a calculator and round each solution to the nearest hundredth.

55. $x^2 = 1.78$ $x = \pm 1.33$ **56.** $y^2 = 9.86$ $y = \pm 3.14$

57. $(x - 1.37)^2 = 5.71$ **58.** $(z + 10.68)^2 = 16.61$

59. $(2y + 1.58)^2 = 21.11$ **60.** $(5z - 5.95)^2 = 14.19$

57. $x = -1.02, 3.76$ **58.** $z = -14.76, -6.60$

Solve each quadratic equation by first factoring the perfect square trinomial on the left side. Then apply the square root property.

61. $x^2 + 4x + 4 = 16$ **62.** $z^2 - 6z + 9 = 25$

63. $y^2 - 10y + 25 = 11$ **64.** $x^2 + 14x + 49 = 31$

65. The area of a circle is found by the equation $A = \pi r^2$. If the area A of a certain circle is 36π square inches, find its radius, r. $r = 6$ in.

45. no real solution
46. no real solution
47. $y = \pm\dfrac{\sqrt{22}}{2}$
48. $y = \pm\dfrac{\sqrt{39}}{3}$

49. $p = -3, 8$ **50.** $p = -\dfrac{1}{3}, 1$ **51.** $x = \dfrac{1 \pm \sqrt{7}}{3}$ **52.** $x = \dfrac{-3 \pm \sqrt{3}}{5}$

53. $x = \dfrac{7 \pm 4\sqrt{2}}{3}$ **54.** $x = \dfrac{11 \pm 3\sqrt{6}}{5}$ **59.** $y = -3.09, 1.51$ **60.** $z = 0.44, 1.94$ **61.** $x = -6, 2$

62. $z = -2, 8$ **63.** $y = 5 \pm \sqrt{11}$ **64.** $x = -7 \pm \sqrt{31}$

66. Neglecting air resistance, the distance d in feet that an object falls in t seconds is given by the equation

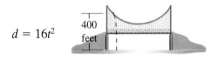

$$d = 16t^2$$

If a sandblaster drops his goggles from a bridge 400 feet from the water below, find how long it takes for the goggles to hit the water. 5 sec.

Review Exercises

Factor each perfect square trinomial. See Section 5.2.

67. $x^2 + 6x + 9$ $(x + 3)^2$ **68.** $y^2 + 10y + 25$ $(y + 5)^2$

69. $x^2 - 4x + 4$ $(x - 2)^2$ **70.** $x^2 - 20x + 100$ $(x - 10)^2$

The following graph shows the number of U.S. households with home offices—those used by corporate workers for their job and those for home business. Use this graph for Exercises 71 through 73. See Section 7.2.

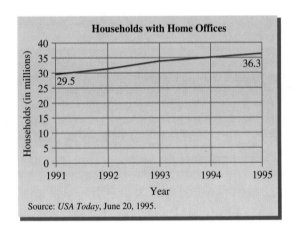

71. Estimate the number of home offices in 1994.

72. The growth of home offices has been almost linear since 1991. Approximate this growth with a linear equation. To do so, find an equation of the line through the ordered pairs $(0, 29.5)$ and $(4, 36.3)$ where x is the number of years since 1991 and y is the number of households (in millions) that have home offices. Write the equation in slope-intercept form. $y = 1.7x + 29.5$

73. Use the equation found in Exercise 72 to predict the number of home offices in the year 2000.

71. 35 million **73.** 44.8 million

10.2 SOLVING QUADRATIC EQUATIONS BY COMPLETING THE SQUARE

O B J E C T I V E S

 Find perfect square trinomials.

Solve quadratic equations by completing the square.

TAPE
BA 10.2

In the last section, we solved equations such as

$$(x + 1)^2 = 8 \quad \text{and} \quad (5x - 2)^2 = 3$$

Thus, if we can write a quadratic equation in the same form as these equations, we can then solve using the square root property. Notice that one side of each equation is a quantity squared and that the other side is a constant. For example, to solve $x^2 + 2x - 4 = 0$, add 4 to both sides and get

$$x^2 + 2x = 4$$

Next, notice that if we add 1 to both sides of the equation the left side is a perfect square trinomial that can be factored.

$$x^2 + 2x + 1 = 4 + 1$$
$$(x + 1)^2 = 5$$

Now solve this equation as we did in the previous section by using the square root property.

$$(x + 1)^2 = 5$$
$$x + 1 = \pm\sqrt{5}$$
$$x = -1 \pm \sqrt{5}$$

Adding a number to $x^2 + 2x$ to form a perfect square trinomial is called **completing the square** on $x^2 + 2x$.

In general, we have the following:

> To complete the square on $x^2 + bx$, add $\left(\dfrac{b}{2}\right)^2$, which is **the square of half of the coefficient of x.**

EXAMPLE 1   Complete the square for each expression and then factor the resulting perfect square trinomial.

a. $x^2 + 10x$ 　　　　　　**b.** $m^2 - 6m$ 　　　　　　**c.** $x^2 + x$

Solution: **a.** The coefficient of the x-term is 10. Half of 10 is 5, and $5^2 = 25$. Add 25.

$$x^2 + 10x + 25 = (x + 5)^2$$

b. Half the coefficient of m is -3, and $(-3)^2$ is 9. Add 9.

$$m^2 - 6m + 9 = (m - 3)^2$$

c. Half the coefficient of x is $\dfrac{1}{2}$ and $\left(\dfrac{1}{2}\right)^2 = \dfrac{1}{4}$. Add $\dfrac{1}{4}$.

$$x^2 + x + \frac{1}{4} = \left(x + \frac{1}{2}\right)^2$$

By completing the square, a quadratic equation can be solved using the square root property.

EXAMPLE 2   Complete the square to solve the quadratic equation $x^2 + 6x + 3 = 0$.

Solution: Isolate the variable terms by subtracting 3 from both sides of the equation.

$$x^2 + 6x + 3 = 0$$
$$x^2 + 6x = -3$$

Next, add the square of half the coefficient of the x-term to *both sides* of the equation so that the left side becomes a perfect square trinomial. The coefficient of x is 6, and half of 6 is 3. Add 3^2 or 9 to both sides.

$$x^2 + 6x + 9 = -3 + 9 \qquad \text{Complete the square.}$$
$$(x + 3)^2 = 6 \qquad \text{Factor the trinomial } x^2 + 6x + 9.$$
$$x + 3 = \pm\sqrt{6} \qquad \text{Apply the square root property.}$$
$$x = -3 \pm \sqrt{6} \qquad \text{Subtract 3 from both sides.}$$

The solution set is $\{-3 - \sqrt{6}, -3 + \sqrt{6}\}$.

> **REMINDER**   Remember, when solving a quadratic equation by completing the square, add the number that completes the square to **both sides of the equation.**

EXAMPLE 3   Complete the square to solve the quadratic equation $y^2 - 10y = -14$.

Solution: The variable terms are already isolated on one side of the equation. The coefficient of y is -10. Half of -10 is -5, and $(-5)^2 = 25$. Add 25 to both sides.

$$y^2 - 10y = -14$$
$$y^2 - 10y + 25 = -14 + 25$$
$$(y - 5)^2 = 11 \qquad \text{Factor the trinomial and simplify } -14 + 25.$$
$$y - 5 = \pm\sqrt{11} \qquad \text{Apply the square root property.}$$
$$y = 5 \pm \sqrt{11} \qquad \text{Add 5 to both sides.}$$

The solution set is $\{5 - \sqrt{11}, 5 + \sqrt{11}\}$.

When the coefficient of the squared variable is not 1, first divide both sides of the equation by the coefficient of the squared variable so that the coefficient is 1. Then complete the square.

EXAMPLE 4   Solve the quadratic equation $4x^2 - 8x - 5 = 0$ by completing the square.

Solution: $4x^2 - 8x - 5 = 0$

$$x^2 - 2x - \frac{5}{4} = 0 \qquad \text{Divide by 4.}$$

$$x^2 - 2x = \frac{5}{4} \qquad \text{Isolate variable terms.}$$

The coefficient of x is -2. Half of -2 is -1, and $(-1)^2 = 1$. Add 1 to both sides.

$$x^2 - 2x + 1 = \frac{5}{4} + 1$$

$$(x - 1)^2 = \frac{9}{4} \qquad \text{Factor } x^2 - 2x + 1 \text{ and simplify } \frac{5}{4} + 1.$$

$$x - 1 = \pm\sqrt{\frac{9}{4}} \qquad \text{Apply the square root property.}$$

$$x = 1 \pm \frac{3}{2} \qquad \text{Add 1 to both sides and simplify the radical.}$$

$$x = 1 + \frac{3}{2} = \frac{5}{2} \quad \text{or} \quad x = 1 - \frac{3}{2} = -\frac{1}{2}.$$

The solution set is $\left\{-\frac{1}{2}, \frac{5}{2}\right\}$.

The following steps may be used to solve a quadratic equation in x by completing the square.

TO SOLVE A QUADRATIC EQUATION IN x BY COMPLETING THE SQUARE

Step 1.   If the coefficient of x^2 is 1, go to step 2. If not, divide both sides of the equation by the coefficient of x^2.

Step 2.   Isolate all terms with variables on one side of the equation.

Step 3.   Complete the square on the resulting binomial expression by adding the square of half of the coefficient of x to both sides of the equation.

Step 4.   Factor the resulting perfect square trinomial.

Step 5.   Apply the square root property to solve the equation.

EXAMPLE 5   Solve the quadratic equation $2y^2 + 6y = -7$ by completing the square.

Solution: The coefficient of y^2 is not 1. Divide both sides by 2, the coefficient of y^2.

$$2y^2 + 6y = -7$$

$$y^2 + 3y = -\frac{7}{2} \qquad \text{Divide by 2.}$$

$$y^2 + 3y + \frac{9}{4} = -\frac{7}{2} + \frac{9}{4} \qquad \text{Add } \left(\frac{3}{2}\right)^2 \text{ or } \frac{9}{4} \text{ to both sides.}$$

$$\left(y + \frac{3}{2}\right)^2 = -\frac{5}{4} \qquad \text{Factor the left side and simplify the right.}$$

There is no real solution to this equation, since the square of a real number cannot be negative.

EXAMPLE 6   Solve the quadratic equation $2x^2 = 10x + 1$ by completing the square.

Solution: First, divide both sides of the equation by 2, the coefficient of x^2.

$$2x^2 = 10x + 1$$

$$x^2 = 5x + \frac{1}{2} \qquad \text{Divide by 2.}$$

Next, isolate the variable terms by subtracting $5x$ from both sides.

$$x^2 - 5x = \frac{1}{2}$$

$$x^2 - 5x + \frac{25}{4} = \frac{1}{2} + \frac{25}{4} \qquad \text{Add } \left(\frac{-5}{2}\right)^2 \text{ or } \frac{25}{4} \text{ to both sides.}$$

$$\left(x - \frac{5}{2}\right)^2 = \frac{27}{4} \qquad \text{Factor the left side and simplify the right.}$$

$$x - \frac{5}{2} = \pm\sqrt{\frac{27}{4}} \qquad \text{Apply the square root property.}$$

$$x - \frac{5}{2} = \pm\frac{3\sqrt{3}}{2} \qquad \text{Simplify.}$$

$$x = \frac{5}{2} \pm \frac{3\sqrt{3}}{2} = \frac{5 \pm 3\sqrt{3}}{2}$$

The solution set is $\left\{\dfrac{5 - 3\sqrt{3}}{2}, \dfrac{5 + 3\sqrt{3}}{2}\right\}$.

MENTAL MATH

Determine the number to add to make each expression a perfect square trinomial. See Example 1.

1. $p^2 + 8p$ 16

2. $p^2 + 6p$ 9

3. $x^2 + 20x$ 100

4. $x^2 + 18x$ 81

5. $y^2 + 14y$ 49

6. $y^2 + 2y$ 1

27. $x = 1 \pm \sqrt{2}$ **28.** $x = 2 \pm \sqrt{2}$ **29.** $y = -4, -1$

30. $y = 2, 3$ **32.** $x = -5, -4$ **33.** $y = \dfrac{-4 \pm \sqrt{6}}{2}$

EXERCISE SET 10.2

5. $\left(x - \dfrac{3}{2}\right)^2$ **6.** $\left(x - \dfrac{5}{2}\right)^2$

Complete the square for each expression and then factor the resulting perfect square trinomial. See Example 1.

1. $x^2 + 4x$ $(x + 2)^2$

2. $x^2 + 6x$ $(x + 3)^2$

3. $k^2 - 12k$ $(k - 6)^2$

4. $k^2 - 16k$ $(k - 8)^2$

5. $x^2 - 3x$

6. $x^2 - 5x$

7. $m^2 - m$ $\left(m - \dfrac{1}{2}\right)^2$

8. $y^2 + y$ $\left(y + \dfrac{1}{2}\right)^2$

Solve each quadratic equation by completing the square. See Examples 2 and 3.   **11.** $x = -6, -2$ **12.** $x = 4, 6$

9. $x^2 - 6x = 0$ $x = 0, 6$

10. $y^2 + 4y = 0$ $y = -4, 0$

11. $x^2 + 8x = -12$

12. $x^2 - 10x = -24$

13. $x^2 + 2x - 5 = 0$

14. $z^2 + 6z - 9 = 0$

15. Find a value of k that will make $x^2 + kx + 16$ a perfect square trinomial. $k = 8$ or $k = -8$

16. Find a value of k that will make $x^2 + kx + 25$ a perfect square trinomial. $k = 10$ or $k = -10$

13. $x = -1 \pm \sqrt{6}$ **14.** $z = -3 \pm 3\sqrt{2}$

Solve each quadratic equation by completing the square. See Examples 4 through 6.   **18.** $x = -5, 1$

17. $4x^2 - 24x = 13$

18. $2x^2 + 8x = 10$

19. $5x^2 + 10x + 6 = 0$

20. $3x^2 - 12x + 14 = 0$

21. $2x^2 = 6x + 5$

22. $4x^2 = -20x + 3$

Solve each quadratic equation by completing the square.

23. $x^2 + 6x - 25 = 0$

24. $x^2 - 6x + 7 = 0$

25. $z^2 + 5z = 7$

26. $x^2 - 7x = 5$

27. $x^2 - 2x - 1 = 0$

28. $x^2 - 4x + 2 = 0$

29. $y^2 + 5y + 4 = 0$

30. $y^2 - 5y + 6 = 0$

31. $3x^2 - 6x = 24$ $x = -2, 4$ **32.** $2x^2 + 18x = -40$

33. $2y^2 + 8y + 5 = 0$

34. $3z^2 + 6z + 4 = 0$

35. $2y^2 - 3y + 1 = 0$

36. $2y^2 - y - 1 = 0$

17. $x = -\dfrac{1}{2}, \dfrac{13}{2}$ **19.** no real solution   **20.** no real solution   **21.** $x = \dfrac{3 \pm \sqrt{19}}{2}$

22. $x = -\dfrac{5}{2} \pm \sqrt{7}$ **23.** $x = -3 \pm \sqrt{34}$ **24.** $x = 3 \pm \sqrt{2}$ **25.** $z = \dfrac{-5 \pm \sqrt{53}}{2}$ **26.** $x = \dfrac{7 \pm \sqrt{69}}{2}$

34. no real solution   **35.** $y = \dfrac{1}{2}, 1$ **36.** $y = -\dfrac{1}{2}, 1$

37. $y = \dfrac{1 \pm \sqrt{13}}{3}$ **38.** $y = \dfrac{1 \pm \sqrt{13}}{4}$ **40.** $y = -2, 5$

37. $3y^2 - 2y - 4 = 0$

38. $4y^2 - 2y - 3 = 0$

39. $y^2 = 5y + 14$ $y = -2, 7$ **40.** $y^2 = 3y + 10$

41. $x(x + 3) = 18$ $x = -6, 3$ **42.** $x(x - 3) = 18$

43. In your own words, describe a perfect square trinomial.   answers may vary   **42.** $x = -3, 6$

44. Describe how to find the number to add to $x^2 - 7x$ to make a perfect square trinomial.
answers may vary

Recall that a graphing calculator may be used to solve an equation. For example, to solve $x^2 + 8x = -12$ (Exercise 11), graph

$$y_1 = x^2 + 8x \quad \text{(left side of equation) and}$$
$$y_2 = -12 \quad \text{(right side of equation)}$$

The x-coordinate of the point of intersection of the graphs is the solution. Use a graphing calculator and solve each equation. Round solutions to the nearest hundredth.   **45.** $x = -6, -2$ **47.** $x \approx -0.68, 3.68$

45. Exercise 11

46. Exercise 12 $x = 4, 6$

47. Exercise 21

48. Exercise 26
$x \approx -0.65, 7.65$

Review Exercises

Simplify each expression. See Section 9.2.

49. $\dfrac{3}{4} - \sqrt{\dfrac{25}{16}}$ $-\dfrac{1}{2}$

50. $\dfrac{3}{5} + \sqrt{\dfrac{16}{25}}$ $\dfrac{7}{5}$

51. $\dfrac{1}{2} - \sqrt{\dfrac{9}{4}}$ -1

52. $\dfrac{9}{10} - \sqrt{\dfrac{49}{100}}$ $\dfrac{1}{5}$

Simplify each expression. See Section 9.4.

53. $\dfrac{6 + 4\sqrt{5}}{2}$ $3 + 2\sqrt{5}$

54. $\dfrac{10 - 20\sqrt{3}}{2}$ $5 - 10\sqrt{3}$

55. $\dfrac{3 - 9\sqrt{2}}{6}$ $\dfrac{1 - 3\sqrt{2}}{2}$

56. $\dfrac{12 - 8\sqrt{7}}{16}$ $\dfrac{3 - 2\sqrt{7}}{4}$

10.3 | SOLVING QUADRATIC EQUATIONS BY THE QUADRATIC FORMULA

TAPE
BA 10.3

O B J E C T I V E

1 Use the quadratic formula to solve quadratic equations.

We can use the technique of completing the square to develop a formula to find solutions of any quadratic equation. We develop and use the **quadratic formula** in this section.

Recall that a quadratic equation in **standard form** is

$$ax^2 + bx + c = 0, \text{ providing } a \neq 0$$

To develop and use the quadratic formula, we need to practice identifying the values of a, b, and c in a quadratic equation.

QUADRATIC EQUATIONS IN STANDARD FORM	
$5x^2 - 6x + 2 = 0$	$a = 5, b = -6, c = 2$
$4y^2 - 9 = 0$	$a = 4, b = 0, c = -9$
$x^2 + x = 0$	$a = 1, b = 1, c = 0$
$\sqrt{2}x^2 + \sqrt{5}x + \sqrt{3} = 0$	$a = \sqrt{2}, b = \sqrt{5}, c = \sqrt{3}$

To derive the quadratic formula, we complete the square for the general quadratic equation in standard form.

$$ax^2 + bx + c = 0$$

First, divide both sides of the equation by the coefficient of x^2 and then isolate the variable terms.

$$x^2 + \frac{b}{a}x + \frac{c}{a} = 0 \qquad \text{Divide by } a; \text{ recall that } a \text{ cannot be 0.}$$

$$x^2 + \frac{b}{a}x = -\frac{c}{a} \qquad \text{Isolate the variable terms.}$$

The coefficient of x is $\frac{b}{a}$. Half of $\frac{b}{a}$ is $\frac{b}{2a}$ and $\left(\frac{b}{2a}\right)^2 = \frac{b^2}{4a^2}$. Add $\frac{b^2}{4a^2}$ to both sides of the equation.

$$x^2 + \frac{b}{a}x + \frac{b^2}{4a^2} = -\frac{c}{a} + \frac{b^2}{4a^2} \qquad \text{Add } \frac{b^2}{4a^2} \text{ to both sides.}$$

$$\left(x + \frac{b}{2a}\right)^2 = -\frac{c}{a} + \frac{b^2}{4a^2} \qquad \text{Factor the left side.}$$

$$\left(x + \frac{b}{2a}\right)^2 = -\frac{4ac}{4a^2} + \frac{b^2}{4a^2} \qquad \text{Multiply } -\frac{c}{a} \text{ by } \frac{4a}{4a} \text{ so that both terms on the right side have a common denominator.}$$

$$\left(x + \frac{b}{2a}\right)^2 = \frac{b^2 - 4ac}{4a^2}$$ Simplify the right side.

Now use the square root property.

$$x + \frac{b}{2a} = \pm \sqrt{\frac{b^2 - 4ac}{4a^2}}$$ Apply the square root property.

$$x + \frac{b}{2a} = \frac{\pm\sqrt{b^2 - 4ac}}{2a}$$ Simplify the radical.

$$x = -\frac{b}{2a} \pm \frac{\sqrt{b^2 - 4ac}}{2a}$$ Subtract $\frac{b}{2a}$ from both sides.

$$= \frac{-b \pm \sqrt{b^2 - 4ac}}{2a}$$ Simplify.

This final equation is called the **quadratic formula** and gives the solutions of any quadratic equation.

Quadratic Formula

If a, b, and c are real numbers and $a \neq 0$, a quadratic equation written in the form $ax^2 + bx + c = 0$ has solutions

$$x = \frac{-b \pm \sqrt{b^2 - 4ac}}{2a}$$

EXAMPLE 1   Use the quadratic formula to solve $3x^2 + x - 3 = 0$.

Solution: This equation is in standard form with $a = 3$, $b = 1$, and $c = -3$. By the quadratic formula,

$$x = \frac{-b \pm \sqrt{b^2 - 4ac}}{2a}$$

$$x = \frac{-1 \pm \sqrt{1^2 - 4 \cdot 3 \cdot (-3)}}{2 \cdot 3}$$ Let $a = 3$, $b = 1$, and $c = -3$.

$$= \frac{-1 \pm \sqrt{1 + 36}}{6}$$ Simplify.

$$= \frac{-1 \pm \sqrt{37}}{6}$$

The solution set is $\left\{ \dfrac{-1 + \sqrt{37}}{6}, \dfrac{-1 - \sqrt{37}}{6} \right\}$.

EXAMPLE 2   Use the quadratic formula to solve $2x^2 - 9x = 5$.

Solution: First, write the equation in standard form by subtracting 5 from both sides.

$$2x^2 - 9x = 5$$
$$2x^2 - 9x - 5 = 0$$

Next, $a = 2$, $b = -9$, and $c = -5$. Substitute these values into the quadratic formula.

$$x = \frac{-b \pm \sqrt{b^2 - 4ac}}{2a}$$

$$x = \frac{-(-9) \pm \sqrt{(-9)^2 - 4 \cdot 2 \cdot (-5)}}{2 \cdot 2} \qquad \text{Substitute in the formula.}$$

$$= \frac{9 \pm \sqrt{81 + 40}}{4} \qquad \text{Simplify.}$$

$$= \frac{9 \pm \sqrt{121}}{4} = \frac{9 \pm 11}{4}$$

Then, $x = \dfrac{9 - 11}{4} = -\dfrac{1}{2}$ or $x = \dfrac{9 + 11}{4} = 5$

Check by substituting $-\frac{1}{2}$ and 5 into the original equation. The solution set is $\left\{ -\frac{1}{2}, 5 \right\}$.

In the example above, the radicand of $\sqrt{121}$ is a perfect square, so the square root is rational.

TO SOLVE A QUADRATIC EQUATION BY THE QUADRATIC FORMULA

Step 1.   Write the quadratic equation in standard form: $ax^2 + bx + c = 0$.

Step 2.   If necessary, clear the equation of fractions to simplify calculations.

Step 3   Identify a, b, and c.

Step 4.   Replace a, b, and c in the quadratic formula by known values, and simplify.

EXAMPLE 3   Use the quadratic formula to solve $7x^2 = 1$.

Solution: Write the equation in standard form by subtracting 1 from both sides.

$$7x^2 = 1$$
$$7x^2 - 1 = 0$$

Next, replace a, b, and c with values: $a = 7, b = 0, c = -1$.

$$x = \frac{0 \pm \sqrt{0^2 - 4 \cdot 7 \cdot (-1)}}{2 \cdot 7}$$ Substitute in the formula.

$$= \frac{\pm\sqrt{28}}{14}$$ Simplify.

$$= \frac{\pm 2\sqrt{7}}{14}$$

$$= \pm\frac{\sqrt{7}}{7}$$

The solution set is $\left\{-\dfrac{\sqrt{7}}{7}, \dfrac{\sqrt{7}}{7}\right\}$.

EXAMPLE 4 Use the quadratic formula to solve $x^2 = -x - 1$.

Solution: First, write the equation in standard form.

$$x^2 + x + 1 = 0$$

Next, replace a, b, and c in the quadratic formula by $a = 1, b = 1$, and $c = 1$.

$$x = \frac{-1 \pm \sqrt{1^2 - 4 \cdot 1 \cdot 1}}{2 \cdot 1}$$ Substitute in the formula.

$$= \frac{-1 \pm \sqrt{-3}}{2}$$ Simplify.

There is no real number solution, because the radicand is negative.

EXAMPLE 5 Use the quadratic formula to solve $\dfrac{1}{2}x^2 - x = 2$.

Solution: Write the equation in standard form and then clear the equation of fractions by multiplying both sides by the LCD 2.

$$\frac{1}{2}x^2 - x = 2$$

$$\frac{1}{2}x^2 - x - 2 = 0$$ Write in standard form.

$$x^2 - 2x - 4 = 0$$ Multiply both sides by 2.

Here, $a = 1$, $b = -2$, and $c = -4$. Substitute these values into the quadratic formula.

$$x = \frac{-(-2) \pm \sqrt{(-2)^2 - 4 \cdot 1 \cdot (-4)}}{2 \cdot 1}$$

$$= \frac{2 \pm \sqrt{20}}{2} = \frac{2 \pm 2\sqrt{5}}{2} \qquad \text{Simplify.}$$

$$= \frac{2(1 \pm \sqrt{5})}{2} = 1 \pm \sqrt{5} \qquad \text{Factor and simplify.}$$

The solution set is $\{1 - \sqrt{5}, 1 + \sqrt{5}\}$.

R E M I N D E R   When simplifying expressions such as

$$\frac{3 \pm 6\sqrt{2}}{6}$$

first factor out a common factor from the terms of the numerator and then simplify.

$$\frac{3 \pm 6\sqrt{2}}{6} = \frac{3(1 \pm 2\sqrt{2})}{2 \cdot 3} = \frac{1 \pm 2\sqrt{2}}{2}$$

---

1. $a = 2, b = 5, c = 3$ **2.** $a = 5, b = -7, c = 1$
3. $a = 10, b = -13, c = -2$ **4.** $a = 1, b = 3, c = -7$

MENTAL MATH

Identify the value of a, b, and c in each quadratic equation.

1. $2x^2 + 5x + 3 = 0$

2. $5x^2 - 7x + 1 = 0$

3. $10x^2 - 13x - 2 = 0$

4. $x^2 + 3x - 7 = 0$

5. $x^2 - 6 = 0$ $a = 1, b = 0, c = -6$

6. $9x^2 - 4 = 0$ $a = 9, b = 0, c = -4$

---

7. $x = 1, 2$ **8.** $x = -1, 6$ **9.** $k = \dfrac{-7 \pm \sqrt{37}}{6}$ **10.** $k = \dfrac{-3 \pm \sqrt{37}}{14}$ **11.** $x = \pm\dfrac{2}{7}$ **12.** $x = \pm\dfrac{\sqrt{15}}{5}$

13. no real solution   **14.** no real solution   **15.** $y = -3, 10$ **16.** $y = -4, 9$ **19.** $m = -3, 4$ **20.** $m = -2, 7$

21. $x = -2 \pm \sqrt{7}$ **22.** $x = -1 \pm \sqrt{11}$

EXERCISE SET 10.3

Simplify the following.

1. $\dfrac{-1 \pm \sqrt{1^2 - 4(1)(-2)}}{2(1)}$ $-2, 1$

2. $\dfrac{-(-5) \pm \sqrt{(-5)^2 - 4(2)(3)}}{2(2)}$ $1, \dfrac{3}{2}$

3. $\dfrac{-5 \pm \sqrt{5^2 - 4(1)(2)}}{2(1)}$ $\dfrac{-5 \pm \sqrt{17}}{2}$

4. $\dfrac{-7 \pm \sqrt{7^2 - 4(2)(1)}}{2(2)}$ $\dfrac{-7 \pm \sqrt{41}}{4}$

5. $\dfrac{-(-4) \pm \sqrt{(-4)^2 - 4(2)(1)}}{2(2)}$ $\dfrac{2 \pm \sqrt{2}}{2}$

6. $\dfrac{-6 \pm \sqrt{6^2 - 4(3)(1)}}{2(3)}$ $-1 \pm \dfrac{\sqrt{6}}{3}$

Use the quadratic formula to solve each quadratic equation. See Examples 1 through 4.

7. $x^2 - 3x + 2 = 0$ **8.** $x^2 - 5x - 6 = 0$

9. $3k^2 + 7k + 1 = 0$ **10.** $7k^2 + 3k - 1 = 0$

44. $y = -1, \dfrac{4}{5}$ **45.** $y = -\dfrac{3}{4}, \dfrac{1}{5}$ **46.** $z = -1, \dfrac{3}{2}$ **47.** no real solution   **48.** no real solution

11. $49x^2 - 4 = 0$ **12.** $25x^2 - 15 = 0$

13. $5z^2 - 4z + 3 = 0$ **14.** $3z^2 + 2x + 1 = 0$

15. $y^2 = 7y + 30$ **16.** $y^2 = 5y + 36$

17. $2x^2 = 10$ $x = \pm\sqrt{5}$ **18.** $5x^2 = 15$ $x = \pm\sqrt{3}$

19. $m^2 - 12 = m$ **20.** $m^2 - 14 = 5m$

21. $3 - x^2 = 4x$ **22.** $10 - x^2 = 2x$

23. no real solution   **24.** no real solution

25. $m = 1 \pm \sqrt{2}$ **26.** $m = 3 \pm \sqrt{7}$

Use the quadratic formula to solve each quadratic equation. See Example 5.

23. $3p^2 - \dfrac{2}{3}p + 1 = 0$ **24.** $\dfrac{5}{2}p^2 - p + \dfrac{1}{2} = 0$

25. $\dfrac{m^2}{2} = m + \dfrac{1}{2}$ **26.** $\dfrac{m^2}{2} = 3m - 1$

27. $4p^2 + \dfrac{3}{2} = -5p$ **28.** $4p^2 + \dfrac{3}{2} = 5p$

27. $p = -\dfrac{3}{4}, -\dfrac{1}{2}$ **28.** $p = \dfrac{1}{2}, \dfrac{3}{4}$ **29.** $a = \dfrac{1}{2}, 3$

Use the quadratic formula to solve each quadratic equation.

29. $2a^2 - 7a + 3 = 0$ **30.** $3a^2 - 7a + 2 = 0$

31. $x^2 - 5x - 2 = 0$ **32.** $x^2 - 2x - 5 = 0$

33. $3x^2 - x - 14 = 0$ **34.** $5x^2 - 13x - 6 = 0$

35. $6x^2 + 9x = 2$ **36.** $3x^2 - 9x = 8$

37. $7p^2 + 2 = 8p$ **38.** $11p^2 + 2 = 10p$

39. $a^2 - 6a + 2 = 0$ **40.** $a^2 - 10a + 19 = 0$

41. $2x^2 - 6x + 3 = 0$ **42.** $5x^2 - 8x + 2 = 0$

43. $3x^2 = 1 - 2x$ **44.** $5y^2 = 4 - x$

45. $20y^2 = 3 - 11y$ **46.** $2z^2 = z + 3$

47. $x^2 + x + 1 = 0$ **48.** $k^2 + 2k + 5 = 0$

49. $4y^2 = 6y + 1$ **50.** $6z^2 + 3z + 2 = 0$

51. $5x^2 = \dfrac{7}{2}x + 1$ **52.** $2x^2 = \dfrac{5}{2}x + \dfrac{7}{2}$

53. $28x^2 + 5x + \dfrac{11}{4} = 0$ **54.** $\dfrac{2}{3}x^2 - 2x - \dfrac{2}{3} = 0$

55. $5z^2 - 2z = \dfrac{1}{5}$ **56.** $9z^2 + 12z = -1$

57. $x^2 + 3\sqrt{2}x - 5 = 0$ **58.** $y^2 - 2\sqrt{5}x - 1 = 0$

32. $x = 1 \pm \sqrt{6}$ **33.** $x = -2, \dfrac{7}{3}$ **34.** $x = -\dfrac{2}{5}, 3$

Use the quadratic formula and a calculator to solve each equation. Round solutions to the nearest tenth.

59. $x^2 + x = 15$ **60.** $y^2 - y = 11$

35. $x = \dfrac{-9 \pm \sqrt{129}}{12}$ **36.** $x = \dfrac{9 \pm \sqrt{177}}{6}$ **37.** $p = \dfrac{4 \pm \sqrt{2}}{7}$ **38.** $p = \dfrac{5 \pm \sqrt{3}}{11}$ **39.** $a = 3 \pm \sqrt{7}$

40. $a = 5 \pm \sqrt{6}$ **41.** $x = \dfrac{3 \pm \sqrt{3}}{2}$ **42.** $x = \dfrac{4 \pm \sqrt{6}}{5}$ **43.** $x = -1, \dfrac{1}{3}$

61. $1.2x^2 - 5.2x - 3.9 = 0$ $x = -0.7, 5.0$

62. $7.3z^2 + 5.4z - 1.1 = 0$ $z = -0.9, 0.2$

A rocket is launched from the top of an 80-foot cliff with an initial velocity of 120 feet per second. The height of the rocket h after t seconds is given by the equation

$$h = -16t^2 + 120t + 80$$

63. How long after the rocket is launched will it be 30 feet from the ground? Round to the nearest tenth of a second.   7.9 sec.

64. How long after the rocket is launched will it strike the ground? Round to the nearest tenth of a second. (*Hint:* The rocket will strike the ground when its height $h = 0$.)   8.1 sec.

49. $y = \dfrac{3 \pm \sqrt{13}}{4}$

50. no real solution

51. $x = \dfrac{7 \pm \sqrt{129}}{20}$

52. $x = \dfrac{5 \pm \sqrt{137}}{8}$

53. no real solution

54. $x = \dfrac{3 \pm \sqrt{13}}{2}$

55. $z = \dfrac{1 \pm \sqrt{2}}{5}$

56. $z = \dfrac{-2 \pm \sqrt{3}}{3}$

57. $x = \dfrac{-3\sqrt{2} \pm \sqrt{38}}{2}$

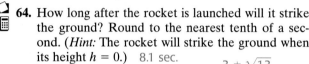

80 feet

65. Explain how the quadratic formula is derived and why it is useful.   answers may vary

58. $y = \sqrt{5} \pm \sqrt{6}$ **59.** $x = -4.4, 3.4$ **60.** $y = -2.9, 3.9$

Review Exercises

Solve the following linear equations. See Section 2.4.

66. $\dfrac{7x}{2} = 3$ $x = \dfrac{6}{7}$ **67.** $\dfrac{5x}{3} = 1$ $x = \dfrac{3}{5}$

68. $\dfrac{5}{7}x - \dfrac{2}{3} = 0$ $x = \dfrac{14}{15}$ **69.** $\dfrac{6}{11}x + \dfrac{1}{5} = 0$ $x = -\dfrac{11}{30}$

70. $\dfrac{3}{4}z + 3 = 0$ $z = -4$ **71.** $\dfrac{5}{2}z + 10 = 0$ $z = -4$

TAPE
BA 10.4

10.4 | SUMMARY OF METHODS FOR SOLVING QUADRATIC EQUATIONS

O B J E C T I V E

 Review methods for solving quadratic equations.

 An important skill in mathematics is learning when to use one technique in favor of another. We now practice this by deciding which method to use when solving quadratic equations. Although both the quadratic formula and completing the square can be used to solve any quadratic equation, the quadratic formula is usually less tedious and thus preferred. The following steps may be used to solve a quadratic equation.

TO SOLVE A QUADRATIC EQUATION

Step 1.   If the equation is in the form $(ax + b)^2 = c$, use the square root property and solve. If not, go to *step 2.*

Step 2.   Write the equation in standard form: $ax^2 + bx + c = 0$.

Step 3.   Try to solve the equation by the factoring method. If not possible, go to *step 4.*

Step 4.   Solve the equation by the quadratic formula.

EXAMPLE 1   Solve $m^2 - 2m - 7 = 0$.

Solution: The equation is in standard form, but the quadratic expression $m^2 - 2m - 7$ is not factorable, so use the quadratic formula with $a = 1$, $b = -2$, and $c = -7$.

$$m^2 - 2m - 7 = 0$$

$$m = \frac{-(-2) \pm \sqrt{(-2)^2 - 4 \cdot 1 \cdot (-7)}}{2 \cdot 1} = \frac{2 \pm \sqrt{32}}{2}$$

$$m = \frac{2 \pm 4\sqrt{2}}{2} = \frac{2(1 \pm 2\sqrt{2})}{2} = 1 \pm 2\sqrt{2}$$

The solution set is $\{1 - 2\sqrt{2}, 1 + 2\sqrt{2}\}$.

EXAMPLE 2   Solve $(3x + 1)^2 = 20$.

Solution: This equation is in a form that makes the square root property easy to apply.

$$(3x + 1)^2 = 20$$
$$3x + 1 = \pm\sqrt{20} \qquad \text{Apply the square root property.}$$
$$3x + 1 = \pm 2\sqrt{5} \qquad \text{Simplify } \sqrt{20}.$$
$$3x = -1 \pm 2\sqrt{5}$$
$$x = \frac{-1 \pm 2\sqrt{5}}{3}$$

The solution set is $\left\{ \dfrac{-1 - 2\sqrt{5}}{3}, \dfrac{-1 + 2\sqrt{5}}{3} \right\}$.

EXAMPLE 3 Solve $x^2 - \dfrac{11}{2}x = -\dfrac{5}{2}$.

Solution: The fractions make factoring more difficult and also complicate the calculations for using the quadratic formula. Clear the equation of fractions by multiplying both sides of the equation by the LCD 2.

$$x^2 - \frac{11}{2}x = -\frac{5}{2}$$

$$x^2 - \frac{11}{2}x + \frac{5}{2} = 0 \qquad \text{Write in standard form.}$$

$$2x^2 - 11x + 5 = 0 \qquad \text{Multiply both sides by 2.}$$
$$(2x - 1)(x - 5) = 0 \qquad \text{Factor.}$$
$$2x - 1 = 0 \quad \text{or} \quad x - 5 = 0 \qquad \text{Apply the zero factor theorem.}$$
$$2x = 1 \quad \text{or} \quad x = 5$$
$$x = \frac{1}{2} \quad \text{or} \quad x = 5$$

The solution set is $\left\{ \dfrac{1}{2}, 5 \right\}$.

1. $x = \dfrac{1}{5}, 2$ **2.** $x = -3, \dfrac{2}{5}$ **4.** $x = 3 \pm \sqrt{2}$ **7.** no real solution **8.** no real solution **9.** $x = 2$ **10.** $x = 3$

12. $p = \dfrac{7}{2}$ **15.** $x = 0, 1, 2$ **16.** $x = -4, -3, 0$ **18.** $z = 0, \dfrac{8}{3}$

EXERCISE SET 10.4

Choose and use a method to solve each equation.

1. $5x^2 - 11x + 2 = 0$ **2.** $5x^2 + 13x - 6 = 0$ **11.** $9 - 6p + p^2 = 0$ $p = 3$ **12.** $49 - 28p + 4p^2 = 0$

3. $x^2 - 1 = 2x$ $x = 1 \pm \sqrt{2}$ **4.** $x^2 + 7 = 6x$ **13.** $4y^2 - 16 = 0$ $y = \pm 2$ **14.** $3y^2 - 27 = 0$ $y = \pm 3$

5. $a^2 = 20$ $a = \pm 2\sqrt{5}$ **6.** $a^2 = 72$ $a = \pm 6\sqrt{2}$ **15.** $x^4 - 3x^3 + 2x^2 = 0$ **16.** $x^3 + 7x^2 + 12x = 0$

7. $x^2 - x + 4 = 0$ **8.** $x^2 - 2x + 7 = 0$ **17.** $(2z + 5)^2 = 25z = -5, 0$ **18.** $(3z - 4)^2 = 16$

9. $3x^2 - 12x + 12 = 0$ **10.** $5x^2 - 30x + 45 = 0$ **19.** $30x = 25x^2 + 2$ **20.** $12x = 4x^2 + 4$

19. $x = \dfrac{3 \pm \sqrt{7}}{5}$ **20.** $x = \dfrac{3 \pm \sqrt{5}}{2}$

21. $\frac{2}{3}m^2 - \frac{1}{3}m - 1 = 0$ **22.** $\frac{5}{8}m^2 + m - \frac{1}{2} = 0$

23. $x^2 - \frac{1}{2}x - \frac{1}{5} = 0$ **24.** $x^2 + \frac{1}{2}x - \frac{1}{8} = 0$

25. $4x^2 - 27x + 35 = 0$ **26.** $9x^2 - 16x + 7 = 0$

27. $(7 - 5x)^2 = 18$ **28.** $(5 - 4x)^2 = 75$

29. $3z^2 - 7z = 12$ **30.** $6z^2 + 7z = 6$

31. $x = x^2 - 110$ **32.** $x = 56 - x^2$

33. $\frac{3}{4}x^2 - \frac{5}{2}x - 2 = 0$ **34.** $x^2 - \frac{6}{5}x - \frac{8}{5} = 0$

35. $x^2 - 0.6x + 0.05 = 0$ **36.** $x^2 - 0.1x - 0.06 = 0$

37. $10x^2 - 11x + 2 = 0$ **38.** $20x^2 - 11x + 1 = 0$

39. $\frac{1}{2}z^2 - 2z + \frac{3}{4} = 0$ **40.** $\frac{1}{5}z^2 - \frac{1}{2}z - 2 = 0$

41. If a line segment AB is divided by a point C into two segments, AC and CB, such that the proportion $\frac{AB}{AC} = \frac{AC}{CB}$ is true, this ratio $\frac{AB}{AC}\left(\text{or }\frac{AC}{CB}\right)$ is called the golden ratio. If AC is 1 unit, find the length of AB. (*Hint:* Let x be the unknown length as shown and substitute into the given proportion.)

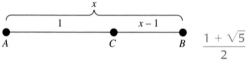

$\dfrac{1 + \sqrt{5}}{2}$

42. The formula $A = P(1 + r)^2$ is used to find the amount of money A in an account after P dollars have been invested in the account paying r annual interest rate for 2 years. Find the interest rate r if $1000 grows to $1690 in 2 years.   30%

21. $m = -1, \frac{3}{2}$ **22.** $m = -2, \frac{2}{5}$ **23.** $x = \dfrac{5 \pm \sqrt{105}}{20}$

24. $x = \dfrac{-1 \pm \sqrt{3}}{4}$

25. $x = \frac{7}{4}, 5$

43. Explain how you will decide what method to use when solving quadratic equations.

answers may vary

Review Exercises

Simplify each expression. See Section 9.2.

44. $\sqrt{48}$ $4\sqrt{3}$ **45.** $\sqrt{104}$ $2\sqrt{26}$

46. $\sqrt{50}$ $5\sqrt{2}$ **47.** $\sqrt{80}$ $4\sqrt{5}$

Solve the following. See Section 2.6.

48. The height of a triangle is 4 times the length of the base. The area of the triangle is 18 square feet. Find the height and base of the triangle.

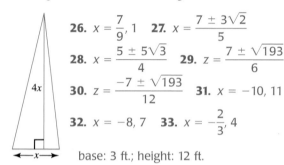

26. $x = \frac{7}{9}, 1$ **27.** $x = \dfrac{7 \pm 3\sqrt{2}}{5}$

28. $x = \dfrac{5 \pm 5\sqrt{3}}{4}$ **29.** $z = \dfrac{7 \pm \sqrt{193}}{6}$

30. $z = \dfrac{-7 \pm \sqrt{193}}{12}$ **31.** $x = -10, 11$

32. $x = -8, 7$ **33.** $x = -\frac{2}{3}, 4$

base: 3 ft.; height: 12 ft.

49. The length of a rectangle is 6 inches more than its width. The area of the rectangle is 391 square inches. Find the dimensions of the rectangle.

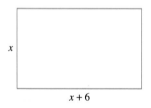

width: 17 in.; length: 23 in.

34. $x = -\frac{4}{5}, 2$ **35.** $x = 0.1, 0.5$ **36.** $x = -0.2, 0.3$

37. $x = \dfrac{11 \pm \sqrt{41}}{20}$ **38.** $x = \dfrac{11 \pm \sqrt{41}}{40}$

39. $z = \dfrac{4 \pm \sqrt{10}}{2}$ **40.** $z = \dfrac{5 \pm \sqrt{185}}{4}$

10.5 COMPLEX SOLUTIONS OF QUADRATIC EQUATIONS

OBJECTIVES

TAPE
BA 10.5

1 Write complex numbers using i notation.

2 Add and subtract complex numbers.

3 Multiply complex numbers.

4 Divide complex numbers.

5 Solve quadratic equations that have complex solutions.

In Chapter 9, we learned that $\sqrt{-4}$, for example, is not a real number because there is no real number whose square is -4. However, our real number system can be extended to include numbers like $\sqrt{-4}$. This extended number system is called the **complex number** system. The complex number system includes the **imaginary unit *i*,** which is defined next.

IMAGINARY UNIT *i*

The imaginary unit, written i, is the number whose square is -1. That is,

$$i^2 = -1 \quad \text{and} \quad i = \sqrt{-1}$$

---

1   We use i to write numbers like $\sqrt{-6}$ as the product of a real number and i. Since $i = \sqrt{-1}$, we have

$$\sqrt{-6} = \sqrt{-1 \cdot 6} = \sqrt{-1} \cdot \sqrt{6} = i\sqrt{6}$$

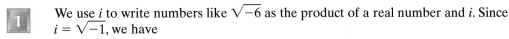

EXAMPLE 1   Write each radical as the product of a real number and i.

    a. $\sqrt{-4}$ **b.** $\sqrt{-11}$ **c.** $\sqrt{-20}$

Solution: Write each negative radicand as a product of a positive number and -1. Then write $\sqrt{-1}$ as i.

    a. $\sqrt{-4} = \sqrt{-1 \cdot 4} = \sqrt{-1} \cdot \sqrt{4} = i \cdot 2 = 2i$

    b. $\sqrt{-11} = \sqrt{-1 \cdot 11} = \sqrt{-1} \cdot \sqrt{11} = i\sqrt{11}$

    c. $\sqrt{-20} = \sqrt{-1 \cdot 20} = \sqrt{-1} \cdot \sqrt{20} = i \cdot 2\sqrt{5} = 2i\sqrt{5}$

The numbers $2i$, $i\sqrt{11}$, and $2i\sqrt{5}$ are called **imaginary numbers.** Both real numbers and imaginary numbers are complex numbers.

COMPLEX NUMBERS AND IMAGINARY NUMBERS

A **complex number** is a number that can be written in the form

    $a + bi$

where a and b are real numbers. A complex number that can be written in the form

    $0 + bi$

$b \neq 0$, is also called an **imaginary number.**

A complex number written in the form $a + bi$ is in **standard form.** We call a the real part and bi the imaginary part of the complex number $a + bi$.

EXAMPLE 2 Identify each number as a complex number by writing it in standard form $a + bi$.

         a. 7   **b.** 0   **c.** $\sqrt{20}$ **d.** $\sqrt{-27}$ **e.** $2 + \sqrt{-4}$

Solution: **a.** 7 is a complex number since $7 = 7 + 0i$

         b. 0 is a complex number since $0 = 0 + 0i$

         c. $\sqrt{20}$ is a complex number since $\sqrt{20} = 2\sqrt{5} = 2\sqrt{5} + 0i$

         d. $\sqrt{-27}$ is a complex number since $\sqrt{-27} = i \cdot 3\sqrt{3} = 0 + 3\sqrt{3}i$

         e. $2 + \sqrt{-4}$ is a complex number since $2 + \sqrt{-4} = 2 + 2i$

2   We now present arithmetic operations—addition, subtraction, multiplication, and division—for the complex number system. Complex numbers are added and subtracted in the same way as we add and subtract polynomials.

EXAMPLE 3 Simplify the sum or difference. Write the result in standard form.

         a. $(2 + 3i) + (-6 - i)$ **b.** $-i + (3 + 7i)$ **c.** $(5 - i) - 4$

Solution: Add the real parts and then add the imaginary parts.

         a. $(2 + 3i) + (-6 - i) = 2 + (-6) + (3i - i) = -4 + 2i$

         b. $-i + (3 + 7i) = 3 + (-i + 7i) = 3 + 6i$

         c. $(5 - i) - 4 = (5 - 4) - i = 1 - i$

EXAMPLE 4 Subtract $(11 - i)$ from $(1 + i)$.

Solution: $(1 + i) - (11 - i) = 1 + i - 11 + i = (1 - 11) + (i + i) = -10 + 2i$

3   Use the distributive property and the FOIL method to multiply complex numbers.

EXAMPLE 5 Find the following products and write in standard form.

         a. $5i(2 - i)$ **b.** $(7 - 3i)(4 + 2i)$ **c.** $(2 + 3i)(2 - 3i)$

Solution: **a.** By the distributive property, we have

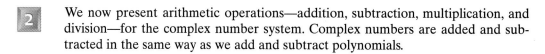

$$5i(2 - i) = 5i \cdot 2 - 5i \cdot i \qquad \text{Apply the distributive property.}$$
$$= 10i - 5i^2$$
$$= 10i - 5(-1) \qquad \text{Write } i^2 \text{ as } -1.$$
$$= 10i + 5$$
$$= 5 + 10i \qquad \text{Write in standard form.}$$

b.

$$\qquad\qquad\qquad\text{F}\quad\text{O}\quad\text{I}\quad\text{L}$$
$$(7 - 3i)(4 + 2i) = 28 + 14i - 12i - 6i^2$$
$$= 28 + 2i - 6(-1) \qquad \text{Write } i^2 \text{ as } -1.$$
$$= 28 + 2i + 6$$
$$= 34 + 2i$$

c.

$$(2 + 3i)(2 - 3i) = 4 - 6i + 6i - 9i^2$$
$$= 4 - 9(-1)$$
$$= 13$$

The product in part (**c**) is the real number 13. Notice that one factor is the sum of 2 and $3i$, and the other factor is the difference of 2 and $3i$. When complex number factors are related as these two are, their product is a real number. In general,

$$(a + bi)(a - bi) = a^2 + b^2$$
$$\text{sum}\quad\text{difference}\ \ \text{real number}$$

The complex numbers $a + bi$ and $a - bi$ are called **complex conjugates** of each other. For example, $2 - 3i$ is the conjugate of $2 + 3i$, and $2 + 3i$ is the conjugate of $2 - 3i$. Also,

The conjugate of $3 - 10i$ is $3 + 10i$.
The conjugate of 5 is 5. (Note that $5 = 5 + 0i$ and its conjugate is $5 - 0i = 5$.)
The conjugate of $4i$ is $-4i$. ($0 - 4i$ is the conjugate of $0 + 4i$.)

4 The fact that the product of a complex number and its conjugate is a real number provides a method for dividing by a complex number and for simplifying fractions whose denominators are complex numbers.

EXAMPLE 6   Write $\dfrac{4 + i}{3 - 4i}$ in standard form.

Solution: To write this quotient as a complex number in the standard form $a + bi$, we need to find an equivalent fraction whose denominator is a real number. By multiplying both numerator and denominator by the denominator's conjugate, we obtain a new fraction that is an equivalent fraction with a real number denominator.

$$\frac{4 + i}{3 - 4i} = \frac{(4 + i)}{(3 - 4i)} \cdot \frac{(3 + 4i)}{(3 + 4i)} \qquad \begin{array}{l}\text{Multiply numerator and denominator}\\ \text{by } 3 + 4i.\end{array}$$

$$= \frac{12 + 16i + 3i + 4i^2}{9 - 16i^2}$$

$$= \frac{12 + 19i + 4(-1)}{9 - 16(-1)}$$

$$= \frac{12 + 19i - 4}{9 + 16} = \frac{8 + 19i}{25}$$

$$= \frac{8}{25} + \frac{19}{25}i \qquad\qquad \text{Write in standard form.}$$

Note that our last step was to write $\dfrac{4 + i}{3 - 4i}$ in standard form $a + bi$, where a and b are real numbers.

> ▣ 5   Some quadratic equations have complex solutions.

EXAMPLE 7   Solve $(x + 2)^2 = -25$ for x.

Solution: Begin by applying the square root property.

$$(x + 2)^2 = -25$$
$$x + 2 = \pm\sqrt{-25}$$
$$x + 2 = \pm 5i \qquad \text{Write } \sqrt{-25} \text{ as } 5i.$$
$$x = -2 \pm 5i$$

The solution set is $\{-2 + 5i, -2 - 5i\}$.

EXAMPLE 8   Solve $m^2 = 4m - 5$.

Solution: Write the equation in standard form and use the quadratic formula to solve.

$$m^2 = 4m - 5$$
$$m^2 - 4m + 5 = 0 \qquad \text{Write the equation in standard form.}$$

Apply the quadratic formula with $a = 1, b = -4,$ and $c = 5$.

$$m = \frac{4 \pm \sqrt{16 - 4 \cdot 1 \cdot 5}}{2 \cdot 1}$$
$$= \frac{4 \pm \sqrt{-4}}{2}$$
$$= \frac{4 \pm 2i}{2}$$
$$= \frac{2(2 \pm i)}{2} = 2 \pm i$$

The solution set is $\{2 - i, 2 + i\}$.

EXAMPLE 9   Solve $x^2 + x = -1$.

Solution:

$$x^2 + x = -1$$

$$x^2 + x + 1 = 0 \qquad \text{Write in standard form.}$$

$$x = \frac{-1 \pm \sqrt{1 - 4 \cdot 1 \cdot 1}}{2 \cdot 1} \qquad \text{Apply the quadratic formula with } a = 1, b = 1, \text{ and } c = 1.$$

$$= \frac{-1 \pm \sqrt{-3}}{2}$$

$$= \frac{-1 \pm i\sqrt{3}}{2}$$

The solution set is $\left\{ \dfrac{-1 - i\sqrt{3}}{2}, \dfrac{-1 + i\sqrt{3}}{2} \right\}$.

9. $-3 + 9i$ **10.** $-2 - i$ **11.** $1 - 3i$ **12.** -9 **15.** $26 - 2i$ **16.** $26 + 2i$ **23.** $x = -1 \pm 3i$ **24.** $y = 2 \pm 5i$

25. $z = \dfrac{3 \pm 2i\sqrt{3}}{2}$ **26.** $p = \dfrac{-5 \pm 3i\sqrt{2}}{3}$ **27.** $y = -3 \pm 2i$ **28.** $y = 1 \pm 2i$ **29.** $x = \dfrac{-7 \pm i\sqrt{15}}{8}$

30. $x = \dfrac{7 \pm i\sqrt{15}}{16}$ **31.** $m = \dfrac{2 \pm i\sqrt{6}}{2}$ **32.** $m = \dfrac{3 \pm i\sqrt{26}}{5}$

33. $15 - 7i$ **34.** $-14 + 8i$ **36.** $-40 - 10i$

EXERCISE SET 10.5

Write each expression in i notation. See Example 1.

1. $\sqrt{-9}$ $3i$

2. $\sqrt{-64}$ $8i$

3. $\sqrt{-100}$ $10i$

4. $\sqrt{-16}$ $4i$

5. $\sqrt{-50}$ $5i\sqrt{2}$

6. $\sqrt{-98}$ $7i\sqrt{2}$

7. $\sqrt{-63}$ $3i\sqrt{7}$

8. $\sqrt{-44}$ $2i\sqrt{11}$

Add or subtract as indicated. See Examples 2 through 4.

9. $(2 - i) + (-5 + 10i)$

10. $(-7 + 2i) + (5 - 3i)$

11. $(3 - 4i) - (2 - i)$

12. $(-6 + i) - (3 + i)$

Multiply. See Example 5.

13. $4i(3 - 2i)$ $8 + 12i$

14. $-2i(5 + 4i)$ $8 - 10i$

15. $(6 - 2i)(4 + i)$

16. $(6 + 2i)(4 - i)$

☐ 17. Earlier in this text, we learned that $\sqrt{-4}$ is not a real number. Explain what that means and explain what type of number $\sqrt{-4}$ is.   answers may vary

☐ 18. Describe how to find the conjugate of a complex number.   answers may vary

Divide. Write each of the following in standard form. See Example 6.

19. $\dfrac{8 - 12i}{4}$ $2 - 3i$

20. $\dfrac{14 + 28i}{-7}$ $-2 - 4i$

21. $\dfrac{7 - i}{4 - 3i}$ $\dfrac{31}{25} + \dfrac{17}{25}i$

22. $\dfrac{4 - 3i}{7 - i}$ $\dfrac{31}{50} - \dfrac{17}{50}i$

Solve the following quadratic equations for complex solutions. See Example 7.

23. $(x + 1)^2 = -9$

24. $(y - 2)^2 = -25$

25. $(2z - 3)^2 = -12$

26. $(3p + 5)^2 = -18$

Solve the following quadratic equations for complex solutions. See Examples 8 and 9.

27. $y^2 + 6y + 13 = 0$

28. $y^2 - 2y + 5 = 0$

29. $4x^2 + 7x + 4 = 0$

30. $8x^2 - 7x + 2 = 0$

31. $2m^2 - 4m + 5 = 0$

32. $5m^2 - 6m + 7 = 0$

37. $-1 + 3i$

38. 9

Perform the indicated operations. Write results in standard form.

33. $3 + (12 - 7i)$

34. $(-14 + 5i) + 3i$

35. $-9i(5i - 7)$ $45 + 63i$

36. $10i(4i - 1)$

37. $(2 - i) - (3 - 4i)$

38. $(3 + i) - (-6 + i)$

39. $\dfrac{15 + 10i}{5i}$ $2 - 3i$

40. $\dfrac{-18 + 12i}{-6i}$ $-2 - 3i$

41. Subtract $2 + 3i$ from $-5 + i$. $-7 - 2i$

42. Subtract $-8 - i$ from $7 - 4i$. $15 - 3i$

43. $(4 - 3i)(4 + 3i)$ 25

44. $(12 - 5i)(12 + 5i)$

44. 169

45. $\dfrac{4-i}{1+2i}$ $\dfrac{2}{5} - \dfrac{9}{5}i$

46. $\dfrac{9-2i}{-3+i}$ $-\dfrac{29}{10} - \dfrac{3}{10}i$

47. $(5 + 2i)^2$ $21 + 20i$

48. $(9 - 7i)^2$ $32 - 126i$

Solve the following quadratic equations for complex solutions.

49. $(y - 4)^2 = -64$

50. $(x + 7)^2 = -1$

51. $4x^2 = -100$ $z = \pm 5i$

52. $7x^2 = -28$ $x = \pm 2i$

53. $z^2 + 6z + 10 = 0$

54. $z^2 + 4z + 13 = 0$

55. $2a^2 - 5a + 9 = 0$

56. $4a^2 + 3a + 2 = 0$

57. $(2x + 8)^2 = -20$

58. $(6z - 4)^2 = -24$

59. $3m^2 + 108 = 0$

60. $5m^2 + 80 = 0$

61. $x^2 + 14x + 50 = 0$

62. $x^2 + 8x + 25 = 0$

Answer the following true or false.

63. Every real number is a complex number.   true

64. Every complex number is a real number.   false

65. If a complex number such as $2 + 3i$ is a solution of a quadratic equation, then its conjugate $2 - 3i$ is also a solution.   true

66. Some imaginary numbers are real numbers.   false

Review Exercises

Graph the following linear equations in two variables. See Section 3.2.   graphical answers in App. F

67. $y = -3$

68. $x = 4$

69. $y = 3x - 2$

70. $y = 2x + 3$

Find the length of the unknown side of each triangle.

71. 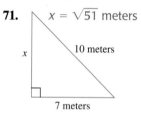 $x = \sqrt{51}$ meters

72. 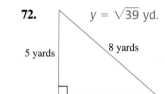 $y = \sqrt{39}$ yd.

49. $y = 4 \pm 8i$ **50.** $x = -7 \pm i$ **53.** $z = -3 \pm i$ **54.** $z = -2 \pm 3i$ **55.** $a = \dfrac{5 \pm i\sqrt{47}}{4}$

56. $a = \dfrac{-3 \pm i\sqrt{23}}{8}$ **57.** $x = -4 \pm i\sqrt{5}$ **58.** $z = \dfrac{2 \pm i\sqrt{6}}{3}$ **59.** $m = \pm 6i$ **60.** $m = \pm 4i$ **61.** $x = -7 \pm i$

62. $x = -4 \pm 3i$

10.6 | GRAPHING QUADRATIC EQUATIONS

TAPE
BA 10.6

O B J E C T I V E S

1 Identify the graph of a quadratic equation as a parabola.

2 Graph quadratic equations of the form $y = ax^2 + bx + c$.

3 Find the intercept points of a parabola.

4 Determine the vertex of a parabola.

1 Recall from Section 3.2 that the graph of a linear equation in two variables $Ax + By = C$ is a straight line. Also recall from Section 7.3 that the graph of a quadratic equation in two variables $y = ax^2 + bx + c$ is a parabola. In this section, we further investigate the graph of a quadratic equation.

To graph the quadratic equation $y = x^2$, select a few values for x and find the corresponding y-values. Make a table of values to keep track. Then plot the points corresponding to these solutions.

If $x = 0$, then $y = 0^2 = 0$.

If $x = -2$, then $y = (-2)^2 = 4$. And so on.

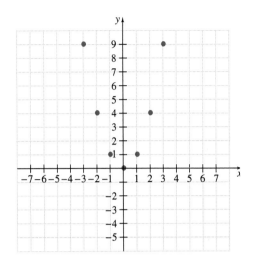

$$y = x^2$$

x	y
0	0
1	1
2	4
3	9
-1	1
-2	4
-3	9

Clearly, these points are not on one straight line. As we saw in Chapter 7, the graph of $y = x^2$ is a smooth curve through the plotted points. This curve is called a **parabola.** The lowest point on a parabola opening upward is called the **vertex.** The vertex is $(0, 0)$ for the parabola $y = x^2$. If we fold the graph paper along the y-axis, the two pieces of the parabola match perfectly. For this reason, we say the graph is **symmetric about the y-axis,** and we call the y-axis the **axis of symmetry.**

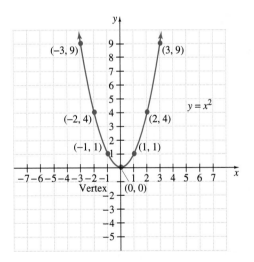

Notice that the parabola that corresponds to the equation $y = x^2$ opens upward. This happens when the coefficient of x^2 is positive. In the equation $y = x^2$, the coefficient of x^2 is 1. Example 1 shows the graph of a quadratic equation whose coefficient of x^2 is negative.

EXAMPLE 1   Graph $y = -2x^2$.

Solution: Select x-values and calculate the corresponding y-values. Plot the ordered pairs found. Then draw a smooth curve through those points. When the coefficient of x^2 is negative, the corresponding parabola opens downward. When a parabola opens downward, the vertex is the highest point of the parabola. The vertex of this parabola is $(0, 0)$ and the axis of symmetry is again the y-axis.

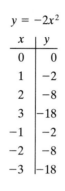

$$y = -2x^2$$

x	y
0	0
1	−2
2	−8
3	−18
−1	−2
−2	−8
−3	−18

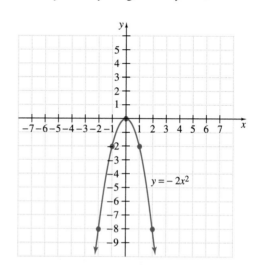

---

3   Just as for linear equations, we can use x- and y-intercepts to help graph quadratic equations. Recall from Chapter 3 that an x-intercept is the x-coordinate of the point where the graph intersects the x-axis. A y-intercept is the y-coordinate of the point where the graph intersects the y-axis.

> R E M I N D E R   Recall that:
>
> To find x-intercepts, let $y = 0$ and solve for x.
> To find y-intercepts, let $x = 0$ and solve for y.

EXAMPLE 2   Graph $y = x^2 - 4$.

Solution: First, find intercepts. To find the y-intercept, let $x = 0$. Then

$$y = 0^2 - 4 = -4$$

To find x-intercepts, let $y = 0$.

$$0 = x^2 - 4$$
$$0 = (x - 2)(x + 2)$$
$$x - 2 = 0 \quad \text{or} \quad x + 2 = 0$$
$$x = 2 \quad \text{or} \quad x = -2$$

Thus far, we have the y-intercept point $(0, -4)$ and the x-intercept points $(2, 0)$ and $(-2, 0)$. Next, select additional x-values, find y-values, plot the points, and draw a smooth curve through the points. The vertex is $(0, -4)$, and the axis of symmetry is the y-axis. This graph has the same shape as the graph of $y = x^2$. It is different from the graph of $y = x^2$ in that the vertex is 4 units lower on the y-axis.

$y = x^2 - 4$

x	y
0	−4
1	−3
2	0
3	5
−1	−3
−2	0
−3	5

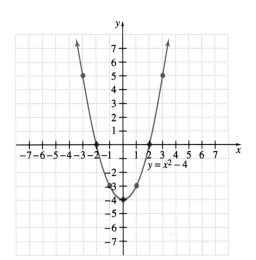

Notice that the graph above passes the vertical line test and is the graph of a function. Since the graph of $y = ax^2 + bx + c, a \neq 0$, is always a parabola opening upward or downward, by the vertical line test, it is always the graph of a function.

EXAMPLE 3   Graph $y = (x + 2)^2$.

Solution: Find the intercepts. To find x-intercepts, let $y = 0$.

$$0 = (x + 2)^2, \quad \text{so } x = -2$$

The x-intercept point is $(-2, 0)$. To find any y-intercepts, let $x = 0$.

$$y = (0 + 2)^2 = 4$$

The y-intercept point is $(0, 4)$. Plot the points $(-2, 0)$ and $(0, 4)$ and then select other values for x to obtain more ordered pairs.

$y = (x + 2)^2$

x	y
-2	0
0	4
1	9
2	16
-1	1
-3	1
-4	4

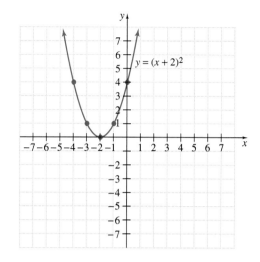

Notice that the graph of $y = (x + 2)^2$ is the same as the graph of $y = x^2$, except the vertex is shifted 2 units to the left, to $(-2, 0)$. The axis of symmetry is not the y-axis, but the vertical line through the vertex $(-2, 0)$. The axis of symmetry is the line $x = -2$.

4 So far we have located the vertex by graphing the equation. However, by writing the equation in a particular form, we can determine the vertex, the axis of symmetry, and whether the parabola opens upward or downward. This can be stated as follows.

THE GRAPH OF A QUADRATIC EQUATION IN THE FORM $y = a(x - h)^2 + k$

The graph of the quadratic equation $y = a(x - h)^2 + k$ is a parabola whose:

- Vertex is (h, k).
- Axis of symmetry is the line $x = h$.
- Direction of opening is upward if $a > 0$ and downward if $a < 0$.

EXAMPLE 4 Graph $y = 3(x - 5)^2 + 2$.

Solution: The equation is written in the form $y = a(x - h)^2 + k$, with $a = 3$, $h = 5$, and $k = 2$. The vertex is then $(5, 2)$, the axis of symmetry is the line $x = 5$, and the graph opens upward since $a = 3$. By letting $x = 0$, we find that the y-intercept is 77, but our grid does not show coordinates as large as 77. Instead, take a point or two on each side of the vertex to determine the shape of the parabola. The x-coordinate of the vertex is 5, so we select $x = 4$ and $x = 6$, for example, to find two other points, as shown in the table.

$y = 3(x - 5)^2 + 2$

x	y
5	2
4	5
6	5

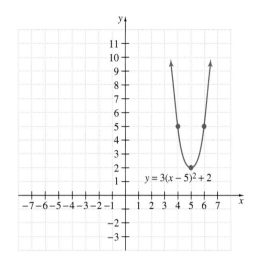

As this parabola demonstrates, some parabolas do not have an x-intercept. The vertex of this parabola is above the x-axis and the graph opens upward, so clearly the graph cannot intersect the x-axis.

If we try to find the x-intercept by replacing y with 0 in the equation, we solve $3(x - 5)^2 + 2 = 0$ or $3(x - 5)^2 = -2$ for x. But this equation has no real number solution. Since the x- and y-axes are real number lines, there is no x-intercept.

The figures below illustrate the possible x-intercepts of a parabola and solutions of the related quadratic equation.

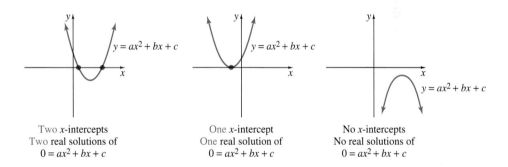

Two x-intercepts
Two real solutions of
$0 = ax^2 + bx + c$

One x-intercept
One real solution of
$0 = ax^2 + bx + c$

No x-intercepts
No real solutions of
$0 = ax^2 + bx + c$

EXAMPLE 5   Graph $y = -(x - 2)^2 + 1$.

Solution: The vertex is $(2, 1)$, the axis of symmetry is the line $x = 2$, and the parabola opens downward since $a = -1$. Again, selecting an x-value on each side of the vertex helps to determine the shape of the parabola. Finding the intercepts often yields a more accurate sketch.

To find x-intercepts, let $y = 0$.

$$0 = -(x - 2)^2 + 1$$
$$(x - 2)^2 = 1$$
$$x - 2 = \pm 1$$
$$x = 2 \pm 1$$
$$x = 3 \quad \text{or} \quad x = 1$$

There are two x-intercept points: $(3, 0)$ and $(1, 0)$. Plot $(3, 0)$ and $(1, 0)$.
To find y-intercepts, let $x = 0$.

$$y = -(0 - 2)^2 + 1 = -3$$

Plot the y-intercept point, $(0, -3)$

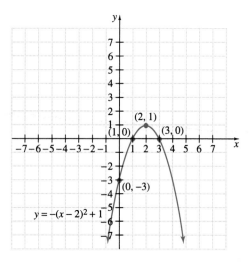

If a quadratic equation is not in the form $y = a(x - h)^2 + k$, then we cannot so easily identify the vertex. Notice that the x-coordinate of the parabola above is halfway between its x-intercepts. We can use this fact to find a formula for the vertex.

Recall that the x-intercepts of a parabola may be found by solving $0 = ax^2 + bx + c$. These solutions, by the quadratic formula, are

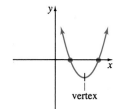

vertex

$$x = \frac{-b - \sqrt{b^2 - 4ac}}{2a}, \quad x = \frac{-b + \sqrt{b^2 - 4ac}}{2a}$$

The x-coordinate of the vertex of a parabola is halfway between its x-intercepts, so the x-value of the vertex may be found by computing the average, or $\frac{1}{2}$ of the sum of the intercepts.

$$x = \frac{1}{2}\left(\frac{-b - \sqrt{b^2 - 4ac}}{2a} + \frac{-b + \sqrt{b^2 - 4ac}}{2a}\right)$$

$$= \frac{1}{2}\left(\frac{-b - \sqrt{b^2 - 4ac} - b + \sqrt{b^2 - 4ac}}{2a}\right)$$

$$= \frac{1}{2}\left(\frac{-2b}{2a}\right)$$

$$= \frac{-b}{2a}$$

This formula may be used to find the x-value of the vertex of a parabola described by the equation $y = ax^2 + bx + c$. The corresponding y-value is found by substituting $\dfrac{-b}{2a}$ for x in the equation.

VERTEX FORMULA

The vertex of the parabola $y = ax^2 + bx + c$ has x-coordinate

$$\frac{-b}{2a}$$

EXAMPLE 6   Graph $y = x^2 - 6x + 8$.

Solution: In the equation $y = x^2 - 6x + 8$, $a = 1$, $b = -6$, and $c = 8$.
The x-coordinate of the vertex is

$$\frac{-b}{2a} = \frac{-(-6)}{2 \cdot 1} = 3.$$

To find the corresponding y-coordinate, let $x = 3$.

$$y = 3^2 - 6 \cdot 3 + 8 = -1.$$

The vertex is $(3, -1)$ and the axis of symmetry is the line $x = 3$. The parabola opens upward since $a = 1$. We also plot intercepts.
To find x-intercepts, let $y = 0$.

$$0 = x^2 - 6x + 8$$

Factor the expression $x^2 - 6x + 8$ to find $(x - 4)(x - 2) = 0$. The x-intercepts are 4 and 2.

If we let $x = 0$ in the original equation, then $y = 8$, the y-intercept. Plot the vertex $(3, -1)$ and the intercept points $(4, 0)$, $(2, 0)$, and $(0, 8)$. Then sketch the parabola.

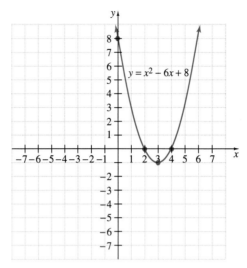

$y = x^2 - 6x + 8$

EXAMPLE 7   Graph $y = x^2 + 2x + 5$.

Solution: In the equation $y = x^2 + 2x + 5$, $a = 1$, $b = 2$, and $c = 5$.

The x-value of the vertex is

$$x = \frac{-b}{2a} = \frac{-2}{2 \cdot 1} = -1$$

The y-value is

$$y = (-1)^2 + 2(-1) + 5 = 4$$

Vertex: $(-1, 4)$

Axis of symmetry: $x = -1$

Opens: Upward

To find x-intercepts, let $y = 0$.

$$0 = (x + 1)^2 + 4$$
$$-4 = (x + 1)^2$$

The equation has no real number solution. Therefore, the graph has no x-intercepts.

To find y-intercepts, let $x = 0$ in the original equation and find that $y = 5$, and the resulting y-intercept point is $(0, 5)$.

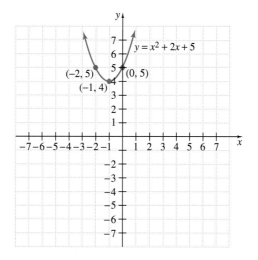

GRAPHING CALCULATOR EXPLORATIONS

Recall that a graphing calculator may be used to solve quadratic equations. The x-intercepts of the graph of $y = ax^2 + bx + c$ are solutions of $0 = ax^2 + bx + c$. To solve $x^2 - 7x - 3 = 0$, for example, graph $y_1 = x^2 - 7x - 3$. The x-intercepts of the graph are the solutions of the equation.

Use a grapher to solve each quadratic equation. Round solutions to two decimal places.

1. $x^2 - 7x - 3 = 0$ $x = -0.41, 7.41$ **2.** $2x^2 - 11x - 1 = 0$ $x = -0.09, 5.59$
3. $-1.7x^2 + 5.6x - 3.7 = 0$ $x = 0.91, 2.38$ **4.** $-5.8x^2 + 2.3x - 3.9 = 0$ no real solution
5. $5.8x^2 - 2.6x - 1.9 = 0$ $x = -0.39, 0.84$ **6.** $7.5x^2 - 3.7x - 1.1 = 0$ $x = -0.21, 0.70$

1. x-intercept: (0, 0),
   y-intercept: (0, 0),
   vertex: (0, 0)

2. x-intercept: (0, 0),
   y-intercept: (0, 0),
   vertex: (0, 0)

3. x-intercept: (1, 0),
   y-intercept: (0, 1),
   vertex: (1, 0)

4. x-intercept: (−2, 0),
   y-intercept: (0, 4),
   vertex: (−2, 0)

graphical answers in App. F

EXERCISE SET 10.6

Graph each quadratic equation. Find the vertex and intercepts. See Examples 1 through 5.

7. $y = \dfrac{1}{3}x^2$ **8.** $y = -\dfrac{1}{2}x^2$

1. $y = 2x^2$ **2.** $y = -2x^2$ **9.** $y = (x - 2)^2 + 1$ **10.** $y = -(x - 2)^2 - 1$

3. $y = (x - 1)^2$ **4.** $y = (x + 2)^2$ **11.** $y = -(x + 1)^2 + 4$ **12.** $y = (x - 1)^2 - 4$

5. $y = -x^2 + 4$ **6.** $y = x^2 - 4$ **13.** $y = -4x^2 + 1$ **14.** $y = 4x^2 - 1$

5. x-intercepts: (−2, 0), (2, 0),
   y-intercept: (0, 4),
   vertex: (0, 4)

6. x-intercepts: (−2, 0), (2, 0),
   y-intercept: (0, −4),
   vertex: (0, −4)

7. x-intercept: (0, 0),
   y-intercept: (0, 0),
   vertex: (0, 0)

8. x-intercept: (0, 0),
   y-intercept: (0, 0),
   vertex: (0, 0)

See p. 592 for additional Exercise Set 10.6 answers.

Write the letter of the graph corresponding to each equation. See Examples 1 through 5.

15. $y = -x^2$ F

16. $y = x^2$ I

17. $y = (x - 2)^2$ A

18. $y = -(x - 1)^2$ E

19. $y = (x + 3)^2 - 1$ H

20. $y = -(x - 2)^2 + 3$ C

21. $y = 2(x + 3)^2$ B

22. $y = -3(x + 1)^2$ G

23. $y = -\dfrac{1}{2}x^2 + 1$ D

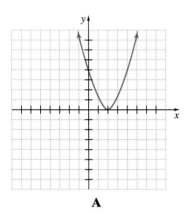

A

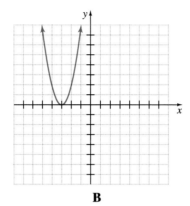

B

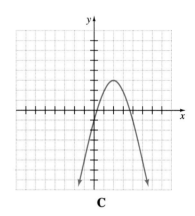

C

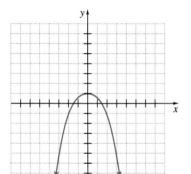

D

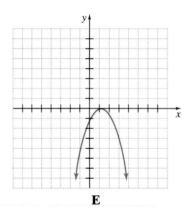

E

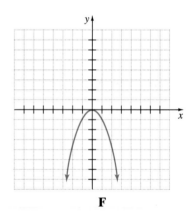

F

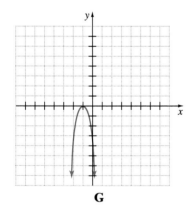

G

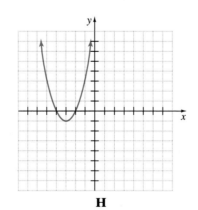

H

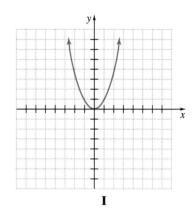

I

graphical answers in App. F

Sketch the graph of each equation. Identify the vertex and the intercepts. See Examples 6 and 7.

24. $y = x^2 + 6x$

25. $y = x^2 - 4x$

26. $y = x^2 + 2x - 8$

27. $y = x^2 - 2x - 3$

28. $y = x^2 - x - 2$

29. $y = x^2 + 2x + 1$

30. $y = x^2 + 5x + 4$

31. $y = x^2 + 7x + 10$

32. $y = x^2 - 4x + 3$

33. $y = x^2 - 6x + 8$

26. $y = (x + 1)^2 - 9$; vertex: $(-1, -9)$;
   x-intercepts: $(-4, 0)$, $(2, 0)$; y-intercept: $(0, -8)$;

The graph of a quadratic equation that takes the form $y = ax^2 + bx + c$ is the graph of a function. Write the domain and the range of each of the functions graphed.

34.

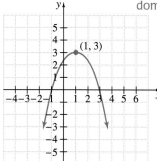

domain: all real numbers;
range: $y \le 3$

35.

domain: all real numbers;
range: $y \le -4$

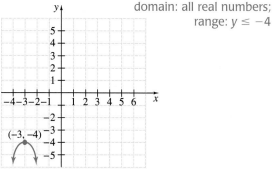

36.

domain: all real numbers;
range: $y \le 1$

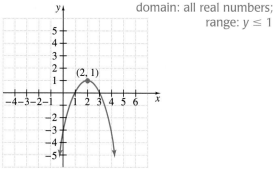

24. $y = (x + 3)^2 - 9$; vertex: $(-3, -9)$;
   x-intercepts: $(-6, 0)$, $(0, 0)$;
   y-intercept: $(0, 0)$

25. $y = (x - 2)^2 - 4$; vertex: $(2, -4)$;
   x-intercepts: $(0, 0)$, $(4, 0)$; y-intercept: $(0, 0)$

37.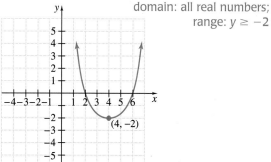

domain: all real numbers;
range: $y \ge -2$

38. The height h of a fireball launched from a Roman candle with an initial velocity of 128 feet per second is given by the equation

$$h = -16t^2 + 128t$$

where t is time in seconds after launch.

Use the graph of this function to answer the questions. **38. a.** 256 ft.

a. Estimate the maximum height of the fireball.

b. Estimate the time when the fireball is at its maximum height. $t = 4$ sec.

c. Estimate the time when the fireball returns to the ground. $t = 8$ sec.

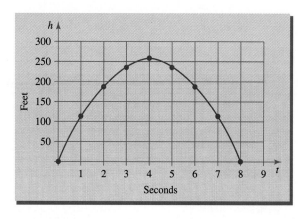

39. C   **41.** A

Match the graph of each quadratic equation of the form $y = a(x - h)^2 + k$ as shown on the next page.

39. $a > 0, h > 0, k > 0$ **40.** $a < 0, h > 0, k > 0$ B

41. $a > 0, h > 0, k < 0$ **42.** $a < 0, h > 0, k < 0$ D

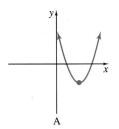

A

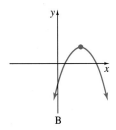

B

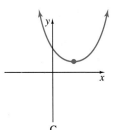

C

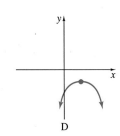

D

Review Exercises

Simplify the following complex fractions. See Section 6.5.

43. $\dfrac{\frac{1}{7}}{\frac{2}{5}} \quad \frac{5}{14}$

44. $\dfrac{\frac{3}{8}}{\frac{1}{7}} \quad \frac{21}{8}$

45. $\dfrac{\frac{1}{x}}{\frac{2}{x^2}} \quad \frac{x}{2}$

46. $\dfrac{\frac{x}{5}}{\frac{2}{x}} \quad \frac{x^2}{10}$

47. $\dfrac{2x}{1 - \frac{1}{x}} \quad \frac{2x^2}{x - 1}$

48. $\dfrac{x}{x - \frac{1}{x}} \quad \frac{x^2}{x^2 - 1}$

49. $\dfrac{\frac{a - b}{2b}}{\frac{b - a}{8b^2}} \quad -4b$

50. $\dfrac{\frac{2a^2}{a - 3}}{\frac{a}{3 - a}} \quad -2a$

GROUP ACTIVITY

MODELING A PHYSICAL SITUATION

MATERIALS:
- Metric ruler
- Grapher with curve-fitting capabilities (optional)

FIGURE 1

FIGURE 2

Model the physical situation of the parabolic path of water from a water fountain. For simplicity, use the given Figure 2 to investigate the following questions.

DATA TABLE

	x	y
Point A	0	0
Point B		
Point V		

1. Collect data for the x-intercepts of the parabolic path. Let points A and B in Figure 2 be on the x-axis and let the coordinates of point A be $(0, 0)$. Use a ruler to measure the distance between points A and B **on Figure 2** to the nearest even one-tenth centimeter, and use this information to determine the coordinates of point B. Record this data in the data table. (*Hint:* If the distance from A to B measures 8 one-tenth centimeters, then the coordinates of point B are $(8, 0)$.)

(continued)

2. Next, collect data for the vertex V of the parabolic path. What is the relationship between the x-coordinate of the vertex and the x-intercepts found in Question 1? What is the axis of symmetry? To locate point V in Figure 2, find the midpoint of the line segment joining points A and B and mark point V on the path of water directly above the midpoint. To approximate the y-coordinate of the vertex, use a ruler to measure its distance from the x-axis to the nearest one-tenth centimeter. Record this data in the data table on the previous page.

3. Plot the points from the data table on a rectangular coordinate system. Sketch the parabola through your points A, B, and V.

4. Which of the following models best fits the data you collected? Explain your reasoning.

(a) $y = 16x + 18$

(b) $y = -13x^2 + 20x$

(c) $y = 0.13x^2 - 2.6x$

(d) $y = -0.13x^2 + 2.6x$

5. (Optional) Enter your data into a grapher and use the quadratic curve-fitting feature to find a model for your data. How does the model compare with your selection from Question 4?

See App. F for Group Activity answers and suggestions.

CHAPTER 10   HIGHLIGHTS

DEFINITIONS AND CONCEPTS	EXAMPLES
SECTION 10.1 SOLVING QUADRATIC EQUATIONS BY THE SQUARE ROOT METHOD	

Square Root Property

If $X^2 = a$ for $X \geq 0$, then $X = \pm\sqrt{a}$

Solve the equation.

$$(x - 1)^2 = 15$$
$$x - 1 = \pm\sqrt{15}$$
$$x = 1 \pm\sqrt{15}$$

SECTION 10.2   SOLVING QUADRATIC EQUATIONS BY COMPLETING THE SQUARE

To solve a quadratic equation by completing the square,

Step 1. If the coefficient of x^2 is not 1, divide both sides of the equation by the coefficient.

Step 2. Isolate all terms with variables on one side.

Step 3. Complete the square by adding the square of half of the coefficient of x to both sides.

Step 4. Factor the perfect square trinomial.

Step 5. Apply the square root property to solve.

Solve $2x^2 + 12x - 10 = 0$ by completing the square.

$$\frac{2x^2}{2} + \frac{12x}{2} - \frac{10}{2} = \frac{0}{2} \quad \text{Divide by 2.}$$

$$x^2 + 6x - 5 = 0 \quad \text{Simplify.}$$
$$x^2 + 6x = 5 \quad \text{Add 5.}$$

The coefficient of x is 6. Half of 6 is 3 and $3^2 = 9$. Add 9 to both sides.

$$x^2 + 6x + 9 = 5 + 9$$
$$(x + 3)^2 = 14 \quad \text{Factor.}$$
$$x + 3 = \pm\sqrt{14}$$
$$x = -3 \pm\sqrt{14}$$

(continued)

DEFINITIONS AND CONCEPTS	EXAMPLES

SECTION 10.3 SOLVING QUADRATIC EQUATIONS BY THE QUADRATIC FORMULA

Quadratic formula

If a, b, and c are real numbers and $a \neq 0$, the quadratic equation $ax^2 + bx + c = 0$ has solutions

$$x = \frac{-b \pm \sqrt{b^2 - 4ac}}{2a}$$

To solve a quadratic equation by the quadratic formula

Step 1. Write the equation in standard form: $ax^2 + bx + c = 0$.

Step 2. If necessary, clear the equation of fractions.

Step 3. Identify a, b, and c.

Step 4. Replace a, b, and c in the quadratic formula by known values, and simplify.

Identify a, b, and c in the quadratic equation

$$4x^2 - 6x = 5$$

First, subtract 5 from both sides.

$$4x^2 - 6x - 5 = 0$$
$$a = 4, b = -6, \quad \text{and} \quad c = -5$$

Solve $3x^2 - 2x - 2 = 0$

In this equation, $a = 3$, $b = -2$, and $c = -2$.

$$x = \frac{-(-2) \pm \sqrt{(-2)^2 - 4(3)(-2)}}{2 \cdot 3}$$
$$= \frac{2 \pm \sqrt{4 - (-24)}}{6}$$
$$= \frac{2 \pm \sqrt{28}}{6} = \frac{2 \pm \sqrt{4 \cdot 7}}{6} = \frac{2 \pm 2\sqrt{7}}{6}$$
$$= \frac{2(1 \pm \sqrt{7})}{2 \cdot 3} = \frac{1 \pm \sqrt{7}}{3}$$

SECTION 10.4 SUMMARY OF METHODS FOR SOLVING QUADRATIC EQUATIONS

To solve a quadratic equation

Step 1. If the equation is in the form $(ax + b)^2 = c$, use the square root property and solve. If not, go to step 2.

Step 2. Write the equation in standard form: $ax^2 + bx + c = 0$.

Step 3. Try to solve by factoring. If not, go to *step 4.*

Step 4. Solve by the quadratic formula.

Solve $(3x - 1)^2 = 10$.

$$3x - 1 = \pm\sqrt{10} \qquad \text{Square root property.}$$
$$3x = 1 \pm \sqrt{10} \qquad \text{Add 1.}$$
$$x = \frac{1 \pm \sqrt{10}}{3} \qquad \text{Divide by 3.}$$

Solve $x(2x + 9) = 5$.

$$2x^2 + 9x - 5 = 0$$
$$(2x - 1)(x + 5) = 0$$
$$2x - 1 = 0 \quad \text{or} \quad x + 5 = 0$$
$$2x = 1$$
$$x = \frac{1}{2} \quad \text{or} \quad x = -5$$

SECTION 10.5 COMPLEX SOLUTIONS OF QUADRATIC EQUATIONS

The **imaginary unit,** written i, is the number whose square is -1. That is,

$$i^2 = -1 \quad \text{and} \quad i = \sqrt{-1}.$$

Write $\sqrt{-10}$ as the product of a real number and i.

$$\sqrt{-10} = \sqrt{-1 \cdot 10} = \sqrt{-1} \cdot \sqrt{10} = i\sqrt{10}$$

(continued)

DEFINITIONS AND CONCEPTS	EXAMPLES

SECTION 10.5   COMPLEX SOLUTIONS OF QUADRATIC EQUATIONS

A **complex number** is a number that can be written in the form $$a + bi$$ where a and b are real numbers. A complex number that can be written in the form $0 + bi$, $b \neq 0$, is also called an **imaginary number.**	Identify each number as a complex number by writing it in **standard form** $a + bi$. $$7 = 7 + 0i$$ $$\sqrt{-5} = 0 + \sqrt{5}i \qquad \text{also an imaginary number}$$ $$1 - \sqrt{-9} = 1 - 3i$$
Complex numbers are added and subtracted in the same way as polynomials are added and subtracted.	Simplify the sum or difference. $$(2 + 3i) - (1 - 6i) = 2 + 3i - 1 + 6i$$ $$= 1 + 9i$$ $$2i + (5 - 3i) = 2i + 5 - 3i$$ $$= 5 - i$$
Use the distributive property to multiply complex numbers.	Multiply: $(4 - i)(-2 + 3i)$ $$= -8 + 12i + 2i - 3i^2$$ $$= -8 + 14i + 3$$ $$= -5 + 14i$$
The complex numbers $a + bi$ and $a - bi$ are called **complex conjugates.**	The conjugate of $(5 - 6i)$ is $(5 + 6i)$.
To write a quotient of complex numbers in standard form $a + bi$, multiply numerator and denominator by the denominator's conjugate.	Write $\dfrac{3 - 2i}{2 + i}$ in standard form. $$\frac{(3 - 2i)}{(2 + i)} \cdot \frac{(2 - i)}{(2 - i)} = \frac{6 - 7i + 2i^2}{4 - i^2}$$ $$= \frac{6 - 7i + 2(-1)}{4 - (-1)}$$ $$= \frac{4 - 7i}{5} \quad \text{or} \quad \frac{4}{5} - \frac{7}{5}i$$
Some quadratic equations have complex solutions.	Solve $x^2 - 3x = -3$ $$x^2 - 3x + 3 = 0$$ $$a = 1, b = -3, c = 3$$ $$x = \frac{-(-3) \pm \sqrt{(-3)^2 - 4(1)(3)}}{2(1)}$$ $$= \frac{3 \pm \sqrt{-3}}{2} = \frac{3 \pm i\sqrt{3}}{2}$$

(continued)

DEFINITIONS AND CONCEPTS	EXAMPLES

SECTION 10.6 GRAPHING QUADRATIC EQUATIONS

The graph of a quadratic equation $y = ax^2 + bx + c$, $a \neq 0$, is called a **parabola.** The lowest point on a parabola opening upward or the highest point on a parabola opening downward is called the **vertex.** The vertical line through the vertex is the **axis of symmetry.**

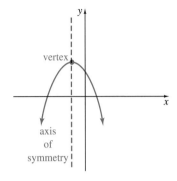

The vertex of the parabola $y = ax^2 + bx + c$ has x-value $\dfrac{-b}{2a}$.

Graph $y = 2x^2 - 6x + 4$

The x-value of the vertex is

$$x = \frac{-b}{2a} = \frac{-(-6)}{2(2)} = \frac{6}{4} = \frac{3}{2}$$

The y-value is

$$y = 2\left(\frac{3}{2}\right)^2 - 6\left(\frac{3}{2}\right) + 4$$

$$y = 2\left(\frac{9}{4}\right) - 9 + 4$$

$$y = \frac{9}{2} - 5 = \frac{9}{2} - \frac{10}{2} = -\frac{1}{2}$$

The vertex is $\left(\dfrac{3}{2}, -\dfrac{1}{2}\right)$

The y-intercept is

$$y = 2 \cdot 0^2 - 6 \cdot 0 + 4 = 4$$

The x-intercepts are found by

$$0 = 2x^2 - 6x + 4$$
$$0 = 2(x^2 - 3x + 2)$$
$$0 = 2(x - 2)(x - 1)$$
$$x - 2 = 0 \quad \text{or} \quad x - 1 = 0$$
$$x = 2 \qquad \text{or} \quad x = 1$$

(continued)

DEFINITIONS AND CONCEPTS	EXAMPLES
SECTION 10.6 GRAPHING QUADRATIC EQUATIONS	

Find more ordered pair solutions as needed.

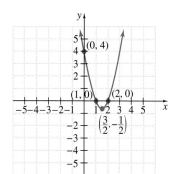

1. $x = -\dfrac{3}{5}, 4$ **2.** $x = -7, -\dfrac{4}{3}$ **3.** $m = -\dfrac{1}{3}, 2$ **4.** $m = \dfrac{5}{7}, -1$

7. $x = -1, 7$ **8.** $x = -3, 4$ **10.** $x = -13, 7$ **13.** $p = -3, 2$

14. $p = -5, 1$ **15.** $x^2 - 10x + 25 = (x - 5)^2$ **16.** $x^2 + 16x + 64 = (x + 8)^2$ **17.** $a^2 + 4a + 4 = (a + 2)^2$

18. $a^2 - 12a + 36 = (a - 6)^2$ **19.** $m^2 - 3m + \dfrac{9}{4} = \left(m - \dfrac{3}{2}\right)^2$ **20.** $m^2 + 5m + \dfrac{25}{4} = \left(m + \dfrac{5}{2}\right)^2$ **21.** $x = 3 \pm \sqrt{2}$

22. $x = -3 \pm \sqrt{2}$ **23.** $y = -1, \dfrac{1}{2}$ **24.** $y = \dfrac{-3 \pm \sqrt{13}}{2}$ **25.** $x = 5 \pm 3\sqrt{2}$ **26.** $x = -2 \pm \sqrt{11}$

27. $x = -1, \dfrac{1}{2}$ **28.** $x = \dfrac{-3 \pm \sqrt{13}}{2}$ **29.** $x = -\dfrac{5}{3}$ **30.** $x = \dfrac{9}{4}$ **31.** $x = \dfrac{2}{5}, \dfrac{1}{3}$ **32.** $x = \dfrac{2}{3}, \dfrac{1}{5}$

33. $x = \dfrac{-1 \pm i\sqrt{39}}{4}$ **34.** no real solution **35.** $z = \dfrac{-1 \pm \sqrt{21}}{10}$ **36.** $z = \dfrac{-7 \pm \sqrt{65}}{8}$

CHAPTER 10 REVIEW

(10.1) *Solve each quadratic equation by factoring or using the square root property.*

1. $(x - 4)(5x + 3) = 0$
2. $(x + 7)(3x + 4) = 0$
3. $3m^2 - 5m = 2$
4. $7m^2 + 2m = 5$
5. $k^2 = 50$ $k = \pm 5\sqrt{2}$
6. $k^2 = 45$ $k = \pm 3\sqrt{5}$
7. $(x - 5)(x - 1) = 12$
8. $(x - 3)(x + 2) = 6$
9. $(x - 11)^2 = 49$ $x = 4, 18$
10. $(x + 3)^2 = 100$
11. $6x^3 - 54x = 0$ $x = 0, \pm 3$
12. $2x^2 - 8 = 0$ $x = \pm 2$
13. $(4p + 2)^2 = 100$
14. $(3p + 6)^2 = 81$

(10.2) *Complete the square for the following expressions and then factor the resulting perfect square trinomial.*

15. $x^2 - 10x$
16. $x^2 + 16x$
17. $a^2 + 4a$
18. $a^2 - 12a$
19. $m^2 - 3m$
20. $m^2 + 5m$

Solve each quadratic equation by completing the square.

21. $x^2 - 6x + 7 = 0$
22. $x^2 + 6x + 7 = 0$
23. $2y^2 + y - 1 = 0$
24. $y^2 + 3y - 1 = 0$

(10.3) *Solve each quadratic equation by using the quadratic formula.*

25. $x^2 - 10x + 7 = 0$
26. $x^2 + 4x - 7 = 0$
27. $2x^2 + x - 1 = 0$
28. $x^2 + 3x - 1 = 0$
29. $9x^2 + 30x + 25 = 0$
30. $16x^2 - 72x + 81 = 0$
31. $15x^2 + 2 = 11x$
32. $15x^2 + 2 = 13x$
33. $2x^2 + x + 5 = 0$
34. $7x^2 - 3x + 1 = 0$

(10.4) *Solve the following equations by using the most appropriate method.*

35. $5z^2 + z - 1 = 0$
36. $4z^2 + 7z - 1 = 0$

See p. 592 for additional Review answers.

37. $4x^4 = x^2$ $x = 0, \pm\dfrac{1}{2}$ **38.** $9x^3 = x$ $x = 0, \pm\dfrac{1}{3}$

39. $2x^2 - 15x + 7 = 0$

40. $x^2 - 6x - 7 = 0$

41. $(3x - 1)^2 = 0$ $x = \dfrac{1}{3}$ **42.** $(2x - 3)^2 = 0$ $x = \dfrac{3}{2}$

43. $x^2 = 6x - 9$ $x = 3$ **44.** $x^2 = 10x - 25$ $x = 5$

45. $\left(\dfrac{1}{2}x - 3\right)^2 = 64$ **46.** $\left(\dfrac{1}{3}x + 1\right)^2 = 49$

47. $x^2 - 0.3x + 0.01 = 0$ **48.** $x^2 + 0.6x - 0.16 = 0$

49. $\dfrac{1}{10}x^2 + x - \dfrac{1}{2} = 0$ **50.** $\dfrac{1}{12}x^2 - \dfrac{1}{2}x + \dfrac{1}{3} = 0$

(10.5)   *Perform the indicated operations. Write the resulting complex number in standard form.*

51. $\sqrt{-144}$ $12i$ **52.** $\sqrt{-36}$ $6i$

53. $\sqrt{-108}$ $6i\sqrt{3}$ **54.** $\sqrt{-500}$ $10i\sqrt{5}$

55. $(7 - i) + (14 - 9i)$ **56.** $(10 - 4i) + (9 - 21i)$

57. $3 - (11 + 2i)$ $-8 - 2i$ **58.** $(-4 - 3i) + 5i$

59. $(2 - 3i)(3 - 2i)$ $-13i$ **60.** $(2 + 5i)(5 - i)$

61. $(3 - 4i)(3 + 4i)$ 25 **62.** $(7 - 2i)(7 - 2i)$

63. $\dfrac{2 - 6i}{4i}$ $\dfrac{3}{2} - \dfrac{1}{2}i$ **64.** $\dfrac{5 - i}{2i}$ $-\dfrac{1}{2} - \dfrac{5}{2}i$

65. $\dfrac{4 - i}{1 + 2i}$ $\dfrac{2}{5} - \dfrac{9}{5}i$ **66.** $\dfrac{1 + 3i}{2 - 7i}$ $-\dfrac{19}{53} + \dfrac{13}{53}i$

Solve each quadratic equation.

67. $3x^2 = -48$ $x = \pm 4i$ **68.** $5x^2 = -125$ $x = \pm 5i$

69. $x^2 - 4x + 13 = 0$ **70.** $x^2 + 4x + 11 = 0$

(10.6)   *Identify the vertex, axis of symmetry, and whether the parabola opens upward or downward for the given quadratic equation.*

71. $y = -3x^2$ **72.** $y = -\dfrac{1}{2}x^2$

73. $y = (x - 3)^2$ **74.** $y = (x - 5)^2$

75. $y = 3x^2 - 7$ **76.** $y = -2x^2 + 25$

77. $y = -5(x - 72)^2 + 14$ **78.** $y = 2(x - 35)^2 - 21$
graphical answers in App. F

Graph the following quadratic equations. Label the vertex and the intercept points with their coordinates.

79. $y = -(x + 1)^2$ **80.** $y = -(x - 2)^2$

81. $y = (x - 2)^2 + 3$ **82.** $y = (x + 3)^2 - 1$

83. $y = x^2 + 5x + 6$ **84.** $y = x^2 - 4x - 8$

85. $y = 2x^2 - 11x - 6$ **86.** $y = 3x^2 - x - 2$

71. vertex: (0, 0); axis of symmetry: $x = 0$; opens downward
72. vertex: (0, 0); axis of symmetry: $x = 0$; opens downward
73. vertex: (3, 0); axis of symmetry: $x = 3$; opens upward
74. vertex: (5, 0); axis of symmetry: $x = 5$; opens upward

Quadratic equations $y = ax^2 + bx + c$ are graphed below. Determine the number of real solutions for the related equation $0 = ax^2 + bx + c$ from each graph. List the solutions.

☐ 87.

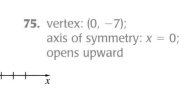

87. 1 solution; $x = -2$

75. vertex: (0, −7);
axis of symmetry: $x = 0$;
opens upward

76. vertex: (0, 25);
axis of symmetry: $x = 0$;
opens downward

☐ 88.

88. 2 solutions; $x = 3$,
$x = -1$

77. vertex: (72, 14);
axis of symmetry: $x = 72$;
opens downward

78. vertex: (35, −21);
of symmetry: $x = 35$;
opens upward

☐ 89.

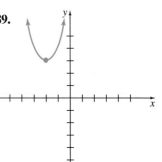

89. no real solutions

☐ 90.

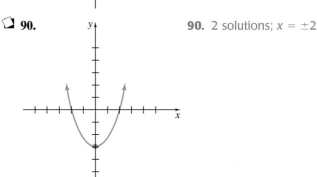

90. 2 solutions; $x = \pm 2$

CHAPTER 10 TEST

Solve by factoring.

1. $2x^2 - 11x = 21$ $x = -\dfrac{3}{2}, 7$ **2.** $x^4 + x^3 - 2x^2 = 0$
$x = -2, 0, 1$

Solve using the square root property.

3. $5k^2 = 80$ $k = \pm 4$ **4.** $(3m - 5)^2 = 8$ $m = \dfrac{5 \pm 2\sqrt{2}}{3}$

Solve by completing the square.

5. $x^2 - 26x + 160 = 0$ **6.** $5x^2 + 9x = 2$
$x = 10, 16$ $x = -2, \dfrac{1}{5}$

Solve using the quadratic formula.

7. $x^2 - 3x - 10 = 0$ **8.** $p^2 - \dfrac{5}{3}p - \dfrac{1}{3} = 0$
$x = -2, 5$ $p = \dfrac{5 \pm \sqrt{37}}{6}$

Solve by the most appropriate method.

9. $(3x - 5)(x + 2) = -6$ **10.** $(3x - 1)^2 = 16$ $x = -1, \dfrac{5}{3}$

11. $3x^2 - 7x - 2 = 0$ **12.** $x^2 - 4x + 5 = 0$

13. $3x^2 - 7x + 2 = 0$ **14.** $2x^2 - 6x + 1 = 0$

9. $x = -\dfrac{4}{3}, 1$ **11.** $x = \dfrac{7 \pm \sqrt{73}}{6}$ **12.** $x = 2 \pm i$

13. $x = \dfrac{1}{3}, 2$ **14.** $x = \dfrac{3 \pm \sqrt{7}}{2}$

15. $x = -3, 0, \dfrac{1}{2}$

15. $2x^5 + 5x^4 - 3x^3 = 0$ **16.** $9x^3 = x$ $x = 0, \pm\dfrac{1}{3}$

Perform the indicated operations. Write the resulting complex number in standard form. **20.** $7 - 6i$

17. $\sqrt{-25}$ $5i$ **18.** $\sqrt{-200}$ $10i\sqrt{2}$

19. $(3 + 2i) + (5 - i)$ $8 + i$ **20.** $(4 - i) - (-3 + 5i)$

21. $(3 + 2i) - (3 - 2i)$ $4i$ **22.** $(3 + 2i) + (3 - 2i)$ 6

23. $(3 + 2i)(3 - 2i)$ 13 **24.** $\dfrac{3 - i}{1 + 2i}$ $\dfrac{1}{5} - \dfrac{7}{5}i$

Graph the quadratic equations. Label the vertex and the intercept points with their coordinates.

25. $y = (x - 1)^2 + 2$ **26.** $y = -2(x + 3)^2$
27. $y = x^2 - 7x + 10$ graphical answers in App. F

CHAPTER 10 CUMULATIVE REVIEW

1. Solve $y + 0.6 = -1.0$. **2.** Solve $8(2 - t) = -5t$.

3. A chemist working on his doctorate degree at Massachusetts Institute of Technology needs 12 liters of a 50% acid solution for a lab experiment. The stockroom has only 40% and 70% solutions. How much of each solution should be mixed together to form 12 liters of a 50% solution?

4. Complete the following ordered pair solutions for the equation $3x + y = 12$. *(Sec. 3.1, Ex. 3)*

    a. (0,) (0, 12) **b.** (, 6) (2, 6) **c.** (−1,)

5. Find the slope of a line perpendicular to the line passing through the points $(-1, 7)$ and $(2, 2)$.

6. Simplify the following expressions. *(Sec. 4.1, Ex. 8)*

    a. 3^0 1 **b.** $(ab)^0$ 1 **c.** $(-5)^0$ 1 **d.** -5^0 −1

7. Multiply: $(3a + b)^3$.

8. Divide $6x^2 + 10x - 5$ by $3x - 1$.

9. Factor $r^2 - r - 42$. **10.** Factor $x^2 + 12x + 36$.

11. Factor $y^3 - 27$. **12.** Factor $64x^3 + 1$.

13. Solve $(5x - 1)(2x^2 + 15x + 18) = 0$.

14. Simplify $\dfrac{2x^2 - 2xy + 3x - 3y}{2x + 3}$.

15. A supermarket offers a 14-ounce box of cereal for \$3.79 and an 18-ounce box of the same brand of cereal for \$4.99. Which is the better buy?

16. Find the domain of each function. *(Sec. 7.4, Ex. 10)*

    a. $g(x) = \dfrac{1}{x}$ **b.** $f(x) = 2x + 1$ all real numbers

17. Solve the system *(Sec. 8.2, Ex. 1)*
$\begin{cases} 2x + y = 10 \\ x = y + 2 \end{cases}$ (4, 2)

18. Find each square root. *(Sec. 9.1, Ex. 1)*

    a. $\sqrt{36}$ 6 **b.** $\sqrt{64}$ 8 **c.** $-\sqrt{25}$ −5

    d. $\sqrt{\dfrac{9}{100}}$ $\dfrac{3}{10}$ **e.** $\sqrt{0}$ 0

19. Find the product and simplify. *(Sec. 9.4, Ex. 3)*

    a. $(\sqrt{5} - 7)(\sqrt{5} + 7)$ −44

    b. $(\sqrt{7x} + 2)^2$ $7x + 4\sqrt{7x} + 4$

(Sec. 9.4, Ex. 7)

20. Rationalize each denominator and simplify.

a. $\dfrac{2}{1+\sqrt{3}}$ $-1+\sqrt{3}$ b. $\dfrac{\sqrt{5}+4}{\sqrt{5}-1}$ $\dfrac{9+5\sqrt{5}}{4}$

21. Use the square root property to solve $(x-3)^2 = 16$. $x = -1, 7$ *(Sec. 10.1, Ex. 4)*

22. Find real solutions of the quadratic equation $2y^2 + 6y = -7$ by completing the square.

23. Solve $m^2 = 4m - 5$. $m = 2 \pm i$ *(Sec. 10.5, Ex. 8)*

22. no real solution *(Sec. 10.2, Ex. 5)*

Exercise Set 10.6 Answers

9. no x-intercepts, y-intercept: $(0, 5)$, vertex: $(2, 1)$

10. no x-intercepts, y-intercept: $(0, -5)$, vertex: $(2, -1)$

11. x-intercepts: $(-3, 0)$, $(1, 0)$, y-intercept: $(0, 3)$, vertex: $(-1, 4)$

12. x-intercept: $(-1, 0)$, $(3, 0)$, y-intercept: $(0, -3)$, vertex: $(1, -4)$

13. x-intercept: $\left(-\dfrac{1}{2}, 0\right), \left(\dfrac{1}{2}, 0\right)$, y-intercepts: $(0, 1)$, vertex: $(0, 1)$

14. x-intercepts: $\left(-\dfrac{1}{2}, 0\right), \left(\dfrac{1}{2}, 0\right)$, y-intercept: $(0, -1)$, vertex: $(0, -1)$

27. $y = (x - 1)^2 - 4$; vertex: $(1, -4)$; x-intercepts: $(-1, 0)$, $(3, 0)$; y-intercept: $(0, -3)$

28. $y = \left(x - \dfrac{1}{2}\right)^2 - \dfrac{9}{4}$; vertex: $\left(\dfrac{1}{2}, -\dfrac{9}{4}\right)$ x-intercepts: $(-1, 0)$, $(2, 0)$; y-intercept: $(0, -2)$

29. $y = (x + 1)^2$; vertex: $(-1, 0)$; x-intercept: $(-1, 0)$, y-intercept: $(0, 1)$

30. $y = \left(x + \dfrac{5}{2}\right)^2 - \dfrac{9}{4}$; vertex: $\left(-\dfrac{5}{2}, -\dfrac{9}{4}\right)$; x-intercepts: $(-4, 0)$, $(-1, 0)$; y-intercept: $(0, 4)$

31. $y = \left(x + \dfrac{7}{2}\right)^2 - \dfrac{9}{4}$; vertex: $\left(-\dfrac{7}{2}, -\dfrac{9}{4}\right)$; x-intercepts: $(-5, 0)$, $(-2, 0)$; y-intercept: $(0, 10)$

32. $y = (x - 2)^2 - 1$; vertex: $(2, -1)$; x-intercepts: $(3, 0)$, $(1, 0)$; y-intercept: $(0, 3)$

33. $y = (x - 3)^2 - 1$; vertex: $(3, -1)$; x-intercepts: $(2, 0)$, $(4, 0)$; y-intercept: $(0, 8)$

Chapter 10 Review Answers

39. $x = \dfrac{1}{2}, 7$ **40.** $x = -1, 7$ **45.** $x = -10, 22$ **46.** $x = -24, 18$ **47.** $x = \dfrac{3 \pm \sqrt{5}}{20}$ **48.** $x = -\dfrac{4}{5}, \dfrac{1}{5}$

49. $x = -5 \pm \sqrt{30}$ **50.** $x = 3 \pm \sqrt{5}$ **55.** $21 - 10i$ **56.** $19 - 25i$ **58.** $-4 + 2i$ **60.** $15 + 23i$

62. $45 - 28i$ **69.** $x = 2 \pm 3i$ **70.** $x = -2 \pm i\sqrt{7}$

Chapter 10 Cumulative Review Answers

1. $y = -1.6$ *(Sec. 2.2, Ex. 2)* **2.** $t = \dfrac{16}{3}$ *(Sec. 2.4, Ex. 2)* **3.** 8 liters @ 40%; 4 liters @ 70% *(Sec. 2.8, Ex. 3)*

4c. $(-1, 15)$ **5.** $\dfrac{3}{5}$ *(Sec. 3.4, Ex. 8)* **7.** $27a^3 + 27a^2b + 9ab^2 + b^3$ *(Sec. 4.3, Ex. 5)*

8. $2x + 4 - \dfrac{1}{3x - 1}$ *(Sec. 4.6, Ex. 6)* **9.** $(r - 7)(r + 6)$ *(Sec. 5.2, Ex. 4)* **10.** $(x + 6)^2$ *(Sec. 5.3, Ex. 7)*

11. $(y - 3)(y^2 + 3y + 9)$ *(Sec. 5.4, Ex. 6)* **12.** $(4x + 1)(16x^2 - 4x + 1)$ *(Sec. 5.4, Ex. 7)*

13. $x = \dfrac{1}{5}, -\dfrac{3}{2}, -6$ *(Sec. 5.6, Ex. 6)* **14.** $x - y$ *(Sec. 6.1, Ex. 9)* **15.** 14 oz. for $3.79 *(Sec. 6.7, Ex. 6)*

16a. all real numbers except 0

OPERATIONS ON DECIMALS

To **add** or **subtract** decimals, write the numbers vertically with decimal points lined up. Add or subtract as with whole numbers and place the decimal point in the answer directly below the decimal points in the problem.

EXAMPLE 1   Add $5.87 + 23.279 + 0.003$.

Solution:

$$\begin{array}{r} 5.87 \\ 23.279 \\ + 0.003 \\ \hline 29.152 \end{array}$$

EXAMPLE 2   Subtract $32.15 - 11.237$.

Solution

$$\begin{array}{r} \overset{1}{}\ \overset{11}{2}.\overset{4}{1}\ \overset{10}{5}\ \overset{}{0} \\ 3\ \cancel{2}.\cancel{1}\ \cancel{5}\ \cancel{0} \\ -\ 1\ 1\ .\ 2\ 3\ 7 \\ \hline 2\ 0\ .\ 9\ 1\ 3 \end{array}$$

To **multiply** decimals, multiply the numbers as if they were whole numbers. The decimal point in the product is placed so that the number of decimal places in the product is the same as the sum of the number of decimal places in the factors.

EXAMPLE 3   Multiply 0.072×3.5.

Solution:

$$\begin{array}{rl} 0.072 & \text{3 decimal places} \\ \times\quad 3.5 & \text{1 decimal place} \\ \hline 360 \\ 216 \\ \hline 0.2520 & \text{4 decimal places} \end{array}$$

To **divide** decimals, move the decimal point in the divisor to the right of the last digit. Move the decimal point in the dividend the same number of places that the decimal point in the divisor was moved. The decimal point in the quotient lies directly above the decimal point in the dividend.

EXAMPLE 4 Divide $9.46 \div 0.04$.

Solution:
$$
\begin{array}{r}
236.5 \\
04.\overline{)946.0} \\
\underline{-8} \\
14 \\
\underline{-12} \\
26 \\
\underline{-24} \\
20 \\
\underline{-20} \\
\end{array}
$$

APPENDIX A EXERCISE SET

Perform the indicated operations.

1. $9.076 + 8.004$ 17.08

2.
$$\begin{array}{r} 6.3 \\ \times\ 0.05 \end{array}$$ 0.315

3.
$$\begin{array}{r} 27.004 \\ -14.2 \\ \hline 12.806 \end{array}$$

4.
$$\begin{array}{r} 0.0036 \\ 7.12 \\ 32.502 \\ +\ 0.05 \\ \hline \end{array}$$ 39.6756

5.
$$\begin{array}{r} 107.92 \\ +\ 3.04 \end{array}$$ 110.96

6. $7.2 \div 4$ 1.8

7. $10 - 7.6$ 2.4

8. $40 \div 0.25$ 160

9. $126.32 - 97.89$ 28.43

10.
$$\begin{array}{r} 3.62 \\ 7.11 \\ 12.36 \\ 4.15 \\ +\ 2.29 \end{array}$$ 29.53

11.
$$\begin{array}{r} 3.25 \\ \times\ 70 \end{array}$$ 227.5

12.
$$\begin{array}{r} 26.014 \\ -\ 7.8 \end{array}$$ 18.214

13. $8.1 \div 3$ 2.7

14.
$$\begin{array}{r} 1.2366 \\ 0.005 \\ 15.17 \\ +\ 0.97 \end{array}$$ 17.3816

15. $55.405 - 6.1711$ 49.2339

16. $8.09 + 0.22$ 8.31

17. $60 \div 0.75$ 80

18. $20 - 12.29$ 7.71

19. $7.612 \div 100$ 0.07612

20.
$$\begin{array}{r} 8.72 \\ 1.12 \\ 14.86 \\ 3.98 \\ +\ 1.99 \end{array}$$ 30.67

21. $12.312 \div 2.7$ 4.56

22. $0.443 \div 100$ 0.00443

23.
$$\begin{array}{r} 569.2 \\ 71.25 \\ +\ 8.01 \end{array}$$ 648.46

24. $3.706 - 2.91$ 0.796

25. $768 - 0.17$ 767.83

26. $63 \div 0.28$ 225

27. $12 + 0.062$ 12.062

28. $0.42 + 18$ 18.42

29. $76 - 14.52$ 61.48

30. $1.1092 \div 0.47$ 2.36

31. $3.311 \div 0.43$ 7.7

32. $7.61 + 0.0004$ 7.6104

33.
$$\begin{array}{r} 762.12 \\ 89.7 \\ +\ 11.55 \end{array}$$ 863.37

34. $444 \div 0.6$ 740

35. $23.4 - 0.821$ 22.579

36. $3.7 + 5.6$ 9.3

37. $476.12 - 112.97$

38. $19.872 \div 0.54$ 36.8

39. $0.007 + 7$ 7.007

40.
$$\begin{array}{r} 51.77 \\ +\ 3.6 \end{array}$$ 55.37

37. 363.15

Review of Angles, Lines, and Special Triangles

The word **geometry** is formed from the Greek words, **geo,** meaning earth, and **metron,** meaning measure. Geometry literally means to measure the earth.

This section contains a review of some basic geometric ideas. It will be assumed that fundamental ideas of geometry such as point, line, ray, and angle are known. In this appendix, the notation $\angle 1$ is read "angle 1" and the notation $m\angle 1$ is read "the measure of angle 1."

We first review types of angles.

ANGLES

A **right angle** is an angle whose measure is 90°. A right angle can be indicated by a square drawn at the vertex of the angle, as shown below.

An angle whose measure is more than 0° but less than 90° is called an **acute angle**.

An angle whose measure is greater than 90° but less than 180° is called an **obtuse angle**.

An angle whose measure is 180° is called a **straight angle**.

Two angles are said to be **complementary** if the sum of their measures is 90°. Each angle is called the **complement** of the other.

Two angles are said to be **supplementary** if the sum of their measures is 180°. Each angle is called the **supplement** of the other.

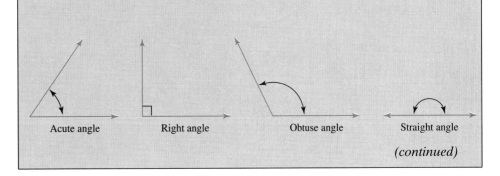

| Acute angle | Right angle | Obtuse angle | Straight angle |

(continued)

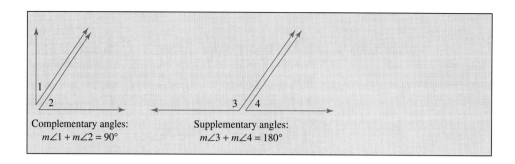

Complementary angles:
$m\angle 1 + m\angle 2 = 90°$

Supplementary angles:
$m\angle 3 + m\angle 4 = 180°$

EXAMPLE 1 If an angle measures 28°, find its complement.

Solution: Two angles are complementary if the sum of their measures is 90°. The complement of a 28° angle is an angle whose measure is $90° - 28° = 62°$. To check, notice that $28° + 62° = 90°$.

Plane is an undefined term that we will describe. A plane can be thought of as a flat surface with infinite length and width, but no thickness. A plane is two dimensional. The arrows in the following diagram indicate that a plane extends indefinitely and has no boundaries.

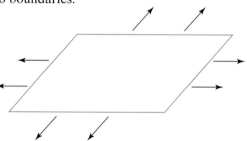

Figures that lie on a plane are called **plane figures.** (See the description of common plane figures in Appendix C.) Lines that lie in the same plane are called **coplanar.**

LINES

Two lines are **parallel** if they lie in the same plane but never meet.

Intersecting lines meet or cross in one point.

Two lines that form right angles when they intersect are said to be **perpendicular.**

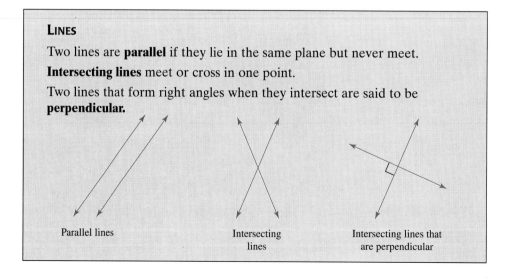

Parallel lines | Intersecting lines | Intersecting lines that are perpendicular

Two intersecting lines form **vertical angles.** Angles 1 and 3 are vertical angles. Also angles 2 and 4 are vertical angles. It can be shown that **vertical angles have equal measures.**

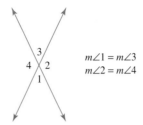

$$m\angle 1 = m\angle 3$$
$$m\angle 2 = m\angle 4$$

Adjacent angles have the same vertex and share a side. Angles 1 and 2 are adjacent angles. Other pairs of adjacent angles are angles 2 and 3, angles 3 and 4, and angles 4 and 1.

A **transversal** is a line that intersects two or more lines in the same plane. Line l is a transversal that intersects lines m and n. The eight angles formed are numbered and certain pairs of these angles are given special names.

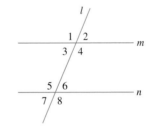

Corresponding angles: $\angle 1$ and $\angle 5$, $\angle 3$ and $\angle 7$, $\angle 2$ and $\angle 6$, and $\angle 4$ and $\angle 8$.

Exterior angles: $\angle 1$, $\angle 2$, $\angle 7$, and $\angle 8$.

Interior angles: $\angle 3$, $\angle 4$, $\angle 5$, and $\angle 6$.

Alternate interior angles: $\angle 3$ and $\angle 6$, $\angle 4$ and $\angle 5$.

These angles and parallel lines are related in the following manner.

PARALLEL LINES CUT BY A TRANSVERSAL

1. If two parallel lines are cut by a transversal, then
   a. **corresponding angles are equal** and
   b. **alternate interior angles are equal.**
2. If corresponding angles formed by two lines and a transversal are equal, then the lines are parallel.
3. If alternate interior angles formed by two lines and a transversal are equal, then the lines are parallel.

EXAMPLE 2   Given that lines m and n are parallel and that the measure of angle 1 is 100°, find the measures of angles 2, 3, and 4.

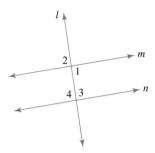

Solution: $m\angle 2 = 100°$,   since angles 1 and 2 are vertical angles.

$m\angle 4 = 100°$,   since angles 1 and 4 are alternate interior angles.

$m\angle 3 = 180° - 100° = 80°$, since angles 4 and 3 are supplementary angles.

A **polygon** is the union of three or more coplanar line segments that intersect each other only at each end point, with each end point shared by exactly two segments.

A **triangle** is a polygon with three sides. The sum of the measures of the three angles of a triangle is 180°. In the following figure, $m\angle 1 + m\angle 2 + m\angle 3 = 180°$.

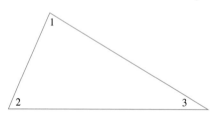

EXAMPLE 3   Find the measure of the third angle of the triangle shown.

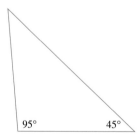

Solution: The sum of the measures of the angles of a triangle is 180°. Since one angle measures 45° and the other angle measures 95°, the third angle measures 180° − 45° − 95° = 40°.

Two triangles are **congruent** if they have the same size and the same shape. In congruent triangles, the measures of corresponding angles are equal and the lengths of corresponding sides are equal. The following triangles are congruent.

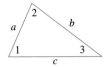 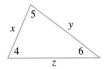

Corresponding angles are equal: $m\angle 1 = m\angle 4, m\angle 2 = m\angle 5$, and $m\angle 3 = m\angle 6$. Also, lengths of corresponding sides are equal: $a = x, b = y$, and $c = z$.

Any one of the following may be used to determine whether two triangles are congruent.

CONGRUENT TRIANGLES

1. If the measures of two angles of a triangle equal the measures of two angles of another triangle and the lengths of the sides between each pair of angles are equal, the triangles are congruent.

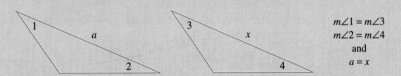

$m\angle 1 = m\angle 3$
$m\angle 2 = m\angle 4$
and
$a = x$

2. If the lengths of the three sides of a triangle equal the lengths of corresponding sides of another triangle, the triangles are congruent.

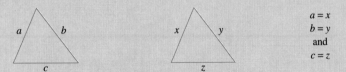

$a = x$
$b = y$
and
$c = z$

3. If the lengths of two sides of a triangle equal the lengths of corresponding sides of another triangle, and the measures of the angles between each pair of sides are equal, the triangles are congruent.

$a = x$
$b = y$
and
$m\angle 1 = m\angle 2$

Two triangles are **similar** if they have the same shape. In similar triangles, the measures of corresponding angles are equal and corresponding sides are in proportion. The following triangles are similar. (All similar triangles drawn in this appendix will be oriented the same.)

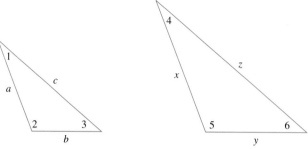

Corresponding angles are equal: $m\angle 1 = m\angle 4$, $m\angle 2 = m\angle 5$, and $m\angle 3 = m\angle 6$.

Also, corresponding sides are proportional: $\dfrac{a}{x} = \dfrac{b}{y} = \dfrac{c}{z}$.

Any one of the following may be used to determine whether two triangles are similar.

SIMILAR TRIANGLES

1. If the measures of two angles of a triangle equal the measures of two angles of another triangle, the triangles are similar.

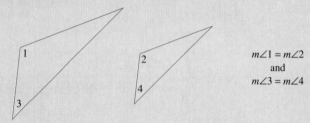

$$m\angle 1 = m\angle 2$$
and
$$m\angle 3 = m\angle 4$$

2. If three sides of one triangle are proportional to three sides of another triangle, the triangles are similar.

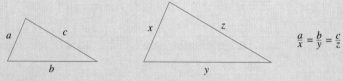

$$\frac{a}{x} = \frac{b}{y} = \frac{c}{z}$$

3. If two sides of a triangle are proportional to two sides of another triangle and the measures of the included angles are equal, the triangles are similar.

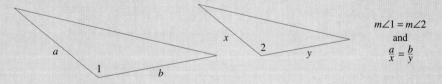

$$m\angle 1 = m\angle 2$$
and
$$\frac{a}{x} = \frac{b}{y}$$

EXAMPLE 4   Given that the following triangles are similar, find the missing length x.

Solution: Since the triangles are similar, corresponding sides are in proportion. Thus, $\frac{2}{3} = \frac{10}{x}$. To solve this equation for x, we multiply both sides by the LCD $3x$.

$$3x\left(\frac{2}{3}\right) = 3x\left(\frac{10}{x}\right)$$
$$2x = 30$$
$$x = 15$$

The missing length is 15 units.

A **right triangle** contains a right angle. The side opposite the right angle is called the **hypotenuse,** and the other two sides are called the **legs.** The **Pythagorean theorem** gives a formula that relates the lengths of the three sides of a right triangle.

THE PYTHAGOREAN THEOREM

If a and b are the lengths of the legs of a right triangle, and c is the length of the hypotenuse, then $a^2 + b^2 = c^2$.

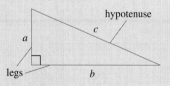

EXAMPLE 5   Find the length of the hypotenuse of a right triangle whose legs have lengths of 3 centimeters and 4 centimeters.

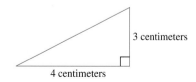

Solution: Because we have a right triangle, we use the Pythagorean theorem. The legs are 3 centimeters and 4 centimeters, so let $a = 3$ and $b = 4$ in the formula.

$$a^2 + b^2 = c^2$$
$$3^2 + 4^2 = c^2$$
$$9 + 16 = c^2$$
$$25 = c^2$$

Since c represents a length, we assume that c is positive. Thus, if c^2 is 25, c must be 5. The hypotenuse has a length of 5 centimeters.

Appendix B Exercise Set

Find the complement of each angle. See Example 1.

1. $19°$ $71°$ **2.** $65°$ $25°$ **3.** $70.8°$ $19.2°$

4. $45\frac{2}{3}°$ $44\frac{1}{3}°$ **5.** $11\frac{1}{4}°$ $78\frac{3}{4}°$ **6.** $19.6°$ $70.4°$

Find the supplement of each angle.

7. $150°$ $30°$ **8.** $90°$ $90°$ **9.** $30.2°$ $149.8°$

10. $81.9°$ $98.1°$ **11.** $79\frac{1}{2}°$ $100\frac{1}{2}°$ **12.** $165\frac{8}{9}°$ $14\frac{1}{9}°$

13. If lines m and n are parallel, find the measures of angles 1 through 7. See Example 2.

$$m\angle 1 = m\angle 5 = m\angle 7 = 110°$$
$$m\angle 2 = m\angle 3 = m\angle 4 = m\angle 6 = 70°$$

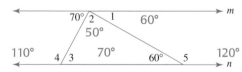

14. If lines m and n are parallel, find the measures of angles 1 through 5. See Example 2.

In each of the following, the measures of two angles of a triangle are given. Find the measure of the third angle. See Example 3.

15. $11°, 79°$ $90°$ **16.** $8°, 102°$ $70°$ **17.** $25°, 65°$ $90°$

18. $44°, 19°$ $117°$ **19.** $30°, 60°$ $90°$ **20.** $67°, 23°$ $90°$

In each of the following, the measure of one angle of a right triangle is given. Find the measures of the other two angles.

21. $45°$ $45°, 90°$ **22.** $60°$ $30°, 90°$ **23.** $17°$ $73°, 90°$

24. $30°$ $60°, 90°$ **25.** $39\frac{3}{4}°$ $50\frac{1}{4}°, 90°$ **26.** $72.6°$ $17.4°, 90°$

Given that each of the following pairs of triangles is similar, find the missing lengths. See Example 4.

27. 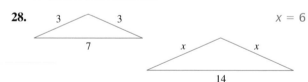 $x = 6$

28. 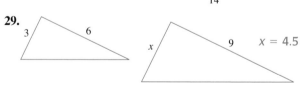 $x = 6$

29. 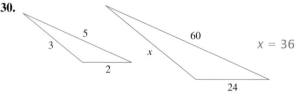 $x = 4.5$

30. $x = 36$

Use the Pythagorean Theorem to find the missing lengths in the right triangles. See Example 5.

31.

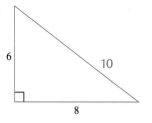

32.

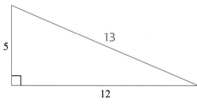

33.

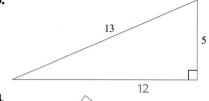

34.

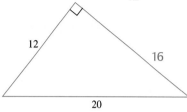

C

REVIEW OF GEOMETRIC FIGURES

PLANE FIGURES HAVE LENGTH AND WIDTH BUT NO THICKNESS OR DEPTH.		
NAME	**DESCRIPTION**	**FIGURE**
POLYGON	Union of three or more coplanar line segments that intersect with each other only at each end point, with each end point shared by two segments.	
TRIANGLE	Polygon with three sides (sum of measures of three angles is 180°).	
SCALENE TRIANGLE	Triangle with no sides of equal length.	
ISOSCELES TRIANGLE	Triangle with two sides of equal length.	
EQUILATERAL TRIANGLE	Triangle with all sides of equal length.	
RIGHT TRIANGLE	Triangle that contains a right angle.	leg, hypotenuse, leg
QUADRILATERAL	Polygon with four sides (sum of measures of four angles is 360°).	

PLANE FIGURES HAVE LENGTH AND WIDTH BUT NO THICKNESS OR DEPTH.		
NAME	**DESCRIPTION**	**FIGURE**
TRAPEZOID	Quadrilateral with exactly one pair of opposite sides parallel.	
ISOSCELES TRAPEZOID	Trapezoid with legs of equal length.	
PARALLELOGRAM	Quadrilateral with both pair of opposite sides parallel and equal in length.	
RHOMBUS	Parallelogram with all sides of equal length.	
RECTANGLE	Parallelogram with four right angles.	
SQUARE	Rectangle with all sides of equal length.	
CIRCLE	All points in a plane the same distance from a fixed point called the **center.**	

SOLID FIGURES HAVE LENGTH, WIDTH, AND HEIGHT OR DEPTH.		
NAME	**DESCRIPTION**	**FIGURE**
RECTANGULAR SOLID	A solid with six sides, all of which are rectangles.	
CUBE	A rectangular solid whose six sides are squares.	
SPHERE	All points the same distance from a fixed point, called the **center.**	
RIGHT CIRCULAR CYLINDER	A cylinder consisting of two circular bases that are perpendicular to its altitude.	
RIGHT CIRCULAR CONE	A cone with a circular base that is perpendicular to its altitude.	

MEAN, MEDIAN, AND MODE

It is sometimes desirable to be able to describe a set of data, or a set of numbers, by a single "middle" number. Three such **measures of central tendency** are the mean, the median, and the mode.

The most common measure of central tendency is the mean (sometimes called the arithmetic mean or the average). The **mean** of a set of data items, denoted by $\bar{x}$, is the sum of the items divided by the number of items.

EXAMPLE 1   Seven students in a psychology class conducted an experiment on mazes. Each student was given a pencil and asked to successfully complete the same maze. The timed results are below.

STUDENT	Ann	Thanh	Carlos	Jesse	Melinda	Ramzi	Dayni
TIME (SECONDS)	13.2	11.8	10.7	16.2	15.9.	13.8	18.5

(a) Who completed the maze in the shortest time? Who completed the maze in the longest time?

(b) Find the mean.

(c) How many students took longer than the mean time? How many students took shorter than the mean time?

Solution: **(a)** Carlos completed the maze in 10.7 seconds, the shortest time. Dayni completed the maze in 18.5 seconds, the longest time.

(b) To find the mean, $\bar{x}$, find the sum of the data items and divide by 7, the number of items.

$$\bar{x} = \frac{13.2 + 11.8 + 10.7 + 16.2 + 15.9 + 13.8 + 18.5}{7} = \frac{100.1}{7} = 14.3$$

(c) Three students, Jesse, Melinda, and Dayni, had times longer than the mean time. Four students, Ann, Thanh, Carlos, and Ramzi, had times shorter than the mean time.

Two other measures of central tendency are the median and the mode.

The **median** of an ordered set of numbers is the middle number. If the number of items is even, the median is the mean of the two middle numbers. The **mode** of a set of numbers is the number that occurs most often. It is possible for a data set to have no mode or more than one mode.

EXAMPLE 2   Find the median and the mode of the following set of numbers. These numbers were high temperatures for fourteen consecutive days in a city in Montana.

$$76, 80, 85, 86, 89, 87, 82, 77, 76, 79, 82, 89, 89, 92$$

Solution: First, write the numbers in order.

$$76, 76, 77, 79, 80, 82, \underbrace{82, 85}_{\text{two}}, 86, 87, \underbrace{89, 89, 89}_{\text{mode}}, 92$$

two
middle numbers

Since there are an even number of items, the median is the mean of the two middle numbers.

$$\text{median} = \frac{82 + 85}{2} = 83.5$$

The mode is 89, since 89 occurs most often.

APPENDIX D EXERCISE SET

For each of the following data sets, find the mean, the median, and the mode. If necessary, round the mean to one decimal place.

1. 21, 28, 16, 42, 38   **2.** 42, 35, 36, 40, 50

3. 7.6, 8.2, 8.2, 9.6, 5.7, 9.1

4. 4.9, 7.1, 6.8, 6.8, 5.3, 4.9

5. 0.2, 0.3, 0.5, 0.6, 0.6, 0.9, 0.2, 0.7, 1.1

6. 0.6, 0.6, 0.8, 0.4, 0.5, 0.3, 0.7, 0.8, 0.1

7. 231, 543, 601, 293, 588, 109, 334, 268

8. 451, 356, 478, 776, 892, 500, 467, 780

1. mean: 29, median: 28, no mode
2. mean: 40.6, median: 40, no mode
3. mean: 8.1, median: 8.2, mode: 8.2
4. mean: 6.0, median: 6.05, mode: 6.8 and 4.9
5. mean: 0.6, median: 0.6, mode: 0.2 and 0.6
6. mean: 0.5, median: 0.6, mode: 0.6 and 0.8
7. mean: 370.9, median: 313.5, no mode
8. mean: 587.5, median: 489, no mode

The ten tallest buildings in the United States are listed below. Use this table for Exercises 9–12.

BUILDING	HEIGHT (FEET)
Sears Tower, Chicago, IL	1454
One World Trade Center (1972), New York, NY	1368
One World Trade Center (1973), New York, NY	1362
Empire State, New York, NY	1250
Amoco, Chicago IL	1136
John Hancock Center, Chicago, IL	1127
First Interstate World Center, Los Angeles, CA	1107
Chrysler, New York, NY	1046
NationsBank Tower, Atlanta, GA	1023
Texas Commerce Tower, Houston, TX	1002

9. Find the mean height for the five tallest buildings.

10. Find the median height for the five tallest buildings.   1362 feet

11. Find the median height for the ten tallest buildings.   1131.5 feet

12. Find the mean height for the ten tallest buildings.

During an experiment, the following times (in seconds) were recorded: 7.8, 6.9, 7.5, 4.7, 6.9, 7.0.

13. Find the mean. Round to the nearest tenth.   6.8

14. Find the median.   **15.** Find the mode.   6.9

In a mathematics class, the following test scores were recorded for a student: 86, 95, 91, 74, 77, 85.

16. Find the mean. Round to the nearest hundredth.

17. Find the median.   **18.** Find the mode.

The following pulse rates were recorded for a group of fifteen students: 78, 80, 66, 68, 71, 64, 82, 71, 70, 65, 70, 75, 77, 86, 72

19. Find the mean.   73

20. Find the median.   71   **21.** Find the mode.

22. How many rates were higher than the mean?   6

23. How many rates were lower than the mean?   9

24. Have each student in your algebra class take his/her pulse rate. Record the data and find the mean, the median, and the mode.

Find the missing numbers in each set of numbers. (These numbers are not necessarily in numerical order.)

25. __, __, 16, 18, __   **26.** __, __, __, __, 40
The mode is 21.   The mode is 35.
The median is 20.   The median is 37.
                             The mean is 38.

9. 1314 feet   **12.** 1187.5 feet   **14.** 6.95   **16.** 84.67   **17.** 85.5   **18.** no mode   **21.** 70 and 71
24. answers will vary   **25.** 21, 21, 24   **26.** 35, 35, 37, 43

E

TABLE OF SQUARES AND SQUARE ROOTS

n	n^2	$\sqrt{n}$	n	n^2	$\sqrt{n}$
1	1	1.000	51	2,601	7.141
2	4	1.414	52	2,704	7.211
3	9	1.732	53	2,809	7.280
4	16	2.000	54	2,916	7.348
5	25	2.236	55	3,025	7.416
6	36	2.449	56	3,136	7.483
7	49	2.646	57	3,249	7.550
8	64	2.828	58	3,364	7.616
9	81	3.000	59	3,481	7.681
10	100	3.162	60	3,600	7.746
11	121	3.317	61	3,721	7.810
12	144	3.464	62	3,844	7.874
13	169	3.606	63	3,969	7.937
14	196	3.742	64	4,096	8.000
15	225	3.873	65	4,225	8.062
16	256	4.000	66	4,356	8.124
17	289	4.123	67	4,489	8.185
18	324	4.243	68	4,624	8.246
19	361	4.359	69	4,761	8.307
20	400	4.472	70	4,900	8.367
21	441	4.583	71	5,041	8.426
22	484	4.690	72	5,184	8.485
23	529	4.796	73	5,329	8.544
24	576	4.899	74	5,476	8.602
25	625	5.000	75	5,625	8.660
26	676	5.099	76	5,776	8.718
27	729	5.196	77	5,929	8.775
28	784	5.292	78	6,084	8.832
29	841	5.385	79	6,241	8.888
30	900	5.477	80	6,400	8.944

n	n^2	$\sqrt{n}$	n	n^2	$\sqrt{n}$
31	961	5.568	81	6,561	9.000
32	1,024	5.657	82	6,724	9.055
33	1,089	5.745	83	6,889	9.110
34	1,156	5.831	84	7,056	9.165
35	1,225	5.916	85	7,225	9.220
36	1,296	6.000	86	7,396	9.274
37	1,369	6.083	87	7,569	9.327
38	1,444	6.164	88	7,744	9.381
39	1,521	6.245	89	7,921	9.434
40	1,600	6.325	90	8,100	9.487
41	1,681	6.403	91	8,281	9.539
42	1,764	6.481	92	8,464	9.592
43	1,849	6.557	93	8,649	9.644
44	1,936	6.633	94	8,836	9.695
45	2,025	6.708	95	9,025	9.747
46	2,116	6.782	96	9,216	9.798
47	2,209	6.856	97	9,409	9.849
48	2,304	6.928	98	9,604	9.899
49	2,401	7.000	99	9,801	9.950
50	2,500	7.071	100	10,000	10.000

F

INSTRUCTOR'S ANSWERS

- Answers to Group Activities
- Answers to Exercises
  Requiring Graphical Solutions
- Notes to the Instructor

Answers to Group Activities

CHAPTER 1
Group Activity (page 62)

Suggestions: You may want to divide your class up into groups of three and ask each member to be responsible for one of the three survey questions.

- For efficiency, one way to implement this activity is conducting the survey orally during class and asking for a show of hands for each category. You could then tally and display the results for the groups to use. Or you could go around the room and have each student give an oral response. Another way to implement this activity and save time is to give a written survey during class one day, collect and compile the raw data yourself, and then pass out the tallied results to the groups when they begin this activity.

- If you choose to have your students do their circle graphs by hand, you may wish to do an example first. Remind students that they can more easily graph the fractions of responses in categories if they divide their circles into a number of (equal) parts equal to the total number of responses. For that reason you may wish to restrict the number of "survey respondents" to a convenient number such as 8 or 16.

- If your students have already been introduced to angle measurement in your class or a prior class, for the circle graphs you might ask them to multiply a category's fraction of total responses by 360° to find the angle measure they should use to plot that category's sector on the circle graph. If you choose to have your students deal with angle measures, remember to ask them to bring a protractor to class as well as the other materials.

- Encourage students who finish early to help others in their group. Encourage group members to check each others' work as time permits.

Answers

For Questions 1–4 and 6, answers will vary depending on survey results.

5. Answers will vary, but look for conclusions such as more people ate Kellogg's cereals, more females than males responded to the survey, there were more people over age 30 than 20 or under, and so on.

CHAPTER 2
Group Activity (page 149)

Suggestions: To save time in class and encourage student involvement, you might consider asking students to find the necessary prices for the fruit salad ingredients outside of class ahead of time. Alternatively, you might supply the necessary circulars and advertisements to each group rather than asking them to bring their own. You might also consider collecting all of the price information yourself in advance and either handing out this information to each group or displaying the information with an overhead projector.

- Emphasize to your students that the ingredient unit prices they use from the circulars *must match the units given in the recipe.* Otherwise, they will need to convert units.

Answers

1. Descriptions will vary. One example is: "The fruit salad cost per person is the sum of the total weight of grapes times the price per pound, the total weight of watermelon times the price per pound, the total number of pints of blueberries times the price per pint, the total weight of nectarines times the price per pound, and the total number of cans of juice times the price per can, all divided by the total number of people to be served."

2. Let c = fruit salad cost per person
   g = price per pound of grapes
   w = price per pound of watermelon
   b = price per pint of blueberries
   n = price per pound of nectarines
   j = price per can of juice

$$c = \frac{4g + 6w + 2b + 2.5n + 2j}{20}$$

3. Answers will vary depending on the prices students use. For instance, students might report prices such as the following:

INGREDIENT	PRICE
grapes	$1.49/lb.
watermelon	$0.25/lb.
blueberries	$1.99/pint
nectarines	$1.29/lb.
frozen juice	$0.89/can

$$c = \frac{4(1.49) + 6(0.25) + 2(1.99) + 2.5(1.29) + 2(0.89)}{20}$$

$$\approx 0.82$$

4. Answers will vary depending on the prices students use. For instance, if students collected the above prices, the additional amount that Sabrina must charge to receive a total of $90 for the event may be found by solving the equation $20(0.82 + x) = 90$ for

x. In this case, she should charge an additional $3.68 per person above the ingredient cost per person. This gives a total price of $4.50 per person, which represents a 449% $\left(\frac{3.68}{0.82} \approx 4.49\right)$ increase over her actual cost per person of $0.82.

CHAPTER 3
Group Activity (page 217)

Suggestions: Groups of four students would work well for this activity. Each pair of students could be responsible for the analysis of four of the eight companies in Questions 1 and 2. Students could then compile their results and discuss Question 3 together to reach a conclusion for Question 4.

- Consider asking each group to write and hand in a brief recommendation report as a business analyst might do.

- If you choose to do optional Question 6 and wish to save time during class, you might supply the necessary references yourself for students to use during class or do the research yourself in advance and supply each group with a different set of data to analyze.

- Optional Question 6 could be assigned as a project, allowing sufficient time for completion.

Answers

1. and **2.**

COMPANY	1993	1994	ORDERED PAIRS (#1)	SLOPE (#2)
Boeing	25,300	21,900	(1993, 25,300) (1994, 21,900)	−3400
United Technologies	20,700	21,200	(1993, 20,700) (1994, 21,200)	500
McDonnell-Douglas	14,500	13,200	(1993, 14,500) (1994, 13,200)	−1300
Lockheed	13,100	13,100	(1993, 13,100) (1994, 13,100)	0
Allied-Signal	11,800	12,800	(1993, 11,800) (1994, 12,800)	1000
Martin Marietta	9,400	9,900	(1993, 9,400) (1994, 9,900)	500
Textron	8,700	9,700	(1993, 8,700) (1994, 9,700)	1000
General Dynamics	4,700	3,700	(1993, 4,700) (1994, 3,700)	−1000

3. Negative slopes: Boeing, McDonnell-Douglas, General Dynamics; based on the data points for these companies, they are losing sales at rates of $3400, $1300, and $1000 million dollars per year, respectively. Zero slopes: Lockheed; based on the data points for this company, it is neither increasing nor losing sales. Parallel lines: United Technologies & Martin Marietta; Textron & Allied-Signal; based on the given data these two pairs of companies are increasing their sales at the same rate per year. Textron is increasing sales at the same rate (+1000) that General Dynamics is losing sales (−1000).

4. Two good choices are Allied-Signal and Textron because they have increased their sales by the greatest amounts.

5. Probably not because two years does not reveal a historical trend. Answers will vary; some possibilities include factors such as debt, stock price history, profit margin, and so on. Additionally, you would want to consider the future prospects for these companies. For example, what orders do they have? What new products are they introducing?

6. Answers will vary depending on the data researched.

CHAPTER 4
Group Activity (page 273)

Suggestions: When your students are completing the tables in this activity, encourage them to double-check their calculations. For example, two students could make the computations simultaneously and then compare their results.

- If your students are comfortable with using a graphing calculator, you might show them how to enter the polynomials with the $\boxed{Y =}$ key and evaluate the polynomials at different values from the home screen. You could also show your students how to create a table with their graphing calculators (if their graphing utilities have this capability).

Answers

1. In both cases, the number of deaths initially decreases through 1992 and then begins to increase again.

YEAR	x	NUMBER OF ACCIDENTAL DEATHS DUE TO MOTOR VEHICLE ACCIDENTS	NUMBER OF ACCIDENTAL DEATHS DUE TO ALL OTHER TYPES OF ACCIDENTS
1990	0	46,426	33,139
1991	1	42,938	31,329
1992	2	41,236	30,979
1993	3	41,320	32,089
1994	4	43,190	34,659

2. $79,565 - 6921x + 1623x^2$; 79,565; 74,267; 72,215; 73,409; 77,849

3.

YEAR	x	NUMBER OF ACCIDENTAL DEATHS DUE TO MOTOR VEHICLE ACCIDENTS	NUMBER OF ACCIDENTAL DEATHS DUE TO ALL OTHER TYPES OF ACCIDENTS	TOTAL NUMBER OF ACCIDENTAL DEATHS DUE TO ALL CAUSES
1990	0	46,426	33,139	79,565
1991	1	42,938	31,329	74,267
1992	2	41,236	30,979	72,215
1993	3	41,320	32,089	73,409
1994	4	43,190	34,659	77,849

4. 172,655; 230,225

5. The graph implies that the total number of accidental deaths will increase dramatically in the future.

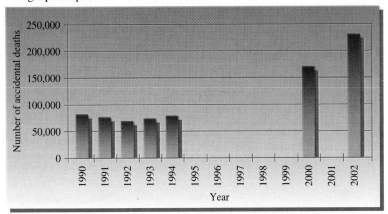

6. The bar graph for this question should be the same as for Question 5.

CHAPTER 5
Group Activity (page 333)

Suggestions: Groups of size six should work well for this activity. Consider asking each pair of group members to be in charge of investigating one of the three options and then sharing his or her findings with the rest of the group.

- Alternatively, divide the entire class into three groups and assign each group to work out one of the three options. After a few minutes, together as a class discuss the findings of each option. Let each group continue with the remaining questions using a correct set of findings.

- You might consider asking groups to actually give their presentations (Question 5) to the rest of the class or having each group give their presentation to one other group.

Answers

1. Diagrams will vary. For example, one possibility is

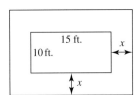

2. Option A: $5[(15 + 2x)(10 + 2x) - (10)(15)]$;
   Option B: $7.5[(15 + 2x)(10 + 2x) - (10)(15)] + 30$;

Option C: $4.5[(15 + 2x)(10 + 2x) - (10)(15)] + 10.86(50)$

3. Option A: 7.5 feet wide; Option B: 5.5 feet wide; Option C: 7 feet wide

4. Answers may vary. If Sam wants the widest patio possible for $3000, he should choose Option A. However, the other materials may be better for his situation. For instance, if he has children, outdoor carpeting may be better (more likely to prevent injuries). If aesthetics are considered, he may prefer the brick patio.

5. Presentations will vary.

CHAPTER 6
Group Activity (page 399)

Suggestions: You might consider reminding your students that a graphing calculator with the capability to make tables would be helpful in Question 1.

Answers

1.

AGE	2	3	4	5	6
YOUNG'S	142.9	200	250	294.1	333.3
COWLING'S	125	166.7	208.3	250	291.7
DIFFERENCE	17.9	33.3	41.7	44.1	41.6

7	8	9	10	11	12	13
368.4	400	428.6	454.5	478.3	500	520
333.3	375	416.7	458.3	500	541.7	583.3
35.1	25	11.9	3.8	21.7	41.7	63.3

2. The dosages for Cowling's Rule fall along a straight (increasing) line; the dosages for Young's Rule are also increasing but fall along a curve.

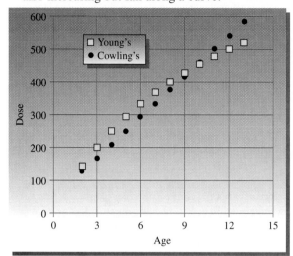

3. No; Yes, between ages 9 and 10.

4. Age 13

5. Yes, at age 23, found by solving $C = \dfrac{D(A + 1)}{24}$ for

A after setting $C = D$; No, because the adult dose is always multiplied by a factor of $\dfrac{A}{A + 12}$ which is never equal to 1, the necessary requirement for the rule to predict exactly the adult dose.

6. Answers will vary. One possibility might be: because children of the same age can differ widely in characteristics such as height, weight, and metabolism, age is probably not the most important factor in determining a child's dose.

CHAPTER 7
Group Activity (page 444)

Suggestions: This activity will give your students practice in matching graphs to equations (note that each equation is matched to *exactly one* graph and vice versa) and the descriptions of several real-life situations to pairs of equations and graphs. You might consider allowing the groups to use a graphing calculator as a tool to verify their answers to match up equations to graphs. For instance, one member of the group could have the responsibility of checking graphs with the graphing calculator. When trying to interpret the real-life situations, encourage your students to try to translate the situation into an algebraic equation and

then match it to one of the available choices or use the clues in the description to picture the graph or use a combination of these techniques.

- Another way to use this activity is as a "partner quiz." Alternatively, a quiz similar to this activity could be given to the class the following day and each individual could receive his or her own score. If every member of the original group scores above a set minimum percent on the quiz, the group can be declared a success and receive some type of reward such as extra credit points.

Answers

a. (iv)
b. (vii)
c. (ii)
d. (i)
e. (viii)
f. (vi)
g. (v)
h. (iii)
i. (ix)
j. (x)

A. Equation: (c); Graph: (ii)
B. Equation: (a); Graph: (iv)
C. Equation: (j); Graph: (x)
D. Equation: (h); Graph: (iii)
E. Equation: (e); Graph: (viii)

CHAPTER 8
Group Activity (page 487)

Suggestions: Make sure that your students are measuring in inches in this group activity. Be sure to point out that the scale given on the grid is $\dfrac{1''}{4} = 10$ miles.

Answers

1. Answers will vary, but coordinates should be in the vicinity of (46, 90).

2. Ship A: $y = \dfrac{5}{2}x - 25$; Ship B: $y = -\dfrac{10}{3}x + 243\dfrac{1}{3}$

3. Coordinates should be approximately (46, 90).

4. (46, 90)

5. Answers will depend on the accuracy of students' measurement. Using the scale of the figure, students should measure a distance close to 80 miles from the collision point for Ship A and close to 68 miles from the collision point for Ship B. If that is the case and if $r_1 =$ Ship A's rate and $r_2 =$ Ship B's rate, to meet at the collision point at the same time,

$$r_1 = \frac{80}{68}r_2 = \frac{20}{17}r_2.$$

CHAPTER 9
Group Activity (page 537)

Suggestions: You could consider supplying the same set of cans for all groups or ask students to bring in their own cans to explore.

- For efficiency, consider using an empty frozen juice can that has its volume marked in milliliters on the outside of the can. Rather than measuring the volume of the can with water and measuring cup, use the volume marked on the can.

- For measuring the radius of a can, students should try to measure the diameter across the top (or bottom) of the can and take half of this measurement for the radius. Point out that the diameter measurement should be taken from the inner side of the can wall to the opposite inner side of the can wall.

- Another way students could find the radius of the can is to have them measure the circumference of the can and solve $C = 2\pi r$ for the radius. Students can measure the circumference of the can by wrapping a piece of string snugly around the can, marking the length of the string that represents the circumference, and then measuring the length of the string with a ruler. Remind students that this circumference is for the outside of the can. If the walls of the can are very thick, the outer circumference will be somewhat different from the inner circumference.

- You might want to use a soda can to start a discussion about how even slight variations from the shape of a right circular cylinder could throw off the calculations and measurements they are making. For instance, some soda cans have a substantial "dimple" in the bottom of the can. How would you measure the height of a can with such a dimple? Some soda cans now have a can top that is smaller in diameter than the rest of the can (and the can is tapered up to this top). How would you measure the radius of such a can? You might want to measure the actual volume of a soda can and then compare it to the volume of a regular right cylinder of the same basic dimensions to show how the different soda can top and bottom changes its volume.

Answers

1. Answers will depend on the can used.
2. Answers will depend on the can used.
3. Answers will depend on the can used. Students can use the formula $r = \sqrt{\dfrac{V}{\pi h}}$ to calculate the radius.
4. Answers will depend on the can used.
5. Answers will vary; however, most of the error can be accounted for by error in measurement. Error could occur in measuring the volume of water with the measuring cup (through spillage or estimation error), in measuring the height of the can, or in measuring the radius of the can.

CHAPTER 10
Group Activity (page 584)

Suggestions: Instead of or in addition to modeling the path of water from the photograph given in the Group Activity, you might consider modeling the path of water using an actual water fountain. If so, students could measure the water's path directly (although this could become a little messy), or you or your students could bend wire into a shape matching the path of the water and measure the wire's shape.

Answers

1. Answers will vary, but the coordinates of point B should be in the vicinity of $(20, 0)$.
2. The x-coordinate of the vertex is halfway between the x-intercepts. The axis of symmetry is the vertical line whose x-intercept is the same as the x-coordinate of the vertex. The coordinates of the vertex should be in the vicinity of $(10, 13)$.
3. Graphs will vary depending on ordered pairs.
4. (d)
5. Answers will vary.

ANSWERS TO EXERCISES REQUIRING GRAPHICAL SOLUTIONS

Exercise Set 3.1

1. quadrant I

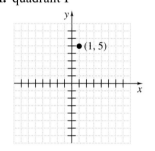

2. quadrant III

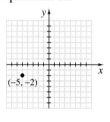

3. no quadrant, *x*-axis

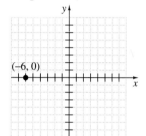

4. no quadrant, *y*-axis

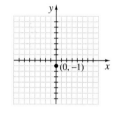

5. quadrant IV

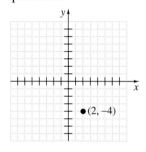

6. quadrant II

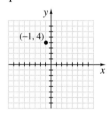

7. no quadrant, *x*-axis

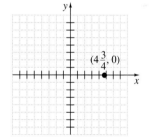

8. no quadrant, *y*-axis

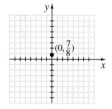

9. no quadrant, origin

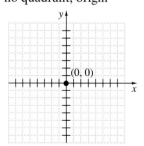

10. no quadrant, *x*-axis

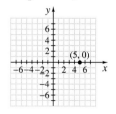

11. no quadrant, *y*-axis

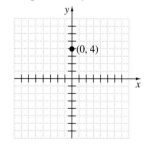

12. quadrant III

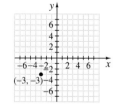

43. $(0, 2), (6, 0), (3, 1)$

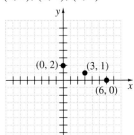

44. $(0, 4), (2, 0), (1, 2)$

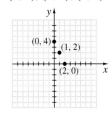

45. $(0, -12), (5, -2), (-3, -18)$

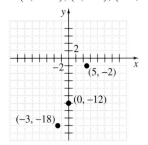

46. $(-2, 0), (-1, 5), (2, 20)$

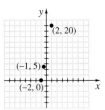

47. $\left(0, \dfrac{5}{7}\right), \left(\dfrac{5}{2}, 0\right), (-1, 1)$

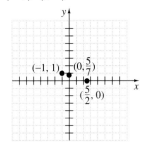

48. $\left(0, -\dfrac{1}{2}\right), \left(1, -\dfrac{1}{3}\right), (-3, -1)$

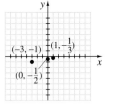

49. $(3, 0), (3, -0.5), \left(3, \dfrac{1}{4}\right)$

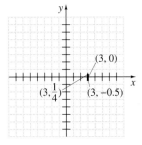

50. $(-2, -1), (0, -1), (-1, -1)$

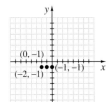

51. $(0, 0), (-5, 1), (10, -2)$

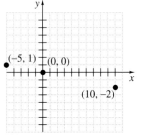

52. $(0, 0), (-2, 6), (-3, 9)$

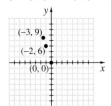

55.

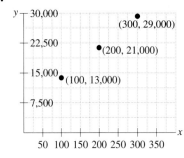

56.

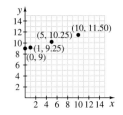

Graphing Calculator Explorations 3.2

1.
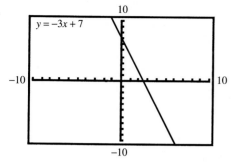
$y = -3x + 7$

2.
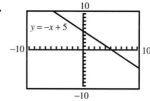
$y = -x + 5$

3.
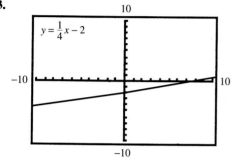
$y = \frac{1}{4}x - 2$

4.
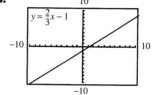
$y = \frac{2}{3}x - 1$

5.
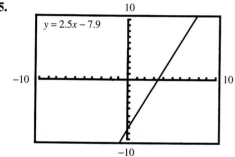
$y = 2.5x - 7.9$

6.
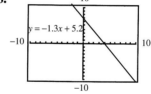
$y = -1.3x + 5.2$

7.
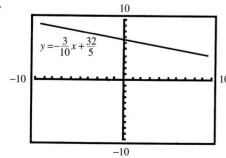
$y = -\frac{3}{10}x + \frac{32}{5}$

8.
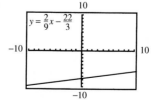
$y = \frac{2}{9}x - \frac{22}{3}$

Exercise Set 3.2

9.

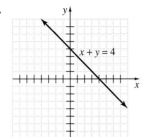

$x + y = 4$

10.

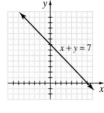

$x + y = 7$

11.

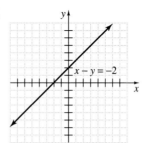

$x - y = -2$

12.

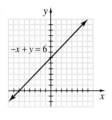

$-x + y = 6$

13.

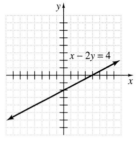

$x - 2y = 4$

14.

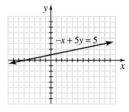

$-x + 5y = 5$

15.

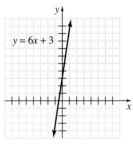

$y = 6x + 3$

16.

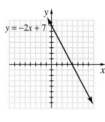

$y = -2x + 7$

17.

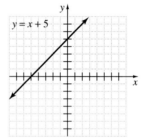

$y = x + 5$

18.

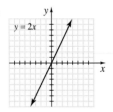

$y = 2x$

19.

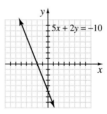

$2x + 3y = 6$

20.

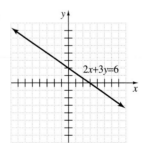

$5x + 2y = -10$

21.

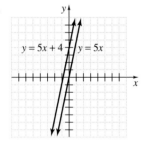

$y = 5x + 4$ $y = 5x$

22.

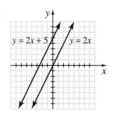

$y = 2x + 5$ $y = 2x$

23.

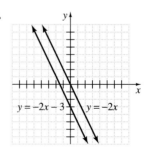

$y = -2x - 3$ $y = -2x$

24.
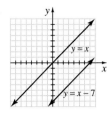

25.
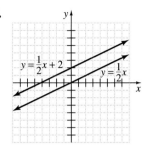

26.
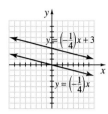

27.
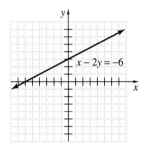

28.
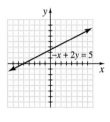

29.
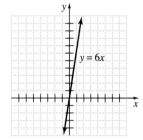

30.
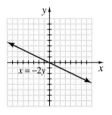

31.
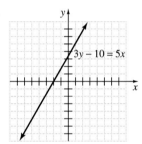

32.
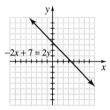

33.
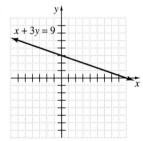

34.
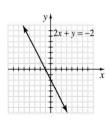

35.
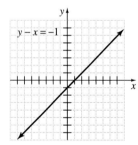

36.
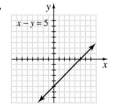

37.
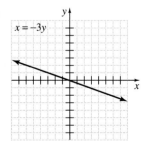

38.
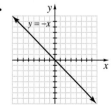

39.

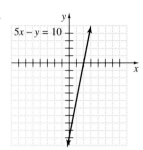

$5x - y = 10$

40.

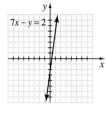

$7x - y = 2$

41.

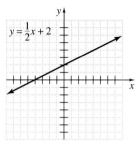

$y = \frac{1}{2}x + 2$

42.

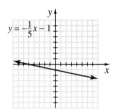

$y = -\frac{1}{5}x - 1$

43.

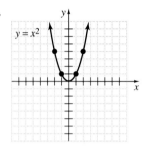

$y = x^2$

44.

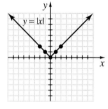

$y = |x|$

Graphing Calculator Explorations 3.3

1.

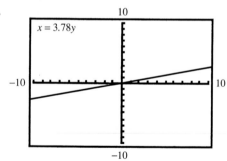

$x = 3.78y$

2.

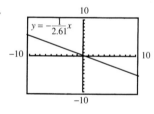

$y = -\frac{1}{2.61}x$

3.

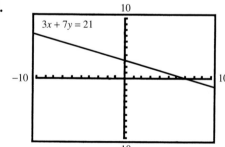

$3x + 7y = 21$

4.

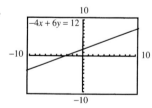

$-4x + 6y = 12$

5.

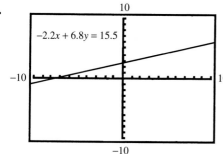

$-2.2x + 6.8y = 15.5$

6. $5.9x - 0.8y = -10.4$

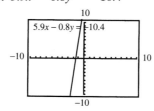

$5.9x - 0.8y = -10.4$

Exercise Set 3.3

11.

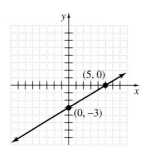

$(5, 0)$
$(0, -3)$

12.

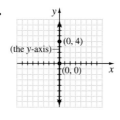

(the y-axis)
$(0, 4)$
$(0, 0)$

13.

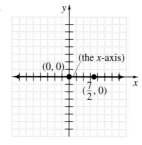

$(0, 0)$
(the x-axis)
$\left(\frac{7}{2}, 0\right)$

14.

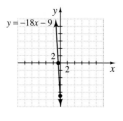

$y = -18x - 9$
2
2

15.

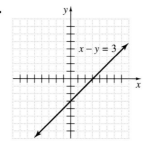

$x - y = 3$

16.

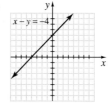

$x - y = -4$

17.

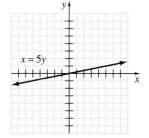

$x = 5y$

18.

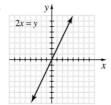

$2x = y$

19.

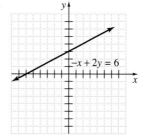

$-x + 2y = 6$

20.

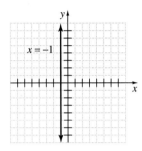

21.

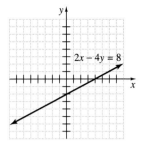

22.

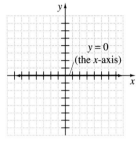

23.

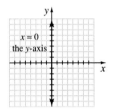

24.

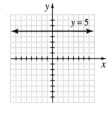

25.

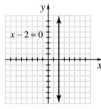

26.

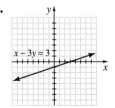

27.

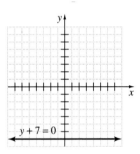

28.

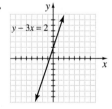

29.

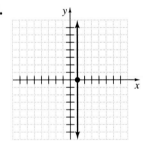

30.

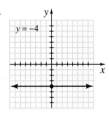

31.

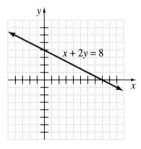

32.

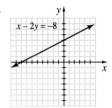

33.

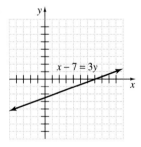

34.

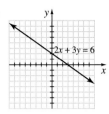

35.
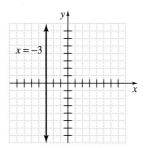

36.
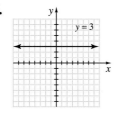

37.
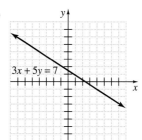

38.
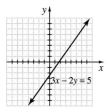

39.
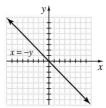

40.
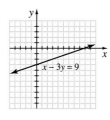

41.
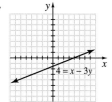

42.
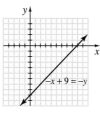

43.
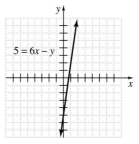

44.
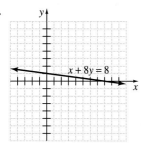

45.
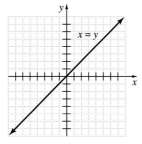

46.
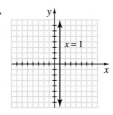

47.
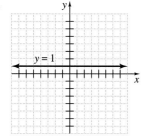

48.
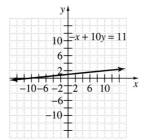

49.
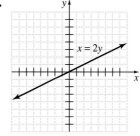

50.

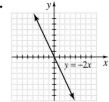

$y = -2x$

51.

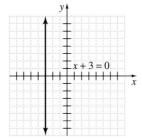

$x + 3 = 0$

52.

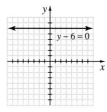

$y - 6 = 0$

53.

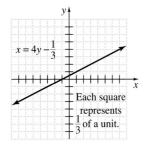

$x = 4y - \frac{1}{3}$

Each square represents $\frac{1}{3}$ of a unit.

54.

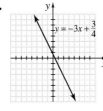

$y = -3x + \frac{3}{4}$

55.

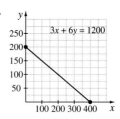

$2x + 3y = 6$

56.

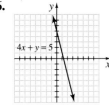

$4x + y = 5$

63.

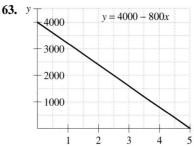

$y = 4000 - 800x$

64.

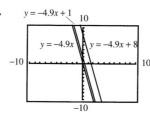

$3x + 6y = 1200$

Graphing Calculator Explorations 3.4

1.

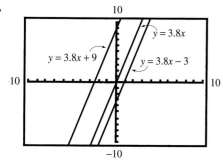

$y = 3.8x$

$y = 3.8x + 9$

$y = 3.8x - 3$

2.

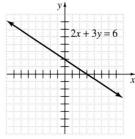

$y = -4.9x + 1$

$y = -4.9x$

$y = -4.9x + 8$

3.

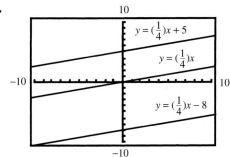

$$y = (\tfrac{1}{4})x + 5$$

$$y = (\tfrac{1}{4})x$$

$$y = (\tfrac{1}{4})x - 8$$

4.

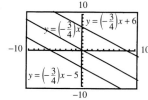

$$y = \left(-\tfrac{3}{4}\right)x$$

$$y = \left(-\tfrac{3}{4}\right)x + 6$$

$$y = \left(-\tfrac{3}{4}\right)x - 5$$

Exercise Set 3.4

31.

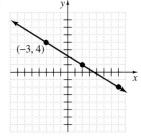

(2, 3)

32.

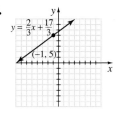

$$y = \tfrac{2}{3}x + \tfrac{17}{3}$$

(−1, 5)

33.

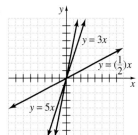

(0, −4)

34.

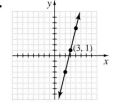

(3, 1)

35.

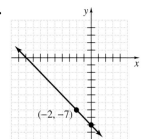

(−2, −7)

36.

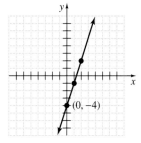

(−4, −3)

37.

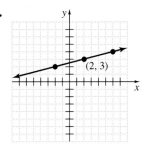

(−3, 4)

38.

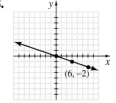

(6, −2)

93.

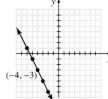

$$y = 3x$$

$$y = (\tfrac{1}{2})x$$

$$y = 5x$$

94.

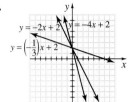

$y = -2x + 2$ $y = -4x + 2$

$y = \left(-\frac{1}{3}\right)x + 2$

99.

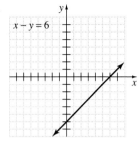

$x - y = 6$

100.

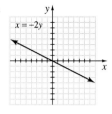

$x = -2y$

101.

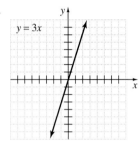

$y = 3x$

102.

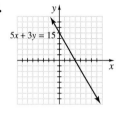

$5x + 3y = 15$

103.

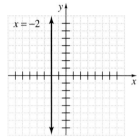

$x = -2$

104.

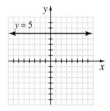

$y = 5$

Exercise Set 3.5

7.

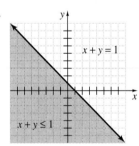

$x + y = 1$

$x + y \leq 1$

8.

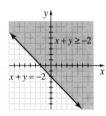

$x + y \geq -2$

$x + y = -2$

9.

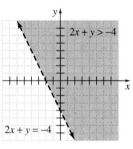

$2x + y > -4$

$2x + y = -4$

10.

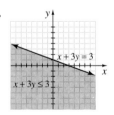

$x + 3y = 3$

$x + 3y \leq 3$

11.

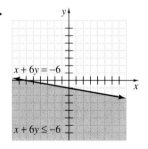

$x + 6y = -6$

$x + 6y \leq -6$

12.

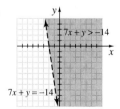

$7x + y > -14$

$7x + y = -14$

13.

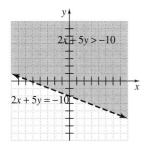

14.

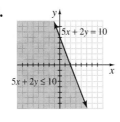

15.

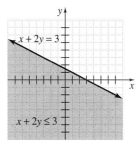

16.

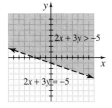

17.

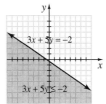

18.

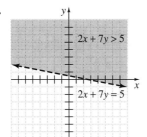

19.

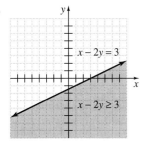

20.

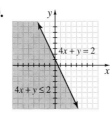

21.

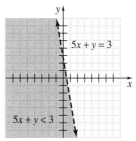

22.

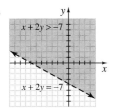

23.

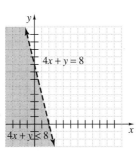

24.

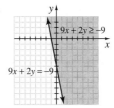

25.

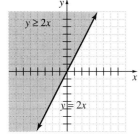

26.

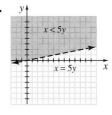

27.

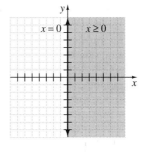

28.
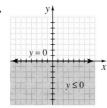

29.
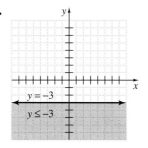

30.
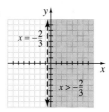

31.
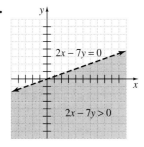

32.
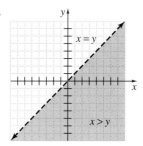

33.
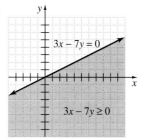

34.
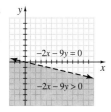

35.
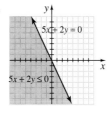

36.
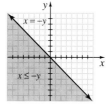

37.
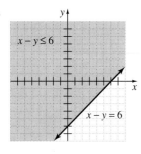

38.
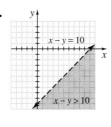

39.
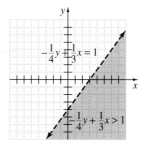

40.
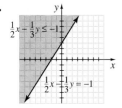

41.
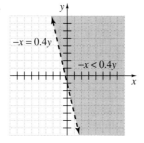

42.
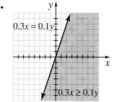

43.

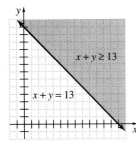

44.

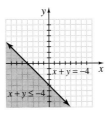

Chapter 3 Review

1.

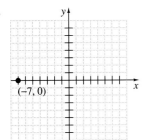

2.

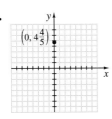

3.

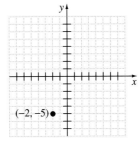

4.

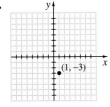

5.

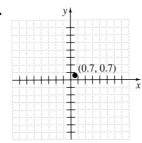

6.

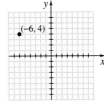

13. $(-3, 0), (1, 3), (9, 9)$

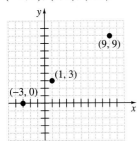

14. $(7, 5), (-7, 5), (0, 5)$

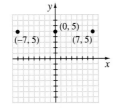

15. $(0, 0), (10, 5), (-10, -5)$

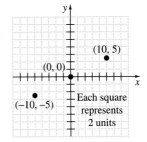

17.
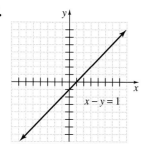
$x - y = 1$

18.
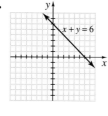
$x + y = 6$

19.
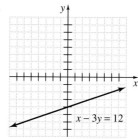
$x - 3y = 12$

20.
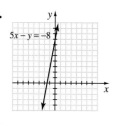
$5x - y = -8$

21.
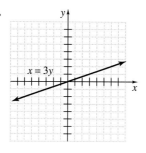
$x = 3y$

22.
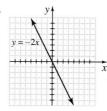
$y = -2x$

23.
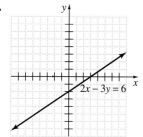
$2x - 3y = 6$

24.
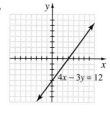
$4x - 3y = 12$

29.
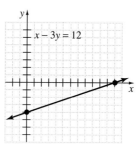
$x - 3y = 12$

30.
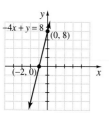
$-4x + y = 8$
$(0, 8)$
$(-2, 0)$

31.
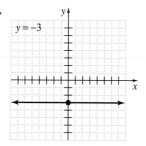
$y = -3$

32.
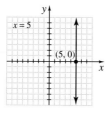
$x = 5$
$(5, 0)$

33.
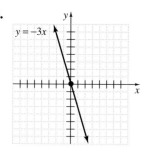
$y = -3x$

34.
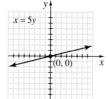
$x = 5y$
$(0, 0)$

35.
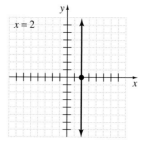
$x = 2$

36.

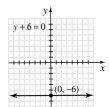

52.

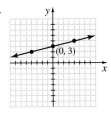

53.

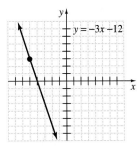

58.

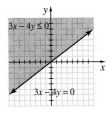

59.

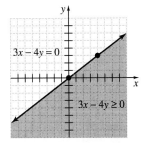

60.

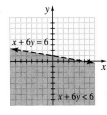

61.

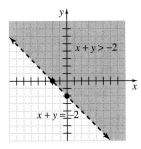

62.

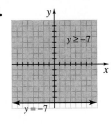

63.

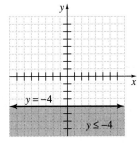

64.

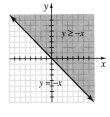

65.

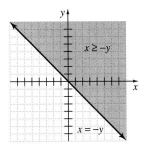

Chapter 3 Test

13.

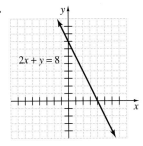

14.

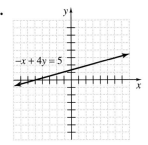

15.

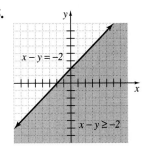

16.

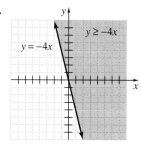

17.

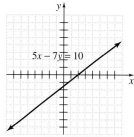

18.

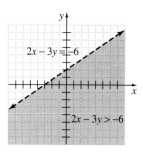

19.

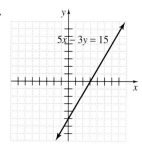

20.

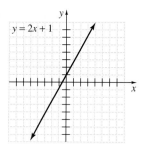

21.

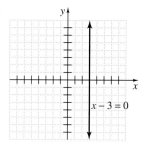

22.

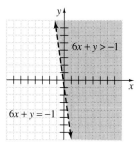

23.

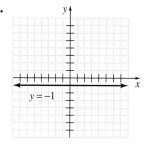

24.

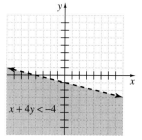

Chapter 3 Cumulative Review

15. (a) quadrant I **(b)** quadrant III **(c)** quadrant IV **(d)** quadrant II **(e)** no quadrant; origin **(f)** no quadrant; y-axis **(g)** no quadrant; x-axis **(h)** no quadrant; y-axis; *Sec. 3.1, Ex. 1*

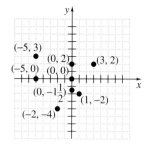

16. *Sec. 3.2, Ex. 2*

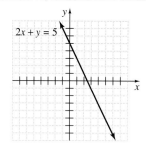

18. *Sec. 3.3, Ex. 6*

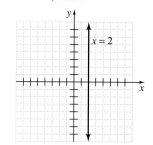

19. $-\dfrac{8}{3}$; *Sec. 3.4, Ex. 1*

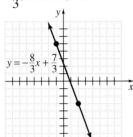

20. *Sec. 3.4, Ex. 3*

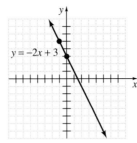

22. *Sec. 3.5, Ex. 2*

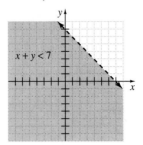

CHAPTER 4

Exponents and Polynomials

Exercise Set 4.2

93.

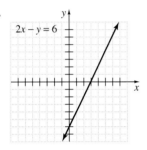

94.

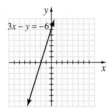

95.

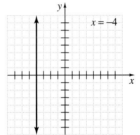

96.

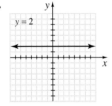

Chapter 4 Cumulative Review

13. *Sec. 3.2, Ex. 3*

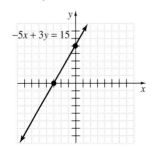

14. *Sec. 3.3, Ex. 2*

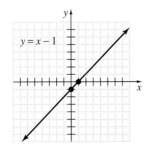

15. $m = 2$; *Sec. 3.4, Ex. 2*

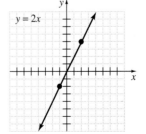

CHAPTER 5
Factoring Polynomials

Exercise Set 5.2

79.

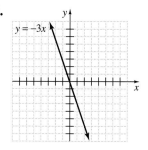

80.

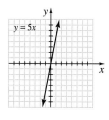

81.

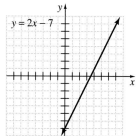

82.

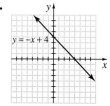

Exercise Set 5.6

83. **(a)** 300; 304; 276; 216; 124; 0; −156   **(b)** 5 sec.   **(c)** 304 ft.   **(d)** Answers may vary.

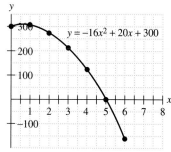

84. **(a)** 0; 84; 136; 156; 144; 100; 24; −84   **(b)** 6.3 sec.   **(c)** 156 ft.   **(d)** Answers may vary.

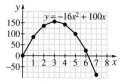

Chapter 5 Cumulative Review

9. *Sec. 3.2, Ex. 5*

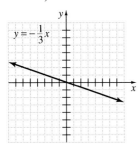

12. *Sec. 3.5, Ex. 3*

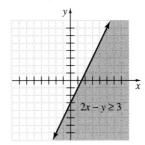

CHAPTER 6

Rational Expressions

Exercise Set 6.1

83. $y = \dfrac{x^2 - 25}{x + 5}$

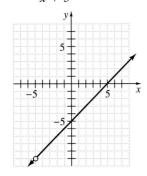

84. $y = \dfrac{x^2 - 16}{x - 4}$

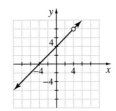

85. $y = \dfrac{x^2 + x - 12}{x + 4}$

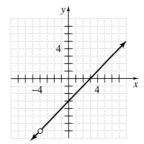

86. $y = \dfrac{x^2 - 6x + 8}{x - 2}$

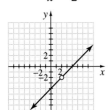

Exercise Set 6.2

75.

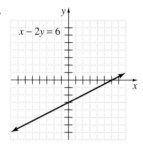

76.

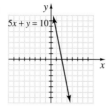

Exercise Set 6.6

75.

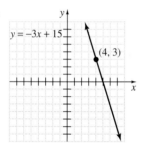

76.

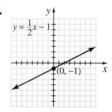

77.

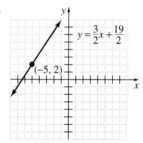

78.

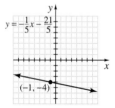

Exercise Set 6.8

44.

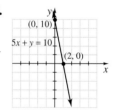

45.

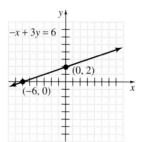

46.

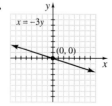

47.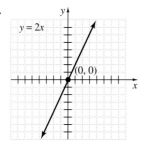

48.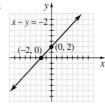

49.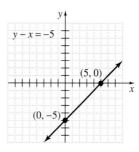

Chapter 6 Cumulative Review

8. *Sec. 3.2, Ex. 4*

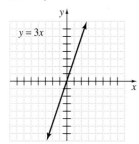

CHAPTER 7

Further Graphing

Graphing Calculator Explorations 7.1

1.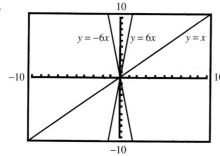

2.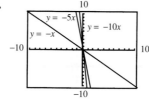

3.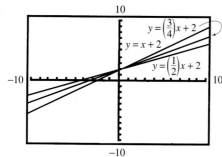

4.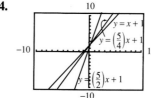

5.

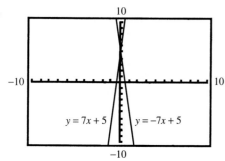

$y = 7x + 5$ $y = -7x + 5$

6.

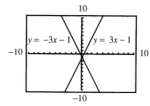

$y = -3x - 1$ $y = 3x - 1$

Exercise Set 7.1

33.

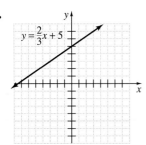

$y = \frac{2}{3}x + 5$

34.

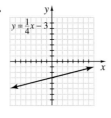

$y = \frac{1}{4}x - 3$

35.

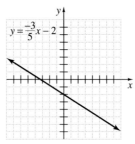

$y = \frac{-3}{5}x - 2$

36.

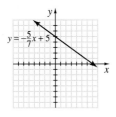

$y = -\frac{5}{7}x + 5$

37.

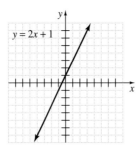

$y = 2x + 1$

38.

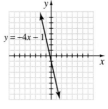

$y = -4x - 1$

39.

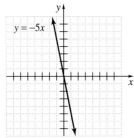

$y = -5x$

40.

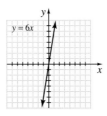

$y = 6x$

41.

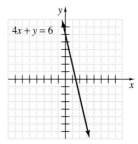

$4x + y = 6$

42.

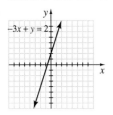

$-3x + y = 2$

43.

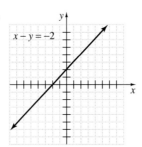

$x - y = -2$

44.

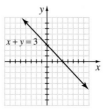

$x + y = 3$

45.

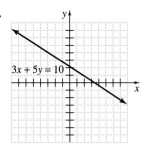

$3x + 5y = 10$

46.

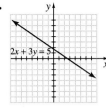

$2x + 3y = 5$

47.

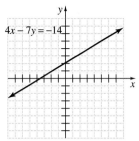

$4x - 7y = -14$

48.

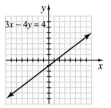

$3x - 4y = 4$

52.

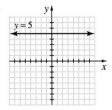

$y = 5$

53.

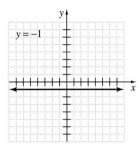

$y = -1$

54.

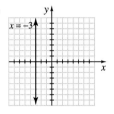

$x = -3$

55.

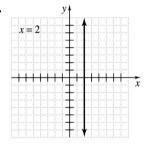

$x = 2$

Exercise Set 7.2

63.

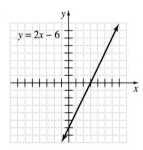

$y = 2x - 6$

64.

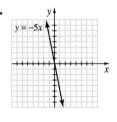

$y = -5x$

65.

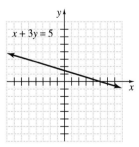

$x + 3y = 5$

66.

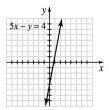

$5x - y = 4$

67.

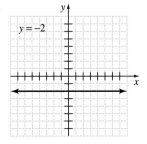

$y = -2$

68.

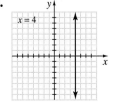

$x = 4$

Exercise Set 7.3

1.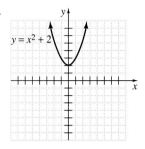

2.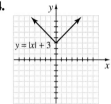

3.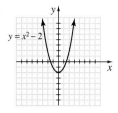

4.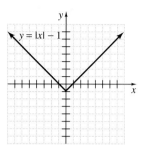

5.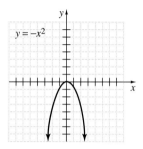

6.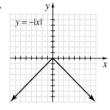

7.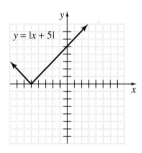

8.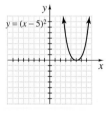

25. linear

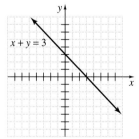

26. linear

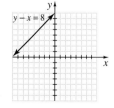

27. linear

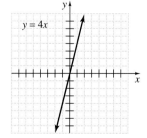

28. linear

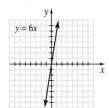

$y = 6x$

29. linear

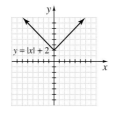

$y = 4x - 2$

30. linear

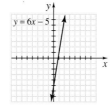

$y = 6x - 5$

31. not linear

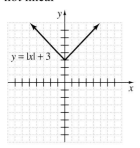

$y = |x| + 3$

32. not linear

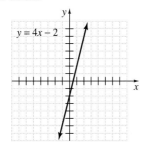

$y = |x| + 2$

33. linear

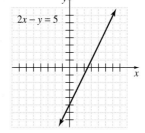

$2x - y = 5$

34. linear

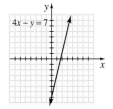

$4x - y = 7$

35. not linear

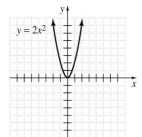

$y = 2x^2$

36. not linear

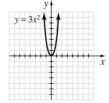

$y = 3x^2$

37. not linear

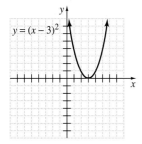

$y = (x - 3)^2$

38. not linear

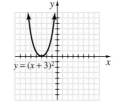

$y = (x + 3)^2$

39. linear

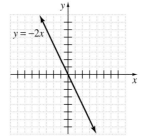

$y = -2x$

40. linear

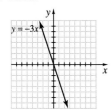

41. linear

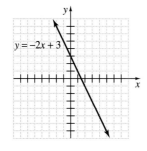

42. linear

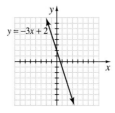

43. not linear

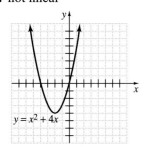

44. not linear

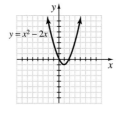

45. $(-3, -27), (-2, -8), (-1, -1), (0, 0), (1, 1), (2, 8), (3, 27)$

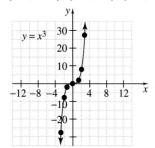

46. $(-3, 81), (-2, 16), (-1, 1), (0, 0), (1, 1), (2, 16), (3, 81)$

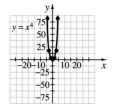

Exercise Set 7.4

49.
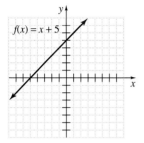
$f(x) = x + 5$

50.
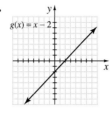
$g(x) = x - 2$

51.
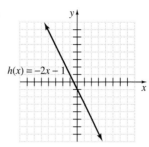
$h(x) = -2x - 1$

52.
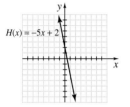
$H(x) = -5x + 2$

53.
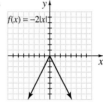
$g(x) = -|x|$

54.
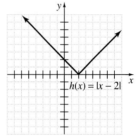
$f(x) = -2|x|$

55.
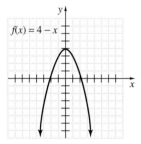
$f(x) = 4 - x$

56.
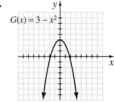
$G(x) = 3 - x^2$

57.
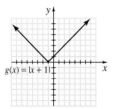
$h(x) = |x - 2|$

58.
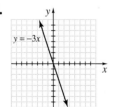
$g(x) = |x + 1|$

Chapter 7 Review

10.
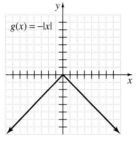
$y = -3x$

11.
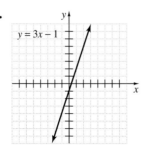
$y = 3x - 1$

12.
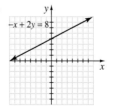
$-x + 2y = 8$

13.

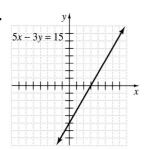

$5x - 3y = 15$

32.

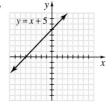

$y = x + 5$

33.

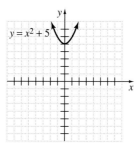

$y = x^2 + 5$

34.

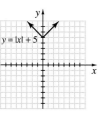

$y = |x| + 5$

35.

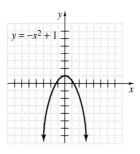

$y = -x^2 + 1$

36.

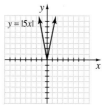

$y = |5x|$

37.

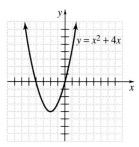

$y = x^2 + 4x$

Chapter 7 Test

8.

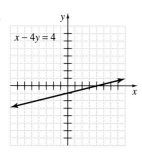

$x - 4y = 4$

9.

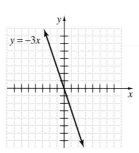

$y = -3x$

10.

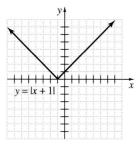

$y = |x + 1|$

11.

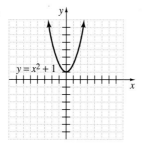

$y = x^2 + 1$

Chapter 7 Cumulative Review

6. *Sec. 3.2, Ex. 6*

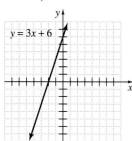

18. *Sec. 7.3, Ex. 1*

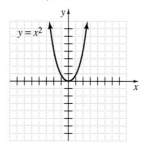

CHAPTER 8

Solving Systems of Linear Equations

Exercise Set 8.1

9. $(2, 3)$; consistent; independent

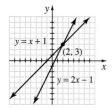

10. $(3, 5)$; consistent; independent

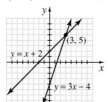

11. $(1, -2)$ consistent; independent

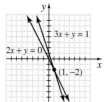

12. $(-1, 3)$; consistent; independent

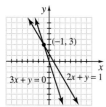

13. $(-2, 1)$; consistent; independent

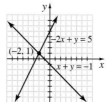

14. $(-1, -2)$; consistent; independent

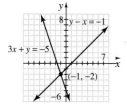

15. $(4, 2)$; consistent; independent

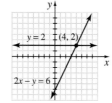

16. $(4, 1)$; consistent; independent

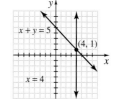

17. no solution; inconsistent; independent

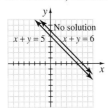

18. $(2, 0)$; consistent; independent

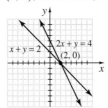

19. infinite number of solutions; consistent; dependent

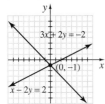

20. no solution; inconsistent; independent

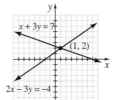

21. $(0, -1)$; consistent; independent

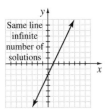

22. $(1, 2)$; consistent; independent

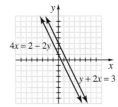

23. $(4, -3)$; consistent; independent

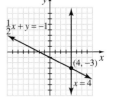

24. $(-1, 4)$; consistent; independent

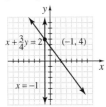

25. $(-5, -7)$; consistent; independent

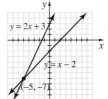

26. $(-3, 2)$; consistent; independent

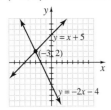

27. $(5, 2)$; consistent; independent

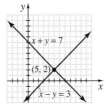

28. $(-1, -3)$; consistent; independent

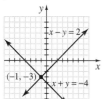

49. (a) $(4, 9)$ **(b)** See graph; yes

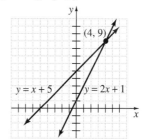

Exercise Set 8.4

35.

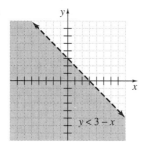

36.

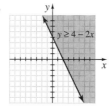

37.

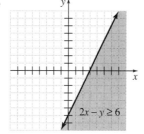

38.

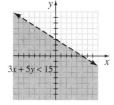

Exercise Set 8.5

1.

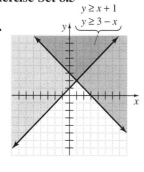

2.

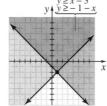

3.

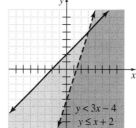

4.
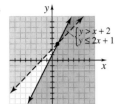
$\begin{cases} y > x + 2 \\ y \le 2x + 1 \end{cases}$

5.
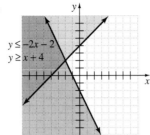
$y \le -2x - 2$
$y \ge x + 4$

6.
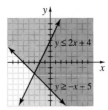
$y \le 2x + 4$
$y \ge -x - 5$

7.
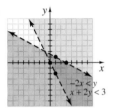
$y \ge -x + 2$
$y \le 2x + 5$

8.
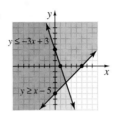
$y \le -3x + 3$
$y \ge x - 5$

9.
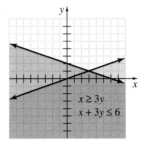
$x \ge 3y$
$x + 3y \le 6$

10.
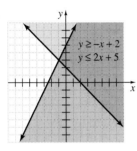
$-2x < y$
$x + 2y < 3$

11.
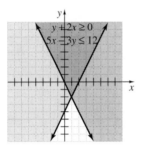
$y + 2x \ge 0$
$5x - 3y \le 12$

12.
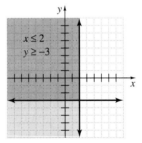
$\begin{cases} y + 2x \le 0 \\ 5x + 3y \ge -2 \end{cases}$

13.
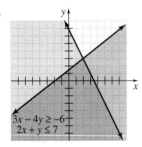
$3x - 4y \ge -6$
$2x + y \le 7$

14.
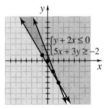
$\begin{cases} 4x - y \ge -2 \\ 2x + 3y \le -8 \end{cases}$

15.
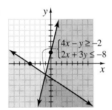
$x \le 2$
$y \ge -3$

16.
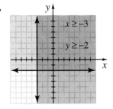
$x \ge -3$
$y \ge -2$

17.
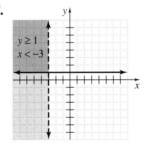
$y \ge 1$
$x < -3$

18.
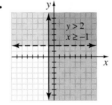
$y > 2$
$x \ge -1$

19.

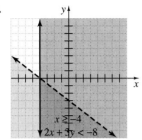

$x \geq -4$
$2x + 3y < -8$

20.

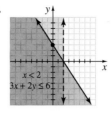

$x < 2$
$3x + 2y \leq 6$

21.

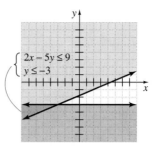

$\begin{cases} 2x - 5y \leq 9 \\ y \leq -3 \end{cases}$

22.

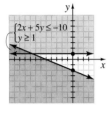

$\begin{cases} 2x + 5y \leq -10 \\ y \geq 1 \end{cases}$

23.

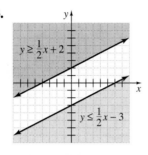

$y \geq \frac{1}{2}x + 2$

$y \leq \frac{1}{2}x - 3$

24.

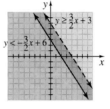

$y \geq \frac{3}{2}x + 3$

$y < -\frac{3}{2}x + 6$

29. (a) $x + y \leq 8, x < 3, x \geq 0, y \geq 0$
(b) See graph.

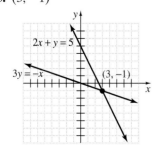

30. (a) $x + y \leq 1000, y \geq 300, x \geq 0$
(b) See graph.

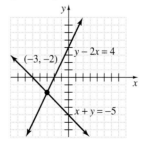

Chapter 8 Review

5. $(3, -1)$

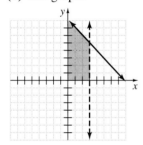

$2x + y = 5$

$3y = -x$

$(3, -1)$

6. $(-1, 1)$

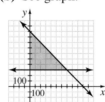

$(-1, 1)$

$2x - y = -3$ $3x + y = -2$

7. $(-3, -2)$

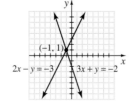

$y - 2x = 4$

$(-3, -2)$

$x + y = -5$

8. No solution

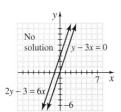

9. $\left(\dfrac{1}{2}, \dfrac{1}{2}\right)$

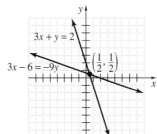

10. $(4, -1)$

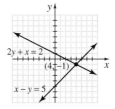

45.

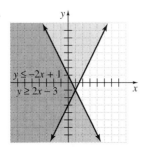

46.

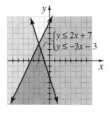

47.

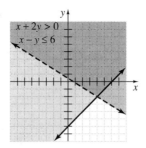

48.

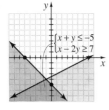

49.

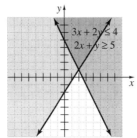

50.

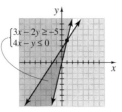

51.

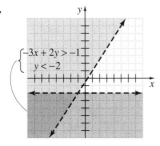

52.

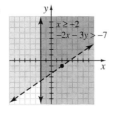

Chapter 8 Test

3. $(-4, 2)$

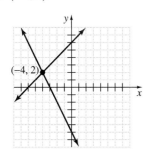

13.

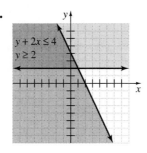

14.

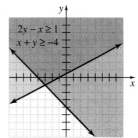

Chapter 8 Cumulative Review

8. *Sec. 3.3, Ex. 3*

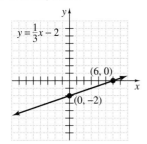

9. *Sec. 3.5, Ex. 4*

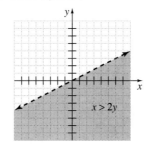

22. *Sec. 8.5, Ex. 3*

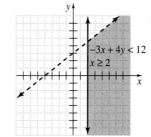

CHAPTER 9
Roots and Radicals

Exercise Set 9.1

80. $(0, 0), (1, 1), (3, 1.7), (4, 2), (9, 3)$

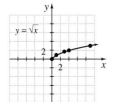

81. $(-8, -2), (-2, -1.3), (-1, -1), (0, 0), (1, 1), (2, 1.3), (8, 2)$

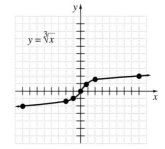

82. See graph; $(2, 0)$

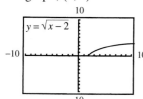

83. See graph; $(-3, 0)$

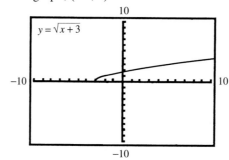

84. See graph; $(-4, 0)$

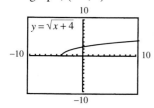

85. See graph; $(5, 0)$

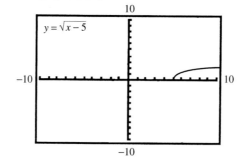

Exercise Set 9.7

77.

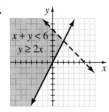

78.

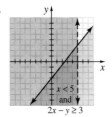

Chapter 9 Cumulative Review

5. *Sec. 3.3, Ex. 7*

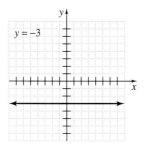

6. *Sec. 3.5, Ex. 5*

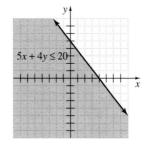

16. *Sec. 8.5, Ex. 2*

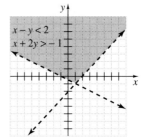

CHAPTER 10

Solving Quadratic Equations

Exercise Set 10.5

67.
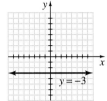

68.
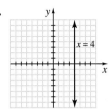

69.
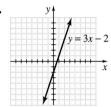

70.
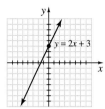

Exercise Set 10.6

1. $y = 2x^2$
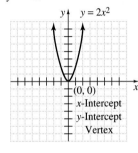

2. $y = -2x^2$
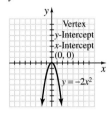

3. $y = (x - 1)^2$
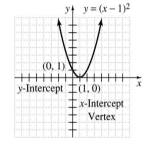

4. $y = (x + 2)^2$
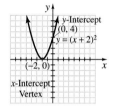

5. $y = -x^2 + 4$
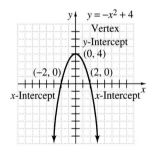

6. $y = x^2 - 4$
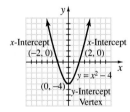

7. $y = \dfrac{1}{3}x^2$
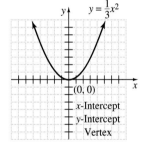

8. $y = -\dfrac{1}{2}x^2$
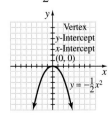

9. $y = (x - 2)^2 + 1$
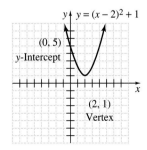

10. $y = -(x - 2)^2 - 1$
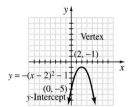

11. $y = -(x + 1)^2 + 4$

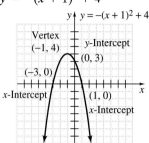

12. $y = (x - 1)^2 - 4$

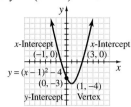

13. $y = -4x^2 + 1$

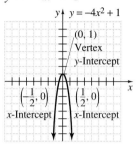

14. $y = 4x^2 - 1$

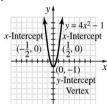

24. $y = x^2 + 6x$

$y = (x + 3)^2 - 9$;
vertex $(-3, -9)$;
y-intercept $(0, 0)$;
x-intercept $(-6, 0)$,
$(0, 0)$

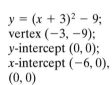

25. $y = x^2 - 4x$

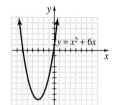

26. $y = x^2 + 2x - 8$

$y = (x + 1)^2 - 9$;
vertex $(-1, -9)$;
y-intercept $(0, -8)$;
x-intercept $(-4, 0)$,
$(2, 0)$

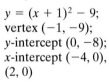

27. $y = x^2 - 2x - 3$

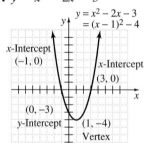

28. $y = x^2 - x - 2$

$y = \left(x - \frac{1}{2}\right)^2 - \frac{9}{4}$;

vertex $\left(\frac{1}{2}, -\frac{9}{4}\right)$;

y-intercept $(0, -2)$;
x-intercept $(-1, 0)$,
$(2, 0)$

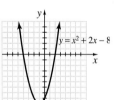

29. $y = x^2 + 2x + 1$

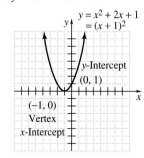

30. $y = x^2 + 5x + 4$

$y = \left(x + \frac{5}{2}\right)^2 - \frac{9}{4}$;

vertex $\left(-\frac{5}{2}, -\frac{9}{4}\right)$;

x-intercept $(-4, 0)$,
$(-1, 0)$;
y-intercept $(0, 4)$

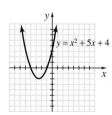

31. $y = x^2 + 7x + 10$

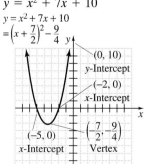

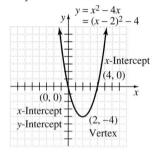
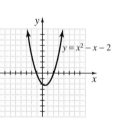
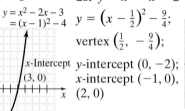

32. $y = x^2 - 4x + 3$

$y = (x - 2)^2 - 1$;
vertex $(2, -1)$;
y-intercept $(0, 3)$;
x-intercept $(3, 0)$,
$(1, 0)$

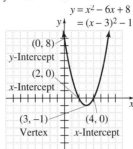

33. $y = x^2 - 6x + 8$

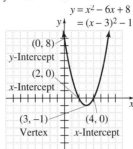

Chapter 10 Review

79. $y = -(x + 1)^2$

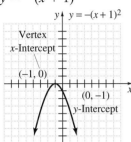

80. $y = -(x - 2)^2$

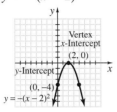

81. $y = (x - 2)^2 + 3$

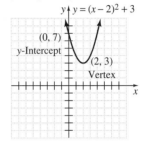

82. $y = (x + 3)^2 - 1$

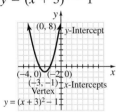

83. $y = x^2 + 5x + 6$

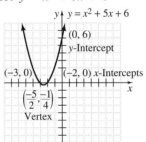

84. $y = x^2 - 4x - 8$

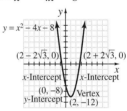

85. $y = 2x^2 - 11x - 6$

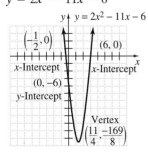

86. $y = 3x^2 - x - 2$

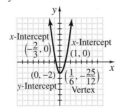

Chapter 10 Test

25.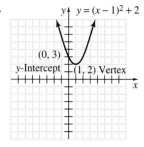

26.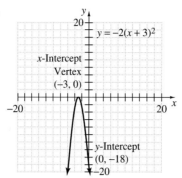

27.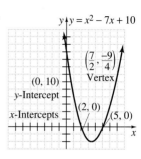

Notes to the Instructor

An Overview

Research on how students learn and regular advances in technology continue to affect how and what we teach. Many of these changes are reflected in the AMATYC Crossroads Guidelines and the NCTM Standards. The material that follows is intended to give a quick overview of several important related topics together with some suggestions to try with your students. Wherever possible, we have drawn on the experiences of and suggestions from instructors who have used these approaches successfully. The topics included in these Notes to the Instructor are:

■ Interactive and Cooperative Learning

■ Interpreting Graphs and Data

■ Alternative Assessment

■ Using Technology

■ Helping Students Succeed

If you have suggestions for other topics, ideas that should be included, or want to share your own successful approaches or instructional strategies, please call 1-800-435-3499, ext. 7404, and ask for a developmental mathematics editor, or e-mail ann_marie_jones@prenhall.com.

Interactive and Cooperative Learning

Many students absorb concepts well when working and learning together in small groups. For this reason you may wish to consider incorporating some interactive and cooperative learning strategies in your classroom. Types of interactive and cooperative learning activities are varied. They include discussion, skill practice and review, concept investigation, concept synthesis, and concept extension. *Beginning Algebra*, Second Edition, includes end-of-section exercises, many of which are appropriate for group work. In addition, chapter-ending Group Activities combine several of the aspects listed above. No matter what type of activity you choose to use, here are some general points to keep in mind.

1. Clearly explain the goals of the cooperative learning activity to students. Encourage active participation by all members of the group.

2. As appropriate, describe how teams will be rewarded for the success of their group.

3. Emphasize that students should share equally in the responsibility for completing the tasks involved in the activity.

4. Stress that each member of the team is accountable for the other team members' understanding of concepts involved in the activity.

5. Highlight how the gains in competence of each individual will contribute to the success of the team.

One way to implement cooperative learning strategies is to have the instructor structure the groups rather than allowing groups to be self-selecting. Before assigning a cooperative learning activity, decide on the optimal group size for the activity and who should be in each group. Consider the following strategies.

■ Try to establish groups with an equal likelihood of success.

■ Predetermine groups with similar female-to-male ratios, ethnic make-up, and low-to-high achiever ratios.

■ Groups can be formed with students' majors as a factor, with each group having a business major, nursing major, and so on.

Methods for random assignment include:

■ Groups can be formed by counting off; for instance, if the class size is 32 and groups of size four are desired, then each of the first eight students are assigned numbers from 1 to 8, and repeat assigning numbers 1 to 8 in this fashion to the remaining students. Then all the students assigned the number 1 create a group, all the 2's form a group, all the 3's form a group, and so on.

■ Groups can be formed by using an appropriate subset of a deck of cards; all the aces in one group, kings in another, and so on.

■ Groups can be created following alphabetical order of first or last names.

■ Partners can be pointed out by the instructor based on "neighbors"—you two, you two, . . .

■ Larger groups can be quickly selected based on location in the classroom; for instance, by rows, or by proximity to each corner of the room.

The instructor's role in interactive and cooperative learning can also include physical setup for the activity, such as rearranging the room, if necessary. Be sure to either provide any necessary special materials or ask students in advance to bring their own.

You should be available for assistance with any portion of the activity, but remind students that their primary resource for aid should be the group itself. You may also suggest guidelines for working together and help create a positive atmosphere for learning. These may include the following.

GUIDELINES FOR WORKING TOGETHER

1. Agree on what your group must do and how you will get it done.

2. Be courteous and listen carefully to other group members.

3. Create an atmosphere that allows group members to be comfortable in asking for help when needed.

4. Remember that not everyone works at the same pace. Be patient. Offer your help. Receive help graciously.

5. Keep in mind that all contributions made by group members are valuable.

6. Ask for your instructor's help only when all other sources of assistance have been exhausted.

7. If you finish early, double-check your work. As appropriate, reflect on your work. Can you make a general rule about the solution or describe a real-world use for what you learned?

Some classroom strategies that have been used successfully during group activities also include the following.

■ Circulate around the room from group to group to monitor progress, steer students in the right direction, and give encouragement.

■ Consider giving each group a blank acetate/transparency and pens at the start of the group activity. Each group can write its results on the transparency. The instructor or students can then share some of these with the class on an overhead projector.

■ If the group is working with a calculator, have everyone discuss and agree which keys to push and why. During the course, have students of all abilities take turns being in charge of using the calculator.

■ Consider selecting some group activities with the goal of previewing and exploring a new topic, while selecting other group activities to reinforce what has already been discussed.

■ "Partner Quizzes" have been used successfully to try to lower test-taking anxiety and to encourage cooperation. (See Alternate Assessment for more information.)

■ Consider assigning roles for members of a group of three or four, such as recorder, facilitator, checker, and reporter.

Individual accountability can be key to the success of an interactive or cooperative learning activity. To guard against "freeloading," you might consider asking each student in the class (or one randomly selected student from each group) to summarize briefly the findings of his or her group or the investigation process used by the group in a sentence or two. You might also consider judging the success of the group based on each individual's improvement. For instance, suppose you give a quiz on factoring and then incorporate a cooperative learning activity for practicing factoring skills. A follow-up quiz on factoring can be used to assess each individual's understanding of factoring. If each member of a group is able to achieve a predetermined goal, such as improving his or her score on the follow-up quiz as compared to the original quiz, then the entire group is rewarded.

You might consider asking your students to report the results of an extensive group activity in the form of a lab report. A lab report might consist of three parts: (1) an introduction in which the title of the lab or group activity is given, the goals of the activity are explained, the necessary materials are listed, and the group members are recorded; (2) an organized description of the process the group used to accomplish the goals of the group activity, including any research performed, tables and/or graphs of data collected, mathematical work completed; and (3) a summary of the group's findings and/or conclusions and, if appropriate, generalizations or conjectures to which the group was led during the course of the activity. If you choose to use this lab reporting method, be sure to emphasize that work done to complete the lab must be done cooperatively. You might divide the activity into sections for which each group member is responsible. If so, the lab report should indicate who played which role and/or who wrote which sections of the lab report. Remind students that they are all responsible for the results given in their group's lab report. Suggest that after each student has made his or her written contribution to the report, the group should review all the contributions together to increase their understanding of the activity.

INTERPRETING GRAPHS AND DATA

Both the NCTM Standards and Addenda and the AMATYC Crossroads Guidelines encourage the inclusion of graphical and data interpretation in mathematics courses. This reflects changing needs in the workplace. Consider the following facts showing a trend toward an increased need to analyze data and information.

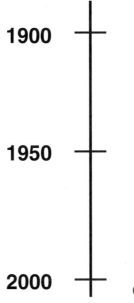

1900

"During the early 1900's, 85% of our workers were in agriculture. Now agriculture involves less than 3% of the workforce."

1950

"In 1950, 73% of U.S. employees worked in production or manufacturing. Now less than 15% do so."

2000

"The Department of Labor estimates that by the year 2000 at least 44% of all workers will be in data services, for example, gathering, processing, retrieving, or analyzing information."
(*Source: The Employee Handbook of New Work Habits for a Radically Changing World* by Price Pritchett. Used with full permission of Pritchett & Associates. All rights reserved.)

Beginning Algebra, Second Edition, provides ample opportunity for several types of interpretation throughout the text. Circle graphs, line graphs, and bar graphs are used often to convey statistical information and are accompanied by questions that reinforce understanding.

Consider bringing into your classroom (or having your students bring in) other examples of current real data from newspapers, magazines, or the Internet. Note that one supplement available with this text is *The New York Times/Themes of the Times* that is created new each year. Another supplement is an *Internet Guide.* In addition, a great source of real data is your students. Consider having students list information such as age, height, shoe size, number of hours worked at a job during the last week, and birth dates. This could be used to develop a range of graphs and charts in which they have a personal interest.

There are many key ideas that could be discussed when creating or analyzing graphs and data. Many such ideas will be specific to the particular graph and the information it conveys. However, some more general ideas you may choose to keep in mind include the following.

READING AND USING GRAPHS

- To show trends over time, a line graph is often useful. The horizontal axis usually indicates the period of time. Look for relationships, if any, between the two quantities. (When reading a line graph, sometimes there may be a break in the axes indicating that some numbers have been omitted. This break is usually shown by a jagged line near the origin.)

- To compare total amounts, such as results from a quantitative survey, a horizontal or vertical bar graph can be a good way to represent the information. One of the axes could be labeled with word phrases rather than numbers.

- A circle graph (pie graph or pie chart) is shown in this text to visually display information using a circle. The circle is divided into wedges called sectors. Often the sectors are labeled with percents.

Encourage your students to study the graphs in this text and to become acquainted with the power of communicating complex data graphically.

ALTERNATIVE ASSESSMENT

Alternative assessment provides a means for evaluating students' progress and/or understanding with tools other than the traditional testing methods. When used with traditional assessment methods, alternative methods can help give a truer picture of a student's comprehension of mathematical concepts. For instance, a student who experiences test anxiety may perform poorly on quizzes or exams but still fully understand the concepts being tested. An alternative assessment method should uncover that understanding. Several common forms of alternative assessment include: topic interviews, self-evaluation, partner quizzes, homework notebooks, journals, portfolios, conceptual and writing exercises, research projects, and demonstrations.

Topic Interviews In this type of assessment, the instructor schedules a short, one-on-one "interview" with each student to informally discuss particular mathematics concepts. For example, the student and instructor could discuss the graph of a particular linear equation, including whether the slope of the line is positive or negative, approximately where the x- and y-intercepts are located, a possible situation that the graph might describe, and so on.

Self-Evaluation Students are asked to respond confidentially in writing to several questions asked by the instructor. These questions could allow students to give their own summaries of the day's class, list concepts they do not fully understand, and suggest ways in which material could be covered more clearly. Students could be asked to rate their own understanding of a topic and supply an explanation or demonstration of why they chose their ratings. Students could also be asked to contribute a quiz or exam question, with the accompanying answer, that they think does a fair job of gauging understanding.

As an alternative, students could be asked only one short question. This can provide the instructor with feedback quickly.

Partner Quizzes This type of assessment may take several forms. In one form, partners are asked to write a short quiz that they think adequately and fairly tests knowledge from a particular section of material. Partners then exchange their quizzes, take their partners' quiz, and then grade the quiz they wrote. This activity requires students to write easily solvable quiz questions by working backwards and requires them to be the "answer authority" when grading the quizzes they wrote. Analyzing the logic another student used in solving a problem can be quite educational to the quiz creator/grader. Another form of partner quizzes asks two students to work together to take a quiz provided by the instructor. Students then help each other understand the concepts required to answer the quiz questions correctly. These may or may not be "open-book" partner quizzes.

Homework Notebook Students are asked to do homework assignments regularly in a notebook. Occasionally, a few selected exercises could be checked for effort and correctness. Evaluation could include awarding some credit for effort (tried the exercise), partial credit for correct process, and credit for correct results. Some instructors may choose to award some credit or bonus points for overall completeness and neatness of homework assignments. Using notebooks provides feedback and can encourage students to attend class and try the exercises.

Ways to organize a notebook that you might suggest to students include writing the text section number or title at the top of each page. Another suggestion can be to leave room for class notes to be written beside the attempted exercises. That is, if a homework exercise is discussed during the next class, the student can make notes or modifications to their solution, if needed. When reviewing for a test, the notes can help students study more effectively. (One way for students to plan for notes is to draw a vertical line on each notebook page such that the page is divided into two unequal parts (columns). The smaller part can represent 1/3 page and the larger part 2/3 of a page. Exercises can be written in the larger area, and any notes can be added in the smaller column.)

Journals Students are asked to keep an ongoing notebook or journal of thoughts about mathematics. Journal writing assignments might include such items as listing the main topics of the day's class, describing the concept that was easiest or hardest to understand, writing a "note" to a friend explaining how to solve a particular problem, listing questions that remain unanswered or writing a question that the student would like to ask the text's author about the topic, describing a real-life situation in which the concepts learned would be useful, creating a "crib sheet" of formulas that could be used on the next test, summarizing certain properties (such as the properties of inequalities or the properties of exponents) and giving an example of each one, or giving advice to students who will be taking this course during the next term.

Portfolios This type of assessment allows students to collect examples of their mathematical work just as one would assemble an art portfolio. Portfolio assignments might include such items as selecting what the student thinks is the best example of his or her work during a week, creating an "original piece" such as an application problem written by the student, choosing a particularly well-drawn graph, picking an interesting piece of graphing technology output, adding examples of effective problem solving or student-corrected homework, or collecting newspaper/magazine articles or comic strips that have a connection to math. You might consider asking the students to supply written explanations of why they chose each piece to add to their portfolios.

Conceptual and Writing Exercises This text contains conceptual and writing exercises. These exercises often ask students to use two or more concepts together. Some require students to stop, think, and explain in their own words the concepts used in the exercises just completed. These exercises can be used alone or as part of a portfolio to assess understanding.

Research Projects Projects that require library research or collecting data outside of the classroom provide opportunities for students to synthesize what they have been learning. Projects may be assigned to groups or individuals. Some of the chapter-ending Group Activities in this text incorporate aspects of independent research.

Demonstrations There are a variety of ways to gauge students' understanding through demonstrations that could be made either to the entire class or a small group. Students might be asked to report on the results of a cooperative learning activity or the findings of a research project to the class, prepare an example or the solution to an exercise that will be demonstrated on the blackboard or overhead during the next class, learn a new graphing calculator function and teach its use to the class, or illustrate a mathematical concept with the use of manipulatives.

Face-to-face meetings or written submissions can address all of these methods of alternative assessment. However, if e-mail is available to you and your students, it can offer another way for them to submit written work and ideas to you, and for you to respond.

USING TECHNOLOGY

The use of a graphing calculator with *Beginning Algebra,* Second Edition, is optional. Graphing Calculator Explorations and Scientific Calculator Explorations give instruction, at point of use, on how to use a calculator as a problem-solving tool in conjunction with the material of the section and allow students to practice particular calculator skills on a short set of exercises. Before the course starts, de-

cide on the role you wish technology to play in your course. Clearly explain your policies on the use of scientific and/or graphing calculators with homework, group work, quizzes, and tests.

Demonstration Graphing technology provides a means for exploring mathematical concepts. You might consider using a graphing calculator as a demonstration tool. It is generally easier to show the effect of different values of b in the equation $y = mx + b$ with a graphing calculator or computer graphing tool than to graph the equations of many lines by hand. Similarly, graphical solutions to systems of two linear equations in two variables may be discovered by demonstrating graphs with a graphing utility. To involve students in the demonstration, you may occasionally want to have a student hold the calculator and press the keys under your direction as you circulate around the room leading a discussion of the concept being illustrated.

Using Graphing Calculators During Class If you have a mix of graphing calculator models being used by students in your classroom, you might consider physically grouping students with the same models together so they may answer one another's questions about the details of using a certain model. Also encourage your students to refer to the owner's manuals that came with their calculators for help.

Rules for Rounding When students use a calculator, they often need to know how to round an answer to present it in the decimal form. Advise students that in this text, numbers are rounded *up* if the digit to the right of the given place value is a 5 or greater and rounded *down* if the digit to the right is a 4 or less. For example, 82.6259 rounded to two decimal places is rounded *up* to 82.63. Also, 82.6259 rounded to the nearest tenth is rounded *down* to 82.6.

You may also want to inform students that many calculators actually carry internally more digits than shown on the display. Encourage students not to round during intermediate steps of a calculation. Each rounding during an intermediate step of a calculation may increase the error of the final calculation.

HELPING STUDENTS SUCCEED

Convey a positive attitude, and encourage your students to have a positive attitude. Also, consider taking the first step to help your students know how to find help, if needed. In addition to your name, office location, and office hours, provide students with your office phone number, e-mail as appropriate, and where to obtain help if you are not available. Consider encouraging students to exchange names and phone numbers with at least one other person in the class. This contact from class can be someone with whom a student can discuss the algebra concepts and exercises, so both students improve their understanding. Depending on the size of your class, spending a little time on student introductions, why they are taking the course, their major and their interests may assist them in finding study partners.

Resources Consider letting students know what other resources are available such as a resource center or library with tutorial software, videotapes, or tutors. At the bookstore, print supplements, such as a student solutions manual or study guide, may also be available. Encourage students to seek help as soon as they have questions. There are many resources they can use to their advantage.

Study Strategies Having good study skills and strategies can be very helpful for students' success at the college level. Study skills particularly appropriate for mathematics are described in this text's Student Solutions Manual and Study Guide to help students. Referring students to these lists may prove useful. Topics covered include suggestions on preparing for class, attending class, taking notes, allocating study time, doing homework, checking exercises, preparing for tests, and taking tests.

In addition, you may choose to elaborate on time management. Having students prepare bar charts or lists for how their time was spent on one or several days can be helpful. Students could allocate time into categories such as sleeping, eating, showering and dressing, traveling, attending classes, studying, working at a job, watching television, exercising, performing household tasks, spending time with friends, and other. Examining the charts can help them understand some of the conflicting demands on their time, and some trade-offs they may have to make. Encourage students to write down on something they will consult often (such as a calendar) how much time they plan to spend on studying algebra. This will help them follow through on their plan.

Some students experience difficulty with math classes not because the material itself is difficult for them but due to the way in which the information is presented. Two possible areas of difficulty include differences in learning styles and proficiency in language.

Learning Styles Society is becoming more and more visually oriented. It shouldn't be surprising that more and more students are, to some degree, visual learners. *Beginning Algebra,* Second Edition, presents a wide variety of tools for learning that students of many different learning styles should find useful. For instance, the steps for problem solving taught by *Beginning Algebra* include "illustrate the problem" as a step in the problem solving process. Other such tools include clear, worked-out examples, visual emphasis of important concepts and definitions, use of color in the explanations, and numerous graphs, including bar graphs and circle graphs, as well as diagrams. Kinesthetic learners benefit from "hands-on" activities and can benefit from working in groups. Real-world examples can make it easier for these students to relate to the underlying mathematics, as it becomes something real that they may have experienced.

Language Issues For students whose first language is not English, mathematics may become difficult due only to the lack of understanding of the language of instruction. For these students, care must be used when introducing new mathematical terms; write out definitions and clearly label each with its term. Try to write down any terms or statements that require emphasis. Students may find it mutually beneficial when a student whose grasp of the English language is good but whose math skills are weak is paired with a student for whom English is a second language but who is proficient in mathematics. A picture is worth a thousand words; drawing a diagram when explaining a concept will be extremely beneficial to any student who is struggling with the English explanation of the concept.

This is just a brief review of some key suggestions that have worked for other instructors around the country. We hope that you will find some of the ideas useful, or that we have stimulated you to think of new approaches appropriate for your students.

INDEX

Photo Credits